THIRD EDITION

Introduction to

LINEAR ALGEBRA

Lee W. Johnson
R. Dean Riess
Jimmy T. Arnold

THIRD EDITION

Introduction to

LINEAR ALGEBRA

Lee W. Johnson
R. Dean Riess
Jimmy T. Arnold

Virginia Polytechnic Institute and State University

Addison-Wesley Publishing Company

Reading, Massachusetts • Menlo Park, California • New York •
Don Mills, Ontario • Wokingham, England • Amsterdam • Bonn •
Sydney • Singapore • Madrid • San Juan • Milan • Paris

Sponsoring Editor: Jerome Grant

Production Supervisor: Mona Zeftel

Technical Art Consultant: Alena B. Konecny

Text Designer: Sally Steele

Cover Designer: Dick Hannus

Illustrator: Intergraphics

Copy Editor: Martha Morong

Manufacturing Manager: Roy Logan

Library of Congress Cataloging-in-Publication Data

Johnson, Lee W.
   Introduction to linear algebra / by Lee W. Johnson, R. Dean Riess,
Jimmy T. Arnold.—3rd ed.
      p.   cm.
   Includes index.
   ISBN 0-201-56801-2
   1. Algebras, Linear.   I. Riess, R. Dean (Ronald Dean), 1940–
II. Arnold, Jimmy (Jimmy Thomas), 1941–   III. Title.
   QA184.J63   1993
   512'.5—dc20   92-26263
                                                    CIP

5 6 7 8 9 10 DO 969594

To our wives
Rochelle, Jan, and Linda

Linear algebra is an important component of undergraduate mathematics, particularly for students majoring in the scientific, engineering, and social science disciplines. At the practical level, matrix theory and the related vector-space concepts provide a language and a powerful computational framework for posing and solving important problems. Beyond this, elementary linear algebra is a valuable introduction to mathematical abstraction and logical reasoning because the theoretical development is self-contained, consistent, and accessible to most students.

Therefore, this book stresses both practical computation and theoretical principles and centers on the principal topics of the first three chapters:

> matrix theory and systems of linear equations,
> elementary vector-space concepts, and
> the eigenvalue problem.

This core material can be used for a brief (10-week) course at the late-freshman/ sophomore level. There is enough additional material in Chapters 4–7 either for a more advanced or a more leisurely paced course.

Features

Our experience in teaching sophomore linear algebra has led us to include the following features in this text.

A gradual increase in the level of abstraction. In a typical linear algebra course, the students find the techniques of Gaussian elimination and matrix operations fairly easy. Then, the ensuing abstract material relating to vector spaces is suddenly much harder. We have done two things to lessen this abrupt midterm jump in difficulty: We have introduced linear independence early in Section 1.8. Also, in Chapter 2, we first study vector-space properties in the familiar geometrical setting of R^n.

An early introduction to eigenvalues. In Chapter 2, elementary vector-space ideas (subspace, basis, dimension, and so on) are introduced in the familiar setting of R^n. Therefore, it is possible to cover the eigenvalue problem very early and in much greater depth than is usually possible. A brief introduction to determinants is given in Section 3.2 to facilitate the early treatment of eigenvalues.

Clarity of exposition.   For many students, linear algebra is the most rigorous and abstract mathematical course they have taken since high school geometry. We have tried to write the text so that it is accessible, but also so that it reveals something of the power of mathematical abstraction. To this end, the topics have been organized so that they flow logically and naturally from the concrete and computational to the more abstract. Numerous examples, many presented in extreme detail, have been included in order to illustrate the concepts. The sections are divided into subsections with boldface headings. This device allows the reader to develop a mental outline of the material and to see how the pieces fit together.

Extensive exercise sets.   We have provided a large number of exercises, ranging from routine drill exercises to interesting applications and exercises of a theoretical nature. The more difficult theoretical exercises have fairly substantial hints. The computational exercises are designed so that the numbers are "nice" and do not obscure the point with a mass of cumbersome arithmetic details.

Applications of general appeal.   Some applications are drawn from difference equations and differential equations. Other applications involve interpolation of data and least-squares approximations. In particular, students from a wide variety of disciplines have encountered problems of drawing curves that fit experimental or empirical data. Hence, they can appreciate techniques from linear algebra that can be applied to such problems.

Computer awareness.   The increased accessibility of computers (especially personal computers) is beginning to affect linear algebra courses in much the same way as it has calculus courses. Accordingly, this text has somewhat of a numerical flavor, and (when it is appropriate) we comment on various aspects of solving linear algebra problems in a computer environment.

Trustworthy answer key.   Except for the theoretical exercises, solutions to the odd-numbered exercises are given at the back of the text. We have expended considerable effort to ensure that these solutions are correct.

Spiraling exercises.   Many sections contain a few exercises that hint at ideas that will be developed later. Such exercises help to get the student involved in thinking about extensions of the material that has just been covered. Thus the student can anticipate a bit of the shape of things to come. This feature helps to lend unity and cohesion to the material.

An early introduction to linear combinations. In Section 1.6, we observe that the matrix-vector product $A\mathbf{x}$ can be expressed as a linear combination of the columns of A, $A\mathbf{x} = x_1\mathbf{A}_1 + x_2\mathbf{A}_2 + \cdots + x_n\mathbf{A}_n$. This viewpoint leads to a simple and natural development for the theory associated with systems of linear equations. For instance, the equation $A\mathbf{x} = \mathbf{b}$ is consistent if and only if $\mathbf{b}$ is expressible as a linear combination of the columns of A. Similarly, a consistent equation $A\mathbf{x} = \mathbf{b}$ has a unique solution if and only if the columns of A are linearly independent. This approach gives some early motivation for the vector-space concepts (introduced in Chapter 2) such as subspace, basis, and dimension. The approach also simplifies ideas such as rank and nullity (which are then naturally given in terms of dimension of appropriate subspaces).

Simple and reliable computer programs. In Chapter 7, listings are given for some simple FORTRAN programs. These listings are in the form of sub-routines and include, for example, programs that solve $(n \times n)$ systems $A\mathbf{x} = \mathbf{b}$, where A and $\mathbf{b}$ may contain complex entries, programs that use Householder transformations to reduce A to Hessenberg form, and programs that use the Givens algorithm to find the eigenvalues and eigenvectors for a symmetric matrix.

An instructor's manual and a student solutions manual are available. The odd-numbered computational exercises have answers at the back of the book. The student solutions manual includes detailed solutions for these exercises. The instructor's manual contains solutions to all the exercises.

Changes in the Third Edition

Some of the changes made in this edition are listed below:

Supplementary exercises. We now include, at the end of each chapter, a set of supplementary exercises. These exercises, some of which are true-false questions, are designed to test the student's understanding of important concepts. They often require the student to use ideas from several different sections.

MATLAB and Derive. New easy-to-use software is beginning to have an impact on the teaching of linear algebra. Furthermore, it seems likely that this new technology will become even more important in the future. Therefore, in each each chapter, we include one or two examples (and some corresponding exercises) that illustrate the use of linear algebra software (such as MATLAB) and computer algebra systems (such as Derive). For instance, MATLAB is used in Example 8 of Section 1.3 and Example 6 of Section 3.6; Derive is used in Example 9 of Section 1.9 and Example 7 of Section 4.6.

More emphasis on applications. This edition has a number of new examples and exercises that illustrate applications of linear algebra. A new section on network flows has also been included in Chapter 1.

Historical notes.   We have added a number of historical notes. These will assist the student in gaining a historical and mathematical perspective of the ideas and concepts of linear algebra.

Organization

To provide greater flexibility, Chapters 3, 4, and 5 are essentially independent. These chapters can be taken in any order once Chapters 1 and 2 are covered. Chapter 6 is a mélange of topics related to the eigenvalue problem: quadratic forms, differential equations, QR factorizations, Householder transformations, generalized eigenvectors, and so on. The sections in Chapter 6 can be covered in various orders. A schematic diagram illustrating the chapter dependencies is given below:

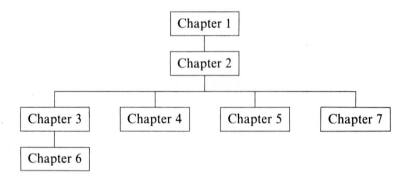

We especially note that Chapter 5 (Determinants) can be covered before Chapter 3 (Eigenvalues). However, Chapter 3 contains a brief introduction to determinants that should prove sufficient to users who do not wish to cover Chapter 5.

A very short but useful course at the beginning level can be built around the following sections:

Section 1.1–1.4, 1.6–1.8, 1.10–1.11
Sections 2.1–2.6
Sections 3.1–3.2, 3.4–3.5

A syllabus that integrates abstract vector spaces.   Chapter 2 introduces elementary vector-space ideas in the familiar setting of R^n. We designed Chapter 2 in this way so that it is possible to cover the eigenvalue problem much earlier and in greater depth than is generally possible. Many instructors, however, prefer an integrated approach to vector spaces, one that combines R^n and abstract vector spaces. The following syllabus, similar to ones used successfully at several universities, allows for a course that integrates abstract vector spaces into Chapter 2. This syllabus also allows for a detailed treatment of determinants:

>   Sections 1.1–1.4, 1.6–1.8, 1.10–1.11
>   Sections 2.1–2.3, 4.1–4.3, 2.4–2.5, 4.4–4.5
>   Sections 3.1–3.3, 5.4–5.5, 3.4–3.7

Augmenting the core sections.   As time and interest permit, the core of Sections 1.1–1.4, 1.6–1.8, 1.10–1.11, 2.1–2.6, 3.1–3.2, and 3.4–3.5 can be augmented by including various combinations of the following sections:

a) *Data fitting and approximation:* 1.9, 2.8–2.9, 6.5–6.6.

b) *Eigenvalue applications:* 3.8, 6.1, 6.2.

c) *More depth in vector space theory:* 2.7, Chapter 4.

d) *More depth in eigenvalue theory:* 3.6–3.7, 6.3–6.4, 6.7–6.8.

e) *Numerical methods:* Chapter 7.

f) *Determinant theory:* Chapter 5.

To allow the possibility of getting quickly to eigenvalues, Chapter 3 contains a brief introduction to determinants. If the time is available and if it is desirable, Chapter 5 (Determinants) can be taken after Chapter 2. In such a course, Section 3.1 can be covered quickly and Sections 3.2–3.3 can be skipped.

Finally, in the interest of developing the students' mathematical sophistication, we have provided proofs for almost every theorem. However, some of the more technical proofs [such as the demonstration that $\det(AB) = \det(A)\det(B)$] are deferred to the end of the sections. As always, constraints of time and class maturity will dictate which proofs should be omitted.

Acknowledgments

A great many valuable contributions to the Third Edition were made by those who reviewed the manuscript as it developed through various stages:

Francis Conlan, Santa Clara University
Michael Ecker, Pennsylvania State University, Wilkes-Barre Campus
George Habetler, Rensselaer Polytechnic Institute
Norman Lee, Ball State University
Joseph Stephen, Northern Illinois University
Theodore Vessey, Saint Olaf College

Blacksburg, Virginia   L.W.J.
                                                                                R.D.R.
                                                                                J.T.A.

CONTENTS

THIRD EDITION

Introduction to

LINEAR
ALGEBRA

Lee W. Johnson
R. Dean Riess
Jimmy T. Arnold

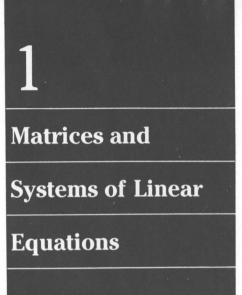

1

Matrices and

Systems of Linear

Equations

Introduction and Gaussian Elimination

In science, engineering, and the social sciences, one of the most important and frequently occurring mathematical problems is finding a simultaneous solution to a set of linear equations involving several unknowns.

The equation

$$x_1 + 2x_2 + x_3 = 1$$

is an example of a linear equation, and $x_1 = 2$, $x_2 = 1$, $x_3 = -3$ is one solution for the equation. In general a ***linear equation*** in n unknowns has the form

$$a_1 x_1 + a_2 x_2 + \cdots + a_n x_n = b, \tag{1}$$

where the coefficients $a_1, a_2, \ldots, a_n$ and the constant b are known, and x_1, $x_2, \ldots, x_n$ denote the unknowns. A ***solution*** to Eq. (1) is any sequence $s_1, s_2, \ldots, s_n$ of numbers such that the substitution $x_1 = s_1$, $x_2 = s_2, \ldots, x_n = s_n$ satisfies the equation.

Equation (1) is called linear because each term has degree one in the variables $x_1, x_2, \ldots, x_n$.

EXAMPLE 1 Determine which of the following equations are linear.

    i) $x_1 + 2x_1 x_2 + 3x_2 = 4$ ii) $x_1^{1/2} + 3x_2 = 4$

    iii) $2x_1^{-1} + \sin x_2 = 0$ iv) $3x_1 - x_2 = x_3 + 1$

SOLUTION   Only Eq. (iv) is linear. The terms $x_1 x_2$, $x_1^{1/2}$, x_1^{-1}, and $\sin x_2$ are all nonlinear. $\qquad\square$

Our concern is to obtain simultaneous solutions to a set of one or more linear equations. The usual approach is to simplify the given system of equations by eliminating variables. The reader is undoubtedly familiar with simple examples such as the following.

EXAMPLE 2 Solve

$$x_1 + x_2 = 3$$
$$x_1 - x_2 = 1.$$

SOLUTION   Eliminate x_2 from the second equation by adding the first equation to the second. This yields

$$x_1 + x_2 = 3$$
$$2x_1 \quad\;\; = 4.$$

Divide the second equation by 2 to obtain

$$x_1 + x_2 = 3$$
$$x_1 \quad\;\; = 2.$$

Substitute for x_1 in the first equation:

$$2 + x_2 = 3$$
$$x_1 \quad\;\; = 2.$$

Solving the first equation for x_2, we now get $x_1 = 2$, $x_2 = 1$ as the unique solution. ☐

Whereas simple systems such as the preceding one are quite easy to solve, it is obvious that a systematic procedure is needed to handle the more complicated problems that arise in practice. To answer this need, the first topic of this chapter is the general problem of solving a system of m linear equations in n unknowns; and the goal is to develop a systematic and computationally efficient method for solving such systems. Whenever appropriate, we will also comment on the practical aspects of solving systems of equations on a digital computer.

General Linear Systems

An **($m \times n$) system of linear equations** is the system of m linear equations in n unknowns:

$$a_{11}x_1 + a_{12}x_2 + \cdots + a_{1n}x_n = b_1$$
$$a_{21}x_1 + a_{22}x_2 + \cdots + a_{2n}x_n = b_2$$
$$\vdots \qquad\qquad\qquad \vdots \quad\;\; \vdots$$
$$a_{m1}x_1 + a_{m2}x_2 + \cdots + a_{mn}x_n = b_m.$$

(2)*

* For clarity of presentation, we assume throughout the chapter that the constants a_{ij} and b_i are real numbers, although all statements are equally valid for complex constants. When we consider eigenvalue problems, we will occasionally encounter linear systems having complex coefficients, but the solution technique is no different.

For example, the general form of a (3×3) system of linear equations is

$$a_{11}x_1 + a_{12}x_2 + a_{13}x_3 = b_1$$
$$a_{21}x_1 + a_{22}x_2 + a_{23}x_3 = b_2$$
$$a_{31}x_1 + a_{32}x_2 + a_{33}x_3 = b_3.$$

A solution to system (2) is a sequence $s_1, \ldots, s_n$ of numbers that is simultaneously a solution for each equation in the system. The double subscript notation used for the coefficients is necessary to provide an "address" for each coefficient. For example, a_{32} appears in the third equation as the coefficient of x_2.

EXAMPLE 3 a) Display the system of equations with coefficients $a_{11} = 2$, $a_{12} = -1$, $a_{13} = -3$, $a_{21} = -2$, $a_{22} = 2$, and $a_{23} = 5$, and with constants $b_1 = -1$ and $b_2 = 3$.

b) Verify that $x_1 = 1$, $x_2 = 0$, $x_3 = 1$ is a solution for the system.

SOLUTION

a) The system is

$$2x_1 - x_2 - 3x_3 = -1$$
$$-2x_1 + 2x_2 + 5x_3 = 3.$$

b) Substituting $x_1 = 1$, $x_2 = 0$, and $x_3 = 1$ yields

$$2(1) - (0) - 3(1) = -1$$
$$-2(1) + 2(0) + 5(1) = 3.$$

Elementary Operations

As we will see, two processes are involved in solving the general $(m \times n)$ system (2):

1. Reduction of the system (that is, the elimination of variables).
2. Description of the set of solutions.

The remainder of this section concerns the reduction procedure. The description of solutions will be considered in the following section.

The goal of the reduction process is to simplify the given system by eliminating unknowns. It is, of course, essential that the reduced system of equations have the same set of solutions as the original system.

> **DEFINITION 1** Two systems of linear equations in n unknowns are *equivalent* provided that they have the same set of solutions.

Thus the reduction procedure must yield an equivalent system of equations. The following theorem provides three operations, called *elementary operations*, that may be used in such a procedure.

THEOREM 1	If one of the following elementary operations is applied to a system of linear equations, then the resulting system is equivalent:

1. Interchange two equations.
2. Multiply an equation by a nonzero scalar.
3. Add a constant multiple of one equation to another.

(In part 2 of Theorem 1, the term *scalar* means a constant; that is, a number.) The proof of Theorem 1 is included in the Exercises of Section 1.1.

To facilitate the use of the elementary operations listed above, we adopt the following notation:

Notation	Elementary operation performed
$E_i \leftrightarrow E_j$	The ith and jth equations are interchanged.
kE_i	The ith equation is multiplied by the nonzero scalar k.
$E_i + kE_j$	k times the jth equation is added to the ith equation.

The following example illustrates the use of elementary operations and the notation.

EXAMPLE 4   Use elementary operations to simplify the system

$$2x_2 + x_3 = -1$$
$$3x_1 + 5x_2 - 5x_3 = 1$$
$$2x_1 + 4x_2 - 2x_3 = 2.$$

SOLUTION   Our goal is to have x_1 appear in the first equation, but to eliminate it from the remaining equations. This can be accomplished by the following steps:

$$2x_2 + x_3 = -1$$
$$3x_1 + 5x_2 - 5x_3 = 1$$
$$2x_1 + 4x_2 - 2x_3 = 2$$

$E_1 \leftrightarrow E_3$:

$$2x_1 + 4x_2 - 2x_3 = 2$$
$$3x_1 + 5x_2 - 5x_3 = 1$$
$$2x_2 + x_3 = -1$$

$1/2E_1$:

$$x_1 + 2x_2 - x_3 = 1$$
$$3x_1 + 5x_2 - 5x_3 = 1$$
$$2x_2 + x_3 = -1$$

$E_2 - 3E_1$:

$$x_1 + 2x_2 - x_3 = 1$$
$$-x_2 - 2x_3 = -2$$
$$2x_2 + x_3 = -1.$$

Theorem 1 assures us that the last system of equations has the same set of solutions as the given system.

Gaussian Elimination

We proceed now to describe a systematic procedure for reducing a system of equations. The simplest and best-known procedure is the variable-elimination technique known as Gaussian elimination. This technique is computationally practical and is the procedure most widely used in computer-software packages that are designed to solve systems of linear equations.

A precise description of Gaussian elimination is most easily given in the language of matrix theory, so we defer this description until Section 1.3. However, in order at least to illustrate Gaussian elimination and allow the reader some practice in solving linear systems, we will indicate the basic steps of the procedure. First we demonstrate Gaussian elimination with the following example.

EXAMPLE 5 Use elementary operations to solve the system

$$x_1 - 2x_2 + x_3 = 2$$
$$2x_1 + x_2 - x_3 = 1 \tag{3a}$$
$$-3x_1 + x_2 - 2x_3 = -5.$$

SOLUTION   Perform elementary operations on system (3a) as follows:

$E_2 - 2E_1, E_3 + 3E_1$:

$$x_1 - 2x_2 + x_3 = 2$$
$$5x_2 - 3x_3 = -3$$
$$-5x_2 + x_3 = 1$$

Gaussian Elimination

Gaussian elimination is named after Karl Friedrich Gauss (1777–1855) who is often called the "prince of mathematicians" because of his extensive work in many areas of mathematics. Gauss described this elimination technique in the context of a specific system of six equations in six unknowns whose solution was needed to determine the orbit of an asteroid. The method, however, was known centuries before Gauss.

$E_3 + E_2$:

$$
\begin{aligned}
x_1 - 2x_2 + x_3 &= 2 \\
5x_2 - 3x_3 &= -3 \\
-2x_3 &= -2.
\end{aligned} \tag{3b}
$$

Solving the last equation for x_3 yields $x_3 = 1$. Substituting for x_3 in the second equation and solving for x_2, we get $x_2 = 0$. Finally, the first equation yields $x_1 = 1$. Because system (3b) is equivalent to the original system (3a), we know that $x_1 = 1, x_2 = 0, x_3 = 1$ is the unique solution of (3a).

$\square$

As Example 5 illustrates, the goal of Gaussian elimination is to reduce the given system of equations to an equivalent system in the "triangular" form of (3b). We now describe the steps in the Gaussian elimination process.

Gaussian Elimination (on the ($m \times n$) system (2)):

Step 1. If necessary, interchange the first equation with another so that x_1 appears in the first equation.

Step 2. Eliminate x_1 from every equation but the first by adding the appropriate multiple of the first equation.

Step 3. Temporarily ignoring the first equation, view the remaining equations as a system of $m - 1$ equations in the unknowns $x_2, \ldots, x_n$. Repeat the procedure on this system.

Note that the application of steps 1 and 2 above to the ($m \times n$) system (2) yields an equivalent system of the form

$$
\begin{aligned}
a'_{11}x_1 + a'_{12}x_2 + \cdots + a'_{1n}x_n &= b'_1 \\
a'_{22}x_2 + \cdots + a'_{2n}x_n &= b'_2 \\
\vdots \qquad \vdots \qquad \vdots \\
a'_{m2}x_2 + \cdots + a'_{mn}x_n &= b'_m.
\end{aligned} \tag{4}
$$

We now view the last $m - 1$ equations of system (4) as a linear system in the unknowns $x_2, \ldots, x_n$. Applying steps 1 and 2 of Gaussian elimination to the system yields a linear system of the form

$$
\begin{aligned}
a''_{11}x_1 + a''_{12}x_2 + a''_{13}x_3 + \cdots + a''_{1n}x_n &= b''_1 \\
a''_{22}x_2 + a''_{23}x_3 + \cdots + a''_{2n}x_n &= b''_2 \\
a''_{33}x_3 + \cdots + a''_{3n}x_n &= b''_3 \\
\vdots \qquad \vdots \qquad \vdots \\
a''_{m3}x_3 + \cdots + a''_{mn}x_n &= b''_m.
\end{aligned}
$$

Although Gaussian elimination may not proceed quite so neatly as we are describing it, our objective is eventually to produce a linear system that is equivalent to system (2) and that has the following form (in the case $m \le n$):

$$c_{11}x_1 + c_{12}x_2 + \cdots + c_{1m}x_m + \cdots + c_{1n}x_n = d_1$$
$$c_{22}x_2 + \cdots + c_{2m}x_m + \cdots + c_{2n}x_n = d_2 \tag{5}$$
$$\vdots \qquad \vdots \qquad \vdots$$
$$c_{mm}x_m + \cdots + c_{mn}x_n = d_m.$$

The next example provides another illustration of Gaussian elimination.

EXAMPLE 6 Use Gaussian elimination to solve

$$2x_2 + x_3 = 3$$
$$-2x_1 + 2x_2 + x_3 = 4$$
$$x_1 - x_2 + x_3 = 1.$$

SOLUTION Clearly the first equation cannot be used to eliminate x_1 from the second and third equations. This problem is easily overcome by interchanging two equations. Thus the given system can be reduced in the following manner:

$E_1 \leftrightarrow E_3$:

$$x_1 - x_2 + x_3 = 1$$
$$-2x_1 + 2x_2 + x_3 = 4$$
$$2x_2 + x_3 = 3$$

$E_2 + 2E_1$:

$$x_1 - x_2 + x_3 = 1$$
$$3x_3 = 6$$
$$2x_2 + x_3 = 3$$

$E_2 \leftrightarrow E_3$:

$$x_1 - x_2 + x_3 = 1$$
$$2x_2 + x_3 = 3$$
$$3x_3 = 6.$$

Solving, we obtain $x_3 = 2$, $x_2 = 1/2$, $x_1 = -1/2$. ▢

The final example of this section uses Gaussian elimination to demonstrate that a system of equations has no solution.

EXAMPLE 7 Use Gaussian elimination to show that the (3×3) system of linear equations

$$x_1 - 2x_2 + x_3 = 3$$
$$-x_1 + 4x_2 = -1 \tag{6a}$$
$$2x_1 + 4x_3 = 12$$

has no solution.

SOLUTION   Perform elementary operations on system (6a) as follows:

$E_2 + E_1, E_3 - 2E_1$:

$$x_1 - 2x_2 + x_3 = 3$$
$$2x_2 + x_3 = 2$$
$$4x_2 + 2x_3 = 6$$

$E_3 - 2E_2$:

$$x_1 - 2x_2 + x_3 = 3$$
$$2x_2 + x_3 = 2 \qquad \textbf{(6b)}$$
$$0x_2 + 0x_3 = 2.$$

Clearly there are no values for x_1, x_2, and x_3 that satisfy the third equation of system (6b). Since system (6b) has no solution, it also follows that the equivalent system (6a) has no solution.   □

Exercises 1.1

Which of the equations in Exercises 1–6 are linear?

1. $x_1 + 2x_3 = 3$

2. $x_1 x_2 + x_2 = 1$

3. $x_1 - x_2 = \sin^2 x_1 + \cos^2 x_1$

4. $x_1 - x_2 = \sin^2 x_1 + \cos^2 x_2$

5. $|x_1| - |x_2| = 0$

6. $\pi x_1 + \sqrt{7} x_2 = \sqrt{3}$

In Exercises 7–10, coefficients are given for a system of the form (2). Display the system and verify that the given values constitute a solution.

7. $a_{11} = 1$, $a_{12} = 3$, $a_{21} = 4$, $a_{22} = -1$,
 $b_1 = 7$, $b_2 = 2$; $x_1 = 1$, $x_2 = 2$

8. $a_{11} = 6$, $a_{12} = -1$, $a_{13} = 1$, $a_{21} = 1$,
 $a_{22} = 2$, $a_{23} = 4$, $b_1 = 14$, $b_2 = 4$; $x_1 = 2$,
 $x_2 = -1$, $x_3 = 1$

9. $a_{11} = 1$, $a_{12} = 1$, $a_{21} = 3$, $a_{22} = 4$,
 $a_{31} = -1$, $a_{32} = 2$, $b_1 = 0$, $b_2 = -1$,
 $b_3 = -3$; $x_1 = 1$, $x_2 = -1$

10. $a_{11} = 0$, $a_{12} = 3$, $a_{21} = 4$, $a_{22} = 0$, $b_1 = 9$,
 $b_2 = 8$; $x_1 = 2$, $x_2 = 3$

In Exercises 11–18, use the elementary operations listed in Theorem 1 to eliminate x_1 from all but one equation; that is, obtain an equivalent system of the form (4).

11. $x_1 + x_2 = 3$
   $-2x_1 + 7x_2 = 9$

12. $2x_1 + 3x_2 = 6$
   $4x_1 - x_2 = 7$

13. $x_1 + 2x_2 - x_3 = 1$
   $x_1 + x_2 + 2x_3 = 2$
   $-2x_1 + x_2 \quad\;\; = 4$

14. $x_2 + x_3 = 4$
   $x_1 - x_2 + 2x_3 = 1$
   $2x_1 + x_2 - x_3 = 6$

15. $x_1 + x_2 = 9$
   $x_1 - x_2 = 7$
   $3x_1 + x_2 = 6$

16. $x_1 + x_2 + x_3 - x_4 = 1$
   $-x_1 + x_2 - x_3 + x_4 = 3$
   $-2x_1 + x_2 + x_3 - x_4 = 2$

17. $x_2 + x_3 - x_4 = 3$
   $x_1 + 2x_2 - x_3 + x_4 = 1$
   $-x_1 + x_2 + 7x_3 - x_4 = 0$

18. $x_1 + x_2 = 0$
   $x_1 - x_2 = 0$
   $3x_1 + x_2 = 0$

In Exercises 19–26, solve the systems using Gaussian elimination.

19. $x_1 + 2x_2 = -5$
   $2x_1 - x_2 = 5$

20. $x_1 + 2x_2 = 4$
   $2x_1 + 6x_2 = 12$

21. $2x_1 + x_2 = 0$
   $3x_1 - x_2 = 0$

22. $x_2 = 6$
   $x_1 + x_2 = 4$

23. $x_1 + 2x_2 + 4x_3 = 6$
$x_1 \quad\quad + 2x_3 = 2$
$x_1 + 3x_2 + 7x_3 = 10$

24. $x_1 - x_2 + x_3 = 1$
$x_1 + x_2 - x_3 = 1$
$2x_1 \quad\quad + 3x_3 = 8$

25. $x_2 + x_3 = 1$
$x_1 + x_2 - x_3 = 0$
$-2x_1 - x_2 + x_3 = -1$

26. $2x_2 - x_3 = 2$
$2x_1 + 4x_2 - 6x_3 = 4$
$-2x_1 + x_2 + x_3 = 1$

The systems in Exercises 27–30 do not have solutions. Verify this fact using Gaussian elimination.

27. $x_1 + x_2 = 3$
$-2x_1 - 2x_2 = 4$

28. $2x_1 - 3x_2 = 4$
$6x_1 - 9x_2 = 4$

29. $x_1 + 2x_2 + x_3 = 3$
$x_1 - x_2 + x_3 = 1$
$-2x_1 - 4x_2 - 2x_3 = 4$

30. $x_1 - 2x_2 - x_3 = 1$
$x_1 - x_2 + x_3 = 0$
$2x_1 \quad\quad + 6x_3 = 1$

In Exercises 31–36, find all values a for which the system has no solution.

31. $x_1 + x_2 = 5$
$2x_1 + ax_2 = 4$

32. $x_1 + 2x_2 = -3$
$ax_1 - 2x_2 = 5$

33. $x_1 + 3x_2 = 4$
$2x_1 + 6x_2 = a$

34. $2x_1 + 4x_2 = a$
$3x_1 + 6x_2 = 5$

35. $3x_1 + ax_2 = 3$
$ax_1 + 3x_2 = 5$

36. $x_1 + ax_2 = 6$
$ax_1 + 2ax_2 = 4$

In Exercises 37 and 38, the terms are from a (2×2) system of the form (2). The system has a solution and one of x_1 and x_2 is given. (a) Find the unknown values. (b) Is the system you use to find the unknown values linear?

37. $a_{11} = 2$, $a_{12} = -a_{22}$, $a_{21} = 1$, $x_2 = -2$, $b_1 = 0$, $b_2 = 3$. Find a_{12}, a_{22}, and x_1.

38. $a_{11} = 2$, $a_{12} = a_{21}$, $a_{22} = -1$, $x_1 = 1$, $b_1 = 0$, $b_2 = 3$. Find a_{12}, a_{21}, and x_2.

39. Consider the equation $2x_1 - 3x_2 + x_3 - x_4 = 3$.
a) In the six different possible combinations, set any two of the variables equal to 1 and graph the equation in terms of the other two.
b) What type of graph do you always get when you set two of the variables equal to two fixed constants?
c) What is one possible reason the equation in formula (1) is called "linear"?

40. A certain three-digit number N equals fifteen times the sum of its digits. If its digits are reversed, the resulting number exceeds N by 396. The one's digit is one larger than the sum of the other two. Give a linear system of three equations whose three unknowns are the digits of N. Solve the system and find N.

41. Find the equation of the parabola, $y = ax^2 + bx + c$, that passes through the points $(-1, 6)$, $(1, 4)$, and $(2, 9)$. (*Hint:* For each point, give a linear equation in a, b, and c.)

42. Three people play a game in which there are always two winners and one loser. They have the understanding that the loser gives each winner an amount equal to what the winner already has. After three games, each has lost just once and each has $24. With how much money did each begin?

43. Consider the (2×2) system

$$a_{11}x_1 + a_{12}x_2 = b_1$$
$$a_{21}x_1 + a_{22}x_2 = b_2.$$

Show that if $a_{11}a_{22} - a_{12}a_{21} \neq 0$, then this system is equivalent to a system of the form

$$c_{11}x_1 + c_{12}x_2 = d_1$$
$$c_{22}x_2 = d_2,$$

where $c_{11} \neq 0$ and $c_{22} \neq 0$. Note that the second system always has a solution. (*Hint:* First suppose that $a_{11} \neq 0$, and then consider the special case that $a_{11} = 0$.)

44. In Exercise 43, suppose that $a_{11}a_{22} - a_{12}a_{21} = 0$ and $a_{11} \neq 0$. Give conditions on b_1 and b_2 that ensure the system has a solution. Can solutions ever be unique?

45. In the (2×2) linear systems (A) and (B) below, c is a nonzero scalar. Prove that any solution, $x_1 = s_1$, $x_2 = s_2$, for (A) is also a solution for (B). Conversely, show that any solution, $x_1 = t_1$, $x_2 = t_2$, for (B) is also a solution for (A). Where is the assumption that c is nonzero required?

(A)
$$a_{11}x_1 + a_{12}x_2 = b_1$$
$$a_{21}x_1 + a_{22}x_2 = b_2$$

(B)
$$a_{11}x_1 + a_{12}x_2 = b_1$$
$$ca_{21}x_1 + ca_{22}x_2 = cb_2$$

46. In the (2×2) linear systems below, the system (B) is obtained from (A) by performing the elementary operation $E_2 + cE_1$. Prove that any solution, $x_1 = s_1, x_2 = s_2$, for (A) is a solution for (B). Similarly, prove that any solution, $x_1 = t_1, x_2 = t_2$, for (B) is a solution for (A).

(A)
$$a_{11}x_1 + a_{12}x_2 = b_1$$
$$a_{21}x_1 + a_{22}x_2 = b_2$$

(B)
$$a_{11}x_1 + a_{12}x_2 = b_1$$
$$(a_{21} + ca_{11})x_1 + (a_{22} + ca_{12})x_2 = b_2 + cb_1$$

47. Prove that any of the elementary operations in Theorem 1 applied to system (2) produces an equivalent system. [*Hint:* To simplify this proof, represent the ith equation in system (2) as $f_i(x_1, x_2, \ldots, x_n) = b_i$; so
$$f_i(x_1, x_2, \ldots, x_n) = a_{i1}x_1 + a_{i2}x_2 + \cdots + a_{in}x_n$$

for $i = 1, 2, \ldots, m$. With this notation, system (2) has the form of (A) below. Next, for example, if a multiple of c times the jth equation is added to the kth equation, a new system of the form (B) is produced:

(A)
$$f_1(x_1, x_2, \ldots, x_n) = b_1$$
$$\vdots$$
$$f_j(x_1, x_2, \ldots, x_n) = b_j$$
$$\vdots$$
$$f_k(x_1, x_2, \ldots, x_n) = b_k$$
$$\vdots$$
$$f_m(x_1, x_2, \ldots, x_n) = b_m$$

(B)
$$f_1(x_1, x_2, \ldots, x_n) = b_1$$
$$\vdots$$
$$f_j(x_1, x_2, \ldots, x_n) = b_j$$
$$\vdots$$
$$g(x_1, x_2, \ldots, x_n) = r$$
$$\vdots$$
$$f_m(x_1, x_2, \ldots, x_n) = b_m$$

where $g(x_1, x_2, \ldots, x_n) = f_k(x_1, x_2, \ldots, x_n) + cf_j(x_1, x_2, \ldots, x_n)$, and $r = b_k + cb_j$. To show that the operation gives an equivalent system, show that any solution for (A) is a solution for (B), and vice versa.]

48. Solve the system of two nonlinear equations in two unknowns
$$x_1^2 - 2x_1 + x_2^2 = 3$$
$$x_1^2 \qquad - x_2^2 = 1.$$

1.2   Solution Sets for Linear Systems

In the previous section we developed a procedure for reducing a system of equations to a simpler but equivalent system. In this section we consider the process of describing the solution set for the system. We begin by considering the possible outcomes when solving the (2×2) system:

$$a_{11}x_1 + a_{12}x_2 = b_1$$
$$a_{21}x_1 + a_{22}x_2 = b_2.$$

Geometrically, solutions to each equation are represented by lines. A simultaneous solution corresponds to a point of intersection. Thus there are three possibilities:

1. The two lines are coincident (the same line), so there are infinitely many solutions.

2. The two lines are parallel, so there are no solutions.

3. The two lines intersect at a single point, so there is a unique solution.

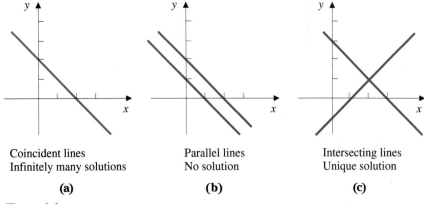

Coincident lines   Parallel lines   Intersecting lines
Infinitely many solutions No solution   Unique solution

(a)   (b)   (c)

Figure 1.1

The (2×2) systems in the following example illustrate each of these three possibilities.

EXAMPLE 1 Give a geometric representation for each of the following systems of equations.

a) $x_1 + x_2 = 2$ b) $x_1 + x_2 = 2$ c) $x_1 + x_2 = 3$
   $2x_1 + 2x_2 = 4$ $x_1 + x_2 = 1$ $x_1 - x_2 = 1$

SOLUTION   The representations are displayed in Fig. 1.1.   □

In view of the following remark, which we will prove in Section 1.4, the examples given above are typical.

REMARK   An $(m \times n)$ system of linear equations has either infinitely many solutions, no solution, or a unique solution.

A system of equations is called **consistent** if it has at least one solution and **inconsistent** if it has no solution. A procedure for solving a system needs to detect the inconsistent systems and, in the consistent systems, to provide a satisfactory algebraic description of the solution set.

The reader is already familiar with examples of consistent systems that have a unique solution. But how can one describe infinitely many solutions? Consider, for example, the (2×3) system

$$\begin{aligned} x_1 \quad\;\; - x_3 &= \;\;\,1 \\ x_2 + x_3 &= -1. \end{aligned} \qquad (1)$$

Note that attempts to simplify the system by eliminating a variable actually result in no further simplification. To illustrate, suppose x_3 is eliminated from the first equation with the operation $E_1 + E_2$. This yields

$$\begin{aligned} x_1 + x_2 \quad\;\;\; &= \quad 0 \\ x_2 + x_3 &= -1. \end{aligned}$$

Now eliminate x_2 from the second equation with the operation $E_2 - E_1$ to obtain

$$\begin{aligned} x_1 + x_2 \quad\ &= \quad 0 \\ -x_1 \quad\ + x_3 &= -1. \end{aligned}$$

Obviously no progress has been made toward describing the solution. Actually system (1), as given, is already satisfactorily reduced according to the Gaussian elimination procedure described in the preceding section. We are unable to eliminate another variable because the system has infinitely many solutions, and at least one free parameter (unconstrained or independent variable) is required in order to describe the solutions. To illustrate, system (1) can be solved for x_1 and x_2 to get

$$\begin{aligned} x_1 &= \quad 1 + x_3 \\ x_2 &= -1 - x_3 \end{aligned} \qquad (2)$$

In this case x_3 is the free parameter, and specific solutions to the system are obtained by assigning values to x_3. For example, setting $x_3 = 1$ yields the solution $x_1 = 2$, $x_2 = -2$, $x_3 = 1$, and setting $x_3 = -1$ gives $x_1 = 0$, $x_2 = 0$, $x_3 = -1$.

Note that the graph of each equation in system (1) is a plane, and each solution for the system corresponds to a point in the intersection of the two planes. Setting $x_3 = t$, where t is now the free parameter, we can rewrite the solution given in (2) as

$$\begin{aligned} x_1 &= \quad 1 + t \\ x_2 &= -1 - t \\ x_3 &= \qquad t, \end{aligned}$$

where these are the parametric equations for the line formed by the two intersecting planes.

Backsolving

More generally, we wish to describe the set of solutions for an $(m \times n)$ system of the following form:

$$\begin{aligned} c_{11}x_1 + c_{12}x_2 + \cdots + c_{1m}x_m + \cdots + c_{1n}x_n &= d_1 \\ c_{22}x_2 + \cdots + c_{2m}x_m + \cdots + c_{2n}x_n &= d_2 \\ \vdots \qquad\quad \vdots \qquad\ \vdots \\ c_{mm}x_m + \cdots + c_{mn}x_n &= d_n. \end{aligned} \qquad (3)$$

The system is already reduced according to the Gaussian elimination procedure. If the coefficients $c_{11}, c_{22}, \ldots, c_{mm}$ are all nonzero, then we can solve system (3) relatively easily. In the special case that $m = n$, the solution will be unique. The process for obtaining the solution to (3) is called **backsolving** and proceeds in the obvious fashion. That is, if we allow $x_{m+1}, \ldots, x_n$ to be free parameters, then x_m is determined from the last equation of (3) by

$$x_m = (d_n - c_{m,m+1}x_{m+1} - \cdots - c_{mn}x_n)/c_{mm}.$$

This formula for x_m is then inserted in the $(m - 1)$st equation of system (3) to determine x_{m-1}. In general the ith equation is used to express x_i in terms of the free parameters $x_{m+1}, \ldots, x_n$.

EXAMPLE 2 Describe the solution set for the (3×5) system

$$\begin{aligned}
2x_1 + x_2 - x_3 - x_4 + 2x_5 &= 3 \\
x_2 - 2x_3 + x_4 + x_5 &= -1 \\
x_3 + 2x_4 - x_5 &= 2.
\end{aligned}$$

SOLUTION Solving the third equation for x_3, we get $x_3 = 2 - 2x_4 + x_5$. Substituting for x_3 in the second equation and solving for x_2 yields $x_2 = 3 - 5x_4 + x_5$. Finally, we substitute for x_2 and x_3 in the first equation and solve for x_1 to obtain $x_1 = 1 + 2x_4 - x_5$. Thus the solution is

$$\begin{aligned}
x_1 &= 1 + 2x_4 - x_5 \\
x_2 &= 3 - 5x_4 + x_5 \\
x_3 &= 2 - 2x_4 + x_5,
\end{aligned}$$

where x_4 and x_5 are independent variables. For example, $x_4 = 0$, $x_5 = 1$, $x_1 = 0$, $x_2 = 4$, $x_3 = 3$ is a solution, as is $x_4 = -1$, $x_5 = 0$, $x_1 = -1$, $x_2 = 8$, $x_3 = 4$. ☐

When backsolving a reduced system of equations such as the system given in (3), there is usually some flexibility in the choice of free variables. (This will be illustrated in later examples.) Our general pattern will be to choose the first variable in each equation as a dependent variable. The remaining variables are then designated as independent variables. In Example 2, for instance, x_1, x_2, and x_3 were chosen as dependent variables, and these were expressed in terms of the independent variables x_4 and x_5. Thus, if the reduced system has m nontrivial equations in n unknowns, then there are m dependent variables and $n - m$ independent variables.

Not every linear system reduces to an equivalent system of the form given in (3) with each $c_{ii} \neq 0$. The general form of a reduced system is best described in matrix terminology, however, so it will be deferred until the next section. We conclude this section with some simple examples that illustrate Gaussian elimination and the backsolving procedure.

EXAMPLE 3 Solve

$$\begin{aligned}
x_1 - 2x_2 - x_3 &= -2 \\
2x_1 + x_2 + 3x_3 &= 1 \\
-3x_1 + x_2 - 2x_3 &= 1.
\end{aligned} \qquad \text{(4a)}$$

SOLUTION   Gaussian elimination proceeds as follows:

$E_2 - 2E_1, E_3 + 3E_1$:

$$
\begin{aligned}
x_1 - 2x_2 - x_3 &= -2 \\
5x_2 + 5x_3 &= 5 \\
-5x_2 - 5x_3 &= -5
\end{aligned}
$$

$E_3 + E_2$:

$$
\begin{aligned}
x_1 - 2x_2 - x_3 &= -2 \\
5x_2 + 5x_3 &= 5 \\
0x_3 &= 0.
\end{aligned}
\qquad \textbf{(4b)}
$$

Since the last equation of system (4b) is satisfied by any value of x_3, we may as well delete it. Thus the given system is equivalent to the (2×3) system

$$
\begin{aligned}
x_1 - 2x_2 - x_3 &= -2 \\
5x_2 + 5x_3 &= 5.
\end{aligned}
\qquad \textbf{(4c)}
$$

Backsolving system (4c), we obtain the solution

$$
\begin{aligned}
x_1 &= -x_3 \\
x_2 &= 1 - x_3,
\end{aligned}
$$

where x_3 is a free parameter. For example, $x_3 = 0$, $x_1 = 0$, $x_2 = 1$ is one solution, and $x_3 = 4$, $x_1 = -4$, $x_2 = -3$ is another.   ☐

In the general $(m \times n)$ linear system given in system (2) of Section 1.1, no restrictions are placed on the relative sizes of m and n. Hence there may be more equations than unknowns $(m > n)$, more unknowns than equations $(m < n)$, or an equal number of equations and unknowns $(m = n)$. The following is an example of a (3×4) "rectangular" system.

EXAMPLE 4 Solve

$$
\begin{aligned}
x_2 + x_3 - x_4 &= 0 \\
x_1 - x_2 + 3x_3 - x_4 &= -2 \\
x_1 + x_2 + x_3 + x_4 &= 2.
\end{aligned}
\qquad \textbf{(5a)}
$$

SOLUTION   System (5a) can be reduced by Gaussian elimination as follows:

$E_1 \leftrightarrow E_2$:

$$
\begin{aligned}
x_1 - x_2 + 3x_3 - x_4 &= -2 \\
x_2 + x_3 - x_4 &= 0 \\
x_1 + x_2 + x_3 + x_4 &= 2
\end{aligned}
$$

$E_3 - E_1$:

$$\begin{aligned}
x_1 - x_2 + 3x_3 - x_4 &= -2 \\
x_2 + x_3 - x_4 &= 0 \\
2x_2 - 2x_3 + 2x_4 &= 4
\end{aligned}$$

$E_3 - 2E_2$:

$$\begin{aligned}
x_1 - x_2 + 3x_3 - x_4 &= -2 \\
x_2 + x_3 - x_4 &= 0 \\
-4x_3 + 4x_4 &= 4.
\end{aligned} \qquad \textbf{(5b)}$$

Backsolving results in the solution

$$\begin{aligned}
x_1 &= 2 - 2x_4, \\
x_2 &= 1 \\
x_3 &= -1 + x_4,
\end{aligned} \qquad \textbf{(5c)}$$

where x_4 is an independent variable. □

Note that system (5b) could just as easily have been backsolved by expressing x_1, x_2, and x_4 in terms of x_3. In this case the solution would be described by

$$\begin{aligned}
x_1 &= -2x_3 \\
x_2 &= 1 \\
x_4 &= 1 + x_3.
\end{aligned} \qquad \textbf{(5d)}$$

Also, the variables in system (5a) could have been rearranged so that Gaussian elimination resulted in solving for x_3 and x_4 in terms of x_1. For this particular example, we will always obtain $x_2 = 1$, so it would not have been possible to solve for x_1, x_3, and x_4 in terms of x_2. What is invariant about these solutions is that $x_2 = 1$, and that any two of the remaining variables can be expressed in terms of the third. Hence there is one independent (unconstrained) variable and there are three dependent (constrained) variables in any case, and we can choose whichever form of the solution best suits our purposes. The next example provides an illustration.

EXAMPLE 5 Exhibit three solutions to system (5a), given in Example 4, such that each of the solutions satisfies one of the following constraints:

a) $x_4 = 3$,   b) $x_3 = 0$,   c) $x_1 = 2$.

SOLUTION The solution given in (5c) designates x_4 as the free variable. Setting $x_4 = 3$ in (5c) yields $x_1 = -4$, $x_2 = 1$, $x_3 = 2$, $x_4 = 3$ as the solution.

The solution given in (5d) designates x_3 as the independent variable. With $x_3 = 0$, it is easy to see from (5d) that $x_1 = 0$, $x_2 = 1$, $x_3 = 0$, $x_4 = 1$ is a solution.

Finally, if system (5b) is backsolved by expressing x_2, x_3, and x_4 in terms of x_1, the solution to system (5a) is described by

$$x_2 = 1$$
$$x_3 = -(1/2)x_1 \tag{5e}$$
$$x_4 = 1 - (1/2)x_1.$$

Setting $x_1 = 2$ gives the solution $x_1 = 2, x_2 = 1, x_3 = -1, x_4 = 0$.

We have now seen examples of linear systems that have a unique solution and examples in which there are infinitely many solutions. The only remaining possibility is that a system may be inconsistent. The next two examples illustrate this case.

EXAMPLE 6 Solve

$$x_1 - 2x_2 - x_3 = -2$$
$$2x_1 + x_2 + 3x_3 = 1 \tag{6a}$$
$$-3x_1 + x_2 - 2x_3 = 1.1.$$

SOLUTION   We proceed with Gaussian elimination as follows:

$E_2 - 2E_1, E_3 + 3E_1$:

$$x_1 - 2x_2 - x_3 = -2$$
$$5x_2 + 5x_3 = 5$$
$$-5x_2 - 5x_3 = -4.9$$

$E_3 + E_2$:

$$x_1 - 2x_2 - x_3 = -2$$
$$5x_2 + 5x_3 = 5 \tag{6b}$$
$$0x_3 = 0.1.$$

Because there is no number x_3 such that $0x_3 = 0.1$, system (6b) is inconsistent. Because (6a) and (6b) are equivalent systems, (6a) has no solution. In a later section we will develop a procedure for finding a "best" approximate solution to (6a), that is, a procedure for finding a set of values x_1, x_2, x_3 that comes closest to solving (or "best fits") the three equations in (6a).

The next example, a (3×2) system, illustrates an **overdetermined system,** that is, a system with more constraints (or equations) than variables. We normally do not expect an overdetermined system to have a solution, but such a system may have a unique solution or even infinitely many solutions (see Exercises 28–30).

EXAMPLE 7 Solve

$$x_1 - x_2 = 3$$
$$2x_1 + x_2 = 6$$
$$x_1 + x_2 = 1.$$

SOLUTION Eliminate x_1 as follows:

$E_2 - 2E_1, E_3 - E_1$:

$$x_1 - x_2 = 3$$
$$3x_2 = 0$$
$$2x_2 = -2.$$

Next, the operation $E_3 - (2/3)E_2$ eliminates x_2 from Eq. (3):

$$x_1 - x_2 = 3$$
$$3x_2 = 0$$
$$0x_2 = -2.$$

Since the third equation cannot be satisfied for any value of x_2, the system is inconsistent. □

Gauss–Jordan Elimination

There is a variation of Gaussian elimination, called *Gauss–Jordan elimination,* that avoids backsolving by using elementary row operations to further reduce the given system. For reasons of accuracy and efficiency, most modern software packages employ a form of Gaussian elimination plus backsolving, rather than the Gauss–Jordan procedure. In the following example, we will illustrate Gauss–Jordan elimination, and after the example, we will comment on the disadvantages of the procedure.

EXAMPLE 8 Solve the system

$$x_1 + x_2 + x_3 + 4x_4 = 4$$
$$2x_1 + 3x_2 + 4x_3 + 9x_4 = 16$$
$$-2x_1 + 3x_3 - 7x_4 = 11.$$

SOLUTION Gauss–Jordan elimination proceeds from left to right as follows:

$E_2 - 2E_1, E_3 + 2E_1$:

$$x_1 + x_2 + x_3 + 4x_4 = 4$$
$$x_2 + 2x_3 + x_4 = 8$$
$$2x_2 + 5x_3 + x_4 = 19$$

$E_1 - E_2, E_3 - 2E_2$:

$$
\begin{aligned}
x_1 \qquad - \ x_3 + 3x_4 &= -4 \\
x_2 + 2x_3 + \ x_4 &= \ \ 8 \\
x_3 - \ x_4 &= \ \ 3
\end{aligned}
$$

$E_1 + E_3, E_2 - 2E_3$:

$$
\begin{aligned}
x_1 \qquad\qquad + 2x_4 &= -1 \\
x_2 \qquad + 3x_4 &= \ \ 2 \\
x_3 - \ x_4 &= \ \ 3.
\end{aligned}
$$

Without backsolving, it is now easy to express x_1, x_2, and x_3 in terms of the unconstrained variable x_4 as

$$
\begin{aligned}
x_1 &= -1 - 2x_4 \\
x_2 &= \ \ \ 2 - 3x_4 \\
x_3 &= \ \ \ 3 + \ x_4.
\end{aligned}
$$

$\square$

The Gauss–Jordan reduction procedure, as illustrated in Example 8, provides a satisfactory approach for solving small systems by hand. For solving larger systems, however, this method substantially increases the number of arithmetic operations performed and thus reduces the efficiency of the procedure. For example, to solve a (10×10) system of equations by Gauss–Jordan reduction (reducing from left to right) requires 165 more multiplications than does Gaussian elimination and backsolving. (*Note:* Operations in which the answer is known in advance to be zero are not counted.)

In either reduction procedure, the most important elementary operation is the third one: Replace E_j by

$$
E_j + mE_i. \tag{7}
$$

When adding a large number to a small number, the computer may be forced essentially to neglect the smaller. Hence, in the worst case with m too large in (7), this operation could replace the jth equation by a multiple of the ith equation and so discard all the information in the jth equation. For slightly smaller values of m, the information in the jth equation can still become severely contaminated. In the Gaussian elimination procedure, pivoting strategies (see Chapter 7) can be used in an attempt to moderate the effects of rounding error. These same strategies cannot be used above the diagonal in Gauss–Jordan elimination, so this procedure may suffer more from the rounding error problem than does Gaussian elimination and backsolving.

In Example 9, we illustrate an alternative version of the Gauss–Jordan elimination procedure. In this procedure Gaussian elimination is used to reduce the system to triangular form. The system is then further reduced by proceeding from right to left. The number of arithmetic operations required is approximately the same as the number required when using Gaussian elimination and backsolving.

EXAMPLE 9 Solve the system given in Example 8 by using Gaussian elimination and then further eliminating variables from right to left.

SOLUTION

$$
\begin{aligned}
x_1 + x_2 + x_3 + 4x_4 &= 4 \\
2x_1 + 3x_2 + 4x_3 + 9x_4 &= 16 \\
-2x_1 \qquad\quad + 3x_3 - 7x_4 &= 11
\end{aligned}
$$

$E_2 - 2E_1, E_3 + 2E_1$:

$$
\begin{aligned}
x_1 + x_2 + x_3 + 4x_4 &= 4 \\
x_2 + 2x_3 + x_4 &= 8 \\
2x_2 + 5x_3 + x_4 &= 19
\end{aligned}
$$

$E_3 - 2E_2$:

$$
\begin{aligned}
x_1 + x_2 + x_3 + 4x_4 &= 4 \\
x_2 + 2x_3 + x_4 &= 8 \\
x_3 - x_4 &= 3
\end{aligned}
$$

$E_1 - E_3, E_2 - 2E_3$:

$$
\begin{aligned}
x_1 + x_2 \qquad\quad + 5x_4 &= 1 \\
x_2 \qquad\quad + 3x_4 &= 2 \\
x_3 - x_4 &= 3
\end{aligned}
$$

$E_1 - E_2$:

$$
\begin{aligned}
x_1 \qquad\qquad\quad + 2x_4 &= -1 \\
x_2 \qquad\quad + 3x_4 &= 2 \\
x_3 - x_4 &= 3.
\end{aligned}
$$

As in Example 8, the solution is now easily obtained without backsolving.

☐

Note that in Examples 8 and 9, the same elementary operations were employed; the only difference is the order in which the operations are used. In the small system given in these examples, the Gauss–Jordan procedure used in Example 8 required one more multiplication than the alternative version demonstrated in Example 9. For large systems, however, the difference in the number of arithmetic operations required becomes a significant factor (approximately 50% more operations).

Exercises 1.2

In Exercises 1–4, determine whether the system has a unique solution, no solution, or infinitely many solutions by sketching a graph for each equation.

1. $2x + y = 5$
$x - y = 1$

2. $2x - y = -1$
$2x - y = 2$

3. $3x + 2y = 6$
$-6x - 4y = -12$

4. $2x + y = 5$
$x - y = 1$
$x + 3y = 9$

In Exercises 5–10, backsolve the reduced system and describe the solution set.

5. $x_1 + x_2 + x_3 = 4$
$x_2 - x_3 = 7$

6. $x_1 - x_2 + 3x_3 = 1$
$2x_2 + 6x_3 = 2$

7. $2x_1 + x_2 - x_3 + x_4 = 7$
$x_2 - x_3 - 2x_4 = 2$

8. $x_1 - x_2 + 3x_3 + x_4 = 3$
$x_2 \qquad + x_4 = 1$

9. $x_1 + 2x_2 - 3x_3 + x_4 = 0$
$x_2 - x_3 + x_4 = 5$
$x_3 - x_4 = 1$

10. $3x_1 - x_2 + x_3 + x_4 + x_5 = 6$
$x_2 - x_3 \qquad = 0$
$x_3 + x_4 - x_5 = 1$

11. In each of parts (a)–(c) below, exhibit a solution for the system in Exercise 5 that satisfies the given constraint.
  a) $x_3 = 0$ b) $x_3 = -1$ c) $x_3 = 1$

12. In each of parts (a)–(d) below, exhibit a solution for the system in Exercise 10 that satisfies the given constraints.
  a) $x_4 = 0$, $x_5 = 0$ b) $x_4 = 1$, $x_5 = 0$
  c) $x_4 = 0$, $x_5 = -1$ d) $x_4 = -1$, $x_5 = 1$

13. Describe the solution set for the systems in Exercises 5 and 6 in terms of x_2. For each system find a solution in which $x_2 = 3$.

14. Describe the solution sets for the systems in Exercises 5 and 6 in terms of x_1. For each system find a solution in which $x_1 = 2$. (*Hint:* To describe the solution sets in terms of x_1, use an elementary operation to eliminate x_2 from the first equation.)

15. Describe the solution sets for the systems in Exercises 7 and 8 in terms of x_2 and x_3. For each system find a solution that satisfies the constraints $x_2 = 4$, $x_3 = 0$; $x_2 = 0$, $x_3 = -8$.

16. Describe the solution sets for the systems in Exercises 7 and 8 in terms of x_1 and x_2. For each system find a solution that satisfies the constraints $x_1 = 3$, $x_2 = 6$; $x_1 = 0$, $x_2 = 3$.

17. Describe the solution set for the system in Exercise 10 in terms of x_3 and x_4. Find a solution in which $x_3 = 3$ and $x_4 = -3$.

In Exercises 18–31, solve the system or state that the system is inconsistent.

18. $2x_1 - 3x_2 = 5$
$-4x_1 + 6x_2 = -10$

19. $x_1 - 2x_2 = 3$
$2x_1 - 4x_2 = 1$

20. $x_1 - x_2 + x_3 = 3$
$2x_1 + x_2 - 4x_3 = -3$

21. $x_1 + x_2 = 2$
$3x_1 + 3x_2 = 6$

22. $x_1 - x_2 + x_3 = 4$
$2x_1 - 2x_2 + 3x_3 = 2$

23. $x_1 + x_2 - x_3 = 2$
$-3x_1 - 3x_2 + 3x_3 = -6$

24. $2x_1 + 3x_2 - 4x_3 = 3$
$x_1 - 2x_2 - 2x_3 = -2$
$-x_1 + 16x_2 + 2x_3 = 16$

25. $x_1 + x_2 - x_3 = 1$
$2x_1 - x_2 + 7x_3 = 8$
$-x_1 + x_2 - 5x_3 = -5$

26. $x_1 + x_2 \qquad - x_5 = 1$
$x_2 + 2x_3 + x_4 + 3x_5 = 1$
$x_1 \qquad - x_3 + x_4 + x_5 = 0$

27. $x_1 \qquad + x_3 + x_4 - 2x_5 = 1$
$2x_1 + x_2 + 3x_3 - x_4 + x_5 = 0$
$3x_1 - x_2 + 4x_3 + x_4 + x_5 = 1$

28. $x_1 + x_2 = 1$
$x_1 - x_2 = 3$
$2x_1 + x_2 = 3$

29. $x_1 + x_2 = 1$
$x_1 - x_2 = 3$
$2x_1 + x_2 = 2$

30. $x_1 + 2x_2 = 1$
$2x_1 + 4x_2 = 2$
$-x_1 - 2x_2 = -1$

31. $x_1 - x_2 - x_3 = 1$
$x_1 \qquad + x_3 = 2$
$x_2 + 2x_3 = 3$

32. Use Gauss–Jordan elimination to solve each of the systems in Exercises 5–10.

In Exercises 33 and 34, find all values α and β where $0 \le \alpha \le 2\pi$ and $0 \le \beta \le 2\pi$.

33. $2\cos\alpha + 4\sin\beta = 3$
$3\cos\alpha - 5\sin\beta = -1$

34. $2\cos^2\alpha - \sin^2\beta = 1$
$12\cos^2\alpha + 8\sin^2\beta = 13$

35. Any circle in the xy-plane has an equation of the form $x^2 + ax + y^2 + by + c = 0$. Find the center and the radius of the circle passing through the points $P(1,0)$, $Q(2,3)$, and $R(1,1)$.

36. Find three numbers whose sum is 34 where the sum of the first and second is 7 and the sum of the second and third is 22.

37. A zoo charges $6 for adults, $3 for students, and $.50 for children. One morning 79 people enter and pay a total of $207. Determine the possible numbers of adults, students, and children.

When an elementary operation is performed on a system of linear equations, the solution set is not changed. For other types of operations, however, the solution set may change. In Exercises 38 and 39, solve the system. Then perform the given nonelementary operation and solve the new system. Are the solution sets the same?

38. $x_1 + x_2 = 3$ Square both sides of the
$x_1 - x_2 = 1$ first equation.

39. $x_1 + 2x_2 = -5$ Multiply both sides of the
$2x_1 - x_2 = 5$ first equation by x_1.

40. The (2×3) system of linear equations

$a_1x + b_1y + c_1z = d_1$

$a_2x + b_2y + c_2z = d_2$

is represented geometrically by two planes. How are the planes related when:
a) The system has no solution?
b) The system has infinitely many solutions?

Is it possible for the system to have a unique solution? Explain.

In Exercises 41–43, determine whether the given (2×3) system of linear equations represents coincident planes (that is, the same plane), two parallel planes, or two planes whose intersection is a line. In the latter case give the parametric equations for the line; that is, give equations of the form $x = at + b$, $y = ct + d$, $z = et + f$.

41. $2x_1 + x_2 + x_3 = 3$ **42.** $x_1 + 2x_2 - x_3 = 2$
$-2x_1 + x_2 - x_3 = 1$ $x_1 + x_2 + x_3 = 3$

43. $x_1 + 3x_2 - 2x_3 = -1$
$2x_1 + 6x_2 - 4x_3 = -2$

44. Describe the solution set of the following system in terms of x_3:

$x_1 + x_2 + x_3 = 3$
$x_1 + 2x_2 \quad\;\; = 5.$

For x_1, x_2, x_3 in the solution set:
a) Find the maximum value of x_3 such that $x_1 \geq 0$ and $x_2 \geq 0$.
b) Find the maximum value of $y = 2x_1 - 4x_2 + x_3$ subject to $x_1 \geq 0$ and $x_2 \geq 0$.
c) Find the minimum value of $y = (x_1 - 1)^2 + (x_2 + 3)^2 + (x_3 + 1)^2$ with no restriction on x_1 or x_2. (*Hint:* Regard y as a function of x_3 and set the derivative equal to 0; then apply the second-derivative test to verify that you have found a minimum.)

1.3 Matrices and Echelon Form

In this section we begin our introduction to matrix theory by relating matrices to the problem of solving systems of linear equations. Initially we show that matrix theory provides a convenient and natural symbolic language to describe linear systems. Later we show that matrix theory is also an appropriate and powerful framework within which to analyze and solve more general linear problems, such as least-squares approximations, the representation of linear operations, and eigenvalue problems.

The rectangular array

$$\begin{bmatrix} 1 & 3 & -1 & 2 \\ 4 & 2 & 1 & -3 \\ 0 & 2 & 0 & 3 \end{bmatrix}$$

is an example of a matrix. More generally, an $(m \times n)$ **matrix** is a rectangular array of numbers of the form

$$A = \begin{bmatrix} a_{11} & a_{12} & \cdots & a_{1n} \\ a_{21} & a_{22} & \cdots & a_{2n} \\ \vdots & & & \vdots \\ a_{m1} & a_{m2} & \cdots & a_{mn} \end{bmatrix}. \tag{1}$$

Thus an $(m \times n)$ matrix has m rows and n columns. The subscripts for the entry a_{ij} indicate that the number appears in the ith row and jth column of A. For example, a_{32} is the entry in the third row and second column of A. We will frequently use the notation $A = (a_{ij})$ to denote a matrix A with entries a_{ij}.

EXAMPLE 1 Display the (2×3) matrix $A = (a_{ij})$, where $a_{11} = 6$, $a_{12} = 3$, $a_{13} = 7$, $a_{21} = 2$, $a_{22} = 1$, and $a_{23} = 4$.

SOLUTION

$$A = \begin{bmatrix} 6 & 3 & 7 \\ 2 & 1 & 4 \end{bmatrix} \qquad \square$$

Matrix Representation of a Linear System

To illustrate the use of matrices to represent linear systems, consider the (3×3) system of equations

$$\begin{align} x_1 + 2x_2 + x_3 &= 4 \\ 2x_1 - x_2 - x_3 &= 1 \\ x_1 + x_2 + 3x_3 &= 0. \end{align} \tag{2}$$

If we display the coefficients and constants for system (2) in matrix form as

$$B = \begin{bmatrix} 1 & 2 & 1 & 4 \\ 2 & -1 & -1 & 1 \\ 1 & 1 & 3 & 0 \end{bmatrix},$$

then we have expressed compactly and naturally all the essential information from (2). The matrix B is called the augmented matrix for (2).

In general, with the $(m \times n)$ system of linear equations

$$\begin{align} a_{11}x_1 + a_{12}x_2 + \cdots + a_{1n}x_n &= b_1 \\ a_{21}x_1 + a_{22}x_2 + \cdots + a_{2n}x_n &= b_2 \\ \vdots \qquad\qquad \vdots \qquad\quad \vdots \\ a_{m1}x_1 + a_{m2}x_2 + \cdots + a_{mn}x_n &= b_m, \end{align} \tag{3a}$$

we associate two matrices. The $(m \times n)$ matrix $A = (a_{ij})$ as given in (1) is called the *coefficient matrix* for system (3a), and the $[m \times (n + 1)]$ matrix

$$B = \begin{bmatrix} a_{11} & a_{12} & \cdots & a_{1n} & b_1 \\ a_{21} & a_{22} & \cdots & a_{2n} & b_2 \\ \vdots & & & & \vdots \\ a_{m1} & a_{m2} & \cdots & a_{mn} & b_m \end{bmatrix}$$ (3b)

is called the *augmented matrix* for (3a). The matrix B is usually denoted as $[A \,|\, \mathbf{b}]$, where A is the coefficient matrix of (3a) and

$$\mathbf{b} = \begin{bmatrix} b_1 \\ b_2 \\ \vdots \\ b_m \end{bmatrix}.$$

EXAMPLE 2 Display the coefficient matrix A and the augmented matrix B for the system

$$x_1 - 2x_2 + x_3 = 2$$
$$2x_1 + x_2 - x_3 = 1$$
$$-3x_1 + x_2 - 2x_3 = -5.$$

SOLUTION The coefficient matrix A and the augmented matrix $[A \,|\, \mathbf{b}]$ are given by

$$A = \begin{bmatrix} 1 & -2 & 1 \\ 2 & 1 & -1 \\ -3 & 1 & -2 \end{bmatrix} \quad \text{and} \quad [A \,|\, \mathbf{b}] = \begin{bmatrix} 1 & -2 & 1 & 2 \\ 2 & 1 & -1 & 1 \\ -3 & 1 & -2 & -5 \end{bmatrix}. \qquad \square$$

Elementary Row Operations

To illustrate the use of the augmented matrix in solving a system of equations, we now solve a linear system by simultaneously performing an elementary operation on the system and the corresponding operation on the rows of the augmented matrix. In the notation for the matrix operations, R_i denotes the ith row of the matrix.

Linear system

$$\begin{aligned} x_1 + 2x_2 - x_3 + 3x_4 &= 2 \\ -x_1 + x_2 + 3x_3 - 2x_4 &= -1 \\ 2x_1 + 7x_2 - x_3 + 9x_4 &= 8 \\ 3x_1 + 3x_2 - 2x_3 + 4x_4 &= -6 \end{aligned}$$

Augmented matrix

$$\begin{bmatrix} 1 & 2 & -1 & 3 & 2 \\ -1 & 1 & 3 & -2 & -1 \\ 2 & 7 & -1 & 9 & 8 \\ 3 & 3 & -2 & 4 & -6 \end{bmatrix}$$

$E_2 + E_1, E_3 - 2E_1, E_4 - 3E_1$:

$$\begin{aligned} x_1 + 2x_2 - x_3 + 3x_4 &= 2 \\ 3x_2 + 2x_3 + x_4 &= 1 \\ 3x_2 + x_3 + 3x_4 &= 4 \\ -3x_2 + x_3 - 5x_4 &= -12 \end{aligned}$$

$R_2 + R_1, R_3 - 2R_1, R_4 - 3R_1$:

$$\begin{bmatrix} 1 & 2 & -1 & 3 & 2 \\ 0 & 3 & 2 & 1 & 1 \\ 0 & 3 & 1 & 3 & 4 \\ 0 & -3 & 1 & -5 & -12 \end{bmatrix}$$

$E_3 - E_2, E_4 + E_2$:

$$\begin{aligned}
x_1 + 2x_2 - x_3 + 3x_4 &= 2 \\
3x_2 + 2x_3 + x_4 &= 1 \\
-x_3 + 2x_4 &= 3 \\
3x_3 - 4x_4 &= -11
\end{aligned}$$

$R_3 - R_2, R_4 + R_2$:

$$\begin{bmatrix}
1 & 2 & -1 & 3 & 2 \\
0 & 3 & 2 & 1 & 1 \\
0 & 0 & -1 & 2 & 3 \\
0 & 0 & 3 & -4 & -11
\end{bmatrix}$$

$E_4 + 3E_3$:

$$\begin{aligned}
x_1 + 2x_2 - x_3 + 3x_4 &= 2 \\
3x_2 + 2x_3 + x_4 &= 1 \\
-x_3 + 2x_4 &= 3 \\
2x_4 &= -2
\end{aligned}$$

$R_4 + 3R_3$:

$$\begin{bmatrix}
1 & 2 & -1 & 3 & 2 \\
0 & 3 & 2 & 1 & 1 \\
0 & 0 & -1 & 2 & 3 \\
0 & 0 & 0 & 2 & -2
\end{bmatrix}$$

Backsolving now yields the solution $x_1 = -8, x_2 = 4, x_3 = -5, x_4 = -1$.

As the preceding example illustrates, the basic idea is to use the augmented matrix as a shorthand notation, thus relieving us of the necessity of keeping track of the unknowns. Since we will now wish to perform elementary operations on the rows of a matrix, we introduce the following terminology.

DEFINITION 2 The following operations, performed on the rows of a matrix, are called *elementary row operations:*

1. Interchange two rows.
2. Multiply a row by a nonzero scalar.
3. Add a constant multiple of one row to another.

As before, we adopt the following notation:

Notation	Elementary row operation
$R_i \leftrightarrow R_j$	The ith and jth rows are interchanged.
kR_i	The ith row is multiplied by the nonzero scalar k.
$R_i + kR_j$	k times the jth row is added to the ith row.

We say that two $(m \times n)$ matrices, B and C, are *row equivalent* if one can be obtained from the other by a sequence of elementary row operations. Now if B is the augmented matrix for a system of linear equations and if C is row equivalent to B, then C is the augmented matrix for an equivalent system. This observation follows because the elementary row operations for matrices exactly duplicate the elementary operations for equations.

Thus we can solve a linear system by forming the augmented matrix, B, for the system and then using elementary row operations to transform the augmented matrix to a row-equivalent matrix, C, which represents a "simpler" system. We will specify this simpler form in the next subsection, but first we illustrate the procedure described above by solving again the system given in Example 4 of Section 1.2.

EXAMPLE 3 Solve

$$x_2 + x_3 - x_4 = 0$$
$$x_1 - x_2 + 3x_3 - x_4 = -2 \quad \textbf{(4a)}$$
$$x_1 + x_2 + x_3 + x_4 = 2.$$

SOLUTION The augmented matrix for (4a) can be reduced as follows:

$$\begin{bmatrix} 0 & 1 & 1 & -1 & 0 \\ 1 & -1 & 3 & -1 & -2 \\ 1 & 1 & 1 & 1 & 2 \end{bmatrix}$$

$R_1 \leftrightarrow R_2$:

$$\begin{bmatrix} 1 & -1 & 3 & -1 & -2 \\ 0 & 1 & 1 & -1 & 0 \\ 1 & 1 & 1 & 1 & 2 \end{bmatrix}$$

$R_3 - R_1$:

$$\begin{bmatrix} 1 & -1 & 3 & -1 & -2 \\ 0 & 1 & 1 & -1 & 0 \\ 0 & 2 & -2 & 2 & 4 \end{bmatrix}$$

$R_3 - 2R_2$:

$$\begin{bmatrix} 1 & -1 & 3 & -1 & -2 \\ 0 & 1 & 1 & -1 & 0 \\ 0 & 0 & -4 & 4 & 4 \end{bmatrix}. \quad \textbf{(4b)}$$

The matrix (4b) is the augmented matrix for the linear system

$$x_1 - x_2 + 3x_3 - x_4 = -2$$
$$x_2 + x_3 - x_4 = 0 \quad \textbf{(4c)}$$
$$-4x_3 + 4x_4 = 4,$$

and system (4c) is equivalent to system (4a). The solution is now found by back-solving as before. ☐

Echelon Form

The matrix in (4b) is an example of a matrix in echelon form. Informally, a matrix C is in echelon form provided that all the nonzero entries appear in a staircase-shaped region in the upper right-hand portion of the matrix. Two such

Matrices and Linear Systems

James Sylvester (1814–1897) is given credit for first using the term "matrix" to denote an array of numbers. His close associate, Arthur Cayley (1821–1895), is usually credited with replacing a system of equations by its augmented matrix. Many centuries earlier, however, a Chinese text solved the system

$$3x_1 + 2x_2 + x_3 = 39$$

$$2x_1 + 3x_2 + x_3 = 34$$

$$x_1 + 2x_2 + 3x_3 = 26$$

by using the following column operations:

$$
\begin{bmatrix} 1 & 2 & 3 \\ 2 & 3 & 2 \\ 3 & 1 & 1 \\ 26 & 34 & 39 \end{bmatrix}
\xrightarrow[\;3C_1 - C_3\;]{3C_2 - 2C_3}
\begin{bmatrix} 0 & 0 & 3 \\ 4 & 5 & 2 \\ 8 & 1 & 1 \\ 39 & 24 & 39 \end{bmatrix}
\xrightarrow{5C_1 - 4C_2}
\begin{bmatrix} 0 & 0 & 3 \\ 0 & 5 & 2 \\ 36 & 1 & 1 \\ 99 & 24 & 39 \end{bmatrix}
$$

matrices in echelon form are illustrated in Fig. 1.2, in which the entries denoted by * are nonzero, whereas the entries denoted by × may or may not be zero.

The purpose of echelon form should be evident. If B is the echelon form of an augmented matrix for a linear system, then either the system is obviously inconsistent or it can be solved easily. In other words, the goal of the Gaussian elimination process should be to reduce an augmented matrix to echelon form. We proceed now with the definition of echelon form.

DEFINITION 3   An $(m \times n)$ matrix C is in *echelon form* if:

1. All rows that consist entirely of zeros are grouped together at the bottom of the matrix.
2. The first (counting from left to right) nonzero entry in the $(i + 1)$st row appears in a column to the right of the first nonzero entry in the ith row.*

$$
C_1 = \begin{bmatrix} * & \times & \times & \times & \times & \times & \times \\ 0 & 0 & * & \times & \times & \times & \times \\ 0 & 0 & 0 & * & \times & \times & \times \\ 0 & 0 & 0 & 0 & 0 & * & \times \\ 0 & 0 & 0 & 0 & 0 & 0 & 0 \end{bmatrix}, \qquad
C_2 = \begin{bmatrix} 0 & * & \times & \times & \times \\ 0 & 0 & * & \times & \times \\ 0 & 0 & 0 & 0 & * \end{bmatrix}
$$

Figure 1.2
Examples of echelon form

---

* Many definitions of echelon form require that the first nonzero entry in each row be a 1, but we do not insist on this. A related concept is that of "reduced echelon form," which we discuss later.

EXAMPLE 4 Determine which of the following matrices are in echelon form:

$$A = \begin{bmatrix} 2 & 1 & 6 & 4 & -2 \\ 0 & 1 & 0 & 2 & 4 \\ 0 & 0 & 3 & 1 & 5 \\ 0 & 0 & 0 & 1 & 4 \end{bmatrix}, \quad B = \begin{bmatrix} 1 & 5 & 0 & 2 & 3 \\ 0 & 0 & 6 & 1 & 1 \\ 0 & 0 & 0 & 2 & 4 \\ 0 & 0 & 0 & 0 & 0 \end{bmatrix}, \quad C = \begin{bmatrix} 3 & 1 & 4 & 6 \\ 0 & 1 & 3 & 6 \\ 0 & 2 & 8 & 15 \\ 0 & 3 & 7 & 19 \end{bmatrix},$$

$$D = \begin{bmatrix} 2 & 1 & 6 & 0 & 3 & 4 \\ 0 & 0 & 2 & 0 & 4 & 3 \\ 0 & 0 & 0 & 0 & 1 & 1 \\ 0 & 0 & 0 & 0 & 0 & -1 \end{bmatrix}, \quad E = \begin{bmatrix} 0 & 5 & 3 & 1 \\ 2 & 6 & 1 & 5 \\ 0 & 0 & 2 & 1 \\ 0 & 0 & 0 & 0 \end{bmatrix}.$$

SOLUTION The matrices A, B, and D are in echelon form, whereas C and E are not. □

In Example 4, the matrix C is not in echelon form. However, we can transform C to echelon form with elementary row operations. In particular, by performing the elementary row operations $R_3 - 2R_2$ and $R_4 - 3R_2$ on matrix C in Example 4, we obtain

$$C_1 = \begin{bmatrix} 3 & 1 & 4 & 6 \\ 0 & 1 & 3 & 6 \\ 0 & 0 & 2 & 3 \\ 0 & 0 & -2 & 1 \end{bmatrix}.$$

The operation $R_4 + R_3$ now yields

$$C_2 = \begin{bmatrix} 3 & 1 & 4 & 6 \\ 0 & 1 & 3 & 6 \\ 0 & 0 & 2 & 3 \\ 0 & 0 & 0 & 4 \end{bmatrix},$$

which is in echelon form.

Note too that matrix E in Example 4 can be put into echelon form with the elementary row operation $R_2 \leftrightarrow R_1$. Indeed, as the next theorem indicates, every matrix can be transformed by elementary row operations to a matrix in echelon form.

THEOREM 2

Let B be an $(m \times n)$ matrix. Then there is an $(m \times n)$ matrix C such that

1. C is in echelon form; and
2. B is row equivalent to C.

If B has only zero entries, then B is already in echelon form. For a non-zero matrix, the reduction steps parallel those for Gaussian elimination given in Section 1.1. The steps are listed below and provide an informal sketch of the proof of Theorem 2.

Reduction to Echelon Form (for an ($m \times n$) matrix)

Step 1. Locate the first (leftmost) column that contains a nonzero entry.

Step 2. If necessary, interchange the first row with another row so that the first nonzero column contains a nonzero entry in the first row.

Step 3. Add appropriate multiples of the first row to each of the succeeding rows so that the first nonzero column has a nonzero entry only in the first row.

Step 4. Temporarily ignore the first row of this matrix and repeat the process on the remaining rows.

The following example illustrates the procedure for transforming a matrix to echelon form.

EXAMPLE 5 Use elementary row operations to find a matrix C such that C is in echelon form and is row equivalent to

$$B = \begin{bmatrix} 0 & 0 & 0 & 1 & 3 & 5 \\ 0 & 1 & 2 & -1 & 2 & 2 \\ 0 & -2 & -4 & 5 & 7 & 14 \\ 0 & 3 & 6 & -4 & 7 & 7 \\ 0 & 0 & 0 & 2 & 4 & 9 \end{bmatrix}.$$

SOLUTION We proceed with elementary row operations as follows:

$R_1 \leftrightarrow R_2$:

$$\begin{bmatrix} 0 & 1 & 2 & -1 & 2 & 2 \\ 0 & 0 & 0 & 1 & 3 & 5 \\ 0 & -2 & -4 & 5 & 7 & 14 \\ 0 & 3 & 6 & -4 & 7 & 7 \\ 0 & 0 & 0 & 2 & 4 & 9 \end{bmatrix}$$

$R_3 + 2R_1, R_4 - 3R_1$:

$$\begin{bmatrix} 0 & 1 & 2 & -1 & 2 & 2 \\ 0 & 0 & 0 & 1 & 3 & 5 \\ 0 & 0 & 0 & 3 & 11 & 18 \\ 0 & 0 & 0 & -1 & 1 & 1 \\ 0 & 0 & 0 & 2 & 4 & 9 \end{bmatrix}$$

$R_3 - 3R_2, R_4 + R_2, R_5 - 2R_2$:

$$\begin{bmatrix} 0 & 1 & 2 & -1 & 2 & 2 \\ 0 & 0 & 0 & 1 & 3 & 5 \\ 0 & 0 & 0 & 0 & 2 & 3 \\ 0 & 0 & 0 & 0 & 4 & 6 \\ 0 & 0 & 0 & 0 & -2 & -1 \end{bmatrix}$$

$R_4 - 2R_3, R_5 + R_3$:

$$\begin{bmatrix} 0 & 1 & 2 & -1 & 2 & 2 \\ 0 & 0 & 0 & 1 & 3 & 5 \\ 0 & 0 & 0 & 0 & 2 & 3 \\ 0 & 0 & 0 & 0 & 0 & 0 \\ 0 & 0 & 0 & 0 & 0 & 2 \end{bmatrix}$$

$R_4 \leftrightarrow R_5$:

$$C = \begin{bmatrix} 0 & 1 & 2 & -1 & 2 & 2 \\ 0 & 0 & 0 & 1 & 3 & 5 \\ 0 & 0 & 0 & 0 & 2 & 3 \\ 0 & 0 & 0 & 0 & 0 & 2 \\ 0 & 0 & 0 & 0 & 0 & 0 \end{bmatrix}.$$

The matrix C is in echelon form and is row equivalent to the matrix B.

The following corollary is an immediate consequence of Theorem 2 and the fact that elementary row operations exactly duplicate elementary operations on equations.

COROLLARY

Any $(m \times n)$ system of linear equations is equivalent to some $(m \times n)$ system of linear equations whose augmented matrix is in echelon form.

Thus the goal of Gaussian elimination is to obtain a system of equations that is equivalent to the given system and that has an augmented matrix in echelon form. We can now summarize the steps for solving a system of equations.

SUMMARY

To solve an $(m \times n)$ system of linear equations:

1. Form the augmented matrix for the system.
2. Use elementary row operations to transform the augmented matrix to a row-equivalent matrix in echelon form.

3. Form the system of equations represented by the reduced matrix.

4. If the system is consistent, complete the solution by using the backsolving techniques described in Section 1.2.

EXAMPLE 6   Solve the (3×4) system of linear equations

$$
\begin{aligned}
x_1 + 2x_2 + x_3 - x_4 &= 1 \\
x_1 + 2x_2 - x_3 + x_4 &= 5 \\
-x_1 + x_2 + 2x_3 \phantom{{}+ x_4} &= 0.
\end{aligned}
\tag{5a}
$$

SOLUTION   We reduce the augmented matrix as follows:

$$
\begin{bmatrix}
1 & 2 & 1 & -1 & 1 \\
1 & 2 & -1 & 1 & 5 \\
-1 & 1 & 2 & 0 & 0
\end{bmatrix}
$$

$R_2 - R_1, R_3 + R_1$:

$$
\begin{bmatrix}
1 & 2 & 1 & -1 & 1 \\
0 & 0 & -2 & 2 & 4 \\
0 & 3 & 3 & -1 & 1
\end{bmatrix}
$$

$R_2 \leftrightarrow R_3$:

$$
\begin{bmatrix}
1 & 2 & 1 & -1 & 1 \\
0 & 3 & 3 & -1 & 1 \\
0 & 0 & -2 & 2 & 4
\end{bmatrix}.
\tag{5b}
$$

The matrix (5b) is in echelon form and is the augmented matrix for the system

$$
\begin{aligned}
x_1 + 2x_2 + x_3 - x_4 &= 1 \\
3x_2 + 3x_3 - x_4 &= 1 \\
-2x_3 + 2x_4 &= 4.
\end{aligned}
\tag{5c}
$$

Furthermore, system (5c) is equivalent to (5a). Backsolving yields

$$
\begin{aligned}
x_1 &= (4x_4 - 5)/3 \\
x_2 &= (-2x_4 + 7)/3 \\
x_3 &= x_4 - 2
\end{aligned}
\tag{5d}
$$

as the solution to (5a).   □

Reduced Echelon Form

In Section 1.2, we presented a reduction technique, called Gauss–Jordan elimination, that avoids backsolving by using elementary operations to reduce the system of equations further. In terms of the augmented matrix, the Gauss–Jordan procedure transforms the matrix for the given system to a matrix in "reduced echelon" form. We close this section with a brief discussion of reduced echelon form, which is a simple extension of the idea of echelon form. In Example 6 above, the reduced system (5c) had augmented matrix

$$A = \begin{bmatrix} 1 & 2 & 1 & -1 & 1 \\ 0 & 3 & 3 & -1 & 1 \\ 0 & 0 & -2 & 2 & 4 \end{bmatrix}$$

in echelon form. As an alternative to backsolving, we can use elementary row operations to transform A into a row-equivalent matrix of the form

$$B = \begin{bmatrix} * & 0 & 0 & \times & \times \\ 0 & * & 0 & \times & \times \\ 0 & 0 & * & \times & \times \end{bmatrix}$$

Clearly it is a trivial matter to solve a linear system for which B is the augmented matrix. B is an example of a matrix in reduced echelon form.

DEFINITION 4   A matrix C that is in echelon form is in *reduced echelon form* provided that the first nonzero element in each nonzero row is the only nonzero entry in its column.

For example, the matrices

$$C = \begin{bmatrix} 2 & 0 & 0 & 1 \\ 0 & 3 & 0 & -1 \\ 0 & 0 & 1 & 2 \end{bmatrix} \quad \text{and} \quad D = \begin{bmatrix} 1 & 2 & 0 & 1 & -1 \\ 0 & 0 & 3 & 2 & 2 \\ 0 & 0 & 0 & 0 & 0 \end{bmatrix}$$

are in reduced echelon form. The next example illustrates the use of the Gauss–Jordan elimination procedure to solve a system by transforming the augmented matrix to a row-equivalent matrix in reduced echelon form.

EXAMPLE 7 Solve the system

$$\begin{aligned} x_1 + x_2 + x_3 + 4x_4 &= 4 \\ 2x_1 + 3x_2 + 4x_3 + 9x_4 &= 16 \\ -2x_1 \qquad\quad + 3x_3 - 7x_4 &= 11 \end{aligned} \tag{6}$$

by transforming the augmented matrix to reduced echelon form.

SOLUTION   The usual Gauss–Jordan reduction proceeds from left to right as follows:

$$\begin{bmatrix} 1 & 1 & 1 & 4 & 4 \\ 2 & 3 & 4 & 9 & 16 \\ -2 & 0 & 3 & -7 & 11 \end{bmatrix}$$

$R_2 - 2R_1, R_3 + 2R_1$:

$$\begin{bmatrix} 1 & 1 & 1 & 4 & 4 \\ 0 & 1 & 2 & 1 & 8 \\ 0 & 2 & 5 & 1 & 19 \end{bmatrix}$$

$R_1 - R_2, R_3 - 2R_2$:

$$\begin{bmatrix} 1 & 0 & -1 & 3 & -4 \\ 0 & 1 & 2 & 1 & 8 \\ 0 & 0 & 1 & -1 & 3 \end{bmatrix}$$

$R_1 + R_3, R_2 - 2R_3$:

$$\begin{bmatrix} 1 & 0 & 0 & 2 & -1 \\ 0 & 1 & 0 & 3 & 2 \\ 0 & 0 & 1 & -1 & 3 \end{bmatrix}.$$

The last matrix above is in reduced echelon form, and the given system is equivalent to the system

$$\begin{aligned} x_1 & & + 2x_4 &= -1 \\ & x_2 & + 3x_4 &= 2 \\ & x_3 - x_4 &= 3. \end{aligned}$$

This system is easily solved, in terms of x_4, without backsolving.

Electronic Aids

One testimony to the practical importance of linear algebra is the wide variety of electronic aids available for linear algebra computations. For instance, many scientific hand calculators can solve systems of linear equations and perform simple matrix operations. For computers there are general-purpose computer algebra systems such as Derive, Mathematica, and Maple that have extensive computational capabilities. Special-purpose linear algebra software such as MATLAB is very easy to use and can perform virtually any type of matrix calculation.

In the following example, we illustrate the use of MATLAB. From time to time, as appropriate, we will include other examples that illustrate the use of electronic aids.

EXAMPLE 8   In certain applications it is necessary to evaluate sums of powers of integers such as

$$1 + 2 + 3 + \cdots + n,$$
$$1^2 + 2^2 + 3^2 + \cdots + n^2,$$
$$1^3 + 2^3 + 3^3 + \cdots + n^3, \quad \text{and so on.}$$

Interestingly, it is possible to derive simple formulas for such sums. For instance, you may be familiar with the formula

$$1 + 2 + 3 + \cdots + n = \frac{n(n+1)}{2}.$$

Such formulas can be derived using the following result: If n and r are positive integers, then there are constants $a_1, a_2, \ldots, a_{r+1}$ such that

$$1^r + 2^r + 3^r + \cdots + n^r = a_1 n + a_2 n^2 + a_3 n^3 + \cdots + a_{r+1} n^{r+1}. \qquad (7)$$

Use Eq. (7) to find the formula for $1^3 + 2^3 + 3^3 + \cdots + n^3$. (*Note:* Equation (7) can be derived from the theory of linear difference equations.)

SOLUTION   From Eq. (7), there are constants a_1, a_2, a_3, and a_4 such that

$$1^3 + 2^3 + 3^3 + \cdots + n^3 = a_1 n + a_2 n^2 + a_3 n^3 + a_4 n^4.$$

If we evaluate the formula above for $n = 1$, $n = 2$, $n = 3$, and $n = 4$, we obtain four equations for a_1, a_2, a_3, and a_4:

$$
\begin{aligned}
a_1 + a_2 + a_3 + a_4 &= 1 & (n = 1)\\
2a_1 + 4a_2 + 8a_3 + 16a_4 &= 9 & (n = 2)\\
3a_1 + 9a_2 + 27a_3 + 81a_4 &= 36 & (n = 3)\\
4a_1 + 16a_2 + 64a_3 + 256a_4 &= 100. & (n = 4)
\end{aligned}
$$

The augmented matrix for this system is

$$A = \begin{bmatrix} 1 & 1 & 1 & 1 & 1 \\ 2 & 4 & 9 & 16 & 9 \\ 3 & 9 & 27 & 81 & 36 \\ 4 & 16 & 64 & 256 & 100 \end{bmatrix}.$$

We used MATLAB to solve the system by transforming A to reduced echelon form. The steps, as they apear on a monitor screen, are shown in Fig. 1.3. The symbol ≫ is a prompt from MATLAB. At the first prompt, we entered the augmented matrix A and then MATLAB displayed A. At the second prompt, we

Adding Integers

Mathematical folklore has it that Gauss discovered the formula $1 + 2 + 3 + \cdots + n = n(n+1)/2$ when he was only ten years old. To occupy time, his teacher asked the students to add the integers from 1 to 100. Gauss immediately wrote an answer and turned his slate over. To his teacher's amazement, Gauss had the only correct answer in the class. Young Gauss had recognized that the numbers could be put in 50 sets of pairs such that the sum of each pair was 101:

$$(50 + 51) + (49 + 52) + (48 + 53) + \cdots + (1 + 100) = 50(101) = 5050.$$

Soon his brilliance was brought to the notice of the Duke of Brunswick, who thereafter sponsored the education of Gauss.

```
>> A = [ 1 , 1 , 1 , 1 , 1 ; 2 , 4 , 8 , 16 , 9 ; 3 , 9 , 27 , 81 , 36 ; 4 , 16 , 64 , 256 , 100 ]

A =
        1        1        1        1        1
        2        4        8       16        9
        3        9       27       81       36
        4       16       64      256      100

>> C = rref (A)

C =
   1.0000        0        0        0        0
        0   1.0000        0        0   0.2500
        0        0   1.0000        0   0.5000
        0        0        0   1.0000   0.2500

>> rat (C, 's' )

ans =
        1        0        0        0        0
        0        1        0        0      1/4
        0        0        1        0      1/2
        0        0        0        1      1/4
```

Figure 1.3
Using MATLAB in Example 8 to row reduce the matrix A to the
matrix C.

entered the MATLAB row-reduction command, $C = \text{rref}(A)$. The new matrix
C, as displayed by MATLAB, is the result of transforming A to reduced echelon
form.

MATLAB normally displays results in decimal form. To obtain a rational
form for the reduced matrix C, we used the command rat(C, 's'), which produced

$$C = \begin{bmatrix} 1 & 0 & 0 & 0 & 0 \\ 0 & 1 & 0 & 0 & 1/4 \\ 0 & 0 & 1 & 0 & 1/2 \\ 0 & 0 & 0 & 1 & 1/4 \end{bmatrix}.$$

From above, we have $a_1 = 0$, $a_2 = 1/4$, $a_3 = 1/2$, and $a_4 = 1/4$. Therefore, the
formula for the sum of the first n cubes is

$$1^3 + 2^3 + 3^3 + \cdots + n^3 = \frac{1}{4}n^2 + \frac{1}{2}n^3 + \frac{1}{4}n^4$$

or, after simplification,

$$1^3 + 2^3 + 3^3 + \cdots + n^3 = \frac{n^2(n+1)^2}{4}.$$

Exercises 1.3

1. Display the (2×3) matrix $A = (a_{ij})$, where $a_{11} = 2$, $a_{12} = 1$, $a_{13} = 6$, $a_{21} = 4$, $a_{22} = 3$, and $a_{23} = 8$.

2. Display the (2×4) matrix $C = (c_{ij})$, where $c_{23} = 4$, $c_{12} = 2$, $c_{21} = 2$, $c_{14} = 1$, $c_{22} = 2$, $c_{24} = 3$, $c_{11} = 1$, and $c_{13} = 7$.

3. Display the (3×3) matrix $Q = (q_{ij})$, where $q_{23} = 1$, $q_{32} = 2$, $q_{11} = 1$, $q_{13} = -3$, $q_{22} = 1$, $q_{33} = 1$, $q_{21} = 2$, $q_{12} = 4$, and $q_{31} = 3$.

4. Suppose the matrix C in Exercise 2 is the augmented matrix for a system of linear equations. Display the system.

5. Repeat Exercise 4 for the matrices in Exercises 1 and 3.

In Exercises 6–11, display the coefficient matrix A and the augmented matrix B for the given system.

6. $\begin{aligned} x_1 - x_2 &= -1 \\ x_1 + x_2 &= 3 \end{aligned}$

7. $\begin{aligned} x_1 + x_2 - x_3 &= 2 \\ 2x_1 \quad\quad - x_3 &= 1 \end{aligned}$

8. $\begin{aligned} x_1 + 3x_2 - x_3 &= 1 \\ 2x_1 + 5x_2 + x_3 &= 5 \\ x_1 + x_2 + x_3 &= 3 \end{aligned}$

9. $\begin{aligned} x_1 + x_2 + 2x_3 &= 6 \\ 3x_1 + 4x_2 - x_3 &= 5 \\ -x_1 + x_2 + x_3 &= 2 \end{aligned}$

10. $\begin{aligned} x_1 + x_2 - 3x_3 &= -1 \\ x_1 + 2x_2 - 5x_3 &= -2 \\ -x_1 - 3x_2 + 7x_3 &= 3 \end{aligned}$

11. $\begin{aligned} x_1 + x_2 + x_3 &= 1 \\ 2x_1 + 3x_2 + x_3 &= 2 \\ x_1 - x_2 + 3x_3 &= 2 \end{aligned}$

Consider the matrices in Exercises 12–22.

a) Either state that the matrix is in echelon form or use elementary row operations to transform it to echelon form.

b) If the matrix is in echelon form, transform it to reduced echelon form.

12. $\begin{bmatrix} 1 & 2 \\ 2 & 1 \end{bmatrix}$

13. $\begin{bmatrix} 1 & 2 \\ 0 & 1 \end{bmatrix}$

14. $\begin{bmatrix} 1 & 2 & -1 \\ 0 & 1 & 3 \end{bmatrix}$

15. $\begin{bmatrix} 2 & 3 & 1 \\ 4 & 1 & 0 \end{bmatrix}$

16. $\begin{bmatrix} 0 & 1 & 1 \\ 1 & 2 & 3 \end{bmatrix}$

17. $\begin{bmatrix} 0 & 0 & 2 & 3 \\ 2 & 0 & 1 & 4 \end{bmatrix}$

18. $\begin{bmatrix} 2 & 0 & 3 & 1 \\ 0 & 0 & 1 & 2 \end{bmatrix}$

19. $\begin{bmatrix} 1 & 3 & 2 & 1 \\ 0 & 1 & 4 & 2 \\ 0 & 0 & 1 & 1 \end{bmatrix}$

20. $\begin{bmatrix} 2 & -1 & 3 \\ 0 & 1 & 1 \\ 0 & 0 & -3 \end{bmatrix}$

21. $\begin{bmatrix} 1 & 2 & -1 & -2 \\ 0 & 2 & -2 & -3 \\ 0 & 0 & 0 & 1 \end{bmatrix}$

22. $\begin{bmatrix} -1 & 4 & -3 & 4 & 6 \\ 0 & 2 & 1 & -3 & -3 \\ 0 & 0 & 0 & 1 & 2 \end{bmatrix}$

23. Solve the systems in Exercises 7, 9, and 11 by reducing the augmented matrix to echelon form.

24. Repeat Exercise 23 for the systems in Exercises 6, 8, and 10.

In Exercises 25–28, solve the system by transforming the augmented matrix to reduced echelon form.

25. $\begin{aligned} 2x_1 - 3x_2 - 3x_3 &= -2 \\ -4x_1 + 3x_2 + 4x_3 &= 3 \\ 2x_1 + 3x_2 + x_3 &= 0 \end{aligned}$

26. $\begin{aligned} x_1 + 2x_2 &= 0 \\ 2x_1 + 5x_2 &= -1 \\ x_1 - x_2 &= 3 \end{aligned}$

27. $\begin{aligned} 2x_1 + x_2 + 3x_3 &= 5 \\ -2x_1 \quad\quad + 3x_3 &= -4 \\ 6x_1 + x_2 \quad\quad &= 14 \end{aligned}$

28. $\begin{aligned} x_1 + 2x_2 + 4x_3 + 3x_4 &= 7 \\ x_1 + 2x_2 + 6x_3 + 6x_4 &= 10 \\ 2x_1 + 4x_2 + 6x_3 + 6x_4 &= 13 \end{aligned}$

In Exercises 29–39, each of the given matrices represents the augmented matrix for a system of linear equations. In each exercise, display the solution set or state that the system is inconsistent.

29. $\begin{bmatrix} 1 & 1 & 0 \\ 0 & 1 & 0 \end{bmatrix}$

30. $\begin{bmatrix} 1 & 1 & 0 \\ 0 & 0 & 2 \end{bmatrix}$

31. $\begin{bmatrix} 1 & 2 & 1 & 0 \\ 0 & 1 & 3 & 1 \end{bmatrix}$

32. $\begin{bmatrix} 1 & 2 & 2 & 1 \\ 0 & 1 & 0 & 0 \end{bmatrix}$

33. $\begin{bmatrix} 1 & 1 & 1 & 0 \\ 0 & 1 & 0 & 0 \\ 0 & 0 & 0 & 1 \end{bmatrix}$

34. $\begin{bmatrix} 1 & 2 & 0 & 1 \\ 0 & 1 & 1 & 0 \\ 0 & 0 & 2 & 0 \end{bmatrix}$

35. $\begin{bmatrix} 1 & 0 & 1 & 0 & 0 \\ 0 & 0 & 1 & 1 & 0 \\ 0 & 0 & 0 & 1 & 0 \end{bmatrix}$ 36. $\begin{bmatrix} 1 & 2 & 1 & 3 \\ 0 & 0 & 0 & 2 \\ 0 & 0 & 0 & 0 \end{bmatrix}$

37. $\begin{bmatrix} 1 & 0 & 0 & 1 \\ 0 & 1 & 0 & 1 \\ 0 & 0 & 0 & 1 \end{bmatrix}$ 38. $\begin{bmatrix} 1 & 1 & 2 & 0 & 2 & 0 \\ 0 & 1 & 1 & 1 & 0 & 0 \\ 0 & 0 & 1 & 2 & 1 & 2 \end{bmatrix}$

39. $\begin{bmatrix} 2 & 1 & 3 & 2 & 0 & 1 \\ 0 & 0 & 1 & 1 & 2 & 1 \\ 0 & 0 & 0 & 0 & 3 & 0 \end{bmatrix}$

In Exercises 40–43, use Eq. (7) to find the formula for the sum. If available, use linear algebra software for Exercises 42 and 43.

40. $1 + 2 + 3 + \cdots + n$

41. $1^2 + 2^2 + 3^2 + \cdots + n^2$

42. $1^4 + 2^4 + 3^4 + \cdots + n^4$

43. $1^5 + 2^5 + 3^5 + \cdots + n^5$

44. Let A and I be as follows:

$$A = \begin{bmatrix} 1 & d \\ c & b \end{bmatrix}, \qquad I = \begin{bmatrix} 1 & 0 \\ 0 & 1 \end{bmatrix}.$$

Prove that if $b - cd \neq 0$, then A is row equivalent to I.

45. As in Fig. 1.2, display all the possible configurations for a (2×3) matrix that is in echelon form. (*Hint:* There are seven such configurations. Consider the various positions that can be occupied by one, two, or none of the symbols.)

46. Repeat Exercise 45 for a (3×2) matrix, for a (3×3) matrix, and for a (3×4) matrix.

47. Consider the matrices B and C:

$$B = \begin{bmatrix} 1 & 2 \\ 2 & 3 \end{bmatrix}, \qquad C = \begin{bmatrix} 1 & 2 \\ 3 & 4 \end{bmatrix}.$$

By Exercise 44, B and C are both row equivalent to matrix I in Exercise 44. Determine elementary row operations that demonstrate that B is row equivalent to C.

48. Repeat Exercise 47 for the matrices

$$B = \begin{bmatrix} 1 & 4 \\ 3 & 7 \end{bmatrix}, \qquad C = \begin{bmatrix} 1 & 2 \\ 2 & 1 \end{bmatrix}.$$

49. Find a cubic polynomial, $p(x) = a + bx + cx^2 + dx^3$, such that $p(1) = 5$, $p'(1) = 5$, $p(2) = 17$, and $p'(2) = 21$.

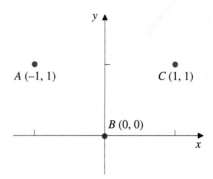

Figure 1.4

50. We want to draw a smooth curve through the points $A(-1, 1)$, $B(0, 0)$, and $C(1, 1)$, shown in Fig. 1.4. The curve is to coincide with the graph of the piecewise-cubic polynomial S defined by

$$S(x) = \begin{cases} p(x), & -1 \leq x \leq 0 \\ q(x), & 0 \leq x \leq 1 \end{cases}$$

where $p(x) = a + bx + cx^2 + dx^3$ and $q(x) = r + sx + tx^2 + ux^3$.

a) Determine the coefficients of p and q so that

$$p(-1) = 1, \qquad p'(-1) = -1, \qquad p(0) = 0,$$
$$q(1) = 1, \qquad q'(1) = 1, \qquad q(0) = 0,$$
$$p'(0) = q'(0), \qquad p''(0) = q''(0).$$

(*Hint:* First use the conditions at $x = 0$ to reduce the size of the system.) *Note:* The four conditions $p(0) = q(0) = 0$, $p'(0) = q'(0)$, and $p''(0) = q''(0)$ guarantee that S is smooth in the sense that S, S', and S'' are continuous on $[-1, 1]$.

b) Sketch the graph of S. *Note:* A piecewise-cubic polynomial that is twice continuously differentiable (such as S) is called a **cubic spline.** Whereas S is made up of two cubic polynomials, the typical cubic spline consists of many cubic polynomial pieces. These pieces are patched together so that the resulting piecewise-cubic polynomial is continuous and has continuous derivatives of orders one and two. Cubic splines find extensive application in the common problem of drawing a smooth curve through data. As a graphic example to illustrate a cubic spline, consider the problem of designing an overhead conveyor whose track is to be attached to the ceiling of a warehouse corridor at points $A_1, A_2, \ldots, A_7$, as

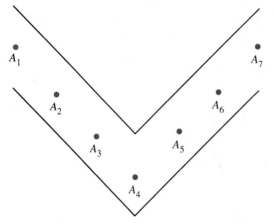

Figure 1.5

shown in Fig. 1.5. In order that the conveyor run smoothly, the track must be shaped so that there are no corners or abrupt changes in concavity; that is, the track should coincide with the graph of a smooth function. Moreover, besides being smooth, the track should be as straight as possible between the joints $A_1, A_2, \dots, A_7$. It can be shown (see a numerical analysis text) that a curve that is both as smooth as possible and as flat (straight) as possible coincides with the graph of a cubic spline.

1.4   Consistent Systems of Linear Equations

We saw in Section 1.2 that a system of linear equations may have a unique solution, infinitely many solutions, or no solution. In this section and in later sections, it will be shown that with certain added bits of information we can, without solving the system, either eliminate one of the three possible outcomes or determine precisely what the outcome will be. This will be important later when situations will arise in which we are not interested in obtaining a specific solution, but we need to know only how many solutions there are.

To illustrate, consider the general (2×3) linear system

$$a_{11}x_1 + a_{12}x_2 + a_{13}x_3 = b_1$$
$$a_{21}x_1 + a_{22}x_2 + a_{23}x_3 = b_2.$$

Geometrically, the system is represented by two planes, and a solution corresponds to a point in the intersection of the planes. The two planes may be parallel, they may be coincident (the same plane), or they may intersect in a line. Thus the system is either inconsistent or has infinitely many solutions; the existence of a unique solution is impossible. We will generalize this situation in the corollary to Theorem 3.

Solution Sets for Consistent Linear Systems

Numerous examples and exercises in the preceding sections have illustrated that the presence of unconstrained variables indicates when a consistent system of linear equations has infinitely many solutions. The next theorem provides a precise formulation, albeit a somewhat technical one, for the number of independent variables that arise when solving a consistent system.

THEOREM 3

Let $[A \,|\, \mathbf{b}]$ be the augmented matrix for a consistent system of linear equations in n unknowns. Further assume that $[A \,|\, \mathbf{b}]$ is row equivalent to a matrix $[C \,|\, \mathbf{d}]$ that is in echelon form and has r nonzero rows. Then $r \leq n$, and in the solution to the given system there are $n - r$ variables that can be assigned arbitrary values.

The remark in Section 1.2 that a system of linear equations has either infinitely many solutions, no solution, or a unique solution is an immediate consequence of Theorem 3. To see this, note, by Theorem 3, that for a given system the following three possibilities exist:

1. The system is inconsistent.
2. The system is consistent and, in the notation of Theorem 3, $r < n$. In this case there are $n - r$ unconstrained variables, so the system has infinitely many solutions.
3. The system is consistent and $r = n$. In this case there are no unconstrained variables, so the system has a unique solution.

The proof of Theorem 3 will be postponed to the end of this section, but the main idea of the proof can be illustrated by considering an example. The (4×5) system of equations

$$
\begin{aligned}
2x_1 - 3x_2 + x_3 - 3x_4 + 2x_5 &= 6 \\
2x_1 - 3x_2 + 5x_3 - x_4 + x_5 &= 8 \\
4x_3 + 3x_4 + 2x_5 &= 3 \\
-2x_1 + 3x_2 + 3x_3 + 3x_4 - 9x_5 &= -6
\end{aligned}
$$

has augmented matrix given by

$$
B = \begin{bmatrix}
2 & -3 & 1 & -3 & 2 & 6 \\
2 & -3 & 5 & -1 & 1 & 8 \\
0 & 0 & 4 & 3 & 2 & 3 \\
-2 & 3 & 3 & 3 & -9 & -6
\end{bmatrix}.
$$

It is a straightforward exercise to reduce B to the matrix

$$
C = \begin{bmatrix}
2 & -3 & 1 & -3 & 2 & 6 \\
0 & 0 & 4 & 2 & -1 & 2 \\
0 & 0 & 0 & 1 & 3 & 1 \\
0 & 0 & 0 & 0 & 0 & 0
\end{bmatrix},
$$

where C is in echelon form. Now the number of nonzero rows in C is three, so, in the notation of Theorem 3, $r = 3$ and $n = 5$. Since $n - r = 5 - 3 = 2$, Theorem 3 asserts that in the solution to the original system, there are two free variables. This is evident in the special case above, because the given system of equations is equivalent to the system represented by C:

$$2x_1 - 3x_2 + x_3 - 3x_4 + 2x_5 = 6$$
$$4x_3 + 2x_4 - x_5 = 2$$
$$x_4 + 3x_5 = 1.$$

In the reduced system above, the unknowns x_1, x_3, and x_4 may be chosen as the dependent variables; that is, the first variable appearing in each equation may be selected as a dependent variable. Thus the number of dependent variables coincides with the number, r, of equations. The remaining $n - r$ variables (x_2 and x_5 in our example) then become independent variables.

The preceding example, in fact, illustrates the following corollary to Theorem 3.

COROLLARY

Consider an $(m \times n)$ system of linear equations. If $m < n$, then either the system is inconsistent or it has infinitely many solutions.

PROOF   Consider an $(m \times n)$ system of linear equations where $m < n$. If the system is inconsistent, there is nothing to prove. If the system is consistent, then Theorem 3 applies. For a consistent system, suppose that the augmented matrix $[A \,|\, \mathbf{b}]$ is row equivalent to a matrix $[C \,|\, \mathbf{d}]$ that is in echelon form and has r nonzero rows. Because the given system has m equations, the augmented matrix $[A \,|\, \mathbf{b}]$ has m rows. Therefore the matrix $[C \,|\, \mathbf{d}]$ also has m rows. Because r is the number of nonzero rows for $[C \,|\, \mathbf{d}]$, it is clear that $r \leq m$. But $m < n$, so it follows that $r < n$. By Theorem 3, there are $n - r$ independent variables. Because $n - r > 0$, the system has infinitely many solutions.

EXAMPLE 1 What are the possibilities for the solution set of a (3×4) system of linear equations? If the system is consistent, what are the possibilities for the number of independent variables?

SOLUTION   By the corollary to Theorem 3, the system either has no solution or has infinitely many solutions. If the system reduces to a system with r equations, then $r \leq 3$. Thus r must be 1, 2, or 3. (The case $r = 0$ can occur only when the original system is the trivial system in which all coefficients and all constants are zero.) If the system is consistent, the number of free parameters is $4 - r$, so the possibilities are 3, 2, and 1.

EXAMPLE 2 What are the possibilities for the solution set of the following (3×4) system?

$$2x_1 - x_2 + x_3 - 3x_4 = 0$$
$$x_1 + 3x_2 - 2x_3 + x_4 = 0 \qquad \textbf{(1)}$$
$$-x_1 - 2x_2 + 4x_3 - x_4 = 0$$

SOLUTION   First note that $x_1 = x_2 = x_3 = x_4 = 0$ is a solution, so the system is consistent. By the corollary to Theorem 3, the system must have infinitely many solutions. That is, $m = 3$ and $n = 4$, so $m < n$.

Homogeneous Systems

System (1) above is an example of a "homogeneous" system of equations. More generally, the $(m \times n)$ system of linear equations given in (2) is called a **homogeneous** system of linear equations:

$$a_{11}x_1 + a_{12}x_2 + \cdots + a_{1n}x_n = 0$$
$$a_{21}x_1 + a_{22}x_2 + \cdots + a_{2n}x_n = 0$$
$$\vdots \qquad\qquad \vdots \quad\; \vdots$$
$$a_{m1}x_1 + a_{m2}x_2 + \cdots + a_{mn}x_n = 0. \tag{2}$$

Thus system (2) is the special case of the general $(m \times n)$ system given earlier (recall system (2) in Section 1.1) in which $b_1 = b_2 = \cdots = b_m = 0$. Note that a homogeneous system is always consistent, because $x_1 = x_2 = \cdots = x_n = 0$ is a solution to system (2). This solution is called the **trivial solution** or **zero solution,** and any other solution is called a **nontrivial solution.** A homogeneous system of equations, therefore, either has the trivial solution as the unique solution or also has nontrivial (and hence infinitely many) solutions. With these observations, the following important theorem is an immediate consequence of the corollary to Theorem 3.

THEOREM 4

A homogeneous $(m \times n)$ system of linear equations always has infinitely many nontrivial solutions when $m < n$.

EXAMPLE 3 What are the possibilities for the solution set of

$$x_1 + 2x_2 + x_3 + 3x_4 = 0$$
$$2x_1 + 4x_2 + 3x_3 + x_4 = 0$$
$$3x_1 + 6x_2 + 6x_3 + 2x_4 = 0?$$

Solve the system.

SOLUTION   By Theorem 4, the system has infinitely many nontrivial solutions. We solve by reducing the augmented matrix:

$$\begin{bmatrix} 1 & 2 & 1 & 3 & 0 \\ 2 & 4 & 3 & 1 & 0 \\ 3 & 6 & 6 & 2 & 0 \end{bmatrix}$$

$R_2 - 2R_1, R_3 - 3R_1$:

$$\begin{bmatrix} 1 & 2 & 1 & 3 & 0 \\ 0 & 0 & 1 & -5 & 0 \\ 0 & 0 & 3 & -7 & 0 \end{bmatrix}$$

$R_3 - 3R_2$:

$$\begin{bmatrix} 1 & 2 & 1 & 3 & 0 \\ 0 & 0 & 1 & -5 & 0 \\ 0 & 0 & 0 & 8 & 0 \end{bmatrix}$$

Note that the last column of zeros is maintained under elementary row operations, so the given system is equivalent to the homogeneous system

$$x_1 + 2x_2 + x_3 + 3x_4 = 0$$
$$x_3 - 5x_4 = 0$$
$$8x_4 = 0.$$

Backsolving gives

$$x_1 = -2x_2$$
$$x_3 = 0$$
$$x_4 = 0$$

as the solution. □

EXAMPLE 4 What are the possibilities for the solution set of

$$2x_1 + 3x_2 - x_3 = 0$$
$$-2x_1 - 2x_2 + 3x_3 = 0$$
$$2x_1 + 6x_2 + 9x_3 = 0?$$

Solve the system.

SOLUTION Theorem 4 no longer applies, because $m = n = 3$. However, because the system is homogeneous, either the trivial solution is the unique solution or there are infinitely many nontrivial solutions. To solve, we reduce the augmented matrix:

$$\begin{bmatrix} 2 & 3 & -1 & 0 \\ -2 & -2 & 3 & 0 \\ 2 & 6 & 9 & 0 \end{bmatrix}$$

$R_2 + R_1, R_3 - R_1$:

$$\begin{bmatrix} 2 & 3 & -1 & 0 \\ 0 & 1 & 2 & 0 \\ 0 & 3 & 10 & 0 \end{bmatrix}$$

$R_3 - 3R_2$:

$$\begin{bmatrix} 2 & 3 & -1 & 0 \\ 0 & 1 & 2 & 0 \\ 0 & 0 & 4 & 0 \end{bmatrix}.$$

Thus the given system is equivalent to the system

$$2x_1 + 3x_2 - x_3 = 0$$
$$x_2 + 2x_3 = 0$$
$$4x_3 = 0,$$

and we can easily see that $x_1 = x_2 = x_3 = 0$ is the only solution for this system.

☐

Proof of Theorem 3 (Optional)*

Suppose that a system of linear equations in n unknowns is consistent and assume that the augmented matrix $[A\,|\,\mathbf{b}]$ for the system reduces to the matrix $[C\,|\,\mathbf{d}]$ in echelon form. Thus $[C\,|\,\mathbf{d}]$ has the form

$$[C\,|\,\mathbf{d}] = \begin{bmatrix} c_{11} & \cdots & c_{1k_2} & \cdots & c_{1k_r} & \cdots & c_{1n} & d_1 \\ 0 & \cdots & c_{2k_2} & \cdots & c_{2k_r} & \cdots & c_{2n} & d_2 \\ \vdots & & \vdots & & \vdots & & \vdots & \vdots \\ 0 & \cdots & 0 & \cdots & c_{rk_r} & \cdots & c_{rn} & d_r \\ 0 & \cdots & 0 & \cdots & 0 & \cdots & 0 & d_{r+1} \\ \vdots & & \vdots & & \vdots & & \vdots & \vdots \\ 0 & \cdots & 0 & \cdots & 0 & \cdots & 0 & d_n \end{bmatrix}.$$

In the matrix above, each of the entries $c_{11}, c_{2k_2}, \ldots, c_{rk_r}$ is the first nonzero entry in its row. If $d_{r+1} \neq 0$, then clearly the system represented by the matrix $[C\,|\,\mathbf{d}]$ is inconsistent, and hence so is the original system. But we are assuming that the system is consistent, so it must be the case that $d_{r+1} = 0$. Since the matrix $[C\,|\,\mathbf{d}]$ is in echelon form, it follows that $d_{r+1} = d_{r+2} = \cdots = d_n = 0$. Thus r is the number of nonzero rows for the matrix $[C\,|\,\mathbf{d}]$.

Since each of the entries $c_{11}, c_{2k_2}, \ldots, c_{rk_r}$ appears in a different column of $[C\,|\,\mathbf{d}]$ and since these entries appear in the first n columns, it follows that $r \leq n$. Furthermore, the results of Section 1.3 imply that the given system is equivalent to the reduced system

$$c_{11}x_1 + \cdots + c_{1k_2}x_{k_2} + \cdots + c_{1k_r}x_{k_r} + \cdots + c_{1n}x_n = d_1$$
$$c_{2k_2}x_{k_2} + \cdots + c_{2k_r}x_{k_r} + \cdots + c_{2n}x_n = d_2$$
$$\vdots \qquad\qquad \vdots \qquad \vdots$$
$$c_{rk_r}x_{k_r} + \cdots + c_{rn}x_n = d_r. \tag{3}$$

* Throughout, material designated as optional may be omitted without any loss of continuity. This material is not prerequisite for topics in later sections except occasionally for other material also designated as optional.

In system (3), the first unknown appearing in each equation can be selected as a dependent variable; that is, each of the unknowns $x_1, x_{k_2}, \ldots, x_{k_r}$ can be expressed in terms of the remaining unknowns. Thus in a reduced system of equations such as (3), the number of dependent variables is equal to the number, r, of equations. This leaves $n - r$ independent variables, as the theorem asserts.

□

Exercises 1.4

In Exercises 1–4, reduce the augmented matrix for the given system to echelon form and, in the notation of Theorem 3, determine n, r, and the number, $n - r$, of independent variables. If $n - r > 0$, then identify $n - r$ independent variables.

1.
$$\begin{aligned} 2x_1 + 2x_2 - x_3 &= 1 \\ -2x_1 - 2x_2 + 4x_3 &= 1 \\ 2x_1 + 2x_2 + 5x_3 &= 5 \\ -2x_1 - 2x_2 - 2x_3 &= -3 \end{aligned}$$

2.
$$\begin{aligned} 2x_1 + 2x_2 &= 1 \\ 4x_1 + 5x_2 &= 4 \\ 4x_1 + 2x_2 &= -2 \end{aligned}$$

3.
$$\begin{aligned} -x_2 + x_3 + x_4 &= 2 \\ x_1 + 2x_2 + 2x_3 - x_4 &= 3 \\ x_1 + 3x_2 + x_3 &= 2 \end{aligned}$$

4.
$$\begin{aligned} x_1 + 2x_2 + 3x_3 + 2x_4 &= 1 \\ x_1 + 2x_2 + 3x_3 + 5x_4 &= 2 \\ 2x_1 + 4x_2 + 6x_3 + x_4 &= 1 \\ -x_1 - 2x_2 - 3x_3 + 7x_4 &= 2 \end{aligned}$$

In Exercises 5 and 6, assume that the given system is consistent. For each system determine, in the notation of Theorem 3, all possibilities for r and for the number, $n - r$, of unconstrained variables. Can the system have a unique solution?

5.
$$\begin{aligned} ax_1 + bx_2 &= c \\ dx_1 + ex_2 &= f \\ gx_1 + hx_2 &= i \end{aligned}$$

6.
$$\begin{aligned} a_{11}x_1 + a_{12}x_2 + a_{13}x_3 + a_{14}x_4 &= b_1 \\ a_{21}x_1 + a_{22}x_2 + a_{23}x_3 + a_{24}x_4 &= b_2 \\ a_{31}x_1 + a_{32}x_2 + a_{33}x_3 + a_{34}x_4 &= b_3 \end{aligned}$$

In Exercises 7–20, determine all possibilities for the solution set (from among infinitely many solutions, a unique solution, or no solution) of the system of linear equations described.

7. A system of 3 equations in 4 unknowns

8. A system of 4 equations in 5 unknowns

9. A homogeneous system of 3 equations in 4 unknowns

10. A homogeneous system of 4 equations in 5 unknowns

11. A system of 3 equations in 2 unknowns

12. A system of 4 equations in 3 unknowns

13. A homogeneous system of 3 equations in 2 unknowns

14. A homogeneous system of 4 equations in 3 unknowns

15. A system of 2 equations in 3 unknowns that has $x_1 = 1, x_2 = 2, x_3 = -1$ as a solution

16. A system of 3 equations in 4 unknowns that has $x_1 = -1, x_2 = 0, x_3 = 2, x_4 = -3$ as a solution

17. A homogeneous system of 2 equations in 2 unknowns

18. A homogeneous system of 3 equations in 3 unknowns

19. A homogeneous system of 2 equations in 2 unknowns that has solution $x_1 = 1, x_2 = -1$

20. A homogeneous system of 3 equations in 3 unknowns that has solution $x_1 = 1, x_2 = 3, x_3 = -1$

In Exercises 21–24, determine by inspection whether the given system has nontrivial solutions or only the trivial solution.

21.
$$\begin{aligned} 2x_1 + 3x_2 - x_3 &= 0 \\ x_1 - x_2 + 2x_3 &= 0 \end{aligned}$$

22.
$$\begin{aligned} x_1 + 2x_2 - x_3 + 2x_4 &= 0 \\ 2x_1 + x_2 + x_3 - x_4 &= 0 \\ 3x_1 - x_2 - 2x_3 + 3x_4 &= 0 \end{aligned}$$

23. $x_1 + 2x_2 - x_3 = 0$
$x_2 + 2x_3 = 0$
$4x_3 = 0$

24. $x_1 - x_2 = 0$
$3x_1 \quad = 0$
$2x_1 + x_2 = 0$

25. Let B be a (4×3) matrix in echelon form.
a) If B has three nonzero rows, then determine the form of B. (Using Fig. 1.2 of Section 1.3 as a guide, denote nonzero entries by $*$ and entries that may or may not be zero by $\times$.)
b) Suppose that a system of 4 linear equations in 2 unknowns has augmented matrix A, where A is a (4×3) matrix row equivalent to B. Demonstrate that the system of equations is inconsistent.

<table>
<tr><td>**1.5**</td><td></td></tr>
</table>

1.5   Applications (Optional)

In this brief section we discuss networks and methods for determining flows in networks. An example of a network is the system of one-way streets shown in Fig. 1.6. A typical problem associated with networks estimating the flow of traffic through this network of streets. Another example is the electrica¹ network shown in Fig. 1.7. A typical problem consists of determining the currents flowing through the loops of the circuit.

Note: The network problems we discuss in this section are kept very simple so the computational details do not obscure the ideas.

Flows in Networks

Networks consist of branches and nodes. For the street network shown in Fig. 1.6, the branches are the streets and the nodes are the intersections. We assume for a network that the total flow into a node is equal to the total flow out of the node. For example, Fig. 1.8 shows a flow of 40 into a node and a total flow of $x_1 + x_2 + 5$ out of the node. Since we assume that the flow into

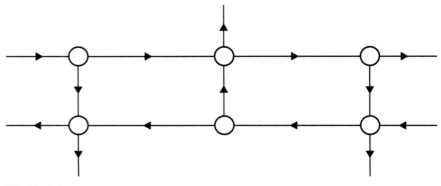

Figure 1.6
A network of one-way streets

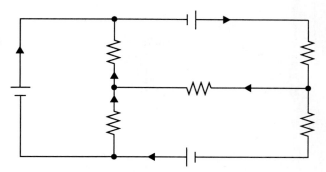

Figure 1.7
An electrical network

a node is equal to the flow out, it follows that the flows x_1 and x_2 must satisfy the linear equation $40 = x_1 + x_2 + 5$, or equivalently,

$$x_1 + x_2 = 35.$$

As an example to illustrate network flow calculations, consider the system of one-way streets in Fig. 1.9, where the flow is given in vehicles per hour. For instance, $x_1 + x_4$ vehicles per hour enter node B while $x_2 + 400$ vehicles per hour leave.

EXAMPLE 1   a) Set up a system of equations that represents traffic flow for the network shown in Fig. 1.9. (The numbers give the average flows into and out of the network at peak traffic hours.)

b) Solve the system of equations. What is the traffic flow if $x_6 = 300$ and $x_7 = 1300$ vehicles per hour?

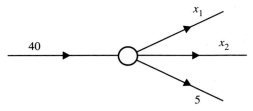

Figure 1.8
Since we assume that the flow into a node
is equal to the flow out, in this case,
$x_1 + x_2 = 35$.

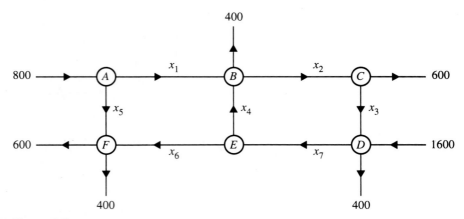

Figure 1.9
The traffic network analyzed in Example 1

SOLUTION

a) Since the flow into a node is equal to the flow out, we obtain the following system of equations:

$$
\begin{aligned}
800 &= x_1 + x_5 & \text{(Node } A) \\
x_1 + x_4 &= 400 + x_2 & \text{(Node } B) \\
x_2 &= 600 + x_3 & \text{(Node } C) \\
1600 + x_3 &= 400 + x_7 & \text{(Node } D) \\
x_7 &= x_4 + x_6 & \text{(Node } E) \\
x_5 + x_6 &= 1000. & \text{(Node } F)
\end{aligned}
$$

b) The augmented matrix for the system above is

$$
\begin{bmatrix}
1 & 0 & 0 & 0 & 1 & 0 & 0 & 800 \\
1 & -1 & 0 & 1 & 0 & 0 & 0 & 400 \\
0 & 1 & -1 & 0 & 0 & 0 & 0 & 600 \\
0 & 0 & 1 & 0 & 0 & 0 & -1 & -1200 \\
0 & 0 & 0 & 1 & 0 & 1 & -1 & 0 \\
0 & 0 & 0 & 0 & 1 & 1 & 0 & 1000
\end{bmatrix}.
$$

Some calculations show that this matrix is row equivalent to

$$
\begin{bmatrix}
1 & 0 & 0 & 0 & 1 & 0 & 0 & 800 \\
0 & 1 & 0 & -1 & 1 & 0 & 0 & 400 \\
0 & 0 & 1 & -1 & 1 & 0 & 0 & -200 \\
0 & 0 & 0 & 1 & -1 & 0 & -1 & -1000 \\
0 & 0 & 0 & 0 & 1 & 1 & 0 & 1000 \\
0 & 0 & 0 & 0 & 0 & 0 & 0 & 0
\end{bmatrix}.
$$

Therefore, the solution is

$$x_1 = x_6 - 200$$
$$x_2 = x_7 - 600$$
$$x_3 = x_7 - 1200$$
$$x_4 = x_7 - x_6$$
$$x_5 = 1000 - x_6.$$

If $x_6 = 300$ and $x_7 = 1300$, then (in vehicles per hour)

$$x_1 = 100, \qquad x_2 = 700, \qquad x_3 = 100, \qquad x_4 = 1000, \qquad x_5 = 700. \quad \square$$

We normally want the flows in a network to be nonnegative. For instance, consider the traffic network in Fig. 1.9. If x_6 were negative, it would indicate that traffic was flowing from F to E rather than in the prescribed direction from E to F.

EXAMPLE 2   Consider the street network in Example 1 (see Fig. 1.9). Suppose that the streets from A to B and from B to C must be closed (that is, $x_1 = 0$ and $x_2 = 0$). How might the traffic be rerouted?

SOLUTION   By Example 1, the flows are

$$x_1 = x_6 - 200$$
$$x_2 = x_7 - 600$$
$$x_3 = x_7 - 1200$$
$$x_4 = x_7 - x_6$$
$$x_5 = 1000 - x_6.$$

Therefore, if $x_1 = 0$ and $x_2 = 0$, it follows that $x_6 = 200$ and $x_7 = 600$. Using these values, we then obtain $x_3 = -600$, $x_4 = 400$, and $x_5 = 800$. In order to have nonnegative flows, we must reverse directions on the street connecting C and D; this change makes $x_3 = 600$ instead of -600. The network flows are shown in Fig. 1.10. $\quad \square$

Electrical Networks

We now consider current flow in simple electrical networks such as the one illustrated in Fig. 1.11. For such networks, current flow is governed by Ohm's law and Kirchhoff's laws, as follows.

Ohm's Law: The voltage drop across a resistor is the product of the current and the resistance.

Kirchhoff's First Law: The sum of the currents flowing into a node is equal to the sum of the currents flowing out.

Kirchhoff's Second Law: The algebraic sum of the voltage drops around a closed loop is equal to the total voltage in the loop.

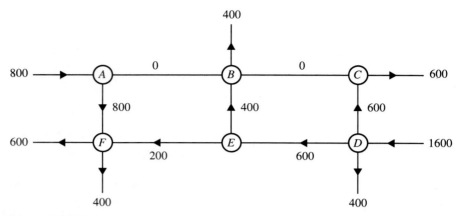

Figure 1.10
The traffic network analyzed in Example 2

(*Note:* With respect to Kirchhoff's second law, two basic closed loops in Fig. 1.11 are the counterclockwise paths *BDCB* and *BCAB*. Also, in each branch, we make a tentative assignment for the direction of current flow. If a current turns out to be negative, we then reverse our assignment for that branch.)

EXAMPLE 3 Determine the currents I_1, I_2, and I_3 for the electrical network shown in Fig. 1.11.

SOLUTION Applying Kirchhoff's second law to the loops *BDCB* and *BCAB*, we obtain equations

$$-10I_2 + 10I_3 = 10 \qquad (BDCB)$$
$$20I_1 + 10I_2 = 5. \qquad (BCAB)$$

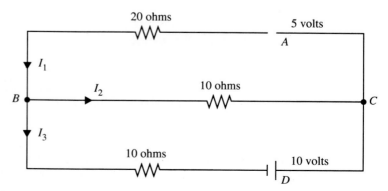

Figure 1.11
The electrical network analyzed in Example 3

Applying Kirchhoff's first law to either of the nodes B or C, we find $I_1 = I_2 + I_3$. Therefore,

$$I_1 - I_2 - I_3 = 0.$$

The augmented matrix for this system of three equations is

$$\begin{bmatrix} 1 & -1 & -1 & 0 \\ 0 & -10 & 10 & 10 \\ 20 & -10 & 0 & 5 \end{bmatrix}.$$

This matrix can be row reduced to

$$\begin{bmatrix} 1 & -1 & -1 & 0 \\ 0 & 1 & -1 & -1 \\ 0 & 0 & 10 & 7 \end{bmatrix}.$$

Therefore, the currents are

$$I_1 = 0.4, \qquad I_2 = -0.3, \qquad I_3 = 0.7.$$

Since I_2 is negative, the current flow is from C to B rather than from B to C, as tentatively assigned in Fig. 1.11. ☐

Exercises 1.5

In Exercises 1 and 2, (a) set up the system of equations that describes traffic flow; (b) determine the flows x_1, x_2, and x_3 if $x_4 = 100$; and (c) determine the maximum and minimum values for x_1 if all the flows are constrained to be nonnegative.

1.

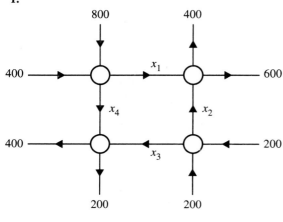

2.

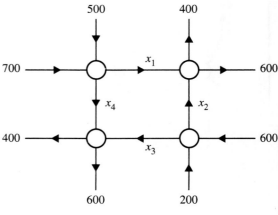

In Exercises 3 and 4, find the flow of traffic in the rotary if $x_1 = 600$.

In Exercises 5–8, determine the currents in the various branches.

3.

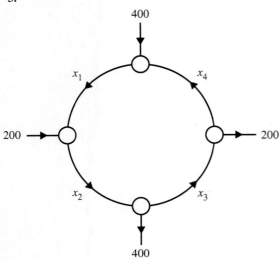

4.

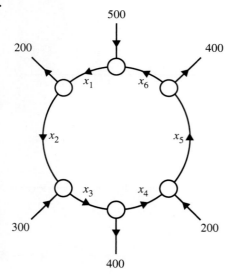

5.

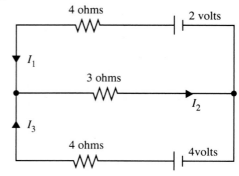

6.

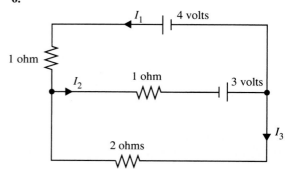

7.

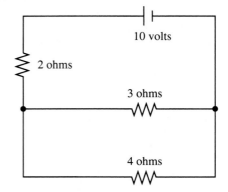

8.

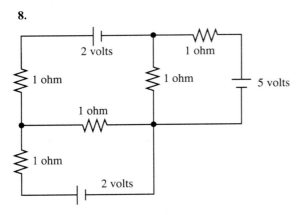

9. a) Set up the system of equations that describes the traffic flow in the accompanying figure.
b) Show that the system is consistent if and only if $a_1 + b_1 + c_1 + d_1 = a_2 + b_2 + c_2 + d_2$.

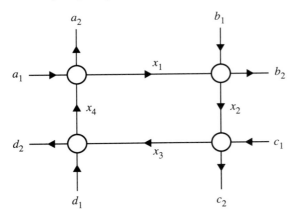

10. The electrical network shown in the accompanying figure is called a **Wheatstone bridge.** In this bridge, R_2 and R_4 are known resistances and R_3 is a known resistance that can be varied. The resistance R_1 is unknown and is to be determined by using the bridge. The resistance R_5 represents the internal resistance of a voltmeter attached between nodes B and D. The bridge is said to be balanced when R_3 is adjusted so that there is no current flowing in the branch between B and D. Show that, when the bridge is balanced, $R_1 R_4 = R_2 R_3$. (In particular, the unknown resistance R_1 can be found from $R_1 = R_2 R_3 / R_4$ when the bridge is balanced.)

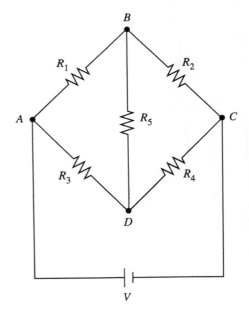

█ 1.6 █

Matrix Operations

In the previous sections, matrices were used as a convenient way of representing systems of equations. But matrices are of considerable interest and importance in their own right, and this section introduces the arithmetic operations that make them a useful computational and theoretical tool.

In this discussion of matrices and matrix operations (and later in the discussion of vectors), it is customary to refer to numerical quantities as **scalars.** For convenience we assume throughout this chapter that all matrix (and vector) entries are real numbers; hence the term *scalar* will mean a real number. In later chapters the term *scalar* will also be applied to complex numbers. We begin with a definition of the equality of two matrices.

> **DEFINITION 5**   Let $A = (a_{ij})$ be an $(m \times n)$ matrix, and let $B = (b_{ij})$ be an $(r \times s)$ matrix. We say that A and B are **equal** (and write $A = B$) if $m = r$, $n = s$, and $a_{ij} = b_{ij}$ for all i and j, $1 \leq i \leq m$, $1 \leq j \leq n$.

Thus two matrices are equal if they have the same number of rows and columns and, moreover, if all their corresponding entries are equal. For example, no two of the matrices

$$A = \begin{bmatrix} 1 & 2 \\ 3 & 4 \end{bmatrix}, \qquad B = \begin{bmatrix} 2 & 1 \\ 4 & 3 \end{bmatrix}, \quad \text{and} \quad C = \begin{bmatrix} 1 & 2 & 0 \\ 3 & 4 & 0 \end{bmatrix}$$

are equal.

Matrix Addition and Scalar Multiplication

The first two arithmetic operations, matrix addition and the multiplication of a matrix by a scalar, are defined quite naturally. In these definitions we use the notation $(Q)_{ij}$ to denote the ijth entry of a matrix Q.

> **DEFINITION 6**   Let $A = (a_{ij})$ and $B = (b_{ij})$ both be $(m \times n)$ matrices. The **sum**, $A + B$, is the $(m \times n)$ matrix defined by
> $$(A + B)_{ij} = a_{ij} + b_{ij}.$$

Note that this definition requires that A and B have the same dimension before their sum is defined. Thus if

$$A = \begin{bmatrix} 1 & 2 & -1 \\ 2 & 3 & 0 \end{bmatrix}, \qquad B = \begin{bmatrix} -3 & 1 & 2 \\ 0 & -4 & 1 \end{bmatrix}, \quad \text{and} \quad C = \begin{bmatrix} 1 & 2 \\ 3 & 1 \end{bmatrix},$$

then

$$A + B = \begin{bmatrix} -2 & 3 & 1 \\ 2 & -1 & 1 \end{bmatrix},$$

while $A + C$ is undefined.

> **DEFINITION 7**   Let $A = (a_{ij})$ be an $(m \times n)$ matrix, and let r be a scalar. The **product**, rA, is the $(m \times n)$ matrix defined by
> $$(rA)_{ij} = ra_{ij}.$$

For example,

$$2\begin{bmatrix} 1 & 3 \\ 2 & -1 \\ 0 & 3 \end{bmatrix} = \begin{bmatrix} 2 & 6 \\ 4 & -2 \\ 0 & 6 \end{bmatrix}.$$

EXAMPLE 1 Let the matrices A, B, and C be given by

$$A = \begin{bmatrix} 1 & 3 \\ -2 & 7 \end{bmatrix}, \quad B = \begin{bmatrix} 6 & 1 \\ 2 & 4 \end{bmatrix}, \quad \text{and} \quad C = \begin{bmatrix} 1 & 2 & -1 \\ 3 & 0 & 5 \end{bmatrix}.$$

Find each of $A + B$, $A + C$, $B + C$, $3C$, and $A + 2B$ or state that the indicated operation is undefined.

SOLUTION   The defined operations yield

$$A + B = \begin{bmatrix} 7 & 4 \\ 0 & 11 \end{bmatrix}, \quad 3C = \begin{bmatrix} 3 & 6 & -3 \\ 9 & 0 & 15 \end{bmatrix}, \quad \text{and} \quad A + 2B = \begin{bmatrix} 13 & 5 \\ 2 & 15 \end{bmatrix},$$

while $A + C$ and $B + C$ are undefined.   ☐

Vectors in R^n

Before proceeding with the definition of matrix multiplication, recall that a point in n-dimensional space is represented by an ordered n-tuple of real numbers $\mathbf{x} = (x_1, x_2, \ldots, x_n)$. Such an n-tuple will be called an **n-dimensional vector** and will be written in the form of a matrix,

$$\mathbf{x} = \begin{bmatrix} x_1 \\ x_2 \\ \vdots \\ x_n \end{bmatrix}.$$

For example, an arbitrary three-dimensional vector has the form

$$\mathbf{x} = \begin{bmatrix} x_1 \\ x_2 \\ x_3 \end{bmatrix},$$

and the vectors

$$\mathbf{x} = \begin{bmatrix} 1 \\ 2 \\ 3 \end{bmatrix}, \quad \mathbf{y} = \begin{bmatrix} 3 \\ 2 \\ 1 \end{bmatrix}, \quad \text{and} \quad \mathbf{z} = \begin{bmatrix} 2 \\ 3 \\ 1 \end{bmatrix}$$

are distinct three-dimensional vectors. The set of all n-dimensional vectors with real components is called **Euclidean n-space** and will be denoted by R^n. Vectors in R^n will be denoted by boldface type. Thus R^n is the set defined by

$$R^n = \left\{ \mathbf{x} \colon \mathbf{x} = \begin{bmatrix} x_1 \\ x_2 \\ \vdots \\ x_n \end{bmatrix} x_1, x_2, \ldots, x_n \text{ real numbers} \right\}.$$

As the notation suggests, an element of R^n can be viewed as an $(n \times 1)$ real matrix, and conversely an $(n \times 1)$ real matrix can be considered an element of R^n. Thus addition and scalar multiplication of vectors is just a special case of these operations for matrices.

In vector calculus the **scalar product** (or **dot product**) of two vectors

$$\mathbf{u} = \begin{bmatrix} u_1 \\ u_2 \\ \vdots \\ u_n \end{bmatrix} \quad \text{and} \quad \mathbf{v} = \begin{bmatrix} v_1 \\ v_2 \\ \vdots \\ v_n \end{bmatrix}$$

in R^n is defined to be the number $u_1 v_1 + u_2 v_2 + \cdots + u_n v_n = \sum_{i=1}^{n} u_i v_i$. For example, if

$$\mathbf{u} = \begin{bmatrix} 2 \\ 3 \\ -1 \end{bmatrix} \quad \text{and} \quad \mathbf{v} = \begin{bmatrix} -4 \\ 2 \\ 3 \end{bmatrix},$$

then the scalar product of $\mathbf{u}$ and $\mathbf{v}$ is $2(-4) + 3(2) + (-1)3 = -5$. The scalar product of two vectors will be considered further in the following section, and in Chapter 2 the properties of R^n will be more fully developed. For the present we return to the introduction of matrix multiplication.

Matrix Multiplication

Matrix multiplication is defined in such a way as to provide a convenient mechanism for describing a linear correspondence between vectors. To illustrate, let the variables $x_1, x_2, \ldots, x_n$ and the variables $y_1, y_2, \ldots, y_m$ be related by the linear equations

$$\begin{aligned}
a_{11}x_1 + a_{12}x_2 + \cdots + a_{1n}x_n &= y_1 \\
a_{21}x_1 + a_{22}x_2 + \cdots + a_{2n}x_n &= y_2 \\
&\vdots \\
a_{m1}x_1 + a_{m2}x_2 + \cdots + a_{mn}x_n &= y_m.
\end{aligned} \tag{1}$$

If we set

$$\mathbf{x} = \begin{bmatrix} x_1 \\ x_2 \\ \vdots \\ x_n \end{bmatrix} \quad \text{and} \quad \mathbf{y} = \begin{bmatrix} y_1 \\ y_2 \\ \vdots \\ y_m \end{bmatrix},$$

then (1) defines a correspondence $\mathbf{x} \to \mathbf{y}$ from vectors in R^n to vectors in R^m. The ith equation of (1) is

$$a_{i1}x_1 + a_{i2}x_2 + \cdots + a_{in}x_n = y_i,$$

and this can be written in a briefer form as

$$\sum_{j=1}^{n} a_{ij}x_j = y_i. \tag{2}$$

If A is the coefficient matrix of system (1),

$$A = \begin{bmatrix} a_{11} & a_{12} & \cdots & a_{1n} \\ a_{21} & a_{22} & \cdots & a_{2n} \\ \vdots & & & \vdots \\ a_{m1} & a_{m2} & \cdots & a_{mn} \end{bmatrix},$$

then the left-hand side of Eq. (2) is precisely the scalar product of the ith row of A with the vector $\mathbf{x}$. Thus if we define the product of A and $\mathbf{x}$ to be the $(m \times 1)$ vector $A\mathbf{x}$ whose ith component is the scalar product of the ith row of A with $\mathbf{x}$, then $A\mathbf{x}$ is given by

$$A\mathbf{x} = \begin{bmatrix} \displaystyle\sum_{j=1}^{n} a_{1j}x_j \\ \displaystyle\sum_{j=1}^{n} a_{2j}x_j \\ \vdots \\ \displaystyle\sum_{j=1}^{n} a_{mj}x_j \end{bmatrix}.$$

Using the definition of equality (Definition 5), we see that the simple matrix equation

$$A\mathbf{x} = \mathbf{y} \tag{3}$$

is equivalent to system (1).

In a natural fashion, we can extend the idea of the product of a matrix and a vector to the product, AB, of an $(m \times n)$ matrix A and an $(n \times s)$ matrix B by defining the ijth entry of AB to be the scalar product of the ith row of A with the jth column of B. Formally, we have the following definition.

DEFINITION 8 Let $A = (a_{ij})$ be an $(m \times n)$ matrix, and let $B = (b_{ij})$ be an $(r \times s)$ matrix. If $n = r$, then the **product** AB is the $(m \times s)$ matrix defined by

$$(AB)_{ij} = \sum_{k=1}^{n} a_{ik}b_{kj}.$$

If $n \neq r$, then the product AB is not defined.

The definition can be visualized by referring to Fig. 1.12.

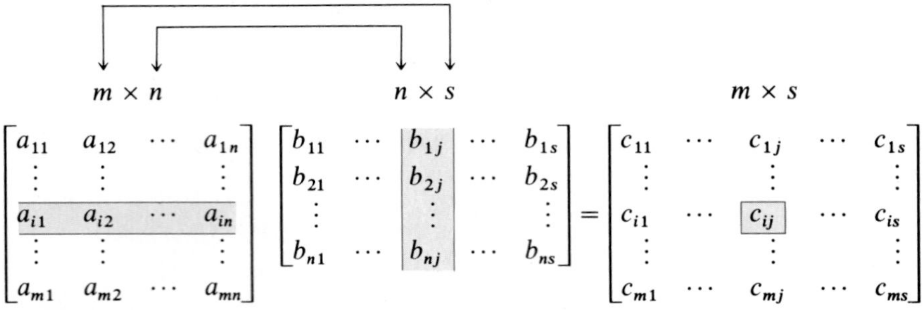

Figure 1.12
The ijth entry of AB is the scalar product of the ith row of A and
the jth column of B.

Thus the product AB is defined only when the inside dimensions of A and B are equal. In this case the outside dimensions, m and s, give the size of AB. Furthermore, the ijth entry of AB is the scalar product of the ith row of A with the jth column of B. For example,

$$\begin{bmatrix} 2 & 1 & -3 \\ -2 & 2 & 4 \end{bmatrix} \begin{bmatrix} -1 & 2 \\ 0 & -3 \\ 2 & 1 \end{bmatrix} = \begin{bmatrix} 2(-1) + 1(0) + (-3)2 & 2(2) + 1(-3) + (-3)1 \\ (-2)(-1) + 2(0) + 4(2) & (-2)2 + 2(-3) + 4(1) \end{bmatrix}$$

$$= \begin{bmatrix} -8 & -2 \\ 10 & -6 \end{bmatrix},$$

whereas the product

$$\begin{bmatrix} 2 & 1 & -3 \\ -2 & 2 & 4 \end{bmatrix} \begin{bmatrix} -1 & 0 & 2 \\ 2 & -3 & 1 \end{bmatrix}$$

is undefined.

EXAMPLE 2 Let the matrices A, B, C, and D be given by

$$A = \begin{bmatrix} 1 & 2 \\ 2 & 3 \end{bmatrix}, \quad B = \begin{bmatrix} -3 & 2 \\ 1 & -2 \end{bmatrix},$$

$$C = \begin{bmatrix} 1 & 0 & -2 \\ 0 & 1 & 1 \end{bmatrix}, \quad \text{and} \quad D = \begin{bmatrix} 3 & 1 \\ -1 & -2 \\ 1 & 1 \end{bmatrix}.$$

Find each of AB, BA, AC, CA, CD, and DC or state that the indicated product is undefined.

SOLUTION The definition of matrix multiplication yields

$$AB = \begin{bmatrix} -1 & -2 \\ -3 & -2 \end{bmatrix}, \quad BA = \begin{bmatrix} 1 & 0 \\ -3 & -4 \end{bmatrix}, \quad \text{and} \quad AC = \begin{bmatrix} 1 & 2 & 0 \\ 2 & 3 & -1 \end{bmatrix}.$$

The product CA is undefined, and

$$CD = \begin{bmatrix} 1 & -1 \\ 0 & -1 \end{bmatrix} \quad \text{and} \quad DC = \begin{bmatrix} 3 & 1 & -5 \\ -1 & -2 & 0 \\ 1 & 1 & -1 \end{bmatrix}.$$

Example 2 illustrates that matrix multiplication is not commutative; that is, normally AB and BA are different matrices. Indeed, the product AB may be defined while the product BA is undefined, or both may be defined but have different dimensions. Even when AB and BA have the same size, they usually are not equal.

EXAMPLE 3 Express each of the linear systems

$$\begin{aligned} x_1 &= 2y_1 - y_2 \\ x_2 &= -3y_1 + 2y_2 \quad \text{and} \\ x_3 &= y_1 + 3y_2 \end{aligned} \qquad \begin{aligned} y_1 &= -4z_1 + 2z_2 \\ y_2 &= 3z_1 + z_2 \end{aligned}$$

as a matrix equation and use matrix multiplication to express x_1, x_2, and x_3 in terms of z_1 and z_2.

SOLUTION We have

$$\begin{bmatrix} x_1 \\ x_2 \\ x_3 \end{bmatrix} = \begin{bmatrix} 2 & -1 \\ -3 & 2 \\ 1 & 3 \end{bmatrix} \begin{bmatrix} y_1 \\ y_2 \end{bmatrix} \quad \text{and} \quad \begin{bmatrix} y_1 \\ y_2 \end{bmatrix} = \begin{bmatrix} -4 & 2 \\ 3 & 1 \end{bmatrix} \begin{bmatrix} z_1 \\ z_2 \end{bmatrix}.$$

Substituting for $\begin{bmatrix} y_1 \\ y_2 \end{bmatrix}$ in the left-hand equation gives

$$\begin{bmatrix} x_1 \\ x_2 \\ x_3 \end{bmatrix} = \begin{bmatrix} 2 & -1 \\ -3 & 2 \\ 1 & 3 \end{bmatrix} \begin{bmatrix} -4 & 2 \\ 3 & 1 \end{bmatrix} \begin{bmatrix} z_1 \\ z_2 \end{bmatrix} = \begin{bmatrix} -11 & 3 \\ 18 & -4 \\ 5 & 5 \end{bmatrix} \begin{bmatrix} z_1 \\ z_2 \end{bmatrix}.$$

Therefore,

$$\begin{aligned} x_1 &= -11z_1 + 3z_2 \\ x_2 &= 18z_1 - 4z_2 \\ x_3 &= 5z_1 + 5z_2. \end{aligned}$$

The use of the matrix equation (3) to represent the linear system (1) provides a convenient notational device for representing the $(m \times n)$ system

$$\begin{aligned} a_{11}x_1 + a_{12}x_2 + \cdots + a_{1n}x_n &= b_1 \\ a_{21}x_1 + a_{22}x_2 + \cdots + a_{2n}x_n &= b_2 \\ \vdots \qquad \vdots \qquad\qquad \vdots \qquad \vdots \\ a_{m1}x_1 + a_{m2}x_2 + \cdots + a_{mn}x_n &= b_m \end{aligned}$$

(4)

of linear equations with unknowns $x_1, \ldots, x_n$. Specifically, if $A = (a_{ij})$ is the coefficient matrix of (4), and if the unknown $(n \times 1)$ matrix $\mathbf{x}$ and the constant $(m \times 1)$ matrix $\mathbf{b}$ are defined by

$$\mathbf{x} = \begin{bmatrix} x_1 \\ x_2 \\ \vdots \\ x_n \end{bmatrix} \quad \text{and} \quad \mathbf{b} = \begin{bmatrix} b_1 \\ b_2 \\ \vdots \\ b_m \end{bmatrix},$$

then the system (4) is equivalent to the matrix equation

$$A\mathbf{x} = \mathbf{b}. \tag{5}$$

EXAMPLE 4 Solve the matrix equation $A\mathbf{x} = \mathbf{b}$, where

$$A = \begin{bmatrix} 1 & 3 & -1 \\ 2 & 5 & -1 \\ 2 & 8 & -2 \end{bmatrix}, \quad \mathbf{x} = \begin{bmatrix} x_1 \\ x_2 \\ x_3 \end{bmatrix}, \quad \text{and} \quad \mathbf{b} = \begin{bmatrix} 2 \\ 6 \\ 6 \end{bmatrix}.$$

SOLUTION   The matrix equation $A\mathbf{x} = \mathbf{b}$ is equivalent to the (3×3) linear system

$$\begin{aligned} x_1 + 3x_2 - x_3 &= 2 \\ 2x_1 + 5x_2 - x_3 &= 6 \\ 2x_1 + 8x_2 - 2x_3 &= 6. \end{aligned}$$

This system can be solved in the usual way—that is, by reducing the augmented matrix—to obtain $x_1 = 2$, $x_2 = 1$, $x_3 = 3$. Therefore,

$$\mathbf{s} = \begin{bmatrix} 2 \\ 1 \\ 3 \end{bmatrix}$$

is the unique solution to $A\mathbf{x} = \mathbf{b}$.

Other Formulations of Matrix Multiplication

It is frequently convenient and useful to express an $(m \times n)$ matrix $A = (a_{ij})$ in the form

$$A = [\mathbf{A}_1, \mathbf{A}_2, \ldots, \mathbf{A}_n], \tag{6}$$

where for each j, $1 \leq j \leq n$, $\mathbf{A}_j$ denotes the jth column of A. That is, $\mathbf{A}_j$ is the $(m \times 1)$ column vector

$$\mathbf{A}_j = \begin{bmatrix} a_{1j} \\ a_{2j} \\ \vdots \\ a_{mj} \end{bmatrix}.$$

For example, if A is the (2×3) matrix

$$A = \begin{bmatrix} 1 & 3 & 6 \\ 2 & 4 & 0 \end{bmatrix}, \qquad (7)$$

then $A = [\mathbf{A}_1, \mathbf{A}_2, \mathbf{A}_3]$, where

$$\mathbf{A}_1 = \begin{bmatrix} 1 \\ 2 \end{bmatrix}, \qquad \mathbf{A}_2 = \begin{bmatrix} 3 \\ 4 \end{bmatrix}, \quad \text{and} \quad \mathbf{A}_3 = \begin{bmatrix} 6 \\ 0 \end{bmatrix}.$$

The next two theorems use (6) to provide alternative ways of expressing the matrix products $A\mathbf{x}$ and AB; these methods will be extremely useful in our later development of matrix theory.

THEOREM 5

Let $A = [\mathbf{A}_1, \mathbf{A}_2, \ldots, \mathbf{A}_n]$ be an $(m \times n)$ matrix whose jth column is $\mathbf{A}_j$, and let $\mathbf{x}$ be the $(n \times 1)$ column vector

$$\mathbf{x} = \begin{bmatrix} x_1 \\ x_2 \\ \vdots \\ x_n \end{bmatrix}.$$

Then the product $A\mathbf{x}$ can be expressed as

$$A\mathbf{x} = x_1 \mathbf{A}_1 + x_2 \mathbf{A}_2 + \cdots + x_n \mathbf{A}_n.$$

The proof of this theorem is not difficult and uses only Definitions 5, 6, 7, and 8; the proof is left as an exercise for the reader. To illustrate Theorem 5, let A be the matrix

$$A = \begin{bmatrix} 1 & 3 & 6 \\ 2 & 4 & 0 \end{bmatrix},$$

and let $\mathbf{x}$ be the vector in R^3,

$$\mathbf{x} = \begin{bmatrix} x_1 \\ x_2 \\ x_3 \end{bmatrix}.$$

Then

$$A\mathbf{x} = \begin{bmatrix} 1 & 3 & 6 \\ 2 & 4 & 0 \end{bmatrix} \begin{bmatrix} x_1 \\ x_2 \\ x_3 \end{bmatrix}$$

$$= \begin{bmatrix} x_1 + 3x_2 + 6x_3 \\ 2x_1 + 4x_2 + 0x_3 \end{bmatrix}$$

$$= x_1 \begin{bmatrix} 1 \\ 2 \end{bmatrix} + x_2 \begin{bmatrix} 3 \\ 4 \end{bmatrix} + x_3 \begin{bmatrix} 6 \\ 0 \end{bmatrix};$$

so that $A\mathbf{x} = x_1\mathbf{A}_1 + x_2\mathbf{A}_2 + x_3\mathbf{A}_3$. In particular, if we set

$$\mathbf{x} = \begin{bmatrix} 2 \\ 2 \\ -3 \end{bmatrix},$$

then $A\mathbf{x} = 2\mathbf{A}_1 + 2\mathbf{A}_2 - 3\mathbf{A}_3$.

From Theorem 5, we see that the matrix equation $A\mathbf{x} = \mathbf{b}$ corresponding to the $(m \times n)$ system (4) can be expressed as

$$x_1\mathbf{A}_1 + x_2\mathbf{A}_2 + \cdots + x_n\mathbf{A}_n = \mathbf{b}. \tag{8}$$

Thus Eq. (8) says that solving $A\mathbf{x} = \mathbf{b}$ amounts to showing that $\mathbf{b}$ can be written in terms of the columns of A.

EXAMPLE 5 Solve

$$x_1\begin{bmatrix} 1 \\ 2 \\ 2 \end{bmatrix} + x_2\begin{bmatrix} 3 \\ 5 \\ 8 \end{bmatrix} + x_3\begin{bmatrix} -1 \\ -1 \\ -2 \end{bmatrix} = \begin{bmatrix} 2 \\ 6 \\ 6 \end{bmatrix}.$$

SOLUTION By Theorem 5, the given equation is equivalent to the matrix equation $A\mathbf{x} = \mathbf{b}$, where

$$A = \begin{bmatrix} 1 & 3 & -1 \\ 2 & 5 & -1 \\ 2 & 8 & -2 \end{bmatrix}, \qquad \mathbf{x} = \begin{bmatrix} x_1 \\ x_2 \\ x_3 \end{bmatrix}, \quad \text{and} \quad \mathbf{b} = \begin{bmatrix} 2 \\ 6 \\ 6 \end{bmatrix}.$$

This equation was solved in Example 4, giving $x_1 = 2, x_2 = 1, x_3 = 3$, so we have

$$2\begin{bmatrix} 1 \\ 2 \\ 2 \end{bmatrix} + \begin{bmatrix} 3 \\ 5 \\ 8 \end{bmatrix} + 3\begin{bmatrix} -1 \\ -1 \\ -2 \end{bmatrix} = \begin{bmatrix} 2 \\ 6 \\ 6 \end{bmatrix}. \qquad \square$$

Although Eq. (8) is not particularly efficient as a computational tool, it is useful for understanding how the internal structure of the coefficient matrix affects the possible solutions of the linear system $A\mathbf{x} = \mathbf{b}$.

Another important observation, which we will use later, is an alternative way of expressing the product of two matrices, as given in Theorem 6.

T H E O R E M   6

Let A be an $(m \times n)$ matrix, and let $B = [\mathbf{B}_1, \mathbf{B}_2, \ldots, \mathbf{B}_s]$ be an $(n \times s)$ matrix whose kth column is $\mathbf{B}_k$. Then the jth column of AB is $A\mathbf{B}_j$, so that

$$AB = [A\mathbf{B}_1, A\mathbf{B}_2, \ldots, A\mathbf{B}_s].$$

PROOF If $A = (a_{ij})$ and $B = (b_{ij})$, then the jth column of AB contains the entries

$$\sum_{k=1}^{n} a_{1k}b_{kj}$$

$$\sum_{k=1}^{n} a_{2k}b_{kj}$$

$$\vdots$$

$$\sum_{k=1}^{n} a_{mk}b_{kj};$$

and these are precisely the components of the column vector $A\mathbf{B}_j$, where

$$\mathbf{B}_j = \begin{bmatrix} b_{1j} \\ b_{2j} \\ \vdots \\ b_{nj} \end{bmatrix}.$$

It follows that we can write AB in the form $AB = [A\mathbf{B}_1, A\mathbf{B}_2, \ldots, A\mathbf{B}_s]$.

To illustrate Theorem 6, let A and B be given by

$$A = \begin{bmatrix} 2 & 6 \\ 0 & 4 \\ 1 & 2 \end{bmatrix} \quad \text{and} \quad B = \begin{bmatrix} 1 & 3 & 0 & 1 \\ 4 & 5 & 2 & 3 \end{bmatrix}.$$

Thus the column vectors for B are

$$\mathbf{B}_1 = \begin{bmatrix} 1 \\ 4 \end{bmatrix}, \quad \mathbf{B}_2 = \begin{bmatrix} 3 \\ 5 \end{bmatrix}, \quad \mathbf{B}_3 = \begin{bmatrix} 0 \\ 2 \end{bmatrix}, \quad \text{and} \quad \mathbf{B}_4 = \begin{bmatrix} 1 \\ 3 \end{bmatrix}$$

and

$$A\mathbf{B}_1 = \begin{bmatrix} 26 \\ 16 \\ 9 \end{bmatrix}, \quad A\mathbf{B}_2 = \begin{bmatrix} 36 \\ 20 \\ 13 \end{bmatrix}, \quad A\mathbf{B}_3 = \begin{bmatrix} 12 \\ 8 \\ 4 \end{bmatrix}, \quad \text{and} \quad A\mathbf{B}_4 = \begin{bmatrix} 20 \\ 12 \\ 7 \end{bmatrix}.$$

Calculating AB, we see immediately that AB is a (3×4) matrix with columns $A\mathbf{B}_1$, $A\mathbf{B}_2$, $A\mathbf{B}_3$, and $A\mathbf{B}_4$; that is,

$$AB = \begin{bmatrix} 26 & 36 & 12 & 20 \\ 16 & 20 & 8 & 12 \\ 9 & 13 & 4 & 7 \end{bmatrix}.$$

Exercises 1.6

The (2×2) matrices listed in (9) are used in several of the exercises that follow.

$$A = \begin{bmatrix} 2 & 1 \\ 1 & 3 \end{bmatrix}, \quad B = \begin{bmatrix} 0 & -1 \\ 1 & 3 \end{bmatrix},$$

$$C = \begin{bmatrix} -2 & 3 \\ 1 & 1 \end{bmatrix}, \quad Z = \begin{bmatrix} 0 & 0 \\ 0 & 0 \end{bmatrix} \tag{9}$$

Exercises 1–6 refer to the matrices in (9).

1. Find (a) $A + B$; (b) $A + C$; (c) $6B$; and (d) $B + 3C$.

2. Find (a) $B + C$; (b) $3A$; (c) $A + 2C$; and (d) $C + 8Z$.

3. Find a matrix D such that $A + D = B$.

4. Find a matrix D such that $A + 2D = C$.

5. Find a matrix D such that $A + 2B + 2D = 3B$.

6. Find a matrix D such that $2A + 5B + D = 2B + 3A$.

The vectors listed in (10) are used in several of the exercises that follow.

$$\mathbf{r} = \begin{bmatrix} 1 \\ 0 \end{bmatrix}, \quad \mathbf{s} = \begin{bmatrix} 2 \\ -3 \end{bmatrix},$$

$$\mathbf{t} = \begin{bmatrix} 1 \\ 4 \end{bmatrix}, \quad \mathbf{u} = \begin{bmatrix} -4 \\ 6 \end{bmatrix} \tag{10}$$

In Exercises 7–12, perform the indicated computation, using the vectors in (10) and the matrices in (9).

7. a) $\mathbf{r} + \mathbf{s}$
   b) $2\mathbf{r} + \mathbf{t}$
   c) $2\mathbf{s} + \mathbf{u}$

8. a) $\mathbf{t} + \mathbf{s}$
   b) $\mathbf{r} + 3\mathbf{u}$
   c) $2\mathbf{u} + 3\mathbf{t}$

9. a) $A\mathbf{r}$
   b) $B\mathbf{r}$
   c) $C(\mathbf{s} + 3\mathbf{t})$

10. a) $B\mathbf{t}$
    b) $C(\mathbf{r} + \mathbf{s})$
    c) $B(\mathbf{r} + \mathbf{s})$

11. a) $(A + 2B)\mathbf{r}$
    b) $(B + C)\mathbf{u}$

12. a) $(A + C)\mathbf{r}$
    b) $(2B + 3C)\mathbf{s}$

Exercises 13–20 refer to the vectors in (10). In each exercise, find scalars a_1 and a_2 that satisfy the given equation or state that the equation has no solution.

13. $a_1\mathbf{r} + a_2\mathbf{s} = \mathbf{t}$

14. $a_1\mathbf{r} + a_2\mathbf{s} = \mathbf{u}$

15. $a_1\mathbf{s} + a_2\mathbf{t} = \mathbf{u}$

16. $a_1\mathbf{s} + a_2\mathbf{t} = \mathbf{r} + \mathbf{t}$

17. $a_1\mathbf{s} + a_2\mathbf{u} = 2\mathbf{r} + \mathbf{t}$

18. $a_1\mathbf{s} + a_2\mathbf{u} = \mathbf{t}$

19. $a_1\mathbf{t} + a_2\mathbf{u} = 3\mathbf{s} + 4\mathbf{t}$

20. $a_1\mathbf{t} + a_2\mathbf{u} = 3\mathbf{r} + 2\mathbf{s}$

Exercises 21–24 refer to the matrices in (9) and the vectors in (10).

21. Find $\mathbf{w}_2$, where $\mathbf{w}_1 = B\mathbf{r}$ and $\mathbf{w}_2 = A\mathbf{w}_1$. Calculate $Q = AB$. Calculate $Q\mathbf{r}$ and verify that $\mathbf{w}_2$ is equal to $Q\mathbf{r}$.

22. Find $\mathbf{w}_2$, where $\mathbf{w}_1 = C\mathbf{s}$ and $\mathbf{w}_2 = A\mathbf{w}_1$. Calculate $Q = AC$. Calculate $Q\mathbf{s}$ and verify that $\mathbf{w}_2$ is equal to $Q\mathbf{s}$.

23. Find $\mathbf{w}_3$, where $\mathbf{w}_1 = C\mathbf{r}$, $\mathbf{w}_2 = B\mathbf{w}_1$, and $\mathbf{w}_3 = A\mathbf{w}_2$. Calculate $Q = A(BC)$ and verify that $\mathbf{w}_3$ is equal to $Q\mathbf{r}$.

24. Find $\mathbf{w}_3$, where $\mathbf{w}_1 = A\mathbf{r}$, $\mathbf{w}_2 = C\mathbf{w}_1$, and $\mathbf{w}_3 = B\mathbf{w}_2$. Calculate $Q = B(CA)$ and verify that $\mathbf{w}_3$ is equal to $Q\mathbf{r}$.

Exercises 25–30 refer to the matrices in (9). Find each of the following.

25. $(A + B)C$

26. $(A + 2B)A$

27. $(A + C)B$

28. $(B + C)Z$

29. $A(BZ)$

30. $Z(AB)$

The matrices and vectors listed in (11) are used in several of the exercises that follow.

$$A = \begin{bmatrix} 2 & 3 \\ 1 & 4 \end{bmatrix}, \quad B = \begin{bmatrix} 1 & 2 \\ 1 & 4 \end{bmatrix}, \quad \mathbf{u} = \begin{bmatrix} 1 \\ 3 \end{bmatrix},$$

$$\mathbf{v} = [2, 4], \quad C = \begin{bmatrix} 2 & 1 \\ 4 & 0 \\ 8 & -1 \\ 3 & 2 \end{bmatrix},$$

$$D = \begin{bmatrix} 2 & 1 & 3 & 6 \\ 2 & 0 & 0 & 4 \\ 1 & -1 & 1 & -1 \\ 1 & 3 & 1 & 2 \end{bmatrix}, \quad \mathbf{w} = \begin{bmatrix} 2 \\ 3 \\ 1 \\ 1 \end{bmatrix}. \tag{11}$$

Exercises 31–41 refer to the matrices and vectors in (11). Find each of the following.

31. AB and BA

32. DC

33. $A\mathbf{u}$ and $\mathbf{v}A$

34. $\mathbf{u}\mathbf{v}$ and $\mathbf{v}\mathbf{u}$

35. $\mathbf{v}(B\mathbf{u})$

36. $B\mathbf{u}$

37. CA

38. CB

39. $C(B\mathbf{u})$

40. $(AB)\mathbf{u}$ and $A(B\mathbf{u})$

41. $(BA)\mathbf{u}$ and $B(A\,\mathbf{u})$

42. In Exercise 40, the calculations $(AB)\mathbf{u}$ and $A(B\mathbf{u})$ produce the same result. Which calculation requires fewer multiplications of individual matrix entries? [For example, it takes two multiplications to get the $(1, 1)$ entry of AB.]

43. The next section will show that all the following calculations produce the same result:

$$C[A(B\mathbf{u})] = (CA)(B\mathbf{u}) = [C(AB)]\mathbf{u} = C[(AB)\mathbf{u}].$$

Convince yourself that the first expression requires the fewest individual multiplications. (*Hint:* Forming $B\mathbf{u}$ takes four multiplications, and thus $A(B\mathbf{u})$ takes eight multiplications, and so on.) Count the number of multiplications required for each of the four calculations above.

44. Refer to the matrices and vectors in (11).
   a) Identify the column vectors in $A = [\mathbf{A}_1, \mathbf{A}_2]$ and $D = [\mathbf{D}_1, \mathbf{D}_2, \mathbf{D}_3, \mathbf{D}_4]$.
   b) In part (a), is $\mathbf{A}_1$ in R^2, R^3, or R^4? Is $\mathbf{D}_1$ in R^2, R^3, or R^4?
   c) Form the (2×2) matrix with columns $[A\mathbf{B}_1, A\mathbf{B}_2]$, and verify that this matrix is the product AB.
   d) Verify that the vector $D\mathbf{w}$ is the same as $2\mathbf{D}_1 + 3\mathbf{D}_2 + \mathbf{D}_3 + \mathbf{D}_4$.

45. Determine whether the following matrix products are defined. When the product is defined, give the size of the product.
   a) AB and BA, where A is (2×3) and B is (3×4)
   b) AB and BA, where A is (2×3) and B is (2×4)
   c) AB and BA, where A is (3×7) and B is (6×3)
   d) AB and BA, where A is (2×3) and B is (3×2)
   e) AB and BA, where A is (3×3) and B is (3×1)
   f) $A(BC)$ and $(AB)C$, where A is (2×3), B is (3×5), and C is (5×4)
   g) AB and BA, where A is (4×1) and B is (1×4)

46. What is the size of the product $(AB)(CD)$, where A is (2×3), B is (3×4), C is (4×4), and D is (4×2)? Also calculate the size of $A[B(CD)]$ and $[(AB)C]D$.

47. If A is a matrix, what should the symbol A^2 mean? What restrictions on A are required in order that A^2 be defined?

48. Set

$$O = \begin{bmatrix} 0 & 0 \\ 0 & 0 \end{bmatrix}, \quad A = \begin{bmatrix} 2 & 0 \\ 0 & 2 \end{bmatrix}, \quad \text{and}$$

$$B = \begin{bmatrix} 1 & b \\ b^{-1} & 1 \end{bmatrix},$$

where $b \neq 0$. Show that O, A, and B are solutions to the matrix equation $X^2 - 2X = O$. Conclude that this "quadratic" equation has infinitely many solutions.

49. Two newspapers compete for subscriptions in a region with 300,000 households. Assume that no household subscribes to both newspapers and that the table below gives the probabilities that a household will change its subscription status during the year.

	From A	*From B*	*From None*
To A	.70	.15	.30
To B	.20	.80	.20
To None	.10	.05	.50

For example, an interpretation of the first column of the table is that during a given year, newspaper A can expect to keep 70% of its current subscribers while losing 20% to newspaper B and 10% to no subscription.

At the beginning of a particular year, suppose that 150,000 households subscribe to newspaper A, 100,000 subscribe to newspaper B, and 50,000 have no subscription. Let P and $\mathbf{x}$ be defined by

$$P = \begin{bmatrix} .70 & .15 & .30 \\ .20 & .80 & .20 \\ .10 & .05 & .50 \end{bmatrix} \quad \text{and} \quad \mathbf{x} = \begin{bmatrix} 150{,}000 \\ 100{,}000 \\ 50{,}000 \end{bmatrix}.$$

The vector $\mathbf{x}$ is called the *state vector* for the beginning of the year.
   a) Calculate $P\mathbf{x}$ and $P^2\mathbf{x}$ and interpret the resulting vectors.
   b) Give a formula for the state vector after n years.

50. Let $A = \begin{bmatrix} 1 & 2 \\ 3 & 4 \end{bmatrix}$.
   a) Find all matrices $B = \begin{bmatrix} a & b \\ c & d \end{bmatrix}$ such that $AB = BA$.
   b) Use the results of part (a) to exhibit (2×2) matrices B and C such that $AB = BA$ and $AC \neq CA$.

51. Let A and B be matrices such that the product AB is defined and is a square matrix. Argue that the product BA is also defined and is a square matrix.

52. Let A and B be matrices such that the product AB is defined. Use Theorem 6 to prove each of the following.
   a) If B has a column of zeros, then so does AB.
   b) If B has two identical columns, then so does AB.

53. a) Express each of the linear systems (i) and (ii) in the form $A\mathbf{x} = \mathbf{b}$.

   i) $2x_1 - x_2 = 3$ ii) $x_1 - 3x_2 + x_3 = 1$
   $\ x_1 + x_2 = 3$ $\ x_1 - 2x_2 + x_3 = 2$
   $\ x_2 - x_3 = -1$

   b) Express systems (i) and (ii) in the form of (8).

   c) Solve systems (i) and (ii) by Gaussian elimination. For each system $A\mathbf{x} = \mathbf{b}$, represent $\mathbf{b}$ as a linear combination of the columns of the coefficient matrix.

54. Solve $A\mathbf{x} = \mathbf{b}$, where A and $\mathbf{b}$ are given by

$$A = \begin{bmatrix} 1 & 1 \\ 1 & 2 \end{bmatrix}, \qquad \mathbf{b} = \begin{bmatrix} 2 \\ 3 \end{bmatrix}.$$

55. Let A and I be the matrices

$$A = \begin{bmatrix} 1 & 1 \\ 1 & 2 \end{bmatrix}, \qquad I = \begin{bmatrix} 1 & 0 \\ 0 & 1 \end{bmatrix}.$$

   a) Find a (2×2) matrix B such that $AB = I$. (*Hint:* Use Theorem 6 to determine the column vectors of B.)

   b) Show that $AB = BA$ for the matrix B found in part (a).

56. Prove Theorem 5 by showing that the ith component of $A\mathbf{x}$ is equal to the ith component of $x_1\mathbf{A}_1 + x_2\mathbf{A}_2 + \cdots + x_n\mathbf{A}_n$, where $1 \le i \le m$.

57. For A and C given below, find a matrix B (if possible) such that $AB = C$.

   a) $A = \begin{bmatrix} 1 & 3 \\ 1 & 4 \end{bmatrix}, \qquad C = \begin{bmatrix} 2 & 6 \\ 3 & 6 \end{bmatrix}$

   b) $A = \begin{bmatrix} 1 & 1 & 1 \\ 0 & 2 & 1 \\ 2 & 4 & 3 \end{bmatrix}, \qquad C = \begin{bmatrix} 1 & 0 & 0 \\ 1 & 2 & 0 \\ 1 & 3 & 5 \end{bmatrix}$

   c) $A = \begin{bmatrix} 1 & 2 \\ 2 & 4 \end{bmatrix}, \qquad C = \begin{bmatrix} 0 & 0 \\ 0 & 0 \end{bmatrix}$, where $B \neq C$.

58. A (3×3) matrix $T = (t_{ij})$ is called an ***upper-triangular*** matrix if T has the form

$$T = \begin{bmatrix} t_{11} & t_{12} & t_{13} \\ 0 & t_{22} & t_{23} \\ 0 & 0 & t_{33} \end{bmatrix}.$$

Formally, T is upper triangular if $t_{ij} = 0$ whenever $i > j$. If A and B are upper-triangular (3×3) matrices, verify that the product AB is also upper triangular.

59. An $(n \times n)$ matrix $T = (t_{ij})$ is called upper triangular if $t_{ij} = 0$ whenever $i > j$. Suppose that A and B are $(n \times n)$ upper-triangular matrices. Use Definition 8 to prove that the product AB is upper triangular. That is, show that the ijth entry of AB is zero when $i > j$.

1.7   Algebraic Properties of Matrix Operations

In the last section we defined the matrix operations of addition, multiplication, and the multiplication of a matrix by a scalar. For these operations to be useful, the basic rules they obey must be determined. As we will presently see, many of the familiar algebraic properties of real numbers also hold for matrices. There are, however, important exceptions. We have already noted, for example, that matrix multiplication is not commutative. Another property of real numbers that does not carry over to matrices is the cancellation law for multiplication. That is, if a, b, and c are real numbers such that $ab = ac$ and $a \neq 0$, then $b = c$. By contrast, consider the three matrices

$$A = \begin{bmatrix} 1 & 1 \\ 1 & 1 \end{bmatrix}, \qquad B = \begin{bmatrix} 1 & 4 \\ 2 & 1 \end{bmatrix}, \quad \text{and} \quad C = \begin{bmatrix} 2 & 2 \\ 1 & 3 \end{bmatrix}.$$

Note that $AB = AC$ but $B \neq C$. This example shows that the familiar cancellation law for real numbers does not apply to matrix multiplication.

Properties of Matrix Operations

The next three theorems list algebraic properties that do hold for matrix operations. In some cases, although the rule seems obvious and the proof simple, certain subtleties should be noted. For example, Theorem 9 asserts that $(r + s)A = rA + sA$, where r and s are scalars and A is an $(m \times n)$ matrix. Although the same addition symbol, $+$, appears on both sides of the equation, two different addition operations are involved; $r + s$ is the sum of two scalars, and $rA + sA$ is the sum of two matrices.

Our first theorem lists some of the properties satisfied by matrix addition.

THEOREM 7

If A, B, and C are $(m \times n)$ matrices, then the following are true:

1. $A + B = B + A$.
2. $(A + B) + C = A + (B + C)$.
3. There exists a unique $(m \times n)$ matrix $\mathcal{O}$ (called the **zero matrix**) such that $A + \mathcal{O} = A$ for every $(m \times n)$ matrix A.
4. Given an $(m \times n)$ matrix A, there exists a unique $(m \times n)$ matrix P such that $A + P = \mathcal{O}$.

These properties are easily established and the proofs of 2–4 are left as exercises. Regarding properties 3 and 4, we note that the zero matrix, $\mathcal{O}$, is the $(m \times n)$ matrix, all of whose entries are zero. Also the matrix P of property 4 is usually called the additive inverse for A, and the reader can show that $P = (-1)A$. The matrix $(-1)A$ is also denoted as $-A$, and the notation $A - B$ means $A + (-B)$. Thus property 4 states that $A - A = \mathcal{O}$.

PROOF OF PROPERTY 1. If $A = (a_{ij})$ and $B = (b_{ij})$ are $(m \times n)$ matrices, then, by Definition 6,

$$(A + B)_{ij} = a_{ij} + b_{ij}.$$

Similarly, by Definition 6,

$$(B + A)_{ij} = b_{ij} + a_{ij}.$$

Since addition of real numbers is commutative, $a_{ij} + b_{ij}$ and $b_{ij} + a_{ij}$ are equal. Therefore, $A + B = B + A$. ▫

Three associative properties involving scalar and matrix multiplication are given in Theorem 8.

THEOREM 8

1. If A, B, and C are $(m \times n)$, $(n \times p)$, and $(p \times q)$ matrices, respectively, then $(AB)C = A(BC)$.
2. If r and s are scalars, then $r(sA) = (rs)A$.
3. $r(AB) = (rA)B = A(rB)$.

The proof is again left to the reader, but we will give one example to illustrate the theorem.

EXAMPLE 1 Demonstrate that $(AB)C = A(BC)$, where

$$A = \begin{bmatrix} 1 & 2 \\ -1 & 3 \end{bmatrix}, \quad B = \begin{bmatrix} 2 & -1 & 3 \\ 1 & -1 & 1 \end{bmatrix}, \quad \text{and} \quad C = \begin{bmatrix} 3 & 1 & 2 \\ -2 & 1 & -1 \\ 4 & -2 & -1 \end{bmatrix}.$$

SOLUTION   Forming the products AB and BC yields

$$AB = \begin{bmatrix} 4 & -3 & 5 \\ 1 & -2 & 0 \end{bmatrix} \quad \text{and} \quad BC = \begin{bmatrix} 20 & -5 & 2 \\ 9 & -2 & 2 \end{bmatrix}.$$

Therefore, $(AB)C$ is the product of a (2×3) matrix with a (3×3) matrix, whereas $A(BC)$ is the product of a (2×2) matrix with a (2×3) matrix. Forming these products, we find

$$(AB)C = \begin{bmatrix} 38 & -9 & 6 \\ 7 & -1 & 4 \end{bmatrix} \quad \text{and} \quad A(BC) = \begin{bmatrix} 38 & -9 & 6 \\ 7 & -1 & 4 \end{bmatrix}. \qquad \square$$

Finally the distributive properties connecting addition and multiplication are given in Theorem 9.

THEOREM 9

1. If A and B are $(m \times n)$ matrices and C is an $(n \times p)$ matrix, then $(A+B)C = AC + BC$.
2. If A is an $(m \times n)$ matrix and B and C are $(n \times p)$ matrices, then $A(B+C) = AB + AC$.
3. If r and s are scalars and A is an $(m \times n)$ matrix, then $(r + s)A = rA + sA$.
4. If r is a scalar and A and B are $(m \times n)$ matrices, then $r(A + B) = rA + rB$.

PROOF   We will prove property 1 and leave the others to the reader. First observe that $(A + B)C$ and $AC + BC$ are both $(m \times p)$ matrices. To show that the components of these two matrices are equal, let $Q = A + B$, where $Q = (q_{ij})$. Then $(A + B)C = QC$, and the rsth component of QC is given by

$$\sum_{k=1}^{n} q_{rk}c_{ks} = \sum_{k=1}^{n} (a_{rk} + b_{rk})c_{ks} = \sum_{k=1}^{n} a_{rk}c_{ks} + \sum_{k=1}^{n} b_{rk}c_{ks}.$$

Because

$$\sum_{k=1}^{n} a_{rk}c_{ks} + \sum_{k=1}^{n} b_{rk}c_{ks}$$

is precisely the rsth entry of $AC + BC$, it follows that $(A + B)C = AC + BC$.

$\square$

The Transpose of a Matrix

The concept of the transpose of a matrix is important in applications. Stated informally, the transpose operation, applied to a matrix A, interchanges the rows and columns of A. The formal definition of transpose is as follows.

> **DEFINITION 9** If $A = (a_{ij})$ is an $(m \times n)$ matrix, then the **transpose** of A, denoted A^T, is the $(n \times m)$ matrix $A^T = (b_{ij})$, where $b_{ij} = a_{ji}$ for all i and j, $1 \leq j \leq m$, and $1 \leq i \leq n$.

The following example illustrates the definition of the transpose of a matrix:

__EXAMPLE 2__ Find the transpose of $A = \begin{bmatrix} 1 & 3 & 7 \\ 2 & 1 & 4 \end{bmatrix}$.

SOLUTION By Definition 9, A^T is the (3×2) matrix

$$A^T = \begin{bmatrix} 1 & 2 \\ 3 & 1 \\ 7 & 4 \end{bmatrix}.$$

In the example above, note that the first row of A becomes the first column of A^T and the second row of A becomes the second column of A^T. Similarly, the columns of A become the rows of A^T. Thus A^T is obtained by interchanging the rows and columns of A.

Three important properties of the transpose are given in Theorem 10.

THEOREM 10

If A and B are $(m \times n)$ matrices and C is an $(n \times p)$ matrix, then:

1. $(A + B)^T = A^T + B^T$.
2. $(AC)^T = C^T A^T$.
3. $(A^T)^T = A$.

PROOF We will leave properties 1 and 3 to the reader and prove property 2. Note first that $(AC)^T$ and $C^T A^T$ are both $(p \times m)$ matrices, so we have only to show that their corresponding entries are equal. From Definition 9, the ijth entry of $(AC)^T$ is the jith entry of AC. Thus the ijth entry of $(AC)^T$ is given by

$$\sum_{k=1}^{n} a_{jk} c_{ki}.$$

Next the ijth entry of $C^T A^T$ is the scalar product of the ith row of C^T with the jth column of A^T. In particular, the ith row of C^T is $[c_{1i}, c_{2i}, \ldots, c_{ni}]$ (the

ith column of C), whereas the jth column of A^T is

$$\begin{bmatrix} a_{j1} \\ a_{j2} \\ \vdots \\ a_{jn} \end{bmatrix}$$

(the jth row of A). Therefore, the ijth entry of $C^T A^T$ is given by

$$c_{1i}a_{j1} + c_{2i}a_{j2} + \cdots + c_{ni}a_{jn} = \sum_{k=1}^{n} c_{ki}a_{jk}.$$

Finally, since

$$\sum_{k=1}^{n} c_{ki}a_{jk} = \sum_{k=1}^{n} a_{jk}c_{ki},$$

the ijth entries of $(AC)^T$ and $C^T A^T$ agree and the matrices are equal.

The transpose operation is used to define certain important types of matrices, such as positive-definite matrices, normal matrices, and symmetric matrices. We will consider these in detail later and give only the definition of a symmetric matrix in this section.

DEFINITION 10   A matrix A is *symmetric* if $A = A^T$.

If A is an $(m \times n)$ matrix, then A^T is an $(n \times m)$ matrix, so we can have $A = A^T$ only if $m = n$. An $(n \times n)$ matrix is called a *square matrix;* thus if a matrix is symmetric, then it must be a square matrix. Furthermore, Definition 9 implies that if $A = (a_{ij})$ is an $(n \times n)$ symmetric matrix, then $a_{ij} = a_{ji}$ for all i and j, $1 \le i, j \le n$. Conversely, if A is square and $a_{ij} = a_{ji}$ for all i and j, then A is symmetric.

EXAMPLE 3   Determine which of the matrices

$$A = \begin{bmatrix} 1 & 2 \\ 2 & 3 \end{bmatrix}, \quad B = \begin{bmatrix} 1 & 2 \\ 1 & 2 \end{bmatrix}, \quad \text{and} \quad C = \begin{bmatrix} 1 & 6 \\ 3 & 1 \\ 2 & 0 \end{bmatrix}$$

is symmetric. Also show that $B^T B$ and $C^T C$ are symmetric.

SOLUTION   By Definition 9,

$$A^T = \begin{bmatrix} 1 & 2 \\ 2 & 3 \end{bmatrix}, \quad B^T = \begin{bmatrix} 1 & 1 \\ 2 & 2 \end{bmatrix}, \quad \text{and} \quad C^T = \begin{bmatrix} 1 & 3 & 2 \\ 6 & 1 & 0 \end{bmatrix}.$$

Thus A is symmetric since $A^T = A$. However, $B^T \ne B$ and $C^T \ne C$. Therefore, B and C are not symmetric. As can be seen, the matrices $B^T B$ and $C^T C$ are symmetric:

$$B^T B = \begin{bmatrix} 2 & 4 \\ 4 & 8 \end{bmatrix} \quad \text{and} \quad C^T C = \begin{bmatrix} 14 & 9 \\ 9 & 37 \end{bmatrix}.$$ □

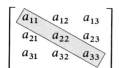

Figure 1.13
Main diagonal

In Exercise 49, the reader is asked to show that $Q^T Q$ is always a symmetric matrix whether or not Q is symmetric.

In the $(n \times n)$ matrix $A = (a_{ij})$, the entries $a_{11}, a_{22}, \ldots, a_{nn}$ are called the **main diagonal** of A. For example, the main diagonal of a (3×3) matrix is illustrated in Fig. 1.13. Since the entries a_{ij} and a_{ji} are "symmetric partners" relative to the main diagonal, symmetric matrices are easily recognizable as those in which the entries form a symmetric array relative to the main diagonal. For example, if

$$A = \begin{bmatrix} 2 & 3 & -1 \\ 3 & 4 & 2 \\ -1 & 2 & 0 \end{bmatrix} \quad \text{and} \quad B = \begin{bmatrix} 1 & 2 & 2 \\ -1 & 3 & 0 \\ 5 & 2 & 6 \end{bmatrix},$$

then, by inspection, A is symmetric whereas B is not.

Scalar Products and Vector Norms

The transpose operation can be used to represent scalar products and vector norms. As we will see, a vector norm provides a method for measuring the size of a vector.

To illustrate the connection between transposes and the scalar product, let $\mathbf{x}$ and $\mathbf{y}$ be vectors in R^3 given by

$$\mathbf{x} = \begin{bmatrix} 1 \\ -3 \\ 2 \end{bmatrix} \quad \text{and} \quad \mathbf{y} = \begin{bmatrix} 1 \\ 2 \\ 1 \end{bmatrix},$$

Then $\mathbf{x}^T$ is the (1×3) vector

$$\mathbf{x}^T = [1, -3, 2],$$

and $\mathbf{x}^T \mathbf{y}$ is the scalar (or (1×1) matrix) given by

$$\mathbf{x}^T \mathbf{y} = [1, -3, 2] \begin{bmatrix} 1 \\ 2 \\ 1 \end{bmatrix} = 1 - 6 + 2 = -3.$$

More generally, if $\mathbf{x}$ and $\mathbf{y}$ are vectors in R^n,

$$\mathbf{x} = \begin{bmatrix} x_1 \\ x_2 \\ \vdots \\ x_n \end{bmatrix}, \quad \mathbf{y} = \begin{bmatrix} y_1 \\ y_2 \\ \vdots \\ y_n \end{bmatrix},$$

then

$$\mathbf{x}^T\mathbf{y} = \sum_{i=1}^{n} x_i y_i;$$

that is, $\mathbf{x}^T\mathbf{y}$ is the scalar product or dot product of $\mathbf{x}$ and $\mathbf{y}$. Also note that $\mathbf{y}^T\mathbf{x} = \sum_{i=1}^{n} y_i x_i = \sum_{i=1}^{n} x_i y_i = \mathbf{x}^T\mathbf{y}$.

One of the basic concepts in computational work is that of the length or norm of a vector. If

$$\mathbf{x} = \begin{bmatrix} a \\ b \end{bmatrix}$$

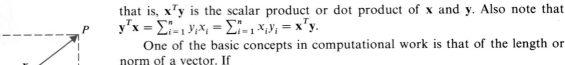

Figure 1.14
Geometric vector in
two-space

is in R^2, then $\mathbf{x}$ can be represented geometrically in the plane as the directed line segment $\overrightarrow{OP}$ from the origin O to the point P, which has coordinates (a, b), as illustrated in Fig. 1.14. By the Pythagorean theorem, the length of the line segment $\overrightarrow{OP}$ is $\sqrt{a^2 + b^2}$.

A similar idea is used in R^n. For a vector $\mathbf{x}$ in R^n,

$$\mathbf{x} = \begin{bmatrix} x_1 \\ x_2 \\ \vdots \\ x_n \end{bmatrix},$$

it is natural to define the *Euclidean length,* or *Euclidean norm* of $\mathbf{x}$, denoted by $\|\mathbf{x}\|$, to be

$$\|\mathbf{x}\| = \sqrt{x_1^2 + x_2^2 + \cdots + x_n^2}.$$

(The quantity $\|\mathbf{x}\|$ gives us a way to measure the size of the vector $\mathbf{x}$.)

Noting that the scalar product of $\mathbf{x}$ with itself is

$$\mathbf{x}^T\mathbf{x} = x_1^2 + x_2^2 + \cdots + x_n^2,$$

we have

$$\|\mathbf{x}\| = \sqrt{\mathbf{x}^T\mathbf{x}}. \tag{1}$$

For vectors $\mathbf{x}$ and $\mathbf{y}$ in R^n, we define the Euclidean distance between $\mathbf{x}$ and $\mathbf{y}$ to be $\|\mathbf{x} - \mathbf{y}\|$. Thus the distance between $\mathbf{x}$ and $\mathbf{y}$ is given by

$$\|\mathbf{x} - \mathbf{y}\| = \sqrt{(\mathbf{x} - \mathbf{y})^T(\mathbf{x} - \mathbf{y})}$$
$$= \sqrt{(x_1 - y_1)^2 + (x_2 - y_2)^2 + \cdots + (x_n - y_n)^2}. \tag{2}$$

<u>EXAMPLE 4</u> If $\mathbf{x}$ and $\mathbf{y}$ in R^3 are given by

$$\mathbf{x} = \begin{bmatrix} -2 \\ 3 \\ 2 \end{bmatrix} \quad \text{and} \quad \mathbf{y} = \begin{bmatrix} 1 \\ 2 \\ -1 \end{bmatrix},$$

then find $\mathbf{x}^T\mathbf{y}$, $\|\mathbf{x}\|$, $\|\mathbf{y}\|$, and $\|\mathbf{x} - \mathbf{y}\|$.

SOLUTION   We have

$$\mathbf{x}^T\mathbf{y} = [-2 \quad 3 \quad 2]\begin{bmatrix} 1 \\ 2 \\ -1 \end{bmatrix} = -2 + 6 - 2 = 2.$$

Also, $\|\mathbf{x}\| = \sqrt{\mathbf{x}^T\mathbf{x}} = \sqrt{4 + 9 + 4} = \sqrt{17}$, and $\|\mathbf{y}\| = \sqrt{\mathbf{y}^T\mathbf{y}} = \sqrt{1 + 4 + 1} = \sqrt{6}$.
Subtracting $\mathbf{y}$ from $\mathbf{x}$ gives

$$\mathbf{x} - \mathbf{y} = \begin{bmatrix} -3 \\ 1 \\ 3 \end{bmatrix},$$

so $\|\mathbf{x} - \mathbf{y}\| = \sqrt{(\mathbf{x} - \mathbf{y})^T(\mathbf{x} - \mathbf{y})} = \sqrt{9 + 1 + 9} = \sqrt{19}$.   ☐

Exercises 1.7

The matrices and vectors listed in (3) are used in several of the exercises that follow.

$$A = \begin{bmatrix} 3 & 1 \\ 4 & 7 \\ 2 & 6 \end{bmatrix}, \qquad B = \begin{bmatrix} 1 & 2 & 1 \\ 7 & 4 & 3 \\ 6 & 0 & 1 \end{bmatrix},$$

$$C = \begin{bmatrix} 2 & 1 & 4 & 0 \\ 6 & 1 & 3 & 5 \\ 2 & 4 & 2 & 0 \end{bmatrix}, \qquad D = \begin{bmatrix} 2 & 1 \\ 1 & 4 \end{bmatrix}, \tag{3}$$

$$E = \begin{bmatrix} 3 & 6 \\ 2 & 3 \end{bmatrix}, \qquad F = \begin{bmatrix} 1 & 1 \\ 1 & 1 \end{bmatrix},$$

$$\mathbf{u} = \begin{bmatrix} 1 \\ -1 \end{bmatrix}, \qquad \mathbf{v} = \begin{bmatrix} -3 \\ 3 \end{bmatrix}$$

Exercises 1–25 refer to the matrices and vectors in (3). In Exercises 1–6, perform the multiplications to verify the given equality or nonequality.

1. $(DE)F = D(EF)$

2. $(FE)D = F(ED)$

3. $DE \neq ED$

4. $EF \neq FE$

5. $F\mathbf{u} = F\mathbf{v}$

6. $3F\mathbf{u} = 7F\mathbf{v}$

In Exercises 7–12, find the matrices.

7. A^T

8. D^T

9. E^TF

10. A^TC

11. $(F\mathbf{v})^T$

12. $(EF)\mathbf{v}$

In Exercises 13–25, calculate the scalars.

13. $\mathbf{u}^T\mathbf{v}$

14. $\mathbf{v}^TF\mathbf{u}$

15. $\mathbf{v}^TD\mathbf{v}$

16. $\mathbf{v}^TF\mathbf{v}$

17. $\mathbf{u}^T\mathbf{u}$

18. $\mathbf{v}^T\mathbf{v}$

19. $\|\mathbf{u}\|$

20. $\|D\mathbf{v}\|$

21. $\|A\mathbf{u}\|$

22. $\|\mathbf{u} - \mathbf{v}\|$

23. $\|F\mathbf{u}\|$

24. $\|F\mathbf{v}\|$

25. $\|(D - E)\mathbf{u}\|$

26. Let A and B be (2×2) matrices. Prove or find a counterexample for the statement: $(A - B)(A + B) = A^2 - B^2$.

27. Let A and B be (2×2) matrices such $A^2 = AB$ and $A \neq \mathcal{O}$. Can we assert that, by cancellation, $A = B$? Explain.

28. Let A and B be as in Exercise 27 above. Find the flaw in the following "proof" that $A = B$.
    Since $A^2 = AB$, $A^2 - AB = \mathcal{O}$. Factoring yields $A(A - B) = \mathcal{O}$. Since $A \neq \mathcal{O}$, it follows that $A - B = \mathcal{O}$. Therefore, $A = B$.

29. Two of the six matrices listed in (3) are symmetric. Identify these matrices.

30. Find (2×2) matrices A and B such that A and B are symmetric, but AB is not symmetric. (*Hint*: $(AB)^T = B^TA^T = BA$.)

31. Let A and B be $(n \times n)$ symmetric matrices. Give a necessary and sufficient condition for AB to be symmetric. (*Hint*: Recall Exercise 30.)

32. Let G be the (2×2) matrix given below, and consider any vector $\mathbf{x}$ in R^2 where both entries are not simultaneously zero:

$$G = \begin{bmatrix} 2 & 1 \\ 1 & 1 \end{bmatrix}, \qquad \mathbf{x} = \begin{bmatrix} x_1 \\ x_2 \end{bmatrix}; \qquad |x_1| + |x_2| > 0.$$

Show that $\mathbf{x}^T G \mathbf{x} > 0$. (*Hint:* Write $\mathbf{x}^T G \mathbf{x}$ as a sum of squares.)

33. Repeat Exercise 32 using the matrix D in (3) in place of G.

34. For F in (3), show that $\mathbf{x}^T F \mathbf{x} \geq 0$ for all $\mathbf{x}$ in R^2. Classify those vectors $\mathbf{x}$ such that $\mathbf{x}^T F \mathbf{x} = 0$.

If $\mathbf{x}$ and $\mathbf{y}$ are vectors in R^n, then the product $\mathbf{x}^T \mathbf{y}$ is often called an inner product. Similarly, the product $\mathbf{x}\mathbf{y}^T$ is often called an outer product. Exercises 35–40 concern outer products; the matrices and vectors are given in (3). In Exercises 35–40, form the outer products.

35. $\mathbf{u}\mathbf{v}^T$ **36.** $\mathbf{u}(F\mathbf{u})^T$ **37.** $\mathbf{v}(E\mathbf{v})^T$

38. $\mathbf{u}(E\mathbf{v})^T$ **39.** $(A\mathbf{u})(A\mathbf{v})^T$ **40.** $(A\mathbf{v})(A\mathbf{u})^T$

41. Let $\mathbf{a}$ and $\mathbf{b}$ be given by

$$\mathbf{a} = \begin{bmatrix} 1 \\ 2 \end{bmatrix} \quad \text{and} \quad \mathbf{b} = \begin{bmatrix} 3 \\ 4 \end{bmatrix}.$$

a) Find $\mathbf{x}$ in R^2 that satisfies both $\mathbf{x}^T \mathbf{a} = 6$ and $\mathbf{x}^T \mathbf{b} = 2$.
b) Find $\mathbf{x}$ in R^2 that satisfies both $\mathbf{x}^T(\mathbf{a} + \mathbf{b}) = 12$ and $\mathbf{x}^T \mathbf{a} = 2$.

42. Let A be a (2×2) matrix, and let B and C be given by

$$B = \begin{bmatrix} 1 & 3 \\ 1 & 4 \end{bmatrix} \quad \text{and} \quad C = \begin{bmatrix} 2 & 3 \\ 4 & 5 \end{bmatrix}.$$

a) If $A^T + B = C$, what is A?
b) If $A^T B = C$, what is A? and
c) Calculate $B C_1$, $\mathbf{B}_1^T C$, $(BC_1)^T C_2$, and $\|C\mathbf{B}_2\|$.

43. Let

$$A = \begin{bmatrix} 4 & -2 & 2 \\ 2 & 4 & -4 \\ 1 & 1 & 0 \end{bmatrix} \quad \text{and} \quad \mathbf{u} = \begin{bmatrix} 1 \\ 3 \\ 2 \end{bmatrix}.$$

a) Verify that $A\mathbf{u} = 2\mathbf{u}$.
b) Without forming A^5, calculate the vector $A^5\mathbf{u}$.
c) Give a formula for $A^n\mathbf{u}$, where n is a positive integer. What property from Theorem 8 is required to derive the formula?

44. Let A, B, and C be $m \times n$ matrices such that $A + C = B + C$. The following statements are the steps

in a proof that $A = B$. Using Theorem 7, provide justification for each of the assertions.

a) There exists an $m \times n$ matrix $\mathcal{O}$ such that $A = A + \mathcal{O}$.
b) There exists an $m \times n$ matrix D such that $A = A + (C + D)$.
c) $A = (A + C) + D = (B + C) + D$.
d) $A = B + (C + D)$.
e) $A = B + \mathcal{O}$.
f) $A = B$.

45. Let A, B, C, and D be matrices such that $AB = D$. and $AC = D$. The following statements are steps in a proof that if r and s are scalars, then $A(rB + sC) = (r + s)D$. Use Theorems 8 and 9 to provide reasons for each of the steps.

a) $A(rB + sC) = A(rB) + A(sC)$.
b) $A(rB + sC) = r(AB) + s(AC) = rD + sD$.
c) $A(rB + sC) = (r + s)D$.

46. Let $\mathbf{x}$ and $\mathbf{y}$ be vectors in R^n such that $\|\mathbf{x}\| = \|\mathbf{y}\| = 1$ and $\mathbf{x}^T \mathbf{y} = 0$. Use (1) to show that $\|\mathbf{x} - \mathbf{y}\| = \sqrt{2}$.

47. Use Theorem 10 to show that $A + A^T$ is symmetric for any square matrix A.

48. Let A be the (2×2) matrix

$$A = \begin{bmatrix} 1 & 2 \\ 3 & 6 \end{bmatrix}.$$

Choose some vector $\mathbf{b}$ in R^2 such that the equation $A\mathbf{x} = \mathbf{b}$ is inconsistent. Verify that the associated equation $A^T A\mathbf{x} = A^T \mathbf{b}$ is consistent for your choice of $\mathbf{b}$. Let $\mathbf{x}^*$ be a solution to $A^T A\mathbf{x} = A^T \mathbf{b}$, and select some vectors $\mathbf{x}$ at random from R^2. Verify that $\|A\mathbf{x}^* - \mathbf{b}\| \leq \|A\mathbf{x} - \mathbf{b}\|$ for any of these random choices for $\mathbf{x}$. (In Chapter 2, we will show that $A^T A\mathbf{x} = A^T \mathbf{b}$ is always consistent for any $(m \times n)$ matrix A regardless of whether $A\mathbf{x} = \mathbf{b}$ is consistent or not. We also show that any solution $\mathbf{x}^*$ of $A^T A\mathbf{x} = A^T \mathbf{b}$ satisfies $\|A\mathbf{x}^* - \mathbf{b}\| \leq \|A\mathbf{x} - \mathbf{b}\|$ for all $\mathbf{x}$ in R^n; that is, such a vector $\mathbf{x}^*$ minimizes the length of the residual vector $\mathbf{r} = A\mathbf{x} - \mathbf{b}$.)

49. Use Theorem 10 to prove each of the following:

a) If Q is any $(m \times n)$ matrix, then $Q^T Q$ and QQ^T are symmetric.
b) If A, B, and C are matrices such that the product ABC is defined, then $(ABC)^T = C^T B^T A^T$. (*Hint:* Set $BC = D$.) *Note:* These proofs can be done quickly without considering the entries in the matrices.

50. Let Q be an $(m \times n)$ matrix and $\mathbf{x}$ any vector in R^n. Prove that $\mathbf{x}^T Q^T Q \mathbf{x} \geq 0$. (*Hint:* Observe that $Q\mathbf{x}$ is a vector in R^m.)

51. Prove properties 2, 3, and 4 of Theorem 7.

52. Prove property 1 of Theorem 8. (*Note:* This is a long exercise, but the proof is similar to the proof of part 2 of Theorem 10.)

53. Prove properties 2 and 3 of Theorem 8.

54. Prove properties 2, 3, and 4 of Theorem 9.

55. Prove properties 1 and 3 of Theorem 10.

1.8   Linear Independence and Nonsingular Matrices

Section 1.6 demonstrated how the general linear system

$$a_{11}x_1 + a_{12}x_2 + \cdots + a_{1n}x_n = b_1$$
$$a_{21}x_1 + a_{22}x_2 + \cdots + a_{2n}x_n = b_2$$
$$\vdots \qquad\qquad\qquad \vdots \quad \vdots \tag{1}$$
$$a_{m1}x_1 + a_{m2}x_2 + \cdots + a_{mn}x_n = b_m$$

can be expressed as a matrix equation $A\mathbf{x} = \mathbf{b}$. We observed in Section 1.2 that system (1) may have a unique solution, infinitely many solutions, or no solution. The material in Section 1.4 illustrates that, with appropriate additional information, we can know which of the three possibilities will occur. The case in which $m = n$ is of particular interest, and in this and later sections we determine conditions on the matrix A in order that an $(n \times n)$ system have a unique solution.

Linear Independence

If $A = [\mathbf{A}_1, \mathbf{A}_2, \ldots, \mathbf{A}_n]$, then, by Theorem 5 of Section 1.6, the equation $A\mathbf{x} = \mathbf{b}$ can be written in terms of the columns of A as

$$x_1\mathbf{A}_1 + x_2\mathbf{A}_2 + \cdots + x_n\mathbf{A}_n = \mathbf{b}. \tag{2}$$

From Eq. (2), it follows that system (1) is consistent if, and only if, $\mathbf{b}$ can be written as a sum of scalar multiples of the column vectors of A. We call a sum such as $x_1\mathbf{A}_1 + x_2\mathbf{A}_2 + \cdots + x_n\mathbf{A}_n$ a *linear combination* of the vectors $\mathbf{A}_1, \mathbf{A}_2, \ldots, \mathbf{A}_n$. Thus $A\mathbf{x} = \mathbf{b}$ is consistent if, and only if, $\mathbf{b}$ is a linear combination of the columns of A.

EXAMPLE 1 If the vectors $\mathbf{A}_1, \mathbf{A}_2, \mathbf{A}_3, \mathbf{b}_1$, and $\mathbf{b}_2$ are given by

$$\mathbf{A}_1 = \begin{bmatrix} 1 \\ 2 \\ -1 \end{bmatrix}, \quad \mathbf{A}_2 = \begin{bmatrix} 1 \\ 3 \\ 1 \end{bmatrix}, \quad \mathbf{A}_3 = \begin{bmatrix} 1 \\ 4 \\ 3 \end{bmatrix}, \quad \mathbf{b}_1 = \begin{bmatrix} 3 \\ 8 \\ 1 \end{bmatrix}, \quad \text{and} \quad \mathbf{b}_2 = \begin{bmatrix} 2 \\ 5 \\ -1 \end{bmatrix},$$

then express each of $\mathbf{b}_1$ and $\mathbf{b}_2$ as a linear combination of the vectors $\mathbf{A}_1, \mathbf{A}_2, \mathbf{A}_3$.

SOLUTION   If $A = [\mathbf{A}_1, \mathbf{A}_2, \mathbf{A}_3]$, that is,

$$A = \begin{bmatrix} 1 & 1 & 1 \\ 2 & 3 & 4 \\ -1 & 1 & 3 \end{bmatrix},$$

then expressing $\mathbf{b}_1$ as a linear combination of $\mathbf{A}_1, \mathbf{A}_2, \mathbf{A}_3$ is equivalent to solving the (3×3) linear system with matrix equation $A\mathbf{x} = \mathbf{b}_1$. The augmented matrix for the system is

$$\begin{bmatrix} 1 & 1 & 1 & 3 \\ 2 & 3 & 4 & 8 \\ -1 & 1 & 3 & 1 \end{bmatrix},$$

and solving in the usual manner yields

$$\begin{aligned} x_1 &= 1 + x_3 \\ x_2 &= 2 - 2x_3, \end{aligned}$$

where x_3 is an unconstrained variable. Thus $\mathbf{b}_1$ can be expressed as a linear combination of $\mathbf{A}_1, \mathbf{A}_2, \mathbf{A}_3$ in infinitely many ways. Taking $x_3 = 2$, for example, yields $x_1 = 3$, $x_2 = -2$, so

$$3\mathbf{A}_1 - 2\mathbf{A}_2 + 2\mathbf{A}_3 = \mathbf{b}_1;$$

that is,

$$3\begin{bmatrix} 1 \\ 2 \\ -1 \end{bmatrix} - 2\begin{bmatrix} 1 \\ 3 \\ 1 \end{bmatrix} + 2\begin{bmatrix} 1 \\ 4 \\ 3 \end{bmatrix} = \begin{bmatrix} 3 \\ 8 \\ 1 \end{bmatrix}.$$

If we attempt to follow the same procedure to express $\mathbf{b}_2$ as a linear combination of $\mathbf{A}_1, \mathbf{A}_2, \mathbf{A}_3$, we discover that the system of equations $A\mathbf{x} = \mathbf{b}_2$ is inconsistent. Therefore, $\mathbf{b}_2$ cannot be expressed as a linear combination of $\mathbf{A}_1, \mathbf{A}_2, \mathbf{A}_3$.

It is convenient at this point to introduce a special symbol, θ, to denote the m-dimensional *zero vector*. Thus θ is the vector in R^m, all of whose components are zero:

$$\theta = \begin{bmatrix} 0 \\ 0 \\ \vdots \\ 0 \end{bmatrix}.$$

We will use θ throughout to designate zero vectors in order to avoid any possible confusion between a zero vector and the scalar zero. With this notation, the $(m \times n)$ homogeneous system

$$
\begin{aligned}
a_{11}x_1 + a_{12}x_2 + \cdots + a_{1n}x_n &= 0 \\
a_{21}x_1 + a_{22}x_2 + \cdots + a_{2n}x_n &= 0 \\
&\vdots \qquad\qquad \vdots \quad \vdots \\
a_{m1}x_1 + a_{m2}x_2 + \cdots + a_{mn}x_n &= 0
\end{aligned}
\tag{3}
$$

has the matrix equation $A\mathbf{x} = \boldsymbol{\theta}$, which can be written as

$$
x_1\mathbf{A}_1 + x_2\mathbf{A}_2 + \cdots + x_n\mathbf{A}_n = \boldsymbol{\theta}.
\tag{4}
$$

In Section 1.4, we observed that the homogeneous system (3) always has the trivial solution $x_1 = x_2 = \cdots = x_n = 0$. Thus in (4), $\boldsymbol{\theta}$ can always be expressed as a linear combination of the columns $\mathbf{A}_1, \mathbf{A}_2, \ldots, \mathbf{A}_n$ of A by taking $x_1 = x_2 = \cdots = x_n = 0$. There may, however, be nontrivial solutions, and this leads to the following definition.

DEFINITION 11 A set of m-dimensional vectors $\{\mathbf{v}_1, \mathbf{v}_2, \ldots, \mathbf{v}_p\}$ is said to be *linearly dependent* if there are scalars $a_1, a_2, \ldots, a_p$, not all of which are zero, such that

$$
a_1\mathbf{v}_1 + a_2\mathbf{v}_2 + \cdots + a_p\mathbf{v}_p = \boldsymbol{\theta}.
$$

A set $\{\mathbf{v}_1, \mathbf{v}_2, \ldots, \mathbf{v}_p\}$ of m-dimensional vectors is said to be *linearly independent* if it is not linearly dependent; that is, the only scalars for which $a_1\mathbf{v}_1 + a_2\mathbf{v}_2 + \cdots + a_p\mathbf{v}_p = \boldsymbol{\theta}$ are the scalars $a_1 = a_2 = \cdots = a_p = 0$.

It follows from Definition 11 that a set $\{\mathbf{v}_1, \mathbf{v}_2, \ldots, \mathbf{v}_p\}$ of m-dimensional vectors is either linearly independent or linearly dependent. The set is linearly independent if the vector equation

$$
x_1\mathbf{v}_1 + x_2\mathbf{v}_2 + \cdots + x_p\mathbf{v}_p = \boldsymbol{\theta}
\tag{5}
$$

has only the trivial solution, and the set is linearly dependent if there is a nontrivial solution. If V is the $(m \times p)$ matrix

$$
V = [\mathbf{v}_1, \mathbf{v}_2, \ldots, \mathbf{v}_p],
$$

then Eq. (5) is equivalent to the matrix equation

$$
V\mathbf{x} = \boldsymbol{\theta}.
\tag{6}
$$

Thus to determine whether the set $\{\mathbf{v}_1, \mathbf{v}_2, \ldots, \mathbf{v}_p\}$ is linearly independent or dependent, we solve the homogeneous system of equations (6) by forming the augmented matrix $[V \,|\, \boldsymbol{\theta}]$ and reducing $[V \,|\, \boldsymbol{\theta}]$ to echelon form. If the system has nontrivial solutions, then $\{\mathbf{v}_1, \mathbf{v}_2, \ldots, \mathbf{v}_p\}$ is a linearly dependent set. If the trivial solution is the only solution, then $\{\mathbf{v}_1, \mathbf{v}_2, \ldots, \mathbf{v}_p\}$ is a linearly independent set.

EXAMPLE 2 Determine whether the set $\{\mathbf{v}_1, \mathbf{v}_2, \mathbf{v}_3\}$ is linearly independent or linearly dependent, where

$$\mathbf{v}_1 = \begin{bmatrix} 1 \\ 2 \\ 3 \end{bmatrix}, \quad \mathbf{v}_2 = \begin{bmatrix} 2 \\ -1 \\ 4 \end{bmatrix}, \quad \text{and} \quad \mathbf{v}_3 = \begin{bmatrix} 0 \\ 5 \\ 2 \end{bmatrix}.$$

SOLUTION To determine whether the set is linearly dependent or not, we must determine whether the vector equation

$$x_1\mathbf{v}_1 + x_2\mathbf{v}_2 + x_3\mathbf{v}_3 = \boldsymbol{\theta} \tag{7}$$

has a nontrivial solution. But Eq. (7) is equivalent to the (3×3) homogeneous system of equations $V\mathbf{x} = \boldsymbol{\theta}$, where $V = [\mathbf{v}_1, \mathbf{v}_2, \mathbf{v}_3]$. The augmented matrix, $[V \mid \boldsymbol{\theta}]$, for this system is

$$\begin{bmatrix} 1 & 2 & 0 & 0 \\ 2 & -1 & 5 & 0 \\ 3 & 4 & 2 & 0 \end{bmatrix}.$$

This matrix reduces to the matrix

$$\begin{bmatrix} 1 & 2 & 0 & 0 \\ 0 & -5 & 5 & 0 \\ 0 & 0 & 0 & 0 \end{bmatrix}$$

in echelon form. Backsolving the resulting system of equations yields the solution $x_1 = -2x_3$, $x_2 = x_3$, where x_3 is arbitrary. In particular, Eq. (7) has nontrivial solutions, so $\{\mathbf{v}_1, \mathbf{v}_2, \mathbf{v}_3\}$ is a linearly dependent set. Setting $x_3 = 1$, for example, gives $x_1 = -2$, $x_2 = 1$. Therefore,

$$-2\mathbf{v}_1 + \mathbf{v}_2 + \mathbf{v}_3 = \boldsymbol{\theta}.$$

Note that from this equation we can express $\mathbf{v}_3$ as a linear combination of $\mathbf{v}_1$ and $\mathbf{v}_2$:

$$\mathbf{v}_3 = 2\mathbf{v}_1 - \mathbf{v}_2.$$

Similarly, of course, $\mathbf{v}_1$ can be expressed as a linear combination of $\mathbf{v}_2$ and $\mathbf{v}_3$, and $\mathbf{v}_2$ can be expressed as a linear combination of $\mathbf{v}_1$ and $\mathbf{v}_3$.

EXAMPLE 3 Determine whether or not the set $\{\mathbf{v}_1, \mathbf{v}_2, \mathbf{v}_3\}$ is linearly dependent, where

$$\mathbf{v}_1 = \begin{bmatrix} 1 \\ 2 \\ -3 \end{bmatrix}, \quad \mathbf{v}_2 = \begin{bmatrix} -2 \\ 1 \\ 1 \end{bmatrix}, \quad \text{and} \quad \mathbf{v}_3 = \begin{bmatrix} 1 \\ -1 \\ -2 \end{bmatrix}.$$

SOLUTION If $V = [\mathbf{v}_1, \mathbf{v}_2, \mathbf{v}_3]$, then the augmented matrix $[V \mid \boldsymbol{\theta}]$ is row equivalent to

$$\begin{bmatrix} 1 & -2 & 1 & 0 \\ 0 & 5 & -3 & 0 \\ 0 & 0 & -2 & 0 \end{bmatrix}.$$

Thus the only solution of $x_1\mathbf{v}_1 + x_2\mathbf{v}_2 + x_3\mathbf{v}_3 = \boldsymbol{\theta}$ is the trivial solution $x_1 = x_2 = x_3 = 0$; so the set $\{\mathbf{v}_1, \mathbf{v}_2, \mathbf{v}_3\}$ is linearly independent.

In contrast to the preceding example, note that $\mathbf{v}_3$ cannot be expressed as a linear combination of $\mathbf{v}_1$ and $\mathbf{v}_2$. If there were scalars a_1 and a_2 such that

$$\mathbf{v}_3 = a_1\mathbf{v}_1 + a_2\mathbf{v}_2,$$

then there would be a nontrivial solution to $x_1\mathbf{v}_1 + x_2\mathbf{v}_2 + x_3\mathbf{v}_3 = \boldsymbol{\theta}$; namely, $x_1 = -a_1, x_2 = -a_2, x_3 = 1.$ ☐

The **unit vectors** $\mathbf{e}_1, \mathbf{e}_2, \ldots, \mathbf{e}_n$ in R^n are defined by

$$\mathbf{e}_1 = \begin{bmatrix} 1 \\ 0 \\ 0 \\ \vdots \\ 0 \end{bmatrix}, \qquad \mathbf{e}_2 = \begin{bmatrix} 0 \\ 1 \\ 0 \\ \vdots \\ 0 \end{bmatrix}, \qquad \mathbf{e}_3 = \begin{bmatrix} 0 \\ 0 \\ 1 \\ \vdots \\ 0 \end{bmatrix}, \ldots, \qquad \mathbf{e}_n = \begin{bmatrix} 0 \\ 0 \\ 0 \\ \vdots \\ 1 \end{bmatrix}. \tag{8}$$

It is easy to see that $\{\mathbf{e}_1, \mathbf{e}_2, \ldots, \mathbf{e}_n\}$ is linearly independent. To illustrate, consider the unit vectors

$$\mathbf{e}_1 = \begin{bmatrix} 1 \\ 0 \\ 0 \end{bmatrix}, \qquad \mathbf{e}_2 = \begin{bmatrix} 0 \\ 1 \\ 0 \end{bmatrix}, \quad \text{and} \quad \mathbf{e}_3 = \begin{bmatrix} 0 \\ 0 \\ 1 \end{bmatrix}$$

in R^3. If $V = [\mathbf{e}_1, \mathbf{e}_2, \mathbf{e}_3]$, then

$$[V \mid \boldsymbol{\theta}] = \begin{bmatrix} 1 & 0 & 0 & 0 \\ 0 & 1 & 0 & 0 \\ 0 & 0 & 1 & 0 \end{bmatrix},$$

so clearly the only solution of $V\mathbf{x} = \boldsymbol{\theta}$ (or equivalently, of $x_1\mathbf{e}_1 + x_2\mathbf{e}_2 + x_3\mathbf{e}_3 = \boldsymbol{\theta}$) is the trivial solution $x_1 = 0, x_2 = 0, x_3 = 0$.

The next example illustrates that in some cases the linear dependence of a set of vectors can be determined by inspection. The example is a special case of Theorem 11, which follows.

EXAMPLE 4 Let $\{\mathbf{v}_1, \mathbf{v}_2, \mathbf{v}_3\}$ be the set of vectors in R^2 given by

$$\mathbf{v}_1 = \begin{bmatrix} 1 \\ 2 \end{bmatrix}, \qquad \mathbf{v}_2 = \begin{bmatrix} 3 \\ 1 \end{bmatrix}, \quad \text{and} \quad \mathbf{v}_3 = \begin{bmatrix} 2 \\ 3 \end{bmatrix}.$$

The Vector Space R^n, $n > 3$

The extension of vectors and their corresponding algebra into more than three dimensions was an extremely important step in the development of mathematics. This advancement is attributed largely to Hermann Grassmann (1809–1877) in his *Ausdehnungslehre*. In this work Grassmann discussed linear independence and dependence and many concepts dealing with the algebraic structure of R^n (such as dimension and subspaces), which we will study in Chapter 2. Unfortunately, Grassmann's work was so difficult to read that it went almost unnoticed for a long period of time and he did not receive as much credit as he deserved.

Without solving the corresponding homogeneous system of equations, show that the set is linearly dependent.

SOLUTION The vector equation $x_1\mathbf{v}_1 + x_2\mathbf{v}_2 + x_3\mathbf{v}_3 = \boldsymbol{\theta}$ is equivalent to the homogeneous system of equations $V\mathbf{x} = \boldsymbol{\theta}$, where $V = [\mathbf{v}_1, \mathbf{v}_2, \mathbf{v}_3]$. But this is the homogeneous system

$$x_1 + 3x_2 + 2x_3 = 0$$
$$2x_1 + x_2 + 3x_3 = 0,$$

consisting of two equations in three unknowns. By Theorem 4 of Section 1.4, the system has nontrivial solutions; hence the set $\{\mathbf{v}_1, \mathbf{v}_2, \mathbf{v}_3\}$ is linearly dependent.

Example 4 is a particular case of the following general result.

THEOREM 11

Let $\{\mathbf{v}_1, \mathbf{v}_2, \ldots, \mathbf{v}_p\}$ be a set of vectors in R^m. If $p > m$, then this set is linearly dependent.

PROOF The set $\{\mathbf{v}_1, \mathbf{v}_2, \ldots, \mathbf{v}_p\}$ is linearly dependent if the equation $V\mathbf{x} = \boldsymbol{\theta}$ has a nontrivial solution, where $V = [\mathbf{v}_1, \mathbf{v}_2, \ldots, \mathbf{v}_p]$. But $V\mathbf{x} = \boldsymbol{\theta}$ represents a homogeneous $(m \times p)$ system of linear equations with $m < p$. By Theorem 4 of Section 1.4, $V\mathbf{x} = \boldsymbol{\theta}$ has nontrivial solutions.

Note that Theorem 11 does not say that if $p \leq m$, then the set $\{\mathbf{v}_1, \mathbf{v}_2, \ldots, \mathbf{v}_p\}$ is linearly independent. Indeed Examples 2 and 3 illustrate that if $p \leq m$, then the set may be either linearly independent or linearly dependent.

Nonsingular Matrices

The concept of linear independence allows us to state precisely which $(n \times n)$ systems of linear equations always have a unique solution. We begin with the following definition.

DEFINITION 12   An $(n \times n)$ matrix A is ***nonsingular*** if the only solution to $A\mathbf{x} = \boldsymbol{\theta}$ is $\mathbf{x} = \boldsymbol{\theta}$. Furthermore, A is said to be ***singular*** if A is not nonsingular.

If $A = [\mathbf{A}_1, \mathbf{A}_2, \ldots, \mathbf{A}_n]$, then $A\mathbf{x} = \boldsymbol{\theta}$ can be written as

$$x_1\mathbf{A}_1 + x_2\mathbf{A}_2 + \cdots + x_n\mathbf{A}_n = \boldsymbol{\theta},$$

so it is an immediate consequence of Definition 12 that A is nonsingular if and only if the column vectors of A form a linearly independent set. This observation is important enough to be stated as a theorem.

THEOREM 12

The $(n \times n)$ matrix $A = [\mathbf{A}_1, \mathbf{A}_2, \ldots, \mathbf{A}_n]$ is nonsingular if and only if $\{\mathbf{A}_1, \mathbf{A}_2, \ldots, \mathbf{A}_n\}$ is a linearly independent set.

<u>EXAMPLE 5</u> Determine whether each of the matrices

$$A = \begin{bmatrix} 1 & 3 \\ 2 & 2 \end{bmatrix} \quad \text{and} \quad B = \begin{bmatrix} 1 & 2 \\ 2 & 4 \end{bmatrix}$$

is singular or nonsingular.

SOLUTION   The augmented matrix $[A \mid \boldsymbol{\theta}]$ for the system $A\mathbf{x} = \boldsymbol{\theta}$ is row equivalent to

$$\begin{bmatrix} 1 & 3 & 0 \\ 0 & -4 & 0 \end{bmatrix},$$

so the trivial solution $x_1 = 0$, $x_2 = 0$ (or $\mathbf{x} = \boldsymbol{\theta}$) is the unique solution. Thus A is nonsingular. On the other hand, B is singular because the vector

$$\mathbf{x} = \begin{bmatrix} -2 \\ 1 \end{bmatrix}$$

is a nontrivial solution of $B\mathbf{x} = \boldsymbol{\theta}$. Equivalently, the columns of B are linearly dependent because

$$-2\mathbf{B}_1 + \mathbf{B}_2 = \boldsymbol{\theta}.$$

The next theorem demonstrates the importance of nonsingular matrices with respect to linear systems.

THEOREM 13

Let A be an $(n \times n)$ matrix. The equation $A\mathbf{x} = \mathbf{b}$ has a unique solution for every $(n \times 1)$ column vector $\mathbf{b}$ if and only if A is nonsingular.

PROOF   Suppose first that $A\mathbf{x} = \mathbf{b}$ has a unique solution no matter what choice we make for $\mathbf{b}$. Choosing $\mathbf{b} = \boldsymbol{\theta}$ implies, by Definition 12, that A is nonsingular.

Conversely, suppose that $A = [\mathbf{A}_1, \mathbf{A}_2, \ldots, \mathbf{A}_n]$ is nonsingular, and let $\mathbf{b}$ be any $(n \times 1)$ column vector. We first show that $A\mathbf{x} = \mathbf{b}$ has a solution. To see this, observe first that.

$$\{\mathbf{A}_1, \mathbf{A}_2, \ldots, \mathbf{A}_n, \mathbf{b}\}$$

is a set of $(n + 1)$ vectors in R^n; so by Theorem 11 this set is linearly dependent. Thus there are scalars $a_1, a_2, \ldots, a_n, a_{n+1}$ such that

$$a_1\mathbf{A}_1 + a_2\mathbf{A}_2 + \cdots + a_n\mathbf{A}_n + a_{n+1}\mathbf{b} = \boldsymbol{\theta}; \tag{9}$$

and moreover not all these scalars are zero. In fact, if $a_{n+1} = 0$ in Eq. (9), then

$$a_1\mathbf{A}_1 + a_2\mathbf{A}_2 + \cdots + a_n\mathbf{A}_n = \boldsymbol{\theta},$$

and it follows that $\{\mathbf{A}_1, \mathbf{A}_2, \ldots, \mathbf{A}_n\}$ is a linearly dependent set. Since this contradicts the assumption that A is nonsingular, we know that a_{n+1} is nonzero. It follows from Eq. (9) that

$$s_1\mathbf{A}_1 + s_2\mathbf{A}_2 + \cdots + s_n\mathbf{A}_n = \mathbf{b},$$

where

$$s_1 = \frac{-a_1}{a_{n+1}}, \qquad s_2 = \frac{-a_2}{a_{n+1}}, \cdots, \qquad s_n = \frac{-a_n}{a_{n+1}}.$$

Thus $A\mathbf{x} = \mathbf{b}$ has a solution $\mathbf{s}$ given by

$$\mathbf{s} = \begin{bmatrix} s_1 \\ s_2 \\ \vdots \\ s_n \end{bmatrix}.$$

This shows that $A\mathbf{x} = \mathbf{b}$ is always consistent when A is nonsingular.

To show that the solution is unique, suppose that the $(n \times 1)$ vector $\mathbf{u}$ is any solution whatsoever to $A\mathbf{x} = \mathbf{b}$; that is, $A\mathbf{u} = \mathbf{b}$. Then $A\mathbf{s} - A\mathbf{u} = \mathbf{b} - \mathbf{b}$, or

$$A(\mathbf{s} - \mathbf{u}) = \boldsymbol{\theta};$$

therefore, $\mathbf{y} = \mathbf{s} - \mathbf{u}$ is a solution to $A\mathbf{x} = \boldsymbol{\theta}$. But A is nonsingular, so $\mathbf{y} = \boldsymbol{\theta}$; that is $\mathbf{s} = \mathbf{u}$. Thus $A\mathbf{x} = \mathbf{b}$ has one, and only one, solution.

In closing we note that for a specific system $A\mathbf{x} = \mathbf{b}$, it is usually easier to demonstrate the existence and/or uniqueness of a solution by using Gaussian elimination and actually solving the system. There are many instances, however, in which theoretical information about existence and uniqueness is extremely valuable to practical computations. A specific instance of this is provided in the next section.

Exercises 1.8

The vectors listed in (10) are used in several of the exercises that follow.

$$\mathbf{v}_1 = \begin{bmatrix} 1 \\ 2 \end{bmatrix}, \qquad \mathbf{v}_2 = \begin{bmatrix} 2 \\ 3 \end{bmatrix}, \qquad \mathbf{v}_3 = \begin{bmatrix} 2 \\ 4 \end{bmatrix},$$

$$\mathbf{v}_4 = \begin{bmatrix} 1 \\ 1 \end{bmatrix}, \qquad \mathbf{v}_5 = \begin{bmatrix} 3 \\ 6 \end{bmatrix},$$

$$\mathbf{u}_0 = \begin{bmatrix} 1 \\ 0 \\ 0 \end{bmatrix}, \qquad \mathbf{u}_1 = \begin{bmatrix} 1 \\ 2 \\ -1 \end{bmatrix}, \qquad \mathbf{u}_2 = \begin{bmatrix} 2 \\ 1 \\ -3 \end{bmatrix}, \qquad \textbf{(10)}$$

$$\mathbf{u}_3 = \begin{bmatrix} -1 \\ 4 \\ 3 \end{bmatrix}, \qquad \mathbf{u}_4 = \begin{bmatrix} 4 \\ 4 \\ 0 \end{bmatrix}, \qquad \mathbf{u}_5 = \begin{bmatrix} 1 \\ 1 \\ 0 \end{bmatrix}$$

In Exercises 1–14, use Eq. (6) to determine whether the given set of vectors is linearly independent or linearly dependent. If the set is linearly dependent, express one vector in the set as a linear combination of the others.

1. $\{\mathbf{v}_1, \mathbf{v}_2\}$ 2. $\{\mathbf{v}_1, \mathbf{v}_3\}$
3. $\{\mathbf{v}_1, \mathbf{v}_5\}$ 4. $\{\mathbf{v}_2, \mathbf{v}_3\}$
5. $\{\mathbf{v}_1, \mathbf{v}_2, \mathbf{v}_3\}$ 6. $\{\mathbf{v}_2, \mathbf{v}_3, \mathbf{v}_4\}$
7. $\{\mathbf{u}_4, \mathbf{u}_5\}$ 8. $\{\mathbf{u}_3, \mathbf{u}_4\}$
9. $\{\mathbf{u}_1, \mathbf{u}_2, \mathbf{u}_5\}$ 10. $\{\mathbf{u}_1, \mathbf{u}_4, \mathbf{u}_5\}$
11. $\{\mathbf{u}_2, \mathbf{u}_4, \mathbf{u}_5\}$ 12. $\{\mathbf{u}_1, \mathbf{u}_2, \mathbf{u}_4\}$
13. $\{\mathbf{u}_0, \mathbf{u}_1, \mathbf{u}_2, \mathbf{u}_4\}$ 14. $\{\mathbf{u}_0, \mathbf{u}_2, \mathbf{u}_3, \mathbf{u}_4\}$

15. Consider the sets of vectors in Exercises 1–14. Using Theorem 11, determine by inspection which of these sets are known to be linearly dependent.

The matrices listed in (11) are used in some of the exercises that follow.

$$A = \begin{bmatrix} 1 & 2 \\ 3 & 4 \end{bmatrix}, \qquad B = \begin{bmatrix} 1 & 2 \\ 2 & 4 \end{bmatrix}, \qquad C = \begin{bmatrix} 1 & 3 \\ 2 & 4 \end{bmatrix},$$

$$D = \begin{bmatrix} 1 & 0 & 0 \\ 0 & 1 & 0 \\ 0 & 1 & 0 \end{bmatrix}, \qquad E = \begin{bmatrix} 0 & 1 & 0 \\ 0 & 0 & 2 \\ 0 & 1 & 3 \end{bmatrix}, \qquad \textbf{(11)}$$

$$F = \begin{bmatrix} 1 & 2 & 1 \\ 0 & 3 & 2 \\ 0 & 0 & 1 \end{bmatrix}$$

In Exercises 16–27, use Definition 12 to determine whether the given matrix is singular or nonsingular. If a matrix M is singular, give all solutions of $M\mathbf{x} = \boldsymbol{\theta}$.

16. A 17. B
18. C 19. AB
20. BA 21. D
22. F 23. $D + F$
24. E 25. EF
26. DE 27. F^T

In Exercises 28–33, determine conditions on the scalars so that the set of vectors is linearly dependent.

28. $\mathbf{v}_1 = \begin{bmatrix} 1 \\ a \end{bmatrix}, \quad \mathbf{v}_2 = \begin{bmatrix} 2 \\ 3 \end{bmatrix}$

29. $\mathbf{v}_1 = \begin{bmatrix} 1 \\ 2 \end{bmatrix}, \quad \mathbf{v}_2 = \begin{bmatrix} 3 \\ a \end{bmatrix}$

30. $\mathbf{v}_1 = \begin{bmatrix} 1 \\ 2 \\ 1 \end{bmatrix}, \quad \mathbf{v}_2 = \begin{bmatrix} 1 \\ 3 \\ 2 \end{bmatrix}, \quad \mathbf{v}_3 = \begin{bmatrix} 0 \\ 1 \\ a \end{bmatrix}$

31. $\mathbf{v}_1 = \begin{bmatrix} 1 \\ 2 \\ 1 \end{bmatrix}, \quad \mathbf{v}_2 = \begin{bmatrix} 1 \\ a \\ 3 \end{bmatrix}, \quad \mathbf{v}_3 = \begin{bmatrix} 0 \\ 2 \\ b \end{bmatrix}$

32. $\mathbf{v}_1 = \begin{bmatrix} a \\ 1 \end{bmatrix}, \quad \mathbf{v}_2 = \begin{bmatrix} b \\ 3 \end{bmatrix}$

33. $\mathbf{v}_1 = \begin{bmatrix} 1 \\ a \end{bmatrix}, \quad \mathbf{v}_2 = \begin{bmatrix} b \\ c \end{bmatrix}$

In Exercises 34–39, the vectors and matrices are from (10) and (11). The equations listed in Exercises 34–39 all have the form $M\mathbf{x} = \mathbf{b}$, and all the equations are consistent. In each exercise, solve the equation and express $\mathbf{b}$ as a linear combination of the columns of M.

34. $A\mathbf{x} = \mathbf{v}_1$ 35. $A\mathbf{x} = \mathbf{v}_3$
36. $C\mathbf{x} = \mathbf{v}_4$ 37. $C\mathbf{x} = \mathbf{v}_2$
38. $F\mathbf{x} = \mathbf{u}_1$ 39. $F\mathbf{x} = \mathbf{u}_3$

In Exercises 40–45, express the given vector **b** as a linear combination of $\mathbf{v}_1$ and $\mathbf{v}_2$, where $\mathbf{v}_1$ and $\mathbf{v}_2$ are in (10).

40. $\mathbf{b} = \begin{bmatrix} 2 \\ 7 \end{bmatrix}$ **41.** $\mathbf{b} = \begin{bmatrix} 3 \\ -1 \end{bmatrix}$

42. $\mathbf{b} = \begin{bmatrix} 0 \\ 4 \end{bmatrix}$ **43.** $\mathbf{b} = \begin{bmatrix} 0 \\ 0 \end{bmatrix}$

44. $\mathbf{b} = \begin{bmatrix} 1 \\ 2 \end{bmatrix}$ **45.** $\mathbf{b} = \begin{bmatrix} 1 \\ 0 \end{bmatrix}$

46. Let $S = \{\mathbf{v}_1, \mathbf{v}_2, \mathbf{v}_3\}$ be a set of vectors in R^3, where $\mathbf{v}_1 = \boldsymbol{\theta}$. Show that S is a linearly dependent set of vectors. (*Hint:* Exhibit a nontrivial solution for either Eq. (5) or Eq. (6).)

47. Let $\{\mathbf{v}_1, \mathbf{v}_2, \mathbf{v}_3\}$ be a set of nonzero vectors in R^m such that $\mathbf{v}_i^T\mathbf{v}_j = 0$ when $i \neq j$. Show that the set is linearly independent. (*Hint:* Set $a_1\mathbf{v}_1 + a_2\mathbf{v}_2 + a_3\mathbf{v}_3 = \boldsymbol{\theta}$ and consider $\boldsymbol{\theta}^T\boldsymbol{\theta}$.)

48. If the set $\{\mathbf{v}_1, \mathbf{v}_2, \mathbf{v}_3\}$ of vectors in R^m is linearly dependent, then argue that the set $\{\mathbf{v}_1, \mathbf{v}_2, \mathbf{v}_3, \mathbf{v}_4\}$ is also linearly dependent for every choice of $\mathbf{v}_4$ in R^m.

49. Suppose that $\{\mathbf{v}_1, \mathbf{v}_2, \mathbf{v}_3\}$ is a linearly independent subset of R^m. Show that the set $\{\mathbf{v}_1, \mathbf{v}_1 + \mathbf{v}_2, \mathbf{v}_1 + \mathbf{v}_2 + \mathbf{v}_3\}$ is also linearly independent.

50. If A and B are $(n \times n)$ matrices such that A is nonsingular and $AB = \mathcal{O}$, then prove that $B = \mathcal{O}$. (*Hint:* Write $B = [\mathbf{B}_1, \ldots, \mathbf{B}_n]$ and consider $AB = [A\mathbf{B}_1, \ldots, A\mathbf{B}_n]$.)

51. If A, B, and C are $(n \times n)$ matrices such that A is nonsingular and $AB = AC$, then prove that $B = C$. (*Hint:* Consider $A(B - C)$ and use the preceding exercise.)

52. Let $A = [\mathbf{A}_1, \ldots, \mathbf{A}_{n-1}]$ be an $(n \times (n-1))$ matrix. Show that $B = [\mathbf{A}_1, \ldots, \mathbf{A}_{n-1}, A\mathbf{b}]$ is singular for every choice of **b** in R^{n-1}.

53. Suppose that C and B are (2×2) matrices and that B is singular. Show that CB is singular. (*Hint:* By Definition 12, there is a vector $\mathbf{x}_1$ in R^2, $\mathbf{x}_1 \neq \boldsymbol{\theta}$, such that $B\mathbf{x}_1 = \boldsymbol{\theta}$.)

54. Let $\{\mathbf{w}_1, \mathbf{w}_2\}$ be a linearly independent set of vectors in R^2. Show that if **b** is any vector in R^2, then **b** is a linear combination of $\mathbf{w}_1$ and $\mathbf{w}_2$. (*Hint:* Consider the (2×2) matrix $A = [\mathbf{w}_1, \mathbf{w}_2]$.)

55. Let A be an $(n \times n)$ nonsingular matrix. Show that A^T is nonsingular as follows:
a) Suppose that **v** is a vector in R^n such that $A^T\mathbf{v} = \boldsymbol{\theta}$. Cite a theorem from this section that guarantees there is a vector **w** in R^n such that $A\mathbf{w} = \mathbf{v}$.
b) By part (a), $A^TA\mathbf{w} = \boldsymbol{\theta}$, and therefore $\mathbf{w}^TA^TA\mathbf{w} = \mathbf{w}^T\boldsymbol{\theta} = 0$. Cite results from Section 1.7 that allow you to conclude that $\|A\mathbf{w}\| = 0$. (*Hint:* What is $(A\mathbf{w})^T$?)
c) Use parts (a) and (b) to conclude that if $A^T\mathbf{v} = \boldsymbol{\theta}$, then $\mathbf{v} = \boldsymbol{\theta}$; this shows that A^T is nonsingular.

56. Let T be an $(n \times n)$ upper-triangular matrix.

$$T = \begin{bmatrix} t_{11} & t_{12} & t_{13} & \cdots & t_{1n} \\ 0 & t_{22} & t_{23} & \cdots & t_{2n} \\ 0 & 0 & t_{33} & \cdots & t_{3n} \\ \vdots & & & & \vdots \\ 0 & 0 & 0 & \cdots & t_{nn} \end{bmatrix}$$

Prove that if $t_{ii} = 0$ for some i, $1 \leq i \leq n$, then T is singular. (*Hint:* If $t_{11} = 0$, find a nonzero vector **v** such that $T\mathbf{v} = \boldsymbol{\theta}$. If $t_{rr} = 0$, but $t_{ii} \neq 0$ for $1, 2, \ldots, r - 1$, use Theorem 4 of Section 1.4 to show that columns $\mathbf{T}_1, \mathbf{T}_2, \ldots, \mathbf{T}_r$ of T are linearly dependent. Then select a nonzero vector **v** such that $T\mathbf{v} = \boldsymbol{\theta}$.)

57. Let T be an $(n \times n)$ upper-triangular matrix as in Exercise 56. Prove that if $t_{ii} \neq 0$ for $i = 1, 2, \ldots, n$, then T is nonsingular. (*Hint:* Let $T = [\mathbf{T}_1, \mathbf{T}_2, \ldots, \mathbf{T}_n]$, and suppose that $a_1\mathbf{T}_1 + a_2\mathbf{T}_2 + \cdots + a_n\mathbf{T}_n = \boldsymbol{\theta}$ for some scalars $a_1, a_2, \ldots, a_n$. First deduce that $a_n = 0$. Next show $a_{n-1} = 0$, and so on.) Note that Exercises 56 and 57 establish that an upper-triangular matrix is singular if and only if one of the entries $t_{11}, t_{22}, \ldots, t_{nn}$ is zero. By Exercise 55 the same result is true for lower-triangular matrices.

58. Suppose that the $(n \times n)$ matrices A and B are row equivalent. Prove that A is nonsingular if and only if B is nonsingular. (*Hint:* The homogeneous systems $A\mathbf{x} = \boldsymbol{\theta}$ and $B\mathbf{x} = \boldsymbol{\theta}$ are equivalent by Theorem 1 of Section 1.1.)

1.9   Data Fitting, Numerical Integration, and Numerical Differentiation (Optional)

In this section we present four applications of matrix theory toward the solution of a practical problem. Three of the applications involve numerical approximation techniques, and the fourth relates to solving certain types of differential equations. In each case, solving the general problem depends on being able to solve a system of linear equations, and the theory of nonsingular matrices will guarantee that a solution exists and is unique.

Polynomial Interpolation

We begin by applying matrix theory to the problem of interpolating data with polynomials. In particular, Theorem 13 of Section 1.8 is used to establish a general existence and uniqueness result for polynomial interpolation. The following example is a simple illustration of polynomial interpolation.

EXAMPLE 1   Find a quadratic polynomial, $q(t)$, such that the graph of $q(t)$ goes through the points $(1, 2)$, $(2, 3)$, and $(3, 6)$ in the ty-plane (see Fig. 1.15).

SOLUTION   A quadratic polynomial $q(t)$ has the form

$$q(t) = a + bt + ct^2, \tag{1a}$$

so our problem reduces to determining constants a, b, and c such that

$$
\begin{aligned}
q(1) &= 2 \\
q(2) &= 3 \\
q(3) &= 6
\end{aligned}
\tag{1b}
$$

The constraints in (1b) are, by (1a), equivalent to

$$
\begin{aligned}
a + b + c &= 2 \\
a + 2b + 4c &= 3 \\
a + 3b + 9c &= 6.
\end{aligned}
\tag{1c}
$$

Clearly (1c) is a system of three linear equations in the three unknowns a, b, and c; so solving (1c) will determine the polynomial $q(t)$. Solving (1c), we find the unique solution $a = 3$, $b = -2$, $c = 1$; therefore, $q(t) = 3 - 2t + t^2$ is the unique quadratic polynomial satisfying the conditions (1b). A portion of the graph of $q(t)$ is shown in Fig. 1.16.

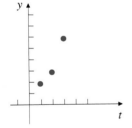

Figure 1.15
Points in the ty-plane

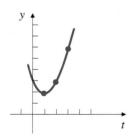

Figure 1.16
Graph of $q(t)$

Frequently polynomial interpolation is used when values of a function $f(t)$ are given in tabular form. For example, given a table of $n + 1$ values of $f(t)$ (see

Table 1.1

t	$f(t)$
t_0	y_0
t_1	y_1
t_2	y_2
$\vdots$	$\vdots$
t_n	y_n

Table 1.1), an *interpolating polynomial* for $f(t)$ is a polynomial, $p(t)$, of the form

$$p(t) = a_0 + a_1 t + a_2 t^2 + \cdots + a_n t^n$$

such that $p(t_i) = y_i = f(t_i)$ for $0 \le i \le n$. Problems of interpolating data in tables are quite common in scientific and engineering work; for example, $y = f(t)$ might describe a temperature distribution as a function of time with $y_i = f(t_i)$ being observed (measured) temperatures. For a time $\hat{t}$ not listed in the table, $p(\hat{t})$ provides an approximation for $f(\hat{t})$.

EXAMPLE 2 Find an interpolating polynomial for the four observations given in Table 1.2. Give an approximation for $f(1.5)$.

Table 1.2

t	$f(t)$
0	3
1	0
2	−1
3	6

SOLUTION In this case, the interpolating polynomial is a polynomial of degree 3 or less,

$$p(t) = a_0 + a_1 t + a_2 t^2 + a_3 t^3,$$

where $p(t)$ satisfies the four constraints $p(0) = 3$, $p(1) = 0$, $p(2) = -1$, and $p(3) = 6$. As in the previous example, these constraints are equivalent to the (4×4) system of equations

$$
\begin{aligned}
a_0 &= 3 \\
a_0 + a_1 + a_2 + a_3 &= 0 \\
a_0 + 2a_1 + 4a_2 + 8a_3 &= -1 \\
a_0 + 3a_1 + 9a_2 + 27a_3 &= 6
\end{aligned}
$$

Solving this system, we find that $a_0 = 3$, $a_1 = -2$, $a_2 = -2$, $a_3 = 1$ is the unique solution. Hence the unique polynomial that interpolates the tabular data for $f(t)$ is

$$p(t) = 3 - 2t - 2t^2 + t^3.$$

The desired approximation for $f(1.5)$ is $p(1.5) = -1.125$. ☐

Note that in each of the two preceding examples, the interpolating polynomial was unique. Theorem 14 below states that this is always the case. The next example considers the general problem of fitting a quadratic polynomial to three data points and illustrates the proof of Theorem 14.

EXAMPLE 3 Given three distinct numbers t_0, t_1, t_2 and any set of three values y_0, y_1, y_2, show that there exists a unique polynomial,

$$p(t) = a_0 + a_1 t + a_2 t^2, \tag{2a}$$

of degree 2 or less such that $p(t_0) = y_0$, $p(t_1) = y_1$, and $p(t_2) = y_2$.

SOLUTION   The given constraints and (2a) define a (3×3) linear system,

$$a_0 + a_1 t_0 + a_2 t_0^2 = y_0$$
$$a_0 + a_1 t_1 + a_2 t_1^2 = y_1 \tag{2b}$$
$$a_0 + a_1 t_2 + a_2 t_2^2 = y_2,$$

where a_0, a_1, and a_2 are the unknowns. The problem is to show that system (2b) has a unique solution. We can write (2b) in matrix form as $T\mathbf{a} = \mathbf{y}$, where

$$T = \begin{bmatrix} 1 & t_0 & t_0^2 \\ 1 & t_1 & t_1^2 \\ 1 & t_2 & t_2^2 \end{bmatrix}, \quad \mathbf{a} = \begin{bmatrix} a_0 \\ a_1 \\ a_2 \end{bmatrix}, \quad \text{and} \quad \mathbf{y} = \begin{bmatrix} y_0 \\ y_1 \\ y_2 \end{bmatrix}. \tag{2c}$$

By Theorem 13, the system is guaranteed to have a unique solution if T is non-singular. To establish that T is nonsingular, it suffices to show that if

$$\mathbf{c} = \begin{bmatrix} c_0 \\ c_1 \\ c_2 \end{bmatrix}$$

is a solution to the homogeneous system $T\mathbf{x} = \boldsymbol{\theta}$, then $\mathbf{c} = \boldsymbol{\theta}$. But $T\mathbf{c} = \boldsymbol{\theta}$ is equivalent to

$$c_0 + c_1 t_0 + c_2 t_0^2 = 0$$
$$c_0 + c_1 t_1 + c_2 t_1^2 = 0 \tag{2d}$$
$$c_0 + c_1 t_2 + c_2 t_2^2 = 0.$$

Let $q(t) = c_0 + c_1 t + c_2 t^2$. Then $q(t)$ has degree at most 2 and, by system (2d), $q(t_0) = q(t_1) = q(t_2) = 0$. Thus $q(t)$ has three distinct real zeros. By Exercise 25, if a quadratic polynomial has three distinct real zeros, then it must be identically zero. That is, $c_0 = c_1 = c_2 = 0$, or $\mathbf{c} = \boldsymbol{\theta}$. Hence T is nonsingular, and so system (2b) has a unique solution.   ☐

The matrix T given in (2c) above is the (3×3) Vandermonde matrix. More generally, for real numbers $t_0, t_1, \ldots, t_n$, the $[(n + 1) \times (n + 1)]$ **Vandermonde matrix** T is defined by

$$T = \begin{bmatrix} 1 & t_0 & t_0^2 & \cdots & t_0^n \\ 1 & t_1 & t_1^2 & \cdots & t_1^n \\ \vdots & & & & \vdots \\ 1 & t_n & t_n^2 & \cdots & t_n^n \end{bmatrix}. \tag{3}$$

Following the argument given in Example 3 and making use of Exercise 26, we can show that if $t_0, t_1, \ldots, t_n$ are distinct, then T is nonsingular. Thus, by Theorem 13, the linear system $T\mathbf{x} = \mathbf{y}$ has a unique solution for each choice of $\mathbf{y}$ in R^{n+1}. As a consequence, we have the following theorem.

THEOREM 14

Given $n + 1$ distinct numbers $t_0, t_1, \ldots, t_n$ and any set of $n + 1$ values y_0, $y_1, \ldots, y_n$, there is one and only one polynomial $p(t)$ of degree n or less, $p(t) = a_0 + a_1 t + \cdots + a_n t^n$, such that $p(t_i) = y_i$, $i = 0, 1, \ldots, n$.

Solutions to Initial Value Problems

The following example provides yet another application of the fact that the Vandermonde matrix T given in (3) is nonsingular when $t_0, t_1, \ldots, t_n$ are distinct. Problems of this sort arise in solving initial value problems in differential equations.

EXAMPLE 4   Given $n + 1$ distinct numbers $t_0, t_1, \ldots, t_n$ and any set of $n + 1$ values y_0, $y_1, \ldots, y_n$, show that there is one, and only one, function that has the form

$$y = a_0 e^{t_0 x} + a_1 e^{t_1 x} + \cdots + a_n e^{t_n x} \tag{4a}$$

and that satisfies the constraints $y(0) = y_0$, $y'(0) = y_1, \ldots, y^{(n)}(0) = y_n$.

SOLUTION   Calculating the first n derivatives of y gives

$$
\begin{aligned}
y &= a_0 e^{t_0 x} & + a_1 e^{t_1 x} & + \cdots + a_n e^{t_n x} \\
y' &= a_0 t_0 e^{t_0 x} & + a_1 t_1 e^{t_1 x} & + \cdots + a_n t_n e^{t_n x} \\
y'' &= a_0 t_0^2 e^{t_0 x} & + a_1 t_1^2 e^{t_1 x} & + \cdots + a_n t_n^2 e^{t_n x} \\
&\ \vdots & \vdots & \qquad\qquad \vdots \\
y^{(n)} &= a_0 t_0^n e^{t_0 x} & + a_1 t_1^n e^{t_1 x} & + \cdots + a_n t_n^n e^{t_n x}.
\end{aligned} \tag{4b}
$$

Substituting $x = 0$ in each equation of system (4b) and setting $y^{(k)}(0) = y_k$ yields the system

$$
\begin{aligned}
y_0 &= a_0 & + a_1 & + \cdots + a_n \\
y_1 &= a_0 t_0 & + a_1 t_1 & + \cdots + a_n t_n \\
y_2 &= a_0 t_0^2 & + a_1 t_1^2 & + \cdots + a_n t_n^2 \\
&\ \vdots & \vdots & \qquad\quad \vdots \\
y_n &= a_0 t_0^n & + a_1 t_1^n & + \cdots + a_n t_n^n
\end{aligned} \tag{4c}
$$

with unknowns $a_0, a_1, \ldots, a_n$. Note that the coefficient matrix for the linear system (4c) is

$$
T^T = \begin{bmatrix}
1 & 1 & \cdots & 1 \\
t_0 & t_1 & \cdots & t_n \\
t_0^2 & t_1^2 & \cdots & t_n^2 \\
\vdots & & & \vdots \\
t_0^n & t_1^n & \cdots & t_n^n
\end{bmatrix}, \tag{4d}
$$

where T is the $[(n + 1) \times (n + 1)]$ Vandermonde matrix given in (3). It is left as an exercise (cf. Exercise 55 of Section 1.8) to show that because T is non-singular, the transpose T^T is also nonsingular. Thus by Theorem 13, the linear system (4c) has a unique solution.

The next example is a specific case of Example 4.

EXAMPLE 5 Find the unique function $y = c_1 e^x + c_2 e^{2x} + c_3 e^{3x}$ that satisfies the constraints $y(0) = 1$, $y'(0) = 2$, and $y''(0) = 0$.

SOLUTION   The given function and its first two derivatives are

$$
\begin{aligned}
y &= c_1 e^x + c_2 e^{2x} + c_3 e^{3x} \\
y' &= c_1 e^x + 2c_2 e^{2x} + 3c_3 e^{3x} \\
y'' &= c_1 e^x + 4c_2 e^{2x} + 9c_3 e^{3x}.
\end{aligned}
\tag{5a}
$$

From (5a) the given constraints are equivalent to

$$
\begin{aligned}
1 &= c_1 + c_2 + c_3 \\
2 &= c_1 + 2c_2 + 3c_3 \\
0 &= c_1 + 4c_2 + 9c_3.
\end{aligned}
\tag{5b}
$$

The augmented matrix for system (5b) is

$$
\begin{bmatrix}
1 & 1 & 1 & 1 \\
1 & 2 & 3 & 2 \\
1 & 4 & 9 & 0
\end{bmatrix},
$$

and solving in the usual manner yields the unique solution $c_1 = -2$, $c_2 = 5$, $c_3 = -2$. Therefore, the function $y = -2e^x + 5e^{2x} - 2e^{3x}$ is the unique function that satisfies the given constraints.

Numerical Integration

The Vandermonde matrix also arises in problems where it is necessary to estimate numerically an integral or a derivative. For example, let $I(f)$ denote the definite integral

$$
I(f) = \int_a^b f(t)\, dt.
$$

If the integrand is fairly complicated or if the integrand is not a standard form that can be found in a table of integrals, then it will be necessary to approximate the value $I(f)$ numerically.

One effective way to approximate $I(f)$ is first to find a polynomial p that approximates f on $[a, b]$,

$$
p(t) \approx f(t), \qquad a \le t \le b.
$$

Next, given that p is a "good" approximation to f, we would expect that the approximation below is also a good one:

$$\int_a^b p(t)\, dt \approx \int_a^b f(t)\, dt. \tag{6}$$

Of course, since p is a polynomial, the integral on the left-hand side of (6) can be easily evaluated and provides a computable estimate to the unknown integral, $I(f)$.

One way to generate a polynomial approximation to f is through interpolation. If we select $n + 1$ points $t_0, t_1, \ldots, t_n$ in $[a, b]$, then the nth-degree polynomial p that satisfies $p(t_i) = f(t_i)$, $0 \le i \le n$, is an approximation to f that can be used in (6) to estimate $I(f)$.

In summary, the numerical integration process outlined above proceeds as follows:

1. Given f, construct the interpolating polynomial, p.
2. Given p, calculate the integral, $\int_a^b p(t)\, dt$.
3. Use $\int_a^b p(t)\, dt$ as the approximation to $\int_a^b f(t)\, dt$.

It turns out that the approximation scheme described above can be simplified considerably and step 1 can be skipped entirely. That is, it is not necessary to construct the actual interpolating polynomial p in order to know the integral of p, $\int_a^b p(t)\, dt$.

We will illustrate the idea with a quadratic interpolating polynomial. Suppose p is the quadratic polynomial that interpolates f at t_0, t_1, and t_2. Next, suppose we can find scalars A_0, A_1, A_2 such that

$$
\begin{aligned}
A_0 \quad + A_1 \quad + A_2 &= \int_a^b 1\, dt \\[4pt]
A_0 t_0 + A_1 t_1 + A_2 t_2 &= \int_a^b t\, dt \\[4pt]
A_0 t_0^2 + A_1 t_1^2 + A_2 t_2^2 &= \int_a^b t^2\, dt.
\end{aligned}
\tag{7}
$$

Now, if the interpolating polynomial p is given by $p(t) = a_0 + a_1 t + a_2 t^2$, then the equations in (7) give

$$
\begin{aligned}
\int_a^b p(t)\, dt &= \int_a^b [a_0 + a_1 t + a_2 t^2]\, dt \\[4pt]
&= a_0 \int_a^b 1\, dt + a_1 \int_a^b t\, dt + a_2 \int_a^b t^2\, dt \\[4pt]
&= a_0 \sum_{i=0}^2 A_i + a_1 \sum_{i=0}^2 A_i t_i + a_2 \sum_{i=0}^2 A_i t_i^2 \\[4pt]
&= \sum_{i=0}^2 A_i [a_0 + a_1 t_i + a_2 t_i^2] \\[4pt]
&= \sum_{i=0}^2 A_i p(t_i).
\end{aligned}
$$

The calculations above demonstrate the following: If we know the values of a quadratic polynomial p at three points t_0, t_1, t_2 and if we can find scalars A_0, A_1, A_2 that satisfy system (7), then we can evaluate the integral of p with the formula

$$\int_a^b p(t)\,dt = \sum_{i=0}^2 A_i p(t_i). \tag{8}$$

Next, since p is the quadratic interpolating polynomial for f, we see that the values of $p(t_i)$ are known to us; that is, $p(t_0) = f(t_0)$, $p(t_1) = f(t_1)$, and $p(t_2) = f(t_2)$. Thus, combining (8) and (6), we obtain

$$\int_a^b p(t)\,dt = \sum_{i=0}^2 A_i p(t_i) = \sum_{i=0}^2 A_i f(t_i) \approx \int_a^b f(t)\,dt,$$

or equivalently,

$$\int_a^b f(t)\,dt \approx \sum_{i=0}^2 A_i f(t_i). \tag{9}$$

The approximation $\sum_{i=0}^2 A_i f(t_i)$ in (9) is known as a numerical integration formula. Observe that once the evaluation points t_0, t_1, t_2 are selected, the scalars A_0, A_1, A_2 are determined by system (7). The coefficient matrix for system (7) has the form

$$A = \begin{bmatrix} 1 & 1 & 1 \\ t_0 & t_1 & t_2 \\ t_0^2 & t_1^2 & t_2^2 \end{bmatrix},$$

and so we see that A is nonsingular since A is the transpose of a Vandermonde matrix (recall matrix (4d)).

In general, if $t_0, t_1, \ldots, t_n$ are $n + 1$ points in $[a, b]$, we can proceed exactly as in the derivation of formula (9) and produce a numerical integration formula of the form

$$\int_a^b f(t)\,dt \approx \sum_{i=0}^n A_i f(t_i). \tag{10}$$

The weights A_i in formula (10) would be determined by solving the "Vandermonde system":

$$A_0 + A_1 + \cdots + A_n = \int_a^b 1\,dt$$

$$A_0 t_0 + A_1 t_1 + \cdots + A_n t_n = \int_a^b t\,dt$$

$$\vdots \qquad\qquad \vdots \qquad \vdots \tag{11}$$

$$A_0 t_0^n + A_1 t_1^n + \cdots + A_n t_n^n = \int_a^b t^n\,dt.$$

The approximation $\sum_{i=0}^n A_i f(t_i)$ is the same number that would be produced by calculating the polynomial p of degree n that interpolates f at $t_0, t_1, \ldots, t_n$ and then evaluating $\int_a^b p(t)\,dt$.

EXAMPLE 6 For an interval $[a, b]$ let $t_0 = a$, $t_1 = (a + b)/2$, and $t_2 = b$. Construct the corresponding numerical integration formula.

SOLUTION   For $t_0 = a$, $t_1 = (a + b)/2$, and $t_2 = b$, the system to be solved is given by (11) with $n = 2$. We write system (11) as $C\mathbf{x} = \mathbf{d}$, where

$$C = \begin{bmatrix} 1 & 1 & 1 \\ a & t_1 & b \\ a^2 & t_1^2 & b^2 \end{bmatrix} \quad \text{and} \quad \mathbf{d} = \begin{bmatrix} b - a \\ (b^2 - a^2)/2 \\ (b^3 - a^3)/3 \end{bmatrix}.$$

It can be shown (Exercise 23) that the augmented matrix $[C \,|\, \mathbf{d}]$ is row equivalent to

$$\begin{bmatrix} 1 & 1 & 1 & b - a \\ 0 & 1 & 2 & b - a \\ 0 & 0 & 6 & b - a \end{bmatrix}.$$

Hence the solution of $C\mathbf{x} = \mathbf{d}$ is $A_0 = (b - a)/6$, $A_1 = 4(b - a)/6$, $A_2 = (b - a)/6$. The corresponding numerical integration formula is

$$\int_a^b f(t)\, dt \approx [(b - a)/6]\{f(a) + 4f[(a + b)/2] + f(b)\}. \tag{12}$$

The reader may be familiar with the approximation above, which is known as Simpson's rule.   ☐

EXAMPLE 7 Use the integration formula (12) to approximate the integral

$$I(f) = \int_0^{1/2} \cos(\pi t^2/2)\, dt.$$

SOLUTION   With $a = 0$ and $b = 1/2$, formula (12) becomes

$$\begin{aligned} I(f) &\approx 1/12[\cos(0) + 4\cos(\pi/32) + \cos(\pi/8)] \\ &= (1/12)[1.0 + 4(0.995184\ldots) + 0.923879\ldots] \\ &= 0.492051\ldots. \end{aligned}$$

☐

Note that in Example 7, the number $I(f)$ is equal to $C(0.5)$, where $C(x)$ denotes the Fresnel integral

$$C(x) = \int_0^x \cos(\pi t^2/2)\, dt.$$

The function $C(x)$ is important in applied mathematics, and extensive tables of the function $C(x)$ are available. The integrand is not a standard form, and $C(x)$ must be evaluated numerically. From a table, $C(0.5) = 0.49223442\ldots$.

Numerical Differentiation

Numerical differentiation formulas can also be derived in the same fashion as numerical integration formulas. In particular, suppose that f is a differentiable function and we wish to estimate the value $f'(a)$, where f is differentiable at $t = a$.

Let p be the polynomial of degree n that interpolates f at $t_0, t_1, \ldots, t_n$, where the interpolation nodes t_i are clustered near $t = a$. Then p provides us with an approximation for f, and we can estimate the value $f'(a)$ by evaluating the derivative of p at $t = a$:

$$f'(a) \approx p'(a).$$

As with a numerical integration formula, it can be shown that the value $p'(a)$ can be expressed as

$$p'(a) = A_0 p(t_0) + A_1 p(t_1) + \cdots + A_n p(t_n). \tag{13}$$

In formula (13), the weights A_i are determined by the system of equations

$$q_0'(a) = A_0 q_0(t_0) + A_1 q_0(t_1) + \cdots + A_n q_0(t_n)$$
$$q_1'(a) = A_0 q_1(t_0) + A_1 q_1(t_1) + \cdots + A_n q_1(t_n)$$
$$\vdots \qquad \vdots \qquad\qquad \vdots$$
$$q_n'(a) = A_0 q_n(t_0) + A_1 q_n(t_1) + \cdots + A_n q_n(t_n),$$

where $q_0(t) = 1$, $q_1(t) = t, \ldots, q_n(t) = t^n$. So if formula (13) holds for the $n + 1$ special polynomials $1, t, \ldots, t^n$, then it holds for every polynomial p of degree n or less.

If p interpolates f at $t_0, t_1, \ldots, t_n$ so that $p(t_i) = f(t_i)$, $0 \le i \le n$, then the approximation $f'(a) \approx p'(a)$ leads to (by formula 13)

$$f'(a) \approx A_0 f(t_0) + A_1 f(t_1) + \cdots + A_n f(t_n). \tag{14}$$

An approximation of the form (14) is called a numerical differentiation formula.

EXAMPLE 8 Derive a numerical differentiation formula of the form

$$f'(a) \approx A_0 f(a - h) + A_1 f(a) + A_2 f(a + h).$$

SOLUTION The weights A_0, A_1, and A_2 are determined by forcing Eq. (13) to hold for $p(t) = 1$, $p(t) = t$, and $p(t) = t^2$. Thus the weights are found by solving the system

$$
\begin{array}{lll}
[p(t) = 1] & 0 = A_0 & + A_1 + A_2 \\
[p(t) = t] & 1 = A_0(a - h) & + A_1(a) + A_2(a + h) \\
[p(t) = t^2] & 2a = A_0(a - h)^2 & + A_1(a)^2 + A_2(a + h)^2.
\end{array}
$$

In matrix form, the system above can be expressed as $C\mathbf{x} = \mathbf{d}$, where

$$C = \begin{bmatrix} 1 & 1 & 1 \\ a - h & a & a + h \\ (a - h)^2 & a^2 & (a + h)^2 \end{bmatrix} \quad \text{and} \quad \mathbf{d} = \begin{bmatrix} 0 \\ 1 \\ 2a \end{bmatrix}.$$

By (4d), the matrix C is nonsingular and (see Exercise 24) the augmented matrix is row equivalent to

$$\begin{bmatrix} 1 & 1 & 1 & 0 \\ 0 & h & 2h & 1 \\ 0 & 0 & 2h^2 & h \end{bmatrix}.$$

Thus the solution is $A_0 = -1/2h$, $A_1 = 0$, $A_2 = 1/2h$. The numerical differentiation formula has the form

$$f'(a) \approx [f(a + h) - f(a - h)]/2h. \tag{15}$$

Note: Formula (15) in this example is known as the centered-difference approximation to $f'(a)$.

These same techniques can be used to derive formulas for estimating higher derivatives.

EXAMPLE 9 Construct a numerical differentiation formula of the form

$$f''(a) \approx A_0 f(a) + A_1 f(a + h) + A_2 f(a + 2h) + A_3 f(a + 3h).$$

SOLUTION The weights A_0, A_1, A_2, and A_3 are determined by forcing the approximation above to be an equality for $p(t) = 1$, $p(t) = t$, $p(t) = t^2$, and $p(t) = t^3$. These constraints lead to the equations

$$\begin{array}{ll} [p(t) = 1] & 0 = A_0 + A_1 + A_2 + A_3 \\ [p(t) = t] & 0 = A_0(a) + A_1(a + h) + A_2(a + 2h) + A_3(a + 3h) \\ [p(t) = t^2] & 2 = A_0(a)^2 + A_1(a + h)^2 + A_2(a + 2h)^2 + A_3(a + 3h)^2 \\ [p(t) = t^3] & 6a = A_0(a)^3 + A_1(a + h)^3 + A_2(a + 2h)^3 + A_3(a + 3h)^3. \end{array}$$

Since the system above is a bit cumbersome to solve by hand, we decided to use the computer algebra system Derive. (Because the coefficient matrix has symbolic rather than numerical entries, we had to use a computer algebra system rather than numerical software such as MATLAB. In particular, Derive is a popular computer algebra system that is menu-driven and very easy to use.) Figure 1.17 shows the results from Derive. Line 2 gives the command to row reduce the augmented matrix for the system. Line 3 gives the results. Therefore, the numerical differentiation formula is

$$f''(a) \approx \frac{1}{h^2} [2f(a) - 5f(a + h) + 4f(a + 2h) - f(a + 3h)].$$

2:    ROW_REDUCE

$$\begin{bmatrix} 1 & 1 & 1 & 1 & 0 \\ a & a+h & a+2h & a+3h & 0 \\ a^2 & (a+h)^2 & (a+2h)^2 & (a+3h)^2 & 2 \\ a^3 & (a+h)^3 & (a+2h)^3 & (a+3h)^3 & 6a \end{bmatrix}$$

3:

$$\begin{bmatrix} 1 & 0 & 0 & 0 & \dfrac{2}{h^2} \\ 0 & 1 & 0 & 0 & -\dfrac{5}{h^2} \\ 0 & 0 & 1 & 0 & \dfrac{4}{h^2} \\ 0 & 0 & 0 & 1 & -\dfrac{1}{h^2} \end{bmatrix}$$

Figure 1.17
Using Derive to solve the system of equations in Example 9

Exercises 1.9

In Exercises 1–6, find the interpolating polynomial for the given table of data. (*Hint:* If the data table has k entries, the interpolating polynomial will be of degree $k - 1$ or less.)

1.
t	0	1	2
y	−1	3	6

2.
t	−1	0	2
y	6	1	−3

3.
t	−1	1	2
y	1	5	7

4.
t	1	3	4
y	5	11	14

5.
t	−1	0	1	2
y	−6	1	4	15

6.
t	−2	−1	1	2
y	−3	1	3	13

In Exercises 7–10, find the constants so that the given function satisfies the given conditions.

7. $y = c_1 e^{2x} + c_2 e^{3x};$ $y(0) = 3,$ $y'(0) = 7$

8. $y = c_1 e^{(x-1)} + c_2 e^{3(x-1)};$ $y(1) = 1,$ $y'(1) = 5$

9. $y = c_1 e^{-x} + c_2 e^{x} + c_3 e^{2x};$ $y(0) = 8,$ $y'(0) = 3,$ $y''(0) = 11$

10. $y = c_1 e^{x} + c_2 e^{2x} + c_3 e^{3x};$ $y(0) = -1,$ $y'(0) = -3,$ $y''(0) = -5$

As in Example 6, find the weights A_i for the numerical integration formulas listed in Exercises 11–16. (*Note:* It can be shown that the special formulas developed in Exercises 11–16 can be "translated" to any interval of the general form $[a, b]$. Similarly, the numerical differentiation formulas in Exercises 17–22 can also be translated.)

11. $\int_0^{3h} f(t)\,dt \approx A_0 f(h) + A_1 f(2h)$

12. $\int_0^{h} f(t)\,dt \approx A_0 f(0) + A_1 f(h)$

13. $\int_0^{3h} f(t)\,dt \approx A_0 f(0) + A_1 f(h) + A_2 f(2h) + A_3 f(3h)$

14. $\int_0^{4h} f(t)\,dt \approx A_0 f(h) + A_1 f(2h) + A_2 f(3h)$

15. $\int_0^{h} f(t)\,dt \approx A_0 f(-h) + A_1 f(0)$

16. $\int_0^{h} f(t)\,dt \approx A_0 f(-h) + A_1 f(0) + A_2 f(h)$

As in Example 8, find the weights for the numerical differentiation formulas in Exercises 17–22. For Exercises 21 and 22, replace $p'(a)$ in formula (13) by $p''(a)$.

17. $f'(0) \approx A_0 f(0) + A_1 f(h)$

18. $f'(0) \approx A_0 f(-h) + A_1 f(0)$

19. $f'(0) \approx A_0 f(0) + A_1 f(h) + A_2 f(2h)$

20. $f'(0) \approx A_0 f(0) + A_1 f(h) + A_2 f(2h) + A_3 f(3h)$

21. $f''(0) \approx A_0 f(-h) + A_1 f(0) + A_2 f(h)$

22. $f''(0) \approx A_0 f(0) + A_1 f(h) + A_2 f(2h)$

23. Complete the calculations in Example 6 by reducing the augmented matrix $[C\,|\,\mathbf{d}]$ to the echelon form given.

24. Complete the calculations in Example 8 by reducing the augmented matrix $[C\,|\,\mathbf{d}]$ to the echelon form given.

25. Let p denote the quadratic polynomial defined by $p(t) = at^2 + bt + c$, where a, b, and c are real numbers. Use Rolle's theorem to prove the following: If t_0, t_1, and t_2 are real numbers such that $t_0 < t_1 < t_2$ and if $p(t_0) = 0$, $p(t_1) = 0$, and $p(t_2) = 0$, then $a = b = c = 0$. (Recall that Rolle's theorem tells us there are values u_1 and u_2 such that u_1 is in (t_0, t_1), u_2 is in (t_1, t_2), $p'(u_1) = 0$, and $p'(u_2) = 0$.)

26. Use mathematical induction to prove that a polynomial of the form $p(t) = a_n t^n + \cdots + a_1 t + a_0$ can have $n+1$ distinct real zeros only if $a_n = a_{n-1} = \cdots =$

$a_1 = a_0 = 0$. (*Hint:* Use Rolle's theorem, as in Exercise 25.)

Exercises 27–33 concern **Hermite interpolation,** where Hermite interpolation means the process of constructing polynomials that match both function values and derivative values.

In Exercises 27–30, find a polynomial p of the form $p(t) = at^3 + bt^2 + ct + d$ that satisfies the given conditions.

27. $p(0) = 2$, $p'(0) = 3$, $p(1) = 8$, $p'(1) = 10$

28. $p(0) = 1$, $p'(0) = 2$, $p(1) = 4$, $p'(1) = 4$

29. $p(-1) = -1$, $p'(-1) = 5$, $p(1) = 9$, $p'(1) = 9$

30. $p(1) = 3$, $p'(1) = 4$, $p(2) = 15$, $p'(2) = 22$

31. Suppose that t_0 and t_1 are distinct real numbers, where $t_0 < t_1$. Prove: If p is any polynomial of the form $p(t) = at^3 + bt^2 + ct + d$, where $p(t_0) = p(t_1) = 0$ and $p'(t_0) = p'(t_1) = 0$, then $a = b = c = d = 0$. (*Hint:* Apply Rolle's theorem.)

32. Suppose t_0 and t_1 are distinct real numbers, where $t_0 < t_1$. Suppose y_0, y_1, s_0, and s_1 are given real numbers. Prove that there is one, and only one, polynomial p of the form $p(t) = at^3 + bt^2 + ct + d$ such that $p(t_0) = y_0$, $p'(t_0) = s_0$, $p(t_1) = y_1$, and $p'(t_1) = s_1$. (*Hint:* Set up a system of four equations corresponding to the four interpolation constraints. Use Exercise 31 to show that the coefficient matrix is nonsingular.)

33. Let $t_0 < t_1 < \cdots < t_n$ be $n+1$ distinct numbers. Let $y_0, y_1, \ldots, y_n$ and $s_0, s_1, \ldots, s_n$ be given real numbers. Show that there is one, and only one, polynomial p of degree $2n+1$ or less such that $p(t_i) = y_i$, $0 \le i \le n$, and $p'(t_i) = s_i$, $0 \le i \le n$. (*Hint:* As in Exercise 31, show that all the coefficients of p are zero if $y_i = s_i = 0$, $0 \le i \le n$. Next, as in Exercise 32, write the system of equations corresponding to the interpolation constraints and verify that the coefficient matrix is nonsingular.)

In Exercises 34 and 35, use linear algebra software to construct the formula.

34. $\int_0^{5h} f(x)\,dx \approx \sum_{j=0}^5 A_j f(jh)$

35. $f'(a) \approx A_0 f(a - 2h) + A_1 f(a - h) + A_2 f(a)$
$\qquad + A_3 f(a + h) + A_4 f(a + 2h)$

1.10 Matrix Inverses and Their Properties

In the preceding sections the matrix equation

$$A\mathbf{x} = \mathbf{b} \tag{1}$$

has been used extensively to represent a system of linear equations. Equation (1) looks, symbolically, like the single linear equation

$$ax = b, \tag{2}$$

where a and b are real numbers. Since Eq. (2) has the unique solution

$$x = a^{-1}b \tag{3}$$

when $a \neq 0$, it is natural to ask whether Eq. (1) can also be solved as

$$\mathbf{x} = A^{-1}\mathbf{b}. \tag{4}$$

In this section we define the inverse of a matrix and develop some of the properties of matrix inverses. In Section 1.11, we prove that a matrix has an inverse if, and only if, it is nonsingular. Thus for a nonsingular matrix A, the matrix equation (1) has the unique solution given by (4).

The Identity Matrix

For a nonzero real number a, the inverse of a is the unique real number a^{-1} with the property that

$$a^{-1}a = aa^{-1} = 1,$$

where 1 is, of course, the multiplicative identity for the set of real numbers. In an analogous manner, if the matrix A has an inverse, A^{-1}, then we require both that $A^{-1}A = AA^{-1}$ and that this product be a multiplicative identity for matrix multiplication.

For each positive integer n, there is an $(n \times n)$ identity element for matrix multiplication called the $(n \times n)$ *identity matrix* and denoted by I_n. The identity matrix I_n is defined to be the $(n \times n)$ matrix that has ones on the main diagonal and zeros elsewhere:

$$I_n = \begin{bmatrix} 1 & 0 & 0 & \cdots & 0 \\ 0 & 1 & 0 & \cdots & 0 \\ 0 & 0 & 1 & \cdots & 0 \\ \vdots & & & & \vdots \\ 0 & 0 & 0 & \cdots & 1 \end{bmatrix}. \tag{5}$$

That is, the ijth entry of I_n is 0 when $i \neq j$ and is 1 when $i = j$. For example, I_2 and I_3 are given by

$$I_2 = \begin{bmatrix} 1 & 0 \\ 0 & 1 \end{bmatrix} \quad \text{and} \quad I_3 = \begin{bmatrix} 1 & 0 & 0 \\ 0 & 1 & 0 \\ 0 & 0 & 1 \end{bmatrix}.$$

The following theorem is an immediate consequence of (5) and the definition of matrix multiplication.

THEOREM 15

If B is an $(m \times n)$ matrix, then $I_m B = B$ and $BI_n = B$. In particular, if A is an $(n \times n)$ matrix, then $I_n A = AI_n = A$.

To illustrate Theorem 15, let

$$A = \begin{bmatrix} 1 & 2 & 0 \\ -1 & 3 & 4 \\ 6 & 1 & 8 \end{bmatrix}, \qquad B = \begin{bmatrix} 2 & 3 & 1 \\ 1 & 5 & 7 \end{bmatrix}, \qquad C = \begin{bmatrix} -2 & 0 \\ 8 & 3 \\ 6 & 1 \end{bmatrix},$$

and

$$\mathbf{x} = \begin{bmatrix} 1 \\ 0 \\ 3 \end{bmatrix}.$$

Note that

$$I_3 A = AI_3 = A$$
$$BI_3 = B$$
$$I_3 C = C$$
$$I_3 \mathbf{x} = \mathbf{x},$$

whereas the products $I_3 B$ and CI_3 are not defined.

Usually the dimension of the identity matrix is clear from the context of the problem under consideration, and it is customary to drop the subscript, n, and denote the $(n \times n)$ identity matrix simply as I. So, for example, if A is an $(n \times n)$ matrix, we will write $IA = AI = A$ instead of $I_n A = AI_n = A$. Finally, note that I_n has the form

$$I_n = [\mathbf{e}_1, \mathbf{e}_2, \ldots, \mathbf{e}_n],$$

where $\mathbf{e}_1, \mathbf{e}_2, \ldots, \mathbf{e}_n$ are the unit vectors introduced in Section 1.8.

Matrix Inverses

Since the various identity matrices play a role in matrix multiplication analogous to that of the number 1 in real number multiplication, it is natural to require that $A^{-1} A = AA^{-1} = I$. The requirement that $A^{-1} A = AA^{-1}$ dictates that A and A^{-1} be square matrices of the same size, and this in turn determines the dimension of I. Thus we have the following definition.

DEFINITION 13   The $(n \times n)$ matrix B is an *inverse* for the $(n \times n)$ matrix A provided that

$$AB = BA = I_n.$$

To illustrate, let

$$A = \begin{bmatrix} 1 & 2 \\ 3 & 4 \end{bmatrix} \quad \text{and} \quad B = \begin{bmatrix} -2 & 1 \\ 3/2 & -1/2 \end{bmatrix}.$$

Then $AB = BA = I$, so B is the inverse of A. We emphasize that nonsquare matrices cannot have inverses and, as the following example illustrates, even a nonzero square matrix need not have an inverse.

EXAMPLE 1   Let A be the (2×2) matrix

$$A = \begin{bmatrix} 1 & 2 \\ 3 & 6 \end{bmatrix}.$$

Show that A has no inverse.

SOLUTION   An inverse for A must be a (2×2) matrix

$$B = \begin{bmatrix} a & b \\ c & d \end{bmatrix}$$

such that $AB = BA = I$. If such a matrix B exists, it must satisfy the following equation:

$$\begin{bmatrix} 1 & 0 \\ 0 & 1 \end{bmatrix} = \begin{bmatrix} 1 & 2 \\ 3 & 6 \end{bmatrix}\begin{bmatrix} a & b \\ c & d \end{bmatrix} = \begin{bmatrix} a + 2c & b + 2d \\ 3a + 6c & 3b + 6d \end{bmatrix}. \tag{6}$$

But (6) requires that $a + 2c = 1$ and $3a + 6c = 0$. This is clearly impossible, so A has no inverse. □

It will be convenient to use the familiar notation A^{-1} to denote an inverse for a matrix A. So far as we know now, however, a single matrix A may have a multitude of inverses, and if this were the case, the symbol A^{-1} would be ambiguous. The next theorem assures us that a matrix can have at most one inverse.

THEOREM 16

If B and C are both inverses for the matrix A, then $B = C$.

PROOF   Since B and C are both inverses for matrix A, we have

$$AB = BA = I \tag{7}$$

and

$$AC = CA = I. \tag{8}$$

Using Eqs. (7) and (8) and the fact that matrix multiplication is associative (part 1 of Theorem 8 in Section 1.7), it follows that

$$B = BI = B(AC) = (BA)C = IC = C. \quad □$$

By Theorem 16, if a square matrix A has an inverse, then it is unique. Thus we may refer to "the" inverse of A and denote it by A^{-1}.

A procedure for calculating the inverse of an $(n \times n)$ nonsingular matrix will be developed in Section 1.11. In the meantime the following remark provides a simple and useful formula for the inverse of a (2×2) matrix.

REMARK   Let A be a (2×2) matrix,

$$A = \begin{bmatrix} a & b \\ c & d \end{bmatrix}, \tag{9}$$

and set $\Delta = ad - bc$.

a) If $\Delta = 0$, then A does not have an inverse.
b) If $\Delta \neq 0$, then A has an inverse given by

$$A^{-1} = \frac{1}{\Delta} \begin{bmatrix} d & -b \\ -c & a \end{bmatrix}. \tag{10}$$

Part (a) of the remark is Exercise 27 of Section 1.11. To verify the formula given in (b), suppose that $\Delta \neq 0$ and define B to be the matrix

$$B = \frac{1}{\Delta} \begin{bmatrix} d & -b \\ -c & a \end{bmatrix}.$$

Then

$$BA = \frac{1}{\Delta} \begin{bmatrix} d & -b \\ -c & a \end{bmatrix} \begin{bmatrix} a & b \\ c & d \end{bmatrix} = \frac{1}{\Delta} \begin{bmatrix} ad - bc & 0 \\ 0 & ad - bc \end{bmatrix} = \begin{bmatrix} 1 & 0 \\ 0 & 1 \end{bmatrix}.$$

Similarly, $AB = I$, so $B = A^{-1}$.

The reader familiar with determinants will recognize the number Δ in the remark as the determinant of the matrix A in (9). We make use of the remark in the following example.

EXAMPLE 2 Let A and B be given by

$$A = \begin{bmatrix} 6 & 8 \\ 3 & 4 \end{bmatrix} \quad \text{and} \quad B = \begin{bmatrix} 1 & 7 \\ 3 & 5 \end{bmatrix}.$$

For each matrix, determine whether an inverse exists and calculate the inverse if it does exist.

SOLUTION   For the matrix A, the number Δ is

$$\Delta = 6(4) - 8(3) = 0,$$

so, by the remark, A cannot have an inverse. For the matrix B, the number Δ is

$$\Delta = 1(5) - 7(3) = -16.$$

According to formula (10),

$$B^{-1} = -\frac{1}{16} \begin{bmatrix} 5 & -7 \\ -3 & 1 \end{bmatrix}.$$

Properties of Matrix Inverses

The following theorem lists some of the properties of matrix inverses.

THEOREM 17

Let A and B be $(n \times n)$ matrices, each of which has an inverse. Then:
1. A^{-1} has an inverse, and $(A^{-1})^{-1} = A$.
2. AB has an inverse, and $(AB)^{-1} = B^{-1}A^{-1}$.
3. If k is a nonzero scalar, then kA has an inverse, and $(kA)^{-1} = (1/k)A^{-1}$.
4. A^T has an inverse, and $(A^T)^{-1} = (A^{-1})^T$.

PROOF

1. Since $AA^{-1} = A^{-1}A = I$, the inverse of A^{-1} is A; that is, $(A^{-1})^{-1} = A$.
2. Note that $(AB)(B^{-1}A^{-1}) = A(BB^{-1})A^{-1} = A(IA^{-1}) = AA^{-1} = I$. Similarly, $(B^{-1}A^{-1})(AB) = I$, so, by Definition 13, $B^{-1}A^{-1}$ is the inverse for AB. Thus $(AB)^{-1} = B^{-1}A^{-1}$.
3. The proof of property 3 is similar to the proofs given for properties 1 and 2 and is left as an exercise.
4. It follows from Theorem 10, property 2, of Section 1.7 that $A^T(A^{-1})^T = (A^{-1}A)^T = I^T = I$. Similarly, $(A^{-1})^TA^T = I$. Therefore, A^T has inverse $(A^{-1})^T$.

Note that the familiar formula $(ab)^{-1} = a^{-1}b^{-1}$ for real numbers is valid only because multiplication of real numbers is commutative. We have already noted that matrix multiplication is not commutative, so, as the following example demonstrates, $(AB)^{-1} \neq A^{-1}B^{-1}$.

EXAMPLE 3 Let A and B be the (2×2) matrices

$$A = \begin{bmatrix} 1 & 3 \\ 2 & 4 \end{bmatrix} \quad \text{and} \quad B = \begin{bmatrix} 3 & -2 \\ 1 & -1 \end{bmatrix}.$$

1. Use formula (10) to calculate A^{-1}, B^{-1}, and $(AB)^{-1}$.
2. Use Theorem 17, property 2, to calculate $(AB)^{-1}$.
3. Show that $(AB)^{-1} \neq A^{-1}B^{-1}$.

SOLUTION   For A the number Δ is $\Delta = 1(4) - 3(2) = -2$, so by formula (10)

$$A^{-1} = \begin{bmatrix} -2 & 3/2 \\ 1 & -1/2 \end{bmatrix}.$$

For B the number Δ is $3(-1) - 1(-2) = -1$, so

$$B^{-1} = \begin{bmatrix} 1 & -2 \\ 1 & -3 \end{bmatrix}.$$

The product AB is given by

$$AB = \begin{bmatrix} 6 & -5 \\ 10 & -8 \end{bmatrix},$$

so by formula (10)

$$(AB)^{-1} = \begin{bmatrix} -4 & 5/2 \\ -5 & 3 \end{bmatrix}.$$

By Theorem 17, property 2,

$$(AB)^{-1} = B^{-1}A^{-1} = \begin{bmatrix} 1 & -2 \\ 1 & -3 \end{bmatrix}\begin{bmatrix} -2 & 3/2 \\ 1 & -1/2 \end{bmatrix} = \begin{bmatrix} -4 & 5/2 \\ -5 & 3 \end{bmatrix}.$$

Finally,

$$A^{-1}B^{-1} = \begin{bmatrix} -2 & 3/2 \\ 1 & -1/2 \end{bmatrix}\begin{bmatrix} 1 & -2 \\ 1 & -3 \end{bmatrix} = \begin{bmatrix} -1/2 & -1/2 \\ 1/2 & -1/2 \end{bmatrix} \neq (AB)^{-1}. \qquad \square$$

Matrix Inverses and Linear Systems

To conclude this section we return to the consideration of an $(n \times n)$ system of linear equations with matrix representation $A\mathbf{x} = \mathbf{b}$. By our next theorem, if A^{-1} exists, then the system of equations has the unique solution given by $\mathbf{x} = A^{-1}\mathbf{b}$.

THEOREM 18

Let

$$A\mathbf{x} = \mathbf{b} \qquad \text{(11a)}$$

be the matrix equation for an $(n \times n)$ system of linear equations, and suppose that A has an inverse. Then the system has a unique solution given by

$$\mathbf{x} = A^{-1}\mathbf{b}. \qquad \text{(11b)}$$

PROOF To see that $A^{-1}\mathbf{b}$ is a solution, substitute it into Eq. (11a) for $\mathbf{x}$. This gives

$$A(A^{-1}\mathbf{b}) = (AA^{-1})\mathbf{b} = I\mathbf{b} = \mathbf{b}.$$

To demonstrate that solution (11b) is unique, let $\mathbf{y}$ be any vector in R^n such that

$$A\mathbf{y} = \mathbf{b}. \qquad \text{(11c)}$$

Multiplying both sides of Eq. (11c) by A^{-1} yields

$$A^{-1}(A\mathbf{y}) = (A^{-1}A)\mathbf{y} = I\mathbf{y} = \mathbf{y} = A^{-1}\mathbf{b}. \qquad \square$$

__EXAMPLE 4__ Use formula (11b) to solve

$$x_1 + 3x_2 = -1$$
$$2x_1 + 4x_2 = 2.$$

SOLUTION   The system given in (12) has the coefficient matrix

$$A = \begin{bmatrix} 1 & 3 \\ 2 & 4 \end{bmatrix}.$$

The inverse of A was calculated in Example 3 and is

$$A^{-1} = \begin{bmatrix} -2 & 3/2 \\ 1 & -1/2 \end{bmatrix}.$$

The matrix equation for (12) is

$$\begin{bmatrix} 1 & 3 \\ 2 & 4 \end{bmatrix} \begin{bmatrix} x_1 \\ x_2 \end{bmatrix} = \begin{bmatrix} -1 \\ 2 \end{bmatrix},$$

so by formula (11b)

$$\begin{bmatrix} x_1 \\ x_2 \end{bmatrix} = \begin{bmatrix} -2 & 3/2 \\ 1 & -1/2 \end{bmatrix} \begin{bmatrix} -1 \\ 2 \end{bmatrix} = \begin{bmatrix} 5 \\ -2 \end{bmatrix}.$$

Thus the system has the unique solution $x_1 = 5, x_2 = -2$.

The reader is cautioned that the system $A\mathbf{x} = \mathbf{b}$ has the solution $\mathbf{x} = A^{-1}\mathbf{b}$ only when A is a square matrix and is invertible. Moreover, even when A is invertible, it is usually more efficient to reduce the augmented matrix $[A|\mathbf{b}]$ than it is to calculate A^{-1}. Thus, from a computational point of view, Gaussian elimination remains the preferred method for solving the system $A\mathbf{x} = \mathbf{b}$.

Ill-Conditioned Matrices

In applications the equation $A\mathbf{x} = \mathbf{b}$ often serves as a mathematical model for a physical problem. In these cases it is important to know whether solutions to $A\mathbf{x} = \mathbf{b}$ are sensitive to small changes in the right-hand side $\mathbf{b}$. If small changes in $\mathbf{b}$ can lead to relatively large changes in the solution $\mathbf{x}$, then the matrix A is called *ill-conditioned.*

The concept of an ill-conditioned matrix is related to the size of A^{-1}. This connection is explained after the next example.

__EXAMPLE 5__ The $(n \times n)$ Hilbert matrix is the matrix whose ijth entry is $1/(i + j - 1)$. For example, the (3×3) Hilbert matrix is

$$\begin{bmatrix} 1 & 1/2 & 1/3 \\ 1/2 & 1/3 & 1/4 \\ 1/3 & 1/4 & 1/5 \end{bmatrix}.$$

Let A denote the (6×6) Hilbert matrix, and consider the vectors $\mathbf{b}$ and $\mathbf{b} + \Delta\mathbf{b}$:

$$\mathbf{b} = \begin{bmatrix} 1 \\ 2 \\ 1 \\ 1.414 \\ 1 \\ 2 \end{bmatrix}, \qquad \mathbf{b} + \Delta\mathbf{b} = \begin{bmatrix} 1 \\ 2 \\ 1 \\ 1.4142 \\ 1 \\ 2 \end{bmatrix}.$$

Note that $\mathbf{b}$ and $\mathbf{b} + \Delta\mathbf{b}$ differ slightly in their fourth components. Compare the solutions of $A\mathbf{x} = \mathbf{b}$ and $A\mathbf{x} = \mathbf{b} + \Delta\mathbf{b}$.

SOLUTION   We used MATLAB to solve these two equations. If $\mathbf{x}_1$ denotes the solution of $A\mathbf{x} = \mathbf{b}$ and $\mathbf{x}_2$ denotes the solution of $A\mathbf{x} = \mathbf{b} + \Delta\mathbf{b}$, the results are (rounded to the nearest integer)

$$\mathbf{x}_1 = \begin{bmatrix} -6538 \\ 185706 \\ -1256237 \\ 3271363 \\ -3616326 \\ 1427163 \end{bmatrix} \quad \text{and} \quad \mathbf{x}_2 = \begin{bmatrix} -6539 \\ 185747 \\ -1256519 \\ 3272089 \\ -3617120 \\ 1427447 \end{bmatrix}.$$

(*Note:* Despite the fact that $\mathbf{b}$ and $\mathbf{b} + \Delta\mathbf{b}$ are nearly equal, $\mathbf{x}_1$ and $\mathbf{x}_2$ differ by almost 800 in their fifth components.)   ☐

Example 5 illustrates that the solutions of $A\mathbf{x} = \mathbf{b}$ and $A\mathbf{x} = \mathbf{b} + \Delta\mathbf{b}$ may be quite different even though $\Delta\mathbf{b}$ is a small vector. In order to explain these differences, let $\mathbf{x}_1$ denote the solution of $A\mathbf{x} = \mathbf{b}$ and $\mathbf{x}_2$ the solution of $A\mathbf{x} = \mathbf{b} + \Delta\mathbf{b}$. Therefore, $A\mathbf{x}_1 = \mathbf{b}$ and $A\mathbf{x}_2 = \mathbf{b} + \Delta\mathbf{b}$. To assess the difference, $\mathbf{x}_2 - \mathbf{x}_1$, we proceed as follows:

$$A\mathbf{x}_2 - A\mathbf{x}_1 = (\mathbf{b} + \Delta\mathbf{b}) - \mathbf{b} = \Delta\mathbf{b}.$$

Therefore, $A(\mathbf{x}_2 - \mathbf{x}_1) = \Delta\mathbf{b}$, or

$$\mathbf{x}_2 - \mathbf{x}_1 = A^{-1}\Delta\mathbf{b}.$$

If A^{-1} contains large entries, then we see from the equation above that $\mathbf{x}_2 - \mathbf{x}_1$ can be large even though $\Delta\mathbf{b}$ is small.

The Hilbert matrices described in Example 5 are well-known examples of ill-conditioned matrices and have large inverses. For example, the inverse of the (6×6) Hilbert matrix is

$$\begin{bmatrix} 36 & -630 & 3360 & -7560 & 7560 & -2772 \\ -630 & 14700 & -88200 & 211680 & -220500 & 83160 \\ 3360 & -88200 & 564480 & -1411200 & 1512000 & -582120 \\ -7560 & 211680 & -1411200 & 3628800 & -3969000 & 1552320 \\ 7560 & -220500 & 1512000 & -3969000 & 4410000 & -1746360 \\ -2772 & 83160 & -582120 & 1552320 & -1746360 & 698544 \end{bmatrix}.$$

Exercises 1.10

In Exercises 1–6, determine n and m so that $I_n A = A$ and $A I_m = A$, where:

1. A is (2×3)
2. A is (5×7)
3. A is (4×4)
4. A is (4×6)
5. A is (4×2)
6. A is (5×5)

As in Example 2, determine whether the (2×2) matrices in Exercises 7–12 have an inverse. If A has an inverse, find A^{-1} and verify that $A^{-1}A = I$.

7. $A = \begin{bmatrix} 2 & 1 \\ 5 & 3 \end{bmatrix}$

8. $A = \begin{bmatrix} -3 & 2 \\ 1 & 1 \end{bmatrix}$

9. $A = \begin{bmatrix} 2 & -2 \\ 2 & 3 \end{bmatrix}$

10. $A = \begin{bmatrix} -1 & 3 \\ 2 & 1 \end{bmatrix}$

11. $A = \begin{bmatrix} 2 & 1 \\ 4 & 2 \end{bmatrix}$

12. $A = \begin{bmatrix} 6 & -2 \\ 9 & -3 \end{bmatrix}$

In Exercises 13–18, solve the given system by forming $\mathbf{x} = A^{-1}\mathbf{b}$, where A is the coefficient matrix for the system.

13. $2x_1 + x_2 = 4$
    $3x_1 + 2x_2 = 2$

14. $x_1 + x_2 = 0$
    $2x_1 + 3x_2 = 4$

15. $x_1 - x_2 = 5$
    $3x_1 - 4x_2 = 2$

16. $2x_1 + 3x_2 = 1$
    $3x_1 + 4x_2 = 7$

17. $3x_1 + x_2 = 10$
    $-x_1 + 3x_2 = 5$

18. $x_1 - x_2 = 10$
    $2x_1 + 3x_2 = 4$

In Exercises 19–22, verify that the given matrix A does not have an inverse. (*Hint:* One of $AB = I$ or $BA = I$ leads to an easy contradiction.)

19. $A = \begin{bmatrix} 0 & 0 & 0 \\ 1 & 2 & 1 \\ 3 & 2 & 1 \end{bmatrix}$

20. $A = \begin{bmatrix} 0 & 4 & 2 \\ 0 & 1 & 7 \\ 0 & 3 & 9 \end{bmatrix}$

21. $A = \begin{bmatrix} 2 & 2 & 4 \\ 1 & 1 & 7 \\ 3 & 3 & 9 \end{bmatrix}$

22. $A = \begin{bmatrix} 1 & 1 & 1 \\ 1 & 1 & 1 \\ 2 & 3 & 2 \end{bmatrix}$

In Exercises 23 and 24, verify that B is the inverse of A by showing that $AB = BA = I$.

23. $A = \begin{bmatrix} -1 & -2 & 11 \\ 1 & 3 & -15 \\ 0 & -1 & 5 \end{bmatrix}$, $B = \begin{bmatrix} 0 & 1 & 3 \\ 5 & 5 & 4 \\ 1 & 1 & 1 \end{bmatrix}$

24. $A = \begin{bmatrix} 1 & 0 & 0 \\ 2 & 1 & 0 \\ 3 & 4 & 1 \end{bmatrix}$, $B = \begin{bmatrix} 1 & 0 & 0 \\ -2 & 1 & 0 \\ 5 & -4 & 1 \end{bmatrix}$

The matrices below are used in Exercises 25–36:

$$A^{-1} = \begin{bmatrix} 3 & 1 \\ 0 & 2 \end{bmatrix}, \quad B = \begin{bmatrix} 1 & 2 \\ 2 & 1 \end{bmatrix}, \quad C^{-1} = \begin{bmatrix} -1 & 1 \\ 1 & 2 \end{bmatrix}. \quad (13)$$

In Exercises 25–36, use Theorem 17 and the matrices in (13) to form Q^{-1}, where Q is the given matrix.

25. $Q = AC$
26. $Q = CA$
27. $Q = A^T$
28. $Q = A^T C$
29. $Q = C^T A^T$
30. $Q = B^{-1}A$
31. $Q = CB^{-1}$
32. $Q = B^{-1}$
33. $Q = 2A$
34. $Q = 10C$
35. $Q = (AC)B^{-1}$
36. $Q = (AB^{-1})C$

37. Consider the (3×3) matrices A, B, and C:

$$A = \begin{bmatrix} 1 & 0 & 0 \\ a & 1 & 0 \\ b & 0 & 1 \end{bmatrix}, \quad B = \begin{bmatrix} 1 & c & 0 \\ 0 & 1 & 0 \\ 0 & d & 1 \end{bmatrix},$$

$$C = \begin{bmatrix} 1 & 0 & e \\ 0 & 1 & f \\ 0 & 0 & 1 \end{bmatrix}. \quad (14)$$

Verify by direct multiplication that the inverse of each matrix in (14) can be obtained by changing the signs of the off-diagonal entries; that is, for example, $A + A^{-1} = 2I$.

38. Show that the matrix

$$A = \begin{bmatrix} \sin\theta & -\cos\theta \\ \cos\theta & \sin\theta \end{bmatrix}$$

has an inverse for all values of θ.

39. Find linear equations expressing x_1 and x_2 in terms of y_1 and y_2 if $-5x_1 + 3x_2 = y_1$ and $2x_1 - x_2 = y_2$. $\left(\text{Hint: Consider the matrix equation } A\mathbf{x} = \mathbf{y}, \right.$

where $A = \begin{bmatrix} -5 & 3 \\ 2 & -1 \end{bmatrix}$, $\mathbf{x} = \begin{bmatrix} x_1 \\ x_2 \end{bmatrix}$, and $\mathbf{y} = \begin{bmatrix} y_1 \\ y_2 \end{bmatrix}.$ $\Big)$

40. If A is a (3×3) matrix such that

$$A^{-1} = \begin{bmatrix} -1 & 4 & 6 \\ 2 & 5 & 7 \\ -9 & 3 & 2 \end{bmatrix},$$

then find a matrix B such that

$$AB = \begin{bmatrix} 3 & 1 \\ 1 & 2 \\ -1 & 1 \end{bmatrix}.$$

41. Simplify the expression $(A^{-1}B)^{-1}(C^{-1}A)^{-1}(B^{-1}C)^{-1}$ for $(n \times n)$ invertible matrices A, B, and C.

42. The equation $x^2 = 1$ can be solved by setting $x^2 - 1 = 0$ and factoring to obtain $(x - 1)(x + 1) = 0$. This yields solutions $x = 1$ and $x = -1$.

  a) Using the factorization technique given above, what (2×2) matrix solutions do you obtain for the matrix equation $X^2 = I$?

  b) Show that

  $$A = \begin{bmatrix} a & 1 - a^2 \\ 1 & -a \end{bmatrix}$$

  is a solution to $X^2 = I$ for every real number a.

  c) Let $b = \pm 1$. Show that

  $$B = \begin{bmatrix} b & 0 \\ c & -b \end{bmatrix}$$

  is a solution to $X^2 = I$ for every real number c.

  d) Explain why the factorization technique used in part (a) did not yield all the solutions to the matrix equation $X^2 = I$.

43. Suppose that A is a (2×2) matrix with columns $\mathbf{u}$ and $\mathbf{v}$, so that $A = [\mathbf{u}, \mathbf{v}]$, $\mathbf{u}$ and $\mathbf{v}$ in R^2. Suppose also that $\mathbf{u}^T\mathbf{u} = 1$, $\mathbf{u}^T\mathbf{v} = 0$, and $\mathbf{v}^T\mathbf{v} = 1$. Prove that $A^TA = I$. (*Hint:* Express the matrix A as

$$A = \begin{bmatrix} u_1 & v_1 \\ u_2 & v_2 \end{bmatrix}, \qquad \mathbf{u} = \begin{bmatrix} u_1 \\ u_2 \end{bmatrix}, \qquad \mathbf{v} = \begin{bmatrix} v_1 \\ v_2 \end{bmatrix}$$

and form the product A^TA.)

44. Let $\mathbf{u}$ be a vector in R^n such that $\mathbf{u}^T\mathbf{u} = 1$. Let $A = I - \mathbf{u}\mathbf{u}^T$, where I is the $(n \times n)$ identity. Verify that $AA = A$. (*Hint:* Write the product $\mathbf{u}\mathbf{u}^T\mathbf{u}\mathbf{u}^T$ as $\mathbf{u}\mathbf{u}^T\mathbf{u}\mathbf{u}^T = \mathbf{u}(\mathbf{u}^T\mathbf{u})\mathbf{u}^T$, and note that $\mathbf{u}^T\mathbf{u}$ is a scalar.)

45. Suppose that A is an $(n \times n)$ matrix such that $AA = A$, as in Exercise 44. Show that if A has an inverse, then $A = I$.

46. Let $A = I - a\mathbf{v}\mathbf{v}^T$, where $\mathbf{v}$ is a nonzero vector in R^n, I is the $(n \times n)$ identity, and a is the scalar given by $a = 2/(\mathbf{v}^T\mathbf{v})$. Show that A is symmetric and that $AA = I$; that is, $A^{-1} = A$.

47. Consider the $(n \times n)$ matrix A defined in Exercise 46. For $\mathbf{x}$ in R^n, show that the product $A\mathbf{x}$ has the form $A\mathbf{x} = \mathbf{x} - \lambda\mathbf{v}$, where λ is a scalar. What is the value of λ for a given $\mathbf{x}$?

48. Suppose that A is an $(n \times n)$ matrix such that $A^TA = I$ (the matrix defined in Exercise 46 is such a matrix). Let $\mathbf{x}$ be any vector in R^n. Show that $\|A\mathbf{x}\| = \|\mathbf{x}\|$; that is, multiplication of $\mathbf{x}$ by A produces a vector $A\mathbf{x}$ having the same length as $\mathbf{x}$.

49. Let $\mathbf{u}$ and $\mathbf{v}$ be vectors in R^n, and let I denote the $(n \times n)$ identity. Let $A = I + \mathbf{u}\mathbf{v}^T$, and suppose $\mathbf{v}^T\mathbf{u} \neq -1$. Establish the **Sherman–Woodberry** formula:

$$A^{-1} = I - a\mathbf{u}\mathbf{v}^T, \qquad a = 1/(1 + \mathbf{v}^T\mathbf{u}). \tag{15}$$

(*Hint:* Form AA^{-1}, where A^{-1} is given by formula (15).)

50. If A is a square matrix, we define the powers A^2, A^3, and so on, as follows: $A^2 = AA$, $A^3 = A(A^2)$, etc. Suppose A is an $(n \times n)$ matrix such that

$$A^3 - 2A^2 + 3A - I = \mathcal{O}.$$

Show that $AB = I$, where $B = A^2 - 2A + 3I$.

51. Suppose that A is $(n \times n)$ and

$$A^2 + b_1A + b_0I = \mathcal{O}, \tag{16}$$

where $b_0 \neq 0$. Show that $AB = I$, where $B = (-1/b_0)[A + b_1I]$.

It can be shown that when A is a (2×2) matrix such that A^{-1} exists, then there are constants b_1 and b_0 such that Eq. (16) holds. Moreover, $b_0 \neq 0$ in Eq. (16) unless A is a multiple of I. In Exercises 52–55, find the constants b_1 and b_0 in Eq. (16) for the given (2×2) matrix. Also, verify that $A^{-1} = (-1/b_0)[A + b_1I]$.

52. A in Exercise 7   **53.** A in Exercise 8

54. A in Exercise 9   **55.** A in Exercise 10

56. a) If linear algebra software is available, solve the systems $A\mathbf{x} = \mathbf{b}_1$ and $A\mathbf{x} = \mathbf{b}_2$, where

$$A = \begin{bmatrix} 0.932 & 0.443 & 0.417 \\ 0.712 & 0.915 & 0.887 \\ 0.632 & 0.514 & 0.493 \end{bmatrix},$$

$$\mathbf{b}_1 = \begin{bmatrix} 1 \\ 1 \\ -1 \end{bmatrix}, \qquad \mathbf{b}_2 = \begin{bmatrix} 1.01 \\ 1.01 \\ -1.01 \end{bmatrix}.$$

Note the large difference between the two solutions.

  b) Calculate A^{-1} and use it to explain the results of part (a).

1.11 Finding the Inverse of a Nonsingular Matrix

In this section we show that an $(n \times n)$ matrix A has an inverse if and only if A is nonsingular, and we give an efficient scheme for calculating the inverse. The first theorem of this section is an important step toward both these objectives.

The Existence of an Inverse

THEOREM 19   If A is an $(n \times n)$ nonsingular matrix, then there is a unique $(n \times n)$ matrix B such that $AB = I$.

PROOF   Because A is nonsingular, Theorem 13 of Section 1.8 asserts that the equation $A\mathbf{x} = \mathbf{b}$ always has a unique solution. In particular, each of the linear systems

$$A\mathbf{x} = \mathbf{e}_1, \qquad A\mathbf{x} = \mathbf{e}_2, \ldots, \qquad A\mathbf{x} = \mathbf{e}_n$$

has a unique solution; that is, there exist unique $(n \times 1)$ vectors $\mathbf{b}_1, \mathbf{b}_2, \ldots, \mathbf{b}_n$ such that

$$A\mathbf{b}_1 = \mathbf{e}_1, \qquad A\mathbf{b}_2 = \mathbf{e}_2, \ldots, \qquad A\mathbf{b}_n = \mathbf{e}_n.$$

If B is the $(n \times n)$ matrix

$$B = [\mathbf{b}_1, \mathbf{b}_2, \ldots, \mathbf{b}_n],$$

then by Theorem 6 of Section 1.6,

$$AB = [A\mathbf{b}_1, A\mathbf{b}_2, \ldots, A\mathbf{b}_n] = [\mathbf{e}_1, \mathbf{e}_2, \ldots, \mathbf{e}_n] = I,$$

which establishes the theorem.

It will be an immediate consequence of our next theorem that the matrix B of Theorem 19 is actually A^{-1}. Thus the proof of Theorem 19 describes a procedure for calculating the inverse of a nonsingular matrix. This procedure is illustrated by the following example.

EXAMPLE 1   Let the (3×3) matrix A be given by

$$A = \begin{bmatrix} 1 & 3 & -1 \\ -2 & -5 & 1 \\ 1 & 5 & -2 \end{bmatrix}.$$

Show that A is nonsingular, and find a (3×3) matrix B such that $AB = I$.

SOLUTION   The homogeneous system $A\mathbf{x} = \boldsymbol{\theta}$ has augmented matrix

$$[A \mid \boldsymbol{\theta}] = \begin{bmatrix} 1 & 3 & -1 & 0 \\ -2 & -5 & 1 & 0 \\ 1 & 5 & -2 & 0 \end{bmatrix},$$

which is row equivalent to

$$C = \begin{bmatrix} 1 & 3 & -1 & 0 \\ 0 & 1 & -1 & 0 \\ 0 & 0 & 1 & 0 \end{bmatrix}.$$

The matrix C represents a homogeneous system that has only the trivial solution; so A is nonsingular and the equation $A\mathbf{x} = \mathbf{b}$ has a unique solution for every (3×1) vector $\mathbf{b}$. In particular, each of the linear systems

$$A\mathbf{x} = \mathbf{e}_1, \qquad A\mathbf{x} = \mathbf{e}_2, \quad \text{and} \quad A\mathbf{x} = \mathbf{e}_3$$

has a unique solution. The equation $A\mathbf{x} = \mathbf{e}_1$ has augmented matrix

$$[A \mid \mathbf{e}_1] = \begin{bmatrix} 1 & 3 & -1 & 1 \\ -2 & -5 & 1 & 0 \\ 1 & 5 & -2 & 0 \end{bmatrix},$$

and this matrix is row equivalent to

$$\begin{bmatrix} 1 & 3 & -1 & 1 \\ 0 & 1 & -1 & 2 \\ 0 & 0 & 1 & -5 \end{bmatrix}.$$

Backsolving yields the unique solution $x_1 = 5$, $x_2 = -3$, $x_3 = -5$, so

$$\mathbf{b}_1 = \begin{bmatrix} 5 \\ -3 \\ -5 \end{bmatrix}$$

is the unique solution for $A\mathbf{x} = \mathbf{e}_1$. Similarly, it is easy to solve $A\mathbf{x} = \mathbf{e}_2$ and $A\mathbf{x} = \mathbf{e}_3$; and the solutions are

$$\mathbf{b}_2 = \begin{bmatrix} 1 \\ -1 \\ -2 \end{bmatrix} \quad \text{and} \quad \mathbf{b}_3 = \begin{bmatrix} -2 \\ 1 \\ 1 \end{bmatrix}.$$

The desired matrix B is now given by

$$B = [\mathbf{b}_1, \mathbf{b}_2, \mathbf{b}_3] = \begin{bmatrix} 5 & 1 & -2 \\ -3 & -1 & 1 \\ -5 & -2 & 1 \end{bmatrix}.$$

To check the calculations, note that

$$AB = \begin{bmatrix} 1 & 3 & -1 \\ -2 & -5 & 1 \\ 1 & 5 & -2 \end{bmatrix} \begin{bmatrix} 5 & 1 & -2 \\ -3 & -1 & 1 \\ -5 & -2 & 1 \end{bmatrix} = \begin{bmatrix} 1 & 0 & 0 \\ 0 & 1 & 0 \\ 0 & 0 & 1 \end{bmatrix}.$$

To conclude that the matrix B obtained in the example above is actually A^{-1}, the definition of a matrix inverse requires us to show further that $BA = I$. The next theorem asserts that this is unnecessary and that the matrix B of Theorem 19 is the inverse of A.

THEOREM 20

If A and B are $(n \times n)$ matrices such that $AB = I$, then $BA = I$. In particular, $B = A^{-1}$.

PROOF   We first show that $AB = I$ implies that B is nonsingular. Let $\mathbf{x}_1$ be a solution to the equation

$$B\mathbf{x} = \boldsymbol{\theta}.$$

Then

$$\mathbf{x}_1 = I\mathbf{x}_1 = (AB)\mathbf{x}_1 = A(B\mathbf{x}_1) = A\boldsymbol{\theta} = \boldsymbol{\theta},$$

and it follows that B is nonsingular because the only solution of $B\mathbf{x} = \boldsymbol{\theta}$ is $\mathbf{x} = \boldsymbol{\theta}$. By Theorem 19, there exists an $(n \times n)$ matrix C such that $BC = I$. Thus,

$$A = AI = A(BC) = (AB)C = IC = C;$$

that is, $BA = I$.

It is an immediate consequence of Theorems 19 and 20 that a nonsingular matrix A has an inverse. Conversely, if an $(n \times n)$ matrix A has an inverse, then, by Theorem 18 of Section 1.10, A is nonsingular. Thus we have proved the following theorem:

THEOREM 21

An $(n \times n)$ matrix A has an inverse if, and only if, A is nonsingular.

The next theorem summarizes some of the properties of nonsingular matrices.

THEOREM 22

Let A be an $(n \times n)$ matrix. Then the following are equivalent:

1. A is nonsingular; that is, the only solution of $A\mathbf{x} = \boldsymbol{\theta}$ is $\mathbf{x} = \boldsymbol{\theta}$.
2. The column vectors of A are linearly independent.
3. $A\mathbf{x} = \mathbf{b}$ always has a unique solution.
4. A has an inverse.

Procedure for Calculating an Inverse

Example 1 illustrated how to calculate the inverse of a matrix following the steps given in the proof of Theorem 19. Our final procedure for calculating the inverse of a nonsingular matrix essentially follows the same steps, but a more efficient way of organizing the calculations is needed. To demonstrate, consider the (3×3) nonsingular matrix

$$A = \begin{bmatrix} 1 & 2 & 3 \\ 2 & 5 & 4 \\ 1 & -1 & 10 \end{bmatrix}. \qquad \textbf{(1a)}$$

Following the proof of Theorem 19, A^{-1} is the matrix $B = [\mathbf{b}_1, \mathbf{b}_2, \mathbf{b}_3]$, where $\mathbf{b}_1, \mathbf{b}_2, \mathbf{b}_3$ are the unique solutions to the three systems

$$A\mathbf{x} = \mathbf{e}_1, \qquad A\mathbf{x} = \mathbf{e}_2, \quad \text{and} \quad A\mathbf{x} = \mathbf{e}_3, \qquad \textbf{(1b)}$$

respectively. To organize the computation of A^{-1} so that A is reduced to echelon form only once instead of three times, we form the (3×6) matrix

$$[A \mid \mathbf{e}_1, \mathbf{e}_2, \mathbf{e}_3].$$

Since $I = [\mathbf{e}_1, \mathbf{e}_2, \mathbf{e}_3]$, this matrix can be written as $[A \mid I]$ and can be reduced to echelon form by using elementary row operations as follows:

$$\begin{bmatrix} 1 & 2 & 3 & 1 & 0 & 0 \\ 2 & 5 & 4 & 0 & 1 & 0 \\ 1 & -1 & 10 & 0 & 0 & 1 \end{bmatrix}$$

$R_2 - 2R_1, R_3 - R_1$:

$$\begin{bmatrix} 1 & 2 & 3 & 1 & 0 & 0 \\ 0 & 1 & -2 & -2 & 1 & 0 \\ 0 & -3 & 7 & -1 & 0 & 1 \end{bmatrix}$$

$R_3 + 3R_2$:

$$\begin{bmatrix} 1 & 2 & 3 & 1 & 0 & 0 \\ 0 & 1 & -2 & -2 & 1 & 0 \\ 0 & 0 & 1 & -7 & 3 & 1 \end{bmatrix}. \qquad \textbf{(1c)}$$

Although the matrix (1c) is in echelon form, it is customary, especially when calculating A^{-1} by hand, to continue the reduction process on (1c) until the matrix $[A \mid I]$ is transformed to a (3×6) matrix of the form $[I \mid B]$. We will see that $A^{-1} = B$. To illustrate, we continue with the reduction of the matrix (1c):

$$\begin{bmatrix} 1 & 2 & 3 & 1 & 0 & 0 \\ 0 & 1 & -2 & -2 & 1 & 0 \\ 0 & 0 & 1 & -7 & 3 & 1 \end{bmatrix}$$

$R_1 - 3R_3, R_2 + 2R_3$:

$$\begin{bmatrix} 1 & 2 & 0 & 22 & -9 & -3 \\ 0 & 1 & 0 & -16 & 7 & 2 \\ 0 & 0 & 1 & -7 & 3 & 1 \end{bmatrix}$$

$R_1 - 2R_2$:

$$[I \mid B] = \begin{bmatrix} 1 & 0 & 0 & 54 & -23 & -7 \\ 0 & 1 & 0 & -16 & 7 & 2 \\ 0 & 0 & 1 & -7 & 3 & 1 \end{bmatrix}. \tag{1d}$$

To see that $B = A^{-1}$, note that $[A \mid I] = [A \mid e_1, e_2, e_2]$ is row equivalent to $[I \mid B] = [I \mid b_1, b_2, b_3]$, and hence the three systems $Ax = e_1$, $Ax = e_2$, and $Ax = e_3$ of (1b) are equivalent to the systems

$$Ix = b_1, \qquad Ix = b_2, \quad \text{and} \quad Ix = b_3, \tag{1e}$$

respectively. For example, $Ix = b_1$ is the system

$$\begin{array}{rcl} x_1 & = & 54 \\ x_2 & = & -16 \\ x_3 & = & -7. \end{array} \tag{1f}$$

Without backsolving, it is obvious from system (1f) that $Ax = e_1$ has the unique solution

$$x = b_1 = \begin{bmatrix} 54 \\ -16 \\ -7 \end{bmatrix}.$$

Similarly, the solutions for $Ax = e_2$ and $Ax = e_3$ are

$$b_2 = \begin{bmatrix} -23 \\ 7 \\ 3 \end{bmatrix} \quad \text{and} \quad b_3 = \begin{bmatrix} -7 \\ 2 \\ 1 \end{bmatrix},$$

respectively. It follows that

$$A^{-1} = B = \begin{bmatrix} 54 & -23 & -7 \\ -16 & 7 & 2 \\ -7 & 3 & 1 \end{bmatrix}.$$

The procedure described above can now be summarized for a general matrix.

Computation of A^{-1}

To calculate the inverse of a nonsingular $(n \times n)$ matrix A, we can proceed as follows:

Step 1. Form the $(n \times 2n)$ matrix $[A \mid I]$.

Step 2. Use elementary row operations to transform $[A \mid I]$ to the form $[I \mid B]$.

In this final form, $B = A^{-1}$.

EXAMPLE 2 Find A^{-1} for

$$A = \begin{bmatrix} 1 & -2 & 1 \\ -1 & 3 & 2 \\ 2 & -2 & 7 \end{bmatrix}.$$

SOLUTION To find A^{-1}, we form the (3×6) matrix $[A \,|\, I]$ and reduce to $[I \,|\, B]$ as follows:

$$\begin{bmatrix} 1 & -2 & 1 & 1 & 0 & 0 \\ -1 & 3 & 2 & 0 & 1 & 0 \\ 2 & -2 & 7 & 0 & 0 & 1 \end{bmatrix}$$

$R_2 + R_1, R_3 - 2R_1$:

$$\begin{bmatrix} 1 & -2 & 1 & 1 & 0 & 0 \\ 0 & 1 & 3 & 1 & 1 & 0 \\ 0 & 2 & 5 & -2 & 0 & 1 \end{bmatrix}$$

$R_3 - 2R_2$:

$$\begin{bmatrix} 1 & -2 & 1 & 1 & 0 & 0 \\ 0 & 1 & 3 & 1 & 1 & 0 \\ 0 & 0 & -1 & -4 & -2 & 1 \end{bmatrix}$$

$R_1 + R_3, R_2 + 3R_3$:

$$\begin{bmatrix} 1 & -2 & 0 & -3 & -2 & 1 \\ 0 & 1 & 0 & -11 & -5 & 3 \\ 0 & 0 & -1 & -4 & -2 & 1 \end{bmatrix}$$

$R_1 + 2R_2, -R_3$:

$$\begin{bmatrix} 1 & 0 & 0 & -25 & -12 & 7 \\ 0 & 1 & 0 & -11 & -5 & 3 \\ 0 & 0 & 1 & 4 & 2 & -1 \end{bmatrix}.$$

The procedure described above now guarantees that

$$A^{-1} = \begin{bmatrix} -25 & -12 & 7 \\ -11 & -5 & 3 \\ 4 & 2 & -1 \end{bmatrix}.$$

The accuracy of the calculations can be checked by showing that $AA^{-1} = I$.

◻

The scheme for organizing the computation of A^{-1} can also be used in other situations. Suppose we are given a number of $(m \times n)$ linear systems, all of which have the same coefficient matrix. For example:

$$A\mathbf{x} = \mathbf{b}_i, \qquad i = 1, 2, \dots, p. \tag{2a}$$

These p systems can be solved by defining the $(m \times p)$ matrix B by

$$B = [\mathbf{b}_1, \mathbf{b}_2, \ldots, \mathbf{b}_p]$$

and then forming the $[m \times (n + p)]$ matrix $[A \,|\, B]$. Although the bookkeeping details vary slightly from those in the construction of A^{-1}, clearly row operations can be used to transform $[A \,|\, B]$ into a row-equivalent matrix $[P \,|\, Q]$, where P is in echelon form and is row equivalent to A and where Q is row equivalent to B. If $Q = [\mathbf{q}_1, \mathbf{q}_2, \ldots, \mathbf{q}_p]$, then $[P \,|\, \mathbf{q}_i]$ is row equivalent to $[A \,|\, \mathbf{b}_i]$, so each system

$$P\mathbf{x} = \mathbf{q}_i, \qquad i = 1, 2, \ldots, p \qquad \textbf{(2b)}$$

is equivalent to the corresponding system $A\mathbf{x} = \mathbf{b}_i$ in (2a). The systems in (2) that are consistent can be determined by inspection and the solution obtained by backsolving. The next example illustrates this procedure.

EXAMPLE 3 Solve each of the (3×3) linear systems $A\mathbf{x} = \mathbf{b}_1$ and $A\mathbf{x} = \mathbf{b}_2$, where

$$A = \begin{bmatrix} 1 & 2 & -1 \\ 2 & 6 & 2 \\ 3 & 4 & -7 \end{bmatrix}, \quad \mathbf{b}_1 = \begin{bmatrix} 2 \\ 5 \\ 4 \end{bmatrix}, \quad \text{and} \quad \mathbf{b}_2 = \begin{bmatrix} 3 \\ 8 \\ 7 \end{bmatrix},$$

or else determine that the system is inconsistent.

SOLUTION The (3×5) matrix $[A \,|\, \mathbf{b}_1, \mathbf{b}_2]$ can be reduced to the form $[P \,|\, \mathbf{q}_1, \mathbf{q}_2]$, where P is in echelon form, by the following steps:

$$\begin{bmatrix} 1 & 2 & -1 & 2 & 3 \\ 2 & 6 & 2 & 5 & 8 \\ 3 & 4 & -7 & 4 & 7 \end{bmatrix}$$

$R_2 - 2R_1, R_3 - 3R_1$:

$$\begin{bmatrix} 1 & 2 & -1 & 2 & 3 \\ 0 & 2 & 4 & 1 & 2 \\ 0 & -2 & -4 & -2 & -2 \end{bmatrix}$$

$R_3 + R_2$:

$$\begin{bmatrix} 1 & 2 & -1 & 2 & 3 \\ 0 & 2 & 4 & 1 & 2 \\ 0 & 0 & 0 & -1 & 0 \end{bmatrix}. \qquad \textbf{(3a)}$$

Writing (3a) as $[P \,|\, \mathbf{q}_1, \mathbf{q}_2]$, we clearly see that the system $P\mathbf{x} = \mathbf{q}_1$ is inconsistent; thus so is the system $A\mathbf{x} = \mathbf{b}_1$. On the other hand, the system $A\mathbf{x} = \mathbf{b}_2$ is equivalent to $P\mathbf{x} = \mathbf{q}_2$, which is the system

$$\begin{aligned} x_1 + 2x_2 - x_3 &= 3 \\ 2x_2 + 4x_3 &= 2. \end{aligned} \qquad \textbf{(3b)}$$

Backsolving (3b) yields the solution

$$x_1 = 1 + 5x_3$$
$$x_2 = 1 - 2x_3.$$

□

Exercises 1.11

1. Let

$$A = \begin{bmatrix} 1 & 2 \\ 3 & 5 \end{bmatrix}.$$

a) Use the method illustrated in Example 1 to calculate A^{-1}.

b) Use the method illustrated in Example 2 to calculate A^{-1}.

2. Repeat Exercise 1 for

$$A = \begin{bmatrix} 1 & 3 & -1 \\ -1 & -4 & 3 \\ 2 & 4 & 3 \end{bmatrix}.$$

In Exercises 3–11, reduce $[A\,|\,I]$ to find A^{-1}. In each case check your calculations by multiplying the given matrix by the derived inverse.

3. $\begin{bmatrix} 1 & 1 \\ 2 & 3 \end{bmatrix}$

4. $\begin{bmatrix} 2 & 3 \\ 6 & 7 \end{bmatrix}$

5. $\begin{bmatrix} 1 & 2 \\ 2 & 1 \end{bmatrix}$

6. $\begin{bmatrix} -1 & -2 & 11 \\ 1 & 3 & -15 \\ 0 & -1 & 5 \end{bmatrix}$

7. $\begin{bmatrix} 1 & 0 & 0 \\ 2 & 1 & 0 \\ 3 & 4 & 1 \end{bmatrix}$

8. $\begin{bmatrix} 1 & 3 & 5 \\ 0 & 1 & 4 \\ 0 & 2 & 7 \end{bmatrix}$

9. $\begin{bmatrix} 1 & 4 & 2 \\ 0 & 2 & 1 \\ 3 & 5 & 3 \end{bmatrix}$

10. $\begin{bmatrix} 1 & -2 & 2 & 1 \\ 1 & -1 & 5 & 0 \\ 2 & -2 & 11 & 2 \\ 0 & 2 & 8 & 1 \end{bmatrix}$

11. $\begin{bmatrix} 1 & 2 & 3 & 1 \\ -1 & 0 & 2 & 1 \\ 2 & 1 & -3 & 0 \\ 1 & 1 & 2 & 1 \end{bmatrix}$

12. Use the inverse found in Exercise 3 to solve the following system for the values listed:

$$x_1 + x_2 = b_1$$
$$2x_1 + 3x_2 = b_2.$$

a) For $b_1 = 1$ and $b_2 = 2$

b) For $b_1 = 1$ and $b_2 = 1$

c) For $b_1 = 2$ and $b_2 = -2$

13. Using the inverse found in Exercise 4, repeat Exercise 12 for the system

$$2x_1 + 3x_2 = b_1$$
$$6x_1 + 7x_2 = b_2.$$

14. Use the inverse found in Exercise 6 to solve the following system for the values listed:

$$-x_1 - 2x_2 + 11x_3 = b_1$$
$$x_1 + 3x_2 - 15x_3 = b_2$$
$$-x_2 + 5x_3 = b_3.$$

a) For $b_1 = 1$, $b_2 = -1$, and $b_3 = 2$

b) For $b_1 = 3$, $b_2 = -2$, and $b_3 = 1$

c) For $b_1 = 0$, $b_2 = 3$, and $b_3 = -1$

15. Using the inverse found in Exercise 9, repeat Exercise 14 for the system

$$x_1 + 4x_2 + 2x_3 = b_1$$
$$2x_2 + x_3 = b_2$$
$$3x_1 + 5x_2 + 3x_3 = b_3.$$

16. a) Solve the systems $A\mathbf{x} = \mathbf{c}_k$, $k = 1, 2, 3$, by row reduction of $[A\,|\,\mathbf{c}_1, \mathbf{c}_2, \mathbf{c}_3]$, where

$$A = \begin{bmatrix} 1 & 1 \\ 2 & 3 \end{bmatrix}, \quad \mathbf{c}_1 = \begin{bmatrix} 2 \\ 1 \end{bmatrix},$$

$$\mathbf{c}_2 = \begin{bmatrix} 0 \\ 2 \end{bmatrix}, \quad \text{and} \quad \mathbf{c}_3 = \begin{bmatrix} -1 \\ 3 \end{bmatrix}.$$

b) Without doing further calculations, exhibit a matrix B such that $AB = C$, where

$$C = \begin{bmatrix} 2 & 0 & -1 \\ 1 & 2 & 3 \end{bmatrix}.$$

(*Hint:* Use Theorem 6 of Section 1.6.)

17. Repeat Exercise 16 for

$$A = \begin{bmatrix} 1 & 3 & -4 \\ 2 & -1 & 3 \\ 8 & 3 & 1 \end{bmatrix}, \quad \mathbf{c}_1 = \begin{bmatrix} 1 \\ 1 \\ 5 \end{bmatrix},$$

$$\mathbf{c}_2 = \begin{bmatrix} 2 \\ -2 \\ -2 \end{bmatrix}, \quad \text{and} \quad \mathbf{c}_3 = \begin{bmatrix} 4 \\ -1 \\ 5 \end{bmatrix},$$

where C is the matrix

$$C = \begin{bmatrix} 1 & 2 & 4 \\ 1 & -2 & -1 \\ 5 & -2 & 5 \end{bmatrix}.$$

18. Let A be the matrix given in Exercise 3. Use the inverse found in Exercise 3 to obtain matrices B and C such that $AB = D$ and $CA = E$, where

$$D = \begin{bmatrix} -1 & 2 & 3 \\ 1 & 0 & 2 \end{bmatrix} \quad \text{and} \quad E = \begin{bmatrix} 2 & -1 \\ 1 & 1 \\ 0 & 3 \end{bmatrix}.$$

19. Repeat Exercise 18 with A being the matrix given in Exercise 6 and where

$$D = \begin{bmatrix} 2 & -1 \\ 1 & 1 \\ 0 & 3 \end{bmatrix} \quad \text{and} \quad E = \begin{bmatrix} -1 & 2 & 3 \\ 1 & 0 & 2 \end{bmatrix}.$$

20. For what values of a is

$$A = \begin{bmatrix} 1 & 1 & -1 \\ 0 & 1 & 2 \\ 1 & 1 & a \end{bmatrix}$$

nonsingular?

21. Find $(AB)^{-1}, (3A)^{-1}$, and $(A^T)^{-1}$ given that

$$A^{-1} = \begin{bmatrix} 1 & 2 & 5 \\ 3 & 1 & 6 \\ 2 & 8 & 1 \end{bmatrix} \quad \text{and}$$

$$B^{-1} = \begin{bmatrix} 3 & -3 & 4 \\ 5 & 1 & 3 \\ 7 & 6 & -1 \end{bmatrix}.$$

22. Find the (3×3) nonsingular matrix A if $A^2 = AB + 2A$, where

$$B = \begin{bmatrix} 2 & 1 & -1 \\ 0 & 3 & 2 \\ -1 & 4 & 1 \end{bmatrix}.$$

23. a) Give examples of nonsingular (2×2) matrices A and B such that $A + B$ is singular.

b) Give examples of singular (2×2) matrices A and B such that $A + B$ is nonsingular.

24. Let A be an $(n \times n)$ nonsingular symmetric matrix. Show that A^{-1} is also symmetric.

25. a) Suppose that $AB = \mathcal{O}$, where A is nonsingular. Use Theorem 21 to prove that $B = \mathcal{O}$.

b) Find a (2×2) matrix B such that $AB = \mathcal{O}$, where B has nonzero entries and where A is the matrix

$$A = \begin{bmatrix} 1 & 1 \\ 1 & 1 \end{bmatrix}.$$

Why does this example not contradict part (a)?

26. Let A, B, and C be matrices such that A is nonsingular and $AB = AC$. Prove that $B = C$.

27. Let A be the (2×2) matrix

$$A = \begin{bmatrix} a & b \\ c & d \end{bmatrix},$$

and set $\Delta = ad - bc$. Prove that if $\Delta = 0$, then A is singular. Conclude that A has no inverse. (*Hint:* Consider the vector

$$\mathbf{v} = \begin{bmatrix} d \\ -c \end{bmatrix};$$

also treat the special case when $d = c = 0$.)

28. Let A and B be $(n \times n)$ nonsingular matrices. Show that AB is also nonsingular. (*Hint:* Use Theorem 21 and Theorem 17, property 2, of Section 1.10.)

29. What is wrong with the following argument that if AB is nonsingular, then each of A and B is also nonsingular?

Since AB is nonsingular, $(AB)^{-1}$ exists by Theorem 21. But by Theorem 17, property 2, $(AB)^{-1} = B^{-1}A^{-1}$. Therefore, A^{-1} and B^{-1} exists, so A and B are nonsingular.

30. Let A and B be $(n \times n)$ matrices such that AB is nonsingular.

a) Prove that B is nonsingular. (*Hint:* Suppose $\mathbf{v}$ is any vector such that $B\mathbf{v} = \boldsymbol{\theta}$ and write $(AB)\mathbf{v}$ as $A(B\mathbf{v})$.)

b) Prove that A is nonsingular. (*Hint:* By part (a), B^{-1} exists. Apply Exercise 28 to the matrices AB and B^{-1}.)

31. Let A be an $(n \times n)$ singular matrix. Argue that at least one of the systems $A\mathbf{x} = \mathbf{e}_k$, $k = 1, 2, \ldots, n$, must be inconsistent, where $\mathbf{e}_1, \mathbf{e}_2, \ldots, \mathbf{e}_n$ are the n-dimensional unit vectors.

Supplementary Computational Exercises

1. Consider the system of equations

$$x_1 \qquad\qquad = 1$$
$$2x_1 + (a^2 + a - 2)x_2 = a^2 - a - 4.$$

For what values of a does the system have infinitely many solutions? No solutions? A unique solution in which $x_2 = 0$?

2. Let

$$A = \begin{bmatrix} 1 & -1 & -1 \\ 2 & -1 & 1 \\ -3 & 1 & -3 \end{bmatrix}, \quad \mathbf{x} = \begin{bmatrix} x_1 \\ x_2 \\ x_3 \end{bmatrix}, \quad \text{and}$$

$$\mathbf{b} = \begin{bmatrix} b_1 \\ b_2 \\ b_3 \end{bmatrix}.$$

a) Determine conditions on b_1, b_2, and b_3 that are necessary and sufficient for the system of equations $A\mathbf{x} = \mathbf{b}$ to be consistent. (*Hint:* Reduce the augmented matrix $[A\,|\,\mathbf{b}]$.)

b) For each of the following choices of $\mathbf{b}$, either show that the system $A\mathbf{x} = \mathbf{b}$ is inconsistent or exhibit the solution.

$$\text{(i) } \mathbf{b} = \begin{bmatrix} 1 \\ 1 \\ 1 \end{bmatrix} \qquad \text{(ii) } \mathbf{b} = \begin{bmatrix} 5 \\ 2 \\ 1 \end{bmatrix}$$

$$\text{(iii) } \mathbf{b} = \begin{bmatrix} 7 \\ 3 \\ 1 \end{bmatrix} \qquad \text{(iv) } \mathbf{b} = \begin{bmatrix} 0 \\ 1 \\ 2 \end{bmatrix}$$

3. Let

$$A = \begin{bmatrix} 1 & -1 & 3 \\ 2 & -1 & 5 \\ -3 & 5 & -10 \\ 1 & 0 & 4 \end{bmatrix} \quad \text{and} \quad \mathbf{x} = \begin{bmatrix} x_1 \\ x_2 \\ x_3 \end{bmatrix}.$$

a) Simultaneously solve each of the systems $A\mathbf{x} = \mathbf{b}_i$, $i = 1, 2, 3$, where

$$\mathbf{b}_1 = \begin{bmatrix} -5 \\ -17 \\ 19 \\ 24 \end{bmatrix}, \quad \mathbf{b}_2 = \begin{bmatrix} 5 \\ 11 \\ -12 \\ 8 \end{bmatrix}, \quad \text{and}$$

$$\mathbf{b}_3 = \begin{bmatrix} 1 \\ 2 \\ -1 \\ 5 \end{bmatrix}.$$

b) Let $B = [\mathbf{b}_1, \mathbf{b}_2, \mathbf{b}_3]$. Use the results of part (a) to exhibit a (3×3) matrix C such that $AC = B$.

4. Let

$$A = \begin{bmatrix} 1 & -1 & 3 \\ 2 & -1 & 4 \end{bmatrix} \quad \text{and} \quad C = \begin{bmatrix} 1 & 2 \\ 3 & 1 \end{bmatrix}.$$

Find a (3×2) matrix B such that $AB = C$.

5. Let A be the nonsingular (5×5) matrix

$$A = [\mathbf{A}_1, \mathbf{A}_2, \mathbf{A}_3, \mathbf{A}_4, \mathbf{A}_5],$$

and let $B = [\mathbf{A}_5, \mathbf{A}_1, \mathbf{A}_4, \mathbf{A}_2, \mathbf{A}_3]$. For a given vector $\mathbf{b}$, suppose that $[1, 3, 5, 7, 9]^T$ is the solution to $B\mathbf{x} = \mathbf{b}$. What is the solution of $A\mathbf{x} = \mathbf{b}$?

6. Let

$$\mathbf{v}_1 = \begin{bmatrix} 1 \\ 1 \\ 3 \end{bmatrix}, \quad \mathbf{v}_2 = \begin{bmatrix} 2 \\ 1 \\ 4 \end{bmatrix}, \quad \text{and} \quad \mathbf{v}_3 = \begin{bmatrix} 5 \\ 2 \\ 9 \end{bmatrix}.$$

a) Solve the vector equation $x_1\mathbf{v}_1 + x_2\mathbf{v}_2 + x_3\mathbf{v}_3 = \mathbf{b}$, where

$$\mathbf{b} = \begin{bmatrix} 8 \\ 5 \\ 18 \end{bmatrix}.$$

b) Show that the set of vectors $\{\mathbf{v}_1, \mathbf{v}_2, \mathbf{v}_3\}$ is linearly dependent by exhibiting a nontrivial solution to the vector equation $x_1\mathbf{v}_1 + x_2\mathbf{v}_2 + x_3\mathbf{v}_3 = \boldsymbol{\theta}$.

7. Let

$$A = \begin{bmatrix} 1 & -1 & 3 \\ 2 & -1 & 5 \\ -1 & 4 & -5 \end{bmatrix}$$

and define a function $T: R^3 \to R^3$ by $T(\mathbf{x}) = A\mathbf{x}$ for each

$$\mathbf{x} = \begin{bmatrix} x_1 \\ x_2 \\ x_3 \end{bmatrix}$$

in R^3.

a) Find a vector $\mathbf{x}$ in R^3 such that $T(\mathbf{x}) = \mathbf{b}$, where

$$\mathbf{b} = \begin{bmatrix} 1 \\ 3 \\ 2 \end{bmatrix}.$$

b) If $\boldsymbol{\theta}$ is the zero vector of R^3, then clearly $T(\boldsymbol{\theta}) = \boldsymbol{\theta}$. Describe all vectors $\mathbf{x}$ in R^3 such that $T(\mathbf{x}) = \boldsymbol{\theta}$.

8. Let

$$\mathbf{v}_1 = \begin{bmatrix} 1 \\ -1 \\ 3 \end{bmatrix}, \quad \mathbf{v}_2 = \begin{bmatrix} 2 \\ -1 \\ 5 \end{bmatrix}, \quad \text{and} \quad \mathbf{v}_3 = \begin{bmatrix} -1 \\ 4 \\ -5 \end{bmatrix}.$$

Find

$$\mathbf{x} = \begin{bmatrix} x_1 \\ x_2 \\ x_3 \end{bmatrix}$$

so that $\mathbf{x}^T\mathbf{v}_1 = 2$, $\mathbf{x}^T\mathbf{v}_2 = 3$, and $\mathbf{x}^T\mathbf{v}_3 = -4$.

9. Find A^{-1} for each of the following matrices A

a) $A = \begin{bmatrix} 1 & 2 & 1 \\ 2 & 5 & 4 \\ 1 & 1 & 0 \end{bmatrix}$

b) $A = \begin{bmatrix} \cos\theta & -\sin\theta \\ \sin\theta & \cos\theta \end{bmatrix}$

10. For what values of λ is the matrix

$$A = \begin{bmatrix} \lambda - 4 & -1 \\ 2 & \lambda - 1 \end{bmatrix}$$

singular? Find A^{-1} if A is nonsingular.

11. Find A if A is (2×2) and $(4A)^{-1} = \begin{bmatrix} 3 & 1 \\ 5 & 2 \end{bmatrix}$.

12. Find A and B if they are (2×2) and

$$A + B = \begin{bmatrix} 4 & 6 \\ 8 & 10 \end{bmatrix} \quad \text{and} \quad A - B = \begin{bmatrix} 2 & 2 \\ 4 & 6 \end{bmatrix}.$$

13. Let

$$A = \begin{bmatrix} 1 & 0 & 0 \\ 0 & -1 & 0 \\ 0 & 0 & -1 \end{bmatrix}.$$

Calculate A^{99} and A^{100}.

In Exercises 14–18, A and B are (3×3) matrices such that

$$A^{-1} = \begin{bmatrix} 2 & 3 & 5 \\ 7 & 2 & 1 \\ 4 & -4 & 3 \end{bmatrix} \quad \text{and} \quad B^{-1} = \begin{bmatrix} -6 & 4 & 3 \\ 7 & -1 & 5 \\ 2 & 3 & 1 \end{bmatrix}.$$

14. Without calculating A, solve the system of equations $A\mathbf{x} = \mathbf{b}$, where

$$\mathbf{x} = \begin{bmatrix} x_1 \\ x_2 \\ x_3 \end{bmatrix} \quad \text{and} \quad \mathbf{b} = \begin{bmatrix} -1 \\ 0 \\ 1 \end{bmatrix}.$$

15. Without calculating A or B, find $(AB)^{-1}$.

16. Without calculating A, find $(3A)^{-1}$.

17. Without calculating A or B, find $(A^TB)^{-1}$.

18. Without calculating A or B, find $[(A^{-1}B^{-1})^{-1}A^{-1}B]^{-1}$.

Supplementary Conceptual Exercises

In Exercises 1–8, answer true or false. Justify your answer by providing a counterexample if the statement is false or an outline of a proof if the statement is true.

1. If A and B are symmetric $(n \times n)$ matrices, then AB is also symmetric.

2. If A is an $(n \times n)$ matrix, then $A + A^T$ is symmetric.

3. If A and B are nonsingular $(n \times n)$ matrices such that $A^2 = I$ and $B^2 = I$, then $(AB)^{-1} = BA$.

4. If A and B are nonsingular $(n \times n)$ matrices, then $A + B$ is also nonsingular.

5. A consistent (3×2) linear system of equations can never have a unique solution.

6. If A is an $(m \times n)$ matrix such that $A\mathbf{x} = \boldsymbol{\theta}$ for every $\mathbf{x}$ in R^n, then A is the $(m \times n)$ zero matrix.

7. If A is a (2×2) nonsingular matrix and $\mathbf{u}_1$ and $\mathbf{u}_2$ are nonzero vectors in R^2, then $\{A\mathbf{u}_1, A\mathbf{u}_2\}$ is linearly independent.

8. Let A be $(m \times n)$ and B be $(p \times q)$. If AB is defined and square, then BA is also defined and square.

In Exercises 9–16, give a brief answer.

9. Let P, Q, and R be nonsingular $(n \times n)$ matrices such that $PQR = I$. Express Q^{-1} in terms of P and R.

10. Suppose that each of A, B, and AB are symmetric $(n \times n)$ matrices. Show that $AB = BA$.

11. Let $\mathbf{u}_1$, $\mathbf{u}_2$, and $\mathbf{u}_3$ be nonzero vectors in R^n such that $\mathbf{u}_1^T \mathbf{u}_2 = 0$, $\mathbf{u}_1^T \mathbf{u}_3 = 0$, and $\mathbf{u}_2^T \mathbf{u}_3 = 0$. Show that $\{\mathbf{u}_1, \mathbf{u}_2, \mathbf{u}_3\}$ is a linearly independent set.

12. Let $\mathbf{u}_1$ and $\mathbf{u}_2$ be linearly dependent vectors in R^2, and let A be a (2×2) matrix. Show that the vectors $A\mathbf{u}_1$ and $A\mathbf{u}_2$ are also linearly dependent.

13. An $(n \times n)$ matrix A is **orthogonal** provided that $A^T = A^{-1}$, that is, if $AA^T = A^TA = I$. If A is an $(n \times n)$ orthogonal matrix, then prove that $\|\mathbf{x}\| = \|A\mathbf{x}\|$ for every vector $\mathbf{x}$ in R^n.

14. An $(n \times n)$ matrix A is **idempotent** if $A^2 = A$. What can you say about A if it is both idempotent and nonsingular?

15. Let A and B be $(n \times n)$ idempotent matrices such that $AB = BA$. Show that AB is also idempotent.

16. An $(n \times n)$ matrix A is **nilpotent of index k** if $A^k = \mathcal{O}$ but $A^i \neq \mathcal{O}$ for $1 \leq i \leq k - 1$.
   a) Show: If A is nilpotent of index 2 or 3, then A is singular.
   b) (Optional) Show: If A is nilpotent of index k, $k \geq 2$, then A is singular. (*Hint:* Consider a proof by contradiction.)

2

The Vector

Space R^n

Introduction

In mathematics and the physical sciences, the term *vector* is applied to a wide variety of objects. Perhaps the most familiar application of the term is to quantities, such as force and velocity, that have both magnitude and direction. Such vectors can be represented in two-space or in three-space as directed line segments or arrows. (A review of geometric vectors is provided in the appendix.) As we will see in Chapter 4, the term *vector* may also be used to describe objects such as matrices, polynomials, and continuous real-valued functions. In this section we demonstrate that R^n, the set of n-dimensional vectors, provides a natural bridge between the intuitive and natural concept of a geometric vector and that of an abstract vector in a general vector space. The remainder of the chapter is concerned with the algebraic and geometric structure of R^n and subsets of R^n. Some of the concepts fundamental to describing this structure are subspace, basis, and dimension. These concepts are introduced and discussed in the first few sections. Although these ideas are relatively abstract, they are easy to understand in R^n and they also have application to concrete problems. Thus R^n will serve as an example and as a model for the study, in Chapter 4, of general vector spaces.

To make the transition from geometric vectors in two-space and three-space to two-dimensional and three-dimensional vectors in R^2 and R^3, recall that the geometric vector, **v**, can be uniquely represented as a directed line segment OP, with initial point at the origin, O, and with terminal point P. If **v** is in two-space and point P has coordinates (a, b), then it is natural to represent **v** in R^2 as the

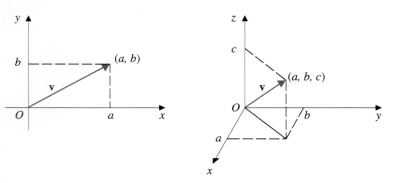

Figure 2.1
Geometric vectors

vector

$$\mathbf{x} = \begin{bmatrix} a \\ b \end{bmatrix}.$$

Similarly, if $\mathbf{v}$ is in three-space and point P has coordinates (a, b, c), then $\mathbf{v}$ can be represented by the vector

$$\mathbf{x} = \begin{bmatrix} a \\ b \\ c \end{bmatrix}$$

in R^3 (see Fig. 2.1). Under the correspondence $\mathbf{v} \to \mathbf{x}$ described above, the usual geometric addition of vectors translates to the standard algebraic addition in R^2 and R^3. Similarly, geometric multiplication by a scalar corresponds precisely to the standard algebraic scalar multiplication (see Fig. 2.2). Thus the study of R^2 and R^3 allows us to translate the geometric properties of vectors to algebraic properties. As we consider vectors from the algebraic viewpoint, it then becomes natural to extend the concept of a vector to other objects that satisfy the same algebraic properties but for which there is no geometric representation. The elements of R^n, $n \geq 4$, are an immediate example.

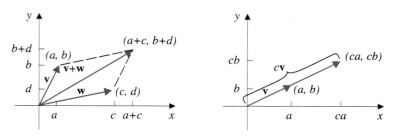

Figure 2.2
Addition and scalar multiplication of vectors

We conclude this section by noting a useful geometric interpretation for vectors in R^2 and R^3. A vector

$$\mathbf{x} = \begin{bmatrix} a \\ b \end{bmatrix}$$

in R^2 can be represented geometrically as the point in the plane that has coordinates (a, b). Similarly, the vector

$$\mathbf{x} = \begin{bmatrix} a \\ b \\ c \end{bmatrix}$$

in R^3 corresponds to the point in three-space that has coordinates (a, b, c). As the next two examples illustrate, this correspondence allows us to interpret geometrically subsets of R^2 and R^3.

EXAMPLE 1 Give a geometric interpretation of the subset W of R^2 defined by

$$W = \{\mathbf{x}: \mathbf{x} = \begin{bmatrix} x_1 \\ x_2 \end{bmatrix}, \quad x_1 + x_2 = 2\}.$$

SOLUTION Geometrically, W is the line in the plane with equation $x + y = 2$ (see Fig. 2.3).

EXAMPLE 2 Let W be the subset of R^3 defined by

$$W = \{\mathbf{x}: \mathbf{x} = \begin{bmatrix} x_1 \\ x_2 \\ 1 \end{bmatrix}, \quad x_1 \text{ and } x_2 \text{ any real numbers}\}.$$

Give a geometric interpretation of W.

SOLUTION Geometrically, W can be viewed as the plane in three-space with equation $z = 1$ (see Fig. 2.4).

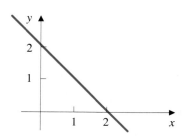

Figure 2.3
The line $x + y = 2$

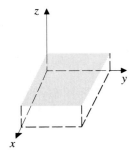

Figure 2.4
The plane $z = 1$

Exercises 2.1

Exercises 1–11 refer to the vectors given in (1).

$$\mathbf{u} = \begin{bmatrix} 3 \\ 1 \end{bmatrix}, \qquad \mathbf{v} = \begin{bmatrix} 1 \\ 2 \end{bmatrix}, \qquad \mathbf{x} = \begin{bmatrix} 0 \\ 1 \\ 3 \end{bmatrix}, \qquad \mathbf{y} = \begin{bmatrix} 2 \\ 1 \\ 0 \end{bmatrix} \quad (1)$$

In Exercises 1–11, sketch the geometric vector (with initial point at the origin) corresponding to each of the vectors given.

1. $\mathbf{u}$ and $-\mathbf{u}$
2. $\mathbf{v}$ and $2\mathbf{v}$
3. $\mathbf{u}$ and $-3\mathbf{u}$
4. $\mathbf{v}$ and $-2\mathbf{v}$
5. $\mathbf{u}$, $\mathbf{v}$, and $\mathbf{u} + \mathbf{v}$
6. $\mathbf{u}$, $2\mathbf{v}$, and $\mathbf{u} + 2\mathbf{v}$
7. $\mathbf{u}$, $\mathbf{v}$, and $\mathbf{u} - \mathbf{v}$
8. $\mathbf{u}$, $\mathbf{v}$, and $\mathbf{v} - \mathbf{u}$
9. $\mathbf{x}$ and $2\mathbf{x}$
10. $\mathbf{y}$ and $-\mathbf{y}$
11. $\mathbf{x}$, $\mathbf{y}$, and $\mathbf{x} + \mathbf{y}$

In Exercises 12–17, interpret the subset W of R^2 geometrically by sketching a graph for W.

12. $W = \{\mathbf{x} \colon \mathbf{x} = \begin{bmatrix} a \\ b \end{bmatrix}, \quad a + b = 1\}$

13. $W = \{\mathbf{x} \colon \mathbf{x} = \begin{bmatrix} x_1 \\ x_2 \end{bmatrix}, \quad x_1 = -3x_2,$
    x_2 any real number$\}$

14. $W = \{\mathbf{w} \colon \mathbf{w} = \begin{bmatrix} 0 \\ b \end{bmatrix}, \quad b$ any real number$\}$

15. $W = \{\mathbf{u} \colon \mathbf{u} = \begin{bmatrix} c \\ d \end{bmatrix}, \quad c + d \geq 0\}$

16. $W = \{\mathbf{x} \colon \mathbf{x} = t\begin{bmatrix} 1 \\ 3 \end{bmatrix}, \quad t$ any real number$\}$

17. $W = \{\mathbf{x} \colon \mathbf{x} = \begin{bmatrix} a \\ b \end{bmatrix}, \quad a^2 + b^2 = 4\}$

In Exercises 18–21, interpret the subset W of R^3 geometrically by sketching a graph for W.

18. $W = \{\mathbf{x} \colon \mathbf{x} = \begin{bmatrix} a \\ 0 \\ 0 \end{bmatrix}, \quad a > 0\}$

19. $W = \{\mathbf{x} \colon \mathbf{x} = \begin{bmatrix} x_1 \\ x_2 \\ x_3 \end{bmatrix}, \quad x_1 = -x_2 - 2x_3\}$

20. $W = \{\mathbf{w} \colon \mathbf{w} = r\begin{bmatrix} 2 \\ 0 \\ 1 \end{bmatrix}, \quad r$ any real number$\}$

21. $W = \{\mathbf{u} \colon \mathbf{u} = \begin{bmatrix} a \\ b \\ c \end{bmatrix}, \quad a^2 + b^2 + c^2 = 1 \quad$ and
    $c \geq 0\}$

In Exercises 22–26, give a set-theoretic description of the given points as a subset W of R^2.

22. The points on the line $x - 2y = 1$
23. The points on the x-axis
24. The points in the upper half-plane

25. The points on the line $y = 2$

26. The points on the parabola $y = x^2$

In Exercises 27–30, give a set-theoretic description of the given points as a subset W of R^3.

27. The points on the plane $x + y - 2z = 0$

28. The points on the line with parametric equations $x = 2t$, $y = -3t$, and $z = t$

29. The points in the yz-plane

30. The points in the plane $y = 2$

2.2 Vector Space Properties of R^n

Recall that R^n is the set of all n-dimensional vectors with real components:

$$R^n = \left\{ \mathbf{x} \colon \mathbf{x} = \begin{bmatrix} x_1 \\ x_2 \\ \vdots \\ x_n \end{bmatrix}, \quad x_1, x_2, \ldots, x_n \text{ real numbers} \right\}.$$

If $\mathbf{x}$ and $\mathbf{y}$ are elements of R^n with

$$\mathbf{x} = \begin{bmatrix} x_1 \\ x_2 \\ \vdots \\ x_n \end{bmatrix} \quad \text{and} \quad \mathbf{y} = \begin{bmatrix} y_1 \\ y_2 \\ \vdots \\ y_n \end{bmatrix},$$

then (see Section 1.6) the vector $\mathbf{x} + \mathbf{y}$ is defined by

$$\mathbf{x} + \mathbf{y} = \begin{bmatrix} x_1 + y_1 \\ x_2 + y_2 \\ \vdots \\ x_n + y_n \end{bmatrix},$$

and if a is a real number, then the vector $a\mathbf{x}$ is defined to be

$$a\mathbf{x} = \begin{bmatrix} ax_1 \\ ax_2 \\ \vdots \\ ax_n \end{bmatrix}.$$

In the context of R^n, scalars are always real numbers. In particular, throughout this chapter, the term *scalar* always means a real number.

The following theorem gives the arithmetic properties of vector addition and scalar multiplication.

As we will see in Chapter 4, any set that satisfies the properties of Theorem 1 is called a ***vector space;*** thus for each positive integer n, R^n is an example of a vector space.

THEOREM 1

If $\mathbf{x}$, $\mathbf{y}$, and $\mathbf{z}$ are vectors in R^n and a and b are scalars, then the following properties hold:

Closure properties:

c1) $\mathbf{x} + \mathbf{y}$ is in R^n.
c2) $a\mathbf{x}$ is in R^n.

Properties of addition:

a1) $\mathbf{x} + \mathbf{y} = \mathbf{y} + \mathbf{x}$.
a2) $\mathbf{x} + (\mathbf{y} + \mathbf{z}) = (\mathbf{x} + \mathbf{y}) + \mathbf{z}$.
a3) R^n contains the zero vector, $\boldsymbol{\theta}$, and $\mathbf{x} + \boldsymbol{\theta} = \mathbf{x}$ for all $\mathbf{x}$ in R^n.
a4) For each vector $\mathbf{x}$ in R^n, there is a vector $-\mathbf{x}$ in R^n such that $\mathbf{x} + (-\mathbf{x}) = \boldsymbol{\theta}$.

Properties of scalar multiplication:

m1) $a(b\mathbf{x}) = (ab)\mathbf{x}$.
m2) $a(\mathbf{x} + \mathbf{y}) = a\mathbf{x} + a\mathbf{y}$.
m3) $(a + b)\mathbf{x} = a\mathbf{x} + b\mathbf{x}$.
m4) $1\mathbf{x} = \mathbf{x}$ for all $\mathbf{x}$ in R^n.

Subspaces of R^n

In this chapter we are interested in subsets, W, of R^n that satisfy all the properties of Theorem 1 (with R^n replaced by W throughout). Such a subset W is called a **subspace** of R^n. For example, consider the subset W of R^3 defined by

$$W = \left\{ \mathbf{x} : \mathbf{x} = \begin{bmatrix} x_1 \\ x_2 \\ 0 \end{bmatrix}, \quad x_1 \text{ and } x_2 \text{ real numbers} \right\}.$$

Viewed geometrically, W is the xy-plane (see Fig. 2.5), so it can be represented by R^2. Therefore, as can be easily shown, W is a subspace of R^3.

The following theorem provides a convenient way of determining when a subset W of R^n is a subspace of R^n.

*O*rigins of Higher-Dimensional Spaces

In addition to Grassmann (see Section 1.8), Sir William Hamilton (1805–1865) also envisioned algebras of n-tuples (which he called "polyplets"). In 1833, Hamilton gave rules for the addition and multiplication of ordered pairs, (a, b), which became the algebra of complex numbers, $z = a + bi$. He searched for years for an extension to 3-tuples. He finally discovered, in a flash of inspiration while crossing a bridge, that the extension was possible if he used 4-tuples $(a, b, c, d) = a + bi + cj + dk$. In this algebra of quaternions, however, multiplication is not commutative; for example, $ij = k$, but $ji = -k$. Hamilton stopped and carved the basic formula, $i^2 = j^2 = k^2 = ijk$, on the bridge. He considered the quaternions his greatest achievement, even though his so-called Hamiltonian principle is considered fundamental to modern physics.

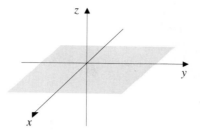

Figure 2.5
W as a subset of R^3

THEOREM 2

A subset W of R^n is a subspace of R^n if and only if the following conditions are met:

s1)* The zero vector, θ, is in W.
s2) $\mathbf{x} + \mathbf{y}$ is in W whenever $\mathbf{x}$ and $\mathbf{y}$ are in W.
s3) $a\mathbf{x}$ is in W whenever $\mathbf{x}$ is in W and a is any scalar.

PROOF   Suppose that W is a subset of R^n that satisfies conditions (s1)–(s3). To show that W is a subspace of R^n, we must show that the 10 properties of Theorem 1 (with R^n replaced by W throughout) are satisfied. But properties (a1), (a2), (m1), (m2), (m3), and (m4) are satisfied by every subset of R^n and so hold in W. Condition (a3) is satisfied by W because the hypothesis (s1) guarantees that θ is in W. Similarly, (c1) and (c2) are given by the hypotheses (s2) and (s3), respectively. The only remaining condition is (a4), and we can easily see that $-\mathbf{x} = (-1)\mathbf{x}$. Thus if $\mathbf{x}$ is in W, then, by (s3), $-\mathbf{x}$ is also in W. Therefore, all the conditions of Theorem 1 are satisfied by W, and W is a subspace of R^n.

For the converse, suppose W is a subspace of R^n. The conditions (a3), (c1), and (c2) of Theorem 1 imply that properties (s1), (s2), and (s3) hold in W.   ☐

The next example illustrates the use of Theorem 2 to verify that a subset W of R^n is a subspace of R^n.

EXAMPLE 1   Let W be the subset of R^3 defined by

$$W = \left\{\mathbf{x}: \mathbf{x} = \begin{bmatrix} x_1 \\ x_2 \\ x_3 \end{bmatrix}, \quad x_1 = x_2 - x_3, \quad x_2 \text{ and } x_3 \text{ any real numbers}\right\}.$$

Verify that W is a subspace of R^3 and give a geometric interpretation of W.

---

* The usual statement of Theorem 2 lists only conditions (s2) and (s3) but assumes that the subset W is nonempty. Thus (s1) replaces the assumption that W is nonempty. The two versions are equivalent (cf. Exercise 34).

SOLUTION   To show that W is a subspace of R^3, we must check that properties (s1)–(s3) of Theorem 2 are satisfied by W. Clearly the zero vector, $\boldsymbol{\theta}$, satisfies the condition $x_1 = x_2 - x_3$. Therefore, $\boldsymbol{\theta}$ is in W, showing that (s1) holds. Now let **u** and **v** be in W, where

$$\mathbf{u} = \begin{bmatrix} u_1 \\ u_2 \\ u_3 \end{bmatrix} \quad \text{and} \quad \mathbf{v} = \begin{bmatrix} v_1 \\ v_2 \\ v_3 \end{bmatrix},$$

and let a be an arbitrary scalar. Since **u** and **v** are in W,

$$u_1 = u_2 - u_3 \quad \text{and} \quad v_1 = v_2 - v_3. \tag{1}$$

The sum $\mathbf{u} + \mathbf{v}$ and the scalar product $a\mathbf{u}$ are given by

$$\mathbf{u} + \mathbf{v} = \begin{bmatrix} u_1 + v_1 \\ u_2 + v_2 \\ u_3 + v_3 \end{bmatrix} \quad \text{and} \quad a\mathbf{u} = \begin{bmatrix} au_1 \\ au_2 \\ au_3 \end{bmatrix}.$$

To see that $\mathbf{u} + \mathbf{v}$ is in W, note that (1) gives

$$u_1 + v_1 = (u_2 - u_3) + (v_2 - v_3) = (u_2 + v_2) - (u_3 + v_3). \tag{2}$$

Thus if the components of **u** and **v** satisfy the condition $x_1 = x_2 - x_3$, then so do the components of the sum $\mathbf{u} + \mathbf{v}$. This argument shows that condition (s2) is met by W. Similarly, from (1),

$$au_1 = a(u_2 - u_3) = au_2 - au_3, \tag{3}$$

so $a\mathbf{u}$ is in W. Therefore, W is a subspace of R^3.

Geometrically, W is the plane whose equation is $x - y + z = 0$ (see Fig. 2.6).

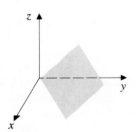

Figure 2.6
The plane $x - y + z = 0$

Verifying That Subsets Are Subspaces

Example 1 illustrates the typical procedure for verifying that a subset W of R^n is a subspace of R^n. In general such a verification proceeds along the following lines:

Steps in Verifying That W Is a Subspace of R^n

1. An algebraic specification for the subset W is given, and this specification serves as a test for determining whether a vector in R^n is or is not in W.
2. Test the zero vector, $\boldsymbol{\theta}$, of R^n to see whether it satisfies the algebraic specification required to be in W. (This shows that W is nonempty.)
3. Choose two arbitrary vectors **x** and **y** from W. Thus **x** and **y** are in R^n, and both vectors satisfy the algebraic specification of W.
4. Test the sum $\mathbf{x} + \mathbf{y}$ to see whether it meets the specification of W.
5. For an arbitrary scalar, a, test the scalar multiple $a\mathbf{x}$ to see whether it meets the specification of W.

The next example illustrates again the use of the procedure described above to verify that a subset W of R^n is a subspace.

EXAMPLE 2 Let W be the subset of R^3 defined by

$$W = \{\mathbf{x}: \mathbf{x} = \begin{bmatrix} x_1 \\ x_2 \\ x_3 \end{bmatrix}, \quad x_2 = 2x_1, \quad x_3 = 3x_1, \quad x_1 \text{ any real number}\}.$$

Verify that W is a subspace of R^3 and give a geometric interpretation of W.

SOLUTION For clarity in this initial example, we explicitly number the five steps used to verify that W is a subspace.

1. The algebraic condition for $\mathbf{x}$ to be in W is

$$x_2 = 2x_1 \quad \text{and} \quad x_3 = 3x_1. \tag{4}$$

In words, $\mathbf{x}$ is in W if and only if the second component of $\mathbf{x}$ is twice the first component and the third component of $\mathbf{x}$ is three times the first.

2. Note that the zero vector, $\boldsymbol{\theta}$, clearly satisfies (4). Therefore, $\boldsymbol{\theta}$ is in W.

3. Next, let $\mathbf{u}$ and $\mathbf{v}$ be two arbitrary vectors in W:

$$\mathbf{u} = \begin{bmatrix} u_1 \\ u_2 \\ u_3 \end{bmatrix} \quad \text{and} \quad \mathbf{v} = \begin{bmatrix} v_1 \\ v_2 \\ v_3 \end{bmatrix}.$$

Because $\mathbf{u}$ and $\mathbf{v}$ are in W, each must satisfy the algebraic specification of W. That is,

$$u_2 = 2u_1 \quad \text{and} \quad u_3 = 3u_1 \tag{5a}$$
$$v_2 = 2v_1 \quad \text{and} \quad v_3 = 3v_1. \tag{5b}$$

4. Next, check whether the sum, $\mathbf{u} + \mathbf{v}$, is in W. (That is, does the vector $\mathbf{u} + \mathbf{v}$ satisfy Eq. (4)?) Now, the sum $\mathbf{u} + \mathbf{v}$ is given by

$$\mathbf{u} + \mathbf{v} = \begin{bmatrix} u_1 + v_1 \\ u_2 + v_2 \\ u_3 + v_3 \end{bmatrix}.$$

By (5a) and (5b), we have

$$u_2 + v_2 = 2(u_1 + v_1) \quad \text{and} \quad (u_3 + v_3) = 3(u_1 + v_1).$$

Thus $\mathbf{u} + \mathbf{v}$ is in W whenever $\mathbf{u}$ and $\mathbf{v}$ are (see Eq. (4)).

5. Similarly, for any scalar a, the scalar multiple $a\mathbf{u}$ is given by

$$a\mathbf{u} = \begin{bmatrix} au_1 \\ au_2 \\ au_3 \end{bmatrix}.$$

Using (5a) gives $au_2 = a(2u_1) = 2(au_1)$ and $au_3 = a(3u_1) = 3(au_1)$. Therefore, $a\mathbf{u}$ is in W whenever $\mathbf{u}$ is (see Eq. (4)).

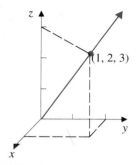

Figure 2.7
A geometric representation of the subspace W; see Example 2

Thus, by Theorem 2, W is a subspace of R^3. Geometrically, W is a line through the origin with parametric equations

$$x = x_1$$
$$y = 2x_1$$
$$z = 3x_1.$$

The graph of the line is given in Fig. 2.7.

Exercise 29 shows that any line in three-space through the origin is a subspace of R^3, and Example 3 of Section 2.3 shows that in three-space any plane through the origin is a subspace. Also note that for each positive integer n, R^n is a subspace of itself and $\{\theta\}$ is a subspace of R^n. We conclude this section with examples of subsets that are not subspaces.

EXAMPLE 3 Let W be the subset of R^3 defined by

$$W = \left\{ \mathbf{x} : \mathbf{x} = \begin{bmatrix} x_1 \\ x_2 \\ 1 \end{bmatrix}, \quad x_1 \text{ and } x_2 \text{ any real numbers} \right\}.$$

Show that W is not a subspace of R^3.

SOLUTION To show that W is not a subspace of R^3, we need only verify that at least one of the properties (s1)–(s3) of Theorem 2 fails. Note that geometrically W can be interpreted as the plane $z = 1$, which does not contain the origin. In other words, the zero vector, θ, is not in W. Because condition (s1) of Theorem 2 is not met, W is not a subspace of R^3. Although it is not necessary to do so, in this example we can also show that both conditions (s2) and (s3) of Theorem 2 fail. To see this, let $\mathbf{x}$ and $\mathbf{y}$ be in W, where

$$\mathbf{x} = \begin{bmatrix} x_1 \\ x_2 \\ 1 \end{bmatrix} \quad \text{and} \quad \mathbf{y} = \begin{bmatrix} y_1 \\ y_2 \\ 1 \end{bmatrix}.$$

Then $\mathbf{x} + \mathbf{y}$ is given by

$$\mathbf{x} + \mathbf{y} = \begin{bmatrix} x_1 + y_1 \\ x_2 + y_2 \\ 2 \end{bmatrix}.$$

In particular, $\mathbf{x} + \mathbf{y}$ is not in W, because the third component of $\mathbf{x} + \mathbf{y}$ does not have the value 1. Similarly,

$$a\mathbf{x} = \begin{bmatrix} ax_1 \\ ax_2 \\ a \end{bmatrix}.$$

So if $a \neq 1$, then $a\mathbf{x}$ is not in W.

EXAMPLE 4 Let W be the subset of R^2 defined by

$$W = \{\mathbf{x}: \mathbf{x} = \begin{bmatrix} x_1 \\ x_2 \end{bmatrix}, \quad x_1 \text{ and } x_2 \text{ any integers}\}.$$

Demonstrate that W is not a subspace of R^2.

SOLUTION   In this case θ is in W, and it is easy to see that if $\mathbf{x}$ and $\mathbf{y}$ are in W, then so is $\mathbf{x} + \mathbf{y}$. If we set

$$\mathbf{x} = \begin{bmatrix} 1 \\ 1 \end{bmatrix}$$

and $a = 1/2$, then $\mathbf{x}$ is in W but $a\mathbf{x}$ is not. Therefore, condition (s3) of Theorem 2 is not met by W, and hence W is not a subspace of R^2.

EXAMPLE 5 Let W be the subspace of R^2 defined by

$$W = \{\mathbf{x}: \mathbf{x} = \begin{bmatrix} x_1 \\ x_2 \end{bmatrix}, \quad \text{where either } x_1 = 0 \text{ or } x_2 = 0\}.$$

Show that W is not a subspace of R^2.

SOLUTION   Let $\mathbf{x}$ and $\mathbf{y}$ be defined by

$$\mathbf{x} = \begin{bmatrix} 1 \\ 0 \end{bmatrix} \quad \text{and} \quad \mathbf{y} = \begin{bmatrix} 0 \\ 1 \end{bmatrix}.$$

Then $\mathbf{x}$ and $\mathbf{y}$ are in W. But

$$\mathbf{x} + \mathbf{y} = \begin{bmatrix} 1 \\ 1 \end{bmatrix}$$

is not in W, so W is not a subspace of R^2. Note that θ is in W, and for any vector $\mathbf{x}$ in W and any scalar a, $a\mathbf{x}$ is again in W. Geometrically, W is the set of points in the plane that lie either on the x-axis or on the y-axis. Either of these axes alone is a subspace of R^2, but, as this example demonstrates, their union is not a subspace.

Exercises 2.2

In Exercises 1–8, W is a subset of R^2 consisting of vectors of the form

$$\mathbf{x} = \begin{bmatrix} x_1 \\ x_2 \end{bmatrix}.$$

In each case determine whether W is a subspace of R^2. If W is a subspace, then give a geometric description of W.

1. $W = \{\mathbf{x}: x_1 = 2x_2\}$

2. $W = \{\mathbf{x}: x_1 - x_2 = 2\}$

3. $W = \{\mathbf{x}: x_1 = x_2 \quad \text{or} \quad x_1 = -x_2\}$

4. $W = \{\mathbf{x}: x_1 \quad \text{and} \quad x_2 \text{ are rational numbers}\}$

5. $W = \{\mathbf{x}: x_1 = 0\}$

6. $W = \{\mathbf{x}: |x_1| + |x_2| = 0\}$

7. $W = \{\mathbf{x}: x_1^2 + x_2 = 1\}$

8. $W = \{\mathbf{x}: x_1 x_2 = 0\}$

In Exercises 9–17, W is a subset of R^3 consisting of vectors of the form

$$\mathbf{x} = \begin{bmatrix} x_1 \\ x_2 \\ x_3 \end{bmatrix}.$$

In each case determine whether W is a subspace of R^3. If W is a subspace, then give a geometric description of W.

9. $W = \{\mathbf{x}: x_3 = 2x_1 - x_2\}$

10. $W = \{\mathbf{x}: x_2 = x_3 + x_1\}$

11. $W = \{\mathbf{x}: x_1 x_2 = x_3\}$

12. $W = \{\mathbf{x}: x_1 = 2x_3\}$

13. $W = \{\mathbf{x}: x_1^2 = x_1 + x_2\}$

14. $W = \{\mathbf{x}: x_2 = 0\}$

15. $W = \{\mathbf{x}: x_1 = 2x_3, \quad x_2 = -x_3\}$

16. $W = \{\mathbf{x}: x_3 = x_2 = 2x_1\}$

17. $W = \{\mathbf{x}: x_2 = x_3 = 0\}$

18. Let $\mathbf{a}$ be a fixed vector in R^3, and define W to be the subset of R^3 given by

$$W = \{\mathbf{x}: \mathbf{a}^T \mathbf{x} = 0\}.$$

Prove that W is a subspace of R^3.

19. Let W be the subspace defined in Exercise 18, where

$$\mathbf{a} = \begin{bmatrix} 1 \\ 2 \\ 3 \end{bmatrix}.$$

Give a geometric description for W.

20. Let W be the subspace defined in Exercise 18, where

$$\mathbf{a} = \begin{bmatrix} 1 \\ 0 \\ 0 \end{bmatrix}.$$

Give a geometric description of W.

21. Let $\mathbf{a}$ and $\mathbf{b}$ be fixed vectors in R^3, and let W be the subset of R^3 defined by

$$W = \{\mathbf{x}: \mathbf{a}^T \mathbf{x} = 0 \quad \text{and} \quad \mathbf{b}^T \mathbf{x} = 0\}.$$

Prove that W is a subspace of R^3.

In Exercises 22–25, W is the subspace of R^3 defined in Exercise 21. For each choice of $\mathbf{a}$ and $\mathbf{b}$, give a geometric description of W.

22. $\mathbf{a} = \begin{bmatrix} 1 \\ -1 \\ 2 \end{bmatrix}, \quad \mathbf{b} = \begin{bmatrix} 2 \\ -1 \\ 3 \end{bmatrix}$

23. $\mathbf{a} = \begin{bmatrix} 1 \\ 2 \\ 2 \end{bmatrix}, \quad \mathbf{b} = \begin{bmatrix} 1 \\ 3 \\ 0 \end{bmatrix}$

24. $\mathbf{a} = \begin{bmatrix} 1 \\ 1 \\ 1 \end{bmatrix}, \quad \mathbf{b} = \begin{bmatrix} 2 \\ 2 \\ 2 \end{bmatrix}$

25. $\mathbf{a} = \begin{bmatrix} 1 \\ 0 \\ -1 \end{bmatrix}, \quad \mathbf{b} = \begin{bmatrix} -2 \\ 0 \\ 2 \end{bmatrix}$

26. In R^4, let $\mathbf{x} = [1, -3, 2, 1]^T$, $\mathbf{y} = [2, 1, 3, 2]^T$, and $\mathbf{z} = [-3, 2, -1, 4]^T$. Set $a = 2$ and $b = -3$. Illustrate that the ten properties of Theorem 1 are satisfied by $\mathbf{x}$, $\mathbf{y}$, $\mathbf{z}$, a, and b.

27. In R^2, suppose that scalar multiplication were defined by

$$a\mathbf{x} = a \begin{bmatrix} x_1 \\ x_2 \end{bmatrix} = \begin{bmatrix} 2ax_1 \\ 2ax_2 \end{bmatrix}$$

for every scalar a. Illustrate with specific examples those properties of Theorem 1 that are not satisfied.

28. Let

$$W = \{\mathbf{x}: \mathbf{x} = \begin{bmatrix} x_1 \\ x_2 \end{bmatrix}, \quad x_2 \geq 0\}.$$

In the statement of Theorem 1, replace each occurrence of R^n with W. Illustrate with specific examples each of the ten properties of Theorem 1 that are not satisfied.

29. In R^3, a line through the origin is the set of all points in R^3 whose coordinates satisfy $x_1 = at$, $x_2 = bt$, and $x_3 = ct$, where t is a variable and a, b, and c are not all zero. Show that a line through the origin is a subspace of R^3.

30. If U and V are subsets of R^n, then the set $U + V$ is defined by

$$U + V = \{\mathbf{x} : \mathbf{x} = \mathbf{u} + \mathbf{v}, \quad \mathbf{u} \text{ in } U \quad \text{and} \quad \mathbf{v} \text{ in } V\}.$$

Prove that if U and V are subspaces of R^n, then $U + V$ is a subspace of R^n.

31. Let U and V be subspaces of R^n. Prove that the intersection, $U \cap V$, is also a subspace of R^n.

32. Let U and V be the subspaces of R^3 defined by

$$U = \{\mathbf{x} : \mathbf{a}^T\mathbf{x} = 0\} \quad \text{and} \quad V = \{\mathbf{x} : \mathbf{b}^T\mathbf{x} = 0\},$$

where

$$\mathbf{a} = \begin{bmatrix} 1 \\ 1 \\ 0 \end{bmatrix} \quad \text{and} \quad \mathbf{b} = \begin{bmatrix} 0 \\ 1 \\ -1 \end{bmatrix}.$$

Demonstrate that the union, $U \cup V$, is not a subspace of R^3 (cf. Exercise 18).

33. Let U and V be subspaces of R^n.
a) Show that the union, $U \cup V$, satisfies properties (s1) and (s3) of Theorem 2.
b) If neither U nor V is a subset of the other, show that $U \cup V$ does not satisfy condition (s2) of Theorem 2. (*Hint:* Choose vectors $\mathbf{u}$ and $\mathbf{v}$ such that $\mathbf{u}$ is in U but not in V and $\mathbf{v}$ is in V but not in U. Assume that $\mathbf{u} + \mathbf{v}$ is in either U or V and reach a contradiction.)

34. Let W be a nonempty subset of R^n that satisfies conditions (s2) and (s3) of Theorem 2. Prove that θ is in W and conclude that W is a subspace of R^n. (Thus property (s1) of Theorem 2 can be replaced with the assumption that W is nonempty.)

2.3 Examples of Subspaces

In this section we introduce several important and particularly useful examples of subspaces of R^n.

The Span of a Subset

To begin, recall that if $\mathbf{v}_1, \ldots, \mathbf{v}_r$ are vectors in R^n, then a vector $\mathbf{y}$ in R^n is a linear combination of $\mathbf{v}_1, \ldots, \mathbf{v}_r$, provided that there exist scalars $a_1, \ldots, a_r$ such that

$$\mathbf{y} = a_1\mathbf{v}_1 + \cdots + a_r\mathbf{v}_r.$$

The next theorem shows that the set of all linear combinations of $\mathbf{v}_1, \ldots, \mathbf{v}_r$ is a subspace of R^n.

THEOREM 3

If $\mathbf{v}_1, \ldots, \mathbf{v}_r$ are vectors in R^n, then the set W consisting of all linear combinations of $\mathbf{v}_1, \ldots, \mathbf{v}_r$ is a subspace of R^n.

PROOF To show that W is a subspace of R^n, we must verify that the three conditions of Theorem 2 are satisfied. Now θ is in W because

$$\theta = 0\mathbf{v}_1 + \cdots + 0\mathbf{v}_r.$$

Next, suppose that $\mathbf{y}$ and $\mathbf{z}$ are in W. Then there exist scalars $a_1, \ldots, a_r, b_1, \ldots, b_r$ such that

$$\mathbf{y} = a_1\mathbf{v}_1 + \cdots + a_r\mathbf{v}_r$$

and

$$\mathbf{z} = b_1\mathbf{v}_1 + \cdots + b_r\mathbf{v}_r.$$

Thus

$$\mathbf{y} + \mathbf{z} = (a_1 + b_1)\mathbf{v}_1 + \cdots + (a_r + b_r)\mathbf{v}_r,$$

so $\mathbf{y} + \mathbf{z}$ is a linear combination of $\mathbf{v}_1, \ldots, \mathbf{v}_r$; that is, $\mathbf{y} + \mathbf{z}$ is in W. Also, for any scalar c,

$$c\mathbf{y} = (ca_1)\mathbf{v}_1 + \cdots + (ca_r)\mathbf{v}_r.$$

In particular, $c\mathbf{y}$ is in W. It follows from Theorem 2 that W is a subspace of R^n.

$\square$

If $S = \{\mathbf{v}_1, \ldots, \mathbf{v}_r\}$ is a subset of R^n, then the subspace W consisting of all linear combinations of $\mathbf{v}_1, \ldots, \mathbf{v}_r$ is called the **subspace spanned by S** and will be denoted by

$$\text{Sp}(S) \quad \text{or} \quad \text{Sp}\{\mathbf{v}_1, \ldots, \mathbf{v}_r\}.$$

For a single vector $\mathbf{v}$ in R^n, $\text{Sp}\{\mathbf{v}\}$ is the subspace

$$\text{Sp}\{\mathbf{v}\} = \{a\mathbf{v}: a \text{ is any real number}\}.$$

Figure 2.8
$\text{Sp}\{\mathbf{v}\}$

If $\mathbf{v}$ is a nonzero vector in R^2 or R^3, then $\text{Sp}\{\mathbf{v}\}$ can be interpreted as the line determined by $\mathbf{v}$ (see Fig. 2.8). As a specific example, consider

$$\mathbf{v} = \begin{bmatrix} 1 \\ 2 \\ 3 \end{bmatrix}.$$

Then

$$\text{Sp}\{\mathbf{v}\} = \{t\begin{bmatrix} 1 \\ 2 \\ 3 \end{bmatrix}: t \text{ is any real number}\}.$$

Thus $\text{Sp}\{\mathbf{v}\}$ is the line with parametric equations

$$x = t$$
$$y = 2t$$
$$z = 3t.$$

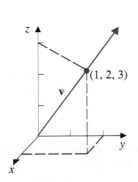

Figure 2.9

$\text{Sp} \left\{ \begin{bmatrix} 1 \\ 2 \\ 3 \end{bmatrix} \right\}$

Equivalently, $\text{Sp}\{\mathbf{v}\}$ is the line that passes through the origin and through the point with coordinates 1, 2, and 3 (see Fig. 2.9).

If $\mathbf{u}$ and $\mathbf{v}$ are noncollinear geometric vectors, then

$$\text{Sp}\{\mathbf{u}, \mathbf{v}\} = \{a\mathbf{u} + b\mathbf{v}: a, b \text{ any real numbers}\}$$

is the plane containing $\mathbf{u}$ and $\mathbf{v}$ (see Fig. 2.10). The following example illustrates this case with a subspace of R^3.

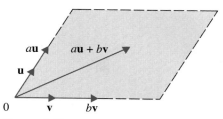

Figure 2.10
Sp{**u**, **v**}

<u>EXAMPLE 1</u> Let **u** and **v** be the three-dimensional vectors

$$\mathbf{u} = \begin{bmatrix} 2 \\ 1 \\ 0 \end{bmatrix} \quad \text{and} \quad \mathbf{v} = \begin{bmatrix} 0 \\ 1 \\ 2 \end{bmatrix}.$$

Determine $W = \text{Sp}\{\mathbf{u}, \mathbf{v}\}$ and give a geometric interpretation of W.

SOLUTION   Let **y** be an arbitrary vector in R^3, where

$$\mathbf{y} = \begin{bmatrix} y_1 \\ y_2 \\ y_3 \end{bmatrix}.$$

Then **y** is in W if and only if there exist scalars x_1 and x_2 such that

$$\mathbf{y} = x_1\mathbf{u} + x_2\mathbf{v}. \tag{1}$$

That is, **y** is in W if and only if there exist scalars x_1 and x_2 such that

$$\begin{aligned} y_1 &= 2x_1 \\ y_2 &= x_1 + x_2 \\ y_3 &= \qquad 2x_2. \end{aligned} \tag{2}$$

The augmented matrix for linear system (2) is

$$\begin{bmatrix} 2 & 0 & y_1 \\ 1 & 1 & y_2 \\ 0 & 2 & y_3 \end{bmatrix},$$

*P*hysical Representations of Vectors

The vector space work of Grassmann and Hamilton was distilled and popularized for the case of R^3 by a Yale University physicist, Josiah Willard Gibbs (1839–1903). Gibbs produced a pamphlet, "Elements of Vector Analysis," mainly for the use of his students. In it, and subsequent articles, Gibbs simplified and improved Hamilton's work in multiple algebras with regard to three-dimensional space. This led to the familiar geometrical representation of vector algebra in terms of operations on directed line segments.

and this matrix is row equivalent to the matrix

$$\begin{bmatrix} 2 & 0 & y_1 \\ 0 & 2 & y_3 \\ 0 & 0 & -(1/2)y_1 + y_2 - (1/2)y_3 \end{bmatrix} \qquad (3)$$

in echelon form. Therefore, linear system (2) is consistent if and only if $-(1/2)y_1 + y_2 - (1/2)y_3 = 0$, or equivalently, if and only if

$$y_1 - 2y_2 + y_3 = 0. \qquad (4)$$

Thus W is the subspace given by

$$W = \{ \mathbf{y} = \begin{bmatrix} y_1 \\ y_2 \\ y_3 \end{bmatrix} : y_1 - 2y_2 + y_3 = 0 \}. \qquad (5)$$

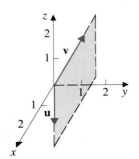

Figure 2.11
The plane
$x - 2y + z = 0$

It also follows from (5) that geometrically W is the plane in three-space with equation $x - 2y + z = 0$ (see Fig. 2.11). □

The Null Space of a Matrix

We now introduce two subspaces that have particular relevance to the linear system of equations $A\mathbf{x} = \mathbf{b}$, where A is an $(m \times n)$ matrix. The first of these subspaces is called the **null space** of A (or the **kernel** of A) and consists of all solutions of $A\mathbf{x} = \boldsymbol{\theta}$.

DEFINITION 1   Let A be an $(m \times n)$ matrix. The **null space** of A [denoted $\mathcal{N}(A)$] is the set of vectors in R^n defined by
$$\mathcal{N}(A) = \{ \mathbf{x}: A\mathbf{x} = \boldsymbol{\theta}, \quad \mathbf{x} \text{ in } R^n \}.$$

In words, the null space consists of all those vectors $\mathbf{x}$ such that $A\mathbf{x}$ is the zero vector. The next theorem shows that the null space of an $(m \times n)$ matrix A is a subspace of R^n.

THEOREM 4

If A is an $(m \times n)$ matrix, then $\mathcal{N}(A)$ is a subspace of R^n.

PROOF   To show that $\mathcal{N}(A)$ is a subspace of R^n, we must verify that the three conditions of Theorem 2 hold. Let $\boldsymbol{\theta}$ be the zero vector in R^n. Then

$$A\boldsymbol{\theta} = \boldsymbol{\theta}, \qquad (6)$$

and so $\boldsymbol{\theta}$ is in $\mathcal{N}(A)$. (*Note:* In Eq. (6), the left $\boldsymbol{\theta}$ is in R^n but the right $\boldsymbol{\theta}$ is in R^m.) Now let $\mathbf{u}$ and $\mathbf{v}$ be vectors in $\mathcal{N}(A)$. Then $\mathbf{u}$ and $\mathbf{v}$ are in R^n and

$$A\mathbf{u} = \boldsymbol{\theta} \quad \text{and} \quad A\mathbf{v} = \boldsymbol{\theta}. \qquad (7)$$

To see that $\mathbf{u} + \mathbf{v}$ is in $\mathcal{N}(A)$, we must test $\mathbf{u} + \mathbf{v}$ against the algebraic specification of $\mathcal{N}(A)$; that is, we must show that $A(\mathbf{u} + \mathbf{v}) = \boldsymbol{\theta}$. But it follows from (7) that

$$A(\mathbf{u} + \mathbf{v}) = A\mathbf{u} + A\mathbf{v} = \boldsymbol{\theta} + \boldsymbol{\theta} = \boldsymbol{\theta},$$

and therefore $\mathbf{u} + \mathbf{v}$ is in $\mathcal{N}(A)$. Similarly, for any scalar a, it follows from (7) that

$$A(a\mathbf{u}) = aA\mathbf{u} = a\boldsymbol{\theta} = \boldsymbol{\theta}.$$

Therefore, $a\mathbf{u}$ is in $\mathcal{N}(A)$. By Theorem 2, $\mathcal{N}(A)$ is a subspace of R^n.

EXAMPLE 2 Describe $\mathcal{N}(A)$, where A is the (3×4) matrix

$$A = \begin{bmatrix} 1 & 1 & 3 & 1 \\ 2 & 1 & 5 & 4 \\ 1 & 2 & 4 & -1 \end{bmatrix}.$$

SOLUTION $\mathcal{N}(A)$ is determined by solving the homogeneous system

$$A\mathbf{x} = \boldsymbol{\theta}. \tag{8}$$

This is accomplished by reducing the augmented matrix $[A \,|\, \boldsymbol{\theta}]$ to echelon form. It is easy to verify that $[A \,|\, \boldsymbol{\theta}]$ is row equivalent to

$$\begin{bmatrix} 1 & 1 & 3 & 1 & 0 \\ 0 & -1 & -1 & 2 & 0 \\ 0 & 0 & 0 & 0 & 0 \end{bmatrix}.$$

Backsolving the corresponding reduced system yields

$$\begin{aligned} x_1 &= -2x_3 - 3x_4 \\ x_2 &= -x_3 + 2x_4 \end{aligned}$$

as the solution to (8). Thus a vector $\mathbf{x}$ in R^4,

$$\mathbf{x} = \begin{bmatrix} x_1 \\ x_2 \\ x_3 \\ x_4 \end{bmatrix},$$

is in $\mathcal{N}(A)$ if and only if $\mathbf{x}$ can be written in the form

$$\mathbf{x} = \begin{bmatrix} -2x_3 - 3x_4 \\ -x_3 + 2x_4 \\ x_3 \\ x_4 \end{bmatrix},$$

where x_3 and x_4 are arbitrary; that is,

$$\mathcal{N}(A) = \left\{ \mathbf{x} : \mathbf{x} = \begin{bmatrix} -2x_3 - 3x_4 \\ -x_3 + 2x_4 \\ x_3 \\ x_4 \end{bmatrix}, \; x_3 \text{ and } x_4 \text{ any real numbers} \right\}.$$

As the next example demonstrates, the fact that $\mathcal{N}(A)$ is a subspace can be used to show that in three-space every plane through the origin is a subspace.

EXAMPLE 3 Verify that any plane through the origin in R^3 is a subspace of R^3.

SOLUTION   The equation of a plane in three-space through the origin is

$$ax + by + cz = 0, \tag{9}$$

where a, b, and c are specified constants not all of which are zero. Now, Eq. (9) can be written as

$$A\mathbf{x} = \boldsymbol{\theta},$$

where A is a (1×3) matrix and $\mathbf{x}$ is in R^3:

$$A = [a \quad b \quad c] \quad \text{and} \quad \mathbf{x} = \begin{bmatrix} x \\ y \\ z \end{bmatrix}.$$

Thus $\mathbf{x}$ is on the plane defined by (9) if and only if $\mathbf{x}$ is in $\mathcal{N}(A)$. Since $\mathcal{N}(A)$ is a subspace of R^3 by Theorem 4, any plane through the origin is a subspace of R^3.

The Range of a Matrix

Another important subspace associated with an $(m \times n)$ matrix A is the *range* of A, defined as follows.

DEFINITION 2   Let A be an $(m \times n)$ matrix. The *range* of A [denoted $\mathcal{R}(A)$] is the set of vectors in R^m defined by

$$\mathcal{R}(A) = \{\mathbf{y}: \mathbf{y} = A\mathbf{x} \text{ for some } \mathbf{x} \text{ in } R^n\}.$$

In words, the range of A consists of the set of all vectors $\mathbf{y}$ in R^m such that the linear system

$$A\mathbf{x} = \mathbf{y}$$

is consistent. As another way to view $\mathcal{R}(A)$, suppose that A is an $(m \times n)$ matrix. We can regard multiplication by A as defining a function from R^n to R^m. In this sense, as $\mathbf{x}$ varies through R^n, the set of all vectors

$$\mathbf{y} = A\mathbf{x}$$

produced in R^m constitutes the "range" of the function.

We saw in Section 1.6 (cf. Theorem 5) that if the $(m \times n)$ matrix A has columns $\mathbf{A}_1, \mathbf{A}_2, \ldots, \mathbf{A}_n$ and if

$$\mathbf{x} = \begin{bmatrix} x_1 \\ x_2 \\ \vdots \\ x_n \end{bmatrix},$$

then the matrix equation

$$A\mathbf{x} = \mathbf{y}$$

is equivalent to the vector equation

$$x_1\mathbf{A}_1 + x_2\mathbf{A}_2 + \cdots + x_n\mathbf{A}_n = \mathbf{y}.$$

Therefore, it follows that

$$\mathscr{R}(A) = \mathrm{Sp}\{\mathbf{A}_1, \mathbf{A}_2, \ldots, \mathbf{A}_n\}.$$

By Theorem 3, $\mathrm{Sp}\{\mathbf{A}_1, \mathbf{A}_2, \ldots, \mathbf{A}_n\}$ is a subspace of R^m. (This subspace is also called the **column space** of matrix A.) Consequently, $\mathscr{R}(A)$ is a subspace of R^m, and we have proved the following theorem.

THEOREM 5

If A is an $(m \times n)$ matrix and if $\mathscr{R}(A)$ is the range of A, then $\mathscr{R}(A)$ is a subspace of R^m.

The next example illustrates a way to give an algebraic specification for $\mathscr{R}(A)$.

EXAMPLE 4 Describe the range of A, where A is the (3×4) matrix

$$A = \begin{bmatrix} 1 & 1 & 3 & 1 \\ 2 & 1 & 5 & 4 \\ 1 & 2 & 4 & -1 \end{bmatrix}.$$

SOLUTION Let $\mathbf{b}$ be an arbitrary vector in R^3,

$$\mathbf{b} = \begin{bmatrix} b_1 \\ b_2 \\ b_3 \end{bmatrix}.$$

Then $\mathbf{b}$ is in $\mathscr{R}(A)$ if and only if the system of equations

$$A\mathbf{x} = \mathbf{b}$$

is consistent. The augmented matrix for the system is

$$[A \,|\, \mathbf{b}] = \begin{bmatrix} 1 & 1 & 3 & 1 & b_1 \\ 2 & 1 & 5 & 4 & b_2 \\ 1 & 2 & 4 & -1 & b_3 \end{bmatrix},$$

which is equivalent to

$$\begin{bmatrix} 1 & 1 & 3 & 1 & b_1 \\ 0 & -1 & -1 & 2 & b_2 - 2b_1 \\ 0 & 0 & 0 & 0 & -3b_1 + b_2 + b_3 \end{bmatrix}.$$

It follows that $A\mathbf{x} = \mathbf{b}$ has a solution [or equivalently, $\mathbf{b}$ is in $\mathscr{R}(A)$] if and only if $-3b_1 + b_2 + b_3 = 0$, or $b_3 = 3b_1 - b_2$, where b_1 and b_2 are arbitrary. Thus

$$\mathscr{R}(A) = \left\{ \mathbf{b}: \mathbf{b} = \begin{bmatrix} b_1 \\ b_2 \\ 3b_1 - b_2 \end{bmatrix}, \quad b_1 \text{ and } b_2 \text{ any real numbers} \right\}. \qquad \square$$

The Row Space of a Matrix

If A is an $(m \times n)$ matrix with columns $\mathbf{A}_1, \mathbf{A}_2, \dots, \mathbf{A}_n$, then we have already defined the **column space** of A to be

$$\text{Sp}\{\mathbf{A}_1, \mathbf{A}_2, \dots, \mathbf{A}_n\}.$$

In a similar fashion, the rows of A can be regarded as vectors $\mathbf{a}_1, \mathbf{a}_2, \dots, \mathbf{a}_m$ in R^n, and the **row space** of A is defined to be

$$\text{Sp}\{\mathbf{a}_1, \mathbf{a}_2, \dots, \mathbf{a}_m\}.$$

For example, if

$$A = \begin{bmatrix} 1 & 2 & 3 \\ 1 & 0 & 1 \end{bmatrix},$$

then the row space of A is $\text{Sp}\{\mathbf{a}_1, \mathbf{a}_2\}$, where

$$\mathbf{a}_1 = [1 \quad 2 \quad 3] \quad \text{and} \quad \mathbf{a}_2 = [1 \quad 0 \quad 1].$$

The following theorem shows that row-equivalent matrices have the same row space.

THEOREM 6	

Let A be an $(m \times n)$ matrix, and suppose that A is row equivalent to the $(m \times n)$ matrix B. Then A and B have the same row space.

The proof of Theorem 6 is given at the end of this section. To illustrate Theorem 6, let A be the (3×3) matrix

$$A = \begin{bmatrix} 1 & -1 & 1 \\ 2 & -1 & 4 \\ 1 & 1 & 5 \end{bmatrix}.$$

By performing the elementary row operations $R_2 - 2R_1$, $R_3 - R_1$, and $R_3 - 2R_2$, we obtain the matrix

$$B = \begin{bmatrix} 1 & -1 & 1 \\ 0 & 1 & 2 \\ 0 & 0 & 0 \end{bmatrix}.$$

By Theorem 6, matrices A and B have the same row space. Clearly the zero row of B contributes nothing as an element of the spanning set, so the row space of B is $\text{Sp}\{\mathbf{b}_1,\mathbf{b}_2\}$, where

$$\mathbf{b}_1 = [1 \ \ -1 \ \ 1] \quad \text{and} \quad \mathbf{b}_2 = [0 \ \ 1 \ \ 2].$$

If the rows of A are denoted by $\mathbf{a}_1, \mathbf{a}_2$, and $\mathbf{a}_3$, then

$$\text{Sp}\{\mathbf{a}_1,\mathbf{a}_2,\mathbf{a}_3\} = \text{Sp}\{\mathbf{b}_1,\mathbf{b}_2\}.$$

More generally, given a subset $S = \{\mathbf{v}_1,\ldots,\mathbf{v}_m\}$ of R^n, Theorem 6 allows us to obtain a "nicer" subset $T = \{\mathbf{w}_1,\ldots,\mathbf{w}_k\}$ of R^n such that $\text{Sp}(S) = \text{Sp}(T)$. The next example illustrates this.

EXAMPLE 5 Let $S = \{\mathbf{v}_1,\mathbf{v}_2,\mathbf{v}_3,\mathbf{v}_4\}$ be a subset of R^3, where

$$\mathbf{v}_1 = \begin{bmatrix} 1 \\ 2 \\ 1 \end{bmatrix}, \quad \mathbf{v}_2 = \begin{bmatrix} 2 \\ 3 \\ 5 \end{bmatrix}, \quad \mathbf{v}_3 = \begin{bmatrix} 1 \\ 4 \\ -5 \end{bmatrix}, \quad \text{and} \quad \mathbf{v}_4 = \begin{bmatrix} 2 \\ 5 \\ -1 \end{bmatrix}.$$

Show that there exists a set $T = \{\mathbf{w}_1,\mathbf{w}_2\}$ consisting of two vectors in R^3 such that $\text{Sp}(S) = \text{Sp}(T)$.

SOLUTION   Let A be the (3×4) matrix

$$A = [\mathbf{v}_1,\mathbf{v}_2,\mathbf{v}_3,\mathbf{v}_4];$$

that is,

$$A = \begin{bmatrix} 1 & 2 & 1 & 2 \\ 2 & 3 & 4 & 5 \\ 1 & 5 & -5 & -1 \end{bmatrix}.$$

The matrix A^T is the (4×3) matrix

$$A^T = \begin{bmatrix} 1 & 2 & 1 \\ 2 & 3 & 5 \\ 1 & 4 & -5 \\ 2 & 5 & -1 \end{bmatrix},$$

and the row vectors of A^T are precisely the vectors $\mathbf{v}_1^T, \mathbf{v}_2^T, \mathbf{v}_3^T$, and $\mathbf{v}_4^T$. It is straightforward to see that A^T reduces to the matrix

$$B^T = \begin{bmatrix} 1 & 2 & 1 \\ 0 & -1 & 3 \\ 0 & 0 & 0 \\ 0 & 0 & 0 \end{bmatrix}.$$

So, by Theorem 6, A^T and B^T have the same row space. Thus A and B have the same column space, where

$$B = \begin{bmatrix} 1 & 0 & 0 & 0 \\ 2 & -1 & 0 & 0 \\ 1 & 3 & 0 & 0 \end{bmatrix}.$$

In particular, $\text{Sp}(S) = \text{Sp}(T)$, where $T = \{\mathbf{w}_1, \mathbf{w}_2\}$,

$$\mathbf{w}_1 = \begin{bmatrix} 1 \\ 2 \\ 1 \end{bmatrix} \quad \text{and} \quad \mathbf{w}_2 = \begin{bmatrix} 0 \\ -1 \\ 3 \end{bmatrix}.$$

Proof of Theorem 6 (Optional)

Assume that A and B are row-equivalent $(m \times n)$ matrices. Then there is a sequence of matrices

$$A = A_1, A_2, \ldots, A_{k-1}, A_k = B$$

such that for $2 \leq j \leq k$, A_j is obtained by performing a single elementary row operation on A_{j-1}. It suffices, then, to show that A_{j-1} and A_j have the same row space for each j, $2 \leq j \leq k$. This means that it is sufficient to consider only the case in which B is obtained from A by a single elementary row operation.

Let A have rows $\mathbf{a}_1, \ldots, \mathbf{a}_m$; that is, A is the $(m \times n)$ matrix

$$A = \begin{bmatrix} \mathbf{a}_1 \\ \vdots \\ \mathbf{a}_j \\ \vdots \\ \mathbf{a}_k \\ \vdots \\ \mathbf{a}_m \end{bmatrix},$$

where each $\mathbf{a}_i$ is a $(1 \times n)$ row vector;

$$\mathbf{a}_i = [a_{i1} \quad a_{i2} \quad \cdots \quad a_{in}].$$

Clearly the order of the rows is immaterial; that is, if B is obtained by interchanging the jth and kth rows of A,

$$B = \begin{bmatrix} \mathbf{a}_1 \\ \vdots \\ \mathbf{a}_k \\ \vdots \\ \mathbf{a}_j \\ \vdots \\ \mathbf{a}_m \end{bmatrix},$$

then A and B have the same row space because

$$\text{Sp}\{\mathbf{a}_1, \ldots, \mathbf{a}_j, \ldots, \mathbf{a}_k, \ldots, \mathbf{a}_m\} = \text{Sp}\{\mathbf{a}_1, \ldots, \mathbf{a}_k, \ldots, \mathbf{a}_j, \ldots, \mathbf{a}_m\}.$$

Next, suppose that B is obtained by performing the row operation $R_k + cR_j$ on

A; thus,

$$B = \begin{bmatrix} \mathbf{a}_1 \\ \vdots \\ \mathbf{a}_j \\ \vdots \\ \mathbf{a}_k + c\mathbf{a}_j \\ \vdots \\ \mathbf{a}_m \end{bmatrix}.$$

If the vector $\mathbf{x}$ is in the row space of A, then there exist scalars $b_1, \ldots, b_m$ such that

$$\mathbf{x} = b_1\mathbf{a}_1 + \cdots + b_j\mathbf{a}_j + \cdots + b_k\mathbf{a}_k + \cdots + b_m\mathbf{a}_m. \tag{10}$$

The vector equation (10) can be rewritten as

$$\mathbf{x} = b_1\mathbf{a}_1 + \cdots + (b_j - cb_k)\mathbf{a}_j + \cdots + b_k(\mathbf{a}_k + c\mathbf{a}_j) + \cdots + b_m\mathbf{a}_m, \tag{11}$$

and hence $\mathbf{x}$ is in the row space of B. Conversely, if the vector $\mathbf{y}$ is in the row space of B, then there exist scalars $d_1, \ldots, d_m$ such that

$$\mathbf{y} = d_1\mathbf{a}_1 + \cdots + d_j\mathbf{a}_j + \cdots + d_k(\mathbf{a}_k + c\mathbf{a}_j) + \cdots + d_m\mathbf{a}_m. \tag{12}$$

But Eq. (12) can be rearranged as

$$\mathbf{y} = d_1\mathbf{a}_1 + \cdots + (d_j + cd_k)\mathbf{a}_j + \cdots + d_k\mathbf{a}_k + \cdots + d_m\mathbf{a}_m, \tag{13}$$

so $\mathbf{y}$ is in the row space of A. Therefore, A and B have the same row space.

The remaining case is the one in which B is obtained from A by multiplying the jth row by the nonzero scalar c. This case is left as Exercise 54 at the end of this section.

Exercises 2.3

Exercises 1–11 refer to the vectors in (14).

$$\mathbf{a} = \begin{bmatrix} 1 \\ -1 \end{bmatrix}, \quad \mathbf{b} = \begin{bmatrix} 2 \\ -3 \end{bmatrix}, \quad \mathbf{c} = \begin{bmatrix} -2 \\ 2 \end{bmatrix}. \tag{14}$$

$$\mathbf{d} = \begin{bmatrix} 1 \\ 0 \end{bmatrix}, \quad \mathbf{e} = \begin{bmatrix} 0 \\ 0 \end{bmatrix}$$

In Exercises 1–11, either show that $Sp(S) = R^2$ or give an algebraic specification for $Sp(S)$. If $Sp(S) \neq R^2$, then give a geometric description of $Sp(S)$.

1. $S = \{\mathbf{a}\}$
2. $S = \{\mathbf{b}\}$
3. $S = \{\mathbf{e}\}$
4. $S = \{\mathbf{a}, \mathbf{b}\}$
5. $S = \{\mathbf{a}, \mathbf{d}\}$
6. $S = \{\mathbf{a}, \mathbf{c}\}$
7. $S = \{\mathbf{b}, \mathbf{e}\}$
8. $S = \{\mathbf{a}, \mathbf{b}, \mathbf{d}\}$
9. $S = \{\mathbf{b}, \mathbf{c}, \mathbf{d}\}$
10. $S = \{\mathbf{a}, \mathbf{b}, \mathbf{e}\}$
11. $S = \{\mathbf{a}, \mathbf{c}, \mathbf{e}\}$

Exercises 12–19 refer to the vectors in (15).

$$\mathbf{v} = \begin{bmatrix} 1 \\ 2 \\ 0 \end{bmatrix}, \quad \mathbf{w} = \begin{bmatrix} 0 \\ -1 \\ 1 \end{bmatrix}, \quad \mathbf{x} = \begin{bmatrix} 1 \\ 1 \\ -1 \end{bmatrix},$$

$$\mathbf{y} = \begin{bmatrix} -2 \\ -2 \\ 2 \end{bmatrix}, \quad \mathbf{z} = \begin{bmatrix} 1 \\ 0 \\ 2 \end{bmatrix} \tag{15}$$

In Exercises 12–19, either show that $Sp(S) = R^3$ or give an algebraic specification for $Sp(S)$. If $Sp(S) \neq R^3$, then give a geometric description of $Sp(S)$.

12. $S = \{\mathbf{v}\}$

13. $S = \{\mathbf{w}\}$

14. $S = \{\mathbf{v}, \mathbf{w}\}$

15. $S = \{\mathbf{v}, \mathbf{x}\}$

16. $S = \{\mathbf{v}, \mathbf{w}, \mathbf{x}\}$

17. $S = \{\mathbf{w}, \mathbf{x}, \mathbf{z}\}$

18. $S = \{\mathbf{v}, \mathbf{w}, \mathbf{z}\}$

19. $S = \{\mathbf{w}, \mathbf{x}, \mathbf{y}\}$

20. Let S be the set given in Exercise 14. For each vector given below, determine whether the vector is in $Sp(S)$. Express those vectors that are in $Sp(S)$ as a linear combination of $\mathbf{v}$ and $\mathbf{w}$.

a) $\begin{bmatrix} 1 \\ 1 \\ 1 \end{bmatrix}$ b) $\begin{bmatrix} 1 \\ 1 \\ -1 \end{bmatrix}$ c) $\begin{bmatrix} 1 \\ 2 \\ 0 \end{bmatrix}$

d) $\begin{bmatrix} 2 \\ 3 \\ 1 \end{bmatrix}$ e) $\begin{bmatrix} -1 \\ 2 \\ 4 \end{bmatrix}$ f) $\begin{bmatrix} 1 \\ 1 \\ 3 \end{bmatrix}$

21. Repeat Exercise 20 for the set S given in Exercise 15.

22. Determine which of the vectors listed in (14) is in the null space of the matrix

$$A = \begin{bmatrix} 2 & 2 \\ 3 & 3 \end{bmatrix}.$$

23. Determine which of the vectors listed in (14) is in the null space of the matrix

$$A = \begin{bmatrix} 0 & 1 \\ 0 & 2 \\ 0 & 3 \end{bmatrix}.$$

24. Determine which of the vectors listed in (15) is in the null space of the matrix

$$A = [-2 \quad 1 \quad 1].$$

25. Determine which of the vectors listed in (15) is in the null space of the matrix

$$A = \begin{bmatrix} 1 & -1 & 0 \\ 2 & -1 & 1 \\ 3 & -5 & -2 \end{bmatrix}.$$

In Exercises 26–37, give an algebraic specification for the null space and the range of the given matrix A.

26. $A = \begin{bmatrix} 1 & -2 \\ -3 & 6 \end{bmatrix}$ 27. $A = \begin{bmatrix} -1 & 3 \\ 2 & -6 \end{bmatrix}$

28. $A = \begin{bmatrix} 1 & 1 \\ 1 & 2 \end{bmatrix}$ 29. $A = \begin{bmatrix} 1 & 1 \\ 2 & 5 \end{bmatrix}$

30. $A = \begin{bmatrix} 1 & -1 & 2 \\ 2 & -1 & 5 \end{bmatrix}$ 31. $A = \begin{bmatrix} 1 & 2 & 1 \\ 3 & 6 & 4 \end{bmatrix}$

32. $A = \begin{bmatrix} 1 & 3 \\ 2 & 7 \\ 1 & 5 \end{bmatrix}$ 33. $A = \begin{bmatrix} 0 & 1 \\ 0 & 2 \\ 0 & 3 \end{bmatrix}$

34. $A = \begin{bmatrix} 1 & -2 & 1 \\ 2 & -3 & 5 \\ 1 & 0 & 7 \end{bmatrix}$ 35. $A = \begin{bmatrix} 1 & 2 & 3 \\ 1 & 3 & 1 \\ 2 & 2 & 10 \end{bmatrix}$

36. $A = \begin{bmatrix} 1 & 0 & -1 \\ -1 & 1 & 2 \\ 1 & 2 & 2 \end{bmatrix}$ 37. $A = \begin{bmatrix} 1 & 2 & 1 \\ 2 & 5 & 4 \\ 1 & 3 & 4 \end{bmatrix}$

38. Let A be the matrix given in Exercise 26.
   a) For each vector $\mathbf{b}$ given below, determine whether $\mathbf{b}$ is in $\mathcal{R}(A)$.
   b) If $\mathbf{b}$ is in $\mathcal{R}(A)$, then exhibit a vector $\mathbf{x}$ in R^2 such that $A\mathbf{x} = \mathbf{b}$.
   c) If $\mathbf{b}$ is in $\mathcal{R}(A)$, then write $\mathbf{b}$ as a linear combination of the columns of A.

   i) $\mathbf{b} = \begin{bmatrix} 1 \\ -3 \end{bmatrix}$ ii) $\mathbf{b} = \begin{bmatrix} -1 \\ 2 \end{bmatrix}$

   iii) $\mathbf{b} = \begin{bmatrix} 1 \\ 1 \end{bmatrix}$ iv) $\mathbf{b} = \begin{bmatrix} -2 \\ 6 \end{bmatrix}$

   v) $\mathbf{b} = \begin{bmatrix} 3 \\ -6 \end{bmatrix}$ vi) $\mathbf{b} = \begin{bmatrix} 0 \\ 0 \end{bmatrix}$

39. Repeat Exercise 38 for the matrix A given in Exercise 27.

40. Let A be the matrix given in Exercise 34.
   a) For each vector $\mathbf{b}$ given below, determine whether $\mathbf{b}$ is in $\mathcal{R}(A)$.
   b) If $\mathbf{b}$ is in $\mathcal{R}(A)$, then exhibit a vector $\mathbf{x}$ in R^3 such that $A\mathbf{x} = \mathbf{b}$.
   c) If $\mathbf{b}$ is in $\mathcal{R}(A)$, then write $\mathbf{b}$ as a linear combination of the columns of A.

   i) $\mathbf{b} = \begin{bmatrix} 1 \\ 2 \\ 0 \end{bmatrix}$ ii) $\mathbf{b} = \begin{bmatrix} 1 \\ 1 \\ -1 \end{bmatrix}$

   iii) $\mathbf{b} = \begin{bmatrix} 4 \\ 7 \\ 2 \end{bmatrix}$ iv) $\mathbf{b} = \begin{bmatrix} 0 \\ 1 \\ 2 \end{bmatrix}$

   v) $\mathbf{b} = \begin{bmatrix} 0 \\ 1 \\ -2 \end{bmatrix}$ vi) $\mathbf{b} = \begin{bmatrix} 0 \\ 0 \\ 0 \end{bmatrix}$

41. Repeat Exercise 40 for the matrix A given in Exercise 35.

42. Let

$$W = \left\{ \mathbf{x} = \begin{bmatrix} 2x_1 - 3x_2 + x_3 \\ -x_1 + 4x_2 - 2x_3 \\ 2x_1 + x_2 + 4x_3 \end{bmatrix} : x_1, x_2, x_3 \text{ real} \right\}.$$

Exhibit a (3×3) matrix A such that $W = \mathcal{R}(A)$. Conclude that W is a subspace of R^3.

43. Let

$$W = \left\{ \mathbf{x} = \begin{bmatrix} x_1 \\ x_2 \\ x_3 \end{bmatrix} : 3x_1 - 4x_2 + 2x_3 = 0 \right\}.$$

Exhibit a (1×3) matrix A such that $W = \mathcal{N}(A)$. Conclude that W is a subspace of R^3.

44. Let S be the set of vectors given in Exercise 16. Exhibit a matrix A such that $\text{Sp}(S) = \mathcal{R}(A)$.

45. Let S be the set of vectors given in Exercise 17. Exhibit a matrix A such that $\text{Sp}(S) = \mathcal{R}(A)$.

In Exercises 46–49, use the technique illustrated in Example 5 to find a set $T = \{\mathbf{w}_1, \mathbf{w}_2\}$ consisting of two vectors such that $\text{Sp}(S) = \text{Sp}(T)$.

46. $S = \left\{ \begin{bmatrix} 1 \\ 0 \\ -1 \end{bmatrix}, \begin{bmatrix} 2 \\ 2 \\ 1 \end{bmatrix}, \begin{bmatrix} 1 \\ 2 \\ 2 \end{bmatrix} \right\}$

47. $S = \left\{ \begin{bmatrix} -2 \\ 1 \\ 3 \end{bmatrix}, \begin{bmatrix} 2 \\ 2 \\ -1 \end{bmatrix}, \begin{bmatrix} -2 \\ 7 \\ 7 \end{bmatrix} \right\}$

48. $S = \left\{ \begin{bmatrix} 1 \\ 0 \\ 1 \end{bmatrix}, \begin{bmatrix} -2 \\ 0 \\ -2 \end{bmatrix}, \begin{bmatrix} 1 \\ 1 \\ 2 \end{bmatrix}, \begin{bmatrix} -2 \\ 3 \\ 1 \end{bmatrix} \right\}$

49. $S = \left\{ \begin{bmatrix} 1 \\ 2 \\ 2 \end{bmatrix}, \begin{bmatrix} 1 \\ 5 \\ 3 \end{bmatrix}, \begin{bmatrix} 0 \\ 6 \\ 2 \end{bmatrix}, \begin{bmatrix} 1 \\ -1 \\ 1 \end{bmatrix} \right\}$

50. Identify the range and the null space for each of the following.
a) The $(n \times n)$ identity matrix
b) The $(n \times n)$ zero matrix
c) Any $(n \times n)$ nonsingular matrix A

51. Let A and B be $(n \times n)$ matrices. Verify that $\mathcal{N}(A) \cap \mathcal{N}(B) \subseteq \mathcal{N}(A + B)$.

52. Let A be an $(m \times r)$ matrix and B an $(r \times n)$ matrix.
a) Show that $\mathcal{N}(B) \subseteq \mathcal{N}(AB)$.
b) Show that $\mathcal{R}(AB) \subseteq \mathcal{R}(A)$.

53. Let W be a subspace of R^n, and let A be an $(m \times n)$ matrix. Let V be the subset of R^m defined by

$$V = \{\mathbf{y} : \mathbf{y} = A\mathbf{x} \text{ for some } \mathbf{x} \text{ in } W\}.$$

Prove that V is a subspace of R^m.

54. Let A be an $(m \times n)$ matrix, and let B be obtained by multiplying the kth row of A by the nonzero scalar c. Prove that A and B have the same row space.

2.4 Bases for Subspaces

Two of the most fundamental concepts of geometry are those of dimension and the use of coordinates to locate a point in space. In this section and the next, we extend these notions to an arbitrary subspace of R^n by introducing the idea of a basis for a subspace. The first part of this section is devoted to developing the definition of a basis, and in the latter part of the section, we present techniques for obtaining bases for the subspaces introduced in Section 2.3. We will consider the concept of dimension in Section 2.5.

An example from R^2 will serve to illustrate the transition from geometry to algebra. We have already seen that each vector $\mathbf{v}$ in R^2,

$$\mathbf{v} = \begin{bmatrix} a \\ b \end{bmatrix}, \tag{1}$$

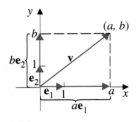

Figure 2.12
$\mathbf{v} = a\mathbf{e}_1 + b\mathbf{e}_2$

can be interpreted geometrically as the point with coordinates a and b. Recall that in R^2 the vectors $\mathbf{e}_1$ and $\mathbf{e}_2$ are defined by

$$\mathbf{e}_1 = \begin{bmatrix} 1 \\ 0 \end{bmatrix} \quad \text{and} \quad \mathbf{e}_2 = \begin{bmatrix} 0 \\ 1 \end{bmatrix}.$$

Clearly the vector $\mathbf{v}$ in (1) can be expressed uniquely as a linear combination of $\mathbf{e}_1$ and $\mathbf{e}_2$:

$$\mathbf{v} = a\mathbf{e}_1 + b\mathbf{e}_2. \tag{2}$$

As we will see later, the set $\{\mathbf{e}_1, \mathbf{e}_2\}$ is an example of a *basis* for R^2 (indeed, it is called the *natural basis* for R^2). In Eq. (2), the vector $\mathbf{v}$ is determined by the coefficients a and b (see Fig. 2.12). Thus the geometric concept of characterizing a point by its coordinates can be interpreted algebraically as determining a vector by its coefficients when the vector is expressed as a linear combination of "basis" vectors. (In fact, the coefficients obtained are often referred to as the coordinates of the vector. This idea will be developed further in Chapter 4.) We turn now to the task of making these ideas precise in the context of an arbitrary subspace W of R^n.

Spanning Sets

Let W be a subspace of R^n, and let S be a subset of W. The discussion above suggests that the first requirement for S to be a basis for W is that each vector in W be expressible as a linear combination of the vectors in S. This leads to the following definition.

DEFINITION 3   Let W be a subspace of R^n, and let $S = \{\mathbf{w}_1, \ldots, \mathbf{w}_m\}$ be a subset of W. We say that S is a *spanning set* for W, or simply that S *spans* W, if every vector $\mathbf{w}$ in W can be expressed as a linear combination of vectors in S;

$$\mathbf{w} = a_1\mathbf{w}_1 + \cdots + a_m\mathbf{w}_m.$$

A restatement of Definition 3 in the notation of the previous section is that S is a spanning set for W provided that $\text{Sp}(S) = W$. It is evident that the set $S = \{\mathbf{e}_1, \mathbf{e}_2, \mathbf{e}_3\}$, consisting of the unit vectors in R^3, is a spanning set for R^3. Specifically, if $\mathbf{v}$ is in R^3,

$$\mathbf{v} = \begin{bmatrix} a \\ b \\ c \end{bmatrix}, \tag{3}$$

then $\mathbf{v} = a\mathbf{e}_1 + b\mathbf{e}_2 + c\mathbf{e}_3$. The next two examples consider other subsets of R^3.

EXAMPLE 1 In R^3, let $S = \{\mathbf{u}_1, \mathbf{u}_2, \mathbf{u}_3\}$, where

$$\mathbf{u}_1 = \begin{bmatrix} 1 \\ -1 \\ 0 \end{bmatrix}, \quad \mathbf{u}_2 = \begin{bmatrix} -2 \\ 3 \\ 1 \end{bmatrix}, \quad \text{and} \quad \mathbf{u}_3 = \begin{bmatrix} 1 \\ 2 \\ 4 \end{bmatrix}.$$

Determine whether S is a spanning set for R^3.

SOLUTION We must determine whether an arbitrary vector $\mathbf{v}$ in R^3 can be expressed as a linear combination of $\mathbf{u}_1$, $\mathbf{u}_2$, and $\mathbf{u}_3$. In other words, we must decide whether the vector equation

$$x_1\mathbf{u}_1 + x_2\mathbf{u}_2 + x_3\mathbf{u}_3 = \mathbf{v}, \tag{4}$$

where $\mathbf{v}$ is the vector in (3), always has a solution. The vector equation (4) is equivalent to the (3×3) linear system with the matrix equation

$$A\mathbf{x} = \mathbf{v}, \tag{5}$$

where A is the (3×3) matrix $A = [\mathbf{u}_1, \mathbf{u}_2, \mathbf{u}_3]$. The augmented matrix for Eq. (5) is

$$[A \,|\, \mathbf{v}] = \begin{bmatrix} 1 & -2 & 1 & a \\ -1 & 3 & 2 & b \\ 0 & 1 & 4 & c \end{bmatrix},$$

and this matrix is row equivalent to

$$\begin{bmatrix} 1 & -2 & 1 & a \\ 0 & 1 & 3 & a+b \\ 0 & 0 & 1 & -a-b+c \end{bmatrix}.$$

Backsolving the resulting system of equations yields

$$x_1 = 10a + 9b - 7c$$
$$x_2 = 4a + 4b - 3c$$
$$x_3 = -a - b + c$$

as the solution to Eq. (4). In particular, Eq. (4) always has a solution, so S is a spanning set for R^3. □

EXAMPLE 2 Let $S = \{\mathbf{v}_1, \mathbf{v}_2, \mathbf{v}_3\}$ be the subset of R^3 defined by

$$\mathbf{v}_1 = \begin{bmatrix} 1 \\ 2 \\ 3 \end{bmatrix}, \quad \mathbf{v}_2 = \begin{bmatrix} -1 \\ 0 \\ -7 \end{bmatrix}, \quad \text{and} \quad \mathbf{v}_3 = \begin{bmatrix} 2 \\ 7 \\ 0 \end{bmatrix}.$$

Does S span R^3?

SOLUTION   Let **v** be the vector given in (3). As before, the vector equation

$$x_1\mathbf{v}_1 + x_2\mathbf{v}_2 + x_3\mathbf{v}_3 = \mathbf{v} \tag{6}$$

is equivalent to the (3×3) system of equations

$$A\mathbf{x} = \mathbf{v}, \tag{7}$$

where $A = [\mathbf{v}_1, \mathbf{v}_2, \mathbf{v}_3]$. The augmented matrix for (7) is

$$[A\,|\,\mathbf{v}] = \begin{bmatrix} 1 & -1 & 2 & a \\ 2 & 0 & 7 & b \\ 3 & -7 & 0 & c \end{bmatrix},$$

and the matrix $[A\,|\,\mathbf{v}]$ is row equivalent to

$$\begin{bmatrix} 1 & -1 & 2 & a \\ 0 & 2 & 3 & -2a + b \\ 0 & 0 & 0 & -7a + 2b + c \end{bmatrix}.$$

It follows that Eq. (6) has a solution if and only if $-7a + 2b + c = 0$. In particular, S does not span R^3. Indeed,

$$\text{Sp}(S) = \{\mathbf{v}: \mathbf{v} = \begin{bmatrix} a \\ b \\ c \end{bmatrix}, \quad \text{where } -7a + 2b + c = 0\}.$$

For example, the vector

$$\mathbf{w} = \begin{bmatrix} 1 \\ 1 \\ 1 \end{bmatrix}$$

is in R^3 but is not in $\text{Sp}(S)$; that is, **w** cannot be expressed as a linear combination of $\mathbf{v}_1$, $\mathbf{v}_2$, and $\mathbf{v}_3$.   □

The next example illustrates a procedure for constructing a spanning set for the null space, $\mathcal{N}(A)$, of a matrix A.

EXAMPLE 3   Let A be the (3×4) matrix

$$A = \begin{bmatrix} 1 & 1 & 3 & 1 \\ 2 & 1 & 5 & 4 \\ 1 & 2 & 4 & -1 \end{bmatrix}.$$

Exhibit a spanning set for $\mathcal{N}(A)$, the null space of A.

SOLUTION   The first step toward obtaining a spanning set for $\mathcal{N}(A)$ is to obtain an algebraic specification for $\mathcal{N}(A)$ by solving the homogeneous system $A\mathbf{x} = \mathbf{0}$. For the given matrix A, this was done in Example 2 of Section 2.3.

Specifically,

$$\mathcal{N}(A) = \left\{ \mathbf{x}: \mathbf{x} = \begin{bmatrix} -2x_3 - 3x_4 \\ -x_3 + 2x_4 \\ x_3 \\ x_4 \end{bmatrix}, \quad x_3 \text{ and } x_4 \text{ any real numbers} \right\}.$$

Thus a vector $\mathbf{x}$ in $\mathcal{N}(A)$ is totally determined by the unconstrained parameters x_3 and x_4. Separating those parameters gives a decomposition of $\mathbf{x}$:

$$\mathbf{x} = \begin{bmatrix} -2x_3 - 3x_4 \\ -x_3 + 2x_4 \\ x_3 \\ x_4 \end{bmatrix} = \begin{bmatrix} -2x_3 \\ -x_3 \\ x_3 \\ 0 \end{bmatrix} + \begin{bmatrix} -3x_4 \\ 2x_4 \\ 0 \\ x_4 \end{bmatrix} = x_3 \begin{bmatrix} -2 \\ -1 \\ 1 \\ 0 \end{bmatrix} + x_4 \begin{bmatrix} -3 \\ 2 \\ 0 \\ 1 \end{bmatrix}. \quad \textbf{(8)}$$

Let $\mathbf{u}_1$ and $\mathbf{u}_2$ be the vectors

$$\mathbf{u}_1 = \begin{bmatrix} -2 \\ -1 \\ 1 \\ 0 \end{bmatrix} \quad \text{and} \quad \mathbf{u}_2 = \begin{bmatrix} -3 \\ 2 \\ 0 \\ 1 \end{bmatrix}.$$

By setting $x_3 = 1$ and $x_4 = 0$ in Eq. (8), we obtain $\mathbf{u}_1$, so $\mathbf{u}_1$ is in $\mathcal{N}(A)$. Similarly, $\mathbf{u}_2$ can be obtained by setting $x_3 = 0$ and $x_4 = 1$, so $\mathbf{u}_2$ is in $\mathcal{N}(A)$. Moreover, it is an immediate consequence of Eq. (8) that each vector $\mathbf{x}$ in $\mathcal{N}(A)$ is a linear combination of $\mathbf{u}_1$ and $\mathbf{u}_2$. Therefore, $\mathcal{N}(A) = \text{Sp}\{\mathbf{u}_1, \mathbf{u}_2\}$; that is, $\{\mathbf{u}_1, \mathbf{u}_2\}$ is a spanning set for $\mathcal{N}(A)$. □

The remaining subspaces introduced in Section 2.3 were either defined or characterized by a spanning set. If $S = \{\mathbf{v}_1, \ldots, \mathbf{v}_r\}$ is a subset of R^n, for instance, then obviously S is a spanning set for $\text{Sp}(S)$. If A is an $(m \times n)$ matrix,

$$A = [\mathbf{A}_1, \ldots, \mathbf{A}_n],$$

then, as we saw in Section 2.3, $\{\mathbf{A}_1, \ldots, \mathbf{A}_n\}$ is a spanning set for $\mathcal{R}(A)$, the range of A. Finally, if

$$A = \begin{bmatrix} \mathbf{a}_1 \\ \mathbf{a}_2 \\ \vdots \\ \mathbf{a}_m \end{bmatrix},$$

where $\mathbf{a}_i$ is the ith-row vector of A, then, by definition, $\{\mathbf{a}_1, \ldots, \mathbf{a}_m\}$ is a spanning set for the row space of A.

Minimal Spanning Sets

If W is a subspace of R^n, $W \neq \{\boldsymbol{\theta}\}$, then spanning sets for W abound. For example, a vector $\mathbf{v}$ in a spanning set can always be replaced by $a\mathbf{v}$, where a is any nonzero scalar. It is easy to demonstrate, however, that not all spanning sets are equally desirable. For example, define **u** in R^2 by

$$\mathbf{u} = \begin{bmatrix} 1 \\ 1 \end{bmatrix}.$$

The set $S = \{\mathbf{e}_1, \mathbf{e}_2, \mathbf{u}\}$ is a spanning set for R^2. Indeed, for an arbitrary vector **v** in R^2,

$$\mathbf{v} = \begin{bmatrix} a \\ b \end{bmatrix},$$

$\mathbf{v} = (a - c)\mathbf{e}_1 + (b - c)\mathbf{e}_2 + c\mathbf{u}$, where c is any real number whatsoever. But the subset $\{\mathbf{e}_1, \mathbf{e}_2\}$ already spans R^2, so the vector **u** is unnecessary.

Recall that a set $\{\mathbf{v}_1, \ldots, \mathbf{v}_m\}$ of vectors in R^n is linearly independent if the vector equation

$$x_1\mathbf{v}_1 + \cdots + x_m\mathbf{v}_m = \boldsymbol{\theta} \tag{9}$$

has only the trivial solution $x_1 = \cdots = x_m = 0$; if Eq. (9) has a nontrivial solution, then the set is linearly dependent. The set $S = \{\mathbf{e}_1, \mathbf{e}_2, \mathbf{u}\}$ is linearly dependent because

$$\mathbf{e}_1 + \mathbf{e}_2 - \mathbf{u} = \boldsymbol{\theta}.$$

Our next example illustrates that a linearly dependent set is not an efficient spanning set; that is, fewer vectors will span the same space.

EXAMPLE 4 Let $S = \{\mathbf{v}_1, \mathbf{v}_2, \mathbf{v}_3\}$ be a subset of R^3, where

$$\mathbf{v}_1 = \begin{bmatrix} 1 \\ 1 \\ 1 \end{bmatrix}, \quad \mathbf{v}_2 = \begin{bmatrix} 2 \\ 3 \\ 1 \end{bmatrix}, \quad \text{and} \quad \mathbf{v}_3 = \begin{bmatrix} 3 \\ 5 \\ 1 \end{bmatrix}.$$

Show that S is a linearly dependent set, and exhibit a subset T of S such that T contains only two vectors but $\text{Sp}(T) = \text{Sp}(S)$.

SOLUTION The vector equation

$$x_1\mathbf{v}_1 + x_2\mathbf{v}_2 + x_3\mathbf{v}_3 = \boldsymbol{\theta} \tag{10}$$

is equivalent to the (3×3) homogeneous system of equations with augmented matrix

$$A = \begin{bmatrix} 1 & 2 & 3 & 0 \\ 1 & 3 & 5 & 0 \\ 1 & 1 & 1 & 0 \end{bmatrix}.$$

Matrix A is row equivalent to

$$B = \begin{bmatrix} 1 & 2 & 3 & 0 \\ 0 & 1 & 2 & 0 \\ 0 & 0 & 0 & 0 \end{bmatrix}$$

in echelon form. Backsolving the system with augmented matrix B gives

$$x_1 = \quad x_3$$
$$x_2 = -2x_3 .$$

Because Eq. (10) has nontrivial solutions, the set S is linearly dependent. Taking $x_3 = 1$, for example, gives $x_1 = 1$, $x_2 = -2$. Therefore,

$$\mathbf{v}_1 - 2\mathbf{v}_2 + \mathbf{v}_3 = \boldsymbol{\theta} . \tag{11}$$

Equation (11) allows us to express $\mathbf{v}_3$ as a linear combination of $\mathbf{v}_1$ and $\mathbf{v}_2$:

$$\mathbf{v}_3 = -\mathbf{v}_1 + 2\mathbf{v}_2 .$$

(Note that we could just as easily have solved Eq. (11) for either $\mathbf{v}_1$ or $\mathbf{v}_2$.) It now follows that

$$\text{Sp}\{\mathbf{v}_1, \mathbf{v}_2\} = \text{Sp}\{\mathbf{v}_1, \mathbf{v}_2, \mathbf{v}_3\} .$$

To illustrate, let $\mathbf{v}$ be in the subspace $\text{Sp}\{\mathbf{v}_1, \mathbf{v}_2, \mathbf{v}_3\}$:

$$\mathbf{v} = a_1\mathbf{v}_1 + a_2\mathbf{v}_2 + a_3\mathbf{v}_3 .$$

Making the substitution $\mathbf{v}_3 = -\mathbf{v}_1 + 2\mathbf{v}_2$ yields

$$\mathbf{v} = a_1\mathbf{v}_1 + a_2\mathbf{v}_2 + a_3(-\mathbf{v}_1 + 2\mathbf{v}_2) .$$

This expression simplifies to

$$\mathbf{v} = (a_1 - a_3)\mathbf{v}_1 + (a_2 + 2a_3)\mathbf{v}_2 ;$$

in particular, $\mathbf{v}$ is in $\text{Sp}\{\mathbf{v}_1, \mathbf{v}_2\}$. Clearly any linear combination of $\mathbf{v}_1$ and $\mathbf{v}_2$ is in $\text{Sp}(S)$ because

$$b_1\mathbf{v}_1 + b_2\mathbf{v}_2 = b_1\mathbf{v}_1 + b_2\mathbf{v}_2 + 0\mathbf{v}_3 .$$

Thus if $T = \{\mathbf{v}_1, \mathbf{v}_2\}$, then $\text{Sp}(T) = \text{Sp}(S)$. ◻

The lesson to be drawn from Example 4 is that a linearly dependent spanning set contains redundant information. That is, if $S = \{\mathbf{w}_1, \ldots, \mathbf{w}_r\}$ is a linearly dependent spanning set for a subspace W, then at least one vector from S is a linear combination of the other $r - 1$ vectors and can be discarded from S to produce a smaller spanning set. On the other hand, if $B = \{\mathbf{v}_1, \ldots, \mathbf{v}_m\}$ is a linearly independent spanning set for W, then no vector in B is a linear combination of the other $m - 1$ vectors in B. Hence if a vector is removed from B, this smaller set cannot be a spanning set for W (in particular, the vector removed from B is in W but cannot be expressed as a linear combination of the vectors retained). In this sense a linearly independent spanning set is a minimal spanning set and hence represents the most efficient way of characterizing the subspace. This idea leads to the following definition.

DEFINITION 4 Let W be a nonzero subspace of R^n. A *basis* for W is a linearly independent spanning set for W.

Note that the zero subspace of R^n, $W = \{\boldsymbol{\theta}\}$, contains only the vector $\boldsymbol{\theta}$. Although it is the case that $\{\boldsymbol{\theta}\}$ is a spanning set for W, the set $\{\boldsymbol{\theta}\}$ is linearly dependent. Thus the concept of a basis is not meaningful for $W = \{\boldsymbol{\theta}\}$.

Uniqueness of Representation

Let $B = \{\mathbf{v}_1, \mathbf{v}_2, \ldots, \mathbf{v}_p\}$ be a basis for a subspace W of R^n, and let $\mathbf{x}$ be a vector in W. Because B is a spanning set, we know that there are scalars $a_1, a_2, \ldots, a_p$ such that

$$\mathbf{x} = a_1\mathbf{v}_1 + a_2\mathbf{v}_2 + \cdots + a_p\mathbf{v}_p. \tag{12}$$

Because B is also a linearly independent set, we can show that the representation of $\mathbf{x}$ in Eq. (12) is unique. That is, if we have any representation of the form $\mathbf{x} = b_1\mathbf{v}_1 + b_2\mathbf{v}_2 + \cdots + b_p\mathbf{v}_p$, then $a_1 = b_1$, $a_2 = b_2, \ldots, a_p = b_p$. To establish this uniqueness, suppose that $b_1, b_2, \ldots, b_p$ are any scalars such that

$$\mathbf{x} = b_1\mathbf{v}_1 + b_2\mathbf{v}_2 + \cdots + b_p\mathbf{v}_p.$$

Subtracting the equation above from (12), we obtain

$$\boldsymbol{\theta} = (a_1 - b_1)\mathbf{v}_1 + (a_2 - b_2)\mathbf{v}_2 + \cdots + (a_p - b_p)\mathbf{v}_p.$$

Then, using the fact that $\{\mathbf{v}_1, \mathbf{v}_2, \ldots, \mathbf{v}_p\}$ is linearly independent, we see that $a_1 - b_1 = 0$, $a_2 - b_2 = 0, \ldots, a_p - b_p = 0$. This discussion of uniqueness leads to the following remark.

REMARK Let $B = \{\mathbf{v}_1, \mathbf{v}_2, \ldots, \mathbf{v}_p\}$ be a basis for W, where W is a subspace of R^n. If $\mathbf{x}$ is in W, then $\mathbf{x}$ can be represented uniquely in terms of the basis B. That is, there are unique scalars $a_1, a_2, \ldots, a_p$ such that

$$\mathbf{x} = a_1\mathbf{v}_1 + a_2\mathbf{v}_2 + \cdots + a_p\mathbf{v}_p.$$

Examples of Bases

It is easy to show that the unit vectors

$$\mathbf{e}_1 = \begin{bmatrix} 1 \\ 0 \\ 0 \end{bmatrix}, \quad \mathbf{e}_2 = \begin{bmatrix} 0 \\ 1 \\ 0 \end{bmatrix}, \quad \text{and} \quad \mathbf{e}_3 = \begin{bmatrix} 0 \\ 0 \\ 1 \end{bmatrix}$$

constitute a basis for R^3. In general, the n-dimensional vectors $\mathbf{e}_1, \mathbf{e}_2, \ldots, \mathbf{e}_n$ form a basis for R^n, frequently called the natural basis.

In Exercise 30, the reader is asked to use Theorem 13 of Section 1.8 to prove that any linearly independent subset $B = \{\mathbf{v}_1, \mathbf{v}_2, \mathbf{v}_3\}$ of R^3 is actually a basis for

R^3. Thus, for example, the vectors

$$\mathbf{v}_1 = \begin{bmatrix} 1 \\ 0 \\ 0 \end{bmatrix}, \quad \mathbf{v}_2 = \begin{bmatrix} 1 \\ 1 \\ 0 \end{bmatrix}, \quad \text{and} \quad \mathbf{v}_3 = \begin{bmatrix} 1 \\ 1 \\ 1 \end{bmatrix}$$

provide another basis for R^3.

In Example 3, a procedure for determining a spanning set for $\mathcal{N}(A)$, the null space of a matrix A, was illustrated. Note in Example 3 that the spanning set $\{\mathbf{u}_1, \mathbf{u}_2\}$ obtained is linearly independent, so it is a basis for $\mathcal{N}(A)$. More generally, if a subspace W of R^n has an algebraic specification in terms of unconstrained variables, the procedure illustrated in Example 3 yields a basis for W. The next example provides another illustration.

EXAMPLE 5 Let A be the (3×4) matrix given in Example 4 of Section 2.3. Use the algebraic specification of $\mathcal{R}(A)$ derived in that example to obtain a basis for $\mathcal{R}(A)$.

SOLUTION In Example 4 of Section 2.3, the range of A was determined to be

$$\mathcal{R}(A) = \{\mathbf{b}: \mathbf{b} = \begin{bmatrix} b_1 \\ b_2 \\ 3b_1 - b_2 \end{bmatrix}, \quad b_1 \text{ and } b_2 \text{ any real numbers}\}.$$

Thus b_1 and b_2 are unconstrained variables, and a vector $\mathbf{b}$ in $\mathcal{R}(A)$ can be decomposed as

$$\mathbf{b} = \begin{bmatrix} b_1 \\ b_2 \\ 3b_1 - b_2 \end{bmatrix} = \begin{bmatrix} b_1 \\ 0 \\ 3b_1 \end{bmatrix} + \begin{bmatrix} 0 \\ b_2 \\ -b_2 \end{bmatrix} = b_1 \begin{bmatrix} 1 \\ 0 \\ 3 \end{bmatrix} + b_2 \begin{bmatrix} 0 \\ 1 \\ -1 \end{bmatrix}. \tag{13}$$

If $\mathbf{u}_1$ and $\mathbf{u}_2$ are defined by

$$\mathbf{u}_1 = \begin{bmatrix} 1 \\ 0 \\ 3 \end{bmatrix} \quad \text{and} \quad \mathbf{u} = \begin{bmatrix} 0 \\ 1 \\ -1 \end{bmatrix},$$

then $\mathbf{u}_1$ and $\mathbf{u}_2$ are in $\mathcal{R}(A)$. One can easily check that $\{\mathbf{u}_1, \mathbf{u}_2\}$ is a linearly independent set, and it is evident from (13) that $\mathcal{R}(A)$ is spanned by $\mathbf{u}_1$ and $\mathbf{u}_2$. Therefore, $\{\mathbf{u}_1, \mathbf{u}_2\}$ is a basis for $\mathcal{R}(A)$.

The previous example illustrates how to obtain a basis for a subspace W, given an algebraic specification for W. The last two examples of this section illustrate two different techniques for constructing a basis for W from a spanning set.

EXAMPLE 6 Let W be the subspace of R^4 spanned by the set $S = \{v_1, v_2, v_3, v_4, v_5\}$, where

$$v_1 = \begin{bmatrix} 1 \\ 1 \\ 2 \\ -1 \end{bmatrix}, \quad v_2 = \begin{bmatrix} 1 \\ 2 \\ 1 \\ 1 \end{bmatrix}, \quad v_3 = \begin{bmatrix} 1 \\ 4 \\ -1 \\ 5 \end{bmatrix},$$

$$v_4 = \begin{bmatrix} 1 \\ 0 \\ 4 \\ -1 \end{bmatrix}, \quad \text{and} \quad v_5 = \begin{bmatrix} 2 \\ 5 \\ 0 \\ 2 \end{bmatrix}.$$

Find a subset of S that is a basis for W.

SOLUTION The procedure is suggested by Example 4. The idea is to solve the dependence relation

$$x_1 v_1 + x_2 v_2 + x_3 v_3 + x_4 v_4 + x_5 v_5 = \theta \tag{14}$$

and then determine which of the v_j's can be eliminated. If V is the (4×5) matrix

$$V = [v_1, v_2, v_3, v_4, v_5],$$

then the augmented matrix $[V \,|\, \theta]$ reduces to

$$\begin{bmatrix} 1 & 1 & 1 & 1 & 2 & 0 \\ 0 & 1 & 3 & -1 & 3 & 0 \\ 0 & 0 & 0 & 1 & -1 & 0 \\ 0 & 0 & 0 & 0 & 0 & 0 \end{bmatrix} \tag{15}$$

The system of equations with augmented matrix (15) has solution

$$\begin{aligned} x_1 &= 2x_3 - x_5 \\ x_2 &= -3x_3 - 2x_5 \\ x_4 &= x_5, \end{aligned} \tag{16}$$

where x_3 and x_5 are unconstrained variables. In particular, the set S is linearly dependent. Moreover, taking $x_3 = 1$ and $x_5 = 0$ yields $x_1 = 2$, $x_2 = -3$, and $x_4 = 0$. Thus Eq. (14) becomes

$$2v_1 - 3v_2 + v_3 = \theta. \tag{17}$$

Since Eq. (17) can be solved for v_3,

$$v_3 = -2v_1 + 3v_2,$$

it follows that v_3 is redundant and can be removed from the spanning set. Similarly, setting $x_3 = 0$ and $x_5 = 1$ gives $x_1 = -1$, $x_2 = -2$, and $x_4 = 1$. In this case Eq. (14) becomes

$$-v_1 - 2v_2 + v_4 + v_5 = \theta,$$

and hence

$$\mathbf{v}_5 = \mathbf{v}_1 + 2\mathbf{v}_2 - \mathbf{v}_4.$$

Since both $\mathbf{v}_3$ and $\mathbf{v}_5$ are in $\mathrm{Sp}\{\mathbf{v}_1,\mathbf{v}_2,\mathbf{v}_4\}$, it follows (as in Example 4) that $\mathbf{v}_1$, $\mathbf{v}_2$, and $\mathbf{v}_4$ span W.

To see that the set $\{\mathbf{v}_1,\mathbf{v}_2,\mathbf{v}_4\}$ is linearly independent, note that the dependence relation

$$x_1\mathbf{v}_1 + x_2\mathbf{v}_2 + x_4\mathbf{v}_4 = \boldsymbol{\theta} \tag{18}$$

is just Eq. (14) with $\mathbf{v}_3$ and $\mathbf{v}_5$ removed. Thus the augmented matrix, $[\mathbf{v}_1,\mathbf{v}_2,\mathbf{v}_4\,|\,\boldsymbol{\theta}]$, for Eq. (18) reduces to

$$\begin{bmatrix} 1 & 1 & 1 & 0 \\ 0 & 1 & -1 & 0 \\ 0 & 0 & 1 & 0 \\ 0 & 0 & 0 & 0 \end{bmatrix}, \tag{19}$$

which is matrix (15) with the third and fifth columns removed. From (19) it is clear that (18) has only the trivial solution, so $\{\mathbf{v}_1,\mathbf{v}_2,\mathbf{v}_4\}$ is a linearly independent set and therefore a basis for W. ☐

The procedure demonstrated in the preceding example can be outlined as follows:

1. A spanning set $S = \{\mathbf{v}_1,\ldots,\mathbf{v}_m\}$ for a subspace W is given.
2. Solve the vector equation

$$x_1\mathbf{v}_1 + \cdots + x_m\mathbf{v}_m = \boldsymbol{\theta}. \tag{20}$$

3. If Eq. (20) has only the trivial solution $x_1 = \cdots = x_m = 0$, then S is a linearly independent set and hence is a basis for W.
4. If Eq. (20) has nontrivial solutions, then there are unconstrained variables. For each x_j that is designated as an unconstrained variable, delete the vector $\mathbf{v}_j$ from the set S. The remaining vectors constitute a basis for W.

Our final technique for constructing a basis uses Theorem 7.

THEOREM 7

If the nonzero matrix A is row equivalent to the matrix B in echelon form, then the nonzero rows of B form a basis for the row space of A.

PROOF By Theorem 6, A and B have the same row space. It follows that the nonzero rows of B span the row space of A. Since the nonzero rows of an echelon matrix are linearly independent vectors, it follows that the nonzero rows of B form a basis for the row space of A. ☐

EXAMPLE 7   Let W be the subspace of R^4 given in Example 6. Use Theorem 7 to construct a basis for W.

SOLUTION   As in Example 6, let V be the (4×5) matrix

$$V = [\mathbf{v}_1, \mathbf{v}_2, \mathbf{v}_3, \mathbf{v}_4, \mathbf{v}_5].$$

Thus W can be viewed as the row space of the matrix V^T, where

$$V^T = \begin{bmatrix} 1 & 1 & 2 & -1 \\ 1 & 2 & 1 & 1 \\ 1 & 4 & -1 & 5 \\ 1 & 0 & 4 & -1 \\ 2 & 5 & 0 & 2 \end{bmatrix}.$$

Since V^T is row equivalent to the matrix

$$B^T = \begin{bmatrix} 1 & 1 & 2 & -1 \\ 0 & 1 & -1 & 2 \\ 0 & 0 & 1 & 2 \\ 0 & 0 & 0 & 0 \\ 0 & 0 & 0 & 0 \end{bmatrix},$$

in echelon form, it follows from Theorem 7 that the nonzero rows of B^T form a basis for the row space of V^T. Consequently the nonzero columns of

$$B = \begin{bmatrix} 1 & 0 & 0 & 0 & 0 \\ 1 & 1 & 0 & 0 & 0 \\ 2 & -1 & 1 & 0 & 0 \\ -1 & 2 & 2 & 0 & 0 \end{bmatrix}$$

are a basis for W. Specifically, the set $\{\mathbf{u}_1, \mathbf{u}_2, \mathbf{u}_3\}$ is a basis of W, where

$$\mathbf{u}_1 = \begin{bmatrix} 1 \\ 1 \\ 2 \\ -1 \end{bmatrix}, \quad \mathbf{u}_2 = \begin{bmatrix} 0 \\ 1 \\ -1 \\ 2 \end{bmatrix}, \quad \text{and} \quad \mathbf{u}_3 = \begin{bmatrix} 0 \\ 0 \\ 1 \\ 2 \end{bmatrix}. \qquad \square$$

The procedure used in the preceding example can be summarized as follows:

1. A spanning set $S = \{\mathbf{v}_1, \ldots, \mathbf{v}_m\}$ for a subspace W of R^n is given.
2. Let V be the $(n \times n)$ matrix $V = [\mathbf{v}_1, \ldots, \mathbf{v}_m]$. Use elementary row operations to transform V^T to a matrix B^T in echelon form.
3. The nonzero columns of B are a basis for W.

Exercises 2.4

In Exercises 1–8, let W be the subspace of R^4 consisting of vectors of the form

$$\mathbf{x} = \begin{bmatrix} x_1 \\ x_2 \\ x_3 \\ x_4 \end{bmatrix}.$$

Find a basis for W when the components of $\mathbf{x}$ satisfy the given conditions.

1. $\begin{aligned} x_1 + x_2 - x_3 \quad &= 0 \\ x_2 \quad - x_4 &= 0 \end{aligned}$

2. $\begin{aligned} x_1 + x_2 - x_3 + x_4 &= 0 \\ x_2 - 2x_3 - x_4 &= 0 \end{aligned}$

3. $x_1 - x_2 + x_3 - 3x_4 = 0$

4. $x_1 - x_2 + x_3 = 0$

5. $x_1 + x_2 = 0$

6. $\begin{aligned} x_1 - x_2 \quad &= 0 \\ x_2 - 2x_3 \quad &= 0 \\ x_3 - x_4 &= 0 \end{aligned}$

7. $\begin{aligned} -x_1 + 2x_2 \quad - x_4 &= 0 \\ x_2 + x_3 \quad &= 0 \end{aligned}$

8. $\begin{aligned} x_1 - x_2 - x_3 + x_4 &= 0 \\ x_2 + x_3 \quad &= 0 \end{aligned}$

9. Let W be the subspace described in Exercise 1. For each vector $\mathbf{x}$ listed below, determine if $\mathbf{x}$ is in W. If $\mathbf{x}$ is in W, then express $\mathbf{x}$ as a linear combination of the basis vectors found in Exercise 1.

a) $\mathbf{x} = \begin{bmatrix} 1 \\ 1 \\ 2 \\ 1 \end{bmatrix}$
b) $\mathbf{x} = \begin{bmatrix} -1 \\ 2 \\ 3 \\ 2 \end{bmatrix}$

c) $\mathbf{x} = \begin{bmatrix} 3 \\ -3 \\ 0 \\ -3 \end{bmatrix}$
d) $\mathbf{x} = \begin{bmatrix} 2 \\ 0 \\ 2 \\ 0 \end{bmatrix}$

10. Let W be the subspace described in Exercise 2. For each vector $\mathbf{x}$ listed below, determine if $\mathbf{x}$ is in W. If $\mathbf{x}$ is in W, then express $\mathbf{x}$ as a linear combination of the basis vectors found in Exercise 2.

a) $\mathbf{x} = \begin{bmatrix} -3 \\ 3 \\ 1 \\ 1 \end{bmatrix}$
b) $\mathbf{x} = \begin{bmatrix} 0 \\ 3 \\ 2 \\ -1 \end{bmatrix}$

c) $\mathbf{x} = \begin{bmatrix} 7 \\ 8 \\ 3 \\ 2 \end{bmatrix}$
d) $\mathbf{x} = \begin{bmatrix} 4 \\ -2 \\ 0 \\ -2 \end{bmatrix}$

In Exercises 11–16:
a) Find a matrix B in echelon form such that B is row equivalent to the given matrix A.
b) Find a basis for the null space of A.
c) As in Example 6, find a basis for the range of A that consists of columns of A. For each column, $\mathbf{A}_j$, of A that does not appear in the basis, express $\mathbf{A}_j$ as a linear combination of the basis vectors.
d) Exhibit a basis for the row space of A.

11. $A = \begin{bmatrix} 1 & 2 & 3 & -1 \\ 3 & 5 & 8 & -2 \\ 1 & 1 & 2 & 0 \end{bmatrix}$
12. $A = \begin{bmatrix} 1 & 1 & 2 \\ 1 & 1 & 2 \\ 2 & 3 & 5 \end{bmatrix}$

13. $A = \begin{bmatrix} 1 & 2 & 1 & 0 \\ 2 & 5 & 3 & -1 \\ 2 & 2 & 0 & 2 \\ 0 & 1 & 1 & -1 \end{bmatrix}$
14. $A = \begin{bmatrix} 2 & 2 & 0 \\ 2 & 1 & 1 \\ 2 & 3 & 0 \end{bmatrix}$

15. $A = \begin{bmatrix} 1 & 2 & 1 \\ 2 & 4 & 1 \\ 3 & 6 & 2 \end{bmatrix}$
16. $A = \begin{bmatrix} 2 & 1 & 2 \\ 2 & 2 & 1 \\ 2 & 3 & 0 \end{bmatrix}$

17. Use the technique illustrated in Example 7 to obtain a basis for the range of A, where A is the matrix given in Exercise 11.

18. Repeat Exercise 17 for the matrix given in Exercise 12.

19. Repeat Exercise 17 for the matrix given in Exercise 13.

20. Repeat Exercise 17 for the matrix given in Exercise 14.

In Exercises 21–24 for the given set S:
a) Find a subset of S that is a basis for $\mathrm{Sp}(S)$ using the technique illustrated in Example 6.

b) Find a basis for Sp(S) using the technique illustrated in Example 7.

21. $S = \left\{ \begin{bmatrix} 1 \\ 2 \end{bmatrix}, \begin{bmatrix} 2 \\ 4 \end{bmatrix} \right\}$

22. $S = \left\{ \begin{bmatrix} 1 \\ 2 \end{bmatrix}, \begin{bmatrix} 2 \\ 1 \end{bmatrix}, \begin{bmatrix} 3 \\ 2 \end{bmatrix} \right\}$

23. $S = \left\{ \begin{bmatrix} 1 \\ 2 \\ 1 \end{bmatrix}, \begin{bmatrix} 2 \\ 5 \\ 0 \end{bmatrix}, \begin{bmatrix} 3 \\ 7 \\ 1 \end{bmatrix}, \begin{bmatrix} 1 \\ 1 \\ 3 \end{bmatrix} \right\}$

24. $S = \left\{ \begin{bmatrix} 1 \\ 2 \\ -1 \\ 3 \end{bmatrix}, \begin{bmatrix} -2 \\ 1 \\ 2 \\ -1 \end{bmatrix}, \begin{bmatrix} -1 \\ -1 \\ 1 \\ -3 \end{bmatrix}, \begin{bmatrix} -2 \\ 2 \\ 2 \\ 0 \end{bmatrix} \right\}$

25. Find a basis for the null space of each of the following matrices.

a) $\begin{bmatrix} 1 & 0 & 0 \\ 1 & 0 & 1 \end{bmatrix}$ b) $\begin{bmatrix} 1 & 1 & 0 \\ 1 & 1 & 0 \end{bmatrix}$

c) $\begin{bmatrix} 1 & 1 & 0 \\ 1 & 1 & 1 \end{bmatrix}$

26. Find a basis for the range of each matrix in Exercise 25.

27. Let $S = \{v_1, v_2, v_3\}$, where

$$v_1 = \begin{bmatrix} 1 \\ 2 \\ 1 \end{bmatrix}, \quad v_2 = \begin{bmatrix} -1 \\ -1 \\ 1 \end{bmatrix}, \quad \text{and} \quad v_3 = \begin{bmatrix} -1 \\ 1 \\ 5 \end{bmatrix}.$$

Show that S is a linearly dependent set, and verify that $\text{Sp}\{v_1, v_2, v_3\} = \text{Sp}\{v_1, v_2\}$.

28. Let $S = \{v_1, v_2, v_3\}$, where

$$v_1 = \begin{bmatrix} 1 \\ 0 \end{bmatrix}, \quad v_2 = \begin{bmatrix} 0 \\ 1 \end{bmatrix}, \quad \text{and} \quad v_3 = \begin{bmatrix} -1 \\ 1 \end{bmatrix}.$$

Find every subset of S that is a basis for R^2.

29. Let $S = \{v_1, v_2, v_3, v_4\}$, where

$$v_1 = \begin{bmatrix} 1 \\ 2 \\ 1 \end{bmatrix}, \quad v_2 = \begin{bmatrix} -1 \\ -1 \\ 1 \end{bmatrix},$$

$$v_3 = \begin{bmatrix} -1 \\ 1 \\ 7 \end{bmatrix}, \quad \text{and} \quad v_4 = \begin{bmatrix} -2 \\ -4 \\ -4 \end{bmatrix}.$$

Find every subset of S that is a basis for R^3.

30. Let $B = \{v_1, v_2, v_3\}$ be a set of linearly independent

vectors in R^3. Prove that B is a basis for R^3. (*Hint:* Use Theorem 13 of Section 1.8 to show that B is a spanning set for R^3.)

31. Let $B = \{v_1, v_2, v_3\}$ be a subset of R^3 such that $\text{Sp}(B) = R^3$. Prove that B is a basis for R^3. (*Hint:* Use Theorem 13 of Section 1.8 to show that B is a linearly independent set.)

In Exercises 32–35, determine whether the given set S is a basis for R^3.

32. $S = \left\{ \begin{bmatrix} 1 \\ -1 \\ -2 \end{bmatrix}, \begin{bmatrix} 1 \\ 1 \\ 2 \end{bmatrix}, \begin{bmatrix} 2 \\ -3 \\ -3 \end{bmatrix} \right\}$

33. $S = \left\{ \begin{bmatrix} 1 \\ 1 \\ -2 \end{bmatrix}, \begin{bmatrix} 2 \\ 5 \\ 2 \end{bmatrix}, \begin{bmatrix} 1 \\ 3 \\ 2 \end{bmatrix} \right\}$

34. $S = \left\{ \begin{bmatrix} 1 \\ -1 \\ -2 \end{bmatrix}, \begin{bmatrix} 1 \\ 1 \\ 2 \end{bmatrix}, \begin{bmatrix} 2 \\ -3 \\ -3 \end{bmatrix}, \begin{bmatrix} 1 \\ 4 \\ 5 \end{bmatrix} \right\}$

35. $S = \left\{ \begin{bmatrix} 1 \\ 1 \\ -2 \end{bmatrix}, \begin{bmatrix} 2 \\ 5 \\ 2 \end{bmatrix} \right\}$

36. Find a vector w in R^3 such that w is not a linear combination of v_1 and v_2:

$$v_1 = \begin{bmatrix} 1 \\ 2 \\ -1 \end{bmatrix} \quad \text{and} \quad v_2 = \begin{bmatrix} 2 \\ -1 \\ -2 \end{bmatrix}.$$

37. Prove that every basis for R^2 contains exactly two vectors. Proceed by showing the following:
a) A basis for R^2 cannot have more than two vectors.
b) A basis for R^2 cannot have one vector. (*Hint:* Suppose that a basis for R^2 could contain one vector. Represent e_1 and e_2 in terms of the basis and obtain a contradiction.)

38. Show that any spanning set for R^n must contain at least n vectors. Proceed by showing that if $u_1, u_2, \ldots, u_p$ are vectors in R^n, and if $p < n$, then there is a nonzero vector v in R^n such that $v^T u_i = 0, 1 \leq i \leq p$. (*Hint:* Write the constraints as a $(p \times n)$ system and use Theorem 4 of Section 1.4.) Given v as above, can v be a linear combination of $u_1, u_2, \ldots, u_p$?

39. Recalling Exercise 38, prove that every basis for R^n contains exactly n vectors.

2.5   Dimension

In this section we translate the geometric concept of dimension into algebraic terms. Clearly R^2 and R^3 have dimension 2 and 3, respectively, since these vector spaces are simply algebraic interpretations of two-space and three-space. It would be natural to extrapolate from these two cases and declare that R^n has dimension n for each positive integer n; indeed, we have earlier referred to elements of R^n as n-dimensional vectors. But if W is a subspace of R^n, how is the dimension of W to be determined? An examination of the subspace, W, of R^3 defined by

$$W = \left\{ \mathbf{x} \colon \mathbf{x} = \begin{bmatrix} x_2 - 2x_3 \\ x_2 \\ x_3 \end{bmatrix}, \quad x_2 \text{ and } x_3 \text{ any real numbers} \right\}$$

suggests a possibility. Geometrically, W is the plane with equation $x = y - 2z$, so naturally the dimension of W is 2. The techniques of the previous section show that W has a basis $\{\mathbf{v}_1, \mathbf{v}_2\}$ consisting of the two vectors

$$\mathbf{v}_1 = \begin{bmatrix} 1 \\ 1 \\ 0 \end{bmatrix} \quad \text{and} \quad \mathbf{v}_2 = \begin{bmatrix} -2 \\ 0 \\ 1 \end{bmatrix}.$$

Thus in this case the dimension of W is equal to the number of vectors in a basis for W.

The Definition of Dimension

More generally, for any subspace W of R^n, we wish to define the dimension of W to be the number of vectors in a basis for W. We have seen, however, that a subspace W may have many different bases. In fact, Exercise 30 of Section 2.4 shows that any set of three linearly independent vectors in R^3 is a basis for R^3. Therefore, for the concept of dimension, as given above, to make sense, we must show that all bases for a given subspace W contain the same number of vectors. This fact will be an easy consequence of the following theorem.

THEOREM 8

Let W be a subspace of R^n, and let $B = \{\mathbf{w}_1, \mathbf{w}_2, \ldots, \mathbf{w}_p\}$ be a spanning set for W containing p vectors. Then any set of $p + 1$ or more vectors in W is linearly dependent.

PROOF   Let $\{\mathbf{s}_1, \mathbf{s}_2, \ldots, \mathbf{s}_m\}$ be any set of m vectors in W, where $m > p$. To show that this set is linearly dependent, we first express each $\mathbf{s}_i$ in terms of the spanning set B:

$$\begin{aligned}
\mathbf{s}_1 &= a_{11}\mathbf{w}_1 + a_{21}\mathbf{w}_2 + \cdots + a_{p1}\mathbf{w}_p \\
\mathbf{s}_2 &= a_{12}\mathbf{w}_1 + a_{22}\mathbf{w}_2 + \cdots + a_{p2}\mathbf{w}_p \\
&\vdots \qquad \vdots \qquad\qquad \vdots \\
\mathbf{s}_m &= a_{1m}\mathbf{w}_1 + a_{2m}\mathbf{w}_2 + \cdots + a_{pm}\mathbf{w}_p.
\end{aligned} \tag{1}$$

To show that $\{\mathbf{s}_1,\mathbf{s}_2,\ldots,\mathbf{s}_m\}$ is linearly dependent, we must show that there is a nontrivial solution of

$$c_1\mathbf{s}_1 + c_2\mathbf{s}_2 + \cdots + c_m\mathbf{s}_m = \boldsymbol{\theta}. \tag{2}$$

Now using system (1), we can rewrite (2) in terms of the vectors in B as

$$\begin{aligned}
c_1(a_{11}\mathbf{w}_1 + a_{21}\mathbf{w}_2 + \cdots + a_{p1}\mathbf{w}_p) + \\
c_2(a_{12}\mathbf{w}_1 + a_{22}\mathbf{w}_2 + \cdots + a_{p2}\mathbf{w}_p) + \\
\cdots + c_m(a_{1m}\mathbf{w}_1 + a_{2m}\mathbf{w}_2 + \cdots + a_{pm}\mathbf{w}_p) = \boldsymbol{\theta}.
\end{aligned} \tag{3a}$$

Equation (3a) can be regrouped as

$$\begin{aligned}
(c_1a_{11} + c_2a_{12} + \cdots + c_ma_{1m})\mathbf{w}_1 + \\
(c_1a_{21} + c_2a_{22} + \cdots + c_ma_{2m})\mathbf{w}_2 + \\
\cdots + (c_1a_{p1} + c_2a_{p2} + \cdots + c_ma_{pm})\mathbf{w}_p = \boldsymbol{\theta}.
\end{aligned} \tag{3b}$$

Now finding $c_1,c_2,\ldots,c_m$ to satisfy Eq. (2) is the same as finding $c_1,c_2,\ldots,c_m$ to satisfy Eq. (3b). Furthermore, we can clearly satisfy Eq. (3b) if we can choose zero for each coefficient of each $\mathbf{w}_i$. Therefore, to obtain one solution of Eq. (3b), it suffices to solve the system

$$\begin{aligned}
a_{11}c_1 + a_{12}c_2 + \cdots + a_{1m}c_m &= 0 \\
a_{21}c_1 + a_{22}c_2 + \cdots + a_{2m}c_m &= 0 \\
\vdots \qquad\qquad \vdots \qquad \vdots& \\
a_{p1}c_1 + a_{p2}c_2 + \cdots + a_{pm}c_m &= 0.
\end{aligned} \tag{4}$$

[Recall that each a_{ij} is a specified constant determined by system (1), whereas each c_i is an unknown parameter of Eq. (2).] The homogeneous system in (4) has more unknowns than equations, so by Theorem 4 of Section 1.4, there is a nontrivial solution to system (4). But a solution to (4) is also a solution to (2), so (2) has a nontrivial solution, and the theorem is established.

As an immediate corollary of Theorem 8, we can show that all bases for a subspace contain the same number of vectors.

COROLLARY

Let W be a subspace of R^n, and let $B = \{\mathbf{w}_1,\mathbf{w}_2,\ldots,\mathbf{w}_p\}$ be a basis for W containing p vectors. Then every basis for W contains p vectors.

PROOF Let $Q = \{\mathbf{u}_1, \mathbf{u}_2, \ldots, \mathbf{u}_r\}$ be any basis for W. Since Q is a spanning set for W, by Theorem 8 any set of $r + 1$ or more vectors in W is linearly dependent. Since B is a linearly independent set of p vectors in W, we know that $p \le r$. Similarly, since B is a spanning set of p vectors for W, any set of $p + 1$ or more vectors in W is linearly dependent. By assumption, Q is a set of r linearly independent vectors in W; so $r \le p$. Now, since we have $p \le r$ and $r \le p$, it must be that $r = p$. ☐

Given that every basis for a subspace contains the same number of vectors, we can make the following definition without any possibility of ambiguity.

DEFINITION 5 Let W be a subspace of R^n. If W has a basis $B = \{\mathbf{w}_1, \mathbf{w}_2, \ldots, \mathbf{w}_p\}$ of p vectors, then we say that W is a subspace of *dimension p*, and we write $\dim(W) = p$.

In Exercise 30, the reader is asked to show that every nonzero subspace of R^n does have a basis. Thus a value for dimension can be assigned to any subspace of R^n, where for completeness we define $\dim(W) = 0$ if W is the zero subspace.

Since R^3 has a basis $\{\mathbf{e}_1, \mathbf{e}_2, \mathbf{e}_3\}$ containing three vectors, we see that $\dim(R^3) = 3$. In general, R^n has a basis $\{\mathbf{e}_1, \mathbf{e}_2, \ldots, \mathbf{e}_n\}$ that contains n vectors; so $\dim(R^n) = n$. Thus the definition of dimension—the number of vectors in a basis—agrees with the usual terminology; R^3 is three-dimensional, and in general, R^n is n-dimensional.

EXAMPLE 1 Let W be the subspace of R^3 defined by

$$W = \{\mathbf{x}\colon \mathbf{x} = \begin{bmatrix} x_1 \\ x_2 \\ x_3 \end{bmatrix}, \quad x_1 = -2x_3, x_2 = x_3, x_3 \text{ arbitrary}\}.$$

Exhibit a basis for W and determine $\dim(W)$.

SOLUTION A vector $\mathbf{x}$ in W can be written in the form

$$\mathbf{x} = \begin{bmatrix} -2x_3 \\ x_3 \\ x_3 \end{bmatrix} = x_3 \begin{bmatrix} -2 \\ 1 \\ 1 \end{bmatrix}.$$

Therefore, the set $\{\mathbf{u}\}$ is a basis for W, where

$$\mathbf{u} = \begin{bmatrix} -2 \\ 1 \\ 1 \end{bmatrix}.$$

It follows that $\dim(W) = 1$. Geometrically, W is the line through the origin and through the point with coordinates $(-2, 1, 1)$, so again the definition of dimension coincides with our geometric intuition. ☐

The next example illustrates the importance of the corollary to Theorem 8.

EXAMPLE 2 Let W be the subspace of R^3, $W = \text{span}\{\mathbf{u}_1, \mathbf{u}_2, \mathbf{u}_3, \mathbf{u}_4\}$, where

$$\mathbf{u}_1 = \begin{bmatrix} 1 \\ 1 \\ 2 \end{bmatrix}, \quad \mathbf{u}_2 = \begin{bmatrix} 2 \\ 4 \\ 0 \end{bmatrix}, \quad \mathbf{u}_3 = \begin{bmatrix} 3 \\ 5 \\ 2 \end{bmatrix}, \quad \text{and} \quad \mathbf{u}_4 = \begin{bmatrix} 2 \\ 5 \\ -2 \end{bmatrix}.$$

Use the techniques illustrated in Examples 5, 6, and 7 of Section 2.4 to find three different bases for W. Give the dimension of W.

SOLUTION

a) The technique used in Example 5 consisted of finding a basis for W by using the algebraic specification for W. In particular, let $\mathbf{b}$ be a vector in R^3:

$$\mathbf{b} = \begin{bmatrix} a \\ b \\ c \end{bmatrix}.$$

Then b is in W if and only if the vector equation

$$x_1\mathbf{u}_1 + x_2\mathbf{u}_2 + x_3\mathbf{u}_3 + x_4\mathbf{u}_4 = \mathbf{b} \tag{5a}$$

is consistent. The matrix equation for (5a) is $U\mathbf{x} = \mathbf{b}$, where U is the (3×4) matrix $U = [\mathbf{u}_1, \mathbf{u}_2, \mathbf{u}_3, \mathbf{u}_4]$. Now, the augmented matrix $[U \,|\, \mathbf{b}]$ is row equivalent to the matrix

$$\begin{bmatrix} 1 & 2 & 3 & 2 & a \\ 0 & 2 & 2 & 3 & -a + b \\ 0 & 0 & 0 & 0 & -4a + 2b + c \end{bmatrix}. \tag{5b}$$

Thus $\mathbf{b}$ is in W if and only if $-4a + 2b + c = 0$ or, equivalently, $c = 4a - 2b$. The subspace W can then be described by

$$W = \left\{ \mathbf{b} \colon \mathbf{b} = \begin{bmatrix} a \\ b \\ 4a - 2b \end{bmatrix}, \quad a \text{ and } b \text{ any real numbers} \right\}.$$

From this description it follows that W has a basis $\{\mathbf{v}_1, \mathbf{v}_2\}$, where

$$\mathbf{v}_1 = \begin{bmatrix} 1 \\ 0 \\ 4 \end{bmatrix} \quad \text{and} \quad \mathbf{v}_2 = \begin{bmatrix} 0 \\ 1 \\ -2 \end{bmatrix}.$$

b) The technique used in Example 6 consisted of discarding redundant vectors from a spanning set for W. In particular since $\{\mathbf{u}_1, \mathbf{u}_2, \mathbf{u}_3, \mathbf{u}_4\}$ spans W, this technique gives a basis for W that is a subset of $\{\mathbf{u}_1, \mathbf{u}_2, \mathbf{u}_3, \mathbf{u}_4\}$. To obtain such a subset, solve the dependence relation

$$x_1\mathbf{u}_1 + x_2\mathbf{u}_2 + x_3\mathbf{u}_3 + x_4\mathbf{u}_4 = \boldsymbol{\theta}. \tag{5c}$$

Note that Eq. (5c) is just Eq. (5a) with $\mathbf{b} = \boldsymbol{\theta}$. It is easily seen from matrix (5b) that Eq. (5c) is equivalent to the reduced system

$$x_1 + 2x_2 + 3x_3 + 2x_4 = 0$$
$$2x_2 + 2x_3 + 3x_4 = 0. \qquad \textbf{(5d)}$$

Backsolving (5d) yields

$$x_1 = -x_3 + x_4$$
$$x_2 = -x_3 - (3/2)x_4,$$

where x_3 and x_4 are arbitrary. Therefore, the vectors $\mathbf{u}_3$ and $\mathbf{u}_4$ can be deleted from the spanning set for W, leaving $\{\mathbf{u}_1, \mathbf{u}_2\}$ as a basis for W.

c) Let U be the (3×4) matrix whose columns span W, $U = [\mathbf{u}_1, \mathbf{u}_2, \mathbf{u}_3, \mathbf{u}_4]$. Following the technique of Example 7, reduce U^T to the matrix

$$C^T = \begin{bmatrix} 1 & 1 & 2 \\ 0 & 2 & -4 \\ 0 & 0 & 0 \\ 0 & 0 & 0 \end{bmatrix}$$

in echelon form. In this case the nonzero columns of

$$C = \begin{bmatrix} 1 & 0 & 0 & 0 \\ 1 & 2 & 0 & 0 \\ 2 & -4 & 0 & 0 \end{bmatrix}$$

form a basis for W; that is, $\{\mathbf{w}_1, \mathbf{w}_2\}$ is a basis for W, where

$$\mathbf{w}_1 = \begin{bmatrix} 1 \\ 1 \\ 2 \end{bmatrix} \quad \text{and} \quad \mathbf{w}_2 = \begin{bmatrix} 0 \\ 2 \\ -4 \end{bmatrix}.$$

In each case the basis obtained for W contains two vectors, so $\dim(W) = 2$. Indeed, viewed geometrically, W is the plane with equation $-4x + 2y + z = 0$.

Properties of a p-Dimensional Subspace

An important feature of dimension is that a p-dimensional subspace W has many of the same properties as R^p. For example, Theorem 11 of Section 1.8 shows that any set of $p + 1$ or more vectors in R^p is linearly dependent. The following theorem shows that this same property and others hold in W when $\dim(W) = p$.

THEOREM 9	

Let W be a subspace of R^n with $\dim(W) = p$.

1. Any set of $p + 1$ or more vectors in W is linearly dependent.
2. Any set of fewer than p vectors in W does not span W.
3. Any set of p linearly independent vectors in W is a basis for W.
4. Any set of p vectors that spans W is a basis for W.

PROOF   Property 1 follows immediately from Theorem 8, because $\dim(W) = p$ means that W has a basis (and hence a spanning set) of p vectors.

Property 2 is equivalent to the statement that a spanning set for W must contain at least p vectors. Again, this is an immediate consequence of Theorem 8.

To establish property 3, let $\{\mathbf{u}_1, \mathbf{u}_2, \ldots, \mathbf{u}_p\}$ be a set of p linearly independent vectors in W. To see that the given set spans W, let $\mathbf{v}$ be any vector in W. By property 1, the set $\{\mathbf{v}, \mathbf{u}_1, \mathbf{u}_2, \ldots, \mathbf{u}_p\}$ is a linearly dependent set of vectors because the set contains $p + 1$ vectors. Thus there are scalars $a_0, a_1, \ldots, a_p$ (not all of which are zero) such that

$$a_0\mathbf{v} + a_1\mathbf{u}_1 + a_2\mathbf{u}_2 + \cdots + a_p\mathbf{u}_p = \boldsymbol{\theta}. \tag{6}$$

In addition, in Eq. (6), a_0 cannot be zero because $\{\mathbf{u}_1, \mathbf{u}_2, \ldots, \mathbf{u}_p\}$ is a linearly independent set. Therefore, Eq. (6) can be rewritten as

$$\mathbf{v} = (-1/a_0)[a_1\mathbf{u}_1 + a_2\mathbf{u}_2 + \cdots + a_p\mathbf{u}_p]. \tag{7}$$

It is clear from Eq. (7) that any vector in W can be expressed as a linear combination of $\mathbf{u}_1, \mathbf{u}_2, \ldots, \mathbf{u}_p$, so the given linearly independent set also spans W. Therefore, the set is a basis.

The proof of property 4 is left as an exercise. ☐

EXAMPLE 3   Let W be the subspace of R^3 given in Example 2, and let $\{\mathbf{v}_1, \mathbf{v}_2, \mathbf{v}_3\}$ be the subset of W defined by

$$\mathbf{v}_1 = \begin{bmatrix} 1 \\ -1 \\ 6 \end{bmatrix}, \qquad \mathbf{v}_2 = \begin{bmatrix} 1 \\ 2 \\ 0 \end{bmatrix}, \quad \text{and} \quad \mathbf{v}_3 = \begin{bmatrix} 2 \\ 1 \\ 6 \end{bmatrix}.$$

Determine which of the subsets $\{\mathbf{v}_1\}$, $\{\mathbf{v}_2\}$, $\{\mathbf{v}_1, \mathbf{v}_2\}$, $\{\mathbf{v}_1, \mathbf{v}_3\}$, $\{\mathbf{v}_2, \mathbf{v}_3\}$, and $\{\mathbf{v}_1, \mathbf{v}_2, \mathbf{v}_3\}$ is a basis for W.

SOLUTION   In Example 2, the subspace W was described as

$$W = \{\mathbf{b} \colon \mathbf{b} = \begin{bmatrix} a \\ b \\ 4a - 2b \end{bmatrix}, \quad a \text{ and } b \text{ any real numbers}\}. \tag{8}$$

Using Eq. (8), we can easily check that $\mathbf{v}_1$, $\mathbf{v}_2$, and $\mathbf{v}_3$ are in W. We saw further in Example 2 that $\dim(W) = 2$. By Theorem 9, property 2, neither of the sets

$\{v_1\}$ or $\{v_2\}$ spans W. By Theorem 9, property 1, the set $\{v_1, v_2, v_3\}$ is linearly dependent. We can easily check that each of the sets $\{v_1, v_2\}$, $\{v_1, v_3\}$, and $\{v_2, v_3\}$ is linearly independent, so by Theorem 9, property 3, each is a basis for W. ☐

The Rank of a Matrix

In this subsection we use the concept of dimension to characterize nonsingular matrices and to determine precisely when a system of linear equations $Ax = b$ is consistent. For an $(m \times n)$ matrix A, the dimension of the null space is called the **nullity of A,** and the dimension of the range of A is called the **rank of A.** The following example will illustrate the relationship between the rank of A and the nullity of A, as well as the relationship between the rank of A and the dimension of the row space of A.

EXAMPLE 4 Find the rank, nullity, and dimension of the row space for the matrix A, where

$$A = \begin{bmatrix} 1 & 1 & 1 & 2 \\ -1 & 0 & 2 & -3 \\ 2 & 4 & 8 & 5 \end{bmatrix}.$$

SOLUTION To find the dimension of the row space of A, observe that A is row equivalent to the matrix

$$B = \begin{bmatrix} 1 & 1 & 1 & 2 \\ 0 & 1 & 3 & -1 \\ 0 & 0 & 0 & 3 \end{bmatrix},$$

and B is in echelon form. Since the nonzero rows of B form a basis for the row space of A, the row space of A has dimension 3.

To find the nullity of A, we must determine the dimension of the null space. Since the homogeneous system $Ax = \theta$ is equivalent to $Bx = \theta$, the null space of A can be determined by backsolving $Bx = \theta$. This gives

$$x_1 = 2x_3$$
$$x_2 = -3x_3$$
$$x_4 = 0 .$$

Thus $\mathscr{N}(A)$ can be described by

$$\mathscr{N}(A) = \{x: x = \begin{bmatrix} 2x_3 \\ -3x_3 \\ x_3 \\ 0 \end{bmatrix}, \quad x_3 \text{ any real number}\}.$$

It now follows that the nullity of A is 1 because the vector

$$\mathbf{v} = \begin{bmatrix} 2 \\ -3 \\ 1 \\ 0 \end{bmatrix}$$

forms a basis for $\mathcal{N}(A)$.

To find the rank of A, we must determine the dimension of the range of A. Recall that $\mathcal{R}(A)$, the range of A, equals the column space of A, so a basis for $\mathcal{R}(A)$ can be found by reducing A^T to echelon form. It is straightforward to show that A^T is row equivalent to the matrix C^T, where

$$C^T = \begin{bmatrix} 1 & -1 & 2 \\ 0 & 1 & 2 \\ 0 & 0 & 3 \\ 0 & 0 & 0 \end{bmatrix}.$$

The nonzero columns of the matrix C,

$$C = \begin{bmatrix} 1 & 0 & 0 & 0 \\ -1 & 1 & 0 & 0 \\ 2 & 2 & 3 & 0 \end{bmatrix},$$

form a basis for $\mathcal{R}(A)$. Thus the rank of A is 3. ☐

Note in the example above that the row space of A is a subspace of R^4, whereas the column space (or range) of A is a subspace of R^3. Thus they are entirely different subspaces; even so, the dimensions are the same, and the next theorem states that this is always the case.

THEOREM 10

If A is an $(m \times n)$ matrix, then the rank of A is equal to the rank of A^T.

The proof of Theorem 10 will be given at the end of this section. Note that the range of A^T is equal to the column space of A^T. But the column space of A^T is precisely the row space of A, so the following corollary is actually a restatement of Theorem 10.

COROLLARY

If A is an $(m \times n)$ matrix, then the row space and the column space of A have the same dimension.

This corollary provides a useful way to determine the rank of a matrix A. Specifically, if A is row equivalent to a matrix B in echelon form, then the number, r, of nonzero rows in B equals the rank of A.

The null space of an ($m \times n$) matrix A is determined by solving the homogeneous system of equations $A\mathbf{x} = \boldsymbol{\theta}$. Suppose the augmented matrix $[A\,|\,\boldsymbol{\theta}]$ for the system is row equivalent to the matrix $[B\,|\,\boldsymbol{\theta}]$, which is in echelon form. Then clearly A is row equivalent to B, and the number, r, of nonzero rows of B equals the rank of A. But r is also the number of nonzero rows of $[B\,|\,\boldsymbol{\theta}]$. It follows from Theorem 3 of Section 1.4 that there are $n - r$ free variables in a solution for $A\mathbf{x} = \boldsymbol{\theta}$. But the number of vectors in a basis for $\mathcal{N}(A)$ equals the number of free variables in the solution for $A\mathbf{x} = \boldsymbol{\theta}$ (cf. Example 3 of Section 2.4); that is, the nullity of A is $n - r$. Thus we have shown, informally, that the following formula holds.

REMARK If A is an ($m \times n$) matrix, then

$$n = \text{rank}(A) + \text{nullity}(A).$$

This remark will be proved formally in a more general context in Chapter 4.

Example 4 illustrates the argument preceding the remark. If A is the matrix given in Example 4,

$$A = \begin{bmatrix} 1 & 1 & 1 & 2 \\ -1 & 0 & 2 & -3 \\ 2 & 4 & 8 & 5 \end{bmatrix},$$

then the augmented matrix $[A\,|\,\boldsymbol{\theta}]$ is row equivalent to

$$[B\,|\,\boldsymbol{\theta}] = \begin{bmatrix} 1 & 1 & 1 & 2 & 0 \\ 0 & 1 & 3 & -1 & 0 \\ 0 & 0 & 0 & 3 & 0 \end{bmatrix}.$$

Since A is row equivalent to B, the corollary to Theorem 10 implies that A has rank 3. Further, in the notation of Theorem 3 of Section 1.4, the system $A\mathbf{x} = \boldsymbol{\theta}$ has $n = 4$ unknowns, and the reduced matrix $[B\,|\,\boldsymbol{\theta}]$ has $r = 3$ nonzero rows. Therefore, the solution for $A\mathbf{x} = \boldsymbol{\theta}$ has $n - r = 4 - 3 = 1$ independent variables, and it follows that the nullity of A is 1. In particular,

$$\text{rank}(A) + \text{nullity}(A) = 3 + 1 = 4,$$

as is guaranteed by the remark.

The following theorem uses the concept of the rank of a matrix to establish necessary and sufficient conditions for a system of equations, $A\mathbf{x} = \mathbf{b}$, to be consistent.

THEOREM 11

An ($m \times n$) system of linear equations, $A\mathbf{x} = \mathbf{b}$, is consistent if and only if

$$\text{rank}(A) = \text{rank}([A\,|\,\mathbf{b}]).$$

PROOF Suppose that $A = [\mathbf{A}_1, \mathbf{A}_2, \dots, \mathbf{A}_n]$. Then the rank of A is the dimension of the column space of A, that is, the subspace

$$\text{Sp}\{\mathbf{A}_1, \mathbf{A}_2, \dots, \mathbf{A}_n\}. \tag{9}$$

Similarly, the rank of $[A \,|\, \mathbf{b}]$ is the dimension of the subspace

$$\text{Sp}\{\mathbf{A}_1, \mathbf{A}_2, \ldots, \mathbf{A}_n, \mathbf{b}\}. \tag{10}$$

But we already know that $A\mathbf{x} = \mathbf{b}$ is consistent if and only if $\mathbf{b}$ is in the column space of A. It follows that $A\mathbf{x} = \mathbf{b}$ is consistent if and only if the subspaces given in (9) and (10) are equal and consequently have the same dimension.

Our final theorem in this section shows that rank can be used to determine nonsingular matrices.

T H E O R E M   1 2

An $(n \times n)$ matrix A is nonsingular if and only if the rank of A is n.

PROOF   Suppose that $A = [\mathbf{A}_1, \mathbf{A}_2, \ldots, \mathbf{A}_n]$. The proof of Theorem 12 rests on the observation that the range of A is given by

$$\mathscr{R}(A) = \text{Sp}\{\mathbf{A}_1, \mathbf{A}_2, \ldots, \mathbf{A}_n\}. \tag{11}$$

If A is nonsingular then, by Theorem 12 of Section 1.8, the columns of A are linearly independent. Thus $\{\mathbf{A}_1, \mathbf{A}_2, \ldots, \mathbf{A}_n\}$ is a basis for $\mathscr{R}(A)$, and the rank of A is n.

Conversely, suppose that A has rank n; that is, $\mathscr{R}(A)$ has dimension n. It is an immediate consequence of Eq. (11) and Theorem 9, property 4, that $\{\mathbf{A}_1, \mathbf{A}_2, \ldots, \mathbf{A}_n\}$ is a basis for $\mathscr{R}(A)$. In particular, the columns of A are linearly independent, so, by Theorem 12 of Section 1.8, A is nonsingular.

Proof of Theorem 10 (Optional)

To prove Theorem 10, let $A = (a_{ij})$ be an $(m \times n)$ matrix. Denote the rows of A by $\mathbf{a}_1, \mathbf{a}_2, \ldots, \mathbf{a}_m$. Thus,

$$\mathbf{a}_i = [a_{i1}, a_{i2}, \ldots, a_{in}].$$

Similarly, let $\mathbf{A}_1, \mathbf{A}_2, \ldots, \mathbf{A}_n$ be the columns of A, where

$$\mathbf{A}_j = \begin{bmatrix} a_{1j} \\ a_{2j} \\ \vdots \\ a_{mj} \end{bmatrix}.$$

Suppose that A^T has rank k. Since the columns of A^T are $\mathbf{a}_1^T, \mathbf{a}_2^T, \dots, \mathbf{a}_m^T$, it follows that if

$$W = \text{Sp}\{\mathbf{a}_1, \mathbf{a}_2, \dots, \mathbf{a}_m\},$$

then $\dim(W) = k$. Therefore, W has a basis $\{\mathbf{w}_1, \mathbf{w}_2, \dots, \mathbf{w}_k\}$, and, by Theorem 9, property 2, $m \geq k$. For $1 \leq j \leq n$, suppose that $\mathbf{w}_j$ is the $(1 \times n)$ vector

$$\mathbf{w}_j = [w_{j1}, w_{j2}, \dots, w_{jn}].$$

Writing each $\mathbf{a}_i$ in terms of the basis yields

$$
\begin{aligned}
[a_{11}, a_{12}, \dots, a_{1n}] &= \mathbf{a}_1 = c_{11}\mathbf{w}_1 + c_{12}\mathbf{w}_2 + \cdots + c_{1k}\mathbf{w}_k \\
[a_{21}, a_{22}, \dots, a_{2n}] &= \mathbf{a}_2 = c_{21}\mathbf{w}_1 + c_{22}\mathbf{w}_2 + \cdots + c_{2k}\mathbf{w}_k \\
&\ \ \vdots \qquad\qquad \vdots \qquad \vdots \qquad\qquad\qquad \vdots \\
[a_{m1}, a_{m2}, \dots, a_{mn}] &= \mathbf{a}_m = c_{m1}\mathbf{w}_1 + c_{m2}\mathbf{w}_2 + \cdots + c_{mk}\mathbf{w}_k.
\end{aligned}
\tag{12}
$$

Equating the jth component of the left side of system (12) with the jth component of the right side yields

$$
\begin{bmatrix} a_{1j} \\ a_{2j} \\ \vdots \\ a_{mj} \end{bmatrix}
= w_{1j}\begin{bmatrix} c_{11} \\ c_{21} \\ \vdots \\ c_{m1} \end{bmatrix}
+ w_{2j}\begin{bmatrix} c_{12} \\ c_{22} \\ \vdots \\ c_{m2} \end{bmatrix}
+ \cdots + w_{kj}\begin{bmatrix} c_{1k} \\ c_{2k} \\ \vdots \\ c_{mk} \end{bmatrix}
\tag{13}
$$

for $1 \leq j \leq n$. For $1 \leq i \leq k$, define $\mathbf{c}_i$ to be the $(m \times 1)$ column vector

$$
\mathbf{c}_i = \begin{bmatrix} c_{1i} \\ c_{2i} \\ \vdots \\ c_{mi} \end{bmatrix}.
$$

Then system (13) becomes

$$\mathbf{A}_j = w_{1j}\mathbf{c}_1 + w_{2j}\mathbf{c}_2 + \cdots + w_{kj}\mathbf{c}_k, \qquad 1 \leq j \leq n. \tag{14}$$

It follows from the equations in (14) that

$$\mathscr{R}(A) = \text{Sp}\{\mathbf{A}_1, \mathbf{A}_2, \dots, \mathbf{A}_n\} \subseteq \text{Sp}\{\mathbf{c}_1, \mathbf{c}_2, \dots, \mathbf{c}_k\}.$$

It follows from Theorem 8 that the subspace

$$V = \text{Sp}\{\mathbf{c}_1, \mathbf{c}_2, \dots, \mathbf{c}_k\}$$

has dimension k, at most. By Exercise 32, $\dim[\mathscr{R}(A)] \leq \dim(V) \leq k$; that is, $\text{rank}(A) \leq \text{rank}(A^T)$.

Since $(A^T)^T = A$, the same argument implies that $\text{rank}(A^T) \leq \text{rank}(A)$. Thus $\text{rank}(A) = \text{rank}(A^T)$.

Exercises 1–14 refer to the vectors in (15).

$$\mathbf{u}_1 = \begin{bmatrix} 1 \\ 1 \end{bmatrix}, \qquad \mathbf{u}_2 = \begin{bmatrix} 1 \\ 2 \end{bmatrix}, \qquad \mathbf{u}_3 = \begin{bmatrix} -1 \\ 1 \end{bmatrix},$$

$$\mathbf{u}_4 = \begin{bmatrix} 0 \\ 0 \end{bmatrix}, \qquad \mathbf{u}_5 = \begin{bmatrix} 3 \\ 3 \end{bmatrix}, \qquad \mathbf{v}_1 = \begin{bmatrix} 1 \\ -1 \\ 1 \end{bmatrix}, \qquad (15)$$

$$\mathbf{v}_2 = \begin{bmatrix} 0 \\ 1 \\ 2 \end{bmatrix}, \qquad \mathbf{v}_3 = \begin{bmatrix} 1 \\ -1 \\ 0 \end{bmatrix}, \qquad \mathbf{v}_4 = \begin{bmatrix} -1 \\ 3 \\ 3 \end{bmatrix}$$

In Exercises 1–6, determine by inspection why the given set S is not a basis for R^2. (That is, either S is linearly dependent or S does not span R^2.)

1. $S = \{\mathbf{u}_1\}$ **2.** $S = \{\mathbf{u}_2\}$

3. $S = \{\mathbf{u}_1, \mathbf{u}_2, \mathbf{u}_3\}$ **4.** $S = \{\mathbf{u}_2, \mathbf{u}_3, \mathbf{u}_5\}$

5. $S = \{\mathbf{u}_1, \mathbf{u}_4\}$ **6.** $S = \{\mathbf{u}_1, \mathbf{u}_5\}$

In Exercises 7–9, determine by inspection why the given set S is not a basis for R^3. (That is, either S is linearly dependent or S does not span R^3.)

7. $S = \{\mathbf{v}_1, \mathbf{v}_2\}$ **8.** $S = \{\mathbf{v}_1, \mathbf{v}_3\}$

9. $S = \{\mathbf{v}_1, \mathbf{v}_2, \mathbf{v}_3, \mathbf{v}_4\}$

In Exercises 10–14, use Theorem 9, property 3, to determine whether the given set is a basis for the indicated vector space.

10. $S = \{\mathbf{u}_1, \mathbf{u}_2\}$ for R^2

11. $S = \{\mathbf{u}_2, \mathbf{u}_3\}$ for R^2

12. $S = \{\mathbf{v}_1, \mathbf{v}_2, \mathbf{v}_3\}$ for R^3

13. $S = \{\mathbf{v}_1, \mathbf{v}_2, \mathbf{v}_4\}$ for R^3

14. $S = \{\mathbf{v}_2, \mathbf{v}_3, \mathbf{v}_4\}$ for R^3

In Exercises 15–20, W is a subspace of R^4 consisting of vectors of the form

$$\mathbf{x} = \begin{bmatrix} x_1 \\ x_2 \\ x_3 \\ x_4 \end{bmatrix}.$$

Determine $\dim(W)$ when the components of $\mathbf{x}$ satisfy the given conditions.

15. $x_1 - 2x_2 + x_3 - x_4 = 0$

16. $x_1 - 2x_3 = 0$

17. $x_1 = -x_2 + 2x_4$
$x_3 = \qquad -x_4$

18. $x_1 \qquad + x_3 - 2x_4 = 0$
$x_2 + 2x_3 - 3x_4 = 0$

19. $x_1 = -x_4$
$x_2 = 3x_4$
$x_3 = 2x_4$

20. $x_1 - x_2 \qquad\quad = 0$
$x_2 - 2x_3 \quad = 0$
$x_3 - x_4 = 0$

In Exercises 21–24, find a basis for $\mathcal{N}(A)$ and give the nullity and the rank of A.

21. $A = \begin{bmatrix} 1 & 2 \\ -2 & -4 \end{bmatrix}$ **22.** $A = \begin{bmatrix} -1 & 2 & 0 \\ 2 & -5 & 1 \end{bmatrix}$

23. $A = \begin{bmatrix} 1 & -1 & 3 \\ 2 & -1 & 8 \\ -1 & 4 & 3 \end{bmatrix}$ **24.** $A = \begin{bmatrix} 1 & 2 & 0 & 5 \\ 1 & 3 & 1 & 7 \\ 2 & 3 & -1 & 9 \end{bmatrix}$

In Exercises 25 and 26, find a basis for $\mathcal{R}(A)$ and give the nullity and the rank of A.

25. $A = \begin{bmatrix} 1 & 2 & 1 \\ -1 & 0 & 3 \\ 1 & 5 & 7 \end{bmatrix}$ **26.** $A = \begin{bmatrix} 1 & 1 & 2 & 0 \\ 2 & 4 & 2 & 4 \\ 2 & 1 & 5 & -2 \end{bmatrix}$

27. Let W be a subspace, and let S be a spanning set for W. Find a basis for W and calculate $\dim(W)$ for each set S.

a) $S = \left\{ \begin{bmatrix} 1 \\ 1 \\ -2 \end{bmatrix}, \begin{bmatrix} -1 \\ -2 \\ 3 \end{bmatrix}, \begin{bmatrix} 1 \\ 0 \\ -1 \end{bmatrix}, \begin{bmatrix} 2 \\ -1 \\ 0 \end{bmatrix} \right\}$

b) $S = \left\{ \begin{bmatrix} 1 \\ 2 \\ -1 \\ 1 \end{bmatrix}, \begin{bmatrix} 3 \\ 1 \\ 1 \\ 2 \end{bmatrix}, \begin{bmatrix} -1 \\ 1 \\ -2 \\ 2 \end{bmatrix}, \begin{bmatrix} 0 \\ -2 \\ 1 \\ 2 \end{bmatrix} \right\}$

28. Let W be the subspace of R^4 defined by $W = \{\mathbf{x}: \mathbf{v}^T\mathbf{x} = 0\}$. Calculate $\dim(W)$, where

$$\mathbf{v} = \begin{bmatrix} 1 \\ 2 \\ -3 \\ -1 \end{bmatrix}.$$

29. Let W be the subspace of R^4 defined by $W = \{\mathbf{x}: \mathbf{a}^T\mathbf{x} = 0 \text{ and } \mathbf{b}^{Tx} = 0 \text{ and } \mathbf{c}^{Tx} = 0\}$.

Calculate dim(W) for

$$\mathbf{a} = \begin{bmatrix} 1 \\ -1 \\ 0 \\ 0 \end{bmatrix}, \quad \mathbf{b} = \begin{bmatrix} 1 \\ 0 \\ -1 \\ 0 \end{bmatrix}, \quad \text{and} \quad \mathbf{c} = \begin{bmatrix} 0 \\ 1 \\ -1 \\ 0 \end{bmatrix}.$$

30. Let W be a nonzero subspace of R^n. Show that W has a basis. (*Hint:* Let $\mathbf{w}_1$ be any nonzero vector in W. If $\{\mathbf{w}_1\}$ is a spanning set for W, then we are done. If not, there is a vector $\mathbf{w}_2$ in W such that $\{\mathbf{w}_1, \mathbf{w}_2\}$ is linearly independent. Why? Continue by asking whether this is a spanning set for W. Why must this process eventually stop?)

31. Suppose that $\{\mathbf{u}_1, \mathbf{u}_2, \ldots, \mathbf{u}_p\}$ is a basis for a subspace W, and suppose that $\mathbf{x}$ is in W with $\mathbf{x} = a_1\mathbf{u}_1 + a_2\mathbf{u}_2 + \cdots + a_p\mathbf{u}_p$. Show that this representation for $\mathbf{x}$ in terms of the basis is unique—that is, if $\mathbf{x} = b_1\mathbf{u}_1 + b_2\mathbf{u}_2 + \cdots + b_p\mathbf{u}_p$, then $b_1 = a_1, b_2 = a_2, \ldots, b_p = a_p$.

32. Let U and V be subspaces of R^n, and suppose that U is a subset of V. Prove that dim(U) $\leq$ dim(V). If dim(U) = dim(V), prove that V is contained in U, and thus conclude that $U = V$.

33. In each case below, determine the largest possible value for the rank of A and the smallest possible value for the nullity of A.
a) A is (3×3)   b) A is (3×4)
c) A is (5×4)

34. If A is a (3×4) matrix, prove that the columns of A are linearly dependent.

35. If A is a (4×3) matrix, prove that the rows of A are linearly dependent.

36. Let A be an ($m \times n$) matrix. Prove that rank(A) $\leq m$ and rank(A) $\leq n$.

37. Let A be a (2×3) matrix with rank 2. Show that the (2×3) system of equations $A\mathbf{x} = \mathbf{b}$ is consistent for every choice of $\mathbf{b}$ in R^2.

38. Let A be a (3×4) matrix with nullity 1. Prove that the (3×4) system of equations $A\mathbf{x} = \mathbf{b}$ is consistent for every choice of $\mathbf{b}$ in R^3.

39. Prove that an ($n \times n$) matrix A is nonsingular if and only if the nullity of A is zero.

40. Let A be an ($m \times m$) nonsingular matrix, and let B be an ($m \times n$) matrix. Prove that $\mathcal{N}(AB) = \mathcal{N}(B)$ and conclude that rank (AB) = rank (B).

41. Prove property 4 of Theorem 9 as follows: Assume that dim(W) = p and let $S = \{\mathbf{w}_1, \ldots, \mathbf{w}_p\}$ be a set of p vectors that spans W. To see that S is linearly independent, suppose that $c_1\mathbf{w}_1 + \cdots + c_p\mathbf{w}_p = \theta$. If $c_i \neq 0$, show that $W = \text{Sp}\{\mathbf{w}_1, \ldots, \mathbf{w}_{i-1}, \mathbf{w}_{i+1}, \ldots, \mathbf{w}_p\}$. Finally, use Theorem 8 to reach a contradiction.

42. Suppose that $S = \{\mathbf{u}_1, \mathbf{u}_2, \ldots, \mathbf{u}_p\}$ is a set of linearly independent vectors in a subspace W, where dim(W) = m and $m > p$. Prove that there is a vector $\mathbf{u}_{p+1}$ in W such that $\{\mathbf{u}_1, \mathbf{u}_2, \ldots, \mathbf{u}_p, \mathbf{u}_{p+1}\}$ is linearly independent. Use this proof to show that a basis including all the vectors in S can be constructed for W.

2.6 Orthogonal Bases for Subspaces

We have seen that a basis provides a very efficient way to characterize a subspace. Also, given a subspace W, we know that there are many different ways to construct a basis for W. In this section we focus on a particular type of basis called an orthogonal basis.

Orthogonal Bases

The idea of orthogonality is a generalization of the vector geometry concept of perpendicularity. If $\mathbf{u}$ and $\mathbf{v}$ are two vectors in R^2 or R^3, then we know that $\mathbf{u}$ and $\mathbf{v}$ are perpendicular if $\mathbf{u}^T\mathbf{v} = 0$ (see Theorem 1 in the Appendix). For example,

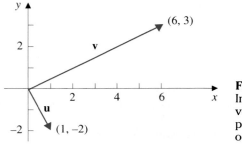

Figure 2.13
In R^2, nonzero vectors **u** and **v** are perpendicular if and only if $\mathbf{u}^T\mathbf{v} = 0$.

consider the vectors **u** and **v** given by

$$\mathbf{u} = \begin{bmatrix} 1 \\ -2 \end{bmatrix} \quad \text{and} \quad \mathbf{v} = \begin{bmatrix} 6 \\ 3 \end{bmatrix}.$$

Clearly $\mathbf{u}^T\mathbf{v} = 0$, and these two vectors are perpendicular when viewed as directed line segments in the plane (see Fig. 2.13).

In general, for vectors in R^n, we use the term "orthogonal" rather than the term "perpendicular." Specifically, if **u** and **v** are vectors in R^n, we say that **u** and **v** are *orthogonal* if

$$\mathbf{u}^T\mathbf{v} = 0.$$

We will also find the concept of an "orthogonal set" of vectors to be useful.

DEFINITION 6   Let $S = \{\mathbf{u}_1, \mathbf{u}_2, \ldots, \mathbf{u}_p\}$ be a set of vectors in R^n. The set S is said to be an *orthogonal set* if each pair of distinct vectors from S is orthogonal; that is, $\mathbf{u}_i^T\mathbf{u}_j = 0$ when $i \neq j$.

<u>EXAMPLE 1</u>   Verify that S is an orthogonal set of vectors, where

$$S = \left\{ \begin{bmatrix} 1 \\ 0 \\ 1 \\ 2 \end{bmatrix}, \begin{bmatrix} 1 \\ 1 \\ -1 \\ 0 \end{bmatrix}, \begin{bmatrix} 1 \\ -2 \\ -1 \\ 0 \end{bmatrix} \right\}.$$

SOLUTION   If we use the notation $S = \{\mathbf{u}_1, \mathbf{u}_2, \mathbf{u}_3\}$, then

$$\mathbf{u}_1^T\mathbf{u}_2 = \begin{bmatrix} 1 & 0 & 1 & 2 \end{bmatrix} \begin{bmatrix} 1 \\ 1 \\ -1 \\ 0 \end{bmatrix} = 1 + 0 - 1 + 0 = 0$$

$$\mathbf{u}_1^T\mathbf{u}_3 = \begin{bmatrix} 1 & 0 & 1 & 2 \end{bmatrix} \begin{bmatrix} 1 \\ -2 \\ -1 \\ 0 \end{bmatrix} = 1 + 0 - 1 + 0 = 0$$

$$\mathbf{u}_2^T\mathbf{u}_3 = \begin{bmatrix} 1 & 1 & -1 & 0 \end{bmatrix} \begin{bmatrix} 1 \\ -2 \\ -1 \\ 0 \end{bmatrix} = 1 - 2 + 1 + 0 = 0.$$

Therefore, $S = \{\mathbf{u}_1, \mathbf{u}_2, \mathbf{u}_3\}$ is an orthogonal set of vectors in R^4.

An important property of an orthogonal set S is that S is necessarily linearly independent (so long as S does not contain the zero vector).

THEOREM 13

Let $S = \{\mathbf{u}_1, \mathbf{u}_2, \ldots, \mathbf{u}_p\}$ be a set of nonzero vectors in R^n. If S is an orthogonal set of vectors, then S is a linearly independent set of vectors.

PROOF   Let $c_1, c_2, \ldots, c_p$ be any scalars that satisfy

$$c_1\mathbf{u}_1 + c_2\mathbf{u}_2 + \cdots + c_p\mathbf{u}_p = \boldsymbol{\theta}. \tag{1}$$

Form the scalar product

$$\mathbf{u}_1^T(c_1\mathbf{u}_1 + c_2\mathbf{u}_2 + \cdots + c_p\mathbf{u}_p) = \mathbf{u}_1^T\boldsymbol{\theta}$$

or

$$c_1(\mathbf{u}_1^T\mathbf{u}_1) + c_2(\mathbf{u}_1^T\mathbf{u}_2) + \cdots + c_p(\mathbf{u}_1^T\mathbf{u}_p) = 0.$$

Since $\mathbf{u}_1^T\mathbf{u}_j = 0$ for $2 \leq j \leq p$, the expression above reduces to

$$c_1(\mathbf{u}_1^T\mathbf{u}_1) = 0. \tag{2}$$

Next, because $\mathbf{u}_1^T\mathbf{u}_1 > 0$ when $\mathbf{u}_1$ is nonzero, we see from Eq. (2) that $c_1 = 0$.

Similarly, forming the scalar product of both sides of Eq. (1) with $\mathbf{u}_i$, we see that $c_i(\mathbf{u}_i^T\mathbf{u}_i) = 0$ or $c_i = 0$ for $1 \leq i \leq p$. Thus S is a linearly independent set of vectors.

By Theorem 13, any orthogonal set S containing p nonzero vectors from a p-dimensional subspace W will be a basis for W [since S is a linearly independent subset of p vectors from W, where $\dim(W) = p$]. Such a basis is called an *orthogonal basis*. In the following definition, recall that the symbol $\|\mathbf{v}\|$ denotes the length of $\mathbf{v}$, $\|\mathbf{v}\| = \sqrt{\mathbf{v}^T\mathbf{v}}$.

DEFINITION 7   Let W be a subspace of R^n, and let $B = \{\mathbf{u}_1, \mathbf{u}_2, \ldots, \mathbf{u}_p\}$ be a basis for W. If B is an orthogonal set of vectors, then B is called an *orthogonal basis* for W.

Furthermore, if $\|\mathbf{u}_i\| = 1$ for $1 \leq i \leq p$, then B is said to be an *orthonormal basis* for W.

The word *orthonormal* suggests both "orthogonal" and "normalized." Thus an orthonormal basis is an orthogonal basis consisting of vectors having length 1, where a vector of length 1 is a "unit" vector or a "normalized" vector. Observe that the unit vectors $e_1, e_2, \ldots, e_n$ form an orthonormal basis for R^n.

EXAMPLE 2   Verify that the set $B = \{v_1, v_2, v_3\}$ is an orthogonal basis for R^3, where

$$v_1 = \begin{bmatrix} 1 \\ 2 \\ 1 \end{bmatrix}, \quad v_2 = \begin{bmatrix} 3 \\ -1 \\ -1 \end{bmatrix}, \quad \text{and} \quad v_3 = \begin{bmatrix} 1 \\ -4 \\ 7 \end{bmatrix}.$$

SOLUTION   We first verify that B is an orthogonal set by calculating

$$v_1^T v_2 = 3 - 2 - 1 = 0$$
$$v_1^T v_3 = 1 - 8 + 7 = 0$$
$$v_2^T v_3 = 3 + 4 - 7 = 0.$$

Now, R^3 has dimension 3. Thus, since B is a set of three vectors and is also a linearly independent set (see Theorem 13), it follows that B is an orthogonal basis for R^3.

The observations made above are stated formally in the following corollary of Theorem 13.

COROLLARY

Let W be a subspace of R^n, where $\dim(W) = p$. If S is an orthogonal set of p nonzero vectors and is also a subset of W, then S is an orthogonal basis for W.

Orthonormal Bases

If $B = \{u_1, u_2, \ldots, u_p\}$ is an orthogonal set, then $C = \{a_1 u_1, a_2 u_2, \ldots, a_p u_p\}$ is also an orthogonal set for any scalars $a_1, a_2, \ldots, a_p$. If B contains only nonzero vectors and if we define the scalars a_i by

$$a_i = \frac{1}{\sqrt{u_i^T u_i}},$$

then C is an **orthonormal** set. That is, we can convert an orthogonal set of nonzero vectors into an orthonormal set by dividing each vector by its length.

EXAMPLE 3   Recall that the set B in Example 2 is an orthogonal basis for R^3. Modify B so that it is an orthonormal basis.

SOLUTION   Given that $B = \{v_1, v_2, v_3\}$ is an orthogonal basis for R^3, we can modify B to be an orthonormal basis by dividing each vector by its length. In particular (see Example 2), the lengths of $\mathbf{v}_1$, $\mathbf{v}_2$, and $\mathbf{v}_3$ are

$$\|\mathbf{v}_1\| = \sqrt{6}, \qquad \|\mathbf{v}_2\| = \sqrt{11}, \quad \text{and} \quad \|\mathbf{v}_3\| = \sqrt{66}.$$

Therefore, the set $C = \{\mathbf{w}_1, \mathbf{w}_2, \mathbf{w}_3\}$ is an orthonormal basis for R^3, where

$$\mathbf{w}_1 = \frac{1}{\sqrt{6}}\mathbf{v}_1 = \begin{bmatrix} 1/\sqrt{6} \\ 2/\sqrt{6} \\ 1/\sqrt{6} \end{bmatrix}, \qquad \mathbf{w}_2 = \frac{1}{\sqrt{11}}\mathbf{v}_2 = \begin{bmatrix} 3/\sqrt{11} \\ -1/\sqrt{11} \\ -1/\sqrt{11} \end{bmatrix}, \quad \text{and}$$

$$\mathbf{w}_3 = \frac{1}{\sqrt{66}}\mathbf{v}_3 = \begin{bmatrix} 1/\sqrt{66} \\ -4/\sqrt{66} \\ 7/\sqrt{66} \end{bmatrix}.$$

Determining Coordinates

Suppose that W is a p-dimensional subspace of R^n and $B = \{\mathbf{w}_1, \mathbf{w}_2, \ldots, \mathbf{w}_p\}$ is a basis for W. If $\mathbf{v}$ is any vector in W, then $\mathbf{v}$ can be written uniquely in the form

$$\mathbf{v} = a_1\mathbf{w}_1 + a_2\mathbf{w}_2 + \cdots + a_p\mathbf{w}_p. \tag{3}$$

(In Eq. (3), the fact that the scalars $a_1, a_2, \ldots, a_p$ are unique is proved in Exercise 31 of Section 2.5.) The scalars $a_1, a_2, \ldots, a_p$ in Eq. (3) are called the *coordinates of v* with respect to the basis B.

As we will see, it is fairly easy to determine the coordinates of a vector with respect to an orthogonal basis. To appreciate the savings in computation, consider how coordinates are found when the basis is not orthogonal. For instance, the set $B_1 = \{\mathbf{v}_1, \mathbf{v}_2, \mathbf{v}_3\}$ is a basis for R^3, where

$$\mathbf{v}_1 = \begin{bmatrix} 1 \\ 1 \\ -1 \end{bmatrix}, \qquad \mathbf{v}_2 = \begin{bmatrix} -1 \\ 2 \\ 1 \end{bmatrix}, \quad \text{and} \quad \mathbf{v}_3 = \begin{bmatrix} 2 \\ -2 \\ 1 \end{bmatrix}.$$

As can be seen, $\mathbf{v}_1^T\mathbf{v}_3 \neq 0$, and so B_1 is not an orthogonal basis. Next, suppose we wish to express some vector $\mathbf{v}$ in R^3, say $\mathbf{v} = [5, -5, -2]^T$, in terms of B_1. We must solve the (3×3) system: $a_1\mathbf{v}_1 + a_2\mathbf{v}_2 + a_3\mathbf{v}_3 = \mathbf{v}$. In matrix terms the coordinates a_1, a_2, and a_3 are found by solving the equation

$$\begin{bmatrix} 1 & -1 & 2 \\ 1 & 2 & -2 \\ -1 & 1 & 1 \end{bmatrix} \begin{bmatrix} a_1 \\ a_2 \\ a_3 \end{bmatrix} = \begin{bmatrix} 5 \\ -5 \\ -2 \end{bmatrix}.$$

(By Gaussian elimination, the solution is $a_1 = 1$, $a_2 = -2$, $a_3 = 1$.)

By contrast, if $B_2 = \{\mathbf{w}_1, \mathbf{w}_2, \mathbf{w}_3\}$ is an orthogonal basis for R^3, it is easy to determine a_1, a_2, and a_3 so that

$$\mathbf{v} = a_1\mathbf{w}_1 + a_2\mathbf{w}_2 + a_3\mathbf{w}_3. \tag{4}$$

To find the coordinate a_1 in Eq. (4), we form the scalar product

$$\mathbf{w}_1^T \mathbf{v} = \mathbf{w}_1^T (a_1 \mathbf{w}_1 + a_2 \mathbf{w}_2 + a_3 \mathbf{w}_3)$$
$$= a_1 (\mathbf{w}_1^T \mathbf{w}_1) + a_2 (\mathbf{w}_1^T \mathbf{w}_2) + a_3 (\mathbf{w}_1^T \mathbf{w}_3)$$
$$= a_1 (\mathbf{w}_1^T \mathbf{w}_1).$$

The last equality follows because $\mathbf{w}_1^T \mathbf{w}_2 = 0$ and $\mathbf{w}_1^T \mathbf{w}_3 = 0$. Therefore, from above,

$$a_1 = \frac{\mathbf{w}_1^T \mathbf{v}}{\mathbf{w}_1^T \mathbf{w}_1}.$$

Similarly,

$$a_2 = \frac{\mathbf{w}_2^T \mathbf{v}}{\mathbf{w}_2^T \mathbf{w}_2} \quad \text{and} \quad a_3 = \frac{\mathbf{w}_3^T \mathbf{v}}{\mathbf{w}_3^T \mathbf{w}_3}.$$

(*Note:* Since B_2 is a basis, $\mathbf{w}_i^T \mathbf{w}_i > 0$, $1 \le i \le 3$.)

EXAMPLE 4   Express the vector $\mathbf{v}$ in terms of the orthogonal basis $B = \{\mathbf{w}_1, \mathbf{w}_2, \mathbf{w}_3\}$, where

$$\mathbf{v} = \begin{bmatrix} 12 \\ -3 \\ 6 \end{bmatrix}, \quad \mathbf{w}_1 = \begin{bmatrix} 1 \\ 2 \\ 1 \end{bmatrix}, \quad \mathbf{w}_2 = \begin{bmatrix} 3 \\ -1 \\ -1 \end{bmatrix}, \quad \text{and} \quad \mathbf{w}_3 = \begin{bmatrix} 1 \\ -4 \\ 7 \end{bmatrix}.$$

SOLUTION   Beginning with the equation

$$\mathbf{v} = a_1 \mathbf{w}_1 + a_2 \mathbf{w}_2 + a_3 \mathbf{w}_3,$$

we form scalar products to obtain

$$\mathbf{w}_1^T \mathbf{v} = a_1 (\mathbf{w}_1^T \mathbf{w}_1), \quad \text{or} \quad 12 = 6a_1$$
$$\mathbf{w}_2^T \mathbf{v} = a_2 (\mathbf{w}_2^T \mathbf{w}_2), \quad \text{or} \quad 33 = 11a_2$$
$$\mathbf{w}_3^T \mathbf{v} = a_3 (\mathbf{w}_3^T \mathbf{w}_3), \quad \text{or} \quad 66 = 66a_3.$$

Thus $a_1 = 2$, $a_2 = 3$, and $a_3 = 1$. Therefore, as can be verified directly, $\mathbf{v} = 2\mathbf{w}_1 + 3\mathbf{w}_2 + \mathbf{w}_3$. ☐

In general, let W be a subspace of R^n and let $B = \{\mathbf{w}_1, \mathbf{w}_2, \ldots, \mathbf{w}_p\}$ be an orthogonal basis for W. If $\mathbf{v}$ is any vector in W, then $\mathbf{v}$ can be expressed uniquely in the form

$$\mathbf{v} = a_1 \mathbf{w}_1 + a_2 \mathbf{w}_2 + \cdots + a_p \mathbf{w}_p, \tag{5a}$$

where

$$a_i = \frac{\mathbf{w}_i^T \mathbf{v}}{\mathbf{w}_i^T \mathbf{w}_i}, \quad 1 \le i \le p. \tag{5b}$$

Constructing an Orthogonal Basis

The next theorem gives a procedure that can be used to generate an orthogonal basis from any given basis. This procedure, called the Gram–Schmidt process, is quite practical from a computational standpoint (although some care must be exercised when programming the procedure for the computer). Generating an orthogonal basis is often the first step in solving problems in least-squares approximation; so Gram–Schmidt orthogonalization is of more than theoretical interest.

THEOREM 14

Gram–Schmidt Let W be a p-dimensional subspace of R^n, and let $\{\mathbf{w}_1, \mathbf{w}_2, \ldots, \mathbf{w}_p\}$ be any basis for W. Then the set of vectors $\{\mathbf{u}_1, \mathbf{u}_2, \ldots, \mathbf{u}_p\}$ is an orthogonal basis for W, where

$$\mathbf{u}_1 = \mathbf{w}_1$$

$$\mathbf{u}_2 = \mathbf{w}_2 - \frac{\mathbf{u}_1^T \mathbf{w}_2}{\mathbf{u}_1^T \mathbf{u}_1} \mathbf{u}_1$$

$$\mathbf{u}_3 = \mathbf{w}_3 - \frac{\mathbf{u}_1^T \mathbf{w}_3}{\mathbf{u}_1^T \mathbf{u}_1} \mathbf{u}_1 - \frac{\mathbf{u}_2^T \mathbf{w}_3}{\mathbf{u}_2^T \mathbf{u}_2} \mathbf{u}_2,$$

and where, in general,

$$\mathbf{u}_i = \mathbf{w}_i - \sum_{k=1}^{i-1} \frac{\mathbf{u}_k^T \mathbf{w}_i}{\mathbf{u}_k^T \mathbf{u}_k} \mathbf{u}_k, \qquad 2 \le i \le p. \tag{6}$$

The proof of Theorem 14 is somewhat technical and we defer it to the end of this section.

In (6) we have explicit expressions that can be used to generate an orthogonal set of vectors $\{\mathbf{u}_1, \mathbf{u}_2, \ldots, \mathbf{u}_p\}$ from a given set of linearly independent vectors. These explicit expressions are especially useful if we have reason to implement the Gram–Schmidt process on a computer.

However, for hand calculations, it is not necessary to memorize formula (6). All we need to remember is the form or the general pattern of the Gram–Schmidt process. In particular, the Gram–Schmidt process starts with a basis $\{\mathbf{w}_1, \mathbf{w}_2, \ldots, \mathbf{w}_p\}$ and generates new vectors $\mathbf{u}_1, \mathbf{u}_2, \mathbf{u}_3, \ldots$ according to the following pattern:

$$\mathbf{u}_1 = \mathbf{w}_1$$
$$\mathbf{u}_2 = \mathbf{w}_2 + a\mathbf{u}_1$$
$$\mathbf{u}_3 = \mathbf{w}_3 + b\mathbf{u}_1 + c\mathbf{u}_2$$
$$\mathbf{u}_4 = \mathbf{w}_4 + d\mathbf{u}_1 + e\mathbf{u}_2 + f\mathbf{u}_3$$
$$\vdots$$
$$\mathbf{u}_i = \mathbf{w}_i + \alpha_1\mathbf{u}_1 + \alpha_2\mathbf{u}_2 + \cdots + \alpha_{i-1}\mathbf{u}_{i-1}$$
$$\vdots$$

In the sequence above, the scalars can be determined in a step-by-step fashion from the orthogonality conditions.

For instance, to determine the scalar a in the definition of $\mathbf{u}_2$, we use the condition $\mathbf{u}_1^T\mathbf{u}_2 = 0$:

$$0 = \mathbf{u}_1^T\mathbf{u}_2 = \mathbf{u}_1^T\mathbf{w}_2 + a\mathbf{u}_1^T\mathbf{u}_1;$$
$$\textit{Therefore:} \quad a = -(\mathbf{u}_1^T\mathbf{w}_2)/(\mathbf{u}_1^T\mathbf{u}_1). \tag{7}$$

To determine the two scalars b and c in the definition of $\mathbf{u}_3$, we use the two conditions $\mathbf{u}_1^T\mathbf{u}_3 = 0$ and $\mathbf{u}_2^T\mathbf{u}_3 = 0$. In particular,

$$
\begin{aligned}
0 = \mathbf{u}_1^T\mathbf{u}_3 &= \mathbf{u}_1^T\mathbf{w}_3 + b\mathbf{u}_1^T\mathbf{u}_1 + c\mathbf{u}_1^T\mathbf{u}_2 \\
&= \mathbf{u}_1^T\mathbf{w}_3 + b\mathbf{u}_1^T\mathbf{u}_1 \quad (\text{since } \mathbf{u}_1^T\mathbf{u}_2 = 0 \quad \text{by (7)})
\end{aligned}
$$
$$\textit{Therefore:} \quad b = -(\mathbf{u}_1^T\mathbf{w}_3)/(\mathbf{u}_1^T\mathbf{u}_1).$$

Similarly,

$$
\begin{aligned}
0 = \mathbf{u}_2^T\mathbf{u}_3 &= \mathbf{u}_2^T\mathbf{w}_3 + b\mathbf{u}_2^T\mathbf{u}_1 + c\mathbf{u}_2^T\mathbf{u}_2 \\
&= \mathbf{u}_2^T\mathbf{w}_3 + c\mathbf{u}_2^T\mathbf{u}_2 \quad (\text{since } \mathbf{u}_2^T\mathbf{u}_1 = 0 \quad \text{by (7)})
\end{aligned}
$$
$$\textit{Therefore:} \quad c = -(\mathbf{u}_2^T\mathbf{w}_3)/(\mathbf{u}_2^T\mathbf{u}_2).$$

Two examples that follow will illustrate the calculations outlined above.

Finally, to use the Gram–Schmidt orthogonalization process to find an orthogonal basis for W, we need some basis for W as a starting point. In many of the applications that require an orthogonal basis for a subspace W, it is relatively easy to produce this initial basis—we will give some examples in a later section. Given a basis for W, the Gram–Schmidt process proceeds in a mechanical fashion using Eq. (6). (*Note:* It was shown in Exercise 30 of Section 2.5 that every nonzero subspace of R^n has a basis. Therefore, by Theorem 14, every nonzero subspace of R^n has an orthogonal basis.)

EXAMPLE 5 Let W be the subspace of R^3 defined by $W = \text{Sp}\{\mathbf{w}_1, \mathbf{w}_2\}$, where

$$\mathbf{w}_1 = \begin{bmatrix} 1 \\ 1 \\ 2 \end{bmatrix} \quad \text{and} \quad \mathbf{w}_2 = \begin{bmatrix} 0 \\ 2 \\ -4 \end{bmatrix}.$$

Use the Gram–Schmidt process to construct an orthogonal basis for W.

SOLUTION We define vectors $\mathbf{u}_1$ and $\mathbf{u}_2$ of the form

$$\mathbf{u}_1 = \mathbf{w}_1$$
$$\mathbf{u}_2 = \mathbf{w}_2 + a\mathbf{u}_1,$$

where the scalar a is found from the condition $\mathbf{u}_1^T\mathbf{u}_2 = 0$. Now, $\mathbf{u}_1 = [1, 1, 2]^T$ and thus $\mathbf{u}_1^T\mathbf{u}_2$ is given by

$$\mathbf{u}_1^T\mathbf{u}_2 = \mathbf{u}_1^T(\mathbf{w}_2 + a\mathbf{u}_1) = \mathbf{u}_1^T\mathbf{w}_2 + a\mathbf{u}_1^T\mathbf{u}_1 = -6 + 6a.$$

Therefore, to have $\mathbf{u}_1^T\mathbf{u}_2 = 0$, we need $a = 1$. With $a = 1$, $\mathbf{u}_2$ is given by $\mathbf{u}_2 = \mathbf{w}_2 + \mathbf{u}_1 = [1, 3, -2]^T$.

In detail, an orthogonal basis for W is $B = \{\mathbf{u}_1, \mathbf{u}_2\}$, where

$$\mathbf{u}_1 = \begin{bmatrix} 1 \\ 1 \\ 2 \end{bmatrix} \quad \text{and} \quad \mathbf{u}_2 = \begin{bmatrix} 1 \\ 3 \\ -2 \end{bmatrix}.$$

For convenience in hand calculations, we can always eliminate fractional components in a set of orthogonal vectors. Specifically, if x and $\mathbf{y}$ are orthogonal, then so are $a\mathbf{x}$ and $\mathbf{y}$ for any scalar a:

$$\text{If} \quad \mathbf{x}^T\mathbf{y} = 0, \quad \text{then} \quad (a\mathbf{x})^T\mathbf{y} = a(\mathbf{x}^T\mathbf{y}) = 0.$$

We will make use of this observation in the following example.

EXAMPLE 6 Use the Gram–Schmidt orthogonalization process to generate an orthogonal basis for $W = \text{Sp}\{\mathbf{w}_1, \mathbf{w}_2, \mathbf{w}_3\}$, where

$$\mathbf{w}_1 = \begin{bmatrix} 0 \\ 1 \\ 2 \\ 1 \end{bmatrix}, \quad \mathbf{w}_2 = \begin{bmatrix} 0 \\ 1 \\ 3 \\ 1 \end{bmatrix}, \quad \text{and} \quad \mathbf{w}_3 = \begin{bmatrix} 1 \\ 1 \\ 1 \\ 0 \end{bmatrix}.$$

SOLUTION First we should check to be sure that $\{\mathbf{w}_1, \mathbf{w}_2, \mathbf{w}_3\}$ is a linearly independent set. A calculation shows that the vectors are linearly independent. (Exercise 27 illustrates what happens when the Gram–Schmidt algorithm is applied to a linearly dependent set.)

To generate an orthogonal basis $\{\mathbf{u}_1, \mathbf{u}_2, \mathbf{u}_3\}$ from $\{\mathbf{w}_1, \mathbf{w}_2, \mathbf{w}_3\}$, we first set

$$\mathbf{u}_1 = \mathbf{w}_1$$
$$\mathbf{u}_2 = \mathbf{w}_2 + a\mathbf{u}_1$$
$$\mathbf{u}_3 = \mathbf{w}_3 + b\mathbf{u}_1 + c\mathbf{u}_2.$$

With $\mathbf{u}_1 = [0, 1, 2, 1]^T$, the orthogonality condition $\mathbf{u}_1^T\mathbf{u}_2 = 0$ leads to $\mathbf{u}_1^T\mathbf{w}_2 + a\mathbf{u}_1^T\mathbf{u}_1 = 0$, or $8 + 6a = 0$. Therefore, $a = -4/3$ and hence

$$\mathbf{u}_2 = \mathbf{w}_2 - (4/3)\mathbf{u}_1 = [0, -1/3, 1/3, -1/3]^T.$$

Next, the conditions $\mathbf{u}_1^T\mathbf{u}_3 = 0$ and $\mathbf{u}_2^T\mathbf{u}_3 = 0$ lead to

$$0 = \mathbf{u}_1^T(\mathbf{w}_3 + b\mathbf{u}_1 + c\mathbf{u}_2) = 3 + 6b$$
$$0 = \mathbf{u}_2^T(\mathbf{w}_3 + b\mathbf{u}_1 + c\mathbf{u}_2) = 0 + (1/3)c.$$

Therefore, $b = -1/2$ and $c = 0$. Having the scalars b and c,

$$\mathbf{u}_3 = \mathbf{w}_3 - (1/2)\mathbf{u}_1 - (0)\mathbf{u}_2 = [1, 1/2, 0, -1/2]^T.$$

For convenience we can eliminate the fractional components in $\mathbf{u}_2$ and $\mathbf{u}_3$, as suggested above, and obtain an orthogonal basis $\{\mathbf{v}_1, \mathbf{v}_2, \mathbf{v}_3\}$, where

$$\mathbf{v}_1 = \begin{bmatrix} 0 \\ 1 \\ 2 \\ 1 \end{bmatrix}, \qquad \mathbf{v}_2 = \begin{bmatrix} 0 \\ -1 \\ 1 \\ -1 \end{bmatrix}, \quad \text{and} \quad \mathbf{v}_3 = \begin{bmatrix} 2 \\ 1 \\ 0 \\ -1 \end{bmatrix}.$$

Note: In Example 6, we could have also eliminated fractional components in the middle of the Gram–Schmidt process. That is, we could have redefined $\mathbf{u}_2$ to be the vector $\mathbf{u}_2 = [0, -1, 1, -1]$ and then calculated $\mathbf{u}_3$ with this new (redefined) multiple of $\mathbf{u}_2$.

As a final example, we use MATLAB to construct orthogonal bases.

EXAMPLE 7   Let A be the (3×5) matrix

$$A = \begin{bmatrix} 1 & 2 & 1 & 3 & 2 \\ 4 & 1 & 0 & 6 & 1 \\ 1 & 1 & 2 & 4 & 5 \end{bmatrix}.$$

Find an orthogonal basis for $\mathscr{R}(A)$ and an orthogonal basis for $\mathscr{N}(A)$.

```
A =

        1        2        1        3        2
        4        1        0        6        1
        1        1        2        4        5

>> orth ( A )

ans =

        0.3841            -0.1173            -0.9158
        0.7682             0.5908             0.2466
        0.5121            -0.7983             0.3170

>> null ( A )

ans =

       -0.7528            -0.0690
       -0.2063             0.1800
       -0.1069            -0.9047
        0.5736            -0.0469
       -0.2243             0.3772
```

Figure 2.14
The MATLAB command orth(A) produces an orthonormal basis for the range of A. The command null(A) gives an orthonormal basis for the null space of A.

SOLUTION   The MATLAB command orth(A) gives an orthonormal basis for the range of A. The command null(A) gives an orthonormal basis for the null space of A. The results are shown in Fig. 2.14. Observe that the basis for $\mathscr{R}(A)$ has three vectors; that is, the dimension of $\mathscr{R}(A)$ is three or, equivalently, A has rank three. The basis for $\mathscr{N}(A)$ has two vectors; that is, the dimension of $\mathscr{N}(A)$ is two, or equivalently, A has nullity two.   ☐

Proof of Theorem 14 (Optional)

We first show that the expression given in Eq. (6) is always defined and that the vectors $\mathbf{u}_1, \mathbf{u}_2, \ldots, \mathbf{u}_p$ are all nonzero. To begin, $\mathbf{u}_1$ is a nonzero vector since $\mathbf{u}_1 = \mathbf{w}_1$. Thus $\mathbf{u}_1^T \mathbf{u}_1 > 0$, and so we can define $\mathbf{u}_2$. Furthermore, we observe that $\mathbf{u}_2$ has the form $\mathbf{u}_2 = \mathbf{w}_2 - b_1 \mathbf{u}_1 = \mathbf{w}_2 - b_1 \mathbf{w}_1$; so $\mathbf{u}_2$ is nonzero since it is a nontrivial linear combination of $\mathbf{w}_1$ and $\mathbf{w}_2$. Proceeding inductively, suppose that $\mathbf{u}_1, \mathbf{u}_2, \ldots, \mathbf{u}_{i-1}$ have been generated by (6); and suppose that each $\mathbf{u}_k$ has the form

$$\mathbf{u}_k = \mathbf{w}_k - c_1 \mathbf{w}_1 - c_2 \mathbf{w}_2 - \cdots - c_{k-1} \mathbf{w}_{k-1}.$$

From this equation, each $\mathbf{u}_k$ is nonzero; and it follows that (6) is a well-defined expression [since $\mathbf{u}_k^T \mathbf{u}_k > 0$ for $1 \le k \le (i-1)$]. Finally, since each $\mathbf{u}_k$ in (6) is a linear combination of $\mathbf{w}_1, \mathbf{w}_2, \ldots, \mathbf{w}_k$, we see that $\mathbf{u}_i$ is a nontrivial linear combination of $\mathbf{w}_1, \mathbf{w}_2, \ldots, \mathbf{w}_i$; and therefore $\mathbf{u}_i$ is nonzero.

All that remains to be proved is that the vectors generated by (6) are orthogonal. Clearly $\mathbf{u}_1^T \mathbf{u}_2 = 0$. Proceeding inductively again, suppose that $\mathbf{u}_j^T \mathbf{u}_k = 0$ for any j and k, where $j \ne k$ and $1 \le j, k \le i-1$. From (6) we have

$$\mathbf{u}_j^T \mathbf{u}_i = \mathbf{u}_j^T \left(\mathbf{w}_i - \sum_{k=1}^{i-1} \frac{\mathbf{u}_k^T \mathbf{w}_i}{\mathbf{u}_k^T \mathbf{u}_k} \mathbf{u}_k \right) = \mathbf{u}_j^T \mathbf{w}_i - \sum_{k=1}^{i-1} \left(\frac{\mathbf{u}_k^T \mathbf{w}_i}{\mathbf{u}_k^T \mathbf{u}_k} \right) (\mathbf{u}_j^T \mathbf{u}_k)$$

$$= \mathbf{u}_j^T \mathbf{w}_i - \left(\frac{\mathbf{u}_j^T \mathbf{w}_i}{\mathbf{u}_j^T \mathbf{u}_j} \right) (\mathbf{u}_j^T \mathbf{u}_j) = 0.$$

Thus $\mathbf{u}_i$ is orthogonal to $\mathbf{u}_j$ for $1 \le j \le i-1$. Having this result, we have shown that $\{\mathbf{u}_1, \mathbf{u}_2, \ldots, \mathbf{u}_p\}$ is an orthogonal set of p nonzero vectors. So, by the corollary of Theorem 13, the vectors $\mathbf{u}_1, \mathbf{u}_2, \ldots, \mathbf{u}_p$ are an orthogonal basis for W.

Exercises 2.6

In Exercises 1–4, verify that $\{\mathbf{u}_1, \mathbf{u}_2, \mathbf{u}_3\}$ is an orthogonal set for the given vectors.

1. $\mathbf{u}_1 = \begin{bmatrix} 1 \\ 1 \\ 1 \end{bmatrix}$, $\mathbf{u}_2 = \begin{bmatrix} -1 \\ 0 \\ 1 \end{bmatrix}$, $\mathbf{u}_3 = \begin{bmatrix} -1 \\ 2 \\ -1 \end{bmatrix}$

2. $\mathbf{u}_1 = \begin{bmatrix} 1 \\ 0 \\ 1 \end{bmatrix}$, $\mathbf{u}_2 = \begin{bmatrix} -1 \\ 0 \\ 1 \end{bmatrix}$, $\mathbf{u}_3 = \begin{bmatrix} 0 \\ 1 \\ 0 \end{bmatrix}$

3. $\mathbf{u}_1 = \begin{bmatrix} 1 \\ 1 \\ 2 \end{bmatrix}$, $\mathbf{u}_2 = \begin{bmatrix} 2 \\ 0 \\ -1 \end{bmatrix}$, $\mathbf{u}_3 = \begin{bmatrix} 1 \\ -5 \\ 2 \end{bmatrix}$

4. $\mathbf{u}_1 = \begin{bmatrix} 2 \\ 1 \\ 2 \end{bmatrix}$, $\mathbf{u}_2 = \begin{bmatrix} 1 \\ 2 \\ -2 \end{bmatrix}$, $\mathbf{u}_3 = \begin{bmatrix} -2 \\ 2 \\ 1 \end{bmatrix}$

In Exercises 5–8, find values a, b, and c such that $\{\mathbf{u}_1, \mathbf{u}_2, \mathbf{u}_3\}$ is an orthogonal set.

5. $\mathbf{u}_1 = \begin{bmatrix} 1 \\ 1 \\ 1 \end{bmatrix}$, $\mathbf{u}_2 = \begin{bmatrix} 2 \\ 2 \\ -4 \end{bmatrix}$, $\mathbf{u}_3 = \begin{bmatrix} a \\ b \\ c \end{bmatrix}$

6. $\mathbf{u}_1 = \begin{bmatrix} 2 \\ 0 \\ 1 \end{bmatrix}$, $\mathbf{u}_2 = \begin{bmatrix} 1 \\ 1 \\ -2 \end{bmatrix}$, $\mathbf{u}_3 = \begin{bmatrix} a \\ b \\ c \end{bmatrix}$

7. $\mathbf{u}_1 = \begin{bmatrix} 1 \\ 1 \\ 1 \end{bmatrix}$, $\mathbf{u}_2 = \begin{bmatrix} -2 \\ -1 \\ a \end{bmatrix}$, $\mathbf{u}_3 = \begin{bmatrix} 4 \\ b \\ c \end{bmatrix}$

8. $\mathbf{u}_1 = \begin{bmatrix} 2 \\ 1 \\ -1 \end{bmatrix}$, $\mathbf{u}_2 = \begin{bmatrix} a \\ 1 \\ -1 \end{bmatrix}$, $\mathbf{u}_3 = \begin{bmatrix} b \\ 3 \\ c \end{bmatrix}$

In Exercises 9–12, express the given vector $\mathbf{v}$ in terms of the orthogonal basis $B = \{\mathbf{u}_1, \mathbf{u}_2, \mathbf{u}_3\}$, where $\mathbf{u}_1$, $\mathbf{u}_2$, and $\mathbf{u}_3$ are as in Exercise 1.

9. $\mathbf{v} = \begin{bmatrix} 1 \\ 1 \\ 0 \end{bmatrix}$ **10.** $\mathbf{v} = \begin{bmatrix} 0 \\ 1 \\ 2 \end{bmatrix}$

11. $\mathbf{v} = \begin{bmatrix} 3 \\ 3 \\ 3 \end{bmatrix}$ **12.** $\mathbf{v} = \begin{bmatrix} 1 \\ 2 \\ 1 \end{bmatrix}$

In Exercises 13–18, use the Gram–Schmidt process to generate an orthogonal set from the given linearly independent vectors.

13. $\begin{bmatrix} 0 \\ 0 \\ 1 \\ 0 \end{bmatrix}$, $\begin{bmatrix} 1 \\ 1 \\ 2 \\ 1 \end{bmatrix}$, $\begin{bmatrix} 1 \\ 0 \\ 1 \\ 1 \end{bmatrix}$ **14.** $\begin{bmatrix} 1 \\ 0 \\ 1 \\ 2 \end{bmatrix}$, $\begin{bmatrix} 2 \\ 1 \\ 0 \\ 2 \end{bmatrix}$, $\begin{bmatrix} 1 \\ -1 \\ 0 \\ 1 \end{bmatrix}$

15. $\begin{bmatrix} 1 \\ 1 \\ 0 \end{bmatrix}$, $\begin{bmatrix} 0 \\ 2 \\ 1 \end{bmatrix}$, $\begin{bmatrix} 1 \\ 1 \\ 6 \end{bmatrix}$ **16.** $\begin{bmatrix} 0 \\ 1 \\ 2 \end{bmatrix}$, $\begin{bmatrix} 3 \\ 6 \\ 2 \end{bmatrix}$, $\begin{bmatrix} 10 \\ -5 \\ 5 \end{bmatrix}$

17. $\begin{bmatrix} 0 \\ 1 \\ 0 \\ 1 \end{bmatrix}$, $\begin{bmatrix} 1 \\ 2 \\ 0 \\ 0 \end{bmatrix}$, $\begin{bmatrix} 0 \\ 2 \\ 1 \\ 0 \end{bmatrix}$ **18.** $\begin{bmatrix} 1 \\ 1 \\ 0 \\ 2 \end{bmatrix}$, $\begin{bmatrix} 0 \\ 2 \\ 1 \\ 2 \end{bmatrix}$, $\begin{bmatrix} 0 \\ 1 \\ 0 \\ 2 \end{bmatrix}$

In Exercises 19 and 20, find a basis for the null space and the range of the given matrix. Then use Gram–Schmidt to obtain orthogonal bases.

19. $\begin{bmatrix} 1 & -2 & 1 & -5 \\ 2 & 1 & 7 & 5 \\ 1 & -1 & 2 & -2 \end{bmatrix}$

20. $\begin{bmatrix} 1 & 3 & 10 & 11 & 9 \\ -1 & 2 & 5 & 4 & 1 \\ 2 & -1 & -1 & 1 & 4 \end{bmatrix}$

21. Argue that any set of four or more nonzero vectors in R^3 cannot be an orthogonal set.

22. Let $S = \{\mathbf{u}_1, \mathbf{u}_2, \mathbf{u}_3\}$ be an orthogonal set of nonzero vectors in R^3. Define the (3×3) matrix A by $A = [\mathbf{u}_1, \mathbf{u}_2, \mathbf{u}_3]$. Show that A is nonsingular and $A^T A = D$, where D is a diagonal matrix. Calculate the diagonal matrix D when A is created from the orthogonal vectors in Exercise 1.

23. Let W be a p-dimensional subspace of R^n. If $\mathbf{v}$ is a vector in W such that $\mathbf{v}^T\mathbf{w} = 0$ for every $\mathbf{w}$ in W, show that $\mathbf{v} = \boldsymbol{\theta}$. (*Hint:* Consider $\mathbf{w} = \mathbf{v}$.)

24. *The Cauchy–Schwarz inequality.* Let $\mathbf{x}$ and $\mathbf{y}$ be vectors in R^n. Prove that $|\mathbf{x}^T\mathbf{y}| \leq \|\mathbf{x}\|\|\mathbf{y}\|$. (*Hint:* Observe that $\|\mathbf{x} - c\mathbf{y}\|^2 \geq 0$ for any scalar c. If $\mathbf{y} \neq \boldsymbol{\theta}$, let $c = \mathbf{x}^T\mathbf{y}/\mathbf{y}^T\mathbf{y}$ and expand $(\mathbf{x} - c\mathbf{y})^T(\mathbf{x} - c\mathbf{y}) \geq 0$. Also treat the case $\mathbf{y} = \boldsymbol{\theta}$.)

25. *The triangle inequality.* Let $\mathbf{x}$ and $\mathbf{y}$ be vectors in R^n. Prove that $\|\mathbf{x} + \mathbf{y}\| \leq \|\mathbf{x}\| + \|\mathbf{y}\|$. (*Hint:* Expand $\|\mathbf{x} + \mathbf{y}\|^2$ and use Exercise 24.)

26. Let $\mathbf{x}$ and $\mathbf{y}$ be vectors in R^n. Prove that $\|\|\mathbf{x}\| - \|\mathbf{y}\|\| \leq \|\mathbf{x} - \mathbf{y}\|$. (*Hint:* For one part consider $\|\mathbf{x} + (\mathbf{y} - \mathbf{x})\|$ and Exercise 25.)

27. If the hypotheses for Theorem 14 were altered so that $\{\mathbf{w}_i\}_{i=1}^{p-1}$ is linearly independent and $\{\mathbf{w}_i\}_{i=1}^{p}$ is linearly dependent, use Exercise 23 to show that Eq. (6) yields $\mathbf{u}_p = \boldsymbol{\theta}$.

28. Let $B = \{\mathbf{u}_1, \mathbf{u}_2, \ldots, \mathbf{u}_p\}$ be an *orthonormal* basis for a subspace W. Let $\mathbf{v}$ be any vector in W, where $\mathbf{v} = a_1\mathbf{u}_1 + a_2\mathbf{u}_2 + \cdots + a_p\mathbf{u}_p$. Show that
$$\|\mathbf{v}\|^2 = a_1^2 + a_2^2 + \cdots + a_p^2.$$

2.7 Linear Transformations from R^n to R^m

In this section we consider a special class of functions, called linear transformations, that map vectors to vectors. As we will presently observe, linear transformations arise naturally as a generalization of matrices. Moreover, linear transformations have important applications in engineering, science, the social sciences, and various branches of mathematics.

The notation for linear transformations follows the usual notation for functions. If V is a subspace of R^n and W is a subspace of R^m, then the notation

$$F: V \to W$$

will denote a function, F, whose domain is the subspace V and whose range is contained in W. Furthermore, for **v** in V we write

$$\mathbf{w} = F(\mathbf{v})$$

to indicate that F maps **v** to **w**. To illustrate, let $F: R^3 \to R^2$ be defined by

$$F(\mathbf{x}) = \begin{bmatrix} x_1 - x_2 \\ x_2 + x_3 \end{bmatrix},$$

where

$$\mathbf{x} = \begin{bmatrix} x_1 \\ x_2 \\ x_3 \end{bmatrix}.$$

In this case if, for example, **v** is the vector

$$\mathbf{v} = \begin{bmatrix} 1 \\ 2 \\ 3 \end{bmatrix},$$

then $F(\mathbf{v}) = \mathbf{w}$, where

$$\mathbf{w} = \begin{bmatrix} -1 \\ 5 \end{bmatrix}.$$

In earlier sections we have seen that an $(m \times n)$ matrix A determines a function from R^n to R^m. Specifically for **x** in R^n, the formula

$$T(\mathbf{x}) = A\mathbf{x} \tag{1}$$

defines a function $T: R^n \to R^m$. To illustrate, let A be the (3×2) matrix

$$A = \begin{bmatrix} 1 & -1 \\ 0 & 2 \\ 3 & 1 \end{bmatrix}.$$

In this case Eq. (1) defines a function $T: R^2 \to R^3$, and the formula for T is

$$T(\mathbf{x}) = T\left(\begin{bmatrix} x_1 \\ x_2 \end{bmatrix}\right) = \begin{bmatrix} 1 & -1 \\ 0 & 2 \\ 3 & 1 \end{bmatrix} \begin{bmatrix} x_1 \\ x_2 \end{bmatrix} = \begin{bmatrix} x_1 - x_2 \\ 2x_2 \\ 3x_1 + x_2 \end{bmatrix};$$

for instance,

$$T\left(\begin{bmatrix} 1 \\ 1 \end{bmatrix}\right) = \begin{bmatrix} 0 \\ 2 \\ 4 \end{bmatrix}.$$

Returning to the general case in which A is an $(m \times n)$ matrix, note that the function T defined by Eq. (1) satisfies the following linearity properties:

$$T(\mathbf{v} + \mathbf{w}) = A(\mathbf{v} + \mathbf{w}) = A\mathbf{v} + A\mathbf{w} = T(\mathbf{v}) + T(\mathbf{w})$$
$$T(c\mathbf{v}) = A(c\mathbf{v}) = cA\mathbf{v} = cT(\mathbf{v}), \tag{2}$$

where $\mathbf{v}$ and $\mathbf{w}$ are any vectors in R^n and c is an arbitrary scalar. We next define a linear transformation to be a function that satisfies the two linearity properties given in (2).

> **DEFINITION 8**   Let V and W be subspaces of R^n and R^m, respectively, and let T be a function from V to W, $T: V \to W$. We say that T is a *linear transformation* if for all $\mathbf{u}$ and $\mathbf{v}$ in V and for all scalars a
>
> $$T(\mathbf{u} + \mathbf{v}) = T(\mathbf{u}) + T(\mathbf{v})$$
>
> and $\tag{3}$
>
> $$T(a\mathbf{u}) = aT(\mathbf{u}).$$

It is apparent from (2) that the function T defined in Eq. (1) by matrix multiplication is a linear transformation. Conversely, if $T: R^n \to R^m$ is a linear transformation, then (see Theorem 15 below) there is an $(m \times n)$ matrix A such that T is defined by Eq. (1). Thus linear transformations from R^n to R^m are precisely those functions that can be defined by matrix multiplication as in Eq. (1). The situation is not so simple for linear transformations on arbitrary vector spaces or even for linear transformations on subspaces of R^n. Thus the concept of a linear transformation is a convenient and useful generalization to arbitrary subspaces of matrix functions defined as in Eq. (1).

Examples of Linear Transformations

Most of the familiar functions from the reals to the reals are not linear transformations. For example, none of the functions

$$f(x) = x + 1, \qquad g(x) = x^2, \qquad h(x) = \sin x, \qquad k(x) = e^x$$

is a linear transformation. Indeed, it will follow from the exercises that a function $f: R \to R$ is a linear transformation if and only if f is defined by $f(x) = ax$ for some scalar a.

We now give several examples to illustrate the use of Definition 8 in verifying whether a function is or is not a linear transformation.

EXAMPLE 1 Let $F: R^3 \to R^2$ be the function defined by

$$F(\mathbf{x}) = \begin{bmatrix} x_1 - x_2 \\ x_2 + x_3 \end{bmatrix}, \quad \text{where} \quad \mathbf{x} = \begin{bmatrix} x_1 \\ x_2 \\ x_3 \end{bmatrix}.$$

Determine whether F is a linear transformation.

SOLUTION We must determine whether the two linearity properties in (3) are satisfied by F. Thus let $\mathbf{u}$ and $\mathbf{v}$ be in R^3,

$$\mathbf{u} = \begin{bmatrix} u_1 \\ u_2 \\ u_3 \end{bmatrix} \quad \text{and} \quad \mathbf{v} = \begin{bmatrix} v_1 \\ v_2 \\ v_3 \end{bmatrix},$$

and let c be a scalar. Then

$$\mathbf{u} + \mathbf{v} = \begin{bmatrix} u_1 + v_1 \\ u_2 + v_2 \\ u_3 + v_3 \end{bmatrix}.$$

Therefore, from the rule defining F,

$$F(\mathbf{u} + \mathbf{v}) = \begin{bmatrix} (u_1 + v_1) - (u_2 + v_2) \\ (u_2 + v_2) + (u_3 + v_3) \end{bmatrix}$$

$$= \begin{bmatrix} u_1 - u_2 \\ u_2 + u_3 \end{bmatrix} + \begin{bmatrix} v_1 - v_2 \\ v_2 + v_3 \end{bmatrix}$$

$$= F(\mathbf{u}) + F(\mathbf{v}).$$

Similarly,

$$F(c\mathbf{u}) = \begin{bmatrix} cu_1 - cu_2 \\ cu_2 + cu_3 \end{bmatrix} = c \begin{bmatrix} u_1 - u_2 \\ u_2 + u_3 \end{bmatrix} = cF(\mathbf{u}),$$

so F is a linear transformation.

Note that F can also be defined as $F(\mathbf{x}) = A\mathbf{x}$, where A is the (2×3) matrix

$$A = \begin{bmatrix} 1 & -1 & 0 \\ 0 & 1 & 1 \end{bmatrix}.$$

EXAMPLE 2 Define $H: R^2 \to R^2$ by

$$H(\mathbf{x}) = \begin{bmatrix} x_1 - x_2 + 1 \\ 3x_2 \end{bmatrix}, \quad \text{where} \quad \mathbf{x} = \begin{bmatrix} x_1 \\ x_2 \end{bmatrix}.$$

Determine whether H is a linear transformation.

SOLUTION   Let **u** and **v** be in R^2:

$$\mathbf{u} = \begin{bmatrix} u_1 \\ u_2 \end{bmatrix} \quad \text{and} \quad \mathbf{v} = \begin{bmatrix} v_1 \\ v_2 \end{bmatrix}.$$

Then

$$H(\mathbf{u} + \mathbf{v}) = \begin{bmatrix} (u_1 + v_1) - (u_2 + v_2) + 1 \\ 3(u_2 + v_2) \end{bmatrix},$$

while

$$H(\mathbf{u}) + H(\mathbf{v}) = \begin{bmatrix} u_1 - u_2 + 1 \\ 3u_2 \end{bmatrix} + \begin{bmatrix} v_1 - v_2 + 1 \\ 3v_2 \end{bmatrix}$$

$$= \begin{bmatrix} (u_1 + v_1) - (u_2 + v_2) + 2 \\ 3(u_2 + v_2) \end{bmatrix}.$$

From above, we see that $H(\mathbf{u} + \mathbf{v}) \neq H(\mathbf{u}) + H(\mathbf{v})$. Therefore, H is not a linear transformation. Although it is not necessary, it can also be verified easily that if $c \neq 1$, then $H(c\mathbf{u}) \neq cH(\mathbf{u})$.   ☐

EXAMPLE 3   Let W be a subspace of R^n such that $\dim(W) = p$, and let $S = \{\mathbf{w}_1, \mathbf{w}_2, \ldots, \mathbf{w}_p\}$ be an orthonormal basis for W. Define $T: R^n \to W$ by

$$T(\mathbf{v}) = (\mathbf{v}^T\mathbf{w}_1)\mathbf{w}_1 + (\mathbf{v}^T\mathbf{w}_2)\mathbf{w}_2 + \cdots + (\mathbf{v}^T\mathbf{w}_p)\mathbf{w}_p. \tag{4}$$

Prove that T is a linear transformation.

SOLUTION   If **u** and **v** are in R^n, then

$$\begin{aligned} T(\mathbf{u} + \mathbf{v}) &= [(\mathbf{u} + \mathbf{v})^T\mathbf{w}_1]\mathbf{w}_1 + [(\mathbf{u} + \mathbf{v})^T\mathbf{w}_2]\mathbf{w}_2 + \cdots + [(\mathbf{u} + \mathbf{v})^T\mathbf{w}_p]\mathbf{w}_p \\ &= [(\mathbf{u}^T + \mathbf{v}^T)\mathbf{w}_1]\mathbf{w}_1 + [(\mathbf{u}^T + \mathbf{v}^T)\mathbf{w}_2]\mathbf{w}_2 + \cdots + [(\mathbf{u}^T + \mathbf{v}^T)\mathbf{w}_p]\mathbf{w}_p \\ &= (\mathbf{u}^T\mathbf{w}_1)\mathbf{w}_1 + (\mathbf{u}^T\mathbf{w}_2)\mathbf{w}_2 + \cdots + (\mathbf{u}^T\mathbf{w}_p)\mathbf{w}_p \\ &\quad + (\mathbf{v}^T\mathbf{w}_1)\mathbf{w}_1 + (\mathbf{v}^T\mathbf{w}_2)\mathbf{w}_2 + \cdots + (\mathbf{v}^T\mathbf{w}_p)\mathbf{w}_p \\ &= T(\mathbf{u}) + T(\mathbf{v}). \end{aligned}$$

It can be shown similarly that $T(c\mathbf{u}) = cT(\mathbf{u})$ for each scalar c, so T is a linear transformation.   ☐

The vector $T(\mathbf{v})$ defined by Eq. (4) is called the orthogonal projection of **v** onto W and will be considered further in Sections 2.8 and 2.9. As a specific illustration of Example 3, let W be the subspace of R^3 consisting of all vectors of the form

$$\mathbf{x} = \begin{bmatrix} x_1 \\ x_2 \\ 0 \end{bmatrix}.$$

Thus W is the xy-plane, and the set $\{\mathbf{e}_1, \mathbf{e}_2\}$ is an orthonormal basis for W. For $\mathbf{x}$ in R^3,

$$\mathbf{x} = \begin{bmatrix} x_1 \\ x_2 \\ x_3 \end{bmatrix},$$

the formula in Eq. (4) yields

$$T(\mathbf{x}) = (\mathbf{x}^T\mathbf{e}_1)\mathbf{e}_1 + (\mathbf{x}^T\mathbf{e}_2)\mathbf{e}_2 = x_1\mathbf{e}_1 + x_2\mathbf{e}_2.$$

Thus,

$$T(\mathbf{x}) = \begin{bmatrix} x_1 \\ x_2 \\ 0 \end{bmatrix}.$$

This transformation is illustrated geometrically by Fig. 2.15.

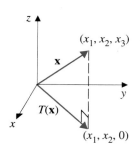

Figure 2.15
Orthogonal projection onto the xy-plane

EXAMPLE 4 Let W be a subspace of R^n, and let a be a scalar. Define $T\colon W \to W$ by $T(\mathbf{w}) = a\mathbf{w}$. Demonstrate that T is a linear transformation.

SOLUTION   If $\mathbf{v}$ and $\mathbf{w}$ are in W, then

$$T(\mathbf{v} + \mathbf{w}) = a(\mathbf{v} + \mathbf{w}) = a\mathbf{v} + a\mathbf{w} = T(\mathbf{v}) + T(\mathbf{w}).$$

Likewise, if c is a scalar, then

$$T(c\mathbf{w}) = a(c\mathbf{w}) = c(a\mathbf{w}) = cT(\mathbf{w}).$$

It follows that T is a linear transformation.

The linear transformation defined in Example 4 is called a ***dilation*** when $a > 1$ and a ***contraction*** when $0 < a < 1$. These cases are illustrated geometrically in Fig. 2.16.

The mapping $I\colon W \to W$ defined by $I(\mathbf{w}) = \mathbf{w}$ is the special case of Example 4 in which $a = 1$. The linear transformation I is called the ***identity transformation.***

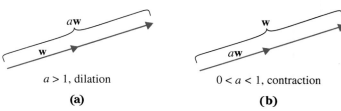

$a > 1$, dilation

(a)

$0 < a < 1$, contraction

(b)

Figure 2.16

<u>EXAMPLE 5</u> Let W be a subspace of R^n, and let θ be the zero vector in R^m. Define T: $W \to R^m$ by $T(\mathbf{w}) = \theta$ for each $\mathbf{w}$ in W. Show that T is a linear transformation.

SOLUTION Let $\mathbf{v}$ and $\mathbf{w}$ be vectors in W, and let c be a scalar. Then

$$T(\mathbf{v} + \mathbf{w}) = \theta = \theta + \theta = T(\mathbf{v}) + T(\mathbf{w})$$

and

$$T(c\mathbf{v}) = \theta = c\theta = cT(\mathbf{v}),$$

so T is a linear transformation.

    The linear transformation T defined in Example 5 is called the ***zero transformation.***
    Later in this section we will consider other examples when we study a particular class of linear transformations from R^2 to R^2. For the present, we turn to further properties of linear transformations.

The Matrix of a Transformation

Let V and W be subspaces, and let $T: V \to W$ be a linear transformation. If $\mathbf{u}$ and $\mathbf{v}$ are vectors in V and if a and b are scalars, then the linearity properties (3) yield

$$T(a\mathbf{u} + b\mathbf{v}) = T(a\mathbf{u}) + T(b\mathbf{v}) = aT(\mathbf{u}) + bT(\mathbf{v}). \tag{5}$$

Inductively we can extend Eq. (5) to any finite subset of V. That is, if $\mathbf{v}_1, \mathbf{v}_2, \ldots, \mathbf{v}_r$ are vectors in V and if $c_1, c_2, \ldots, c_r$ are scalars, then

$$T(c_1\mathbf{v}_1 + c_2\mathbf{v}_2 + \cdots + c_r\mathbf{v}_r) = c_1 T(\mathbf{v}_1) + c_2 T(\mathbf{v}_2) + \cdots + c_r T(\mathbf{v}_r). \tag{6}$$

The following example illustrates an application of Eq. (6).

<u>EXAMPLE 6</u> Let W be the subspace of R^3 defined by

$$W = \left\{ \mathbf{x} : \mathbf{x} = \begin{bmatrix} x_2 + 2x_3 \\ x_2 \\ x_3 \end{bmatrix}, \quad x_2 \text{ and } x_3 \text{ any real numbers} \right\}.$$

Then $\{\mathbf{w}_1, \mathbf{w}_2\}$ is a basis for W, where

$$\mathbf{w}_1 = \begin{bmatrix} 1 \\ 1 \\ 0 \end{bmatrix} \quad \text{and} \quad \mathbf{w}_2 = \begin{bmatrix} 2 \\ 0 \\ 1 \end{bmatrix}.$$

Suppose that $T: W \to R^2$ is a linear transformation such that $T(\mathbf{w}_1) = \mathbf{u}_1$ and $T(\mathbf{w}_2) = \mathbf{u}_2$, where

$$\mathbf{u}_1 = \begin{bmatrix} 1 \\ 1 \end{bmatrix} \quad \text{and} \quad \mathbf{u}_2 = \begin{bmatrix} 1 \\ -1 \end{bmatrix}.$$

Let the vector $\mathbf{w}$ be given by

$$\mathbf{w} = \begin{bmatrix} -1 \\ 3 \\ -2 \end{bmatrix}.$$

Show that $\mathbf{w}$ is in W, express $\mathbf{w}$ as a linear combination of $\mathbf{w}_1$ and $\mathbf{w}_2$, and use Eq. (6) to determine $T(\mathbf{w})$.

SOLUTION It follows from the description of W that $\mathbf{w}$ is in W. Furthermore, it is easy to see that

$$\mathbf{w} = 3\mathbf{w}_1 - 2\mathbf{w}_2.$$

By Eq. (6),

$$T(\mathbf{w}) = 3T(\mathbf{w}_1) - 2T(\mathbf{w}_2) = 3\mathbf{u}_1 - 2\mathbf{u}_2 = 3\begin{bmatrix} 1 \\ 1 \end{bmatrix} - 2\begin{bmatrix} 1 \\ -1 \end{bmatrix}.$$

Thus,

$$T(\mathbf{w}) = \begin{bmatrix} 1 \\ 5 \end{bmatrix}.$$

Example 6 illustrates that the action of a linear transformation T on a subspace W is completely determined once the action of T on a basis for W is known. Our next example provides yet another illustration of this fact.

EXAMPLE 7 Let $T: R^3 \to R^2$ be a linear transformation such that

$$T(\mathbf{e}_1) = \begin{bmatrix} 1 \\ 2 \end{bmatrix}, \qquad T(\mathbf{e}_2) = \begin{bmatrix} -1 \\ 1 \end{bmatrix}, \quad \text{and} \quad T(\mathbf{e}_3) = \begin{bmatrix} 2 \\ 3 \end{bmatrix}.$$

For an arbitrary vector $\mathbf{x}$ in R^3,

$$\mathbf{x} = \begin{bmatrix} x_1 \\ x_2 \\ x_3 \end{bmatrix},$$

give a formula for $T(\mathbf{x})$.

SOLUTION The vector $\mathbf{x}$ can be written in the form

$$\mathbf{x} = x_1\mathbf{e}_1 + x_2\mathbf{e}_2 + x_3\mathbf{e}_3,$$

so by Eq. (6),

$$T(\mathbf{x}) = x_1 T(\mathbf{e}_1) + x_2 T(\mathbf{e}_2) + x_3 T(\mathbf{e}_3). \tag{7}$$

Thus,

$$T(\mathbf{x}) = x_1\begin{bmatrix} 1 \\ 2 \end{bmatrix} + x_2\begin{bmatrix} -1 \\ 1 \end{bmatrix} + x_3\begin{bmatrix} 2 \\ 3 \end{bmatrix} = \begin{bmatrix} x_1 - x_2 + 2x_3 \\ 2x_1 + x_2 + 3x_3 \end{bmatrix}.$$

Continuing with the notation of the preceding example, let A be the (2×3) matrix with columns $T(\mathbf{e}_1)$, $T(\mathbf{e}_2)$, $T(\mathbf{e}_3)$; thus,

$$A = [T(\mathbf{e}_1), T(\mathbf{e}_2), T(\mathbf{e}_3)] = \begin{bmatrix} 1 & -1 & 2 \\ 2 & 1 & 3 \end{bmatrix}.$$

It is an immediate consequence of Eq. (7) and Theorem 5 of Section 1.6 that $T(\mathbf{x}) = A\mathbf{x}$. Thus Example 7 illustrates the following theorem.

THEOREM 15

Let $T: R^n \rightarrow R^m$ be a linear transformation, and let $\mathbf{e}_1, \mathbf{e}_2, \ldots, \mathbf{e}_n$ be the unit vectors in R^n. If A is the $(m \times n)$ matrix defined by

$$A = [T(\mathbf{e}_1), T(\mathbf{e}_2), \ldots, T(\mathbf{e}_n)],$$

then $T(\mathbf{x}) = A\mathbf{x}$ for all $\mathbf{x}$ in R^n.

PROOF   If $\mathbf{x}$ is a vector in R^n,

$$\mathbf{x} = \begin{bmatrix} x_1 \\ x_2 \\ \vdots \\ x_n \end{bmatrix},$$

then $\mathbf{x}$ can be expressed in the form

$$\mathbf{x} = x_1\mathbf{e}_1 + x_2\mathbf{e}_2 + \cdots + x_n\mathbf{e}_n.$$

It now follows from Eq. (6) that

$$T(\mathbf{x}) = x_1 T(\mathbf{e}_1) + x_2 T(\mathbf{e}_2) + \cdots + x_n T(\mathbf{e}_n). \tag{8}$$

If $A = [T(\mathbf{e}_1), T(\mathbf{e}_2), \ldots, T(\mathbf{e}_n)]$, then by Theorem 5 of Section 1.6, the right-hand side of Eq. (8) is simply $A\mathbf{x}$. Thus Eq. (8) is equivalent to $T(\mathbf{x}) = A\mathbf{x}$.

EXAMPLE 8   Let $T: R^2 \rightarrow R^3$ be the linear transformation defined by the formula

$$T\left(\begin{bmatrix} x_1 \\ x_2 \end{bmatrix}\right) = \begin{bmatrix} x_1 + 2x_2 \\ -x_1 + x_2 \\ 2x_1 - x_2 \end{bmatrix}.$$

Find a matrix A such that $T(\mathbf{x}) = A\mathbf{x}$ for each $\mathbf{x}$ in R^2.

SOLUTION   By Theorem 15, A is the (3×2) matrix

$$A = [T(\mathbf{e}_1), T(\mathbf{e}_2)].$$

It is an easy calculation that

$$T(\mathbf{e}_1) = \begin{bmatrix} 1 \\ -1 \\ 2 \end{bmatrix} \quad \text{and} \quad T(\mathbf{e}_2) = \begin{bmatrix} 2 \\ 1 \\ -1 \end{bmatrix}.$$

Therefore,

$$A = \begin{bmatrix} 1 & 2 \\ -1 & 1 \\ 2 & -1 \end{bmatrix}.$$

One can easily verify that $T(\mathbf{x}) = A\mathbf{x}$ for each $\mathbf{x}$ in R^2.

Null Space and Range

Associated with a linear transformation, T, are two important and useful subspaces called the null space and the range of T. These are defined as follows:

DEFINITION 9 Let V and W be subspaces, and let $T: V \rightarrow W$ be a linear transformation. The **null space** of T, denoted by $\mathcal{N}(T)$, is the subset of V given by

$$\mathcal{N}(T) = \{\mathbf{v}: \mathbf{v} \text{ is in } V \text{ and } T(\mathbf{v}) = \boldsymbol{\theta}\}.$$

The **range** of T, denoted by $\mathcal{R}(T)$, is the subset of W defined by

$$\mathcal{R}(T) = \{\mathbf{w}: \mathbf{w} \text{ is in } W \text{ and } \mathbf{w} = T(\mathbf{v}) \text{ for some } \mathbf{v} \text{ in } V\}.$$

That $\mathcal{N}(T)$ and $\mathcal{R}(T)$ are subspaces will be proved in the more general context of Chapter 4. If T maps R^n into R^m, then by Theorem 15 there exists an $(m \times n)$ matrix A such that $T(\mathbf{x}) = A\mathbf{x}$. In this case it is clear that the null space of T is the null space of A and the range of T coincides with the range of A.

As with matrices, the dimension of the null space of a linear transformation T is called the **nullity** of T, and the dimension of the range of T is called the **rank** of T. If T is defined by matrix multiplication, $T(\mathbf{x}) = A\mathbf{x}$, then the transformation T and the matrix A have the same nullity and the same rank. Moreover, if $T: R^n \rightarrow R^m$, then A is an $(m \times n)$ matrix, so it follows from the remark in Section 2.5 that

$$\text{rank}(T) + \text{nullity}(T) = n. \tag{9}$$

Formula (9) will be proved in a more general setting in Chapter 4.

The next two examples illustrate the use of the matrix of T to determine the null space and the range of T.

EXAMPLE 9 Let F be the linear transformation given in Example 1, $F: R^3 \rightarrow R^2$. Describe the null space and the range of F, and determine the nullity and the rank of F.

SOLUTION It follows from Theorem 15 that $F(\mathbf{x}) = A\mathbf{x}$, where A is the (2×3) matrix

$$A = [F(\mathbf{e}_1), F(\mathbf{e}_2), F(\mathbf{e}_3)] = \begin{bmatrix} 1 & -1 & 0 \\ 0 & 1 & 1 \end{bmatrix}.$$

Thus the null space and the range of F coincide, respectively, with the null space and the range of A. Since A is already in echelon form, the null space of A is determined by backsolving the homogeneous system $A\mathbf{x} = \boldsymbol{0}$, where $\mathbf{x}$ is in R^3:

$$\mathbf{x} = \begin{bmatrix} x_1 \\ x_2 \\ x_3 \end{bmatrix}.$$

This gives

$$\mathcal{N}(F) = \mathcal{N}(A) = \{\mathbf{x}: x_1 = -x_3 \quad \text{and} \quad x_2 = -x_3\}.$$

Using the techniques of Section 2.4, we can easily see that the vector

$$\mathbf{u} = \begin{bmatrix} -1 \\ -1 \\ 1 \end{bmatrix}$$

is a basis for $\mathcal{N}(F)$, so F has nullity 1. By Eq. (9),

$$\text{rank}(F) = n - \text{nullity}(F) = 3 - 1 = 2.$$

Thus $\mathcal{R}(F)$ is a two-dimensional subspace of R^2, and hence $\mathcal{R}(F) = R^2$.

Alternatively, note that the system of equations $A\mathbf{x} = \mathbf{b}$ has a solution for each $\mathbf{b}$ in R^2, so $\mathcal{R}(F) = \mathcal{R}(A) = R^2$. $\square$

EXAMPLE 10 Let $T: R^2 \to R^3$ be the linear transformation given in Example 8. Describe the null space and the range of T, and determine the nullity and the rank of T.

SOLUTION In Example 8 it was shown that $T(\mathbf{x}) = A\mathbf{x}$, where A is the (3×2) matrix

$$A = \begin{bmatrix} 1 & 2 \\ -1 & 1 \\ 2 & -1 \end{bmatrix}.$$

If $\mathbf{b}$ is the (3×1) vector,

$$\mathbf{b} = \begin{bmatrix} b_1 \\ b_2 \\ b_3 \end{bmatrix},$$

then the augmented matrix $[A\,|\,\mathbf{b}]$ for the linear system $A\mathbf{x} = \mathbf{b}$ is row equivalent to

$$\begin{bmatrix} 1 & 2 & b_1 \\ 0 & 3 & b_1 + b_2 \\ 0 & 0 & (-1/3)b_1 + (5/3)b_2 + b_3 \end{bmatrix}. \tag{10}$$

Therefore, $T(\mathbf{x}) = A\mathbf{x} = \mathbf{b}$ can be solved if and only if $0 = (-1/3)b_1 + (5/3)b_2 + b_3$.

The range of T can thus be described as

$$\mathcal{R}(T) = \mathcal{R}(A)$$

$$= \left\{ \mathbf{b}: \mathbf{b} = \begin{bmatrix} b_1 \\ b_2 \\ (1/3)b_1 - (5/3)b_2 \end{bmatrix}, \quad b_1 \text{ and } b_2 \text{ any real numbers} \right\}.$$

A basis for $\mathcal{R}(T)$ is $\{\mathbf{u}_1, \mathbf{u}_2\}$ where

$$\mathbf{u}_1 = \begin{bmatrix} 1 \\ 0 \\ 1/3 \end{bmatrix} \quad \text{and} \quad \mathbf{u}_2 = \begin{bmatrix} 0 \\ 1 \\ -5/3 \end{bmatrix}.$$

Thus T has rank 2, and by Eq. (9),

$$\text{nullity}(T) = n - \text{rank}(T) = 2 - 2 = 0.$$

It follows that T has null space $\{\boldsymbol{\theta}\}$. Alternatively, it is clear from matrix (10), with $\mathbf{b} = \boldsymbol{\theta}$, that the homogeneous system of equations $A\mathbf{x} = \boldsymbol{\theta}$ has only the trivial solution. Therefore, $\mathcal{N}(T) = \mathcal{N}(A) = \{\boldsymbol{\theta}\}$.

Orthogonal Transformations on R^2 (Optional)

It is often informative and useful to view linear transformations on either R^2 or R^3 from a geometric point of view. To illustrate this general notion, the remainder of this section is devoted to determining those linear transformations $T: R^2 \to R^2$ that preserve the length of a vector; that is, we are interested in linear transformations T such that

$$\|T(\mathbf{v})\| = \|\mathbf{v}\| \tag{11}$$

for all $\mathbf{v}$ in R^2. Transformations that satisfy Eq. (11) are called *orthogonal transformations.* We begin by giving some examples of orthogonal transformations.

EXAMPLE 11 Let θ be a fixed angle, and let $T: R^2 \to R^2$ be the linear transformation defined by $T(\mathbf{v}) = A\mathbf{v}$, where A is the (2×2) matrix

$$A = \begin{bmatrix} \cos\theta & -\sin\theta \\ \sin\theta & \cos\theta \end{bmatrix}.$$

Give a geometric interpretation of T, and show that T is an orthogonal transformation.

SOLUTION Suppose that $\mathbf{v}$ and $T(\mathbf{v})$ are given by

$$\mathbf{v} = \begin{bmatrix} a \\ b \end{bmatrix} \quad \text{and} \quad T(\mathbf{v}) = \begin{bmatrix} c \\ d \end{bmatrix}.$$

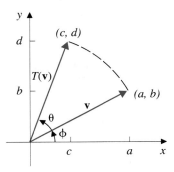

Figure 2.17
Rotation through the angle θ

Then $T(\mathbf{v}) = A\mathbf{v}$, so

$$\begin{bmatrix} c \\ d \end{bmatrix} = \begin{bmatrix} \cos\theta & -\sin\theta \\ \sin\theta & \cos\theta \end{bmatrix} \begin{bmatrix} a \\ b \end{bmatrix} = \begin{bmatrix} a\cos\theta - b\sin\theta \\ a\sin\theta + b\cos\theta \end{bmatrix}. \tag{12}$$

We proceed now to show that $T(\mathbf{v})$ is obtained geometrically by rotating the vector $\mathbf{v}$ through the angle θ. To see this, let ϕ be the angle between $\mathbf{v}$ and the positive x-axis (see Fig. 2.17), and set $r = \|\mathbf{v}\|$. Then the coordinates a and b can be written as

$$a = r\cos\phi, \qquad b = r\sin\phi. \tag{13}$$

Making the substitution (13) for a and b in (12) yields

$$c = r\cos\phi\cos\theta - r\sin\phi\sin\theta = r\cos(\phi + \theta)$$

and

$$d = r\cos\phi\sin\theta + r\sin\phi\cos\theta = r\sin(\phi + \theta).$$

Therefore, c and d are the coordinates of the point obtained by rotating the point (a, b) through the angle θ. Clearly then, $\|T(\mathbf{v})\| = \|\mathbf{v}\|$ and T is an orthogonal linear transformation.

The linear transformation T defined in Example 11 is called a **rotation.** Thus if A is a (2×2) matrix,

$$A = \begin{bmatrix} a & -b \\ b & a \end{bmatrix},$$

where $a^2 + b^2 = 1$, then the linear transformation $T(\mathbf{v}) = A\mathbf{v}$ is the rotation through the angle θ, $0 \le \theta < 2\pi$, where $\cos\theta = a$ and $\sin\theta = b$.

EXAMPLE 12 Define $T: R^2 \to R^2$ by $T(\mathbf{v}) = A\mathbf{v}$, where

$$A = \begin{bmatrix} -1/2 & \sqrt{3}/2 \\ -\sqrt{3}/2 & -1/2 \end{bmatrix}.$$

Give a geometric interpretation of T.

SOLUTION Since $\cos(4\pi/3) = -1/2$ and $\sin(4\pi/3) = -\sqrt{3}/2$, T is the rotation through the angle $4\pi/3$.

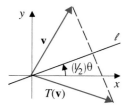

Figure 2.18
Reflection about
a line

Now let l be a line in the plane that passes through the origin, and let $\mathbf{v}$ be a vector in the plane. If we define $T(\mathbf{v})$ to be the symmetric image of $\mathbf{v}$ relative to l (see Fig. 2.18), then clearly T preserves the length of $\mathbf{v}$. It can be shown that T is multiplication by the matrix

$$A = \begin{bmatrix} \cos\theta & \sin\theta \\ \sin\theta & -\cos\theta \end{bmatrix},$$

where $(1/2)\theta$ is the angle between l and the positive x-axis. Any such transformation is called a **reflection.** Note that a reflection T is also an orthogonal linear transformation.

EXAMPLE 13 Let $T: R^2 \to R^2$ be defined by $T(\mathbf{v}) = A\mathbf{v}$, where A is the (2×2) matrix

$$A = \begin{bmatrix} 1/2 & \sqrt{3}/2 \\ \sqrt{3}/2 & -1/2 \end{bmatrix}.$$

Give a geometric interpretation of T.

SOLUTION   Since $\cos(\pi/3) = 1/2$ and $\sin(\pi/3) = \sqrt{3}/2$, T is the reflection about the line l, where l is the line that passes through the origin at an angle of 30 degrees.

The next theorem gives a characterization of orthogonal transformations on R^2. A consequence of this theorem will be that every orthogonal transformation is either a rotation or a reflection.

THEOREM 16

Let $T: R^2 \to R^2$ be a linear transformation. Then T is an orthogonal transformation if and only if $\|T(\mathbf{e}_1)\| = \|T(\mathbf{e}_2)\| = 1$ and $T(\mathbf{e}_1)$ is perpendicular to $T(\mathbf{e}_2)$.

PROOF   If T is an orthogonal transformation, then $\|T(\mathbf{v})\| = \|\mathbf{v}\|$ for every vector $\mathbf{v}$ in R^2. In particular, $\|T(\mathbf{e}_1)\| = \|\mathbf{e}_1\| = 1$, and similarly $\|T(\mathbf{e}_2)\| = 1$. Set $\mathbf{u}_1 = T(\mathbf{e}_1)$, $\mathbf{u}_2 = T(\mathbf{e}_2)$, and $\mathbf{v} = [1, 1]^T = \mathbf{e}_1 + \mathbf{e}_2$. Then

$$2 = \|\mathbf{v}\|^2 = \|T(\mathbf{v})\|^2 = \|T(\mathbf{e}_1 + \mathbf{e}_2)\|^2 = \|T(\mathbf{e}_1) + T(\mathbf{e}_2)\|^2.$$

Thus,

$$\begin{aligned} 2 &= \|\mathbf{u}_1 + \mathbf{u}_2\|^2 \\ &= (\mathbf{u}_1 + \mathbf{u}_2)^T(\mathbf{u}_1 + \mathbf{u}_2) \\ &= (\mathbf{u}_1^T + \mathbf{u}_2^T)(\mathbf{u}_1 + \mathbf{u}_2) \\ &= \mathbf{u}_1^T\mathbf{u}_1 + \mathbf{u}_1^T\mathbf{u}_2 + \mathbf{u}_2^T\mathbf{u}_1 + \mathbf{u}_2^T\mathbf{u}_2 \\ &= \|\mathbf{u}_1\|^2 + 2\mathbf{u}_1^T\mathbf{u}_2 + \|\mathbf{u}_2\|^2 \\ &= 2 + 2\mathbf{u}_1^T\mathbf{u}_2. \end{aligned}$$

It follows that $\mathbf{u}_1^T\mathbf{u}_2 = 0$, so $\mathbf{u}_1$ is perpendicular to $\mathbf{u}_2$.
The proof of the converse is Exercise 47.

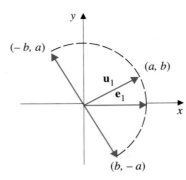

Figure 2.19
Choices for $\mathbf{u}_2$

We can now use Theorem 16 to give a geometric description for any orthogonal linear transformation, T, on R^2. First, suppose that $T(\mathbf{e}_1) = \mathbf{u}_1$ and $T(\mathbf{e}_2) = \mathbf{u}_2$. If

$$\mathbf{u}_1 = \begin{bmatrix} a \\ b \end{bmatrix},$$

then $1 = \|\mathbf{u}_1\|^2 = a^2 + b^2$. Since $\|\mathbf{u}_2\| = 1$ and $\mathbf{u}_2$ is perpendicular to $\mathbf{u}_1$, there are two choices for $\mathbf{u}_2$ (see Fig. 2.19): either

$$\mathbf{u}_2 = \begin{bmatrix} -b \\ a \end{bmatrix} \quad \text{or} \quad \mathbf{u}_2 = \begin{bmatrix} b \\ -a \end{bmatrix}.$$

In either case, it follows from Theorem 15 that T is defined by $T(\mathbf{v}) = A\mathbf{v}$, where A is the (2×2) matrix $A = [\mathbf{u}_1, \mathbf{u}_2]$. Thus if

$$\mathbf{u}_2 = \begin{bmatrix} -b \\ a \end{bmatrix},$$

then

$$A = \begin{bmatrix} a & -b \\ b & a \end{bmatrix},$$

so T is a rotation. If

$$\mathbf{u}_2 = \begin{bmatrix} b \\ -a \end{bmatrix},$$

then

$$A = \begin{bmatrix} a & b \\ b & -a \end{bmatrix},$$

and T is a reflection. In either case note that $A^TA = I$, so $A^T = A^{-1}$ (cf. Exercise 48). An $(n \times n)$ real matrix with the property that $A^TA = I$ is called an ***orthogonal matrix.*** Thus we have shown that an orthogonal transformation on R^2 is defined by

$$T(\mathbf{x}) = A\mathbf{x},$$

where A is an orthogonal matrix.

Exercises 2.7

1. Define $T: R^2 \to R^2$ by

$$T\left(\begin{bmatrix} x_1 \\ x_2 \end{bmatrix}\right) = \begin{bmatrix} 2x_1 - 3x_2 \\ -x_1 + x_2 \end{bmatrix}.$$

Find each of the following.

a) $T\left(\begin{bmatrix} 0 \\ 0 \end{bmatrix}\right)$ b) $T\left(\begin{bmatrix} 1 \\ 1 \end{bmatrix}\right)$

c) $T\left(\begin{bmatrix} 2 \\ 1 \end{bmatrix}\right)$ d) $T\left(\begin{bmatrix} -1 \\ 0 \end{bmatrix}\right)$

2. Define $T: R^2 \to R^2$ by $T(\mathbf{x}) = A\mathbf{x}$, where

$$A = \begin{bmatrix} 1 & -1 \\ -3 & 3 \end{bmatrix}.$$

Find each of the following.

a) $T\left(\begin{bmatrix} 2 \\ 2 \end{bmatrix}\right)$ b) $T\left(\begin{bmatrix} 3 \\ 1 \end{bmatrix}\right)$

c) $T\left(\begin{bmatrix} 2 \\ 0 \end{bmatrix}\right)$ d) $T\left(\begin{bmatrix} 0 \\ 0 \end{bmatrix}\right)$

3. Let $T: R^3 \to R^2$ be the linear transformation defined by

$$T\left(\begin{bmatrix} x_1 \\ x_2 \\ x_3 \end{bmatrix}\right) = \begin{bmatrix} x_1 + 2x_2 + 4x_3 \\ 2x_1 + 3x_2 + 5x_3 \end{bmatrix}.$$

Which of the following vectors are in the null space of T?

a) $\begin{bmatrix} 0 \\ 0 \\ 0 \end{bmatrix}$ b) $\begin{bmatrix} 2 \\ -3 \\ 1 \end{bmatrix}$

c) $\begin{bmatrix} 1 \\ 2 \\ 1 \end{bmatrix}$ d) $\begin{bmatrix} -1 \\ 3/2 \\ -1/2 \end{bmatrix}$

4. Let $T: R^2 \to R^2$ be the function defined in Exercise 1. Find $\mathbf{x}$ in R^2 such that $T(\mathbf{x}) = \mathbf{b}$, where

$$\mathbf{b} = \begin{bmatrix} 2 \\ -2 \end{bmatrix}.$$

5. Let $T: R^2 \to R^2$ be the function given in Exercise 1. Show that for each $\mathbf{b}$ in R^2, there is an $\mathbf{x}$ in R^2 such that $T(\mathbf{x}) = \mathbf{b}$.

6. Let T be the linear transformation given in Exer-

cise 2. Find $\mathbf{x}$ in R^2 such that $T(\mathbf{x}) = \mathbf{b}$, where

$$\mathbf{b} = \begin{bmatrix} -2 \\ 6 \end{bmatrix}.$$

7. Let T be the linear transformation given in Exercise 2. Show that there is no $\mathbf{x}$ in R^2 such that $T(\mathbf{x}) = \mathbf{b}$ for

$$\mathbf{b} = \begin{bmatrix} 1 \\ 1 \end{bmatrix}.$$

In Exercises 8–17, determine whether the function F is a linear transformation.

8. $F: R^2 \to R^2$ defined by $F\left(\begin{bmatrix} x_1 \\ x_2 \end{bmatrix}\right) = \begin{bmatrix} 2x_1 - x_2 \\ x_1 + 3x_2 \end{bmatrix}$

9. $F: R^2 \to R^2$ defined by $F\left(\begin{bmatrix} x_1 \\ x_2 \end{bmatrix}\right) = \begin{bmatrix} x_2 \\ x_1 \end{bmatrix}$

10. $F: R^2 \to R^2$ defined by $F\left(\begin{bmatrix} x_1 \\ x_2 \end{bmatrix}\right) = \begin{bmatrix} x_1 + x_2 \\ 1 \end{bmatrix}$

11. $F: R^2 \to R^2$ defined by $F\left(\begin{bmatrix} x_1 \\ x_2 \end{bmatrix}\right) = \begin{bmatrix} x_1^2 \\ x_1 x_2 \end{bmatrix}$

12. $F: R^3 \to R^2$ defined by

$$F\left(\begin{bmatrix} x_1 \\ x_2 \\ x_3 \end{bmatrix}\right) = \begin{bmatrix} x_1 - x_2 + x_3 \\ -x_1 + 3x_2 - 2x_3 \end{bmatrix}$$

13. $F: R^3 \to R^2$ defined by $F\left(\begin{bmatrix} x_1 \\ x_2 \\ x_3 \end{bmatrix}\right) = \begin{bmatrix} x_1 \\ x_2 \end{bmatrix}$

14. $F: R^2 \to R^3$ defined by $F\left(\begin{bmatrix} x_1 \\ x_2 \end{bmatrix}\right) = \begin{bmatrix} x_1 - x_2 \\ -x_1 + x_2 \\ x_2 \end{bmatrix}$

15. $F: R^2 \to R^3$ defined by $F\left(\begin{bmatrix} x_1 \\ x_2 \end{bmatrix}\right) = \begin{bmatrix} x_1 \\ x_2 \\ 0 \end{bmatrix}$

16. $F: R^2 \to R$ defined by $F\left(\begin{bmatrix} x_1 \\ x_2 \end{bmatrix}\right) = 2x_1 + 3x_2$

17. $F: R^2 \to R$ defined by $F\left(\begin{bmatrix} x_1 \\ x_2 \end{bmatrix}\right) = |x_1| + |x_2|$

18. Let W be the subspace of R^3 defined by

$$W = \{\mathbf{x}: \mathbf{x} = \begin{bmatrix} x_1 \\ x_2 \\ x_3 \end{bmatrix}, \quad x_2 = x_3 = 0\}.$$

Find an orthonormal basis for W, and use Eq. (4) of Example 3 to give a formula for the orthogonal projection $T: R^3 \to W$; that is, determine $T(\mathbf{v})$ for arbitrary $\mathbf{v}$ in R^3:

$$\mathbf{v} = \begin{bmatrix} a \\ b \\ c \end{bmatrix}.$$

Give a geometric interpretation of W, $\mathbf{v}$, and $T(\mathbf{v})$.

19. Let $T: R^2 \to R^3$ be a linear transformation such that $T(\mathbf{e}_1) = \mathbf{u}_1$ and $T(\mathbf{e}_2) = \mathbf{u}_2$, where

$$\mathbf{u}_1 = \begin{bmatrix} 1 \\ 0 \\ -1 \end{bmatrix} \quad \text{and} \quad \mathbf{u}_2 = \begin{bmatrix} 2 \\ 1 \\ 0 \end{bmatrix}.$$

Find each of the following.

a) $T\left(\begin{bmatrix} 1 \\ 1 \end{bmatrix}\right)$ b) $T\left(\begin{bmatrix} 2 \\ -1 \end{bmatrix}\right)$

c) $T\left(\begin{bmatrix} 3 \\ 2 \end{bmatrix}\right)$

20. Let $T: R^2 \to R^2$ be a linear transformation such that $T(\mathbf{v}_1) = \mathbf{u}_1$ and $T(\mathbf{v}_2) = \mathbf{u}_2$, where

$$\mathbf{v}_1 = \begin{bmatrix} 0 \\ 1 \end{bmatrix}, \quad \mathbf{v}_2 = \begin{bmatrix} -1 \\ 1 \end{bmatrix},$$

$$\mathbf{u}_1 = \begin{bmatrix} 0 \\ 2 \end{bmatrix}, \quad \text{and} \quad \mathbf{u}_2 = \begin{bmatrix} 3 \\ 1 \end{bmatrix}.$$

Find each of the following.

a) $T\left(\begin{bmatrix} 1 \\ 1 \end{bmatrix}\right)$ b) $T\left(\begin{bmatrix} 2 \\ -1 \end{bmatrix}\right)$

c) $T\left(\begin{bmatrix} 3 \\ 2 \end{bmatrix}\right)$

In Exercises 21–24, the action of a linear transformation T on a basis for either R^2 or R^3 is given. In each case use Eq. (6) to derive a formula for T.

21. $T\left(\begin{bmatrix} 1 \\ 1 \end{bmatrix}\right) = \begin{bmatrix} 2 \\ -1 \end{bmatrix}$ and $T\left(\begin{bmatrix} 1 \\ -1 \end{bmatrix}\right) = \begin{bmatrix} 0 \\ 3 \end{bmatrix}$

22. $T\left(\begin{bmatrix} 1 \\ 1 \end{bmatrix}\right) = \begin{bmatrix} 1 \\ 2 \\ 1 \end{bmatrix}$ and $T\left(\begin{bmatrix} 1 \\ -1 \end{bmatrix}\right) = \begin{bmatrix} 0 \\ 2 \\ 2 \end{bmatrix}$

23. $T\left(\begin{bmatrix} 1 \\ 0 \\ 1 \end{bmatrix}\right) = \begin{bmatrix} 0 \\ 1 \end{bmatrix}, \quad T\left(\begin{bmatrix} 0 \\ -1 \\ 1 \end{bmatrix}\right) = \begin{bmatrix} 1 \\ 0 \end{bmatrix},$

$T\left(\begin{bmatrix} 1 \\ -1 \\ 0 \end{bmatrix}\right) = \begin{bmatrix} 0 \\ 0 \end{bmatrix}$

24. $T\left(\begin{bmatrix} 1 \\ 0 \\ 1 \end{bmatrix}\right) = \begin{bmatrix} 0 \\ -1 \\ 1 \end{bmatrix}, \quad T\left(\begin{bmatrix} 0 \\ -1 \\ 1 \end{bmatrix}\right) = \begin{bmatrix} 2 \\ 1 \\ 0 \end{bmatrix},$

$T\left(\begin{bmatrix} 1 \\ -1 \\ 0 \end{bmatrix}\right) = \begin{bmatrix} 0 \\ 0 \\ 1 \end{bmatrix}$

In Exercises 25–30, a linear transformation T is given. In each case find a matrix A such that $T(\mathbf{x}) = A\mathbf{x}$. Also describe the null space and the range of T and give the rank and the nullity of T.

25. $T: R^2 \to R^2$ defined by $T\left(\begin{bmatrix} x_1 \\ x_2 \end{bmatrix}\right) = \begin{bmatrix} x_1 + 3x_2 \\ 2x_1 + x_2 \end{bmatrix}$

26. $T: R^2 \to R^3$ defined by $T\left(\begin{bmatrix} x_1 \\ x_2 \end{bmatrix}\right) = \begin{bmatrix} x_1 - x_2 \\ x_1 + x_2 \\ x_2 \end{bmatrix}$

27. $T: R^2 \to R$ defined by $T\left(\begin{bmatrix} x_1 \\ x_2 \end{bmatrix}\right) = 3x_1 + 2x_2$

28. $T: R^3 \to R^3$ defined by $T\left(\begin{bmatrix} x_1 \\ x_2 \\ x_3 \end{bmatrix}\right) = \begin{bmatrix} x_1 + x_2 \\ x_3 \\ x_2 \end{bmatrix}$

29. $T: R^3 \to R^2$ defined by $T\left(\begin{bmatrix} x_1 \\ x_2 \\ x_3 \end{bmatrix}\right) = \begin{bmatrix} x_1 - x_2 \\ x_2 - x_3 \end{bmatrix}$

30. $T: R^3 \to R$ defined by $T\left(\begin{bmatrix} x_1 \\ x_2 \\ x_3 \end{bmatrix}\right) = 2x_1 - x_2 + 4x_3$

31. Let a be a real number and define $f: R \to R$ by $f(x) = ax$ for each x in R. Show that f is a linear transformation.

32. Let $T: R \to R$ be a linear transformation and suppose that $T(1) = a$. Show that $T(x) = ax$ for each x in R.

33. Let $T: R^2 \to R^2$ be the function that maps each point in R^2 to its reflection with respect to the x-axis. Give a formula for T and show that T is a linear transformation.

34. Let $T: R^2 \to R^2$ be the function that maps each point in R^2 to its reflection with respect to the line $y = x$. Give a formula for T and show that T is a linear transformation.

35. Let V and W be subspaces, and let $F: V \to W$ and $G: V \to W$ be linear transformations. Define $F + G$: $V \to W$ by $[F + G](\mathbf{v}) = F(\mathbf{v}) + G(\mathbf{v})$ for each $\mathbf{v}$ in V. Prove that $F + G$ is a linear transformation.

36. Let $F: R^3 \to R^2$ and $G: R^3 \to R^2$ be defined by

$$F\left(\begin{bmatrix} x_1 \\ x_2 \\ x_3 \end{bmatrix}\right) = \begin{bmatrix} 2x_1 - 3x_2 + x_3 \\ 4x_1 + 2x_2 - 5x_3 \end{bmatrix}$$

and

$$G\left(\begin{bmatrix} x_1 \\ x_2 \\ x_3 \end{bmatrix}\right) = \begin{bmatrix} -x_1 + 4x_2 + 2x_3 \\ -2x_1 + 3x_2 + 3x_3 \end{bmatrix}.$$

a) Give a formula for the linear transformation $F + G$ (cf. Exercise 35 above).
b) Find matrices A, B, and C such that $F(\mathbf{x}) = A\mathbf{x}$, $G(\mathbf{x}) = B\mathbf{x}$, and $(F + G)(\mathbf{x}) = C\mathbf{x}$.
c) Verify that $C = A + B$.

37. Let V and W be subspaces, and let $T: V \to W$ be a linear transformation. If a is a scalar, define $aT: V \to W$ by $[aT](\mathbf{v}) = a[T(\mathbf{v})]$ for each $\mathbf{v}$ in V. Show that aT is a linear transformation.

38. Let $T: R^3 \to R^2$ be the linear transformation defined in Exercise 29. The linear transformation $[3T]: R^3 \to R^2$ is defined in Exercise 37 above.
a) Give a formula for the transformation $3T$.
b) Find matrices A and B such that $T(\mathbf{x}) = A\mathbf{x}$ and $[3T](\mathbf{x}) = B\mathbf{x}$.
c) Verify that $B = 3A$.

39. Let U, V, and W be subspaces, and let $F: U \to V$ and $G: V \to W$ be linear transformations. Prove that the composition $G \circ F: U \to W$ of F and G, defined by $[G \circ F](\mathbf{u}) = G(F(\mathbf{u}))$ for each $\mathbf{u}$ in U, is a linear transformation.

40. Let $F: R^3 \to R^2$ and $G: R^2 \to R^3$ be linear transformations defined by

$$F\left(\begin{bmatrix} x_1 \\ x_2 \\ x_3 \end{bmatrix}\right) = \begin{bmatrix} -x_1 + 2x_2 - 4x_3 \\ 2x_1 + 5x_2 + x_3 \end{bmatrix}$$

and

$$G\left(\begin{bmatrix} x_1 \\ x_2 \end{bmatrix}\right) = \begin{bmatrix} x_1 - 2x_2 \\ 3x_1 + 2x_2 \\ -x_1 + x_2 \end{bmatrix}.$$

a) By Exercise 39 above, $G \circ F: R^3 \to R^3$ is a linear transformation. Give a formula for $G \circ F$.
b) Find matrices A, B, and C such that $F(\mathbf{x}) = A\mathbf{x}$, $G(\mathbf{x}) = B\mathbf{x}$, and $[G \circ F](\mathbf{x}) = C\mathbf{x}$.
c) Verify that $C = BA$.

41. Let B be an $(m \times n)$ matrix, and let $T: R^n \to R^m$ be defined by $T(\mathbf{x}) = B\mathbf{x}$ for each $\mathbf{x}$ in R^n. If A is the matrix for T given by Theorem 15, show that $A = B$.

42. Let $F: R^n \to R^p$ and $G: R^p \to R^m$ be linear transformations, and assume that Theorem 15 yields matrices A and B, respectively, for F and G. Show that the matrix for the composition $G \circ F$ (cf. Exercise 39) is BA. (*Hint:* Show that $(G \circ F)(\mathbf{x}) = BA\mathbf{x}$ for $\mathbf{x}$ in R^n and then apply Exercise 41.)

43. Let $I: R^n \to R^n$ be the identity transformation. Determine the matrix A such that $I(\mathbf{x}) = A\mathbf{x}$ for each $\mathbf{x}$ in R^n.

44. Let a be a real number and define $T: R^n \to R^n$ by $T(\mathbf{x}) = a\mathbf{x}$ (cf. Example 4). Determine the matrix A such that $T(\mathbf{x}) = A\mathbf{x}$ for each $\mathbf{x}$ in R^n.

Exercises 45–49 are based on optional material.

45. Let $T: R^2 \to R^2$ be a rotation through the angle θ. In each of the following cases, exhibit the matrix for T. Also represent $\mathbf{v}$ and $T(\mathbf{v})$ geometrically, where

$$\mathbf{v} = \begin{bmatrix} 1 \\ 1 \end{bmatrix}.$$

a) $\theta = \pi/2$ b) $\theta = \pi/3$ c) $\theta = 2\pi/3$

46. Let $T: R^2 \to R^2$ be the reflection with respect to the line l. In each of the following cases, exhibit the matrix for T. Also represent $\mathbf{e}_1, \mathbf{e}_2, T(\mathbf{e}_1)$, and $T(\mathbf{e}_2)$ geometrically.
a) l is the x-axis.   b) l is the y-axis.
c) l is the line with equation $y = x$.
d) l is the line with equation $y = \sqrt{3}x$.

47. Let $T: R^2 \to R^2$ be a linear transformation that satisfies the conditions of Theorem 16. Show that T is orthogonal. (*Hint:* If $\mathbf{v} = [a, b]^T$, then $\mathbf{v} = a\mathbf{e}_1 + b\mathbf{e}_2$. Now use Eq. (6).)

48. Let $T: R^2 \to R^2$ be an orthogonal linear transformation, and let A be the corresponding (2×2) matrix. Show that $A^T A = I$. (*Hint:* Use Theorem 16.)

49. Let $A = [\mathbf{A}_1, \mathbf{A}_2]$ be a (2×2) matrix such that $A^T A = I$, and define $T: R^2 \to R^2$ by $T(\mathbf{x}) = A\mathbf{x}$.
a) Show that $\{\mathbf{A}_1, \mathbf{A}_2\}$ is an orthonormal set.
b) Use Theorem 16 to show that T is an orthogonal transformation.

2.8	**The Least-Squares Problem in R^n (Optional)**

In this section we consider a problem that is important in its own right and is widely used in applications. Before stating the problem, we need to recall some notation. Let $\mathbf{x}$ be a vector in R^n,

$$\mathbf{x} = \begin{bmatrix} x_1 \\ x_2 \\ \vdots \\ x_n \end{bmatrix}.$$

We define the **_length_** or **_magnitude_** of $\mathbf{x}$ to be the number $\|\mathbf{x}\|$, where

$$\|\mathbf{x}\| = \sqrt{\mathbf{x}^T\mathbf{x}} = \sqrt{x_1^2 + x_2^2 + \cdots + x_n^2}.$$

Furthermore, the **_distance_** between two vectors $\mathbf{x}$ and $\mathbf{y}$ is defined to be the number $\|\mathbf{x} - \mathbf{y}\|$:

$$\|\mathbf{x} - \mathbf{y}\| = \sqrt{(\mathbf{x} - \mathbf{y})^T(\mathbf{x} - \mathbf{y})}$$
$$= \sqrt{(x_1 - y_1)^2 + (x_2 - y_2)^2 + \cdots + (x_n - y_n)^2}.$$

The problem we wish to consider is stated below.

The Least-Squares Problem in R^n

Let W be a p-dimensional subspace of R^n. Given a vector $\mathbf{v}$ in R^n, find a vector $\mathbf{w}^*$ in W such that

$$\|\mathbf{v} - \mathbf{w}^*\| \leq \|\mathbf{v} - \mathbf{w}\|, \quad \text{for all } \mathbf{w} \text{ in } W.$$

The vector $\mathbf{w}^*$ is called the **_best least-squares approximation to v_**.

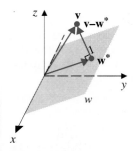

Figure 2.20
$\mathbf{w}^*$ is the closest point in the plane W to $\mathbf{v}$

That is, among all vectors $\mathbf{w}$ in W, we want to find the special vector $\mathbf{w}^*$ in W that is closest to $\mathbf{v}$. Although this problem can be extended to some very complicated and abstract settings, examination of the geometry of a simple special case will exhibit a fundamental principle that extends to all such problems.

Consider the special case where W is a two-dimensional subspace of R^3. Geometrically, we can visualize W as a plane through the origin (see Fig. 2.20). Given a point $\mathbf{v}$ not on W, we wish to find the point in the plane, $\mathbf{w}^*$, that is closest to $\mathbf{v}$. The geometry of this problem seems to insist (see Fig. 2.20) that $\mathbf{w}^*$ is characterized by the fact that the vector $\mathbf{v} - \mathbf{w}^*$ is perpendicular to the plane W.

The next theorem shows that Fig. 2.20 is not atypical. That is, if $\mathbf{v} - \mathbf{w}^*$ is orthogonal to every vector in W, then $\mathbf{w}^*$ is the best least-squares approximation to $\mathbf{v}$.

THEOREM 17

Let W be a p-dimensional subspace of R^n, and let $\mathbf{v}$ be a vector in R^n. Suppose there is a vector $\mathbf{w}^*$ in W such that $(\mathbf{v} - \mathbf{w}^*)^T\mathbf{w} = 0$ for every vector $\mathbf{w}$ in W. Then for any $\mathbf{w}$ in W, $\|\mathbf{v} - \mathbf{w}^*\| \leq \|\mathbf{v} - \mathbf{w}\|$.

PROOF   Let $\mathbf{w}$ be any vector in W and consider the following:

$$\|\mathbf{v} - \mathbf{w}\|^2 = \|(\mathbf{v} - \mathbf{w}^*) + (\mathbf{w}^* - \mathbf{w})\|^2$$
$$= (\mathbf{v} - \mathbf{w}^*)^T(\mathbf{v} - \mathbf{w}^*) + 2(\mathbf{v} - \mathbf{w}^*)^T(\mathbf{w}^* - \mathbf{w})$$
$$+ (\mathbf{w}^* - \mathbf{w})^T(\mathbf{w}^* - \mathbf{w}) \qquad (1)$$
$$= \|\mathbf{v} - \mathbf{w}^*\|^2 + \|\mathbf{w}^* - \mathbf{w}\|^2.$$

[The last equality follows because $\mathbf{w}^* - \mathbf{w}$ is a vector in W, and therefore $(\mathbf{v} - \mathbf{w}^*)^T(\mathbf{w}^* - \mathbf{w}) = 0$.] Since $\|\mathbf{w}^* - \mathbf{w}\|^2 \geq 0$, it follows from Eq. (1) that $\|\mathbf{v} - \mathbf{w}\|^2 \geq \|\mathbf{v} - \mathbf{w}^*\|^2$. Therefore, $\mathbf{w}^*$ is the best approximation to $\mathbf{v}$.   ☐

The equality in (1), $\|\mathbf{v} - \mathbf{w}\|^2 = \|\mathbf{v} - \mathbf{w}^*\|^2 + \|\mathbf{w}^* - \mathbf{w}\|^2$, is reminiscent of the Pythagorean theorem. A schematic view of this connection is sketched in Fig. 2.21.

In a later theorem we will show that there is always one, and only one, vector $\mathbf{w}^*$ in W such that $\mathbf{v} - \mathbf{w}^*$ is orthogonal to every vector in W. Thus it will be established that the best approximation always exists and is always unique. The proof of this fact will be constructive, so we now concentrate on methods for finding $\mathbf{w}^*$.

Finding Best Approximations

Theorem 17 suggests a procedure for finding the best approximation $\mathbf{w}^*$. In particular, we should search for a vector $\mathbf{w}^*$ in W that satisfies the following condition:

$$\text{If } \mathbf{w} \text{ is in } W, \quad \text{then } (\mathbf{v} - \mathbf{w}^*)^T\mathbf{w} = 0.$$

The search for $\mathbf{w}^*$ is simplified if we make the following observation: If $\mathbf{v} - \mathbf{w}^*$ is orthogonal to every vector in W, then $\mathbf{v} - \mathbf{w}^*$ is also orthogonal to every vector in a basis for W. In fact, the condition that $\mathbf{v} - \mathbf{w}^*$ be orthogonal to the basis vectors is both necessary and sufficient for $\mathbf{v} - \mathbf{w}^*$ to be orthogonal to every vector in W.

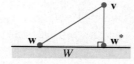

Figure 2.21
A geometric interpretation of the vector $\mathbf{w}^*$ in W closest to $\mathbf{v}$

<table>
<tr><td>THEOREM 18</td><td>

Let W be a p-dimensional subspace of R^n, and let $\{\mathbf{u}_1, \mathbf{u}_2, \ldots, \mathbf{u}_p\}$ be a basis for W. Let $\mathbf{v}$ be a vector in R^n. Then $(\mathbf{v} - \mathbf{w}^*)^T\mathbf{w} = 0$ for all $\mathbf{w}$ in W if and only if

$$(\mathbf{v} - \mathbf{w}^*)^T\mathbf{u}_i = 0, \qquad 1 \le i \le p.$$

</td></tr>
</table>

The proof of Theorem 18 is left as Exercise 21.

As Theorem 18 states, the best approximation $\mathbf{w}^*$ is determined by the p equations:

$$(\mathbf{v} - \mathbf{w}^*)^T\mathbf{u}_1 = 0$$
$$(\mathbf{v} - \mathbf{w}^*)^T\mathbf{u}_2 = 0$$
$$\vdots$$
$$(\mathbf{v} - \mathbf{w}^*)^T\mathbf{u}_p = 0$$

(2)

Suppose we can show that these p equations always have a unique solution. Then, by Theorem 17, it will follow that the best approximation exists and is unique.

Before addressing the question of whether system (2) has a unique solution, we will consider an example that illustrates a procedure for solving (2).

EXAMPLE 1 Let W be the subspace of R^3 defined by

$$W = \{\mathbf{x}: \mathbf{x} = \begin{bmatrix} x_1 \\ x_2 \\ x_3 \end{bmatrix}, \quad x_1 + x_2 - 3x_3 = 0\}.$$

Let $\mathbf{v}$ be the vector $\mathbf{v} = [1, -2, -4]^T$. Use system (2) to find $\mathbf{w}^*$ in W such that $(\mathbf{v} - \mathbf{w}^*)^T\mathbf{w} = 0$ for all $\mathbf{w}$ in W.

SOLUTION A basis for W is $\{\mathbf{u}_1, \mathbf{u}_2\}$, where

$$\mathbf{u}_1 = \begin{bmatrix} 1 \\ -1 \\ 0 \end{bmatrix} \quad \text{and} \quad \mathbf{u}_2 = \begin{bmatrix} 3 \\ 0 \\ 1 \end{bmatrix}.$$

Now the equations $(\mathbf{v} - \mathbf{w}^*)^T\mathbf{u}_1 = 0$ and $(\mathbf{v} - \mathbf{w}^*)^T\mathbf{u}_2 = 0$ can be written as

$$(\mathbf{w}^*)^T\mathbf{u}_1 = \mathbf{v}^T\mathbf{u}_1$$
$$(\mathbf{w}^*)^T\mathbf{u}_2 = \mathbf{v}^T\mathbf{u}_2.$$

(3)

If $\mathbf{w}^*$ is to be in W, then $\mathbf{w}^*$ must be expressible as $\mathbf{w}^* = a\mathbf{u}_1 + b\mathbf{u}_2$ for some scalars a and b.

Thus (3) leads to the conditions

$$(a\mathbf{u}_1 + b\mathbf{u}_2)^T\mathbf{u}_1 = \mathbf{v}^T\mathbf{u}_1$$
$$(a\mathbf{u}_1 + b\mathbf{u}_2)^T\mathbf{u}_2 = \mathbf{v}^T\mathbf{u}_2.$$

Equivalently,

$$a(\mathbf{u}_1^T\mathbf{u}_1) + b(\mathbf{u}_2^T\mathbf{u}_1) = \mathbf{v}^T\mathbf{u}_1$$
$$a(\mathbf{u}_1^T\mathbf{u}_2) + b(\mathbf{u}_2^T\mathbf{u}_2) = \mathbf{v}^T\mathbf{u}_2.$$

Evaluating the scalar products above, we are led to

$$2a + 3b = 3$$
$$3a + 10b = -1.$$

From the conditions above, it follows that $a = 3$ and $b = -1$. Thus the vector $\mathbf{w}^*$ given by $\mathbf{w}^* = 3\mathbf{u}_1 - \mathbf{u}_2$ will satisfy the orthogonality conditions in (3).

In detail, $\mathbf{w}^*$ is the vector

$$\mathbf{w}^* = 3\mathbf{u}_1 - \mathbf{u}_2 = \begin{bmatrix} 0 \\ -3 \\ -1 \end{bmatrix}.$$

Note that the vector $\mathbf{v} - \mathbf{w}^* = [1, 1, -3]^T$ is orthogonal to $\mathbf{u}_1$ and $\mathbf{u}_2$.

The procedure used in Example 1 will be formalized in the next subsection.

The Normal Equations

Let W be a p-dimensional subspace of R^n with a basis $\{\mathbf{u}_1, \mathbf{u}_2, \ldots, \mathbf{u}_p\}$, and let $\mathbf{v}$ be a vector in R^n. If there is a vector $\mathbf{w}^*$ in W such that $(\mathbf{v} - \mathbf{w}^*)^T\mathbf{u}_i = 0$, $1 \le i \le p$, then $\mathbf{w}^*$ is the best approximation to $\mathbf{v}$. This result follows from Theorems 17 and 18 and was illustrated in Example 1.

Thus we focus on the system of equations in (2), which can be rewritten as

$$\mathbf{u}_1^T\mathbf{w}^* = \mathbf{u}_1^T\mathbf{v}$$
$$\mathbf{u}_2^T\mathbf{w}^* = \mathbf{u}_2^T\mathbf{v}$$
$$\vdots \qquad\qquad (4)$$
$$\mathbf{u}_p^T\mathbf{w}^* = \mathbf{u}_p^T\mathbf{v}.$$

Because we want $\mathbf{w}^*$ to be in W, we need scalars $x_1, x_2, \ldots, x_p$ such that $\mathbf{w}^* = x_1\mathbf{u}_1 + x_2\mathbf{u}_2 + \cdots + x_p\mathbf{u}_p$ and system (4) is satisfied. Therefore, the best approximation $\mathbf{w}^*$ can be found if we solve the p equations:

$$\mathbf{u}_1^T(x_1\mathbf{u}_1 + x_2\mathbf{u}_2 + \cdots + x_p\mathbf{u}_p) = \mathbf{u}_1^T\mathbf{v}$$
$$\mathbf{u}_2^T(x_1\mathbf{u}_1 + x_2\mathbf{u}_2 + \cdots + x_p\mathbf{u}_p) = \mathbf{u}_2^T\mathbf{v}$$
$$\vdots$$
$$\mathbf{u}_p^T(x_1\mathbf{u}_1 + x_2\mathbf{u}_2 + \cdots + x_p\mathbf{u}_p) = \mathbf{u}_p^T\mathbf{v}.$$

The system above has the form

$$(\mathbf{u}_1^T\mathbf{u}_1)x_1 + (\mathbf{u}_1^T\mathbf{u}_2)x_2 + \cdots + (\mathbf{u}_1^T\mathbf{u}_p)x_p = \mathbf{u}_1^T\mathbf{v}$$
$$(\mathbf{u}_2^T\mathbf{u}_1)x_1 + (\mathbf{u}_2^T\mathbf{u}_2)x_2 + \cdots + (\mathbf{u}_2^T\mathbf{u}_p)x_p = \mathbf{u}_2^T\mathbf{v}$$
$$\vdots \qquad\qquad\qquad\qquad\qquad\qquad\qquad\qquad (5)$$
$$(\mathbf{u}_p^T\mathbf{u}_1)x_1 + (\mathbf{u}_p^T\mathbf{u}_2)x_2 + \cdots + (\mathbf{u}_p^T\mathbf{u}_p)x_p = \mathbf{u}_p^T\mathbf{v}.$$

The system of equations in (5) can be expressed compactly as follows: Let A denote the $(n \times p)$ matrix whose columns are formed from the basis vectors $\mathbf{u}_1, \mathbf{u}_2, \ldots, \mathbf{u}_p$:

$$A = [\mathbf{u}_1, \mathbf{u}_2, \ldots, \mathbf{u}_p].$$

It can be shown (see Exercise 22) that system (5) is the same as the equation

$$A^TA\mathbf{x} = A^T\mathbf{v}, \qquad\qquad\qquad (6)$$

where $\mathbf{x} = [x_1, x_2, \ldots, x_p]^T$. The equations in (5) are called ***normal equations*** and (6) expresses them in matrix terms.

The next theorem shows that Eq. (6) always has a unique solution and that the solution leads to the best approximation.

THEOREM 19

Let W be a p-dimensional subspace of R^n with basis $\{\mathbf{u}_1, \mathbf{u}_2, \ldots, \mathbf{u}_p\}$. Let A denote the $(n \times p)$ matrix defined by $A = [\mathbf{u}_1, \mathbf{u}_2, \ldots, \mathbf{u}_p]$. Then:

a) The $(p \times p)$ matrix A^TA is nonsingular.

Let $\mathbf{v}$ be any vector in R^n, and let $\mathbf{x}^*$ be the unique vector in R^p that satisfies $A^TA\mathbf{x}^* = A^T\mathbf{v}$. Then:

b) The vector $\mathbf{w}^* = A\mathbf{x}$ is in W.

Let $\mathbf{w}^* = A\mathbf{x}^*$, and let $\mathbf{w}$ be any vector in W, $\mathbf{w} \neq \mathbf{w}^*$. Then:

c) $\|\mathbf{v} - \mathbf{w}^*\| < \|\mathbf{v} - \mathbf{w}\|$.

PROOF   To show that A^TA is nonsingular, let $\mathbf{q}$ be any p-dimensional vector such that $A^TA\mathbf{q} = \boldsymbol{\theta}$. Clearly it then follows that

$$\mathbf{q}^T(A^TA\mathbf{q}) = \mathbf{q}^T\boldsymbol{\theta} = 0.$$

But $\mathbf{q}^T(A^TA\mathbf{q}) = (\mathbf{q}^TA^T)(A\mathbf{q}) = (A\mathbf{q})^T(A\mathbf{q})$. Therefore, if $A^TA\mathbf{q} = \boldsymbol{\theta}$, then $(A\mathbf{q})^T(A\mathbf{q}) = 0$, and this implies that $A\mathbf{q} = \boldsymbol{\theta}$. Now for $\mathbf{q} = [q_1, q_2, \ldots, q_p]^T$,

$$A\mathbf{q} = q_1\mathbf{u}_1 + q_2\mathbf{u}_2 + \cdots + q_p\mathbf{u}_p. \qquad\qquad (7)$$

Since $\{\mathbf{u}_1, \mathbf{u}_2, \ldots, \mathbf{u}_p\}$ is a basis for W, it is a linearly independent set. Therefore, we see from Eq. (7) that $A\mathbf{q} = \boldsymbol{\theta}$ cannot hold unless $\mathbf{q} = \boldsymbol{\theta}$. Thus A^TA is nonsingular.

To prove property (b) of Theorem 19, we need only note (as in Eq. 7) that $A\mathbf{x}$ is a linear combination of $\mathbf{u}_1, \mathbf{u}_2, \ldots, \mathbf{u}_p$ for any $\mathbf{x}$ in R^p.

The proof of property (c) is essentially a recapitulation of the discussion leading to the normal equations given in (6). We include the proof here for completeness. Suppose $A^TA\mathbf{x}^* = A^T\mathbf{v}$ and let $\mathbf{w}^* = A\mathbf{x}^*$. Then it follows that

$$A^T\mathbf{w}^* = A^T\mathbf{v}$$

or

$$A^T(\mathbf{v} - \mathbf{w}^*) = \boldsymbol{\theta}. \qquad (8)$$

Since the ith row of A^T is equal to $\mathbf{u}_i^T$, we can see from Eq. (8) that $\mathbf{u}_i^T(\mathbf{v}-\mathbf{w}^*)=0$ for $i = 1, 2, \ldots, p$. Then, since $\{\mathbf{u}_1, \mathbf{u}_2, \ldots, \mathbf{u}_p\}$ is a basis for W, it follows from Theorem 18 that $\mathbf{w}^T(\mathbf{v} - \mathbf{w}^*) = 0$ for all $\mathbf{w}$ in W. Next, by Theorem 17, we have $\|\mathbf{v} - \mathbf{w}\|^2 = \|\mathbf{v} - \mathbf{w}^*\|^2 + \|\mathbf{w}^* - \mathbf{w}\|^2$; recall Eq. (1) in the proof of Theorem 17. Thus from (1), if $\mathbf{w}$ is in W, then $\|\mathbf{v} - \mathbf{w}\| > \|\mathbf{v} - \mathbf{w}^*\|$, unless $\mathbf{w} = \mathbf{w}^*$.

EXAMPLE 2 Consider the matrix B and the vector $\mathbf{b}$, where

$$B = \begin{bmatrix} 1 & 0 & 2 \\ 0 & 2 & 2 \\ -1 & 1 & -1 \\ -1 & 2 & 0 \end{bmatrix} \quad \text{and} \quad \mathbf{b} = \begin{bmatrix} 3 \\ -3 \\ 0 \\ -3 \end{bmatrix}.$$

Find $\mathbf{w}^*$ in the range of B such that $\|\mathbf{b} - \mathbf{w}^*\| \le \|\mathbf{b} - \mathbf{w}\|$ for all $\mathbf{w}$ in $\mathcal{R}(B)$.

SOLUTION   A basis for $\mathcal{R}(B)$ consists of $\{\mathbf{u}_1, \mathbf{u}_2\}$, where

$$\mathbf{u}_1 = \begin{bmatrix} 1 \\ 0 \\ -1 \\ -1 \end{bmatrix} \quad \text{and} \quad \mathbf{u}_2 = \begin{bmatrix} 0 \\ 2 \\ 1 \\ 2 \end{bmatrix}.$$

Let A be the (4×2) matrix $A = [\mathbf{u}_1, \mathbf{u}_2]$, and consider $A^TA\mathbf{x} = A^T\mathbf{b}$:

$$\begin{bmatrix} 3 & -3 \\ -3 & 9 \end{bmatrix}\begin{bmatrix} x_1 \\ x_2 \end{bmatrix} = \begin{bmatrix} 6 \\ -12 \end{bmatrix}.$$

The solution of $A^TA\mathbf{x} = A^T\mathbf{b}$ is $\mathbf{x}^* = [1, -1]^T$, and therefore $\mathbf{w}^* = A\mathbf{x}^* = \mathbf{u}_1 - \mathbf{u}_2$. In detail,

$$\mathbf{w}^* = \mathbf{u}_1 - \mathbf{u}_2 = \begin{bmatrix} 1 \\ -2 \\ -2 \\ -3 \end{bmatrix}.$$

As a check, we form $\mathbf{b} - \mathbf{w}^* = [2, -1, 2, 0]^T$ and note that $(\mathbf{b} - \mathbf{w}^*)^T\mathbf{u}_1 = 0$ and $(\mathbf{b} - \mathbf{w}^*)^T\mathbf{u}_2 = 0$.

Orthogonal Bases

Suppose that an orthogonal basis is used for the p-dimensional subspace W in Theorem 19. With an orthogonal basis, the matrix $A^T A$ is diagonal, and an explicit formula can be given for the best approximation.

Specifically, let $\{\mathbf{u}_1, \mathbf{u}_2, \ldots, \mathbf{u}_p\}$ be an orthogonal basis for W, where W is a subspace of R^n. Let $\mathbf{v}$ be any vector in R^n. The best approximation $\mathbf{w}^*$ has the form

$$\mathbf{w}^* = x_1 \mathbf{u}_1 + x_2 \mathbf{u}_2 + \cdots + x_p \mathbf{u}_p,$$

where the coefficients $x_1, x_2, \ldots, x_n$ can be found from system (5). Since the basis vectors are orthogonal, we know that $\mathbf{u}_i^T \mathbf{u}_j = 0$, $i \neq j$. Therefore, system (5) reduces to

$$
\begin{aligned}
(\mathbf{u}_1^T \mathbf{u}_1) x_1 & = \mathbf{u}_1^T \mathbf{v} \\
(\mathbf{u}_2^T \mathbf{u}_2) x_2 & = \mathbf{u}_2^T \mathbf{v} \\
& \ \ \vdots \\
(\mathbf{u}_p^T \mathbf{u}_p) x_p & = \mathbf{u}_p^T \mathbf{v}.
\end{aligned}
$$

Thus $x_i = (\mathbf{u}_i^T \mathbf{v})/(\mathbf{u}_i^T \mathbf{u}_i)$, and hence $\mathbf{w}^*$ is given by

$$\mathbf{w}^* = \frac{\mathbf{u}_1^T \mathbf{v}}{\mathbf{u}_1^T \mathbf{u}_1} \mathbf{u}_1 + \frac{\mathbf{u}_2^T \mathbf{v}}{\mathbf{u}_2^T \mathbf{u}_2} \mathbf{u}_2 + \cdots + \frac{\mathbf{u}_p^T \mathbf{v}}{\mathbf{u}_p^T \mathbf{u}_p} \mathbf{u}_p. \tag{9}$$

EXAMPLE 3 Let W be the subspace of R^3 described in Example 1. Find an orthogonal basis for W and calculate the best approximation to the vector $\mathbf{v} = [1, -2, -4]^T$.

SOLUTION In Example 1, we saw that W had a basis consisting of the vectors

$$\begin{bmatrix} 1 \\ -1 \\ 0 \end{bmatrix} \quad \text{and} \quad \begin{bmatrix} 3 \\ 0 \\ 1 \end{bmatrix}.$$

Applying the Gram–Schmidt process, we obtain an orthogonal basis $\{\mathbf{u}_1, \mathbf{u}_2\}$, where

$$\mathbf{u}_1 = \begin{bmatrix} 1 \\ -1 \\ 0 \end{bmatrix} \quad \text{and} \quad \mathbf{u}_2 = \begin{bmatrix} 3 \\ 3 \\ 2 \end{bmatrix}.$$

Having an orthogonal basis, we can calculate the best approximation $\mathbf{w}^*$ from (9):

$$\mathbf{w}^* = \frac{\mathbf{u}_1^T \mathbf{v}}{\mathbf{u}_1^T \mathbf{u}_1} \mathbf{u}_1 + \frac{\mathbf{u}_2^T \mathbf{v}}{\mathbf{u}_2^T \mathbf{u}_2} \mathbf{u}_2 = \frac{3}{2} \begin{bmatrix} 1 \\ -1 \\ 0 \end{bmatrix} + \frac{-11}{22} \begin{bmatrix} 3 \\ 3 \\ 2 \end{bmatrix}$$

$$= \begin{bmatrix} 3/2 \\ -3/2 \\ 0 \end{bmatrix} + \begin{bmatrix} -3/2 \\ -3/2 \\ -1 \end{bmatrix} = \begin{bmatrix} 0 \\ -3 \\ -1 \end{bmatrix}.$$

Note that the vector $\mathbf{w}^* = [0, -3, -1]^T$ is (as it should be) the same as the best approximation computed in Example 1.

☐

The Projection Matrix

Let W be a p-dimensional subspace of R^n with basis $\{\mathbf{u}_1, \mathbf{u}_2, \ldots, \mathbf{u}_p\}$. By Theorem 19, if $\mathbf{v}$ is any vector in R^n, then there is a unique best approximation $\mathbf{w}^*$ to $\mathbf{v}$. Furthermore, $\mathbf{w}^*$ can be found by setting $\mathbf{w}^* = A\mathbf{x}^*$, where $\mathbf{x}^*$ is the unique solution of $A^T A\mathbf{x} = A^T\mathbf{v}$, $A = [\mathbf{u}_1, \mathbf{u}_2, \ldots, \mathbf{u}_p]$. Since $A^T A$ is nonsingular, we can write

$$\mathbf{x}^* = (A^T A)^{-1} A^T \mathbf{v}.$$

Therefore, it follows that $\mathbf{w}^* = A\mathbf{x}^*$ is given by

$$\mathbf{w}^* = A(A^T A)^{-1} A^T \mathbf{v}. \tag{10}$$

Suppose we choose a different basis for W, say $\{\mathbf{v}_1, \mathbf{v}_2, \ldots, \mathbf{v}_p\}$. Let B denote the $(n \times p)$ matrix given by $B = [\mathbf{v}_1, \mathbf{v}_2, \ldots, \mathbf{v}_p]$. Clearly Theorem 19 applies to B as well as to A, and so we can also find the best approximation $\mathbf{w}^*$ from

$$\mathbf{w}^* = B(B^T B)^{-1} B^T \mathbf{v}. \tag{11}$$

To summarize: Theorem 19 states the best approximation is unique. For any $\mathbf{v}$ in R^n, we can calculate the best approximation either from Eq. (10) or from Eq. (11). Thus for all $\mathbf{v}$ in R^n,

$$A(A^T A)^{-1} A^T \mathbf{v} = B(B^T B)^{-1} B^T \mathbf{v}. \tag{12}$$

As can be shown (see Exercise 24), Eq. (12) implies that $A(A^T A)^{-1} A^T = B(B^T B)^{-1} B^T$. (Note that both $A(A^T A)^{-1} A^T$ and $B(B^T B)^{-1} B^T$ are $(n \times n)$ matrices.)

Equation (12) shows that no matter what basis $\{\mathbf{v}_1, \mathbf{v}_2, \ldots, \mathbf{v}_p\}$ is chosen for W, the matrix $B(B^T B)^{-1} B^T$ is equal to $A(A^T A)^{-1} A^T$. Thus with any subspace W of R^n, there is a unique $(n \times n)$ matrix P_W associated with W. The matrix P_W is called the ***projection matrix for W*** and can be calculated as follows:

1. Let $\{\mathbf{q}_1, \mathbf{q}_2, \ldots, \mathbf{q}_p\}$ be any basis for W.
2. Let Q be the $(n \times p)$ matrix $Q = [\mathbf{q}_1, \mathbf{q}_2, \ldots, \mathbf{q}_p]$.
3. Then $P_W = Q(Q^T Q)^{-1} Q^T$.

(*Note:* As we observed in Eqs. (11) and (12), the vector $\mathbf{w}^*$ given by $\mathbf{w}^* = P_W\mathbf{v}$ is the best approximation to $\mathbf{v}$. The vector $P_W\mathbf{v}$ is often called the ***projection of v on W*.**)

EXAMPLE 4   Find the projection matrix P_W for the subspace W defined by

$$W = \left\{ \mathbf{x}: \mathbf{x} = \begin{bmatrix} x_1 \\ x_2 \\ x_3 \end{bmatrix}, \quad x_1 + x_2 - 3x_3 = 0 \right\}.$$

For $\mathbf{v} = [1, -2, -4]^T$, calculate $P_W\mathbf{v}$.

SOLUTION   The subspace W is the same as in Example 1, where we used a basis consisting of $\mathbf{u}_1 = [1, -1, 0]^T$ and $\mathbf{u}_2 = [3, 0, 1]^T$. To illustrate that the projection matrix is independent of the basis, we choose a basis for W consisting of $\mathbf{q}_1 = [1, 2, 1]^T$ and $\mathbf{q}_2 = [2, 1, 1]^T$.

With this choice of basis, $Q = [\mathbf{q}_1, \mathbf{q}_2]$ and

$$Q^T Q = \begin{bmatrix} 6 & 5 \\ 5 & 6 \end{bmatrix}, \qquad (Q^T Q)^{-1} = \frac{1}{11} \begin{bmatrix} 6 & -5 \\ -5 & 6 \end{bmatrix}.$$

Therefore, $P_W = Q(Q^T Q)^{-1} Q^T$ is given by

$$P_W = \frac{1}{11} \begin{bmatrix} 1 & 2 \\ 2 & 1 \\ 1 & 1 \end{bmatrix} \begin{bmatrix} 6 & -5 \\ -5 & 6 \end{bmatrix} \begin{bmatrix} 1 & 2 & 1 \\ 2 & 1 & 1 \end{bmatrix} = \frac{1}{11} \begin{bmatrix} 10 & -1 & 3 \\ -1 & 10 & 3 \\ 3 & 3 & 2 \end{bmatrix}.$$

Next, for $\mathbf{v} = [1, -2, -4]^T$, we have $P_W \mathbf{v} = [0, -3, -1]^T$. Note that $\mathbf{w}^* = P_W \mathbf{v}$ is the best approximation to $\mathbf{v}$. This is the same vector found in Example 1 by using other techniques. ▢

The projection matrix has several interesting properties that are explored in the exercises.

Exercises 2.8

Exercises 1–20 refer to the subspaces listed below:

a) $W = \left\{ \mathbf{x}: \mathbf{x} = \begin{bmatrix} x_1 \\ x_2 \\ x_3 \end{bmatrix}, \quad x_1 - 2x_2 + x_3 = 0 \right\}$

b) $W = \mathscr{R}(B), \quad B = \begin{bmatrix} 1 & 2 \\ 1 & 1 \\ 0 & 1 \end{bmatrix}$

c) $W = \mathscr{R}(B), \quad B = \begin{bmatrix} 1 & 2 & 4 \\ -1 & 0 & -2 \\ 1 & 1 & 3 \end{bmatrix}$

d) $W = \left\{ \mathbf{x}: \mathbf{x} = \begin{bmatrix} x_1 \\ x_2 \\ x_3 \end{bmatrix}, \quad \begin{matrix} x_1 + x_2 + x_3 = 0 \\ x_1 - x_2 - x_3 = 0 \end{matrix} \right\}$

In Exercises 1–10, find a basis for the indicated subspace W. For the given vector $\mathbf{v}$, solve the normal equations (6) and determine the best approximation $\mathbf{w}^*$. Verify that $\mathbf{v} - \mathbf{w}^*$ is orthogonal to the basis vectors.

1. W given by (a), $\mathbf{v} = [1, 2, 6]^T$

2. W given by (a), $\mathbf{v} = [3, 0, 3]^T$

3. W given by (a), $\mathbf{v} = [1, 1, 1]^T$

4. W given by (b), $\mathbf{v} = [1, 1, 6]^T$

5. W given by (b), $\mathbf{v} = [3, 3, 3]^T$

6. W given by (b), $\mathbf{v} = [3, 0, 3]^T$

7. W given by (c), $\mathbf{v} = [2, 0, 4]^T$

8. W given by (c), $\mathbf{v} = [4, 0, -1]^T$

9. W given by (d), $\mathbf{v} = [1, 3, 1]^T$

10. W given by (d), $\mathbf{v} = [3, 4, 0]^T$

In Exercises 11–16, find an orthogonal basis for the indicated subspace W. Use Eq. (9) to determine the best approximation $\mathbf{w}^*$ for the given vector $\mathbf{v}$.

11. W and $\mathbf{v}$ as in Exercise 1

12. W and $\mathbf{v}$ as in Exercise 2

13. W and $\mathbf{v}$ as in Exercise 4

14. W and $\mathbf{v}$ as in Exercise 5

15. W and $\mathbf{v}$ as in Exercise 7

16. W and $\mathbf{v}$ as in Exercise 8

In Exercises 17–20, find the projection matrix P_W for the indicated subspace W. Use P_W to determine the best approximation $\mathbf{w}^*$ for the given vector $\mathbf{v}$.

17. W given by (a), $\mathbf{v} = [6, 0, 3]^T$

18. W given by (b), $\mathbf{v} = [6, 2, 1]^T$

19. W given by (c), $\mathbf{v} = [1, 1, 3]^T$

20. W given by (d), $\mathbf{v} = [4, 4, 6]^T$

21. Prove Theorem 18.

22. Let A denote the $(n \times p)$ matrix $A = [\mathbf{u}_1, \mathbf{u}_2, \ldots, \mathbf{u}_p]$. Verify that system (5) can be expressed as $A^T A \mathbf{x} = A^T \mathbf{v}$.

23. Consider $\mathbf{w}^*$ as given in Eq. (9). Verify directly that $(\mathbf{v} - \mathbf{w}^*)^T \mathbf{u}_1 = 0, (\mathbf{v} - \mathbf{w}^*)^T \mathbf{u}_2 = 0, \ldots, (\mathbf{v} - \mathbf{w}^*)^T \mathbf{u}_p = 0$.

24. Suppose that P and Q are two $(n \times n)$ matrices. Suppose also that $P\mathbf{x} = Q\mathbf{x}$ for all $\mathbf{x}$ in R^n. Show that $P = Q$.

25. Let W be a p-dimensional subspace of R^n with basis $\{\mathbf{u}_1, \mathbf{u}_2, \ldots, \mathbf{u}_p\}$. Let $A = [\mathbf{u}_1, \mathbf{u}_2, \ldots, \mathbf{u}_p]$, and

let P denote the projection matrix $P = A(A^TA)^{-1}A^T$. Show that P is symmetric. (*Hint:* You will have to consider whether the inverse of a symmetric matrix is also symmetric.)

26. Let P be the projection matrix of Exercise 25. Show that $P^2 = P$. (Recall that a square matrix Q such that $Q^2 = Q$ is called ***idempotent***.)

27. Show that the only $(n \times n)$ nonsingular idempotent matrix is the identity matrix.

28. Calculate the projection matrix P in the case that $W = R^n$. (*Hint:* Choose a convenient basis for R^n.)

29. Let P denote the projection matrix in Exercise 25. Note that a vector $\mathbf{w}$ is in W if and only if $\mathbf{w} = A\mathbf{x}$ for some p-dimensional vector $\mathbf{x}$. Show that if $\mathbf{w}$ is in W, then $P\mathbf{w} = \mathbf{w}$. (That is, the projection matrix acts like the identity matrix on W.)

2.9 Fitting Data and Least-Squares Solutions of Overdetermined Systems (Optional)

Let A be an $(m \times n)$ matrix, and let $\mathbf{b}$ be an $(m \times 1)$ vector. Even if the linear system $A\mathbf{x} = \mathbf{b}$ has no solution, we may still wish to find $\mathbf{x}^*$ in R^n such that

$$\|A\mathbf{x}^* - \mathbf{b}\| \le \|A\mathbf{x} - \mathbf{b}\| \quad \text{for all } \mathbf{x} \text{ in } R^n. \tag{1}$$

That is, $\mathbf{x}^*$ minimizes the magnitude of the residual vector, $A\mathbf{x} - \mathbf{b}$, and (in this sense) comes closest to solving the system.

Let $W = \mathcal{R}(A)$, the range space of A. If $\mathbf{w}^* = A\mathbf{x}^*$ and $\mathbf{w} = A\mathbf{x}$, then $\mathbf{w}$ and $\mathbf{w}^*$ are in W. Thus inequality (1) can be solved by finding $\mathbf{w}^*$ in W such that

$$\|\mathbf{w}^* - \mathbf{b}\| \le \|\mathbf{w} - \mathbf{b}\| \quad \text{for all } \mathbf{w} \text{ in } W. \tag{2}$$

As the problem in (1) is restated in (2), we see that a solution can be obtained from the techniques described in the last section. That is, $\mathbf{x}^*$ minimizes $\|A\mathbf{x} - \mathbf{b}\|$, $\mathbf{x}$ in R^n if and only if $A\mathbf{x}^* = \mathbf{w}^*$, where $\mathbf{w}^*$ minimizes $\|\mathbf{w} - \mathbf{b}\|$, $\mathbf{w}$ in $\mathcal{R}(A)$.

For problem (1), note that we are interested in finding $\mathbf{x}^*$ rather than $\mathbf{w}^*$. Also, whereas the best approximation $\mathbf{w}^*$ is unique, there may be many vectors $\mathbf{x}^*$ that minimize $\|A\mathbf{x} - \mathbf{b}\|$. An example will serve to illustrate these distinctions.

EXAMPLE 1 Consider the (4×3) matrix A and the (4×1) vector $\mathbf{b}$, where

$$A = \begin{bmatrix} 1 & 0 & 2 \\ 0 & 2 & 2 \\ -1 & 1 & -1 \\ -1 & 2 & 0 \end{bmatrix} \quad \text{and} \quad \mathbf{b} = \begin{bmatrix} 3 \\ -3 \\ 0 \\ -3 \end{bmatrix}.$$

Find all vectors $\mathbf{x}^*$ in R^3 that minimize the quantity $\|A\mathbf{x} - \mathbf{b}\|$.

SOLUTION   In Example 2 of Section 2.8, we found $\mathbf{w}^*$ in $\mathscr{R}(A)$ such that $\|\mathbf{w}^* - \mathbf{b}\| \le \|\mathbf{w} - \mathbf{b}\|$ for all $\mathbf{w}$ in $\mathscr{R}(A)$. The unique vector $\mathbf{w}^*$ in $\mathscr{R}(A)$ that is closest to $\mathbf{b}$ is

$$\mathbf{w}^* = \begin{bmatrix} 1 \\ -2 \\ -2 \\ -3 \end{bmatrix}.$$

Therefore, if $\mathbf{x}^*$ is any vector in R^3 such that $A\mathbf{x}^* = \mathbf{w}^*$, then we will have $\|A\mathbf{x}^* - \mathbf{b}\| \le \|A\mathbf{x} - \mathbf{b}\|$ for all $\mathbf{x}$ in R^3. The system $A\mathbf{x} = \mathbf{w}^*$ is given by

$$\begin{array}{rcr} x_1 \qquad\quad + 2x_3 &=& 1 \\ 2x_2 + 2x_3 &=& -2 \\ -x_1 + x_2 - x_3 &=& -2 \\ -x_1 + 2x_2 \qquad\quad &=& -3. \end{array}$$

This system is equivalent to

$$\begin{array}{rcr} x_1 \qquad\quad + 2x_3 &=& 1 \\ 2x_2 + 2x_3 &=& -2. \end{array}$$

Thus the general form for a vector $\mathbf{x}^*$ that satisfies $A\mathbf{x}^* = \mathbf{w}^*$ is

$$\mathbf{x}^* = \begin{bmatrix} 1 \\ -1 \\ 0 \end{bmatrix} + a \begin{bmatrix} -2 \\ -1 \\ 1 \end{bmatrix}, \quad a \text{ any scalar.} \qquad \square$$

Overdetermined Systems

Consider a system of equations $A\mathbf{x} = \mathbf{b}$, where A is an $(m \times n)$ matrix with $m > n$ and $\mathbf{b}$ in R^m. Such a system is frequently called an ***overdetermined system*** because there are more equations than unknowns. A vector $\mathbf{x}^*$ in R^n that minimizes $\|A\mathbf{x} - \mathbf{b}\|$ is called a ***least-squares solution*** to $A\mathbf{x} = \mathbf{b}$. Usually an overdetermined system $A\mathbf{x} = \mathbf{b}$ is not consistent, and so finding vectors $\mathbf{x}^*$ that minimize $\|A\mathbf{x} - \mathbf{b}\|$ is a reasonable goal. (If the system $A\mathbf{x} = \mathbf{b}$ is consistent, then a least-squares solution $\mathbf{x}^*$ is a solution in the usual sense of the word because $\|A\mathbf{x}^* - \mathbf{b}\| = 0$.) As we will see later, overdetermined systems arise frequently in data-fitting problems.

The following theorem gives a method for finding least-squares solutions to overdetermined systems. The method is a slight modification of Theorem 19.

THEOREM 20

Let A be an $(m \times n)$ matrix with $m \geq n$, and let $\mathbf{b}$ be a vector in R^m. Then the system $A^TA\mathbf{x} = A^T\mathbf{b}$ is consistent. Moreover,

$$\|A\mathbf{x}^* - \mathbf{b}\| \leq \|A\mathbf{x} - \mathbf{b}\| \quad \text{for all } \mathbf{x} \text{ in } R^n$$

if and only if $\mathbf{x}^*$ is a solution of

$$A^TA\mathbf{x} = A^T\mathbf{b}.$$

Also, the vector $\mathbf{x}^*$ is unique if and only if A has rank n.

The proof of Theorem 20 uses ideas from the last section and is left to the exercises. The main difference between Theorem 19 and Theorem 20 is that Theorem 20 does not insist that the column vectors of A be linearly independent.

EXAMPLE 2 Rework Example 1 by using Theorem 20.

SOLUTION Example 1 asks for the least-squares solutions to $A\mathbf{x} = \mathbf{b}$, where

$$A = \begin{bmatrix} 1 & 0 & 2 \\ 0 & 2 & 2 \\ -1 & 1 & -1 \\ -1 & 2 & 0 \end{bmatrix} \quad \text{and} \quad \mathbf{b} = \begin{bmatrix} 3 \\ -3 \\ 0 \\ -3 \end{bmatrix}.$$

From Theorem 20, these solutions $\mathbf{x}^*$ are found by solving $A^TA\mathbf{x} = A^T\mathbf{b}$.
Now A^TA and $A^T\mathbf{b}$ are given by

$$A^TA = \begin{bmatrix} 3 & -3 & 3 \\ -3 & 9 & 3 \\ 3 & 3 & 9 \end{bmatrix} \quad \text{and} \quad A^T\mathbf{b} = \begin{bmatrix} 6 \\ -12 \\ 0 \end{bmatrix}.$$

The augmented matrix $[A^TA \,|\, A^T\mathbf{b}]$ is row equivalent to

$$\begin{bmatrix} 1 & -1 & 1 & 2 \\ 0 & 1 & 1 & -1 \\ 0 & 0 & 0 & 0 \end{bmatrix}.$$

Thus the least-squares solutions (the solutions of $A^TA\mathbf{x} = A^T\mathbf{b}$) are given by

$$\mathbf{x}^* = \begin{bmatrix} 1 - 2x_3 \\ -1 - x_3 \\ x_3 \end{bmatrix} = \begin{bmatrix} 1 \\ -1 \\ 0 \end{bmatrix} + x_3 \begin{bmatrix} -2 \\ -1 \\ 1 \end{bmatrix}.$$

(*Note:* The solutions in Examples 1 and 2 agree, although they were found with different procedures.)

Linear Least-Squares Fits to Data

Suppose the laws of physics tell us that two measurable quantities, t and y, are related in a linear fashion:

$$y = mt + c. \tag{3}$$

Suppose also that we wish to determine experimentally the values of m and c. If we know that Eq. (3) models the phenomena exactly and that we have made no experimental error, then we can determine m and c with only two experimental observations. For instance, if $y = y_0$ when $t = t_0$ and if $y = y_1$ when $t = t_1$, we can solve for m and c from

$$\begin{array}{c} mt_0 + c = y_0 \\ mt_1 + c = y_1 \end{array} \quad \text{or} \quad \begin{bmatrix} t_0 & 1 \\ t_1 & 1 \end{bmatrix}\begin{bmatrix} m \\ c \end{bmatrix} = \begin{bmatrix} y_0 \\ y_1 \end{bmatrix}.$$

Usually we must be resigned to experimental errors or to imperfections in the model given by Eq. (3). In this case we would probably make a number of experimental observations, (t_i, y_i), for $i = 0, 1, \ldots, k$. Using these observed values in Eq. (3) leads to an overdetermined system of the form

$$mt_0 + c = y_0$$
$$mt_1 + c = y_1$$
$$\vdots$$
$$mt_k + c = y_k.$$

In matrix terms, this overdetermined system can be expressed as $A\mathbf{x} = \mathbf{b}$, where

$$A = \begin{bmatrix} t_0 & 1 \\ t_1 & 1 \\ \vdots & \vdots \\ t_k & 1 \end{bmatrix}, \quad \mathbf{x} = \begin{bmatrix} m \\ c \end{bmatrix}, \quad \text{and} \quad \mathbf{b} = \begin{bmatrix} y_0 \\ y_1 \\ \vdots \\ y_k \end{bmatrix}.$$

In this context a least-squares solution to $A\mathbf{x} = \mathbf{b}$ is a vector $\mathbf{x}^* = [m^*, c^*]^T$ that minimizes $\|A\mathbf{x} - \mathbf{b}\|$, where

$$\|A\mathbf{x} - \mathbf{b}\|^2 = \sum_{i=0}^{k} [(mt_i + c) - y_i]^2.$$

Finally, the function defined by $y = m^*t + c^*$ is called a least-squares linear fit to the data.

EXAMPLE 3 Consider the experimental observations given in the following table:

t	1	4	8	11
y	1	2	4	5

Find the least-squares linear fit to the data.

8. In (a)–(c) below, use the given information to determine the nullity of T.
   a) $T: R^3 \to R^2$ and the rank of T is 2.
   b) $T: R^3 \to R^3$ and the rank of T is 2.
   c) $T: R^3 \to R^3$ and the rank of T is 3.

9. In (a)–(c) below, use the given information to determine the rank of T.
   a) $T: R^3 \to R^2$ and the nullity of T is 2.
   b) $T: R^3 \to R^3$ and the nullity of T is 1.
   c) $T: R^2 \to R^3$ and the nullity of T is 0.

10. Let $B = \{\mathbf{x}_1, \mathbf{x}_2\}$ be a basis for R^2, and let $T: R^2 \to R^2$ be a linear transformation such that
$$T(\mathbf{x}_1) = \begin{bmatrix} 1 \\ 1 \end{bmatrix} \quad \text{and} \quad T(\mathbf{x}_2) = \begin{bmatrix} 2 \\ -1 \end{bmatrix}.$$
If $\mathbf{e}_1 = \mathbf{x}_1 - 2\mathbf{x}_2$ and $\mathbf{e}_2 = 2\mathbf{x}_1 + \mathbf{x}_2$, where $\mathbf{e}_1$ and $\mathbf{e}_2$ are the unit vectors in R^2, then find the matrix of T.

11. Let
$$\mathbf{b} = \begin{bmatrix} a \\ b \end{bmatrix}$$
and suppose that $T: R^3 \to R^2$ is a linear transformation defined by $T(\mathbf{x}) = A\mathbf{x}$, where A is a (2×3) matrix such that the augmented matrix $[A \mid \mathbf{b}]$ reduces to
$$\begin{bmatrix} 1 & 0 & 8 & -5a + 3b \\ 0 & 1 & -3 & 2a - b \end{bmatrix}.$$
   a) Find vectors $\mathbf{x}_1$ and $\mathbf{x}_2$ in R^3 such that $T(\mathbf{x}_1) = \mathbf{e}_1$ and $T(\mathbf{x}_2) = \mathbf{e}_2$, where $\mathbf{e}_1$ and $\mathbf{e}_2$ are the unit vectors in R^2.
   b) Exhibit a nonzero vector $\mathbf{x}_3$ in R^3 such that $\mathbf{x}_3$ is in $\mathcal{N}(T)$.
   c) Show that $B = \{\mathbf{x}_1, \mathbf{x}_2, \mathbf{x}_3\}$ is a basis for R^3.
   d) Express each of the unit vectors $\mathbf{e}_1, \mathbf{e}_2, \mathbf{e}_3$ of R^3 as a linear combination of the vectors in B. Now calculate $T(\mathbf{e}_i)$, $i = 1, 2, 3$, and determine the matrix A.

In Exercises 12–18, $\mathbf{b} = [a, b, c, d]^T$, $T: R^6 \to R^4$ is a linear transformation defined by $T(\mathbf{x}) = A\mathbf{x}$, and A is a (4×6) matrix such that the augmented matrix $[A \mid \mathbf{b}]$ reduces to
$$\begin{bmatrix} 1 & 0 & 2 & 0 & -3 & 1 & 4a + b - 2c \\ 0 & 1 & -1 & 0 & 2 & 2 & 12a + 5b - 7c \\ 0 & 0 & 0 & 1 & -1 & -2 & -5a - 2b + 3c \\ 0 & 0 & 0 & 0 & 0 & 0 & -16a - 7b + 9c + d \end{bmatrix}.$$

12. Exhibit a basis for the row space of A, and determine the rank and the nullity of A.

13. Determine which of the following vectors are in $\mathcal{R}(T)$. Explain how you can tell.
$$\mathbf{w}_1 = \begin{bmatrix} 1 \\ -1 \\ 1 \\ 0 \end{bmatrix}, \quad \mathbf{w}_2 = \begin{bmatrix} 1 \\ 1 \\ 3 \\ 2 \end{bmatrix},$$
$$\mathbf{w}_3 = \begin{bmatrix} 2 \\ -2 \\ 1 \\ 9 \end{bmatrix}, \quad \mathbf{w}_4 = \begin{bmatrix} 2 \\ 1 \\ 4 \\ 3 \end{bmatrix}$$

14. For each vector $\mathbf{w}_i$, $i = 1, 2, 3, 4$, listed in Exercise 13, if the system of equations $A\mathbf{x} = \mathbf{w}_i$ is consistent, then exhibit a solution.

15. For each vector $\mathbf{w}_i$, $i = 1, 2, 3, 4$, listed in Exercise 13, if $\mathbf{w}_i$ is in $\mathcal{R}(T)$, then find a vector $\mathbf{x}$ in R^6 such that $T(\mathbf{x}) = \mathbf{w}_i$.

16. Suppose that $A = [\mathbf{A}_1, \mathbf{A}_2, \mathbf{A}_3, \mathbf{A}_4, \mathbf{A}_5, \mathbf{A}_6]$.
   a) For each vector $\mathbf{w}_i$, $i = 1, 2, 3, 4$, listed in Exercise 13, if $\mathbf{w}_i$ is in the column space of A, then express $\mathbf{w}_i$ as a linear combination of the columns of A.
   b) Find a subset of $\{\mathbf{A}_1, \mathbf{A}_2, \mathbf{A}_3, \mathbf{A}_4, \mathbf{A}_5, \mathbf{A}_6\}$ that is a basis for the column space of A.
   c) For each column, $\mathbf{A}_j$, of A that does not appear in the basis obtained in part (b), express $\mathbf{A}_j$ as a linear combination of the basis vectors.
   d) Let $\mathbf{b} = [1, -2, 1, -7]^T$. Show that $\mathbf{b}$ is in the column space of A, and express $\mathbf{b}$ as a linear combination of the basis vectors found in part (b).
   e) If $\mathbf{x} = [2, 3, 1, -1, 1, 1]^T$, then express $A\mathbf{x}$ as a linear combination of the basis vectors found in part (b).

17. a) Give an algebraic specification for $\mathcal{R}(T)$, and use that specification to determine a basis for $\mathcal{R}(T)$.
   b) Show that $\mathbf{b} = [1, 2, 3, 3]^T$ is in $\mathcal{R}(T)$, and express $\mathbf{b}$ as a linear combination of the basis vectors found in part (a).

18. a) Exhibit a basis for $\mathcal{N}(T)$.
   b) Show that $\mathbf{x} = [6, 1, 1, -2, 2, -2]^T$ is in $\mathcal{N}(T)$, and express $\mathbf{x}$ as a linear combination of the basis vectors found in part (a).

Supplementary Conceptual Exercises

In Exercises 1–12, answer true or false. Justify your answer by providing a counterexample if the statement is false or an outline of a proof if the statement is true.

1. If W is a subspace of R^n and $\mathbf{x}$ and $\mathbf{y}$ are vectors in R^n such that $\mathbf{x} + \mathbf{y}$ is in W, then $\mathbf{x}$ is in W and $\mathbf{y}$ is in W.

2. If W is a subspace of R^n and $a\mathbf{x}$ is in W, where a is a nonzero scalar and $\mathbf{x}$ is in R^n, then $\mathbf{x}$ is in W.

3. If $S = \{\mathbf{x}_1, \ldots, \mathbf{x}_k\}$ is a subset of R^n and $k \leq n$, then S is a linearly independent set.

4. If $S = \{\mathbf{x}_1, \ldots, \mathbf{x}_k\}$ is a subset of R^n and $k > n$, then S is a linearly dependent set.

5. If $S = \{\mathbf{x}_1, \ldots, \mathbf{x}_k\}$ is a subset of R^n and $k < n$, then S is not a spanning set for R^n.

6. If $S = \{\mathbf{x}_1, \ldots, \mathbf{x}_k\}$ is a subset of R^n and $k \geq n$, then S is a spanning set for R^n.

7. If S_1 and S_2 are linearly independent subsets of R^n, then the set $S_1 \cup S_2$ is also linearly independent.

8. If W is a subspace of R^n, then W has exactly one basis.

9. If W is a subspace of R^n, and $\dim(W) = k$, then W contains exactly k vectors.

10. If B is a basis for R^n and W is a subspace of R^n, then some subset of B is a basis for W.

11. If W is a subspace of R^n, and $\dim(W) = n$, then $W = R^n$.

12. Let W_1 and W_2 be subspaces of R^n with bases B_1 and B_2, respectively. Then $B_1 \cap B_2$ is a basis for $W_1 \cap W_2$.

In Exercises 13–23, give a brief answer.

13. Let W be a subspace of R^n, and set $V = \{\mathbf{x}: \mathbf{x}$ is in R^n but $\mathbf{x}$ is not in $W\}$. Determine if V is a subspace of R^n.

14. Explain what is wrong with the following argument: Let W be a subspace of R^n, and let $B = \{\mathbf{e}_1, \ldots, \mathbf{e}_n\}$ be the basis of R^n consisting of the unit vectors. Since B is linearly independent and since every vector $\mathbf{w}$ in W can be written as a linear com-

bination of the vectors in B, it follows that B is a basis for W.

15. If $B = \{\mathbf{x}_1, \mathbf{x}_2, \mathbf{x}_3\}$ is a basis for R^3, show that $B' = \{\mathbf{x}_1, \mathbf{x}_1 + \mathbf{x}_2, \mathbf{x}_1 + \mathbf{x}_2 + \mathbf{x}_3\}$ is also a basis for R^3.

16. Let W be a subspace of R^n, and let $S = \{\mathbf{w}_1, \ldots, \mathbf{w}_k\}$ be a linearly independent subset of W such that $\{\mathbf{w}_1, \ldots, \mathbf{w}_k, \mathbf{w}\}$ is linearly dependent for every $\mathbf{w}$ in W. Prove that S is a basis for W.

17. Let $\{\mathbf{u}_1, \ldots, \mathbf{u}_n\}$ be a linearly independent subset of R^n, and let $\mathbf{x}$ in R^n be such that $\mathbf{u}_1^T\mathbf{x} = \cdots = \mathbf{u}_n^T\mathbf{x} = 0$. Show that $\mathbf{x} = \boldsymbol{\theta}$.

18. Let $\mathbf{u}$ be a nonzero vector in R^n, and let W be the subset of R^n defined by $W = \{\mathbf{x}: \mathbf{u}^T\mathbf{x} = 0\}$.
   a) Prove that W is a subspace of R^n.
   b) Show that $\dim(W) = n - 1$.
   c) If $\boldsymbol{\theta} = \mathbf{w} + c\mathbf{u}$, where $\mathbf{w}$ is in W and c is a scalar, show that $\mathbf{w} = \boldsymbol{\theta}$ and $c = 0$. [*Hint:* Consider $\mathbf{u}^T(\mathbf{w} + c\mathbf{u})$.]
   d) If $\{\mathbf{w}_1, \ldots, \mathbf{w}_{n-1}\}$ is a basis for W, show that $\{\mathbf{w}_1, \ldots, \mathbf{w}_{n-1}, \mathbf{u}\}$ is a basis for R^n. [*Hint:* Suppose that $c_1\mathbf{w}_1 + \cdots + c_{n-1}\mathbf{w}_{n-1} + c\mathbf{u} = \boldsymbol{\theta}$. Now set $\mathbf{w} = c_1\mathbf{w}_1 + \cdots + c_{n-1}\mathbf{w}_{n-1}$ and use part (c).]

19. Let V and W be subspaces of R^n such that $V \cap W = \{\boldsymbol{\theta}\}$ and $\dim(V) + \dim(W) = n$.
   a) If $\mathbf{v} + \mathbf{w} = \boldsymbol{\theta}$, where $\mathbf{v}$ is in V and $\mathbf{w}$ is in W, show that $\mathbf{v} = \boldsymbol{\theta}$ and $\mathbf{w} = \boldsymbol{\theta}$.
   b) If B_1 is a basis for V and B_2 is a basis for W, show that $B_1 \cup B_2$ is a basis for R^n. (*Hint:* Use part (a) to show that $B_1 \cup B_2$ is linearly independent.)
   c) If $\mathbf{x}$ is in R^n, show that $\mathbf{x}$ can be written in the form $\mathbf{x} = \mathbf{v} + \mathbf{w}$, where $\mathbf{v}$ is in V and $\mathbf{w}$ is in W. (*Hint:* First note that $\mathbf{x}$ can be written as a linear combination of the vectors in $B_1 \cup B_2$.)
   d) Show that the representation obtained in part (c) is unique; that is, if $\mathbf{x} = \mathbf{v}_1 + \mathbf{w}_1$, where $\mathbf{v}_1$ is in V and $\mathbf{w}_1$ is in W, then $\mathbf{v} = \mathbf{v}_1$ and $\mathbf{w} = \mathbf{w}_1$.

20. A linear transformation $T: R^n \to R^n$ is **onto** provided that $\mathcal{R}(T) = R^n$. Prove each of the following.
   a) If the rank of T is n, then T is onto.
   b) If the nullity of T is zero, then T is onto.

c) If T is onto, then the rank of T is n and the nullity of T is zero.

21. If $T: R^n \to R^m$ is a linear transformation, then show that $T(\theta_n) = \theta_m$, where θ_n and θ_m are the zero vectors in R^n and R^m, respectively.

22. Let $T: R^n \to R^m$ be a linear transformation, and suppose that $S = \{\mathbf{x}_1, \ldots, \mathbf{x}_k\}$ is a subset of R^n such that $\{T(\mathbf{x}_1), \ldots, T(\mathbf{x}_k)\}$ is a linearly independent subset of R^m. Show that the set S is linearly independent.

23. Let $T: R^n \to R^m$ be a linear transformation with nullity zero. If $S = \{\mathbf{x}_1, \ldots, \mathbf{x}_k\}$ is a linearly independent subset of R^n, then show that $\{T(\mathbf{x}_1), \ldots, T(\mathbf{x}_k)\}$ is a linearly independent subset of R^m.

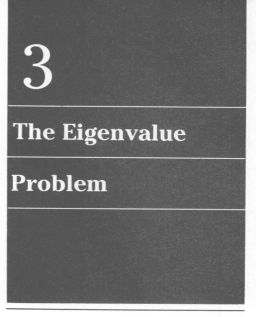

3

The Eigenvalue

Problem

3.1 Introduction

The *eigenvalue problem,* the topic of this chapter, is a problem of considerable theoretical interest and wide-ranging application. For instance, applications found in Sections 3.8 and 4.10 and Chapter 6 include procedures for:

a) solving systems of differential equations;

b) analyzing population growth models;

c) calculating powers of matrices;

d) diagonalizing linear transformations; and

e) simplifying and describing the graphs of quadratic forms in two and three variables.

The eigenvalue problem is formulated as follows.

The Eigenvalue Problem

For an $(n \times n)$ matrix A, find all scalars λ such that the equation

$$A\mathbf{x} = \lambda\mathbf{x} \qquad\qquad (1)$$

has a nonzero solution, $\mathbf{x}$. Such a scalar λ is called an *eigenvalue* of A, and any nonzero $(n \times 1)$ vector $\mathbf{x}$ satisfying Eq. (1) is called an *eigenvector* corresponding to λ.

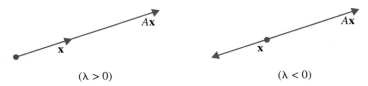

Figure 3.1
Let $A\mathbf{x} = \lambda\mathbf{x}$, where $\mathbf{x}$ is a nonzero vector. Then $\mathbf{x}$ and $A\mathbf{x}$ are parallel vectors.

Let $\mathbf{x}$ be an eigenvector of A corresponding to an eigenvalue λ. Then the vector $A\mathbf{x}$ is a scalar multiple of $\mathbf{x}$ (see Eq. (1)). Represented as geometric vectors, $\mathbf{x}$ and $A\mathbf{x}$ have the same direction if λ is positive and the opposite direction if λ is negative (see Fig. 3.1).

Now, we can rewrite Eq. (1) as

$$A\mathbf{x} - \lambda\mathbf{x} = \boldsymbol{\theta},$$

or

$$(A - \lambda I)\mathbf{x} = \boldsymbol{\theta}, \qquad \mathbf{x} \neq \boldsymbol{\theta}, \tag{2}$$

where I is the $(n \times n)$ identity matrix. If Eq. (2) is to have nonzero solutions, then λ must be chosen so that the $(n \times n)$ matrix $A - \lambda I$ is singular. Therefore, the eigenvalue problem consists of two parts:

1. Find all scalars λ such that $A - \lambda I$ is singular.
2. Given a scalar λ such that $A - \lambda I$ is singular, find all nonzero vectors $\mathbf{x}$ such that $(A - \lambda I)\mathbf{x} = \boldsymbol{\theta}$.

If we know an eigenvalue of A, then the variable-elimination techniques described in Chapter 1 provide an efficient way to find the eigenvectors. The new feature of the eigenvalue problem is in part 1, determining all scalars λ such that the matrix $A - \lambda I$ is singular. In the next subsection, we discuss how such values λ are found.

Eigenvalues for (2 × 2) Matrices

Before discussing how the eigenvalue problem is solved for a general $(n \times n)$ matrix A, we first consider the special case where A is a (2×2) matrix. In particular, suppose we want to solve the eigenvalue problem for a matrix A of the form

$$A = \begin{bmatrix} a & b \\ c & d \end{bmatrix}.$$

As we noted above, the first step is to find all scalars λ such that $A - \lambda I$ is singular. The matrix $A - \lambda I$ is given by

$$A - \lambda I = \begin{bmatrix} a & b \\ c & d \end{bmatrix} - \begin{bmatrix} \lambda & 0 \\ 0 & \lambda \end{bmatrix},$$

or

$$A - \lambda I = \begin{bmatrix} a - \lambda & b \\ c & d - \lambda \end{bmatrix}.$$

Next we recall (Exercise 27 in Section 1.11) that a (2×2) matrix is singular if and only if the product of the diagonal entries is equal to the product of the off-diagonal entries. That is, if B is the (2×2) matrix

$$B = \begin{bmatrix} r & s \\ t & u \end{bmatrix}, \tag{3a}$$

then

$$B \text{ is singular} \Leftrightarrow ru - st = 0. \tag{3b}$$

If we apply the result in (3b) to the matrix $A - \lambda I$, it follows that $A - \lambda I$ is singular if and only if λ is a value such that

$$(a - \lambda)(d - \lambda) - bc = 0. \tag{4}$$

Expanding the equation for λ given above, we obtain the following condition on λ:

$$\lambda^2 - (a + d)\lambda + (ad - bc) = 0.$$

Equivalently, $A - \lambda I$ is singular if and only if λ is a root of the polynomial equation

$$t^2 - (a + d)t + (ad - bc) = 0. \tag{5}$$

An example will serve to illustrate the idea.

EXAMPLE 1 Find all scalars λ such that $A - \lambda I$ is singular, where

$$A = \begin{bmatrix} 5 & -2 \\ 6 & -2 \end{bmatrix}.$$

SOLUTION   The matrix $A - \lambda I$ has the form

$$A - \lambda I = \begin{bmatrix} 5 - \lambda & -2 \\ 6 & -2 - \lambda \end{bmatrix}.$$

As in Eq. (4), $A - \lambda I$ is singular if and only if

$$-(5 - \lambda)(2 + \lambda) + 12 = 0,$$

or $\lambda^2 - 3\lambda + 2 = 0$. Since $\lambda^2 - 3\lambda + 2 = (\lambda - 2)(\lambda - 1)$, it follows that $A - \lambda I$ is singular if and only if $\lambda = 2$ or $\lambda = 1$.   ☐

As a check for the calculations in Example 1, we list the matrices $A - \lambda I$ for $\lambda = 2$ and $\lambda = 1$:

$$A - 2I = \begin{bmatrix} 3 & -2 \\ 6 & -4 \end{bmatrix}, \quad A - I = \begin{bmatrix} 4 & -2 \\ 6 & -3 \end{bmatrix}. \tag{6}$$

Note that the matrices displayed above, $A - 2I$ and $A - I$, are singular.

Eigenvectors for (2 × 2) Matrices

As we observed earlier, the eigenvalue problem consists of two steps: First find the eigenvalues (the scalars λ such that $A - \lambda I$ is singular). Next find the eigenvectors [the nonzero vectors $\mathbf{x}$ such that $(A - \lambda I)\mathbf{x} = \boldsymbol{\theta}$].

In the following example, we find the eigenvectors for matrix A in Example 1.

EXAMPLE 2 For matrix A in Example 1, determine the eigenvectors corresponding to $\lambda = 2$ and to $\lambda = 1$.

SOLUTION   According to Eq. (2), the eigenvectors corresponding to $\lambda = 2$ are the nonzero solutions of $(A - 2I)\mathbf{x} = \boldsymbol{\theta}$. Thus, for the singular matrix $A - 2I$ listed in (6), we need to solve the homogeneous system

$$3x_1 - 2x_2 = 0$$
$$6x_1 - 4x_2 = 0.$$

The solution of this system is given by $3x_1 = 2x_2$, or $x_1 = (2/3)x_2$. Thus all the nonzero solutions of $(A - 2I)\mathbf{x} = \boldsymbol{\theta}$ are of the form

$$\mathbf{x} = \begin{bmatrix} (2/3)x_2 \\ x_2 \end{bmatrix} = x_2 \begin{bmatrix} 2/3 \\ 1 \end{bmatrix}, \qquad x_2 \neq 0.$$

For $\lambda = 1$, the solutions of $(A - I)\mathbf{x} = \boldsymbol{\theta}$ are found by solving

$$4x_1 - 2x_2 = 0$$
$$6x_1 - 3x_2 = 0.$$

The nonzero solutions of this system are all of the form

$$\mathbf{x} = \begin{bmatrix} (1/2)x_2 \\ x_2 \end{bmatrix} = x_2 \begin{bmatrix} 1/2 \\ 1 \end{bmatrix}, \qquad x_2 \neq 0.$$

The results of Examples 1 and 2 provide the solution to the eigenvalue problem for the matrix A, where A is given by

$$A = \begin{bmatrix} 5 & -2 \\ 6 & -2 \end{bmatrix}.$$

In a summary form, the eigenvalues and corresponding eigenvectors are as listed below:

Eigenvalue: $\lambda = 2$;   *Eigenvectors:* $\mathbf{x} = a \begin{bmatrix} 2/3 \\ 1 \end{bmatrix}$, $a \neq 0$.

Eigenvalue: $\lambda = 1$;   *Eigenvectors:* $\mathbf{x} = a \begin{bmatrix} 1/2 \\ 1 \end{bmatrix}$, $a \neq 0$.

Note that for a given eigenvalue λ, there are infinitely many eigenvectors corresponding to λ. Since Eq. (2) is a homogeneous system, it follows that if $\mathbf{x}$ is an eigenvector corresponding to λ, then so is $a\mathbf{x}$ for any nonzero scalar a.

Finally, we make the following observation: If A is a (2×2) matrix, then we have a simple test for determining those values λ such that $A - \lambda I$ is singular. But if A is an $(n \times n)$ matrix with $n > 2$, we do not (as yet) have a test for determining whether $A - \lambda I$ is singular. In the next section a singularity test based on the theory of determinants will be developed.

Exercises 3.1

In Exercises 1–12, find the eigenvalues and the eigenvectors for the given matrix.

1. $A = \begin{bmatrix} 1 & 0 \\ 2 & 3 \end{bmatrix}$

2. $A = \begin{bmatrix} 2 & 1 \\ 0 & -1 \end{bmatrix}$

3. $A = \begin{bmatrix} 2 & -1 \\ -1 & 2 \end{bmatrix}$

4. $A = \begin{bmatrix} 1 & -2 \\ 1 & 4 \end{bmatrix}$

5. $A = \begin{bmatrix} 2 & 1 \\ 1 & 2 \end{bmatrix}$

6. $A = \begin{bmatrix} 3 & -1 \\ 5 & -3 \end{bmatrix}$

7. $A = \begin{bmatrix} 1 & 0 \\ 2 & 1 \end{bmatrix}$

8. $A = \begin{bmatrix} 2 & 3 \\ 0 & 2 \end{bmatrix}$

9. $A = \begin{bmatrix} 2 & 2 \\ 3 & 3 \end{bmatrix}$

10. $A = \begin{bmatrix} 1 & 2 \\ 4 & 8 \end{bmatrix}$

11. $A = \begin{bmatrix} 1 & -1 \\ 1 & 3 \end{bmatrix}$

12. $A = \begin{bmatrix} 2 & -1 \\ 1 & 4 \end{bmatrix}$

Apply the singularity test (4) to the matrices in Exercises 13–16. Show that there is no *real* scalar λ such that $A - \lambda I$ is singular. (*Note:* Complex eigenvalues are discussed in Section 3.6.)

13. $A = \begin{bmatrix} -2 & -1 \\ 5 & 2 \end{bmatrix}$

14. $A = \begin{bmatrix} 3 & -2 \\ 5 & -3 \end{bmatrix}$

15. $A = \begin{bmatrix} 2 & -1 \\ 1 & 2 \end{bmatrix}$

16. $A = \begin{bmatrix} 1 & -1 \\ 1 & 1 \end{bmatrix}$

17. Consider the (2×2) *symmetric* matrix

$$A = \begin{bmatrix} a & b \\ b & d \end{bmatrix}.$$

Show that there are always real scalars λ such that $A - \lambda I$ is singular. (*Hint:* Use the quadratic formula for the roots of Eq. (5).)

18. Consider the (2×2) matrix A given by

$$A = \begin{bmatrix} a & b \\ -b & a \end{bmatrix}, \quad b \neq 0.$$

Show that there are no real scalars λ such that $A - \lambda I$ is singular.

19. Let A be a (2×2) matrix. Show that A and A^T have the same set of eigenvalues by considering the polynomial equation (5).

3.2 Determinants and the Eigenvalue Problem

Now we turn our attention to the eigenvalue problem for a general $(n \times n)$ matrix A. As we observed in the last section, the first task is to determine all scalars λ such that the matrix $A - \lambda I$ is singular.

For a (2×2) matrix $A - \lambda I$ given by

$$A - \lambda I = \begin{bmatrix} a - \lambda & b \\ c & d - \lambda \end{bmatrix},$$

Determinants

Determinants were studied and extensively used long before matrix algebra was developed. In 1693, the co-founder of calculus, Gottfried Wilhelm Leibniz (1646–1716), essentially used determinants to determine if a (3×3) linear system was consistent. (Similar work was done ten years earlier in Japan by Seki-Kowa.) Cramer's Rule (see Section 5.4), which uses determinants to solve linear systems, was developed in 1729 by Colin Maclaurin (1698–1746). Joseph Louis Lagrange (1736–1813) used determinants to express the area of a triangle and the volume of a tetrahedron.

It was Augustin-Louis Cauchy (1789–1857) who first coined the term "determinant" and in 1812 published a unification of the theory of determinants. In subsequent publications Cauchy used determinants in a variety of ways such as the development of the functional determinant commonly called the "Jacobian."

we have a simple test for singularity:

$$A - \lambda I \text{ is singular} \Leftrightarrow (a - \lambda)(d - \lambda) - bc = 0.$$

For a general $(n \times n)$ matrix A, the theory of determinants can be used to discover those values λ such that $A - \lambda I$ is singular.

Determinant theory has long intrigued mathematicians. The reader has probably learned how to calculate determinants, at least for (2×2) and (3×3) matrices. The purpose of this section is to review briefly those aspects of determinant theory that can be used in the eigenvalue problem. A formal development of determinants, including proofs, definitions, and the important properties of determinants, can be found in Chapter 5. In this section we present three basic results: an algorithm for evaluating determinants, a characterization of singular matrices in terms of determinants, and a result concerning determinants of matrix products.

Determinants of (2 × 2) Matrices

We begin with the definition for the determinant of a (2×2) matrix.

DEFINITION 1   Let A be the (2×2) matrix

$$A = \begin{bmatrix} a_{11} & a_{12} \\ a_{21} & a_{22} \end{bmatrix}.$$

The **determinant** of A, denoted by $\det(A)$, is the number

$$\det(A) = \begin{vmatrix} a_{11} & a_{12} \\ a_{21} & a_{22} \end{vmatrix} = a_{11}a_{22} - a_{21}a_{12}.$$

(*Note:* As Definition 1 indicates, the determinant of a (2×2) matrix is simply the difference of the products of the diagonal entries and the off-diagonal entries. Thus, in the context of the singularity test displayed in (3) in the previous section, a (2×2) matrix A is singular if and only if $\det(A) = 0$. Also note that we designate the determinant of A by vertical bars when we wish to exhibit the entries of A.)

EXAMPLE 1 Find $\det(A)$, where

$$A = \begin{bmatrix} 2 & 4 \\ 1 & 3 \end{bmatrix}.$$

SOLUTION By Definition 1,

$$\det(A) = \begin{vmatrix} 2 & 4 \\ 1 & 3 \end{vmatrix} = 2 \cdot 3 - 1 \cdot 4 = 2. \qquad \square$$

EXAMPLE 2 Find $\det(A)$, where

$$A = \begin{bmatrix} 2 & 4 \\ 3 & 6 \end{bmatrix}.$$

SOLUTION By Definition 1,

$$\det(A) = \begin{vmatrix} 2 & 4 \\ 3 & 6 \end{vmatrix} = 2 \cdot 6 - 3 \cdot 4 = 0. \qquad \square$$

Again, Examples 1 and 2 are special instances that reaffirm our earlier observation about the singularity of a (2×2) matrix A. That is, A is singular if and only if $\det(A) = 0$.

Determinants of (3×3) Matrices

In Definition 1, we associated a number, $\det(A)$, with a (2×2) matrix A. This number assignment had the property that $\det(A) = 0$ if and only if A is singular. We now develop a similar association of a number, $\det(A)$, with an $(n \times n)$ matrix A.

We first consider the case in which $n = 3$.

DEFINITION 2 Let A be the (3×3) matrix

$$A = \begin{bmatrix} a_{11} & a_{12} & a_{13} \\ a_{21} & a_{22} & a_{23} \\ a_{31} & a_{32} & a_{33} \end{bmatrix}.$$

The *determinant of* A is the number $\det(A)$, where

$$\det(A) = a_{11} \begin{vmatrix} a_{22} & a_{23} \\ a_{32} & a_{33} \end{vmatrix} - a_{12} \begin{vmatrix} a_{21} & a_{23} \\ a_{31} & a_{33} \end{vmatrix} + a_{13} \begin{vmatrix} a_{21} & a_{22} \\ a_{31} & a_{32} \end{vmatrix}. \qquad (1)$$

(*Note:* The determinant of a (3 × 3) matrix is defined to be the "weighted sum" of three (2 × 2) determinants. Similarly, the determinant of an (*n* × *n*) matrix will be defined as the weighted sum of *n* determinants each of order [(*n* − 1) × (*n* − 1)].)

EXAMPLE 3 Find det(*A*), where

$$A = \begin{bmatrix} 1 & 2 & -1 \\ 5 & 3 & 4 \\ -2 & 0 & 1 \end{bmatrix}.$$

SOLUTION From Definition 2,

$$\det(A) = (1)\begin{vmatrix} 3 & 4 \\ 0 & 1 \end{vmatrix} - (2)\begin{vmatrix} 5 & 4 \\ -2 & 1 \end{vmatrix} + (-1)\begin{vmatrix} 5 & 3 \\ -2 & 0 \end{vmatrix}$$

$$= 1(3 \cdot 1 - 4 \cdot 0) - 2[5 \cdot 1 - 4(-2)] - 1[5 \cdot 0 - 3(-2)]$$

$$= 3 - 26 - 6 = -29.$$

Minors and Cofactors

If we examine the three (2 × 2) determinants that appear in Eq. (1), we can see a pattern. In particular, the entries in the first (2 × 2) determinant can be obtained from the matrix *A* by striking out the first row and column of *A*. Similarly, the entries in the second (2 × 2) determinant can be obtained by striking out the first row and second column of *A*. Finally, striking out the first row and third column yields the third (2 × 2) determinant.

The process of generating "submatrices" by striking out rows and columns is fundamental to the definition of a general (*n* × *n*) determinant. For a general (*n* × *n*) matrix *A*, we will use the notation M_{rs} to designate the [(*n* − 1) × (*n* − 1)] matrix generated by removing row *r* and column *s* from *A* (see Definition 3 below).

> **DEFINITION 3** Let $A = (a_{ij})$ be an (*n* × *n*) matrix. The [(*n* − 1) × (*n* − 1)] matrix that results from removing the *r*th row and *s*th column from *A* is called a *minor matrix of A* and is designated by M_{rs}.

An example will serve to illustrate the idea in Definition 3.

EXAMPLE 4 List the minor matrices M_{21}, M_{23}, M_{42}, and M_{11} for the (4 × 4) matrix *A* given by

$$A = \begin{bmatrix} 1 & 2 & 1 & 3 \\ 0 & 1 & 2 & 0 \\ 4 & 2 & 0 & -1 \\ -2 & 3 & 1 & 1 \end{bmatrix}.$$

SOLUTION The minor matrix M_{21} is obtained from A by removing the second row and the first column from A:

$$M_{21} = \begin{bmatrix} 2 & 1 & 3 \\ 2 & 0 & -1 \\ 3 & 1 & 1 \end{bmatrix}.$$

Similarly, we have

$$M_{23} = \begin{bmatrix} 1 & 2 & 3 \\ 4 & 2 & -1 \\ -2 & 3 & 1 \end{bmatrix}, \quad M_{42} = \begin{bmatrix} 1 & 1 & 3 \\ 0 & 2 & 0 \\ 4 & 0 & -1 \end{bmatrix}, \quad \text{and}$$

$$M_{11} = \begin{bmatrix} 1 & 2 & 0 \\ 2 & 0 & -1 \\ 3 & 1 & 1 \end{bmatrix}.$$

☐

Using the notation for a minor matrix, we can reinterpret the definition of a (3×3) determinant as follows: If $A = (a_{ij})$ is a (3×3) matrix, then from Eq. (1) and Definition 3,

$$\det(A) = a_{11} \det(M_{11}) - a_{12} \det(M_{12}) + a_{13} \det(M_{13}). \tag{2}$$

In determinant theory, the number $\det(M_{ij})$ is called a minor. Precisely, if $A = (a_{ij})$ is an $(n \times n)$ matrix, then the number $\det(M_{ij})$ is the **minor of the (i, j)th entry, a_{ij}.** In addition, the numbers A_{ij} defined by

$$A_{ij} = (-1)^{i+j} \det(M_{ij})$$

are known as **cofactors** (or **signed minors**). Thus the expression for $\det(A)$ in Eq. (2) is known as a **cofactor expansion** corresponding to the first row.

It is natural, then, to wonder about other cofactor expansions of A that parallel the one given in Eq. (2). For instance, what is the cofactor expansion of A corresponding to the second row or even, perhaps, corresponding to the third column?

By analogy, a cofactor expansion along the second row would have the form

$$-a_{21} \det(M_{21}) + a_{22} \det(M_{22}) - a_{23} \det(M_{23}). \tag{3}$$

An expansion along the third column would take the form

$$a_{13} \det(M_{13}) - a_{23} \det(M_{23}) + a_{33} \det(M_{33}). \tag{4}$$

EXAMPLE 5 Let A denote the (3×3) matrix from Example 3,

$$A = \begin{bmatrix} 1 & 2 & -1 \\ 5 & 3 & 4 \\ -2 & 0 & 1 \end{bmatrix}.$$

Calculate the second-row and third-column cofactor expansions defined by Eqs. (3) and (4), respectively.

SOLUTION   According to the pattern in Eq. (3), a second-row expansion has the value

$$-5\begin{vmatrix} 2 & -1 \\ 0 & 1 \end{vmatrix} + 3\begin{vmatrix} 1 & -1 \\ -2 & 1 \end{vmatrix} - 4\begin{vmatrix} 1 & 2 \\ -2 & 0 \end{vmatrix} = -10 - 3 - 16 = -29.$$

Using Eq. (4), we obtain a third-column expansion given by

$$-\begin{vmatrix} 5 & 3 \\ -2 & 0 \end{vmatrix} - 4\begin{vmatrix} 1 & 2 \\ -2 & 0 \end{vmatrix} + \begin{vmatrix} 1 & 2 \\ 5 & 3 \end{vmatrix} = -6 - 16 - 7 = -29. \qquad \square$$

(*Note:* For the (3×3) matrix A in Example 5, there are three possible row expansions and three possible column expansions. It can be shown that each of these six expansions yields exactly the same value, namely, -29. In general, as we observe in the next subsection, all row expansions and all column expansions produce the same value for any $(n \times n)$ matrix.)

The Determinant of an $(n \times n)$ Matrix

We now give an inductive definition for $\det(A)$, the determinant of an $(n \times n)$ matrix. That is, $\det(A)$ is defined in terms of determinants of $[(n - 1) \times (n - 1)]$ matrices. The natural extension of Definition 2 is the following.

DEFINITION 4   Let $A = (a_{ij})$ be an $(n \times n)$ matrix. The ***determinant of A*** is the number $\det(A)$, where

$$\det(A) = a_{11} \det(M_{11}) - a_{12} \det(M_{12}) + \cdots + (-1)^{n+1} a_{1n} \det(M_{1n})$$

$$= \sum_{j=1}^{n} (-1)^{j+1} a_{1j} \det(M_{1j}). \qquad (5)$$

The definition for $\det(A)$ can be stated in a briefer form if we recall the notation A_{ij} for a cofactor. That is, $A_{ij} = (-1)^{i+j} \det(M_{ij})$. Using the cofactor notation, we can rephrase Definition 4 as

$$\det(A) = \sum_{j=1}^{n} a_{1j} A_{1j}. \qquad (6)$$

In the following example we see how Eq. (5) gives the determinant of a (4×4) matrix as the sum of four (3×3) determinants, where each (3×3) determinant is the sum of three (2×2) determinants.

EXAMPLE 6   Use Definition 4 to calculate the $\det(A)$, where

$$A = \begin{bmatrix} 1 & 2 & -1 & 1 \\ -1 & 0 & 2 & -2 \\ 3 & -1 & 1 & 1 \\ 2 & 0 & -1 & 2 \end{bmatrix}.$$

SOLUTION   The determinants of the minor matrices M_{11}, M_{12}, M_{13}, and M_{14} are (3×3) determinants and are calculated as before with Definition 2:

$$\det(M_{11}) = \begin{vmatrix} 0 & 2 & -2 \\ -1 & 1 & 1 \\ 0 & -1 & 2 \end{vmatrix} = 0 \begin{vmatrix} 1 & 1 \\ -1 & 2 \end{vmatrix} - 2 \begin{vmatrix} -1 & 1 \\ 0 & 2 \end{vmatrix} + (-2) \begin{vmatrix} -1 & 1 \\ 0 & -1 \end{vmatrix} = 2$$

$$\det(M_{12}) = \begin{vmatrix} -1 & 2 & -2 \\ 3 & 1 & 1 \\ 2 & -1 & 2 \end{vmatrix} = (-1) \begin{vmatrix} 1 & 1 \\ -1 & 2 \end{vmatrix} - 2 \begin{vmatrix} 3 & 1 \\ 2 & 2 \end{vmatrix} + (-2) \begin{vmatrix} 3 & 1 \\ 2 & -1 \end{vmatrix} = -1$$

$$\det(M_{13}) = \begin{vmatrix} -1 & 0 & -2 \\ 3 & -1 & 1 \\ 2 & 0 & 2 \end{vmatrix} = (-1) \begin{vmatrix} -1 & 1 \\ 0 & 2 \end{vmatrix} - 0 \begin{vmatrix} 3 & 1 \\ 2 & 2 \end{vmatrix} + (-2) \begin{vmatrix} 3 & -1 \\ 2 & 0 \end{vmatrix} = -2$$

$$\det(M_{14}) = \begin{vmatrix} -1 & 0 & 2 \\ 3 & -1 & 1 \\ 2 & 0 & -1 \end{vmatrix} = (-1) \begin{vmatrix} -1 & 1 \\ 0 & -1 \end{vmatrix} - 0 \begin{vmatrix} 3 & 1 \\ 2 & -1 \end{vmatrix} + 2 \begin{vmatrix} 3 & -1 \\ 2 & 0 \end{vmatrix} = 3.$$

Hence, from Eq. (5) with $n = 4$,

$$\det(A) = 1(2) - 2(-1) + (-1)(-2) - 1(3) = 3.$$

EXAMPLE 7   For the (4×4) matrix A in Example 6, calculate the second-column cofactor expansion given by

$$-a_{12} \det(M_{12}) + a_{22} \det(M_{22}) - a_{32} \det(M_{32}) + a_{42} \det(M_{42}).$$

SOLUTION   From Example 6, $\det(M_{12}) = -1$. Since $a_{22} = 0$ and $a_{42} = 0$, we need not calculate $\det(M_{22})$ and $\det(M_{42})$. The only other value needed is $\det(M_{32})$, where

$$\det(M_{32}) = \begin{vmatrix} 1 & -1 & 1 \\ -1 & 2 & -2 \\ 2 & -1 & 2 \end{vmatrix} = 1 \begin{vmatrix} 2 & -2 \\ -1 & 2 \end{vmatrix} - (-1) \begin{vmatrix} -1 & -2 \\ 2 & 2 \end{vmatrix} + 1 \begin{vmatrix} -1 & 2 \\ 2 & -1 \end{vmatrix} = 1.$$

Thus the second-column expansion gives the value

$$-2(-1) + 0 \det(M_{22}) - (-1)(1) + 0 \det(M_{42}) = 3.$$

From Example 6, $\det(A) = 3$. From Example 7, a second-column expansion also produces the same value, 3. The theorem below states that a cofactor expansion along any row or any column always produces the same number, $\det(A)$. The expansions in the theorem are phrased in the same (brief) notation as in Eq. (6). The proof of Theorem 1 is given in Chapter 5.

THEOREM 1

Let $A = (a_{ij})$ be an $(n \times n)$ matrix with minor matrices M_{ij} and cofactors $A_{ij} = (-1)^{i+j} \det(M_{ij})$. Then

$$\det(A) = \sum_{j=1}^{n} a_{ij} A_{ij} \qquad (i\text{th-row expansion})$$

$$= \sum_{i=1}^{n} a_{ij} A_{ij} \qquad (j\text{th-column expansion})$$

Because of Theorem 1, we can always find $\det(A)$ by choosing the row or column of A with the most zeros for the cofactor expansion. (If $a_{ij} = 0$, then $a_{ij} A_{ij} = 0$, and we need not compute A_{ij}.) In the next section we consider how to use elementary row or column operations to create zeros and hence simplify determinant calculations.

Determinants and Singular Matrices

Theorems 2 and 3, which follow, are fundamental to our study of eigenvalues. These theorems are merely stated here; their proofs are given in Chapter 5.

THEOREM 2

Let A and B be $(n \times n)$ matrices. Then

$$\det(AB) = \det(A)\det(B).$$

The following example illustrates Theorem 2.

EXAMPLE 8   Calculate $\det(A)$, $\det(B)$, and $\det(AB)$ for the matrices

$$A = \begin{bmatrix} 1 & 2 \\ -1 & 1 \end{bmatrix} \quad \text{and} \quad B = \begin{bmatrix} 2 & 3 \\ 1 & -1 \end{bmatrix}.$$

SOLUTION   The product, AB, is given by

$$AB = \begin{bmatrix} 4 & 1 \\ -1 & -4 \end{bmatrix}.$$

Clearly $\det(AB) = -15$. We also see that $\det(A) = 3$ and $\det(B) = -5$. Observe, for this special case, that $\det(A)\det(B) = \det(AB)$.

To study the eigenvalue problem for an $(n \times n)$ matrix, we need a test for singularity. The following theorem shows that determinant theory provides a simple and elegant test.

THEOREM 3

Let A be an $(n \times n)$ matrix. Then

$$A \text{ is singular if and only if } \det(A) = 0.$$

Theorem 3 is already familiar for the case in which A is a (2×2) matrix (recall Definition 1 and Examples 1 and 2). An outline for the proof of Theorem 3 is given in the next section. Finally, in Section 3.4 we will be able to use Theorem 3 to devise a procedure for solving the eigenvalue problem.

We conclude this brief introduction to determinants by observing that it is easy to calculate the determinant of a triangular matrix.

THEOREM 4

Let $T = (t_{ij})$ be an $(n \times n)$ triangular matrix. Then

$$\det(T) = t_{11}t_{22}\ldots t_{nn}.$$

The proof of Theorem 4 is left to the exercises. The next example illustrates how a proof for Theorem 4 might be constructed.

EXAMPLE 9   Use a cofactor expansion (as in Definition 4 or Theorem 1) to calculate $\det(T)$:

$$T = \begin{bmatrix} 2 & 1 & 3 & 7 \\ 0 & 4 & 8 & 1 \\ 0 & 0 & 1 & 5 \\ 0 & 0 & 0 & 3 \end{bmatrix}.$$

SOLUTION   By Theorem 1, we can use a cofactor expansion along any row or column to calculate $\det(T)$. Because of the structure of T, an expansion along the first column or the fourth row will be easiest.

Expanding along the first column, we find

$$\det(T) = \begin{vmatrix} 2 & 1 & 3 & 7 \\ 0 & 4 & 8 & 1 \\ 0 & 0 & 1 & 5 \\ 0 & 0 & 0 & 3 \end{vmatrix} = 2 \begin{vmatrix} 4 & 8 & 1 \\ 0 & 1 & 5 \\ 0 & 0 & 3 \end{vmatrix}$$

$$= (2)(4) \begin{vmatrix} 1 & 5 \\ 0 & 3 \end{vmatrix} = 24.$$

This example provides a special case of Theorem 4.

(*Note:* An easy corollary to Theorem 4 is the following: If I is the $(n \times n)$ identity matrix, then $\det(I) = 1$. In the exercises some additional results are derived from the theorems in this section and from the fact that $\det(I) = 1$.)

Exercises 3.2

In Exercises 1–6, list the minor matrix M_{ij}, and calculate the cofactor $A_{ij} = (-1)^{i+j}\det(M_{ij})$ for the matrix A given by

$$A = \begin{bmatrix} 2 & -1 & 3 & 1 \\ 4 & 1 & 3 & -1 \\ 6 & 2 & 4 & 1 \\ 2 & 2 & 0 & -2 \end{bmatrix} \quad (7)$$

1. M_{11} 2. M_{21} 3. M_{31}

4. M_{41} 5. M_{34} 6. M_{43}

7. Use the results of Exercises 1–4 to calculate $\det(A)$ for the matrix A given in (7).

In Exercises 8–19, calculate the determinant of the given matrix. Use Theorem 3 to state whether the matrix is singular or nonsingular.

8. $A = \begin{bmatrix} 2 & 1 \\ -1 & 2 \end{bmatrix}$ 9. $A = \begin{bmatrix} 1 & -1 \\ -2 & 2 \end{bmatrix}$

10. $A = \begin{bmatrix} 2 & 3 \\ 4 & 6 \end{bmatrix}$ 11. $A = \begin{bmatrix} 1 & 1 \\ 2 & 1 \end{bmatrix}$

12. $A = \begin{bmatrix} 1 & 2 & 4 \\ 2 & 3 & 7 \\ 4 & 2 & 10 \end{bmatrix}$ 13. $A = \begin{bmatrix} 2 & -3 & 2 \\ -1 & -2 & 1 \\ 3 & 1 & -1 \end{bmatrix}$

14. $A = \begin{bmatrix} 1 & 2 & 1 \\ 0 & 3 & 2 \\ -1 & 1 & 1 \end{bmatrix}$ 15. $A = \begin{bmatrix} 2 & 0 & 0 \\ 1 & 3 & 2 \\ 2 & 1 & 4 \end{bmatrix}$

16. $A = \begin{bmatrix} 2 & 0 & 0 \\ 3 & 1 & 0 \\ 2 & 4 & 2 \end{bmatrix}$ 17. $A = \begin{bmatrix} 1 & 2 & 1 & 5 \\ 0 & 3 & 0 & 0 \\ 0 & 4 & 1 & 2 \\ 0 & 3 & 1 & 4 \end{bmatrix}$

18. $A = \begin{bmatrix} 0 & 1 & 0 & 0 \\ 0 & 0 & 1 & 0 \\ 1 & 0 & 0 & 0 \\ 0 & 0 & 0 & 1 \end{bmatrix}$ 19. $A = \begin{bmatrix} 0 & 0 & 0 & 2 \\ 0 & 0 & 3 & 1 \\ 0 & 2 & 1 & 2 \\ 3 & 4 & 1 & 4 \end{bmatrix}$

20. Let $A = (a_{ij})$ be a given (3×3) matrix. Form the associated (3×5) matrix B shown below:

$$B = \begin{bmatrix} a_{11} & a_{12} & a_{13} & a_{11} & a_{12} \\ a_{21} & a_{22} & a_{23} & a_{21} & a_{22} \\ a_{31} & a_{32} & a_{33} & a_{31} & a_{32} \end{bmatrix}.$$

a) Subtract the sum of the three upward diagonal products from the sum of the three downward diagonal products and argue that your result is equal to $\det(A)$.

b) Show, by example, that a similar "basketweave algorithm" cannot be used to calculate the determinant of a (4×4) matrix.

In Exercises 21 and 22, find all ordered pairs (x, y) such that A is singular.

21. $A = \begin{bmatrix} x & y & 1 \\ 2 & 3 & 1 \\ 0 & -1 & 1 \end{bmatrix}$ 22. $A = \begin{bmatrix} x & 1 & 1 \\ 2 & 1 & 1 \\ 0 & -1 & y \end{bmatrix}$

23. Let $A = (a_{ij})$ be the $(n \times n)$ matrix specified thus: $a_{ij} = d$ for $i = j$ and $a_{ij} = 1$ for $i \neq j$. For $n = 2, 3$, and 4, show that

$$\det(A) = (d - 1)^{n-1}(d - 1 + n).$$

24. Let A and B be $(n \times n)$ matrices. Use Theorems 2 and 3 to give a quick proof of each of the following.
   a) If either A or B is singular, then AB is singular.
   b) If AB is singular, then either A or B is singular.

25. Suppose that A is an $(n \times n)$ nonsingular matrix, and recall that $\det(I) = 1$, where I is the $(n \times n)$ identity matrix. Show that $\det(A^{-1}) = 1/\det(A)$.

26. If A and B are $(n \times n)$ matrices, then usually $AB \neq BA$. Nonetheless, argue that always $\det(AB) = \det(BA)$.

In Exercises 27–30, use Theorem 2 and Exercise 25 to evaluate the given determinant, where A and B are $(n \times n)$ matrices with $\det(A) = 3$ and $\det(B) = 5$.

27. $\det(ABA^{-1})$ 28. $\det(A^2B)$

29. $\det(A^{-1}B^{-1}A^2)$ 30. $\det(AB^{-1}A^{-1}B)$

31. a) Let A be an $(n \times n)$ matrix. If $n = 3$, $\det(A)$ can be found by evaluating three (2×2) determinants. If $n = 4$, $\det(A)$ can be found by evaluating twelve (2×2) determinants. Give a formula, $H(n)$, for the number of (2×2) determinants necessary to find $\det(A)$ for an arbitrary n.
   b) Suppose you can perform additions, subtractions, multiplications, and divisions each at a rate of one per second. How long does it take to evaluate $H(n)$ (2×2) determinants when $n = 2$, $n = 5$, and $n = 10$?

32. Let U and V be $(n \times n)$ upper-triangular matrices. Prove a special case of Theorem 2: $\det(UV) = \det(U)\det(V)$. (*Hint:* Use the definition for matrix multiplication to calculate the diagonal entries of the product UV and then apply Theorem 4. You will also need to recall from Exercise 59 in Section 1.6 that UV is an upper-triangular matrix.)

33. Let V be an $(n \times n)$ triangular matrix. Use Theorem 4 to prove that $\det(V^T) = \det(V)$.

34. Let $T = (t_{ij})$ be an $(n \times n)$ upper-triangular matrix. Prove that $\det(T) = t_{11}t_{22}\ldots t_{nn}$. (*Hint:* Use mathematical induction, beginning with a (2×2) upper-triangular determinant.)

3.3 Elementary Operations and Determinants (Optional)*

We saw in Section 3.2 that having many zero entries in a matrix simplifies the calculation of its determinant. The ultimate case is given in Theorem 4. If $T = (t_{ij})$ is an $(n \times n)$ triangular matrix, then it is very easy to calculate $\det(T)$:

$$\det(T) = t_{11}t_{22}\ldots t_{nn}.$$

In Chapter 1, we used elementary row operations to create zero entries. We now consider these row operations (along with similar column operations) and describe their effect on the value of the determinant. For instance, consider the (2×2) matrices

$$A = \begin{bmatrix} 1 & 2 \\ 3 & 4 \end{bmatrix} \quad \text{and} \quad B = \begin{bmatrix} 3 & 4 \\ 1 & 2 \end{bmatrix}.$$

Clearly B is the result of interchanging the first and second rows of A (an elementary row operation). Also, we see that $\det(A) = -2$, whereas $\det(B) = 2$. This computation demonstrates that performing an elementary operation may change the value of the determinant. As we will see, however, it is possible to predict in advance the nature of any changes that might be produced by an elementary operation. For example, we will see that a row interchange always reverses the sign of the determinant.

Before studying the effects of elementary row operations on determinants, we consider the following theorem, which is an immediate consequence of Theorem 1.

THEOREM 5

If A is an $(n \times n)$ matrix, then

$$\det(A) = \det(A^T).$$

* The results in this section are not required for a study of the eigenvalue problem. It is included here for the convenience of the reader and because it follows naturally from definitions and theorems in the previous section.

PROOF   The proof is by induction, and we begin with the case $n = 2$. Let $A = (a_{ij})$ be a (2×2) matrix:

$$A = \begin{bmatrix} a_{11} & a_{12} \\ a_{21} & a_{22} \end{bmatrix}, \qquad A^T = \begin{bmatrix} a_{11} & a_{21} \\ a_{12} & a_{22} \end{bmatrix}.$$

Hence it is clear that $\det(A) = \det(A^T)$ when A is a (2×2) matrix.

The inductive step hinges on the following observation about minor matrices: Suppose that B is a square matrix and let $C = B^T$. Next, let M_{rs} and N_{rs} denote minor matrices of B and C, respectively. Then these minor matrices are related by

$$N_{ij} = (M_{ji})^T. \tag{1}$$

(In words, the ijth minor matrix of B^T is equal to the transpose of the jith minor matrix of B.)

To proceed with the induction, suppose that Theorem 5 is valid for all $(k \times k)$ matrices, $2 \le k \le n - 1$. Let A be an $(n \times n)$ matrix, where $n > 2$. Let M_{rs} denote the minor matrices of A, and let N_{rs} denote the minor matrices of A^T. Consider an expansion of $\det(A)$ along the first row and an expansion of $\det(A^T)$ along the first column:

$$\begin{aligned} \det(A) &= a_{11} \det(M_{11}) - a_{12} \det(M_{12}) + \cdots + (-1)^{n+1} a_{1n} \det(M_{1n}) \\ \det(A^T) &= a_{11} \det(N_{11}) - a_{12} \det(N_{21}) + \cdots + (-1)^{n+1} a_{1n} \det(N_{n1}). \end{aligned} \tag{2}$$

(The expansion for $\det(A^T)$ in (2) incorporates the fact that the first-column entries for A^T are the same as the first-row entries of A.)

By Eq. (1), the minor matrices N_{j1} in (2) satisfy $N_{j1} = (M_{1j})^T$, $1 \le j \le n$. By the inductive hypotheses, $\det(M_{1j}^T) = \det(M_{1j})$, since M_{1j} is a matrix of order $n - 1$. Therefore, both expansions in (2) have the same value, showing that $\det(A^T) = \det(A)$.   ☐

One valuable aspect of Theorem 5 is that it tells us an elementary column operation applied to a square matrix A will affect $\det(A)$ in precisely the same way as the corresponding elementary row operation.

Effects of Elementary Operations

We first consider how the determinant changes when rows of a matrix are interchanged.

THEOREM 6

Let A be an $(n \times n)$ matrix, and let B be formed by interchanging any two rows (or columns) of A. Then

$$\det(B) = -\det(A).$$

PROOF   First we consider the case where the two rows to be interchanged are adjacent, say, the ith and $(i + 1)$st rows. Let M_{ij}, $1 \leq j \leq n$, be the minor matrices of A from the ith row, and let $N_{i+1,j}$, $1 \leq j \leq n$, be the minor matrices of B from the $(i + 1)$st row. A bit of reflection will reveal that $N_{i+1,j} = M_{ij}$. Since $a_{i1}, a_{i2}, \ldots, a_{in}$ are the elements of the $(i + 1)$st row of B, we have

$$\det(B) = \sum_{j=1}^{n} (-1)^{i+1+j} a_{ij} \det(N_{i+1,j})$$

$$= - \sum_{j=1}^{n} (-1)^{i+j} a_{ij} \det(N_{i+1,j})$$

$$= - \sum_{j=1}^{n} (-1)^{i+j} a_{ij} \det(M_{ij}), \qquad \text{since } N_{i+1,j} = M_{ij}$$

$$= -\det(A).$$

Thus far we know that interchanging any two *adjacent* rows changes the sign of the determinant. Now suppose that B is formed by interchanging the ith and kth rows of A, where $k \geq i + 1$. The ith row can be moved to the kth row by $(k - i)$ successive interchanges of adjacent rows. The original kth row at this point is now the $(k - 1)$st row. This row can be moved to the ith row by $(k - 1 - i)$ successive interchanges of adjacent rows. At this point all other rows are in their original positions. Hence we have formed B with $2k - 1 - 2i$ successive interchanges of adjacent rows. Thus,

$$\det(B) = (-1)^{(2k-1-2i)} \det(A) = -\det(A). \qquad \square$$

COROLLARY

If A is an $(n \times n)$ matrix with two identical rows (columns), then $\det(A) = 0$.

We leave the proof of the corollary as an exercise.

EXAMPLE 1   Find $\det(A)$, where

$$A = \begin{bmatrix} 0 & 0 & 0 & 4 \\ 0 & 0 & 3 & 2 \\ 0 & 1 & 2 & 5 \\ 2 & 3 & 1 & 3 \end{bmatrix}.$$

SOLUTION   We could calculate $\det(A)$ by using a cofactor expansion, but we also see that we can rearrange the rows of A to produce a triangular matrix. Adopting the latter course of action, we have

$$\det(A) = \begin{vmatrix} 0 & 0 & 0 & 4 \\ 0 & 0 & 3 & 2 \\ 0 & 1 & 2 & 5 \\ 2 & 3 & 1 & 3 \end{vmatrix} = - \begin{vmatrix} 2 & 3 & 1 & 3 \\ 0 & 0 & 3 & 2 \\ 0 & 1 & 2 & 5 \\ 0 & 0 & 0 & 4 \end{vmatrix} = \begin{vmatrix} 2 & 3 & 1 & 3 \\ 0 & 1 & 2 & 5 \\ 0 & 0 & 3 & 2 \\ 0 & 0 & 0 & 4 \end{vmatrix} = 24. \qquad \square$$

Next we consider the effect of another elementary operation.

THEOREM 7

Suppose that B is obtained from the $(n \times n)$ matrix A by multiplying one row (or column) of A by a nonzero scalar c and leaving the other rows (or columns) unchanged. Then

$$\det(B) = c\det(A).$$

PROOF Suppose that $[ca_{i1}, ca_{i2}, \ldots, ca_{in}]$ is the ith row of B. Since the other rows of B are unchanged from A, the minor matrices of B from the ith row are the same as M_{ij}, the minor matrices of A from the ith row. Using a cofactor expansion from the ith row of B to calculate $\det(B)$ gives

$$\det(B) = \sum_{j=1}^{n} (ca_{ij})(-1)^{i+j}\det(M_{ij})$$

$$= c\sum_{j=1}^{n} a_{ij}(-1)^{i+j}\det(M_{ij})$$

$$= c\det(A). \qquad \square$$

As we see in the next theorem, the third elementary operation leaves the determinant unchanged. (*Note:* Theorem 8 is also valid when the word *column* is substituted for the word *row*.)

THEOREM 8

Let A be an $(n \times n)$ matrix. Suppose that B is the matrix obtained from A by replacing the ith row of A by the ith row of A plus a constant multiple of the kth row of A, $k \neq i$. Then

$$\det(B) = \det(A).$$

PROOF Note that the ith row of B has the form

$$[a_{i1} + ca_{k1}, a_{i2} + ca_{k2}, \ldots, a_{in} + ca_{kn}].$$

Since the other rows of B are unchanged from A, the minor matrices taken with respect to the ith row of B are the same as the minor matrices M_{ij} of A. Using a cofactor expansion of $\det(B)$ from the ith row, we have

$$\det(B) = \sum_{j=1}^{n} (a_{ij} + ca_{kj})(-1)^{i+j}\det(M_{ij})$$

$$= \sum_{j=1}^{n} a_{ij}(-1)^{i+j}\det(M_{ij}) + c\sum_{j=1}^{n} a_{kj}(-1)^{i+j}\det(M_{ij})$$

$$= \det(A) + c\sum_{j=1}^{n} a_{kj}(-1)^{i+j}\det(M_{ij}). \qquad (3)$$

Theorem 8 will be proved if we can show that the last summation on the right-hand side of Eq. (3) has the value zero. In order to prove that the summation has the value zero, construct a matrix Q by replacing the ith row of A by the kth row of A. The matrix Q so constructed has two identical rows (the kth row of A appears both as the kth row and the ith row of Q). Therefore, by the corollary to Theorem 6, $\det(Q) = 0$. Next, expanding $\det(Q)$ along the ith row of Q, we obtain (since the ith-row minors of Q are the same as those of A and since the ijth entry of Q is a_{kj})

$$0 = \det(Q) = \sum_{j=1}^{n} a_{kj}(-1)^{i+j}\det(M_{ij}). \tag{4}$$

Substituting Eq. (4) into Eq. (3) establishes the theorem. □

EXAMPLE 2   Evaluate $\det(A)$, where

$$A = \begin{bmatrix} 1 & 2 & 1 \\ 0 & 3 & 2 \\ -2 & 1 & 1 \end{bmatrix}.$$

SOLUTION   The value of $\det(A)$ is unchanged if we add a multiple of 2 times row 1 to row 3. The effect of this row operation will be to introduce another zero entry in the first column. Specifically,

$$\det(A) = \begin{vmatrix} 1 & 2 & 1 \\ 0 & 3 & 2 \\ -2 & 1 & 1 \end{vmatrix} \xrightarrow[=]{(R_3 + 2R_1)} \begin{vmatrix} 1 & 2 & 1 \\ 0 & 3 & 2 \\ 0 & 5 & 3 \end{vmatrix} = -1. \quad □$$

Using Elementary Operations to Simplify Determinants

Clearly it is usually easier to calculate the determinant of a matrix with several zero entries than to calculate one with no zero entries. Therefore, a common strategy in determinant evaluation is to mimic the steps of Gaussian elimination—that is, to use elementary row or column operations to reduce the matrix to triangular form.

EXAMPLE 3   Evaluate $\det(A)$, where

$$A = \begin{bmatrix} 1 & 2 & -1 \\ 5 & 3 & 4 \\ -2 & 0 & 1 \end{bmatrix}.$$

SOLUTION   With Gaussian elimination, we would first form the matrix B by the following operations: Replace R_2 by $R_2 - 5R_1$ and replace R_3 by $R_3 + 2R_1$. From Theorem 8, the matrix B produced by these two row operations has the same determinant as the original matrix A. In detail:

$$\det(A) = \begin{vmatrix} 1 & 2 & -1 \\ 5 & 3 & 4 \\ -2 & 0 & 1 \end{vmatrix} = \begin{vmatrix} 1 & 2 & -1 \\ 0 & -7 & 9 \\ 0 & 4 & -1 \end{vmatrix} = 1 \begin{vmatrix} -7 & 9 \\ 4 & -1 \end{vmatrix} = 7 - 36 = -29. \quad □$$

We could have created a zero in the (2, 1) position of the last (2 × 2) determinant. The formula for (2 × 2) determinants is so simple, however, that it is customary to evaluate a (2 × 2) determinant directly. The next example illustrates that we need not always attempt to go to a triangular form in order to simplify a determinant.

EXAMPLE 4   Evaluate det(A), where

$$A = \begin{bmatrix} 1 & 2 & -1 & 1 \\ -1 & 0 & 2 & -2 \\ 3 & -1 & 1 & 1 \\ 2 & 0 & -1 & 2 \end{bmatrix}.$$

SOLUTION   We can introduce a third zero in the second column if we replace R_1 by $R_1 + 2R_3$:

$$\det(A) = \begin{vmatrix} 1 & 2 & -1 & 1 \\ -1 & 0 & 2 & -2 \\ 3 & -1 & 1 & 1 \\ 2 & 0 & -1 & 2 \end{vmatrix} = \begin{vmatrix} 7 & 0 & 1 & 3 \\ -1 & 0 & 2 & -2 \\ 3 & -1 & 1 & 1 \\ 2 & 0 & -1 & 2 \end{vmatrix} = -(-1) \begin{vmatrix} 7 & 1 & 3 \\ -1 & 2 & -2 \\ 2 & -1 & 2 \end{vmatrix}.$$

(The second equality is from Theorem 8. The third equality is from an expansion along the second column.) Next we replace R_2 by $R_2 - 2R_1$ and R_3 by $R_1 + R_3$. The details are

$$\det(A) = \begin{vmatrix} 7 & 1 & 3 \\ -1 & 2 & -2 \\ 2 & -1 & 2 \end{vmatrix} = \begin{vmatrix} 7 & 1 & 3 \\ -15 & 0 & -8 \\ 9 & 0 & 5 \end{vmatrix}$$

$$= -1 \begin{vmatrix} -15 & -8 \\ 9 & 5 \end{vmatrix} = (75 - 72) = 3. \qquad \square$$

The next example illustrates that if the entries in a determinant are integers, then we can avoid working with fractions until the last step. The technique involves multiplying various rows by constants to make each entry in a column divisible by the pivot entry in the column.

EXAMPLE 5   Find det(A), where

$$A = \begin{bmatrix} 2 & 3 & -2 & 4 \\ 3 & -3 & 5 & 2 \\ 5 & 2 & 4 & 3 \\ -3 & 4 & -3 & 2 \end{bmatrix}.$$

SOLUTION   We first multiply rows 2, 3, and 4 by 2 to make them divisible by 2. The row reduction operations to create zeros in the first column can then proceed without using fractions. The row operations are $R_2 - 3R_1$, $R_3 - 5R_1$, and $R_4 + 3R_1$:

$$\det(A) = \begin{vmatrix} 2 & 3 & -2 & 4 \\ 3 & -3 & 5 & 2 \\ 5 & 2 & 4 & 3 \\ -3 & 4 & -3 & 2 \end{vmatrix} = \frac{1}{8} \begin{vmatrix} 2 & 3 & -2 & 4 \\ 6 & -6 & 10 & 4 \\ 10 & 4 & 8 & 6 \\ -6 & 8 & -6 & 4 \end{vmatrix}$$

$$= \frac{1}{8} \begin{vmatrix} 2 & 3 & -2 & 4 \\ 0 & -15 & 16 & -8 \\ 0 & -11 & 18 & -14 \\ 0 & 17 & -12 & 16 \end{vmatrix} = \frac{2}{8} \begin{vmatrix} -15 & 16 & -8 \\ -11 & 18 & -14 \\ 17 & -12 & 16 \end{vmatrix}$$

$$= \frac{1}{4} \begin{vmatrix} -15 & 2(8) & 2(-4) \\ -11 & 2(9) & 2(-7) \\ 17 & 2(-6) & 2(8) \end{vmatrix} = \frac{2(2)}{4} \begin{vmatrix} -15 & 8 & -4 \\ -11 & 9 & -7 \\ 17 & -6 & 8 \end{vmatrix}.$$

We now multiply the second row by 4 and use $R_2 - 7R_1$ and $R_3 + 2R_1$:

$$\det(A) = \frac{1}{4} \begin{vmatrix} -15 & 8 & -4 \\ -44 & 36 & -28 \\ 17 & -6 & 8 \end{vmatrix} = \frac{1}{4} \begin{vmatrix} -15 & 8 & -4 \\ 61 & -20 & 0 \\ -13 & 10 & 0 \end{vmatrix}$$

$$= \frac{-4}{4}(610 - 260) = -350. \qquad \square$$

The examples above illustrate that there are many strategies that will lead to a simpler determinant calculation. Exactly which choices are made are determined by experience and personal preference.

Proof of Theorem 3

In the last section we stated Theorem 3: An $(n \times n)$ matrix A is singular if and only if $\det(A) = 0$. The results of this section enable us to sketch a proof for Theorem 3.

If A is an $(n \times n)$ matrix, then we know from Chapter 1 that we can use Gaussian elimination to produce a row-equivalent upper-triangular matrix T. This matrix T can be formed by using row interchanges and adding multiples of one row to other rows. Thus, by Theorems 6 and 8,

$$\det(A) = \pm \det(T). \tag{5}$$

An outline for the proof of Theorem 3 is given below. We use t_{ij} to denote the entries of the upper-triangular matrix T:

1. $\det(A) = 0 \Leftrightarrow \det(T) = 0$,   by Eq. (5);
2. $\det(T) = 0 \Leftrightarrow t_{ii} = 0$ for some i,   by Theorem 4;
3. $t_{ii} = 0$ for some $i \Leftrightarrow T$ singular   (see Exercise 56 of Section 1.8);
4. T singular $\Leftrightarrow A$ singular,   since T and A are row equivalent.

Exercises 3.3

In Exercises 1–6, evaluate $\det(A)$ by using row operations to introduce zeros into the second and third entries of the first column.

1. $A = \begin{bmatrix} 1 & 2 & 1 \\ 3 & 0 & 2 \\ -1 & 1 & 3 \end{bmatrix}$ **2.** $A = \begin{bmatrix} 2 & 4 & 6 \\ 3 & 1 & 2 \\ 1 & 2 & 1 \end{bmatrix}$

3. $A = \begin{bmatrix} 3 & 6 & 9 \\ 2 & 0 & 2 \\ 1 & 2 & 0 \end{bmatrix}$ **4.** $A = \begin{bmatrix} 1 & 1 & 2 \\ -2 & 1 & 3 \\ 1 & 4 & 1 \end{bmatrix}$

5. $A = \begin{bmatrix} 2 & 4 & -3 \\ 3 & 2 & 5 \\ 2 & 3 & 4 \end{bmatrix}$ **6.** $A = \begin{bmatrix} 3 & 4 & -2 \\ 2 & 3 & 5 \\ 2 & 4 & 3 \end{bmatrix}$

In Exercises 7–12, use only column interchanges or row interchanges to produce a triangular determinant and then find the value of the original determinant.

7. $\begin{vmatrix} 1 & 0 & 0 & 0 \\ 2 & 0 & 0 & 3 \\ 1 & 1 & 0 & 1 \\ 1 & 4 & 2 & 2 \end{vmatrix}$ **8.** $\begin{vmatrix} 0 & 0 & 3 & 1 \\ 2 & 1 & 0 & 1 \\ 0 & 0 & 0 & 2 \\ 0 & 2 & 2 & 1 \end{vmatrix}$

9. $\begin{vmatrix} 0 & 0 & 2 & 0 \\ 0 & 0 & 1 & 3 \\ 0 & 4 & 1 & 3 \\ 2 & 1 & 5 & 6 \end{vmatrix}$ **10.** $\begin{vmatrix} 0 & 0 & 1 & 0 \\ 1 & 2 & 1 & 3 \\ 0 & 0 & 0 & 5 \\ 0 & 3 & 1 & 2 \end{vmatrix}$

11. $\begin{vmatrix} 0 & 0 & 1 & 0 \\ 0 & 2 & 6 & 3 \\ 2 & 4 & 1 & 5 \\ 0 & 0 & 0 & 4 \end{vmatrix}$ **12.** $\begin{vmatrix} 0 & 1 & 0 & 0 \\ 0 & 2 & 0 & 3 \\ 2 & 1 & 0 & 6 \\ 3 & 2 & 2 & 4 \end{vmatrix}$

In Exercises 13–18, assume that the (3×3) matrix A satisfies $\det(A) = 2$, where A is given by

$$A = \begin{bmatrix} a & b & c \\ d & e & f \\ g & h & i \end{bmatrix}.$$

Calculate $\det(B)$ in each case.

13. $B = \begin{bmatrix} a & b & 3c \\ d & e & 3f \\ g & h & 3i \end{bmatrix}$ **14.** $B = \begin{bmatrix} d & e & f \\ g & h & i \\ a & b & c \end{bmatrix}$

15. $B = \begin{bmatrix} b & a & c \\ e & d & f \\ h & g & i \end{bmatrix}$

16. $B = \begin{bmatrix} a & b & c \\ a+d & b+e & c+f \\ g & h & i \end{bmatrix}$

17. $B = \begin{bmatrix} d & e & f \\ 2a & 2b & 2c \\ g & h & i \end{bmatrix}$ **18.** $B = \begin{bmatrix} d & f & e \\ a & c & b \\ g & i & h \end{bmatrix}$

In Exercises 19–22, evaluate the (4×4) determinants. Theorems 6–8 can be used to simplify the calculations.

19. $\begin{vmatrix} 2 & 4 & 2 & 6 \\ 1 & 3 & 2 & 1 \\ 2 & 1 & 2 & 3 \\ 1 & 2 & 1 & 1 \end{vmatrix}$ **20.** $\begin{vmatrix} 0 & 2 & 1 & 3 \\ 1 & 2 & 1 & 0 \\ 0 & 1 & 1 & 3 \\ 2 & 2 & 1 & 2 \end{vmatrix}$

21. $\begin{vmatrix} 0 & 4 & 1 & 3 \\ 0 & 2 & 2 & 1 \\ 1 & 3 & 1 & 2 \\ 2 & 2 & 1 & 4 \end{vmatrix}$ **22.** $\begin{vmatrix} 2 & 2 & 4 & 4 \\ 1 & 1 & 3 & 3 \\ 1 & 0 & 2 & 1 \\ 4 & 1 & 3 & 2 \end{vmatrix}$

In Exercises 23 and 24, use row operations to obtain a triangular determinant and find the value of the original Vandermonde determinant.

23. $\begin{vmatrix} 1 & a & a^2 \\ 1 & b & b^2 \\ 1 & c & c^2 \end{vmatrix}$ **24.** $\begin{vmatrix} 1 & a & a^2 & a^3 \\ 1 & b & b^2 & b^3 \\ 1 & c & c^2 & c^3 \\ 1 & d & d^2 & d^3 \end{vmatrix}$

25. Let A be an $(n \times n)$ matrix. Use Theorem 7 to argue that $\det(cA) = c^n \det(A)$.

26. Prove the corollary to Theorem 6. (*Hint:* Suppose that the ith and jth rows of A are identical. Interchange these two rows and let B denote the matrix that results. How are $\det(A)$ and $\det(B)$ related?)

27. Find examples of (2×2) matrices A and B such that $\det(A + B) \neq \det(A) + \det(B)$.

28. An $(n \times n)$ matrix A is called *skew symmetric* if $A^T = -A$. Show that if A is skew symmetric, then $\det(A) = (-1)^n \det(A)$. (*Hint:* Use Theorem 5 and Exercise 25.) Now, argue that an $(n \times n)$ skew-symmetric matrix is singular when n is an odd integer.

3.4 Eigenvalues and the Characteristic Polynomial

Having given the brief introduction to determinant theory presented in Section 3.2, we return to the central topic of this chapter, the eigenvalue problem. For reference, recall that the eigenvalue problem for an $(n \times n)$ matrix A has two parts:

1. Find all scalars λ such that $A - \lambda I$ is singular. (Such scalars are the *eigenvalues* of A.)
2. Given an eigenvalue λ, find all nonzero vectors $\mathbf{x}$ such that $(A - \lambda I)\mathbf{x} = \boldsymbol{\theta}$. (Such vectors are the *eigenvectors* corresponding to the eigenvalue λ.)

In this section we focus on part 1, finding the eigenvalues. In the next section we discuss eigenvectors.

In Section 3.1, we were able to determine the eigenvalues of a (2×2) matrix by using a test for singularity given by Eq. (4) in Section 3.1. Knowing Theorem 3 from Section 3.2, we now have a test for singularity that is applicable to any $(n \times n)$ matrix. As applied to the eigenvalue problem, Theorem 3 can be used as follows:

$$A - \lambda I \text{ is singular} \Leftrightarrow \det(A - \lambda I) = 0. \tag{1}$$

An example will illustrate how the singularity test given in Eq. (1) is used in practice.

EXAMPLE 1 Use the singularity test given in Eq. (1) to determine the eigenvalues of the (3×3) matrix A, where

$$A = \begin{bmatrix} 1 & 1 & 1 \\ 0 & 3 & 3 \\ -2 & 1 & 1 \end{bmatrix}.$$

SOLUTION A scalar λ is an eigenvalue of A if and only if $A - \lambda I$ is singular. According to the singularity test in Eq. (1), λ is an eigenvalue of A if and only if λ is a scalar such that

$$\det(A - \lambda I) = 0.$$

Thus we focus on $\det(A - \lambda I)$, where $A - \lambda I$ is the matrix given by

$$A - \lambda I = \begin{bmatrix} 1 & 1 & 1 \\ 0 & 3 & 3 \\ -2 & 1 & 1 \end{bmatrix} - \begin{bmatrix} \lambda & 0 & 0 \\ 0 & \lambda & 0 \\ 0 & 0 & \lambda \end{bmatrix}$$
$$= \begin{bmatrix} 1-\lambda & 1 & 1 \\ 0 & 3-\lambda & 3 \\ -2 & 1 & 1-\lambda \end{bmatrix}.$$

Expanding $\det(A - \lambda I)$ along the first column, we have

$$\det(A - \lambda I) = \begin{vmatrix} 1-\lambda & 1 & 1 \\ 0 & 3-\lambda & 3 \\ -2 & 1 & 1-\lambda \end{vmatrix}$$

$$= (1-\lambda)\begin{vmatrix} 3-\lambda & 3 \\ 1 & 1-\lambda \end{vmatrix} - (0)\begin{vmatrix} 1 & 1 \\ 1 & 1-\lambda \end{vmatrix}$$

$$+ (-2)\begin{vmatrix} 1 & 1 \\ 3-\lambda & 3 \end{vmatrix}$$

$$= (1-\lambda)[(3-\lambda)(1-\lambda) - 3] - 2[3 - (3-\lambda)]$$
$$= (1-\lambda)[\lambda^2 - 4\lambda] - 2[\lambda]$$
$$= [-\lambda^3 + 5\lambda^2 - 4\lambda] - [2\lambda]$$
$$= -\lambda^3 + 5\lambda^2 - 6\lambda$$
$$= -\lambda(\lambda^2 - 5\lambda + 6)$$
$$= -\lambda(\lambda - 3)(\lambda - 2).$$

From the singularity test in Eq. (1), we see that $A - \lambda I$ is singular if and only if $\lambda = 0$, $\lambda = 3$, or $\lambda = 2$. ▢

The ideas developed in Example 1 will be formalized in the next subsection.

The Characteristic Polynomial

From the singularity condition given in Eq. (1), we know that $A - \lambda I$ is singular if and only if $\det(A - \lambda I) = 0$. In Example 1, for a (3×3) matrix A, we saw that the expression $\det(A - \lambda I)$ was a polynomial of degree 3 in λ. In general, it can be shown that $\det(A - \lambda I)$ is a polynomial of degree n in λ when A is $(n \times n)$. Then, since $A - \lambda I$ is singular if and only if $\det(A - \lambda I) = 0$, it follows that the eigenvalues of A are precisely the zeros of the polynomial $\det(A - \lambda I)$.

To avoid any possible confusion between the eigenvalues λ of A and the problem of finding the zeros of this associated polynomial (called the characteristic polynomial of A), we will use the variable t instead of λ in the characteristic polynomial and write $p(t) = \det(A - tI)$. To summarize this discussion, we give Theorems 9 and 10.

THEOREM 9

Let A be an $(n \times n)$ matrix. Then $\det(A - tI)$ is a polynomial of degree n in t.

The proof of Theorem 9 is somewhat tedious and we omit it. The fact that $\det(A - tI)$ is a polynomial leads us to the next definition.

> **DEFINITION 5**   Let A be an $(n \times n)$ matrix. The nth-degree polynomial, $p(t)$, given by
>
> $$p(t) = \det(A - tI)$$
>
> is called the **characteristic polynomial** for A.

Again, in the context of the singularity test in Eq. (1), the roots of $p(t) = 0$ are the eigenvalues of A. This observation is stated formally in the next theorem.

THEOREM 10

Let A be an $(n \times n)$ matrix, and let p be the characteristic polynomial for A. Then the eigenvalues of A are precisely the roots of $p(t) = 0$.

Theorem 10 has the effect of replacing the original problem—determining values λ for which $A - \lambda I$ is singular—by an equivalent problem, finding the roots of a polynomial equation $p(t) = 0$. Since polynomials are familiar and an immense amount of theoretical and computational machinery has been developed for solving polynomial equations, we should feel more comfortable with the eigenvalue problem.

The equation $p(t) = 0$ that must be solved to find the eigenvalues of A is called the **characteristic equation.** Suppose that $p(t)$ has degree n, where $n \geq 1$. Then the equation $p(t) = 0$ can have no more than n distinct roots. From this fact, it follows that:

a) An $(n \times n)$ matrix can have no more than n distinct eigenvalues.

Also, by the fundamental theorem of algebra, the equation $p(t) = 0$ always has

The Fundamental Theorem of Algebra

The eigenvalues of an $(n \times n)$ matrix A are the zeros of $p(t) = \det(A - tI)$, a polynomial of degree n. The fundamental theorem of algebra states that the equation $p(t) = 0$ has a solution, r_1, in the field of complex numbers. Since $q(t) = p(t)/(t - r_1)$ is a polynomial of degree $n - 1$, repeated use of this result allows us to write

$$p(t) = a(t - r_1)(t - r_2) \cdots (t - r_n).$$

A number of famous mathematicians (including Newton, Euler, d'Alembert, and Lagrange) attempted proofs of the fundamental theorem. In 1799, Gauss critiqued these attempts and presented a proof of his own. He admitted that his proof contained an unestablished assertion, but he stated that its validity could not be doubted. Gauss gave three more proofs in his lifetime, but all suffered from an imperfect understanding of the concept of continuity and the structure of the complex number system. These properties were established in 1874 by Weierstrass and not only made the proofs by Gauss rigorous, but a 1746 proof due to d'Alembert as well.

at least one root (possibly complex). Therefore:

b) An $(n \times n)$ matrix always has at least one eigenvalue (possibly complex).

Finally, we recall that any nth-degree polynomial $p(t)$ can be written in the factored form

$$p(t) = a(t - r_1)(t - r_2) \cdots (t - r_n).$$

The zeros of $p, r_1, r_2, \ldots, r_n$, however, need not be distinct or real. The number of times the factor $(t - r)$ appears in the factorization of $p(t)$ given above is called the ***algebraic multiplicity*** of r.

EXAMPLE 2   Find the characteristic polynomial and the eigenvalues for the (2×2) matrix

$$A = \begin{bmatrix} 1 & 5 \\ 3 & 3 \end{bmatrix}.$$

SOLUTION   By Definition 5, the characteristic polynomial is found by calculating $p(t) = \det(A - tI)$, or

$$p(t) = \begin{vmatrix} 1 - t & 5 \\ 3 & 3 - t \end{vmatrix} = (1 - t)(3 - t) - 15$$

$$= t^2 - 4t - 12 = (t - 6)(t + 2).$$

By Theorem 10, the eigenvalues of A are the roots of $p(t) = 0$; thus the eigenvalues are $\lambda = 6$ and $\lambda = -2$.   ☐

EXAMPLE 3   Find the characteristic polynomial and the eigenvalues for the (2×2) matrix

$$A = \begin{bmatrix} 2 & -1 \\ 1 & 2 \end{bmatrix}.$$

SOLUTION   The characteristic polynomial is

$$p(t) = \begin{vmatrix} 2 - t & -1 \\ 1 & 2 - t \end{vmatrix} = t^2 - 4t + 5.$$

By the quadratic formula, the eigenvalues are $\lambda = 2 + i$ and $\lambda = 2 - i$. Therefore, this example illustrates that a matrix with real entries can have eigenvalues that are complex. In Section 3.6, we discuss complex eigenvalues and eigenvectors at length.   ☐

EXAMPLE 4   Find the characteristic polynomial and the eigenvalues for the (3×3) matrix

$$A = \begin{bmatrix} 3 & -1 & -1 \\ -12 & 0 & 5 \\ 4 & -2 & -1 \end{bmatrix}.$$

SOLUTION By Definition 5, the characteristic polynomial is given by $p(t) = \det(A - tI)$, or

$$p(t) = \begin{vmatrix} 3-t & -1 & -1 \\ -12 & -t & 5 \\ 4 & -2 & -1-t \end{vmatrix}.$$

Expanding along the first column, we have

$$p(t) = (3-t)\begin{vmatrix} -t & 5 \\ -2 & -1-t \end{vmatrix} + 12\begin{vmatrix} -1 & -1 \\ -2 & -1-t \end{vmatrix} + 4\begin{vmatrix} -1 & -1 \\ -t & 5 \end{vmatrix}$$

$$= (3-t)[t(1+t)+10] + 12[(1+t)-2] + 4[-5-t]$$
$$= (3-t)[t^2 + t + 10] + 12[t-1] + 4[-t-5]$$
$$= [-t^3 + 2t^2 - 7t + 30] + [12t - 12] + [-4t - 20]$$
$$= -t^3 + 2t^2 + t - 2.$$

By Theorem 10, the eigenvalues of A are the roots of $p(t) = 0$. We can write $p(t)$ as

$$p(t) = -(t-2)(t-1)(t+1),$$

and thus the eigenvalues of A are $\lambda = 2$, $\lambda = 1$, and $\lambda = -1$. $\qquad\square$

(*Note:* Finding or approximating the root of a polynomial equation is a task that is generally best left to the computer. Therefore, so that the theory associated with the eigenvalue problem is not hidden by a mass of computational details, the examples and exercises in this chapter will usually be constructed so that the characteristic equation has integer roots.)

Special Results

If we know the eigenvalues of a matrix A, then we also know the eigenvalues of certain matrices associated with A. A list of such results is found in Theorems 11 and 12.

THEOREM 11

Let A be an $(n \times n)$ matrix, and let λ be an eigenvalue of A. Then:

a) λ^k is an eigenvalue of A^k, $k = 2, 3, \ldots$.
b) If A is nonsingular, then $1/\lambda$ is an eigenvalue of A^{-1}.
c) If α is any scalar, then $\lambda + \alpha$ is an eigenvalue of $A + \alpha I$.

PROOF Property (a) is proved by induction, and we begin with the case $k = 2$. Suppose that λ is an eigenvalue of A with an associated eigenvector, $\mathbf{x}$. That is,

$$A\mathbf{x} = \lambda\mathbf{x}, \qquad \mathbf{x} \neq \boldsymbol{\theta}. \qquad (2)$$

Multiplying both sides of Eq. (2) by the matrix A gives

$$A(A\mathbf{x}) = A(\lambda\mathbf{x})$$
$$A^2\mathbf{x} = \lambda(A\mathbf{x})$$
$$A^2\mathbf{x} = \lambda(\lambda\mathbf{x})$$
$$A^2\mathbf{x} = \lambda^2\mathbf{x}, \qquad \mathbf{x} \neq \boldsymbol{\theta}.$$

Thus λ^2 is an eigenvalue of A^2 with a corresponding eigenvector, $\mathbf{x}$.

In the exercises the reader is asked to finish the proof of property (a) and prove properties (b) and (c) of Theorem 11. (*Note:* As the proof of Theorem 11 will demonstrate, if $\mathbf{x}$ is any eigenvector of A, then $\mathbf{x}$ is also an eigenvector of A^k, A^{-1}, and $A + \alpha I$.)

___

EXAMPLE 5 Let A be the (3×3) matrix in Example 4. Determine the eigenvalues of A^5, A^{-1}, and $A + 2I$.

SOLUTION   From Example 4, the eigenvalues of A are $\lambda = 2$, $\lambda = 1$, and $\lambda = -1$. By Theorem 11, A^5 has eigenvalues $\lambda = 2^5 = 32$, $\lambda = 1^5 = 1$, and $\lambda = (-1)^5 = -1$. Since A^5 is a (3×3) matrix and can have no more than three eigenvalues, those eigenvalues must be 32, 1, and -1.

Similarly, the eigenvalues of A^{-1} are $\lambda = 1/2$, $\lambda = 1$, and $\lambda = -1$. The eigenvalues of $A + 2I$ are $\lambda = 4$, $\lambda = 3$, and $\lambda = 1$.

The proof of the next theorem rests on the following fact (see Section 1.8): If B is a square matrix, then both B and B^T are nonsingular or both B and B^T are singular. (See also Exercise 30.)

THEOREM 12

Let A be an $(n \times n)$ matrix. Then A and A^T have the same eigenvalues.

PROOF   Observe that $(A - \lambda I)^T = A^T - \lambda I$. By our earlier remark, $A - \lambda I$ and $(A - \lambda I)^T$ are either both singular or both nonsingular. Thus λ is an eigenvalue of A if and only if λ is an eigenvalue of A^T.

The next result follows immediately from the definition of an eigenvalue. We write the result as a theorem because it provides another important characterization of singularity.

THEOREM 13

Let A be an $(n \times n)$ matrix. Then A is singular if and only if $\lambda = 0$ is an eigenvalue of A.

(*Note:* If A is singular, then the eigenvectors corresponding to $\lambda = 0$ are in the null space of A.)

Our final theorem treats a class of matrices for which eigenvalues can be determined by inspection.

THEOREM 14

Let $T = (t_{ij})$ be an $(n \times n)$ triangular matrix. Then the eigenvalues of T are the diagonal entries, $t_{11}, t_{22}, \ldots, t_{nn}$.

PROOF Since T is triangular, the matrix $T - tI$ is also triangular. The diagonal entries of $T - tI$ are $t_{11} - t, t_{22} - t, \ldots, t_{nn} - t$. Thus, by Theorem 4, the characteristic polynomial is given by

$$p(t) = \det(T - tI) = (t_{11} - t)(t_{22} - t) \cdots (t_{nn} - t).$$

By Theorem 10, the eigenvalues are $\lambda = t_{11}, \lambda = t_{22}, \ldots, \lambda = t_{nn}$. □

EXAMPLE 6 Find the characteristic polynomial and the eigenvalues for the matrix A given by

$$A = \begin{bmatrix} 1 & 2 & 1 & 0 \\ 0 & 3 & -1 & 1 \\ 0 & 0 & 2 & 1 \\ 0 & 0 & 0 & 3 \end{bmatrix}.$$

SOLUTION By Theorem 4, $p(t) = \det(A - tI)$ has the form

$$p(t) = (1 - t)(3 - t)^2(2 - t).$$

The eigenvalues are $\lambda = 1$, $\lambda = 2$, and $\lambda = 3$. The eigenvalues $\lambda = 1$ and $\lambda = 2$ have algebraic multiplicity 1, whereas the eigenvalue $\lambda = 3$ has algebraic multiplicity 2. □

Computational Considerations

In all the examples we have considered so far, it was possible to factor the characteristic polynomial and thus determine the eigenvalues by inspection. In reality we can rarely expect to be able to factor the characteristic polynomial; so we must solve the characteristic equation by using numerical root-finding methods such as those discussed in Chapter 7. To be more specific about root finding, we recall that there are formulas for the roots of some polynomial equations. For instance, the solution of the linear equation

$$at + b = 0, \qquad a \neq 0,$$

is given by

$$t = -\frac{b}{a};$$

and the roots of the quadratic equation

$$at^2 + bt + c = 0, \qquad a \neq 0,$$

are given by the familiar quadratic formula

$$t = \frac{-b \pm \sqrt{b^2 - 4ac}}{2a}.$$

There are similar (although more complicated) formulas for the roots of third-degree and fourth-degree polynomial equations. Unfortunately there are no such formulas for polynomials of degree 5 or higher [that is, formulas that express the zeros of $p(t)$ as a simple function of the coefficients of $p(t)$]. Moreover, in the mid-nineteenth century Abel *proved* that such formulas cannot exist for polynomials of degree 5 or higher.* This means that in general we cannot expect to find the eigenvalues of a large matrix exactly—the best we can do is to find good approximations to the eigenvalues. The eigenvalue problem differs qualitatively from the problem of solving $A\mathbf{x} = \mathbf{b}$. For a system $A\mathbf{x} = \mathbf{b}$, if we are willing to invest the effort required to solve the system by hand, we can obtain the exact solution in a finite number of steps. On the other hand, we cannot in general expect to find roots of a polynomial equation in a finite number of steps.

Finding roots of the characteristic equation is not the only computational aspect of the eigenvalue problem that must be considered. In fact, it is not hard to see that special techniques must be developed even to find the characteristic polynomial. To see the dimensions of this problem, consider the characteristic polynomial of an $(n \times n)$ matrix A: $p(t) = \det(A - tI)$. The evaluation of $p(t)$ from a cofactor expansion of $\det(A - tI)$ ultimately requires the evaluation of $n!/2$ determinants of order (2×2). Even for modest values of n, the number $n!/2$ is alarmingly large. For instance,

$$10!/2 = 1,814,400,$$

whereas

$$20!/2 > 1.2 \times 10^{18}.$$

The enormous number of calculations required to compute $\det(A - tI)$ means that we cannot find $p(t)$ in any practical sense by expanding $\det(A - tI)$. In Chapter 5, we note that there are relatively efficient ways of finding $\det(A)$, but these techniques (which amount to using elementary row operations to triangularize A) are not useful in our problem of computing $\det(A - tI)$ because of the variable t. In Section 6.3, we resolve this difficulty by using similarity transformations to transform A to a matrix H, where A and H have the same characteristic polynomial, and where it is a trivial matter to calculate the characteristic polynomial for H. Moreover, these transformation methods will give us some other important results as a by-product, results such as the

* For a historical discussion, see J. E. Maxfield and M. W. Maxfield, *Abstract Algebra and Solution by Radicals* (Philadelphia: W. B. Saunders, 1971).

Cayley–Hamilton theorem, which have some practical computational significance. Finally, in Chapter 7, we discuss other numerical procedures that are currently being used in software packages designed to solve the eigenvalue problem. Exercises 31–34 suggest another method for finding the characteristic polynomial, based on evaluating $\det(A - tI)$ for selected values of t.

Exercises 3.4

In Exercises 1–14, find the characteristic polynomial and the eigenvalues for the given matrix. Also, give the algebraic multiplicity of each eigenvalue. (*Note:* In each case the eigenvalues are integers.)

1. $\begin{bmatrix} 1 & 0 \\ 2 & 3 \end{bmatrix}$ 2. $\begin{bmatrix} 2 & 1 \\ 0 & -1 \end{bmatrix}$

3. $\begin{bmatrix} 2 & -1 \\ -1 & 2 \end{bmatrix}$ 4. $\begin{bmatrix} 13 & -16 \\ 9 & -11 \end{bmatrix}$

5. $\begin{bmatrix} 1 & -1 \\ 1 & 3 \end{bmatrix}$ 6. $\begin{bmatrix} 2 & 2 \\ 3 & 3 \end{bmatrix}$

7. $\begin{bmatrix} -6 & -1 & 2 \\ 3 & 2 & 0 \\ -14 & -2 & 5 \end{bmatrix}$ 8. $\begin{bmatrix} -2 & -1 & 0 \\ 0 & 1 & 1 \\ -2 & -2 & -1 \end{bmatrix}$

9. $\begin{bmatrix} 3 & -1 & -1 \\ -12 & 0 & 5 \\ 4 & -2 & -1 \end{bmatrix}$ 10. $\begin{bmatrix} -7 & 4 & -3 \\ 8 & -3 & 3 \\ 32 & -16 & 13 \end{bmatrix}$

11. $\begin{bmatrix} 2 & 4 & 4 \\ 0 & 1 & -1 \\ 0 & 1 & 3 \end{bmatrix}$ 12. $\begin{bmatrix} 6 & 4 & 4 & 1 \\ 4 & 6 & 1 & 4 \\ 4 & 1 & 6 & 4 \\ 1 & 4 & 4 & 6 \end{bmatrix}$

13. $\begin{bmatrix} 5 & 4 & 1 & 1 \\ 4 & 5 & 1 & 1 \\ 1 & 1 & 4 & 2 \\ 1 & 1 & 2 & 4 \end{bmatrix}$ 14. $\begin{bmatrix} 1 & -1 & -1 & -1 \\ -1 & 1 & -1 & -1 \\ -1 & -1 & 1 & -1 \\ -1 & -1 & -1 & 1 \end{bmatrix}$

15. Prove property (b) of Theorem 11. (*Hint:* Begin with $A\mathbf{x} = \lambda\mathbf{x}, \mathbf{x} \neq \mathbf{0}$.)

16. Prove property (c) of Theorem 11.

17. Complete the proof of property (a) of Theorem 11.

18. Let $q(t) = t^3 - 2t^2 - t + 2$; and for any $(n \times n)$ matrix H, define the "matrix polynomial" $q(H)$ given by

$q(H) = H^3 - 2H^2 - H + 2I,$

where I is the $(n \times n)$ identity matrix.
a) Prove that if λ is an eigenvalue of H, then the number $q(\lambda)$ is an eigenvalue of the matrix $q(H)$. (*Hint:* Suppose that $H\mathbf{x} = \lambda\mathbf{x}$, where $\mathbf{x} \neq \mathbf{0}$, and use Theorem 11 to evaluate $q(H)\mathbf{x}$.)
b) Use part (a) to calculate the eigenvalues of $q(A)$ and $q(B)$, where A and B are from Exercises 7 and 8, respectively.

19. With $q(t)$ as in Exercise 18, verify that $q(C)$ is the zero matrix, where C is from Exercise 9. (Note that $q(t)$ is the characteristic polynomial for C. See Exercises 20–23.)

Exercises 20–23 illustrate the *Cayley–Hamilton theorem,* which states that if $p(t)$ is the characteristic polynomial for A, then $p(A)$ is the zero matrix. (As in Exercise 18, $p(A)$ is the $(n \times n)$ matrix that comes from substituting A for t in $p(t)$.) In Exercises 20–23, verify that $p(A) = \mathcal{O}$ for the given matrix A.

20. A in Exercise 3
21. A in Exercise 4
22. A in Exercise 9
23. A in Exercise 13

24. This problem establishes a special case of the Cayley–Hamilton theorem.
a) Prove that if B is a (3×3) matrix, and if $B\mathbf{x} = \mathbf{0}$ for every $\mathbf{x}$ in R^3, then B is the zero matrix. (*Hint:* Consider $B\mathbf{e}_1, B\mathbf{e}_2,$ and $B\mathbf{e}_3$.)
b) Suppose that $\lambda_1, \lambda_2,$ and λ_3 are the eigenvalues of a (3×3) matrix A, and suppose that $\mathbf{u}_1, \mathbf{u}_2,$ and $\mathbf{u}_3$ are corresponding eigenvectors. Prove that if $\{\mathbf{u}_1, \mathbf{u}_2, \mathbf{u}_3\}$ is a linearly independent set, and if $p(t)$ is the characteristic polynomial for A, then $p(A)$ is the zero matrix. (*Hint:* Any vector $\mathbf{x}$ in R^3 can be expressed as a linear combination of $\mathbf{u}_1, \mathbf{u}_2,$ and $\mathbf{u}_3$.)

25. Consider the (2×2) matrix A given by

$$A = \begin{bmatrix} a & b \\ c & d \end{bmatrix}.$$

The characteristic polynomial for A is $p(t) = t^2 - (a+d)t + (ad - bc)$. Verify the Cayley–Hamilton theorem for (2×2) matrices by forming A^2 and showing that $p(A)$ is the zero matrix.

26. Let A be the (3×3) upper-triangular matrix given by

$$A = \begin{bmatrix} a & d & f \\ 0 & b & e \\ 0 & 0 & c \end{bmatrix}.$$

The characteristic polynomial for A is $p(t) = -(t-a)(t-b)(t-c)$. Verify that $p(A)$ has the form $p(A) = -(A - aI)(A - bI)(A - cI)$. (*Hint:* Expand $p(t)$ and $p(A)$; for instance, $(A - bI)(A - cI) = A^2 - (b+c)A + bcI$.) Next, show that $p(A)$ is the zero matrix by forming the product of the matrices $A - aI$, $A - bI$, and $A - cI$. (*Hint:* Form the product $(A - bI)(A - cI)$ first.)

27. Let $q(t) = t^n + a_{n-1}t^{n-1} + \cdots + a_1 t + a_0$, and define the $(n \times n)$ "companion" matrix by

$$A = \begin{bmatrix} -a_{n-1} & -a_{n-2} & \cdots & -a_1 & -a_0 \\ 1 & 0 & \cdots & 0 & 0 \\ 0 & 1 & \cdots & 0 & 0 \\ \vdots & & & & \vdots \\ 0 & 0 & \cdots & 1 & 0 \end{bmatrix}.$$

a) For $n = 2$ and for $n = 3$, show that $\det(A - tI) = (-1)^n q(t)$.

b) Give the companion matrix A for the polynomial $q(t) = t^4 + 3t^3 - t^2 + 2t - 2$. Verify that $q(t)$ is the characteristic polynomial for A.

c) Prove for all n that $\det(A - tI) = (-1)^n q(t)$.

28. The "power method" is a numerical method used to estimate the "dominant" eigenvalue of a matrix A. (See Chapter 7 for the practical details. By the dominant eigenvalue, we mean the one that is largest in absolute value.) The algorithm proceeds as follows:

a) Choose any starting vector x_0, $x_0 \neq \theta$.

b) Let $x_{k+1} = Ax_k$, $k = 0, 1, 2, \ldots$.

c) Let $\beta_k = x_k^T x_{k+1}/x_k^T x_k$, $k = 0, 1, 2, \ldots$.

Under suitable conditions it can be shown that $\{\beta_k\} \to \lambda_1$, where λ_1 is the dominant eigenvalue of A. Use the power method to estimate the dominant eigenvalue of the matrix in Exercise 9. Use the starting vector

$$x = \begin{bmatrix} 1 \\ 1 \\ 1 \end{bmatrix}$$

and calculate β_0, β_1, β_2, β_3, and β_4.

29. This exercise gives a condition under which the power method (see Exercise 28) converges. Suppose that A is an $(n \times n)$ matrix and has real eigenvalues $\lambda_1, \lambda_2, \ldots, \lambda_n$ with corresponding eigenvectors $u_1, u_2, \ldots, u_n$. Furthermore, suppose that $|\lambda_1| > |\lambda_2| \geq \cdots \geq |\lambda_n|$ and the starting vector x_0 satisfies $x_0 = c_1 u_1 + c_2 u_2 + \cdots + c_n u_n$, where $c_1 \neq 0$. Prove that

$$\lim_{k \to \infty} \beta_k = \lambda_1.$$

(*Hint:* Observe that $x_j = A^j x_0$, $j = 1, 2, \ldots$, and use Theorem 11 to calculate x_{k+1} and x_k. Next, factor all powers of λ_1 from the numerator and denominator of $\beta_k = x_k^T x_{k+1}/x_k^T x_k$.)

30. Theorem 12 shows that A and A^T have the same eigenvalues. In Theorem 5 of Section 3.3, it was shown that $\det(A) = \det(A^T)$. Use this result to show that A and A^T have the same characteristic polynomial. (*Note:* Theorem 12 proves that $A - \lambda I$ and $A^T - \lambda I$ are singular or nonsingular together. This exercise shows that the eigenvalues of A and A^T have the same algebraic multiplicity.)

The characteristic polynomial $p(t) = \det(A - tI)$ has the form $p(t) = (-1)^n t^n + a_{n-1}t^{n-1} + \cdots + a_1 t + a_0$. The coefficients of $p(t)$ can be found by evaluating $\det(A - tI)$ at n distinct values of t and solving the resulting Vandermonde system for $a_{n-1}, \ldots, a_1, a_0$. Employ this technique in Exercises 31–34 to find the characteristic polynomial for the indicated matrix A.

31. A in Exercise 5

32. A in Exercise 6

33. A in Exercise 7

34. A in Exercise 8

Eigenvectors and Eigenspaces

As we saw in the previous section, we can find the eigenvalues of a matrix A by solving the characteristic equation $\det(A - tI) = 0$. Once we know the eigenvalues, the familiar technique of Gaussian elimination can be employed to find the eigenvectors that correspond to the various eigenvalues.

In particular, the eigenvectors corresponding to an eigenvalue λ of A are the nonzero solutions of

$$(A - \lambda I)\mathbf{x} = \boldsymbol{\theta}. \tag{1}$$

Given a value for λ, the equations in (1) can be solved for $\mathbf{x}$ by using Gaussian elimination.

EXAMPLE 1 Find the eigenvectors that correspond to the eigenvalues of matrix A in Example 1 of Section 3.4.

SOLUTION For matrix A in Example 1, $A - \lambda I$ is the matrix

$$A - \lambda I = \begin{bmatrix} 1 - \lambda & 1 & 1 \\ 0 & 3 - \lambda & 3 \\ -2 & 1 & 1 - \lambda \end{bmatrix}.$$

Also, from Example 1 we know that the eigenvalues of A are given by $\lambda = 0$, $\lambda = 2$, and $\lambda = 3$.

For each eigenvalue λ, we find the eigenvectors that correspond to λ by solving the system $(A - \lambda I)\mathbf{x} = \boldsymbol{\theta}$. For the eigenvalue $\lambda = 0$, we have $(A - 0I)\mathbf{x} = \boldsymbol{\theta}$, or $A\mathbf{x} = \boldsymbol{\theta}$, to solve:

$$\begin{aligned} x_1 + x_2 + x_3 &= 0 \\ 3x_2 + 3x_3 &= 0 \\ -2x_1 + x_2 + x_3 &= 0. \end{aligned}$$

The solution of this system is $x_1 = 0$, $x_2 = -x_3$, with x_3 arbitrary. Thus the eigenvectors of A corresponding to $\lambda = 0$ are given by

$$\mathbf{x} = \begin{bmatrix} 0 \\ -a \\ a \end{bmatrix} = a \begin{bmatrix} 0 \\ -1 \\ 1 \end{bmatrix}, \qquad a \neq 0,$$

and any such vector $\mathbf{x}$ satisfies $A\mathbf{x} = 0 \cdot \mathbf{x}$. This equation illustrates that the definition of eigenvalues does allow the possibility that $\lambda = 0$ is an eigenvalue. We stress, however, that the zero vector is never considered an eigenvector (after all, $A\mathbf{x} = \lambda \mathbf{x}$ is always satisfied for $\mathbf{x} = \boldsymbol{\theta}$, no matter what value λ has).

The eigenvectors corresponding to the eigenvalue $\lambda = 3$ are found by solving $(A - 3I)\mathbf{x} = \boldsymbol{\theta}$:

$$\begin{aligned} -2x_1 + x_2 + x_3 &= 0 \\ 3x_3 &= 0 \\ -2x_1 + x_2 - 2x_3 &= 0. \end{aligned}$$

The solution of this system is $x_3 = 0$, $x_2 = 2x_1$, with x_1 arbitrary. Thus the nontrivial solutions of $(A - 3I)\mathbf{x} = \boldsymbol{\theta}$ (the eigenvectors of A corresponding to $\lambda = 3$) all have the form

$$\mathbf{x} = \begin{bmatrix} a \\ 2a \\ 0 \end{bmatrix} = a \begin{bmatrix} 1 \\ 2 \\ 0 \end{bmatrix}, \qquad a \neq 0.$$

Finally, the eigenvectors corresponding to $\lambda = 2$ are found from $(A - 2I)\mathbf{x} = \boldsymbol{\theta}$, and the solution is $x_1 = -2x_3$, $x_2 = -3x_3$, with x_3 arbitrary. So the eigenvectors corresponding to $\lambda = 2$ are of the form

$$\mathbf{x} = \begin{bmatrix} -2a \\ -3a \\ a \end{bmatrix} = a \begin{bmatrix} -2 \\ -3 \\ 1 \end{bmatrix}, \qquad a \neq 0. \qquad \square$$

We pause here to make several comments. As Example 1 shows, there are infinitely many eigenvectors that correspond to a given eigenvalue. This comment should be obvious, for if $A - \lambda I$ is a singular matrix, there are infinitely many nontrivial solutions of $(A - \lambda I)\mathbf{x} = \boldsymbol{\theta}$. In particular, if $A\mathbf{x} = \lambda\mathbf{x}$ for some nonzero vector $\mathbf{x}$, then we also have $A\mathbf{y} = \lambda\mathbf{y}$ when $\mathbf{y} = a\mathbf{x}$, with a being any scalar. Thus any nonzero multiple of an eigenvector is again an eigenvector.

Next, we again note that the scalar $\lambda = 0$ may be an eigenvalue of a matrix, as Example 1 showed. In fact, from Theorem 13 of Section 3.4 we know that $\lambda = 0$ is an eigenvalue of A whenever A is singular.

Last, we observe from Example 1 that finding all the eigenvectors corresponding to $\lambda = 0$ is precisely the same as finding the null space of A and then deleting the zero vector, $\boldsymbol{\theta}$. Likewise, the eigenvectors of A corresponding to $\lambda = 2$ and $\lambda = 3$ are the nonzero vectors in the null space of $A - 2I$ and $A - 3I$, respectively.

Eigenspaces and Geometric Multiplicity

In the discussion above we made the following observation: If λ is an eigenvalue of A, then the eigenvectors corresponding to λ are precisely the nonzero vectors in the null space of $A - \lambda I$. It is convenient to formalize this observation.

DEFINITION 6   Let A be an $(n \times n)$ matrix. If λ is an eigenvalue of A, then:

a) The null space of $A - \lambda I$ is denoted by E_λ and is called the ***eigenspace*** of λ.

b) The dimension of E_λ is called the ***geometric multiplicity*** of λ.

(*Note:* Since $A - \lambda I$ is singular, the dimension of E_λ, the geometric multiplicity of λ, is always at least 1 and may be larger. It can be shown that the geometric multiplicity of λ is never larger than the algebraic multiplicity of λ. The next three examples illustrate some of the possibilities.)

EXAMPLE 2 Determine the algebraic and geometric multiplicities for the eigenvalues of A

$$A = \begin{bmatrix} 1 & 1 & 0 \\ 0 & 1 & 1 \\ 0 & 0 & 1 \end{bmatrix}.$$

SOLUTION The characteristic polynomial is $p(t) = (1 - t)^3$, and thus the only eigenvalue of A is $\lambda = 1$. The eigenvalue $\lambda = 1$ has algebraic multiplicity 3. The eigenspace is found by solving $(A - I)\mathbf{x} = \boldsymbol{0}$. The system $(A - I)\mathbf{x} = \boldsymbol{0}$ is

$$x_2 = 0$$
$$x_3 = 0.$$

Thus $\mathbf{x}$ is in the eigenspace E_λ corresponding to $\lambda = 1$ if and only if $\mathbf{x}$ has the form

$$\mathbf{x} = \begin{bmatrix} x_1 \\ 0 \\ 0 \end{bmatrix} = x_1 \begin{bmatrix} 1 \\ 0 \\ 0 \end{bmatrix}. \tag{2}$$

The geometric multiplicity of the eigenvalue $\lambda = 1$ is 1, and $\mathbf{x}$ is an eigenvector if $\mathbf{x}$ has the form (2) with $x_1 \neq 0$. $\qquad\square$

EXAMPLE 3 Determine the algebraic and geometric multiplicities for the eigenvalues of B

$$B = \begin{bmatrix} 1 & 1 & 0 \\ 0 & 1 & 0 \\ 0 & 0 & 1 \end{bmatrix}.$$

SOLUTION The characteristic polynomial is $p(t) = (1 - t)^3$, so $\lambda = 1$ is the only eigenvalue and it has algebraic multiplicity 3.

The corresponding eigenspace is found by solving $(B - I)\mathbf{x} = \boldsymbol{0}$. Now $(B - I)\mathbf{x} = \boldsymbol{0}$ if and only if $\mathbf{x}$ has the form

$$\mathbf{x} = \begin{bmatrix} x_1 \\ 0 \\ x_3 \end{bmatrix} = x_1 \begin{bmatrix} 1 \\ 0 \\ 0 \end{bmatrix} + x_3 \begin{bmatrix} 0 \\ 0 \\ 1 \end{bmatrix}. \tag{3}$$

By (3), the eigenspace has dimension 2, and so the eigenvalue $\lambda = 1$ has geometric multiplicity 2. The eigenvectors of B are the nonzero vectors of the form (3). $\qquad\square$

EXAMPLE 4 Determine the algebraic and geometric multiplicities for the eigenvalues of C

$$C = \begin{bmatrix} 1 & 0 & 0 \\ 0 & 1 & 0 \\ 0 & 0 & 1 \end{bmatrix}.$$

SOLUTION   The characteristic polynomial is $p(t) = (1 - t)^3$, so $\lambda = 1$ has algebraic multiplicity 3.

The eigenspace is found by solving $(C - I)\mathbf{x} = \boldsymbol{\theta}$, and since $C - I$ is the zero matrix, every vector in R^3 is in the null space of $C - I$. The geometric multiplicity of the eigenvalue $\lambda = 1$ is equal to 3.   □

(*Note:* The matrices in Examples 2, 3, and 4 all have the same characteristic polynomial, $p(t) = (1 - t)^3$. However, the respective eigenspaces are different.)

Defective Matrices

For applications (such as diagonalization) it will be important to know whether an $(n \times n)$ matrix A has a set of n linearly independent eigenvectors. As we will see later, if A is an $(n \times n)$ matrix and if some eigenvalue of A has a geometric multiplicity that is less than its algebraic multiplicity, then A will not have a set of n linearly independent eigenvectors. Such a matrix is called defective.

> **DEFINITION 7**   Let A be an $(n \times n)$ matrix. If there is an eigenvalue λ of A such that the geometric multiplicity of λ is less than the algebraic multiplicity of λ, then A is called a ***defective*** matrix.

Note that the matrices in Examples 1 and 4 are not defective. The matrices in Examples 2 and 3 are defective. Example 5 provides another instance of a defective matrix.

EXAMPLE 5   Find all the eigenvalues and eigenvectors of the matrix A:

$$A = \begin{bmatrix} -4 & 1 & 1 & 1 \\ -16 & 3 & 4 & 4 \\ -7 & 2 & 2 & 1 \\ -11 & 1 & 3 & 4 \end{bmatrix}.$$

Also, determine the algebraic and geometric multiplicities of the eigenvalues.

SOLUTION   Omitting the details, a cofactor expansion yields

$$\det(A - tI) = t^4 - 5t^3 + 9t^2 - 7t + 2$$
$$= (t - 1)^3(t - 2).$$

Hence the eigenvalues are $\lambda = 1$ (algebraic multiplicity 3) and $\lambda = 2$ (algebraic multiplicity 1).

In solving $(A - 2I)\mathbf{x} = \boldsymbol{\theta}$, we reduce the augmented matrix $[A - 2I \,|\, \boldsymbol{\theta}]$ as follows, multiplying rows 2, 3, and 4 by constants to avoid working with fractions:

$$[A - 2I \,|\, \theta] = \begin{bmatrix} -6 & 1 & 1 & 1 & 0 \\ -16 & 1 & 4 & 4 & 0 \\ -7 & 2 & 0 & 1 & 0 \\ -11 & 1 & 3 & 2 & 0 \end{bmatrix} \sim \begin{bmatrix} -6 & 1 & 1 & 1 & 0 \\ -48 & 3 & 12 & 12 & 0 \\ -42 & 12 & 0 & 6 & 0 \\ -66 & 6 & 18 & 12 & 0 \end{bmatrix}$$

$$\sim \begin{bmatrix} -6 & 1 & 1 & 1 & 0 \\ 0 & -5 & 4 & 4 & 0 \\ 0 & 5 & -7 & -1 & 0 \\ 0 & -5 & 7 & 1 & 0 \end{bmatrix} \sim \begin{bmatrix} -6 & 1 & 1 & 1 & 0 \\ 0 & -5 & 4 & 4 & 0 \\ 0 & 0 & -3 & 3 & 0 \\ 0 & 0 & 0 & 0 & 0 \end{bmatrix}.$$

Backsolving yields $x_1 = 3x_4/5$, $x_2 = 8x_4/5$, $x_3 = x_4$. Hence **x** is an eigenvector corresponding to $\lambda = 2$ only if

$$\mathbf{x} = \begin{bmatrix} x_1 \\ x_2 \\ x_3 \\ x_4 \end{bmatrix} = \begin{bmatrix} 3x_4/5 \\ 8x_4/5 \\ x_4 \\ x_4 \end{bmatrix} = \frac{x_4}{5}\begin{bmatrix} 3 \\ 8 \\ 5 \\ 5 \end{bmatrix}, \qquad x_4 \neq 0,$$

Thus the algebraic and geometric multiplicities of the eigenvalue $\lambda = 2$ are equal to 1.

In solving $(A - I)\mathbf{x} = \theta$, we reduce the augmented matrix $[A - I \,|\, \theta]$:

$$[A - I \,|\, \theta] = \begin{bmatrix} -5 & 1 & 1 & 1 & 0 \\ -16 & 2 & 4 & 4 & 0 \\ -7 & 2 & 1 & 1 & 0 \\ -11 & 1 & 3 & 3 & 0 \end{bmatrix} \sim \begin{bmatrix} -5 & 1 & 1 & 1 & 0 \\ -80 & 10 & 20 & 20 & 0 \\ -35 & 10 & 5 & 5 & 0 \\ -55 & 5 & 15 & 15 & 0 \end{bmatrix}$$

$$\sim \begin{bmatrix} -5 & 1 & 1 & 1 & 0 \\ 0 & -6 & 4 & 4 & 0 \\ 0 & 3 & -2 & -2 & 0 \\ 0 & -6 & 4 & 4 & 0 \end{bmatrix} \sim \begin{bmatrix} -5 & 1 & 1 & 1 & 0 \\ 0 & 3 & -2 & -2 & 0 \\ 0 & 0 & 0 & 0 & 0 \\ 0 & 0 & 0 & 0 & 0 \end{bmatrix}.$$

Backsolving yields $x_1 = (x_3 + x_4)/3$ and $x_2 = 2(x_3 + x_4)/3$. Thus **x** is an eigenvector corresponding to $\lambda = 1$ only if **x** is a nonzero vector of the form

$$\mathbf{x} = \begin{bmatrix} x_1 \\ x_2 \\ x_3 \\ x_4 \end{bmatrix} = \begin{bmatrix} (x_3 + x_4)/3 \\ 2(x_3 + x_4)/3 \\ x_3 \\ x_4 \end{bmatrix} = \frac{x_3}{3}\begin{bmatrix} 1 \\ 2 \\ 3 \\ 0 \end{bmatrix} + \frac{x_4}{3}\begin{bmatrix} 1 \\ 2 \\ 0 \\ 3 \end{bmatrix}. \tag{4}$$

By (4), the eigenspace E_λ corresponding to $\lambda = 1$ has a basis consisting of the vectors

$$\begin{bmatrix} 1 \\ 2 \\ 3 \\ 0 \end{bmatrix} \quad \text{and} \quad \begin{bmatrix} 1 \\ 2 \\ 0 \\ 3 \end{bmatrix}.$$

Since E_λ has dimension 2, the eigenvalue $\lambda = 1$ has geometric multiplicity 2 and algebraic multiplicity 3. (Matrix A is defective.)   ▢

The next theorem shows that a matrix can be defective only if it has repeated eigenvalues. (As shown in Example 4, however, repeated eigenvalues do not necessarily mean that a matrix is defective.)

THEOREM 15

Let $\mathbf{u}_1, \mathbf{u}_2, \ldots, \mathbf{u}_k$ be eigenvectors of an $(n \times n)$ matrix A corresponding to distinct eigenvalues $\lambda_1, \lambda_2, \ldots, \lambda_k$. That is,

$$A\mathbf{u}_j = \lambda_j\mathbf{u}_j \quad \text{for } j = 1, 2, \ldots, k; \quad k \leq n \tag{5}$$

$$\lambda_i \neq \lambda_j \quad \text{for } i \neq j; \quad 1 \leq i, j \leq k. \tag{6}$$

Then $\{\mathbf{u}_1, \mathbf{u}_2, \ldots, \mathbf{u}_k\}$ is a linearly independent set.

PROOF   Since $\mathbf{u}_1 \neq \boldsymbol{\theta}$, the set $\{\mathbf{u}_1\}$ is trivially linearly independent. If the set $\{\mathbf{u}_1, \mathbf{u}_2, \ldots, \mathbf{u}_k\}$ were linearly dependent, then there would exist an integer m, $2 \leq m \leq k$, such that:

a) $S_1 \equiv \{\mathbf{u}_1, \mathbf{u}_2, \ldots, \mathbf{u}_{m-1}\}$ is linearly independent.

b) $S_2 \equiv \{\mathbf{u}_1, \mathbf{u}_2, \ldots, \mathbf{u}_{m-1}, \mathbf{u}_m\}$ is linearly dependent.

Now since S_2 is linearly dependent, there exist scalars $c_1, c_2, \ldots, c_m$ (not all zero) such that

$$c_1\mathbf{u}_1 + c_2\mathbf{u}_2 + \cdots + c_{m-1}\mathbf{u}_{m-1} + c_m\mathbf{u}_m = \boldsymbol{\theta}. \tag{7}$$

Furthermore, c_m in Eq. (7) cannot be zero. (If $c_m = 0$, then Eq. (7) would imply that S_1 is linearly dependent, contradicting (a).)

Multiplying both sides of (7) by A and using $A\mathbf{u}_j = \lambda_j\mathbf{u}_j$, we obtain

$$c_1\lambda_1\mathbf{u}_1 + c_2\lambda_2\mathbf{u}_2 + \cdots + c_{m-1}\lambda_{m-1}\mathbf{u}_{m-1} + c_m\lambda_m\mathbf{u}_m = \boldsymbol{\theta}. \tag{8}$$

Multiplying both sides of (7) by λ_m yields

$$c_1\lambda_m\mathbf{u}_1 + c_2\lambda_m\mathbf{u}_2 + \cdots + c_{m-1}\lambda_m\mathbf{u}_{m-1} + c_m\lambda_m\mathbf{u}_m = \boldsymbol{\theta}. \tag{9}$$

Subtracting Eq. (8) from Eq. (9), we find that

$$c_1(\lambda_m - \lambda_1)\mathbf{u}_1 + c_2(\lambda_m - \lambda_2)\mathbf{u}_2 + \cdots + c_{m-1}(\lambda_m - \lambda_{m-1})\mathbf{u}_{m-1} = \boldsymbol{\theta}. \tag{10}$$

If we set $\beta_j = c_j(\lambda_m - \lambda_j)$, $1 \leq j \leq m-1$, Eq. (10) becomes

$$\beta_1\mathbf{u}_1 + \beta_2\mathbf{u}_2 + \cdots + \beta_{m-1}\mathbf{u}_{m-1} = \boldsymbol{\theta}.$$

Since S_1 is linearly independent, it then follows that

$$\beta_1 = \beta_2 = \cdots = \beta_{m-1} = 0,$$

or

$$c_j(\lambda_m - \lambda_j) = 0, \quad \text{for } j = 1, 2, \ldots, m-1.$$

Because $\lambda_m \neq \lambda_j$ for $j \neq m$, we must have $c_j = 0$ for $1 \leq j \leq m-1$.

Finally (see Eq. 7), if $c_j = 0$ for $1 \leq j \leq m - 1$, then $c_m\mathbf{u}_m = \mathbf{0}$. Since $c_m \neq 0$, it follows that $\mathbf{u}_m = \mathbf{0}$. But $\mathbf{u}_m$ is an eigenvector, and so $\mathbf{u}_m \neq \mathbf{0}$. Hence we have contradicted the assumption that there is an m, $m \leq k$, such that S_2 is linearly dependent. Thus $\{\mathbf{u}_1, \mathbf{u}_2, \ldots, \mathbf{u}_k\}$ is linearly independent.

An important and useful corollary to Theorem 15 is given below.

COROLLARY

Let A be an $(n \times n)$ matrix. If A has n distinct eigenvalues, then A has a set of n linearly independent eigenvectors.

Exercises 3.5

Given below is a list of matrices and their respective characteristic polynomials. These are referred to in Exercises 1–11.

$$A = \begin{bmatrix} 2 & -1 \\ -1 & 2 \end{bmatrix}, \qquad B = \begin{bmatrix} 1 & -1 \\ 1 & 3 \end{bmatrix},$$

$$p(t) = (t - 3)(t - 1), \qquad p(t) = (t - 2)^2,$$

$$C = \begin{bmatrix} -6 & -1 & 2 \\ 3 & 2 & 0 \\ -14 & -2 & 5 \end{bmatrix}, \qquad D = \begin{bmatrix} -7 & 4 & -3 \\ 8 & -3 & 3 \\ 32 & -16 & 13 \end{bmatrix},$$

$$p(t) = -(t - 1)^2(t + 1), \qquad p(t) = -(t - 1)^3,$$

$$E = \begin{bmatrix} 6 & 4 & 4 & 1 \\ 4 & 6 & 1 & 4 \\ 4 & 1 & 6 & 4 \\ 1 & 4 & 4 & 6 \end{bmatrix}, \qquad F = \begin{bmatrix} 1 & -1 & -1 & -1 \\ -1 & 1 & -1 & -1 \\ -1 & -1 & 1 & -1 \\ -1 & -1 & -1 & 1 \end{bmatrix},$$

$$p(t) = (t + 1)(t - 5)^2(t - 15), \qquad p(t) = (t + 2)(t - 2)^3$$

In Exercises 1–11, find a basis for the eigenspace E_λ for the given matrix and the value of λ. Determine the algebraic and geometric multiplicities of λ.

1. A, $\lambda = 3$
2. A, $\lambda = 1$
3. B, $\lambda = 2$
4. C, $\lambda = 1$
5. C, $\lambda = -1$
6. D, $\lambda = 1$
7. E, $\lambda = -1$
8. E, $\lambda = 5$
9. E, $\lambda = 15$
10. F, $\lambda = -2$
11. F, $\lambda = 2$

In Exercises 12–17, find the eigenvalues and the eigenvectors for the given matrix. Is the matrix defective?

12. $\begin{bmatrix} 1 & 1 & -1 \\ 0 & 2 & -1 \\ 0 & 0 & 1 \end{bmatrix}$
13. $\begin{bmatrix} 2 & 1 & 2 \\ 0 & 3 & 2 \\ 0 & 0 & 2 \end{bmatrix}$
14. $\begin{bmatrix} 1 & 2 & 1 \\ 0 & 1 & 2 \\ 0 & 0 & 1 \end{bmatrix}$
15. $\begin{bmatrix} 2 & 0 & 3 \\ 0 & 2 & 1 \\ 0 & 0 & 1 \end{bmatrix}$
16. $\begin{bmatrix} -1 & 6 & 2 \\ 0 & 5 & -6 \\ 1 & 0 & -2 \end{bmatrix}$
17. $\begin{bmatrix} 3 & -1 & -1 \\ -12 & 0 & 5 \\ 4 & -2 & -1 \end{bmatrix}$

18. If a vector $\mathbf{x}$ is a linear combination of eigenvectors of a matrix A, then it is easy to calculate the product $\mathbf{y} = A^k\mathbf{x}$ for any positive integer k. For instance, suppose that $A\mathbf{u}_1 = \lambda_1\mathbf{u}_1$ and $A\mathbf{u}_2 = \lambda_2\mathbf{u}_2$, where $\mathbf{u}_1$ and $\mathbf{u}_2$ are nonzero vectors. If $\mathbf{x} = a_1\mathbf{u}_1 + a_2\mathbf{u}_2$, then (see Theorem 11 of Section 3.4) $\mathbf{y} = A^k\mathbf{x} = A^k(a_1\mathbf{u}_1 + a_2\mathbf{u}_2) = a_1A^k\mathbf{u}_1 + a_2A^k\mathbf{u}_2 = a_1(\lambda_1)^k\mathbf{u}_1 + a_2(\lambda_2)^k\mathbf{u}_2$. Find $A^{10}\mathbf{x}$, where

$$A = \begin{bmatrix} 4 & -2 \\ 5 & -3 \end{bmatrix} \quad \text{and} \quad \mathbf{x} = \begin{bmatrix} 0 \\ 9 \end{bmatrix}.$$

19. As in Exercise 18, calculate $A^{10}\mathbf{x}$ for

$$A = \begin{bmatrix} 1 & 2 & -1 \\ 0 & 5 & -2 \\ 0 & 6 & -2 \end{bmatrix} \quad \text{and} \quad \mathbf{x} = \begin{bmatrix} 2 \\ 4 \\ 7 \end{bmatrix}.$$

20. Consider a (4×4) matrix H of the form

$$H = \begin{bmatrix} \times & \times & \times & \times \\ a & \times & \times & \times \\ 0 & b & \times & \times \\ 0 & 0 & c & \times \end{bmatrix} \qquad \textbf{(11)}$$

In matrix (11) the entries designated $\times$ may be zero or nonzero. Suppose, in (11), that a, b, and c are nonzero. Let λ be any eigenvalue of H. Show that the geometric multiplicity of λ is equal to 1. (*Hint:* Verify that the rank of $H - \lambda I$ is exactly equal to 3.)

21. An $(n \times n)$ matrix P is called ***idempotent*** if $P^2 = P$. Show that if P is an invertible idempotent matrix, then $P = I$.

22. Let P be an idempotent matrix. Show that the only eigenvalues of P are $\lambda = 0$ and $\lambda = 1$. (*Hint:* Suppose that $P\mathbf{x} = \lambda\mathbf{x}$, $\mathbf{x} \neq \boldsymbol{\theta}$.)

23. Let $\mathbf{u}$ be a vector in R^n such that $\mathbf{u}^T\mathbf{u} = 1$. Show that the $(n \times n)$ matrix $P = \mathbf{u}\mathbf{u}^T$ is an idempotent matrix. (*Hint:* Use the associative properties of matrix multiplication.)

24. Verify that if Q is idempotent, then so is $I - Q$. Also verify that $(I - 2Q)^{-1} = I - 2Q$.

25. Suppose that $\mathbf{u}$ and $\mathbf{v}$ are vectors in R^n such that $\mathbf{u}^T\mathbf{u} = 1$, $\mathbf{v}^T\mathbf{v} = 1$, and $\mathbf{u}^T\mathbf{v} = 0$. Show that $P = \mathbf{u}\mathbf{u}^T + \mathbf{v}\mathbf{v}^T$ is idempotent.

26. Show that any nonzero vector of the form $a\mathbf{u} + b\mathbf{v}$ is an eigenvector corresponding to $\lambda = 1$ for the matrix P in Exercise 25.

27. Let A be an $(n \times n)$ symmetric matrix, with (real) distinct eigenvalues $\lambda_1, \lambda_2, \dots, \lambda_n$. Let the corre-

sponding eigenvectors $\mathbf{u}_1, \mathbf{u}_2, \dots, \mathbf{u}_n$ be chosen so that $\|\mathbf{u}_i\| = 1$ (that is, $\mathbf{u}_i^T\mathbf{u}_i = 1$). Exercise 29 shows that A can be "decomposed" as

$$A = \lambda_1\mathbf{u}_1\mathbf{u}_1^T + \lambda_2\mathbf{u}_2\mathbf{u}_2^T + \cdots + \lambda_n\mathbf{u}_n\mathbf{u}_n^T. \qquad \textbf{(12)}$$

Verify decomposition (12) for each of the following matrices.

a) $B = \begin{bmatrix} 2 & -1 \\ -1 & 2 \end{bmatrix}$ b) $C = \begin{bmatrix} 1 & 2 \\ 2 & 1 \end{bmatrix}$

c) $D = \begin{bmatrix} 3 & 2 \\ 2 & 0 \end{bmatrix}$

28. Let A be a symmetric matrix and suppose that $A\mathbf{u} = \lambda\mathbf{u}$, $\mathbf{u} \neq \boldsymbol{\theta}$ and $A\mathbf{v} = \beta\mathbf{v}$, $\mathbf{v} \neq \boldsymbol{\theta}$. Also suppose that $\lambda \neq \beta$. Show that $\mathbf{u}^T\mathbf{v} = 0$. (*Hint:* Since $A\mathbf{v}$ and $\mathbf{u}$ are vectors, $(A\mathbf{v})^T\mathbf{u} = \mathbf{u}^T(A\mathbf{v})$. Rewrite the term $(A\mathbf{v})^T\mathbf{u}$ by using Theorem 10, property 2, of Section 1.7.)

29. Having A as in Exercise 27, we see from Exercise 28 that $\mathbf{u}_i^T\mathbf{u}_j = 0$, $i \neq j$. By the corollary to Theorem 15, $\{\mathbf{u}_1, \mathbf{u}_2, \dots, \mathbf{u}_n\}$ is an orthonormal basis for R^n. To show that decomposition (12) is valid, let C denote the right-hand side of (12). Then show that $(A - C)\mathbf{u}_i = \boldsymbol{\theta}$ for $1 \leq i \leq n$. Finally, show that $A - C$ is the zero matrix. (*Hint:* Look at Exercise 24 in Section 3.4.)

(*Note:* We will see in the next section that a real symmetric matrix has only real eigenvalues. It can also be shown that the eigenvectors can be chosen to be orthonormal, even when the eigenvalues are not distinct. Thus decomposition (12) is valid for any real symmetric matrix A.)

3.6   Complex Eigenvalues and Eigenvectors

Up to now we have not examined, in detail, the case in which the characteristic equation has complex roots—that is, the case in which a matrix has complex eigenvalues. We will see that the possibility of complex eigenvalues does not pose any additional problems except that the eigenvectors corresponding to complex eigenvalues will have complex components, and complex arithmetic will be required to find these eigenvectors.

EXAMPLE 1   Find the eigenvalues and the eigenvectors for

$$A = \begin{bmatrix} 3 & 1 \\ -2 & 1 \end{bmatrix}.$$

SOLUTION The characteristic polynomial for A is $p(t) = t^2 - 4t + 5$. The eigenvalues of A are the roots of $p(t) = 0$, which we can find from the quadratic formula,

$$\lambda = \frac{4 \pm \sqrt{-4}}{2} = 2 \pm i,$$

where $i = \sqrt{-1}$. Thus despite the fact that A is a real matrix, the eigenvalues of A are complex, $\lambda = 2 + i$ and $\lambda = 2 - i$. To find the eigenvectors of A corresponding to $\lambda = 2 + i$, we must solve $[A - (2 + i)I]\mathbf{x} = \boldsymbol{\theta}$, which leads to the (2×2) homogeneous system

$$\begin{aligned} (1 - i)x_1 + \quad x_2 &= 0 \\ -2x_1 - (1 + i)x_2 &= 0. \end{aligned} \tag{1}$$

At the end of this section, we will discuss the details of how such a system is solved. For the moment, we merely note that if the first equation is multiplied by $1 + i$, then Eq. (1) is equivalent to

$$\begin{aligned} 2x_1 + (1 + i)x_2 &= 0 \\ -2x_1 - (1 + i)x_2 &= 0. \end{aligned}$$

Thus the solutions of (1) are determined by $x_1 = -(1 + i)x_2/2$. The nonzero solutions of (1), the eigenvectors corresponding to $\lambda = 2 + i$, are of the form

$$\mathbf{x} = a \begin{bmatrix} 1 + i \\ -2 \end{bmatrix}, \qquad a \neq 0.$$

Similar calculations show that the eigenvectors of A corresponding to $\lambda = 2 - i$ are all of the form

$$\mathbf{x} = b \begin{bmatrix} 1 - i \\ -2 \end{bmatrix}, \qquad b \neq 0. \qquad \square$$

Complex Numbers

Ancient peoples knew that certain quadratic equations, such as $x^2 + 1 = 0$, had no real solutions. This posed no difficulty, however, because their particular problems (such as finding the intersections of a line and a circle) did not require complex solutions. Hence people paid complex numbers little attention and referred to them as "imaginary" numbers. In 1545, however, Cardano published a formula for finding the roots of a cubic equation that often required algebraic manipulation of $\sqrt{-1}$ in order to find certain real solutions. Guided by Cardano's formula, Bombelli, in 1572, is credited with working out the algebra of complex numbers. However, the important link to geometry, the association of $a + bi$ with the point (a, b), was not developed for another hundred years.

Probably the two people most influential in developing complex numbers into their essential role in describing scientific phenomena were Leonhard Euler (1707–1783) and Augustin-Louis Cauchy (1789–1857). Besides introducing much of the mathematical notation used today, Euler used complex numbers to unify the study of exponential, logarithmic, and trigonometric functions. Cauchy is regarded as the founder of the field of functions of a complex variable. Many terms and results in the extension of calculus to complex variables are due to Cauchy and named after him.

Complex Arithmetic and Complex Vectors

Before giving the major theoretical results of this section, we briefly review several of the details of complex arithmetic. We will usually represent a complex number z in the form $z = a + ib$, where a and b are real numbers and $i^2 = -1$. In the representation $z = a + ib$, a is called the **real part** of z, and b is called the **imaginary part** of z. If $z = a + ib$ and $w = c + id$, then $z + w = (a + c) + i(b + d)$, whereas $zw = (ac - bd) + i(ad + bc)$. Thus, for example, if $z_1 = 2 + 3i$ and $z_2 = 1 - i$, then

$$z_1 + z_2 = 3 + 2i \quad \text{and} \quad z_1 z_2 = 5 + i.$$

If z is the complex number $z = a + ib$, then the **conjugate** of z (denoted by $\bar{z}$) is defined to be $\bar{z} = a - ib$. We list several properties of the conjugate operation:

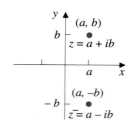

$$\overline{(z + w)} = \bar{z} + \bar{w}$$
$$\overline{(zw)} = \bar{z}\bar{w}$$
$$z + \bar{z} = 2a \qquad \qquad \textbf{(2)}$$
$$z - \bar{z} = 2ib$$
$$z\bar{z} = a^2 + b^2.$$

Figure 3.2
A complex number and its conjugate

From the last equality, we note that $z\bar{z}$ is a positive real quantity when $z \neq 0$. In fact, if we visualize z as the point (a, b) in the coordinate plane (called the **complex plane**), then $\sqrt{a^2 + b^2}$ is the distance from (a, b) to the origin (see Fig. 3.2). Hence we define the **magnitude** of z to be $|z|$, where

$$|z| = \sqrt{\bar{z}z} = \sqrt{a^2 + b^2}.$$

We also note from (2) that if $z = \bar{z}$, then $b = 0$ and so z is a real number.

EXAMPLE 2 Let $z = 4 - 2i$ and $w = 3 + 5i$.

a) Find the values of the real and imaginary parts of w.

b) Calculate $\bar{z}$, $\bar{w}$, and $|z|$.

c) Find $u = 2z + 3\bar{w}$ and $v = \bar{z}w$.

SOLUTION

a) Since $w = 3 + 5i$, the real part of w is 3 and the imaginary part is 5.

b) For $z = 4 - 2i$, $\bar{z} = 4 + 2i$. Similarly, since $w = 3 + 5i$, we have $\bar{w} = 3 - 5i$. Finally, $|z| = \sqrt{(4)^2 + (-2)^2} = \sqrt{20}$.

c) Here, $2z = 2(4 - 2i) = 8 - 4i$, whereas $3\bar{w} = 3(3 - 5i) = 9 - 15i$. Therefore,

$$u = 2z + 3\bar{w} = (8 - 4i) + (9 - 15i) = 17 - 19i.$$

The product $v = \bar{z}w$ is calculated as follows:

$$v = \bar{z}w = (4 + 2i)(3 + 5i)$$
$$= 12 + 6i + 20i + 10i^2$$
$$= 2 + 26i.$$

The conjugate operation is useful when dealing with matrices and vectors that have complex components. We define the conjugate of a vector as follows: If $\mathbf{x} = [x_1, x_2, \ldots, x_n]^T$, then the *conjugate vector* (denoted by $\bar{\mathbf{x}}$) is given by

$$\bar{\mathbf{x}} = \begin{bmatrix} \bar{x}_1 \\ \bar{x}_2 \\ \vdots \\ \bar{x}_n \end{bmatrix}.$$

In (3), an example of a vector $\mathbf{x}$ and its conjugate, $\bar{\mathbf{x}}$, is given:

$$\mathbf{x} = \begin{bmatrix} 2 + 3i \\ 4 \\ 1 - 7i \end{bmatrix}, \qquad \bar{\mathbf{x}} = \begin{bmatrix} 2 - 3i \\ 4 \\ 1 + 7i \end{bmatrix}. \tag{3}$$

The *magnitude* or *norm* of a complex vector $\mathbf{x}$ (denoted by $\|\mathbf{x}\|$) is defined in terms of $\mathbf{x}$ and $\bar{\mathbf{x}}$:

$$\|\mathbf{x}\| = \sqrt{\bar{\mathbf{x}}^T \mathbf{x}}. \tag{4}$$

With respect to Eq. (4), note that

$$\bar{\mathbf{x}}^T \mathbf{x} = \bar{x}_1 x_1 + \bar{x}_2 x_2 + \cdots + \bar{x}_n x_n = |x_1|^2 + |x_2|^2 + \cdots + |x_n|^2.$$

(If $\mathbf{x}$ is a real vector, so that $\bar{\mathbf{x}} = \mathbf{x}$, then the definition for $\|\mathbf{x}\|$ in Eq. (4) agrees with our earlier definition in Section 1.7.)

As the next example illustrates, the scalar product $\mathbf{x}^T \mathbf{y}$ will usually be complex valued if $\mathbf{x}$ and $\mathbf{y}$ are complex vectors.

EXAMPLE 3 Find $\mathbf{x}^T \mathbf{y}$, $\|\mathbf{x}\|$, and $\|\mathbf{y}\|$ for

$$\mathbf{x} = \begin{bmatrix} 2 \\ 1 - i \\ 3 + 2i \end{bmatrix} \quad \text{and} \quad \mathbf{y} = \begin{bmatrix} i \\ 1 + i \\ 2 - i \end{bmatrix}.$$

SOLUTION For $\mathbf{x}^T \mathbf{y}$, we find

$$\begin{aligned} \mathbf{x}^T \mathbf{y} &= (2)(i) + (1 - i)(1 + i) + (3 + 2i)(2 - i) \\ &= (2i) + (2) + (8 + i) \\ &= 10 + 3i. \end{aligned}$$

Similarly, $\|\mathbf{x}\| = \sqrt{\bar{\mathbf{x}}^T \mathbf{x}} = \sqrt{4 + 2 + 13} = \sqrt{19}$, whereas $\|\mathbf{y}\| = \sqrt{\bar{\mathbf{y}}^T \mathbf{y}} = \sqrt{1 + 2 + 5} = \sqrt{8}$. ☐

Eigenvalues of Real Matrices

In a situation where complex numbers might arise, it is conventional to refer to a vector $\mathbf{x}$ as a real vector if all the components of $\mathbf{x}$ are known to be real numbers. Similarly, we use the term *real matrix* to denote a matrix A, all of whose entries are real.

With these preliminaries, we can present two important results. The first result was illustrated in Example 1. We found $\lambda = 2 + i$ to be an eigenvalue with a corresponding eigenvector $\mathbf{x} = [1 + i, -2]^T$. We also found that the conjugates, $\bar{\lambda} = 2 - i$ and $\bar{\mathbf{x}} = [1 - i, -2]^T$, were the other eigenvalue/eigenvector pair. That is, the eigenvalues and eigenvectors occurred in conjugate pairs. The next theorem tells us that Example 1 is typical.

THEOREM 16

Let A be a real $(n \times n)$ matrix with an eigenvalue λ and corresponding eigenvector $\mathbf{x}$. Then $\bar{\lambda}$ is also an eigenvalue of A, and $\bar{\mathbf{x}}$ is an eigenvector corresponding to $\bar{\lambda}$.

PROOF   It can be shown (Exercise 36) that $\overline{\lambda \mathbf{x}} = \bar{\lambda} \bar{\mathbf{x}}$. Furthermore, since A is real, it can be shown (Exercise 36) that

$$\overline{A\mathbf{x}} = \bar{A}\bar{\mathbf{x}} = A\bar{\mathbf{x}}.$$

Using these two results and the assumption $A\mathbf{x} = \lambda\mathbf{x}$, we obtain

$$A\bar{\mathbf{x}} = \overline{A\mathbf{x}} = \overline{\lambda\mathbf{x}} = \bar{\lambda}\bar{\mathbf{x}}, \qquad \bar{\mathbf{x}} \neq \boldsymbol{\theta}.$$

Thus $\bar{\lambda}$ is an eigenvalue corresponding to the eigenvector $\bar{\mathbf{x}}$.   ☐

Finally, as the next theorem shows, there is an important class of matrices for which the possibility of complex eigenvalues is precluded.

THEOREM 17

If A is an $(n \times n)$ real symmetric matrix, then all the eigenvalues of A are real.

PROOF   Let A be any $(n \times n)$ real symmetric matrix, and suppose that $A\mathbf{x} = \lambda\mathbf{x}$, where $\mathbf{x} \neq \boldsymbol{\theta}$ and where we allow the possibility that $\mathbf{x}$ is a complex vector. To isolate λ, we first note that

$$\bar{\mathbf{x}}^T(A\mathbf{x}) = \bar{\mathbf{x}}^T(\lambda\mathbf{x}) = \lambda(\bar{\mathbf{x}}^T\mathbf{x}). \tag{5}$$

Regarding $A\mathbf{x}$ as a vector, we see that $\bar{\mathbf{x}}^T(A\mathbf{x}) = (A\mathbf{x})^T\bar{\mathbf{x}}$ (since, in general, $\mathbf{u}^T\mathbf{v} = \mathbf{v}^T\mathbf{u}$ for complex vectors $\mathbf{u}$ and $\mathbf{v}$). Using this observation in Eq. (5), we obtain

$$\lambda\bar{\mathbf{x}}^T\mathbf{x} = \bar{\mathbf{x}}^T(A\mathbf{x}) = (A\mathbf{x})^T\bar{\mathbf{x}} = \mathbf{x}^TA^T\bar{\mathbf{x}} = \mathbf{x}^TA\bar{\mathbf{x}}, \tag{6}$$

with the last equality above holding because $A = A^T$. Since A is real, we also know that $A\bar{\mathbf{x}} = \overline{\lambda\mathbf{x}}$; hence we deduce from Eq. (6) that

$$\lambda\bar{\mathbf{x}}^T\mathbf{x} = \mathbf{x}^TA\bar{\mathbf{x}} = \mathbf{x}^T(\bar{\lambda}\bar{\mathbf{x}}) = \bar{\lambda}\mathbf{x}^T\bar{\mathbf{x}},$$

or

$$\lambda\bar{\mathbf{x}}^T\mathbf{x} = \bar{\lambda}\mathbf{x}^T\bar{\mathbf{x}}. \tag{7}$$

Because $\mathbf{x} \neq \boldsymbol{\theta}$, $\bar{\mathbf{x}}^T\mathbf{x}$ is nonzero, and so from Eq. (7) we see that $\bar{\lambda} = \lambda$, which means that λ is real.   ☐

Gaussian Elimination for Systems with Complex Coefficients (Optional)

The remainder of this section is concerned with the computational details of solving $(A - \lambda I)\mathbf{x} = \boldsymbol{\theta}$ when λ is complex. We will see that although the arithmetic is tiresome, we can use Gaussian elimination to solve a system of linear equations that has some complex coefficients in exactly the same way that we solve systems of linear equations having real coefficients. For example, consider the (2×2) system

$$a_{11}x_1 + a_{12}x_2 = b_1$$
$$a_{21}x_1 + a_{22}x_2 = b_2,$$

where the coefficients a_{ij} may be complex. Just as before, we can multiply the first equation by $-a_{21}/a_{11}$, add the result to the second equation to eliminate x_1 from the second equation, and then backsolve to find x_2 and x_1. For larger systems with complex coefficients, the principles of Gaussian elimination are exactly the same as they are for real systems; only the computational details are different.

One computational detail that might be unfamiliar is dividing one complex number by another [the first step of Gaussian elimination for the (2×2) system above is to form a_{21}/a_{11}]. To see how a complex division is carried out, let $z = a + ib$ and $w = c + id$, where $w \neq 0$. To form the quotient z/w, we multiply numerator and denominator by $\bar{w}$:

$$\frac{z}{w} = \frac{z\bar{w}}{w\bar{w}}.$$

In detail, we have

$$\frac{z}{w} = \frac{z\bar{w}}{w\bar{w}} = \frac{(a + ib)(c - id)}{c^2 + d^2} = \frac{(ac + bd) + i(bc - ad)}{c^2 + d^2}. \tag{8}$$

Our objective is to express the quotient z/w in the standard form $z/w = r + is$, where r and s are real numbers; from Eq. (8), r and s are given by

$$r = \frac{ac + bd}{c^2 + d^2} \quad \text{and} \quad s = \frac{bc - ad}{c^2 + d^2}.$$

For instance,

$$\frac{2 + 3i}{1 + 2i} = \frac{(2 + 3i)(1 - 2i)}{(1 + 2i)(1 - 2i)} = \frac{8 - i}{5} = \frac{8}{5} - \frac{1}{5}i.$$

EXAMPLE 4 Use Gaussian elimination to solve the system in (1):

$$(1 - i)x_1 + \qquad x_2 = 0$$
$$-2x_1 - (1 + i)x_2 = 0.$$

SOLUTION   The initial step in solving this system is to multiply the first equation by $2/(1 - i)$ and then add the result to the second equation. Following the discussion above, we write $2/(1 - i)$ as

$$\frac{2}{1 - i} = \frac{2(1 + i)}{(1 - i)(1 + i)} = \frac{2 + 2i}{2} = 1 + i.$$

Multiplying the first equation by $1 + i$ and adding the result to the second equation produces the equivalent system

$$(1 - i)x_1 + x_2 = 0$$
$$0 = 0,$$

which leads to $x_1 = -x_2/(1 - i)$. Simplifying, we obtain

$$x_1 = \frac{-x_2}{1 - i} = \frac{-x_2(1 + i)}{(1 - i)(1 + i)} = \frac{-(1 + i)}{2}x_2.$$

With $x_2 = -2a$, the solutions are all of the form

$$\mathbf{x} = a\begin{bmatrix} 1 + i \\ -2 \end{bmatrix}. \qquad \square$$

(*Note:* Since we are allowing the possibility of vectors with complex components, we will also allow the parameter a in Example 4 to be complex. For example, with $a = i$ we see that

$$\mathbf{x} = \begin{bmatrix} -1 + i \\ -2i \end{bmatrix}$$

is also a solution.)

EXAMPLE 5 Find the eigenvalues and the eigenvectors of A, where

$$A = \begin{bmatrix} -2 & -2 & -9 \\ -1 & 1 & -3 \\ 1 & 1 & 4 \end{bmatrix}.$$

SOLUTION   The characteristic polynomial of A is

$$p(t) = -(t - 1)(t^2 - 2t + 2).$$

Thus the eigenvalues of A are $\lambda = 1$, $\lambda = 1 + i$, $\lambda = 1 - i$.

As we noted above, the complex eigenvalues occur in conjugate pairs; and if we find an eigenvector $\mathbf{x}$ for $\lambda = 1 + i$, then we immediately see that $\bar{\mathbf{x}}$ is an eigenvector for $\bar{\lambda} = 1 - i$. In this example we find the eigenvectors for $\lambda = 1 + i$ by reducing the augmented matrix $[A - \lambda I \,|\, \boldsymbol{\theta}]$ to echelon form. Now for $\lambda = 1 + i$,

$$[A - \lambda I \,|\, \boldsymbol{\theta}] = \begin{bmatrix} -3 - i & -2 & -9 & 0 \\ -1 & -i & -3 & 0 \\ 1 & 1 & 3 - i & 0 \end{bmatrix}.$$

To introduce a zero into the $(2, 1)$ position, we use the multiple m, where

$$m = \frac{1}{-3 - i} = \frac{-1}{3 + i} = \frac{-(3 - i)}{(3 + i)(3 - i)} = \frac{-3 + i}{10}.$$

Multiplying the first row by m and adding the result to the second row, and then multiplying the first row by $-m$ and adding the result to the third row, we find that $[A - \lambda I \,|\, \theta]$ is row equivalent to

$$\begin{bmatrix} -3 - i & -2 & -9 & 0 \\ 0 & \dfrac{6 - 12i}{10} & \dfrac{-3 - 9i}{10} & 0 \\ 0 & \dfrac{4 + 2i}{10} & \dfrac{3 - i}{10} & 0 \end{bmatrix}.$$

Multiplying the second and third by 10 in the matrix above, we obtain a row-equivalent matrix:

$$\begin{bmatrix} -3 - i & -2 & -9 & 0 \\ 0 & 6 - 12i & -3 - 9i & 0 \\ 0 & 4 + 2i & 3 - i & 0 \end{bmatrix}.$$

Completing the reduction, we multiply the second row by r and add the result to the third row, where r is the multiple

$$r = \frac{-(4 + 2i)}{6 - 12i} = \frac{-(4 + 2i)(6 + 12i)}{(6 - 12i)(6 + 12i)} = \frac{-60i}{180} = \frac{-i}{3}.$$

We obtain the row-equivalent matrix

$$\begin{bmatrix} -3 - i & -2 & -9 & 0 \\ 0 & 6 - 12i & -3 - 9i & 0 \\ 0 & 0 & 0 & 0 \end{bmatrix};$$

and the eigenvectors of A corresponding to $\lambda = 1 + i$ are found by solving

$$\begin{aligned} -(3 + i)x_1 - 2x_2 &= 9x_3 \\ (6 - 12i)x_2 &= (3 + 9i)x_3, \end{aligned} \tag{9}$$

with x_3 arbitrary, $x_3 \neq 0$. We first find x_2 from

$$x_2 = \frac{3 + 9i}{6 - 12i}x_3 = \frac{(3 + 9i)(6 + 12i)}{180}x_3 = \frac{-90 + 90i}{180}x_3,$$

or

$$x_2 = \frac{-1 + i}{2}x_3.$$

From the first equation in (9), we obtain

$$-(3 + i)x_1 = 2x_2 + 9x_3 = (8 + i)x_3,$$

or

$$x_1 = \frac{-(8+i)}{3+i}x_3 = \frac{-(8+i)(3-i)}{10}x_3 = \frac{-25+5i}{10}x_3 = \frac{-5+i}{2}x_3.$$

Setting $x_3 = 2a$, we have $x_2 = (-1+i)a$ and $x_1 = (-5+i)a$; so the eigenvectors of A corresponding to $\lambda = 1+i$ are all of the form

$$\mathbf{x} = \begin{bmatrix} (-5+i)a \\ (-1+i)a \\ 2a \end{bmatrix} = a\begin{bmatrix} -5+i \\ -1+i \\ 2 \end{bmatrix}, \quad a \neq 0.$$

Furthermore, we know that eigenvectors of A corresponding to $\bar{\lambda} = 1-i$ have the form

$$\bar{\mathbf{x}} = b\begin{bmatrix} -5-i \\ -1-i \\ 2 \end{bmatrix}, \quad b \neq 0.$$

If linear algebra software is available, then finding eigenvalues and eigenvectors is a simple matter.

EXAMPLE 6 Find the eigenvalues and the eigenvectors for the (4×4) matrix

$$A = \begin{bmatrix} 3 & 3 & 6 & 9 \\ 1 & 4 & 3 & 7 \\ 2 & -5 & 8 & 3 \\ 2 & -9 & 7 & 4 \end{bmatrix}.$$

```
A =

     3     3     6     9
     1     4     3     7
     2    -5     8     3
     2    -9     7     4

>> [ V, D ] = eig ( A )

V =

    0.6897 + 0.2800i    0.6897 - 0.2800i    0.8216     0.9609
    0.4761 + 0.2051i    0.4761 - 0.2051i    0.4196    -0.0067
    0.1338 + 0.2255i    0.1338 - 0.2255i    0.3014    -0.2765
   -0.1139 + 0.3090i   -0.1139 - 0.3090i   -0.2409    -0.0160

D =

    6.9014 + 5.3028i         0                 0          0
         0             6.9014 - 5.3028i         0          0
         0                  0            4.0945          0
         0                  0                 0      1.1027
```

Figure 3.3
MATLAB was used to find the eigenvalues and eigenvectors of matrix A in Example 6—that is, $AV = VD$ or $V^{-1}AV = D$, where D is diagonal.

SOLUTION We used MATLAB for this problem. The command $[V, D] = \text{eig(A)}$ produces a diagonal matrix D and a matrix of eigenvectors V. That is, $AV = DV$ or (if A is not defective) $V^{-1}AV = D$. The results from MATLAB are shown in Fig. 3.3. As can be seen from the matrix D in Fig. 3.3, A has two complex eigenvalues, which are (to the places shown) $\lambda = 6.9014 + 5.3028i$ and $\lambda = 6.9014 - 5.3028i$. In addition, A has two real eigenvalues $\lambda = 4.0945$ and $\lambda = 1.1027$. Eigenvectors are found in the corresponding columns of V.

As the examples above indicate, finding eigenvectors that correspond to a complex eigenvalue proceeds exactly as for a real eigenvalue except for the additional details required by complex arithmetic.

Although complex eigenvalues and eigenvectors may seem an undue complication, they are in fact fairly important to applications. For instance, we note (without trying to be precise) that oscillatory and periodic solutions to first-order systems of differential equations correspond to complex eigenvalues; and since many physical systems exhibit such behavior, we need some way to model them.

Exercises 3.6

In Exercises 1–18, $s = 1 + 2i$, $u = 3 - 2i$, $v = 4 + i$, $w = 2 - i$, and $z = 1 + i$. In each exercise perform the indicated calculation and express the result in the form $a + ib$.

1. $\bar{u}$

2. $\bar{z}$

3. $u + \bar{v}$

4. $\bar{z} + w$

5. $u + \bar{u}$

6. $s - \bar{s}$

7. $v\bar{v}$

8. $u\bar{v}$

9. $s^2 - w$

10. $z^2 w$

11. $\bar{u}w^2$

12. $s(u^2 + v)$

13. u/v

14. v/u^2

15. s/z

16. $(w + \bar{v})/u$

17. $w + iz$

18. $s - iw$

Find the eigenvalues and the eigenvectors for the matrices in Exercises 19–24. (For the matrix in Exercise 24, one eigenvalue is $\lambda = 1 + 5i$.)

19. $\begin{bmatrix} 6 & 8 \\ -1 & 2 \end{bmatrix}$

20. $\begin{bmatrix} 2 & 4 \\ -2 & -2 \end{bmatrix}$

21. $\begin{bmatrix} -2 & -1 \\ 5 & 2 \end{bmatrix}$

22. $\begin{bmatrix} 5 & -5 & -5 \\ -1 & 4 & 2 \\ 3 & -5 & -3 \end{bmatrix}$

23. $\begin{bmatrix} 1 & -4 & -1 \\ 3 & 2 & 3 \\ 1 & 1 & 3 \end{bmatrix}$

24. $\begin{bmatrix} 1 & -5 & 0 & 0 \\ 5 & 1 & 0 & 0 \\ 0 & 0 & 1 & -2 \\ 0 & 0 & 2 & 1 \end{bmatrix}$

In Exercises 25 and 26, solve the linear system.

25. $(1 + i)x + iy = 5 + 4i$
$(1 - i)x - 4y = -11 + 5i$

26. $(1 - i)x - (3 + i)y = -5 - i$
$(2 + i)x + (1 + 2i)y = 1 + 6i$

In Exercises 27–30, calculate $\|\mathbf{x}\|$.

27. $\mathbf{x} = \begin{bmatrix} 1 + i \\ 2 \end{bmatrix}$

28. $\mathbf{x} = \begin{bmatrix} 3 + i \\ 2 - i \end{bmatrix}$

29. $\mathbf{x} = \begin{bmatrix} 1 - 2i \\ i \\ 3 + i \end{bmatrix}$

30. $\mathbf{x} = \begin{bmatrix} 2i \\ 1 - i \\ 3 \end{bmatrix}$

In Exercises 31–34, use linear algebra software to find the eigenvalues and the eigenvectors.

31. $\begin{bmatrix} 2 & 2 & 5 \\ 5 & 3 & 7 \\ 1 & 5 & 3 \end{bmatrix}$

32. $\begin{bmatrix} 1 & 2 & 8 \\ 8 & 4 & 9 \\ 2 & 6 & 1 \end{bmatrix}$

33. $\begin{bmatrix} 5 & -1 & 0 & 8 \\ 3 & 6 & 8 & -3 \\ 1 & 1 & 4 & 2 \\ 9 & 7 & 6 & 9 \end{bmatrix}$

34. $\begin{bmatrix} 5 & 5 & 4 & 6 \\ 0 & 8 & 6 & 7 \\ 1 & 2 & 3 & 1 \\ 6 & 3 & 8 & 5 \end{bmatrix}$

35. Establish the five properties of the conjugate operation listed in (2).

36. Let A be an $(m \times n)$ matrix and let B be an $(n \times p)$ matrix, where the entries of A and B may be complex. Use Exercise 35 and the definition of AB to show that $\overline{AB} = \bar{A}\bar{B}$. (By $\bar{A}$, we mean the matrix whose ijth entry is the conjugate of the ijth entry of A.) If A is a real matrix and $\mathbf{x}$ is an $(n \times 1)$ vector, show that $\overline{A\mathbf{x}} = A\bar{\mathbf{x}}$.

37. Let A be an $(m \times n)$ matrix, where the entries of A may be complex. It is customary to use the symbol A^* to denote the matrix

$$A^* = (\bar{A})^T.$$

Suppose that A is an $(m \times n)$ matrix and B is an $(n \times p)$ matrix. Use Exercise 36 and the properties of the transpose operation to give a quick proof that $(AB)^* = B^*A^*$.

38. An $(n \times n)$ matrix A is called **Hermitian** if $A^* = A$.
   a) Prove that a Hermitian matrix A has only real eigenvalues. (*Hint:* Observing that $\bar{\mathbf{x}}^T\mathbf{x} = \mathbf{x}^*\mathbf{x}$, modify the proof of Theorem 17.)
   b) Let $A = (a_{ij})$ be an $(n \times n)$ Hermitian matrix. Show that a_{ii} is real for $1 \le i \le n$.

39. Let $p(t) = a_0 + a_1 t + \cdots + a_n t^n$, where the coefficients $a_0, a_1, \ldots, a_n$ are all real.
   a) Prove that if r is a complex root of $p(t) = 0$, then $\bar{r}$ is also a root of $p(t) = 0$.
   b) If $p(t)$ has degree 3, argue that $p(t)$ must have at least one real root.
   c) If A is a (3×3) real matrix, argue that A must have at least one real eigenvalue.

40. An $(n \times n)$ real matrix A is called **orthogonal** if $A^T A = I$. Let λ be an eigenvalue of an orthogonal matrix A, where $\lambda = r + is$. Prove that $\lambda\bar{\lambda} = r^2 + s^2 = 1$. (*Hint:* First show that $\|A\mathbf{x}\| = \|\mathbf{x}\|$ for any vector $\mathbf{x}$.)

41. A real symmetric $(n \times n)$ matrix A is called **positive definite** if $\mathbf{x}^T A\mathbf{x} > 0$ for all $\mathbf{x}$ in R^n, $\mathbf{x} \ne \boldsymbol{\theta}$. Prove that the eigenvalues of a real symmetric positive-definite matrix A are all positive.

42. An $(n \times n)$ matrix A is called **unitary** if $A^*A = I$. (If A is a real unitary matrix, then A is orthogonal; see Exercise 40.) Show that if A is unitary and λ is an eigenvalue for A, then $|\lambda| = 1$.

3.7   **Similarity Transformations and Diagonalization**

In Chapter 1, we saw that two linear systems of equations have the same solution if their augmented matrices are row equivalent. In this chapter we are interested in identifying classes of matrices that have the same eigenvalues.

As we know, the eigenvalues of an $(n \times n)$ matrix A are the zeros of its characteristic polynomial,

$$p(t) = \det(A - tI).$$

Thus if an $(n \times n)$ matrix B has the same characteristic polynomial as A, then A and B have the same eigenvalues. As we will see, it is fairly simple to find such matrices B.

Similarity

In particular, let A be an $(n \times n)$ matrix and let S be a nonsingular $(n \times n)$ matrix. Then, as the following calculation shows, the matrices A and $B = S^{-1}AS$ have the same characteristic polynomial. To establish this fact, observe that the characteristic polynomial for $S^{-1}AS$ is given by

$$
\begin{aligned}
p(t) &= \det(S^{-1}AS - tI) \\
&= \det(S^{-1}AS - tS^{-1}S) \\
&= \det[S^{-1}(A - tI)S] \\
&= \det(S^{-1})\det(A - tI)\det(S), \text{ by Theorem 2} \\
&= [\det(S^{-1})\det(S)]\det(A - tI) \\
&= \det(A - tI).
\end{aligned}
\tag{1}
$$

(The last equality above follows because $\det(S^{-1})\det(S) = \det(S^{-1}S) = \det(I) = 1$.)

Thus, by (1), the matrices $S^{-1}AS$ and A have the same characteristic polynomial and hence the same set of eigenvalues. The discussion above leads to the next definition.

DEFINITION 8   The $(n \times n)$ matrices A and B are said to be *similar* if there is a nonsingular $(n \times n)$ matrix S such that $B = S^{-1}AS$.

The calculations carried out in (1) show that similar matrices have the same characteristic polynomial. Consequently the following theorem is immediate.

THEOREM 18

If A and B are similar $(n \times n)$ matrices, then A and B have the same eigenvalues. Moreover, these eigenvalues have the same algebraic multiplicity.

Although similar matrices always have the same characteristic polynomial, it is not true that two matrices with the same characteristic polynomial are necessarily similar. As a simple example, consider the two matrices

$$
A = \begin{bmatrix} 1 & 0 \\ 1 & 1 \end{bmatrix} \quad \text{and} \quad I = \begin{bmatrix} 1 & 0 \\ 0 & 1 \end{bmatrix}.
$$

Now $p(t) = (1 - t)^2$ is the characteristic polynomial for both A and I; so A and I have the same set of eigenvalues. If A and I were similar, however, there would be a (2×2) matrix S such that

$$
I = S^{-1}AS.
$$

But the equation $I = S^{-1}AS$ is equivalent to $S = AS$, which is in turn equivalent to $SS^{-1} = A$ or $I = A$. Thus I and A cannot be similar. (A repetition of this argument shows that the only matrix similar to the identity matrix is I itself.) In this respect, similarity is a more fundamental concept for the eigenvalue problem than is the characteristic polynomial; two matrices can have exactly the same characteristic polynomial without being similar; so similarity leads to a more finely detailed way of distinguishing matrices.

Although similar matrices have the same eigenvalues, they do not generally have the same eigenvectors. For example, if $B = S^{-1}AS$ and if $B\mathbf{x} = \lambda\mathbf{x}$, then

$$S^{-1}AS\mathbf{x} = \lambda\mathbf{x} \quad \text{or} \quad A(S\mathbf{x}) = \lambda(S\mathbf{x}).$$

Thus if $\mathbf{x}$ is an eigenvector for B corresponding to λ, then $S\mathbf{x}$ is an eigenvector for A corresponding to λ.

Diagonalization

Computations involving an $(n \times n)$ matrix A can often be simplified if we know that A is similar to a diagonal matrix. To illustrate, suppose $S^{-1}AS = D$, where D is a diagonal matrix. Next, suppose we need to calculate the power A^k, where k is a positive integer. Knowing that $D = S^{-1}AS$, we can proceed as follows:

$$\begin{aligned} D^k &= (S^{-1}AS)^k \\ &= S^{-1}A^kS. \end{aligned} \qquad (2)$$

(The fact that $(S^{-1}AS)^k = S^{-1}A^kS$ is established in Exercise 25.) Note that because D is a diagonal matrix, it is easy to form the power D^k.

Once the matrix D^k is computed, the matrix A^k can be recovered from Eq. (2) by forming SD^kS^{-1}:

$$SD^kS^{-1} = S(S^{-1}A^kS)S^{-1} = A^k.$$

Whenever an $(n \times n)$ matrix A is similar to a diagonal matrix, we say that A is *diagonalizable*. The next theorem gives a characterization of diagonalizable matrices.

THEOREM 19

An $(n \times n)$ matrix A is diagonalizable if and only if A possesses a set of n linearly independent eigenvectors.

PROOF   Suppose that $\{\mathbf{u}_1, \mathbf{u}_2, \ldots, \mathbf{u}_n\}$ is a set of n linearly independent eigenvectors for A:

$$A\mathbf{u}_k = \lambda_k\mathbf{u}_k, \qquad k = 1, 2, \ldots, n.$$

Let S be the $(n \times n)$ matrix whose column vectors are the eigenvectors of A:

$$S = [\mathbf{u}_1, \mathbf{u}_2, \ldots, \mathbf{u}_n].$$

Now S is a nonsingular matrix; so S^{-1} exists where

$$S^{-1}S = [S^{-1}\mathbf{u}_1, S^{-1}\mathbf{u}_2, \ldots, S^{-1}\mathbf{u}_n] = [\mathbf{e}_1, \mathbf{e}_2, \ldots, \mathbf{e}_n] = I. \qquad (3)$$

Furthermore, since $A\mathbf{u}_k = \lambda_k\mathbf{u}_k$, we obtain

$$AS = [A\mathbf{u}_1, A\mathbf{u}_2, \ldots, A\mathbf{u}_n] = [\lambda_1\mathbf{u}_1, \lambda_2\mathbf{u}_2, \ldots, \lambda_n\mathbf{u}_n];$$

and so from Eq. (3),

$$S^{-1}AS = [\lambda_1 S^{-1}\mathbf{u}_1, \lambda_2 S^{-1}\mathbf{u}_2, \ldots, \lambda_n S^{-1}\mathbf{u}_n] = [\lambda_1\mathbf{e}_1, \lambda_2\mathbf{e}_2, \ldots, \lambda_n\mathbf{e}_n].$$

Therefore, $S^{-1}AS$ has the form

$$S^{-1}AS = \begin{bmatrix} \lambda_1 & 0 & 0 & \cdots & 0 \\ 0 & \lambda_2 & 0 & \cdots & 0 \\ 0 & 0 & \lambda_3 & \cdots & 0 \\ \vdots & & & & \vdots \\ 0 & 0 & 0 & \cdots & \lambda_n \end{bmatrix} = D;$$

and we have shown that if A has n linearly independent eigenvectors, then A is similar to a diagonal matrix.

Now suppose that $C^{-1}AC = D$, where C is nonsingular and D is a diagonal matrix. Let us write C and D in column form as

$$C = [\mathbf{C}_1, \mathbf{C}_2, \ldots, \mathbf{C}_n] \quad \text{and} \quad D = [d_1\mathbf{e}_1, d_2\mathbf{e}_2, \ldots, d_n\mathbf{e}_n].$$

From $C^{-1}AC = D$, we obtain $AC = CD$, and we write both of these in column form as

$$AC = [A\mathbf{C}_1, A\mathbf{C}_2, \ldots, A\mathbf{C}_n]$$
$$CD = [d_1 C\mathbf{e}_1, d_2 C\mathbf{e}_2, \ldots, d_n C\mathbf{e}_n].$$

But since $C\mathbf{e}_k = \mathbf{C}_k$ for $k = 1, 2, \ldots, n$, we see that $AC = CD$ implies

$$A\mathbf{C}_k = d_k\mathbf{C}_k, \qquad k = 1, 2, \ldots, n.$$

Since C is nonsingular, the vectors $\mathbf{C}_1, \mathbf{C}_2, \ldots, \mathbf{C}_n$ are linearly independent (and in particular, no $\mathbf{C}_k$ is the zero vector). Thus the diagonal entries of D are the eigenvalues of A, and the column vectors of C are a set of n linearly independent eigenvectors. ☐

Note that the proof of Theorem 19 gives a procedure for diagonalizing an $(n \times n)$ matrix A. That is, if A has n linearly independent eigenvectors $\mathbf{u}_1, \mathbf{u}_2, \ldots, \mathbf{u}_n$, then the matrix $S = [\mathbf{u}_1, \mathbf{u}_2, \ldots, \mathbf{u}_n]$ will diagonalize A.

EXAMPLE 1 Show that A is diagonalizable by finding a matrix S such that $S^{-1}AS = D$:

$$A = \begin{bmatrix} 5 & -6 \\ 3 & -4 \end{bmatrix}.$$

SOLUTION It is easy to verify that A has eigenvalues $\lambda_1 = 2$ and $\lambda_2 = -1$ with corresponding eigenvectors

$$\mathbf{u}_1 = \begin{bmatrix} 2 \\ 1 \end{bmatrix} \quad \text{and} \quad \mathbf{u}_2 = \begin{bmatrix} 1 \\ 1 \end{bmatrix}.$$

Forming $S = [\mathbf{u}_1, \mathbf{u}_2]$, we obtain

$$S = \begin{bmatrix} 2 & 1 \\ 1 & 1 \end{bmatrix}, \qquad S^{-1} = \begin{bmatrix} 1 & -1 \\ -1 & 2 \end{bmatrix}.$$

As a check on the calculations, we form $S^{-1}AS$. The matrix AS is given by

$$AS = \begin{bmatrix} 5 & -6 \\ 3 & -4 \end{bmatrix}\begin{bmatrix} 2 & 1 \\ 1 & 1 \end{bmatrix} = \begin{bmatrix} 4 & -1 \\ 2 & -1 \end{bmatrix}.$$

Next, forming $S^{-1}(AS)$, we obtain

$$S^{-1}(AS) = \begin{bmatrix} 1 & -1 \\ -1 & 2 \end{bmatrix}\begin{bmatrix} 4 & -1 \\ 2 & -1 \end{bmatrix} = \begin{bmatrix} 2 & 0 \\ 0 & -1 \end{bmatrix} = D. \qquad \square$$

EXAMPLE 2 Use the result of Example 1 to calculate A^{10}, where

$$A = \begin{bmatrix} 5 & -6 \\ 3 & -4 \end{bmatrix}.$$

SOLUTION   As was noted in Eq. (2), $D^{10} = S^{-1}A^{10}S$. Therefore, $A^{10} = SD^{10}S^{-1}$. Now by Example 1,

$$D^{10} = \begin{bmatrix} 2^{10} & 0 \\ 0 & (-1)^{10} \end{bmatrix} = \begin{bmatrix} 1024 & 0 \\ 0 & 1 \end{bmatrix}.$$

Hence $A^{10} = SD^{10}S^{-1}$ is given by

$$A^{10} = \begin{bmatrix} 2047 & -2046 \\ 1023 & -1022 \end{bmatrix}. \qquad \square$$

Sometimes complex arithmetic is necessary to diagonalize a real matrix.

EXAMPLE 3 Show that A is diagonalizable by finding a matrix S such that $S^{-1}AS = D$:

$$A = \begin{bmatrix} 1 & 1 \\ -1 & 1 \end{bmatrix}.$$

SOLUTION A has eigenvalues $\lambda_1 = 1 + i$ and $\lambda_2 = 1 - i$, with corresponding eigenvectors

$$\mathbf{u}_1 = \begin{bmatrix} 1 \\ i \end{bmatrix} \quad \text{and} \quad \mathbf{u}_2 = \begin{bmatrix} 1 \\ -i \end{bmatrix}.$$

Forming the matrix $S = [\mathbf{u}_1, \mathbf{u}_2]$, we obtain

$$S = \begin{bmatrix} 1 & 1 \\ i & -i \end{bmatrix}, \qquad S^{-1} = \begin{bmatrix} \dfrac{1}{2} & -\dfrac{i}{2} \\ \dfrac{1}{2} & \dfrac{i}{2} \end{bmatrix}.$$

As a check, note that AS is given by

$$AS = \begin{bmatrix} 1 & 1 \\ -1 & 1 \end{bmatrix}\begin{bmatrix} 1 & 1 \\ i & -i \end{bmatrix} = \begin{bmatrix} 1 + i & 1 - i \\ -1 + i & -1 - i \end{bmatrix}.$$

Next, $S^{-1}(AS)$ is the matrix

$$S^{-1}(AS) = \frac{1}{2}\begin{bmatrix} 1 & -i \\ 1 & i \end{bmatrix}\begin{bmatrix} 1+i & 1-i \\ -1+i & -1-i \end{bmatrix} = \begin{bmatrix} 1+i & 0 \\ 0 & 1-i \end{bmatrix} = D. \quad \square$$

Some types of matrices are known to be diagonalizable. The next theorem lists one such condition. Then, in the last subsection, we prove the important theorem: If A is a real symmetric matrix, then A is diagonalizable.

THEOREM 20

Let A be an $(n \times n)$ matrix with n distinct eigenvalues. Then A is diagonalizable.

PROOF By Theorem 15, if A has n distinct eigenvalues, then A has a set of n linearly independent eigenvectors. Thus by Theorem 19, A is diagonalizable.

$\square$

As the next example shows, a matrix A may be diagonalizable even though it has repeated eigenvalues.

EXAMPLE 4 Show that A is diagonalizable, where

$$A = \begin{bmatrix} 25 & -8 & 30 \\ 24 & -7 & 30 \\ -12 & 4 & -14 \end{bmatrix}.$$

SOLUTION The eigenvalues of A are $\lambda_1 = \lambda_2 = 1$ and $\lambda_3 = 2$. The eigenspace corresponding to $\lambda_1 = \lambda_2 = 1$ has dimension 2, with a basis $\{\mathbf{u}_1, \mathbf{u}_2\}$, where

$$\mathbf{u}_1 = \begin{bmatrix} 1 \\ 3 \\ 0 \end{bmatrix} \quad \text{and} \quad \mathbf{u}_2 = \begin{bmatrix} -4 \\ 3 \\ 4 \end{bmatrix}.$$

An eigenvector corresponding to $\lambda_3 = 2$ is

$$\mathbf{u}_3 = \begin{bmatrix} 4 \\ 4 \\ -2 \end{bmatrix}.$$

Defining S by $S = [\mathbf{u}_1, \mathbf{u}_2, \mathbf{u}_3]$, we can verify that

$$S^{-1}AS = D = \begin{bmatrix} 1 & 0 & 0 \\ 0 & 1 & 0 \\ 0 & 0 & 2 \end{bmatrix}. \quad \square$$

Orthogonal Matrices

A remarkable and useful fact about symmetric matrices is that they are always diagonalizable. Moreover, the diagonalization of a symmetric matrix A can be accomplished with a special type of matrix known as an orthogonal matrix.

DEFINITION 9   A real $(n \times n)$ matrix Q is called an *orthogonal* matrix if Q is invertible and $Q^{-1} = Q^T$.

Definition 9 can be rephrased as follows: A real square matrix Q is orthogonal if and only if

$$Q^TQ = I. \tag{4}$$

Another useful description of orthogonal matrices can be obtained from Eq. (4). In particular, suppose that $Q = [\mathbf{q}_1, \mathbf{q}_2, \ldots, \mathbf{q}_n]$ is an $(n \times n)$ matrix. Since the ith row of Q^T is equal to $\mathbf{q}_i^T$, the definition of matrix multiplication tells us:

The ijth entry of Q^TQ is equal to $\mathbf{q}_i^T\mathbf{q}_j$.

Therefore, by Eq. (4), an $(n \times n)$ matrix $Q = [\mathbf{q}_1, \mathbf{q}_2, \ldots, \mathbf{q}_n]$ is orthogonal if and only if:

The columns of Q, $\{\mathbf{q}_1, \mathbf{q}_2, \ldots, \mathbf{q}_n\}$,
form an orthonormal set of vectors. $\tag{5}$

EXAMPLE 5   Verify that the matrices, Q_1 and Q_2 are orthogonal:

$$Q_1 = \frac{1}{\sqrt{2}} \begin{bmatrix} 1 & 0 & 1 \\ 0 & \sqrt{2} & 0 \\ -1 & 0 & 1 \end{bmatrix} \quad \text{and} \quad Q_2 = \begin{bmatrix} 0 & 0 & 1 \\ 1 & 0 & 0 \\ 0 & 1 & 0 \end{bmatrix}.$$

SOLUTION   We use Eq. (4) to show that Q_1 is orthogonal. Specifically,

$$Q_1^TQ_1 = \frac{1}{2} \begin{bmatrix} 1 & 0 & -1 \\ 0 & \sqrt{2} & 0 \\ 1 & 0 & 1 \end{bmatrix} \begin{bmatrix} 1 & 0 & 1 \\ 0 & \sqrt{2} & 0 \\ -1 & 0 & 1 \end{bmatrix} = \frac{1}{2} \begin{bmatrix} 2 & 0 & 0 \\ 0 & 2 & 0 \\ 0 & 0 & 2 \end{bmatrix} = I.$$

We use condition (5) to show that Q_2 is orthogonal. The column vectors of Q_2 are, in the order they appear, $\{\mathbf{e}_2, \mathbf{e}_3, \mathbf{e}_1\}$. Since these vectors are orthonormal, it follows from (5) that Q_2 is orthogonal.

From the characterization of orthogonal matrices given in (5), the following observation can be made: If $Q = [\mathbf{q}_1, \mathbf{q}_2, \ldots, \mathbf{q}_n]$ is an $(n \times n)$ orthogonal matrix and if $P = [\mathbf{p}_1, \mathbf{p}_2, \ldots, \mathbf{p}_n]$ is formed by rearranging the columns of Q, then P is also an orthogonal matrix.

As a special case of this observation, suppose that P is a matrix formed by rearranging the columns of the identity matrix, I. Then, since I is an orthogonal matrix, it follows that P is orthogonal as well. Such a matrix P, formed by rearranging the columns of I, is called a ***permutation*** matrix. The matrix Q_2 in Example 5 is a specific instance of a (3×3) permutation matrix.

Orthogonal matrices have some special properties that make them valuable tools for applications. These properties were mentioned in Section 2.7 with regard to (2×2) orthogonal matrices. Suppose we think of an $(n \times n)$ matrix Q as defining a function (or linear transformation) from R^n to R^n. That is, for $\mathbf{x}$ in R^n, consider the function defined by

$$\mathbf{y} = Q\mathbf{x}.$$

As the next theorem shows, if Q is orthogonal, then the function $\mathbf{y} = Q\mathbf{x}$ preserves the lengths of vectors and the angles between pairs of vectors.

THEOREM 21

Let Q be an $(n \times n)$ orthogonal matrix.

a) If $\mathbf{x}$ is in R^n, then $\|Q\mathbf{x}\| = \|\mathbf{x}\|$.
b) If $\mathbf{x}$ and $\mathbf{y}$ are in R^n, then $(Q\mathbf{x})^T(Q\mathbf{y}) = \mathbf{x}^T\mathbf{y}$.
c) $\text{Det}(Q) = \pm 1$.

PROOF We will prove property (a) and leave properties (b) and (c) to the exercises. Let $\mathbf{x}$ be a vector in R^n. Then

$$\|Q\mathbf{x}\| = \sqrt{(Q\mathbf{x})^T(Q\mathbf{x})} = \sqrt{\mathbf{x}^T Q^T Q\mathbf{x}} = \sqrt{\mathbf{x}^T I \mathbf{x}} = \sqrt{\mathbf{x}^T \mathbf{x}} = \|\mathbf{x}\|.$$

The fact that $\mathbf{x}^T(Q^TQ)\mathbf{x} = \mathbf{x}^T I \mathbf{x}$ comes from Eq. (4). ☐

Theorem 21 can be illustrated geometrically (see Figs. 3.4 and 3.5). In Fig. 3.4(a), a vector $\mathbf{x}$ in R^2 is shown, where $\|\mathbf{x}\| = 1$. The vector $Q\mathbf{x}$ is shown

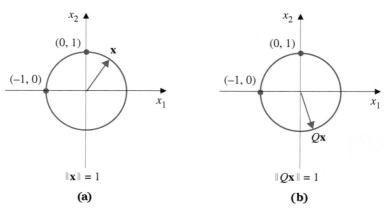

Figure 3.4
The length of $\mathbf{x}$ is equal to the length of $Q\mathbf{x}$.

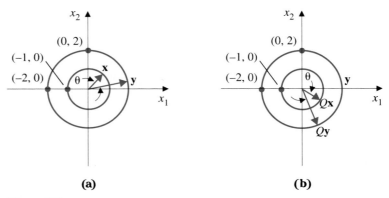

Figure 3.5
The angle between **x** and **y** is equal to the angle between Q**x**
and Q**y**.

in Fig. 3.4(b), where, by Theorem 21, Q**x** also has length 1. In Fig. 3.5(a),
vectors **x** and **y** are shown, where $\|\mathbf{x}\| = 1$ and $\|\mathbf{y}\| = 2$. From vector geometry,
we know that the angle θ between **x** and **y** satisfies the condition

$$\mathbf{x}^T\mathbf{y} = \|\mathbf{x}\|\|\mathbf{y}\|\cos\theta, \qquad 0 \le \theta \le \pi. \tag{6}$$

In Fig. 3.5(b), the vectors Q**x** and Q**y** are shown, where the angle between Q**x**
and Q**y** is also equal to θ. To establish that the angle between **x** and **y** is equal
to the angle between Q**x** and Q**y**, we can argue as follows: Let γ denote the
angle between Q**x** and Q**y**, where $0 \le \gamma \le \pi$. As in (6), the angle γ satisfies
the condition

$$(Q\mathbf{x})^T(Q\mathbf{y}) = \|Q\mathbf{x}\|\|Q\mathbf{y}\|\cos\gamma, \qquad 0 \le \gamma \le \pi. \tag{7}$$

By Theorem 21, $(Q\mathbf{x})^T(Q\mathbf{y}) = \mathbf{x}^T\mathbf{y}$ and $\|Q\mathbf{x}\|\|Q\mathbf{y}\| = \|\mathbf{x}\|\|\mathbf{y}\|$. Thus, from (6) and
(7), $\cos\theta = \cos\gamma$. Since the cosine function is one-to-one on $[0, \pi]$, the condition
$\cos\theta = \cos\gamma$ implies that $\theta = \gamma$.

Diagonalization of Symmetric Matrices

We conclude this section by showing that every symmetric matrix can be diagonalized by an orthogonal matrix. Several approaches can be used to establish
this diagonalization result. We choose to demonstrate it by first stating a special
case of a theorem known as Schur's theorem.

THEOREM 22

Let A be an $(n \times n)$ matrix, where A has only real eigenvalues. Then there is an
$(n \times n)$ orthogonal matrix Q such that

$$Q^TAQ = T,$$

where T is an $(n \times n)$ upper-triangular matrix.

We leave the proof of Theorem 22 as a series of somewhat challenging exercises. It is important to observe that the triangular matrix T in Theorem 22 is similar to A. That is, since $Q^{-1} = Q^T$, it follows that Q^TAQ is a similarity transformation.

Schur's theorem (of which Theorem 22 is a special case) states that any $(n \times n)$ matrix A is "unitarily" similar to a triangular matrix T. The definition of a unitary matrix is given in the exercises of the previous section.

Linear algebra software can be used to find matrices Q and T that satisfy the conclusions of Schur's theorem: $Q^TAQ = T$. Note that we can rewrite $Q^TAQ = T$ as

$$A = QTQ^T.$$

The decomposition $A = QTQ^T$ is called a **Schur decomposition** or a **Schur factorization** of A.

EXAMPLE 6   The (3×3) matrix A has real eigenvalues:

$$A = \begin{bmatrix} 2 & 4 & 3 \\ 7 & 5 & 9 \\ 1 & 3 & 1 \end{bmatrix}.$$

Find an orthogonal matrix Q and an upper-triangular matrix T such that $Q^TAQ = T$.

SOLUTION   We used MATLAB in this example. The MATLAB command $[Q, T] = \text{schur}(A)$ yields appropriate matrices Q and T (see Fig. 3.6). Since A

```
A =
          2        4        3
          7        5        9
          1        3        1

>> [ Q , T ] = schur ( A )

Q =
     - 0.4421    0.7193   - 0.5359
     - 0.8514  - 0.1486     0.5030
     - 0.2822  - 0.6786   - 0.6781


T =
     11.6179     2.1869     6.6488
           0   - 0.3125     0.1033
           0          0   - 3.3055
```

Figure 3.6
MATLAB was used in Example 6 to find matrices Q and T such that $Q^TAQ = T$.

and T are similar, the eigenvalues of A are the diagonal entries of T. Thus, to the places shown in Fig. 3.6, the eigenvalues of A are $\lambda = 11.6179$, $\lambda = -0.3125$, and $\lambda = -3.3055$.

With Theorem 22, it is a simple matter to show that any real symmetric matrix can be diagonalized by an orthogonal matrix. In fact, as the next theorem states, a matrix is orthogonally diagonalizable if and only if the matrix is symmetric. We will use this result in Section 6.1 when we discuss diagonalizing quadratic forms.

THEOREM 23

Let A be a real $(n \times n)$ matrix.

a) If A is symmetric, then there is an orthogonal matrix Q such that $Q^TAQ = D$, where D is diagonal.
b) If $Q^TAQ = D$, where Q is orthogonal and D is diagonal, then A is a symmetric matrix.

PROOF To prove property (a), suppose A is symmetric. Recall, by Theorem 17, that A has only real eigenvalues. Thus, by Theorem 22, there is an orthogonal matrix Q such that $Q^TAQ = M$, where M is an upper-triangular matrix. Using the transpose operation on the equality $M = Q^TAQ$ and also using the fact that $A^T = A$, we obtain

$$M^T = (Q^TAQ)^T = Q^TA^TQ = Q^TAQ = M.$$

Thus, since M is upper triangular and $M^T = M$, it follows that M is a diagonal matrix.

To prove property (b), suppose that $Q^TAQ = D$, where Q is orthogonal and D is diagonal. Since D is diagonal, we know that $D^T = D$. Thus, using the transpose operation on the equality $Q^TAQ = D$, we obtain

$$Q^TAQ = D = D^T = (Q^TAQ)^T = Q^TA^TQ.$$

From this result, we see that $Q^TAQ = Q^TA^TQ$. Multiplying by Q and Q^T, we obtain

$$Q(Q^TAQ)Q^T = Q(Q^TA^TQ)Q^T$$
$$(QQ^T)A(QQ^T) = (QQ^T)A^T(QQ^T)$$
$$A = A^T.$$

Thus, since $A = A^T$, matrix A is symmetric.

Theorem 23 states that every real symmetric matrix A is orthogonally diagonalizable; that is, $Q^TAQ = D$, where Q is orthogonal and D is diagonal. From the proof of Theorem 19 (also see Examples 1, 3, and 4), the eigenvalues of A

are the diagonal entries of D, and eigenvectors of A can be chosen as the columns of Q. Since the columns of Q form an orthonormal set, the following result is a corollary of Theorem 23.

COROLLARY

Let A be a real $(n \times n)$ symmetric matrix. It is possible to choose eigenvectors $\mathbf{u}_1, \mathbf{u}_2, \ldots, \mathbf{u}_n$ for A such that $\{\mathbf{u}_1, \mathbf{u}_2, \ldots, \mathbf{u}_n\}$ is an orthonormal basis for R^n.

The corollary is illustrated in the next example. Before presenting the example, we note the following fact, which is established in Exercise 43:

If $\mathbf{u}$ and $\mathbf{v}$ are eigenvectors of a symmetric matrix and if $\mathbf{u}$ and $\mathbf{v}$ belong to different eigenspaces, then $\mathbf{u}^T\mathbf{v} = 0$. **(8)**

Note that if A is not symmetric, then eigenvectors corresponding to different eigenvalues are not generally orthogonal.

EXAMPLE 7 Find an orthonormal basis for R^4 consisting of eigenvectors of the matrix

$$A = \begin{bmatrix} 1 & -1 & -1 & -1 \\ -1 & 1 & -1 & -1 \\ -1 & -1 & 1 & -1 \\ -1 & -1 & -1 & 1 \end{bmatrix}.$$

SOLUTION Matrix A is a special case of the Rodman matrix (see Exercise 42). The characteristic polynomial for A is given by

$$p(t) = \det(A - tI) = (t - 2)^3(t + 2).$$

Thus the eigenvalues of A are $\lambda_1 = \lambda_2 = \lambda_3 = 2$ and $\lambda_4 = -2$.

It is easy to verify that corresponding eigenvectors are given by

$$\mathbf{w}_1 = \begin{bmatrix} 1 \\ -1 \\ 0 \\ 0 \end{bmatrix}, \quad \mathbf{w}_2 = \begin{bmatrix} 1 \\ 0 \\ -1 \\ 0 \end{bmatrix}, \quad \mathbf{w}_3 = \begin{bmatrix} 1 \\ 0 \\ 0 \\ -1 \end{bmatrix}, \quad \text{and} \quad \mathbf{w}_4 = \begin{bmatrix} 1 \\ 1 \\ 1 \\ 1 \end{bmatrix}.$$

Note that $\mathbf{w}_1$, $\mathbf{w}_2$, and $\mathbf{w}_3$ belong to the eigenspace associated with $\lambda = 2$, whereas $\mathbf{w}_4$ is in the eigenspace associated with $\lambda = -2$. As is promised by condition (8), $\mathbf{w}_1^T\mathbf{w}_4 = \mathbf{w}_2^T\mathbf{w}_4 = \mathbf{w}_3^T\mathbf{w}_4 = 0$. Also note that the matrix S defined by $S = [\mathbf{w}_1, \mathbf{w}_2, \mathbf{w}_3, \mathbf{w}_4]$ will diagonalize A. However, S is not an orthogonal matrix.

To obtain an orthonormal basis for R^4 (and hence an orthogonal matrix Q that diagonalizes A), we first find an orthogonal basis for the eigenspace associated with $\lambda = 2$. Applying the Gram–Schmidt process to the set $\{\mathbf{w}_1, \mathbf{w}_2, \mathbf{w}_3\}$, we produce orthogonal vectors

$$\mathbf{x}_1 = \begin{bmatrix} 1 \\ -1 \\ 0 \\ 0 \end{bmatrix}, \qquad \mathbf{x}_2 = \begin{bmatrix} 1/2 \\ 1/2 \\ -1 \\ 0 \end{bmatrix}, \quad \text{and} \quad \mathbf{x}_3 = \begin{bmatrix} 1/3 \\ 1/3 \\ 1/3 \\ -1 \end{bmatrix}.$$

Thus the set $\{\mathbf{x}_1, \mathbf{x}_2, \mathbf{x}_3, \mathbf{w}_4\}$ is an orthogonal basis for R^4 consisting of eigenvectors of A. This set can then be normalized to determine an orthonormal basis for R^4 and an orthogonal matrix Q that diagonalizes A.

We conclude by mentioning a result that is useful in applications. Let A be an $(n \times n)$ symmetric matrix with eigenvalues $\lambda_1, \lambda_2, \ldots, \lambda_n$. Let $\mathbf{u}_1, \mathbf{u}_2, \ldots, \mathbf{u}_n$ be a corresponding set of orthonormal eigenvectors, where $A\mathbf{u}_i = \lambda_i\mathbf{u}_i$, $1 \le i \le n$. Matrix A can be expressed in the form

$$A = \lambda_1\mathbf{u}_1\mathbf{u}_1^T + \lambda_2\mathbf{u}_2\mathbf{u}_2^T + \cdots + \lambda_n\mathbf{u}_n\mathbf{u}_n^T. \tag{9}$$

In Eq. (9), each $(n \times n)$ matrix $\mathbf{u}_i\mathbf{u}_i^T$ is a rank-one matrix. Expression (9) is called a *spectral decomposition for A.* A proof for (9) can be constructed along the lines of Exercise 29 of Section 3.5.

Exercises 3.7

In Exercises 1–12, determine whether the given matrix A is diagonalizable. If A is diagonalizable, calculate A^5 using the method of Example 2.

1. $A = \begin{bmatrix} 2 & -1 \\ -1 & 2 \end{bmatrix}$

2. $A = \begin{bmatrix} 1 & -1 \\ -1 & 1 \end{bmatrix}$

3. $A = \begin{bmatrix} -3 & 2 \\ -2 & 1 \end{bmatrix}$

4. $A = \begin{bmatrix} 1 & 3 \\ 0 & 1 \end{bmatrix}$

5. $A = \begin{bmatrix} 1 & 0 \\ 10 & 2 \end{bmatrix}$

6. $A = \begin{bmatrix} -1 & 7 \\ 0 & 1 \end{bmatrix}$

7. $A = \begin{bmatrix} 3 & -2 & -4 \\ 8 & -7 & -16 \\ -3 & 3 & 7 \end{bmatrix}$

8. $A = \begin{bmatrix} -1 & -1 & -4 \\ -8 & -3 & -16 \\ 1 & 2 & 7 \end{bmatrix}$

9. $A = \begin{bmatrix} 3 & -1 & -1 \\ -12 & 0 & 5 \\ 4 & -2 & -1 \end{bmatrix}$

10. $A = \begin{bmatrix} 1 & 1 & -1 \\ 0 & 2 & -1 \\ 0 & 0 & 1 \end{bmatrix}$

11. $A = \begin{bmatrix} 1 & 1 & -2 \\ 0 & 2 & -1 \\ 0 & 0 & 1 \end{bmatrix}$

12. $A = \begin{bmatrix} 1 & 3 & 3 \\ 0 & 5 & 4 \\ 0 & 0 & 1 \end{bmatrix}$

In Exercises 13–18, use condition (5) to determine whether the given matrix Q is orthogonal.

13. $Q = \begin{bmatrix} 0 & 1 \\ 1 & 0 \end{bmatrix}$

14. $Q = \dfrac{1}{\sqrt{5}}\begin{bmatrix} 1 & -2 \\ 2 & 1 \end{bmatrix}$

15. $Q = \begin{bmatrix} 2 & -1 \\ 1 & 2 \end{bmatrix}$

16. $Q = \begin{bmatrix} 3 & 2 \\ -2 & 3 \end{bmatrix}$

17. $Q = \dfrac{1}{\sqrt{6}}\begin{bmatrix} \sqrt{3} & 1 & \sqrt{2} \\ 0 & -2 & \sqrt{2} \\ -\sqrt{3} & 1 & \sqrt{2} \end{bmatrix}$

18. $Q = \begin{bmatrix} 1 & 1 & -4 \\ 2 & -2 & 1 \\ 1 & 3 & 2 \end{bmatrix}$

In Exercises 19 and 20, find values α, β, a, b, and c such that matrix Q is orthogonal. Choose positive values for α and β. (*Hint:* Use (5) to determine the values.)

19. $Q = \begin{bmatrix} \alpha & \beta & a \\ 0 & 2\beta & b \\ \alpha & -\beta & c \end{bmatrix}$ **20.** $Q = \begin{bmatrix} \alpha & -\beta & a \\ \alpha & 3\beta & b \\ \alpha & -2\beta & c \end{bmatrix}$

In Exercises 21–24, use linear algebra software to find an orthogonal matrix Q and an upper-triangular matrix T such that $Q^T A Q = T$. (*Note:* In each exercise, the matrix A has only real eigenvalues.)

21. $\begin{bmatrix} 1 & 0 & 1 \\ 3 & 3 & 5 \\ 2 & 6 & 2 \end{bmatrix}$ **22.** $\begin{bmatrix} 3 & 0 & 7 \\ 9 & -6 & 4 \\ 1 & 1 & 4 \end{bmatrix}$

23. $\begin{bmatrix} 4 & 5 & 2 & 8 \\ 0 & 6 & 7 & 5 \\ 2 & 4 & 5 & 3 \\ 9 & 7 & 3 & 6 \end{bmatrix}$ **24.** $\begin{bmatrix} 4 & 7 & 3 & 5 \\ 8 & 5 & 7 & 8 \\ 2 & 4 & 3 & 5 \\ 0 & 5 & 7 & 4 \end{bmatrix}$

25. Let A be an $(n \times n)$ matrix, and let S be a nonsingular $(n \times n)$ matrix.
   a) Verify that $(S^{-1}AS)^2 = S^{-1}A^2S$ and that $(S^{-1}AS)^3 = S^{-1}A^3S$.
   b) Prove by induction that $(S^{-1}AS)^k = S^{-1}A^kS$ for any positive integer k.

26. Show that if A is diagonalizable and if B is similar to A, then B is diagonalizable. (*Hint:* Suppose that $S^{-1}AS = D$ and $W^{-1}AW = B$.)

27. Suppose that B is similar to A. Show each of the following.
   a) $B + \alpha I$ is similar to $A + \alpha I$.
   b) B^T is similar to A^T.
   c) If A is nonsingular, then B is nonsingular and, moreover, B^{-1} is similar to A^{-1}.

28. Prove properties (b) and (c) of Theorem 21. (*Hint:* For property (c), use the fact that $Q^T Q = I$.)

29. Let $\mathbf{u}$ be a vector in R^n such that $\mathbf{u}^T\mathbf{u} = 1$. Let $Q = I - 2\mathbf{u}\mathbf{u}^T$. Show that Q is an orthogonal matrix. Also, calculate the vector $Q\mathbf{u}$. Is $\mathbf{u}$ an eigenvector for Q?

30. Suppose that A and B are orthogonal $(n \times n)$ matrices. Show that AB is an orthogonal matrix.

31. Let $\mathbf{x}$ be a nonzero vector in R^2, $\mathbf{x} = [a, b]^T$. Find a vector $\mathbf{y}$ in R^2 such that $\mathbf{x}^T\mathbf{y} = 0$ and $\mathbf{y}^T\mathbf{y} = 1$.

32. Let A be a real (2×2) matrix with only real eigenvalues. Suppose that $A\mathbf{u} = \lambda\mathbf{u}$, where $\mathbf{u}^T\mathbf{u} = 1$. By Exercise 31, there is a vector $\mathbf{v}$ in R^2 such that $\mathbf{u}^T\mathbf{v} = 0$ and $\mathbf{v}^T\mathbf{v} = 1$. Let Q be the (2×2) matrix given by $Q = [\mathbf{u}, \mathbf{v}]$, and note that Q is an orthogonal matrix. Verify that

$$Q^T A Q = \begin{bmatrix} \lambda & \mathbf{u}^T A\mathbf{v} \\ 0 & \mathbf{v}^T A\mathbf{v} \end{bmatrix}.$$

(Thus Theorem 22 is proved for a (2×2) matrix A.)

In Exercises 33–36, use the procedure outlined in Exercise 32 to find an orthogonal matrix Q such that $Q^T A Q = T$, T upper triangular.

33. $A = \begin{bmatrix} 1 & -1 \\ 1 & 3 \end{bmatrix}$ **34.** $A = \begin{bmatrix} 5 & -2 \\ 6 & -2 \end{bmatrix}$

35. $A = \begin{bmatrix} 2 & -1 \\ -1 & 2 \end{bmatrix}$ **36.** $A = \begin{bmatrix} 2 & 2 \\ 3 & 3 \end{bmatrix}$

37. Let A and R be $(n \times n)$ matrices. Show that the ijth entry of $R^T A R$ is given by $\mathbf{R}_i^T A \mathbf{R}_j$, where $R = [\mathbf{R}_1, \mathbf{R}_2, \ldots, \mathbf{R}_n]$.

38. Let A be a real (3×3) matrix with only real eigenvalues. Suppose that $A\mathbf{u} = \lambda\mathbf{u}$, where $\mathbf{u}^T\mathbf{u} = 1$. By the Gram–Schmidt process, there are vectors $\mathbf{v}$ and $\mathbf{w}$ in R^3 such that $\{\mathbf{u}, \mathbf{v}, \mathbf{w}\}$ is an orthonormal set. Consider the orthogonal matrix Q given by $Q = [\mathbf{u}, \mathbf{v}, \mathbf{w}]$. Verify that

$$Q^T A Q = \begin{bmatrix} \lambda & \mathbf{u}^T A\mathbf{v} & \mathbf{u}^T A\mathbf{w} \\ 0 & \mathbf{v}^T A\mathbf{v} & \mathbf{v}^T A\mathbf{w} \\ 0 & \mathbf{w}^T A\mathbf{v} & \mathbf{w}^T A\mathbf{w} \end{bmatrix} = \begin{bmatrix} \lambda & \mathbf{u}^T A\mathbf{v} & \mathbf{u}^T A\mathbf{w} \\ 0 & & \\ 0 & & A_1 \end{bmatrix}.$$

39. Let $B = Q^T A Q$, where Q and A are as in Exercise 38. Consider the (2×2) submatrix of B given by A_1 in Exercise 38. Show that the eigenvalues of A_1 are real. (*Hint:* Calculate $\det(B - tI)$ and show that every eigenvalue of A_1 is an eigenvalue of B. Then make a statement showing that all the eigenvalues of B are real.)

40. Let $B = Q^T A Q$, where Q and A are as in Exercise 38. By Exercises 32 and 39, there is a (2×2) matrix S such that $S^T S = I$, $S^T A_1 S = T_1$, where T_1 is upper triangular. Form the (3 × 3) matrix R:

$$R = \begin{bmatrix} 1 & 0 & 0 \\ 0 & & \\ 0 & & S \end{bmatrix}$$

Verify each of the following.
a) $R^T R = I$.
b) $R^T Q^T A Q R$ is an upper-triangular matrix.
(Note that this exercise verifies Theorem 22 for a (3 × 3) matrix A.)

41. Following the outline of Exercises 38–40, use induction to prove Theorem 22.

42. Consider the ($n \times n$) symmetric matrix $A = (a_{ij})$ defined as follows:
a) $a_{ii} = 1, 1 \le i \le n$;
b) $a_{ij} = -1, i \ne j, 1 \le i, j \le n$.
(A (4 × 4) version of this matrix is given in Example 7.) Verify that the eigenvalues of A are $\lambda = 2$ (geometric multiplicity $n - 1$) and $\lambda = 2 - n$ (geometric multiplicity 1). (*Hint:* Show that the following are eigenvectors: $\mathbf{u}_i = \mathbf{e}_1 - \mathbf{e}_i$, $2 \le i \le n$ and $\mathbf{u}_1 = [1, 1, \ldots, 1]^T$.)

43. Suppose that A is a real symmetric matrix and that $A\mathbf{u} = \lambda\mathbf{u}$, $A\mathbf{v} = \beta\mathbf{v}$, where $\lambda \ne \beta$, $\mathbf{u} \ne \mathbf{0}$, and $\mathbf{v} \ne \mathbf{0}$. Show that $\mathbf{u}^T\mathbf{v} = 0$. (*Hint:* Consider $\mathbf{u}^T A\mathbf{v}$.)

3.8 Difference Equations; Markov Chains; Systems of Differential Equations (Optional)

In this section we examine how eigenvalues can be used to solve difference equations and systems of differential equations. In Chapter 6, we treat other applications of eigenvalues and also return to a deeper study of systems of differential equations.

Let A be an ($n \times n$) matrix, and let $\mathbf{x}_0$ be a vector in R^n. Consider the sequence of vectors $\{\mathbf{x}_k\}$ defined by

$$\mathbf{x}_1 = A\mathbf{x}_0$$
$$\mathbf{x}_2 = A\mathbf{x}_1$$
$$\mathbf{x}_3 = A\mathbf{x}_2$$
$$\vdots$$

In general, this sequence is given by

$$\mathbf{x}_k = A\mathbf{x}_{k-1}, \qquad k = 1, 2, \ldots. \tag{1}$$

Vector sequences that are generated as in Eq. (1) occur in a variety of applications and serve as mathematical models to describe population growth, ecological systems, radar tracking of airborne objects, digital control of chemical processes, and the like. One of the objectives in such models is to describe the behavior of the sequence $\{\mathbf{x}_k\}$ in qualitative or quantitative terms. In this section we see that the behavior of the sequence $\{\mathbf{x}_k\}$ can be analyzed from the eigenvalues of A.

The following simple example illustrates a typical sequence of the form (1).

EXAMPLE 1 Let $\mathbf{x}_k = A\mathbf{x}_{k-1}$, $k = 1, 2, \ldots$. Calculate $\mathbf{x}_1, \mathbf{x}_2, \mathbf{x}_3, \mathbf{x}_4$, and $\mathbf{x}_5$, where

$$A = \begin{bmatrix} .8 & .2 \\ .2 & .8 \end{bmatrix} \quad \text{and} \quad \mathbf{x}_0 = \begin{bmatrix} 1 \\ 2 \end{bmatrix}.$$

SOLUTION Some routine but tedious calculations show that

$$\mathbf{x}_1 = A\mathbf{x}_0 = \begin{bmatrix} 1.2 \\ 1.8 \end{bmatrix}, \qquad \mathbf{x}_2 = A\mathbf{x}_1 = \begin{bmatrix} 1.32 \\ 1.68 \end{bmatrix}, \qquad \mathbf{x}_3 = A\mathbf{x}_2 = \begin{bmatrix} 1.392 \\ 1.608 \end{bmatrix}$$

$$\mathbf{x}_4 = A\mathbf{x}_3 = \begin{bmatrix} 1.4352 \\ 1.5648 \end{bmatrix}, \quad \text{and} \quad \mathbf{x}_5 = A\mathbf{x}_4 = \begin{bmatrix} 1.46112 \\ 1.53888 \end{bmatrix}. \qquad \square$$

In Example 1, the first six terms of a vector sequence $\{\mathbf{x}_k\}$ are listed. An inspection of these first few terms suggests that the sequence might have some regular pattern of behavior. For instance, the first components of these vectors are steadily increasing, whereas the second components are steadily decreasing. In fact, as shown in Example 3, this monotonic behavior persists for all terms of the sequence $\{\mathbf{x}_k\}$. Moreover, it can be shown that

$$\lim_{k \to \infty} \mathbf{x}_k = \mathbf{x}^*,$$

where the limit vector $\mathbf{x}^*$ is given by

$$\mathbf{x}^* = \begin{bmatrix} 1.5 \\ 1.5 \end{bmatrix}.$$

Difference Equations

Let A be an $(n \times n)$ matrix. The equation

$$\mathbf{x}_k = A\mathbf{x}_{k-1} \tag{2}$$

is called a **difference equation.** A **solution** to the difference equation is any sequence of vectors $\{\mathbf{x}_k\}$ that satisfies Eq. (2). That is, a solution is a sequence $\{\mathbf{x}_k\}$ whose successive terms are related by $\mathbf{x}_1 = A\mathbf{x}_0$, $\mathbf{x}_2 = A\mathbf{x}_1, \dots, \mathbf{x}_k = A\mathbf{x}_{k-1}, \dots$. (Equation 2 is not the most general form of a difference equation.)

The basic challenge posed by a difference equation is to describe the behavior of the sequence $\{\mathbf{x}_k\}$. Some specific questions are:

1. For a given starting vector $\mathbf{x}_0$, is there a vector $\mathbf{x}^*$ such that

$$\lim_{k \to \infty} \mathbf{x}_k = \mathbf{x}^*?$$

2. If the sequence $\{\mathbf{x}_k\}$ does have a limit, $\mathbf{x}^*$, what is the limit vector?
3. Find a "formula" that can be used to calculate $\mathbf{x}_k$ in terms of the starting vector $\mathbf{x}_0$.
4. Given a vector $\mathbf{b}$ and an integer k, determine $\mathbf{x}_0$ so that $\mathbf{x}_k = \mathbf{b}$.
5. Given a vector $\mathbf{b}$, characterize the set of starting vectors $\mathbf{x}_0$ for which $\{\mathbf{x}_k\} \to \mathbf{b}$.

Unlike many equations, difference equation (2) does not raise any interesting questions concerning the existence or uniqueness of solutions. For a given starting vector $\mathbf{x}_0$, we see that a solution to (2) always exists because it can be constructed. For instance, in Example 1 we found the first six terms of the solution to the given difference equation. In terms of uniqueness, suppose $\mathbf{x}_0$ is a given starting vector. It can be shown (see Exercise 21) that if $\{\mathbf{w}_k\}$ is any sequence satisfying Eq. (2) and if $\mathbf{w}_0 = \mathbf{x}_0$, then $\mathbf{w}_k = \mathbf{x}_k$, $k = 1, 2, \ldots$.

The next example shows how a difference equation might serve as a mathematical model for a physical process. The model is kept very simple so that the details do not obscure the ideas. Thus the example should be considered illustrative rather than realistic.

EXAMPLE 2 Suppose that animals are being raised for market, and the grower wishes to determine how the annual rate of harvesting animals will affect the yearly size of the herd.

SOLUTION To begin, let $x_1(k)$ and $x_2(k)$ be the "state variables" that measure the size of the herd in the kth year of operation, where

$x_1(k)$ = number of animals less than one year old at year k

$x_2(k)$ = number of animals more than one year old at year k.

We assume that animals less than one year old do not reproduce, and that animals more than one year old have a reproduction rate of b per year. Thus if the herd has $x_2(k)$ mature animals at year k, we expect to have $x_1(k + 1)$ young animals at year $k + 1$, where

$$x_1(k + 1) = bx_2(k).$$

Next we assume that the young animals have a death rate of d_1 per year, and the mature animals have a death rate of d_2 per year. Furthermore, we assume that the mature animals are harvested at a rate of h per year, and that young animals are not harvested. Thus we expect to have $x_2(k + 1)$ mature animals at year $k + 1$, where

$$x_2(k + 1) = x_1(k) + x_2(k) - d_1 x_1(k) - d_2 x_2(k) - h x_2(k).$$

This equation reflects the following facts: An animal that is young at year k will mature by year $k + 1$; an animal that is mature at year k is still mature at year $k + 1$; a certain percentage of young and mature animals will die during the year; and a certain percentage of mature animals will be harvested during the year. Collecting like terms in the second equation and combining the two equations, we obtain the state equations for the herd:

$$\begin{aligned} x_1(k + 1) &= bx_2(k) \\ x_2(k + 1) &= (1 - d_1)x_1(k) + (1 - d_2 - h)x_2(k). \end{aligned} \tag{3}$$

The state equations give the size and composition of the herd at year $k + 1$ in terms of the size and composition of the herd at year k. For example, if we know the initial composition of the herd at year zero, $x_1(0)$ and $x_2(0)$, we can use (3) to calculate the composition of the herd after one year, $x_1(1)$ and $x_2(1)$.

In matrix form, (3) becomes

$$\mathbf{x}(k) = A\mathbf{x}(k - 1), \qquad k = 1, 2, 3, \ldots.$$

where

$$\mathbf{x}(k) = \begin{bmatrix} x_1(k) \\ x_2(k) \end{bmatrix} \quad \text{and} \quad A = \begin{bmatrix} 0 & b \\ (1 - d_1) & (1 - d_2 - h) \end{bmatrix}.$$

In the context of this example, the growth and composition of the herd are governed by the eigenvalues of A, and these can be controlled by varying the parameter h.

Solving Difference Equations

Consider the difference equation

$$\mathbf{x}_k = A\mathbf{x}_{k-1}, \tag{4}$$

where A is an $(n \times n)$ matrix. The key to finding a useful form for solutions of Eq. (4) is to observe that the sequence $\{\mathbf{x}_k\}$ can be calculated by multiplying powers of A by the starting vector $\mathbf{x}_0$. That is,

$$\mathbf{x}_1 = A\mathbf{x}_0$$
$$\mathbf{x}_2 = A\mathbf{x}_1 = A(A\mathbf{x}_0) = A^2\mathbf{x}_0$$
$$\mathbf{x}_3 = A\mathbf{x}_2 = A(A^2\mathbf{x}_0) = A^3\mathbf{x}_0$$
$$\mathbf{x}_4 = A\mathbf{x}_3 = A(A^3\mathbf{x}_0) = A^4\mathbf{x}_0,$$

and, in general,

$$\mathbf{x}_k = A^k\mathbf{x}_0, \quad k = 1, 2, \ldots. \tag{5}$$

Next, let A have eigenvalues $\lambda_1, \lambda_2, \ldots, \lambda_n$ and corresponding eigenvectors $\mathbf{u}_1, \mathbf{u}_2, \ldots, \mathbf{u}_n$. We now make a critical assumption: Let us suppose that matrix A is not defective. That is, let us suppose that the set of eigenvectors $\{\mathbf{u}_1, \mathbf{u}_2, \ldots, \mathbf{u}_n\}$ is linearly independent.

With the assumption that A is not defective, we can use the set of eigenvectors as a basis for R^n. In particular, any starting vector $\mathbf{x}_0$ can be expressed as a linear combination of the eigenvectors:

$$\mathbf{x}_0 = a_1\mathbf{u}_1 + a_2\mathbf{u}_2 + \cdots + a_n\mathbf{u}_n.$$

Then, using Eq. (5), we can obtain the following expression for $\mathbf{x}_k$:

$$\mathbf{x}_k = A^k\mathbf{x}_0$$
$$= A^k(a_1\mathbf{u}_1 + a_2\mathbf{u}_2 + \cdots + a_n\mathbf{u}_n)$$
$$= a_1 A^k\mathbf{u}_1 + a_2 A^k\mathbf{u}_2 + \cdots + a_n A^k\mathbf{u}_n$$
$$= a_1(\lambda_1)^k\mathbf{u}_1 + a_2(\lambda_2)^k\mathbf{u}_2 + \cdots + a_n(\lambda_n)^k\mathbf{u}_n. \tag{6}$$

(The last equality above comes from Theorem 11 of Section 3.4: If $A\mathbf{u} = \lambda\mathbf{u}$, then $A^k\mathbf{u} = \lambda^k\mathbf{u}$.)

Note that if A does not have a set of n linearly independent eigenvectors, then the expression for $\mathbf{x}_k$ in Eq. (6) must be modified. The modification depends on the idea of a "generalized eigenvector." It can be shown (see Section 6.8) that we can always choose a basis for R^n consisting of eigenvectors and generalized eigenvectors of A.

EXAMPLE 3 Use Eq. (6) to find an expression for $\mathbf{x}_k$, where $\mathbf{x}_k$ is the kth term of the sequence in Example 1. Use your expression to calculate $\mathbf{x}_k$ for $k = 10$ and $k = 20$. Determine whether the sequence $\{\mathbf{x}_k\}$ converges.

SOLUTION The sequence $\{\mathbf{x}_k\}$ in Example 1 is generated by $\mathbf{x}_k = A\mathbf{x}_{k-1}$, $k = 1, 2, \ldots$, where

$$A = \begin{bmatrix} .8 & .2 \\ .2 & .8 \end{bmatrix} \quad \text{and} \quad \mathbf{x}_0 = \begin{bmatrix} 1 \\ 2 \end{bmatrix}.$$

Now the characteristic polynomial for A is

$$p(t) = t^2 - 1.6t + 0.6 = (t-1)(t-0.6).$$

Therefore, the eigenvalues of A are $\lambda_1 = 1$ and $\lambda_2 = 0.6$. Corresponding eigenvectors are

$$\mathbf{u}_1 = \begin{bmatrix} 1 \\ 1 \end{bmatrix} \quad \text{and} \quad \mathbf{u}_2 = \begin{bmatrix} 1 \\ -1 \end{bmatrix}.$$

The starting vector $\mathbf{x}_0$ can be expressed in terms of the eigenvectors as $\mathbf{x}_0 = 1.5\mathbf{u}_1 - 0.5\mathbf{u}_2$:

$$\mathbf{x}_0 = \begin{bmatrix} 1 \\ 2 \end{bmatrix} = 1.5\begin{bmatrix} 1 \\ 1 \end{bmatrix} - 0.5\begin{bmatrix} 1 \\ -1 \end{bmatrix}.$$

Therefore, the terms of the sequence $\{\mathbf{x}_k\}$ are given by

$$\begin{aligned} \mathbf{x}_k = A^k\mathbf{x}_0 &= A^k(1.5\mathbf{u}_1 - 0.5\mathbf{u}_2) \\ &= 1.5A^k\mathbf{u}_1 - 0.5A^k\mathbf{u}_2 \\ &= 1.5(1)^k\mathbf{u}_1 - 0.5(0.6)^k\mathbf{u}_2 \\ &= 1.5\mathbf{u}_1 - 0.5(0.6)^k\mathbf{u}_2. \end{aligned}$$

In detail, the components of $\mathbf{x}_k$ are

$$\mathbf{x}_k = \begin{bmatrix} 1.5 - 0.5(0.6)^k \\ 1.5 + 0.5(0.6)^k \end{bmatrix}, \quad k = 0, 1, 2, \ldots. \tag{7}$$

For $k = 10$ and $k = 20$, we calculate $\mathbf{x}_k$ from Eq. (7), finding

$$\mathbf{x}_{10} = \begin{bmatrix} 1.496976\ldots \\ 1.503023\ldots \end{bmatrix} \quad \text{and} \quad \mathbf{x}_{20} = \begin{bmatrix} 1.499981\ldots \\ 1.500018\ldots \end{bmatrix}.$$

Finally, since $\lim_{k\to\infty}(0.6)^k = 0$, we see from Eq. (7) that

$$\lim_{k\to\infty} \mathbf{x}_k = \mathbf{x}^* = \begin{bmatrix} 1.5 \\ 1.5 \end{bmatrix}.$$

Types of Solutions to Difference Equations

If we reflect about the results of Example 3, the following observations emerge: Suppose a sequence $\{\mathbf{x}_k\}$ is generated by $\mathbf{x}_k = A\mathbf{x}_{k-1}$, $k = 1, 2, \ldots$, where A is the (2×2) matrix

$$A = \begin{bmatrix} .8 & .2 \\ .2 & .8 \end{bmatrix}.$$

Then, no matter what starting vector $\mathbf{x}_0$ is selected, the sequence $\{\mathbf{x}_k\}$ will either converge to the zero vector or the sequence will converge to a multiple of $\mathbf{u}_1 = [1, 1]^T$.

To verify this observation, let $\mathbf{x}_0$ be any given initial vector. We can express $\mathbf{x}_0$ in terms of the eigenvectors:

$$\mathbf{x}_0 = a_1\mathbf{u}_1 + a_2\mathbf{u}_2.$$

Since the eigenvalues of A are $\lambda_1 = 1$ and $\lambda_2 = 0.6$, the vector $\mathbf{x}_k$ is given by

$$\mathbf{x}_k = A^k\mathbf{x}_0 = a_1(1)^k\mathbf{u}_1 + a_2(0.6)^k\mathbf{u}_2 = a_1\mathbf{u}_1 + a_2(0.6)^k\mathbf{u}_2.$$

Given this expression for $\mathbf{x}_k$, there are only two possibilities:

1. If $a_1 \neq 0$, then $\lim_{k \to \infty} \mathbf{x}_k = a_1\mathbf{u}_1$.
2. If $a_1 = 0$, then $\lim_{k \to \infty} \mathbf{x}_k = \boldsymbol{\theta}$.

In general, an analogous description can be given for the possible solutions of any difference equation. Specifically, let A be a nondefective $(n \times n)$ matrix with eigenvalues $\lambda_1, \lambda_2, \ldots, \lambda_n$. For convenience, let us assume the eigenvalues are indexed according to their magnitude, where

$$|\lambda_1| \geq |\lambda_2| \geq \cdots \geq |\lambda_n|.$$

Let $\mathbf{x}_0$ be any initial vector and consider the sequence $\{\mathbf{x}_k\}$, where $\mathbf{x}_k = A\mathbf{x}_{k-1}$, $k = 1, 2, \ldots$. Finally, suppose $\mathbf{x}_0$ is expressed as

$$\mathbf{x}_0 = a_1\mathbf{u}_1 + a_2\mathbf{u}_2 + \cdots + a_n\mathbf{u}_n,$$

where $a_1 \neq 0$.

From Eq. (6), we have the following possibilities for the sequence $\{\mathbf{x}_k\}$:

1. If $|\lambda_1| < 1$, then $\lim_{k \to \infty} \mathbf{x}_k = \boldsymbol{\theta}$.
2. If $|\lambda_1| = 1$, then there is a constant $M > 0$ such that $\|\mathbf{x}_k\| \leq M$, for all k.
3. If $\lambda_1 = 1$ and $|\lambda_2| < 1$, then $\lim_{k \to \infty} \mathbf{x}_k = a_1\mathbf{u}_1$.
4. If $|\lambda_1| > 1$, then $\lim_{k \to \infty} \|\mathbf{x}_k\| = \infty$.

Other possibilities exist that are not listed above. For example, if $\lambda_1 = 1$, $\lambda_2 = 1$, and $|\lambda_3| < 1$, then $\{\mathbf{x}_k\} \to a_1\mathbf{u}_1 + a_2\mathbf{u}_2$.

Also, in listing the possibilities we assumed that A was not defective and that $a_1 \neq 0$. If $a_1 = 0$ but $a_2 \neq 0$, it should be clear that a similar list can be made by using λ_2 in place of λ_1. If matrix A is defective, it can be shown (see Section 6.8) that the list above is still valid, with the following exception (see Exercise 19 for an example): If $|\lambda_1| = 1$ and if the geometric multiplicity of λ_1 is less than the algebraic multiplicity, then it will usually be the case that $\|\mathbf{x}_k\| \to \infty$ as $k \to \infty$.

EXAMPLE 4   For the herd model described in Example 2, let the parameters be given by $b = 0.9$, $d_1 = 0.1$, and $d_2 = 0.2$. Thus $\mathbf{x}_k = A\mathbf{x}_{k-1}$, where

$$A = \begin{bmatrix} 0 & .9 \\ .9 & .8 - h \end{bmatrix}.$$

Determine a harvest rate h so that the herd neither dies out nor grows without bound.

SOLUTION   For any given harvest rate h, the matrix A will have eigenvalues λ_1 and λ_2, where $|\lambda_1| \geq |\lambda_2|$. If $|\lambda_1| < 1$, then $\{\mathbf{x}_k\} \to \boldsymbol{\theta}$, and the herd is dying out. If $|\lambda_1| > 1$, then $\{\|\mathbf{x}_k\|\} \to \infty$, which indicates that the herd is increasing without bound.

Therefore, we want to select h so that $\lambda_1 = 1$. For any given h, λ_1 and λ_2 are roots of the characteristic equation $p(t) = 0$, where

$$p(t) = \det(A - tI) = t^2 - (.8 - h)t - .81.$$

To have $\lambda_1 = 1$, we need $p(1) = 0$, or

$$(1)^2 - (.8 - h)(1) - .81 = 0,$$

or

$$h - .61 = 0.$$

Thus a harvest rate of $h = 0.61$ will lead to $\lambda_1 = 1$ and $\lambda_2 = -0.81$.

Note that to the extent the herd model in Examples 2 and 4 is valid, a harvest rate of less than 0.61 will cause the herd to grow, whereas a rate greater than 0.61 will cause the herd to decrease. A harvest rate of 0.61 will cause the herd to approach a "steady-state" distribution of 9 young animals for every 10 mature animals. That is, for any initial vector $\mathbf{x}_0 = a_1\mathbf{u}_1 + a_2\mathbf{u}_2$, we have (with $h = 0.61$)

$$\mathbf{x}_k = a_1\mathbf{u}_1 + a_2(-0.81)^k\mathbf{u}_2,$$

where the eigenvectors $\mathbf{u}_1$ and $\mathbf{u}_2$ are given by

$$\mathbf{u}_1 = \begin{bmatrix} 9 \\ 10 \end{bmatrix} \quad \text{and} \quad \mathbf{u}_2 = \begin{bmatrix} 10 \\ -9 \end{bmatrix}.$$

Markov Chains

A special type of difference equation arises in the study of "Markov chains" or "Markov processes." We cannot go into the interesting theory of Markov chains, but we will give an example that illustrates some of the ideas.

EXAMPLE 5 An automobile rental company has three locations, which we designate as P, Q, and R. When an automobile is rented at one of the locations, it may be returned to any of the three locations.

Suppose, at some specific time, that there are p cars at location P, q cars at Q, and r cars at R. Experience has shown, in any given week, that the p cars at location P are distributed as follows: 10% are rented and returned to Q, 30% are rented and returned to R, and 60% remain at P (these either are not rented or are rented and returned to P). Similar rental histories are known for locations Q and R, as summarized below.

Weekly Distribution History

Location P: 60% stay at P, 10% go to Q, 30% go to R.
Location Q: 10% go to P, 80% stay at Q, 10% go to R.
Location R: 10% go to P, 20% go to Q, 70% stay at R.

SOLUTION   Let $\mathbf{x}_k$ represent the state of the rental fleet at the beginning of week k:

$$\mathbf{x}_k = \begin{bmatrix} p(k) \\ q(k) \\ r(k) \end{bmatrix}.$$

For the state vector $\mathbf{x}_k$, $p(k)$ denotes the number of cars at location P, $q(k)$ the number at Q, and $r(k)$ the number at R.

From the weekly distribution history, we see that

$$p(k + 1) = .6p(k) + .1q(k) + .1r(k)$$
$$q(k + 1) = .1p(k) + .8q(k) + .2r(k)$$
$$r(k + 1) = .3p(k) + .1q(k) + .7r(k).$$

(For instance, the number of cars at P when week $k + 1$ begins is determined by the 60% that remain at P, the 10% that arrive from Q, and the 10% that arrive from R.)

To the extent that the weekly distribution percentages do not change, the rental fleet is rearranged among locations P, Q, and R according to the rule $\mathbf{x}_{k+1} = A\mathbf{x}_k$, $k = 0, 1, \ldots$, where A is the (3×3) matrix

$$A = \begin{bmatrix} .6 & .1 & .1 \\ .1 & .8 & .2 \\ .3 & .1 & .7 \end{bmatrix}.$$

Example 5 represents a situation in which a fixed population (the rental fleet) is rearranged in stages (week by week) among a fixed number of categories (the locations P, Q, and R). Moreover, in Example 5 the rules governing the rearrangement remain fixed from stage to stage (the weekly distribution percentages stay constant). In general, such problems can be modeled by a difference equation of the form

$$\mathbf{x}_{k+1} = A\mathbf{x}_k, \qquad k = 0, 1, \ldots.$$

For such problems the matrix A is often called a ***transition*** matrix. Such a matrix has two special properties:

The entries of A are all nonnegative.   **(8a)**

In each column of A, the sum of the entries has the value 1.   **(8b)**

It turns out that a matrix having properties (8a) and (8b) always has an eigenvalue of $\lambda = 1$. This fact is established in Exercise 26 and illustrated in the next example.

EXAMPLE 6   Suppose the automobile rental company described in Example 5 has a fleet of 600 cars. Initially an equal number of cars is based at each location, so that $p(0) = 200$, $q(0) = 200$, and $r(0) = 200$. As in Example 5, let the week-by-week distribution of cars be governed by $\mathbf{x}_{k+1} = A\mathbf{x}_k$, $k = 0, 1, \ldots$, where

$$\mathbf{x}_k = \begin{bmatrix} p(k) \\ q(k) \\ r(k) \end{bmatrix}, \qquad A = \begin{bmatrix} .6 & .1 & .1 \\ .1 & .8 & .2 \\ .3 & .1 & .7 \end{bmatrix}, \quad \text{and} \quad \mathbf{x}_0 = \begin{bmatrix} 200 \\ 200 \\ 200 \end{bmatrix}.$$

Find $\lim_{k \to \infty} \mathbf{x}_k$. Determine the number of cars at each location in the kth week, for $k = 1, 5$, and 10.

SOLUTION   If A is not defective, we can use Eq. (6) to express $\mathbf{x}_k$ as

$$\mathbf{x}_k = a_1 (\lambda_1)^k \mathbf{u}_1 + a_2 (\lambda_2)^k \mathbf{u}_2 + a_3 (\lambda_3)^k \mathbf{u}_3,$$

where $\{\mathbf{u}_1, \mathbf{u}_2, \mathbf{u}_3\}$ is a basis for R^3, consisting of eigenvectors of A.

It can be shown that A has eigenvalues $\lambda_1 = 1$, $\lambda_2 = .6$, and $\lambda_3 = .5$. Thus A has three linearly independent eigenvectors:

$$\lambda_1 = 1, \qquad \mathbf{u}_1 = \begin{bmatrix} 4 \\ 9 \\ 7 \end{bmatrix}; \qquad \lambda_2 = .6 \qquad \mathbf{u}_2 = \begin{bmatrix} 0 \\ 1 \\ -1 \end{bmatrix};$$

$$\lambda_3 = .5, \qquad \mathbf{u}_3 = \begin{bmatrix} -1 \\ -1 \\ 2 \end{bmatrix}.$$

The initial vector, $\mathbf{x}_0 = [200, 200, 200]^T$, can be written as

$$\mathbf{x}_0 = 30\mathbf{u}_1 - 150\mathbf{u}_2 - 80\mathbf{u}_3.$$

Thus the vector $\mathbf{x}_k = [p(k), q(k), r(k)]^T$ is given by

$$
\begin{aligned}
\mathbf{x}_k &= A^k \mathbf{x}_0 \\
&= A^k (30\mathbf{u}_1 - 150\mathbf{u}_2 - 80\mathbf{u}_3) \\
&= 30(\lambda_1)^k \mathbf{u}_1 - 150(\lambda_2)^k \mathbf{u}_2 - 80(\lambda_3)^k \mathbf{u}_3 \\
&= 30\mathbf{u}_1 - 150(.6)^k \mathbf{u}_2 - 80(.5)^k \mathbf{u}_3.
\end{aligned}
$$ **(9)**

From the expression above, we see that

$$\lim_{k \to \infty} \mathbf{x}_k = 30\mathbf{u}_1 = \begin{bmatrix} 120 \\ 270 \\ 210 \end{bmatrix}.$$

Therefore, as the weeks proceed, the rental fleet will tend to an equilibrium state with 120 cars at P, 270 cars at Q, and 210 cars at R. To the extent that the model is valid, location Q will require the largest facility for maintenance, parking, and the like.

Finally, using Eq. (9), we can calculate the state of the fleet for the kth week:

$$\mathbf{x}_1 = \begin{bmatrix} 160 \\ 220 \\ 220 \end{bmatrix}, \quad \mathbf{x}_5 = \begin{bmatrix} 122.500 \\ 260.836 \\ 216.664 \end{bmatrix}, \quad \text{and} \quad \mathbf{x}_{10} = \begin{bmatrix} 120.078 \\ 269.171 \\ 210.751 \end{bmatrix}. \qquad \square$$

Note that the components of $\mathbf{x}_{10}$ are rounded to three places. Of course, for an actual fleet the state vectors $\mathbf{x}_k$ must have only integer components. The fact that the sequence defined in Eq. (9) need not have integer components represents a limitation of the assumed distribution model.

Systems of Differential Equations

Difference equations are useful for describing the state of a physical system at discrete values of time. Mathematical models that describe the evolution of a physical system for all values of time are frequently expressed in terms of a differential equation or a system of differential equations. A simple example of a system of differential equations is

$$\begin{aligned} v'(t) &= av(t) + bw(t) \\ w'(t) &= cv(t) + dw(t). \end{aligned} \tag{10}$$

In Eq. (10), the problem is to find functions $v(t)$ and $w(t)$ that simultaneously satisfy these equations and in which initial conditions $v(0)$ and $w(0)$ may also be specified. We can express Eq. (10) in matrix terms if we let

$$\mathbf{x}(t) = \begin{bmatrix} v(t) \\ w(t) \end{bmatrix}.$$

Then Eq. (10) can be written as $\mathbf{x}'(t) = A\mathbf{x}(t)$, where

$$\mathbf{x}'(t) = \begin{bmatrix} v'(t) \\ w'(t) \end{bmatrix} \quad \text{and} \quad A = \begin{bmatrix} a & b \\ c & d \end{bmatrix}.$$

The equation $\mathbf{x}'(t) = A\mathbf{x}(t)$ is reminiscent of the simple scalar differential equation, $y'(t) = \alpha y(t)$, which is frequently used in calculus to model problems such as radioactive decay or bacterial growth. To find a function $y(t)$ that satisfies the identity $y'(t) = \alpha y(t)$, we rewrite the equation as $y'(t)/y(t) = \alpha$. Integrating both sides with respect to t yields $\ln|y(t)| = \alpha t + \beta$, or equivalently $y(t) = y_0 e^{\alpha t}$, where $y_0 = y(0)$.

Using the scalar equation as a guide, we assume the vector equation $\mathbf{x}'(t) = A\mathbf{x}(t)$ has a solution of the form

$$\mathbf{x}(t) = e^{\lambda t}\mathbf{u}, \tag{11}$$

where $\mathbf{u}$ is a constant vector. To see if the function $\mathbf{x}(t)$ in Eq. (11) can be a solution, we differentiate and get $\mathbf{x}'(t) = \lambda e^{\lambda t}\mathbf{u}$. On the other hand, $A\mathbf{x}(t) = e^{\lambda t}A\mathbf{u}$; so (11) will be a solution of $\mathbf{x}'(t) = A\mathbf{x}(t)$ if and only if

$$e^{\lambda t}(A - \lambda I)\mathbf{u} = \mathbf{0}. \tag{12}$$

Now $e^{\lambda t} \neq 0$ for all values of t; so Eq. (12) will be satisfied only if $(A - \lambda I)\mathbf{u} = \mathbf{0}$. Therefore, if λ is an eigenvalue of A and $\mathbf{u}$ is a corresponding eigenvector, then $\mathbf{x}(t)$ given in (11) is a solution to $\mathbf{x}'(t) = A\mathbf{x}(t)$. (*Note:* The choice $\mathbf{u} = \mathbf{0}$ will also give a solution, but it is a trivial solution.)

If the (2×2) matrix A has eigenvalues λ_1 and λ_2 with corresponding eigenvectors $\mathbf{u}_1$ and $\mathbf{u}_2$, then two solutions of $\mathbf{x}'(t) = A\mathbf{x}(t)$ are $\mathbf{x}_1(t) = e^{\lambda_1 t}\mathbf{u}_1$ and $\mathbf{x}_2(t) = e^{\lambda_2 t}\mathbf{u}_2$. It is easy to verify that any linear combination of $\mathbf{x}_1(t)$ and $\mathbf{x}_2(t)$ is also a solution; so

$$\mathbf{x}(t) = a_1\mathbf{x}_1(t) + a_2\mathbf{x}_2(t) \tag{13}$$

will solve $\mathbf{x}'(t) = A\mathbf{x}(t)$ for any choice of scalars a_1 and a_2. Finally, the initial-value problem consists of finding a solution to $\mathbf{x}'(t) = A\mathbf{x}(t)$ that satisfies an initial condition, $\mathbf{x}(0) = \mathbf{x}_0$, where $\mathbf{x}_0$ is some specified vector. Given the form of $\mathbf{x}_1(t)$ and $\mathbf{x}_2(t)$, it is clear from (13) that $\mathbf{x}(0) = a_1\mathbf{u}_1 + a_2\mathbf{u}_2$. If the eigenvectors $\mathbf{u}_1$ and $\mathbf{u}_2$ are linearly independent, we can always choose scalars b_1 and b_2 so that $\mathbf{x}_0 = b_1\mathbf{u}_1 + b_2\mathbf{u}_2$; and therefore $\mathbf{x}(t) = b_1\mathbf{x}_1(t) + b_2\mathbf{x}_2(t)$ is the solution of $\mathbf{x}'(t) = A\mathbf{x}(t)$, $\mathbf{x}(0) = \mathbf{x}_0$.

EXAMPLE 7 Solve the initial-value problem

$$v'(t) = v(t) - 2w(t), \qquad v(0) = 4$$
$$w'(t) = v(t) + 4w(t), \qquad w(0) = -3.$$

SOLUTION In vector form, the given equation can be expressed as $\mathbf{x}'(t) = A\mathbf{x}(t)$, $\mathbf{x}(0) = \mathbf{x}_0$, where

$$\mathbf{x}(t) = \begin{bmatrix} v(t) \\ w(t) \end{bmatrix}, \qquad A = \begin{bmatrix} 1 & -2 \\ 1 & 4 \end{bmatrix}, \qquad \text{and} \qquad \mathbf{x}_0 = \begin{bmatrix} 4 \\ -3 \end{bmatrix}.$$

The eigenvalues of A are $\lambda_1 = 2$ and $\lambda_2 = 3$, with corresponding eigenvectors

$$\mathbf{u}_1 = \begin{bmatrix} 2 \\ -1 \end{bmatrix} \quad \text{and} \quad \mathbf{u}_2 = \begin{bmatrix} 1 \\ -1 \end{bmatrix}.$$

As before, $\mathbf{x}_1(t) = e^{2t}\mathbf{u}_1$ and $\mathbf{x}_2(t) = e^{3t}\mathbf{u}_2$ are solutions of $\mathbf{x}'(t) = A\mathbf{x}(t)$, as is any linear combination, $\mathbf{x}(t) = b_1\mathbf{x}_1(t) + b_2\mathbf{x}_2(t)$. We now need only choose appropriate constants b_1 and b_2 so that $\mathbf{x}(0) = \mathbf{x}_0$, where we know $\mathbf{x}(0) = b_1\mathbf{u}_1 +$

$b_2\mathbf{u}_2$. For $\mathbf{x}_0$ as given, it is routine to find $\mathbf{x}_0 = \mathbf{u}_1 + 2\mathbf{u}_2$. Thus the solution of $\mathbf{x}'(t) = A\mathbf{x}(t)$, $\mathbf{x}(0) = \mathbf{x}_0$ is $\mathbf{x}(t) = \mathbf{x}_1(t) + 2\mathbf{x}_2(t)$, or

$$\mathbf{x}(t) = e^{2t}\mathbf{u}_1 + 2e^{3t}\mathbf{u}_2.$$

In terms of the functions v and w, we have

$$\mathbf{x}(t) = \begin{bmatrix} v(t) \\ w(t) \end{bmatrix} = e^{2t}\begin{bmatrix} 2 \\ -1 \end{bmatrix} + 2e^{3t}\begin{bmatrix} 1 \\ -1 \end{bmatrix} = \begin{bmatrix} 2e^{2t} + 2e^{3t} \\ -e^{2t} - 2e^{3t} \end{bmatrix}. \qquad \square$$

In general, given the problem of solving

$$\mathbf{x}'(t) = A\mathbf{x}(t), \qquad \mathbf{x}(0) = \mathbf{x}_0, \qquad \textbf{(14)}$$

where A is an $(n \times n)$ matrix, we can proceed just as above. We first find the eigenvalues $\lambda_1, \lambda_2, \ldots, \lambda_n$ of A and corresponding eigenvectors $\mathbf{u}_1, \mathbf{u}_2, \ldots, \mathbf{u}_n$. For each i, $\mathbf{x}_i(t) = e^{\lambda_i t}\mathbf{u}_i$ is a solution of $\mathbf{x}'(t) = A\mathbf{x}(t)$, as is the general expression

$$\mathbf{x}(t) = b_1\mathbf{x}_1(t) + b_2\mathbf{x}_2(t) + \cdots + b_n\mathbf{x}_n(t). \qquad \textbf{(15)}$$

As before, $\mathbf{x}(0) = b_1\mathbf{u}_1 + b_2\mathbf{u}_2 + \cdots + b_n\mathbf{u}_n$; so if $\mathbf{x}_0$ can be expressed as a linear combination of $\mathbf{u}_1, \mathbf{u}_2, \ldots, \mathbf{u}_n$, then we can construct a solution to Eq. (14) in the form of (15). If the eigenvectors of A do not form a basis for R^n, we can still get a solution of the form (15); but a more detailed analysis is required.

Exercises 3.8

In Exercises 1–6, consider the vector sequence $\{\mathbf{x}_k\}$, where $\mathbf{x}_k = A\mathbf{x}_{k-1}$, $k = 1, 2, \ldots$. For the given starting vector $\mathbf{x}_0$, calculate $\mathbf{x}_1$, $\mathbf{x}_2$, $\mathbf{x}_3$, and $\mathbf{x}_4$ by using direct multiplication, as in Example 1.

1. $A = \begin{bmatrix} 0 & 1 \\ 1 & 0 \end{bmatrix}$, $\mathbf{x}_0 = \begin{bmatrix} 2 \\ 4 \end{bmatrix}$

2. $A = \begin{bmatrix} .5 & .5 \\ .5 & .5 \end{bmatrix}$, $\mathbf{x}_0 = \begin{bmatrix} 16 \\ 8 \end{bmatrix}$

3. $A = \begin{bmatrix} .5 & .25 \\ .5 & .75 \end{bmatrix}$, $\mathbf{x}_0 = \begin{bmatrix} 128 \\ 64 \end{bmatrix}$

4. $A = \begin{bmatrix} 2 & -1 \\ -1 & 2 \end{bmatrix}$, $\mathbf{x}_0 = \begin{bmatrix} 3 \\ 1 \end{bmatrix}$

5. $A = \begin{bmatrix} 1 & 4 \\ 1 & 1 \end{bmatrix}$, $\mathbf{x}_0 = \begin{bmatrix} -1 \\ 2 \end{bmatrix}$

6. $A = \begin{bmatrix} 3 & 1 \\ 4 & 3 \end{bmatrix}$, $\mathbf{x}_0 = \begin{bmatrix} 2 \\ 0 \end{bmatrix}$

In Exercises 7–14, let $\mathbf{x}_k = A\mathbf{x}_{k-1}$, $k = 1, 2, \ldots$, for the given A and $\mathbf{x}_0$. Find an expression for $\mathbf{x}_k$ by using

Eq. (6), as in Example 3. With a calculator, compute $\mathbf{x}_4$ and $\mathbf{x}_{10}$ from the expression. Comment on $\lim_{k \to \infty}\mathbf{x}_k$.

7. A and $\mathbf{x}_0$ in Exercise 1

8. A and $\mathbf{x}_0$ in Exercise 2

9. A and $\mathbf{x}_0$ in Exercise 3

10. A and $\mathbf{x}_0$ in Exercise 4

11. A and $\mathbf{x}_0$ in Exercise 5

12. A and $\mathbf{x}_0$ in Exercise 6

13. $A = \begin{bmatrix} 3 & -1 & -1 \\ -12 & 0 & 5 \\ 4 & -2 & -1 \end{bmatrix}$, $\mathbf{x}_0 = \begin{bmatrix} 3 \\ -14 \\ 8 \end{bmatrix}$

14. $A = \begin{bmatrix} -6 & 1 & 3 \\ -3 & 0 & 2 \\ -20 & 2 & 10 \end{bmatrix}$, $\mathbf{x}_0 = \begin{bmatrix} 1 \\ 1 \\ -1 \end{bmatrix}$

In Exercises 15–18, solve the initial-value problem.

15. $u'(t) = 5u(t) - 6v(t)$, $u(0) = 4$
    $v'(t) = 3u(t) - 4v(t)$, $v(0) = 1$

16. $u'(t) = u(t) + 2v(t), \quad u(0) = 1$
$v'(t) = 2u(t) + v(t), \quad v(0) = 5$

17. $u'(t) = u(t) + v(t) + w(t), \quad u(0) = 3$
$v'(t) = 3v(t) + 3w(t), \quad v(0) = 3$
$w'(t) = -2u(t) + v(t) + w(t), \quad w(0) = 1$

18. $u'(t) = -2u(t) + 2v(t) - 3w(t), \quad u(0) = 3$
$v'(t) = 2u(t) + v(t) - 6w(t), \quad v(0) = -1$
$w'(t) = -u(t) - 2v(t), \quad w(0) = 3$

19. Consider the matrix A given by

$$A = \begin{bmatrix} 1 & 2 \\ 0 & 1 \end{bmatrix}.$$

Note that $\lambda = 1$ is the only eigenvalue of A.
a) Verify that A is defective.
b) Consider the sequence $\{\mathbf{x}_k\}$ determined by $\mathbf{x}_k = A\mathbf{x}_{k-1}$, $k = 1, 2, \ldots$, where $\mathbf{x}_0 = [1, 1]^T$. Use induction to show that $\mathbf{x}_k = [2k + 1, 1]^T$. (This exercise gives an example of a sequence $\mathbf{x}_k = A\mathbf{x}_{k-1}$, where $\lim_{k\to\infty}\|\mathbf{x}_k\| = \infty$, even though A has no eigenvalue larger than 1 in magnitude.)

In Exercises 20 and 21, choose a value α so that the matrix A has an eigenvalue of $\lambda = 1$. Then, for $\mathbf{x}_0 = [1, 1]^T$, calculate $\lim_{k\to\infty}\mathbf{x}_k$, where $\mathbf{x}_k = A\mathbf{x}_{k-1}$, $k = 1, 2, \ldots$.

20. $A = \begin{bmatrix} .5 & .5 \\ .5 & 1 + \alpha \end{bmatrix}$

21. $A = \begin{bmatrix} 0 & .3 \\ .6 & 1 + \alpha \end{bmatrix}$

22. Suppose that $\{\mathbf{u}_k\}$ and $\{\mathbf{v}_k\}$ are sequences satisfying $\mathbf{u}_k = A\mathbf{u}_{k-1}$, $k = 1, 2, \ldots$, and $\mathbf{v}_k = A\mathbf{v}_{k-1}$, $k = 1, 2, \ldots$. Show that if $\mathbf{u}_0 = \mathbf{v}_0$, then $\mathbf{u}_i = \mathbf{v}_i$ for all i.

23. Let $B = (b_{ij})$ be an $(n \times n)$ matrix. Matrix B is called a stochastic matrix if B contains only nonnegative entries and if $b_{i1} + b_{i2} + \cdots + b_{in} = 1$, $1 \leq i \leq n$. (That is, B is a stochastic matrix if B^T satisfies conditions 8a and 8b.) Show that $\lambda = 1$ is an eigenvalue of B. (*Hint:* Consider the vector $\mathbf{w} = [1, 1, \ldots, 1]^T$.)

24. Suppose that B is a stochastic matrix whose entries are all positive. By Exercise 23, $\lambda = 1$ is an eigenvalue of B. Show that if $B\mathbf{u} = \mathbf{u}$, $\mathbf{u} \neq \boldsymbol{\theta}$, then $\mathbf{u}$ is a multiple of the vector $\mathbf{w}$ defined in Exercise 23. (*Hint:* Define $\mathbf{v} = \alpha\mathbf{u}$ so that $v_i = 1$ and $|v_j| \leq 1$, $1 \leq j \leq n$. Consider the ith equations in $B\mathbf{w} = \mathbf{w}$ and $B\mathbf{v} = \mathbf{v}$.)

25. Let B be a stochastic matrix, and let λ be any eigenvalue of B. Show that $|\lambda| \leq 1$. For simplicity, assume that λ is real. (*Hint:* Suppose that $B\mathbf{u} = \lambda\mathbf{u}$, $\mathbf{u} \neq \boldsymbol{\theta}$. Define a vector $\mathbf{v}$ as in Exercise 24.)

26. Let A be an $(n \times n)$ matrix satisfying conditions (8a) and (8b). Show that $\lambda = 1$ is an eigenvalue of A and that if $A\mathbf{u} = \beta\mathbf{u}$, $\mathbf{u} \neq \boldsymbol{\theta}$, then $|\beta| \leq 1$. (*Hint:* Matrix A^T is stochastic.)

27. Suppose that $(A - \lambda I)\mathbf{u} = \boldsymbol{\theta}$, $\mathbf{u} \neq \boldsymbol{\theta}$, and there is a vector $\mathbf{v}$ such that $(A - \lambda I)\mathbf{v} = \mathbf{u}$. Then $\mathbf{v}$ is called a ***generalized eigenvector***. Show that $\{\mathbf{u}, \mathbf{v}\}$ is a linearly independent set. (*Hint:* Note that $A\mathbf{v} = \lambda\mathbf{v} + \mathbf{u}$. Suppose that $a\mathbf{u} + b\mathbf{v} = \boldsymbol{\theta}$, and multiply this equation by A.)

28. Let A, $\mathbf{u}$, and $\mathbf{v}$ be as in Exercise 27. Show that $A^k\mathbf{v} = \lambda^k\mathbf{v} + k\lambda^{k-1}\mathbf{u}$, $k = 1, 2, \ldots$.

29. Consider matrix A in Exercise 19.
a) Find an eigenvector $\mathbf{u}$ and a generalized eigenvector $\mathbf{v}$ for A.
b) Express $\mathbf{x}_0 = [1, 1]^T$ as $\mathbf{x}_0 = a\mathbf{u} + b\mathbf{v}$.
c) Using the result of Exercise 28, find an expression for $A^k\mathbf{x}_0 = A^k(a\mathbf{u} + b\mathbf{v})$.
d) Verify that $A^k\mathbf{x}_0 = [2k + 1, 1]^T$ as was shown by other means in Exercise 19.

Supplementary Computational Exercises

1. Find all values x such that A is singular, where

$$A = \begin{bmatrix} x & 1 & 2 \\ 3 & x & 0 \\ 0 & -1 & 1 \end{bmatrix}.$$

2. For what values x does A have only real eigenvalues, where

$$A = \begin{bmatrix} 2 & 1 \\ x & 3 \end{bmatrix}?$$

3. Let

$$A = \begin{bmatrix} a & b \\ c & d \end{bmatrix},$$

where $a + b = 2$ and $c + d = 2$. Show that $\lambda = 2$ is an eigenvalue for A. (*Hint:* Guess an eigenvector.)

4. Let A and B be (3×3) matrices such that $\det(A) = 2$ and $\det(B) = 9$. Find the values of each of the following.
   a) $\det(A^{-1}B^2)$
   b) $\det(3A)$
   c) $\det(AB^2A^{-1})$

5. For what values x is A defective, where

$$A = \begin{bmatrix} 2 & x \\ 0 & 2 \end{bmatrix}.$$

In Exercises 6–9, A is a (2×2) matrix such that $A^2 + 3A - I = \mathcal{O}$.

6. Suppose we know that

$$A\mathbf{u} = \begin{bmatrix} 2 \\ 1 \end{bmatrix}, \quad \text{where } \mathbf{u} = \begin{bmatrix} 1 \\ 3 \end{bmatrix}.$$

Find $A^2\mathbf{u}$ and $A^3\mathbf{u}$.

7. Show that A is nonsingular. (*Hint:* Is there a non-zero vector $\mathbf{x}$ such that $A\mathbf{x} = \boldsymbol{\theta}$?)

8. Find $A^{-1}\mathbf{u}$, where $\mathbf{u}$ is as in Exercise 6.

9. Using the fact that $A^2 = I - 3A$, we can find scalars a_k and b_k such that $A^k = a_k A + b_k I$. Find these scalars for $k = 2, 3, 4,$ and 5.

In Exercises 10 and 11, find the eigenvalue λ_i given the corresponding eigenvector $\mathbf{u}_i$. Do not calculate the characteristic polynomial for A.

10. $A = \begin{bmatrix} 2 & -12 \\ 1 & -5 \end{bmatrix}$, $\mathbf{u}_1 = \begin{bmatrix} 4 \\ 1 \end{bmatrix}$, $\mathbf{u}_2 = \begin{bmatrix} 3 \\ 1 \end{bmatrix}$

11. $A = \begin{bmatrix} 1 & 2 \\ -1 & 4 \end{bmatrix}$, $\mathbf{u}_1 = \begin{bmatrix} 2 \\ 1 \end{bmatrix}$, $\mathbf{u}_2 = \begin{bmatrix} 1 \\ 1 \end{bmatrix}$

12. Find x so that $\mathbf{u}$ is an eigenvector. What is the corresponding eigenvalue λ?

$$A = \begin{bmatrix} 2 & x \\ 1 & -5 \end{bmatrix}, \quad \mathbf{u} = \begin{bmatrix} 1 \\ -1 \end{bmatrix}$$

13. Find x and y so that $\mathbf{u}$ is an eigenvector corresponding to the eigenvalue $\lambda = 1$:

$$A = \begin{bmatrix} x & y \\ 2x & -y \end{bmatrix}, \quad \mathbf{u} = \begin{bmatrix} -1 \\ 1 \end{bmatrix}$$

14. Find x and y so that $\mathbf{u}$ is an eigenvector corresponding to the eigenvalue $\lambda = 4$:

$$A = \begin{bmatrix} x + y & y \\ x - 3 & 1 \end{bmatrix}, \quad \mathbf{u} = \begin{bmatrix} -3 \\ 1 \end{bmatrix}$$

Supplementary Conceptual Exercises

In Exercises 1–8, answer true or false. Justify your answer by providing a counterexample if the statement is false or an outline of a proof if the statement is true. In each exercise, A is a real $(n \times n)$ matrix.

1. If A is nonsingular with $A^{-1} = A^T$, then $\det(A) = 1$.

2. If $\mathbf{x}$ is an eigenvector for A, where A is nonsingular, then $\mathbf{x}$ is also an eigenvector for A^{-1}.

3. If A is nonsingular, then $\det(A^4)$ is positive.

4. If A is defective, then A is singular.

5. If A is an orthogonal matrix and if $\mathbf{x}$ is in R^n, then $\|A\mathbf{x}\| = \|\mathbf{x}\|$.

6. If S is $(n \times n)$ and nonsingular, then A and $S^{-1}AS$ have the same eigenvalues.

7. If A and B are diagonal $(n \times n)$ matrices, then $\det(A + B) = \det(A) + \det(B)$.

8. If A is singular, then A is defective.

In Exercises 9–14, give a brief answer.

9. Suppose that A and Q are $(n \times n)$ matrices where Q is orthogonal. Then we know that A and $B = Q^TAQ$ have the same eigenvalues.
   a) If $\mathbf{x}$ is an eigenvector of B corresponding to λ, give an eigenvector of A corresponding to λ.
   b) If $\mathbf{u}$ is an eigenvector of A corresponding to λ, give an eigenvector of B corresponding to λ.

10. Suppose that A is $(n \times n)$ and $A^3 = \mathcal{O}$. Show that 0 is the only eigenvalue of A.

11. Show that if A is $(n \times n)$ and is similar to the $(n \times n)$ identity I, then $A = I$.

12. Let A and B be $(n \times n)$ with A nonsingular. Show that AB and BA are similar. (*Hint:* Consider $S^{-1}ABS = BA$.)

13. Suppose that A and B are $(n \times n)$ and A is similar to B. Show that A^k is similar to B^k for $k = 2, 3,$ and 4.

14. Let $\mathbf{u}$ be a vector in R^n such that $\mathbf{u}^T\mathbf{u} = 1$, and let A denote the $(n \times n)$ matrix $A = I - 2\mathbf{u}\mathbf{u}^T$.
    a) Is A symmetric?
    b) Is A orthogonal?
    c) Calculate $A\mathbf{u}$.
    d) Suppose that $\mathbf{w}$ is in R^n and $\mathbf{u}^T\mathbf{w} = 0$. What is $A\mathbf{w}$?
    e) Give the eigenvalues of A and give the geometric multiplicity for each eigenvalue.

4

Vector Spaces

and Linear

Transformations

Introduction

Chapter 2 illustrated that by passing from a purely geometric view of vectors to an algebraic perspective we could, in a natural way, extend the concept of a vector to include elements of R^n. Using R^n as a model, this chapter extends the notion of a vector even further to include objects such as matrices, polynomials, functions continuous on a given interval, and solutions to certain differential equations. Most of the elementary concepts (such as subspace, basis, and dimension) that are important to understanding vector spaces are immediate generalizations of the same concepts in R^n.

Linear transformations were also introduced in Chapter 2, and we showed in Section 2.7 that a linear transformation, T, from R^n to R^m is always defined by matrix multiplication; that is,

$$T(\mathbf{x}) = A\mathbf{x} \tag{1}$$

for some $(m \times n)$ matrix A. In Sections 4.7–4.10, we will consider linear transformations on arbitrary vector spaces, thus extending the theory of mappings defined as in Eq. (1) to a more general setting. For example, differentiation and integration can be viewed as linear transformations.

Although the theory of vector spaces is relatively abstract, the vector-space structure provides a unifying framework of great flexibility, and many important practical problems fit naturally into a vector-space framework. As examples, the set of all solutions to a differential equation such as

$$a(x)y'' + b(x)y' + c(x)y = 0$$

can be shown to be a two-dimensional vector space. Thus if two "linearly independent" solutions are known, then all the solutions are determined. The previously defined notion of dot product can be extended to more general vector spaces and used to define the distance between two vectors. This notion is essential when one wishes to approximate one object with another (for example, to approximate a function with a polynomial). Linear transformations permit a natural extension of the important concepts of eigenvalues and eigenvectors to arbitrary vector spaces.

A basic feature of vector spaces is that they possess both an algebraic character and a geometric character. In this regard the geometric character frequently gives a pictorial insight into how a particular problem can be solved, whereas the algebraic character is used actually to calculate a solution.

As an example of how we can use this dual geometric/algebraic character of vector spaces, consider the following. In 1811 and 1822, Fourier, in his *Mathematical Theory of Heat,* made extremely important discoveries by using trigonometric series of the form

$$s(x) = \sum_{k=0}^{\infty} (a_k \cos kx + b_k \sin kx)$$

to represent functions, $f(x)$, $-\pi \le x \le \pi$. Today these representations can be visualized and utilized in a simple way using vector-space concepts.

For any positive integer n, let $\mathscr{S}_n$ represent the set of all trigonometric polynomials of degree at most n:

$$\mathscr{S}_n = \left\{ s_n(x): s_n(x) = \sum_{k=0}^{n} (a_k \cos kx + b_k \sin kx), \ a_k \text{ and } b_k \text{ real numbers} \right\}.$$

Now, if $s_n^*(x)$ is the best approximation in $\mathscr{S}_n$ to $f(x)$, then we might hope that

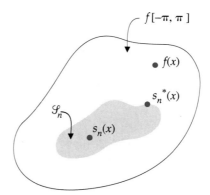

Figure 4.1
Among all $s_n(x)$ in S_n, we are
searching for $s^*(x)$, which best
approximates $f(x)$, $-\pi \le x \le \pi$.

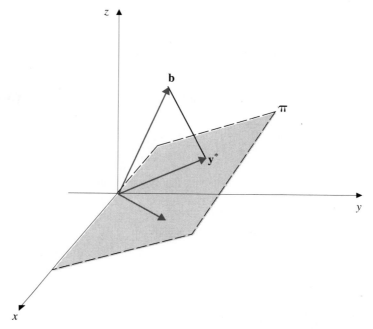

Figure 4.2
The vector **y*** in Π is closer to **b** than is any other vector **y** in Π if and only if **b** – **y*** is perpendicular to all **y** in Π.

$s(x) = \lim_{n \to \infty} s_n^*(x)$. A heuristic picture of this setting is shown in Fig. 4.1, where $\mathscr{F}[-\pi, \pi]$ denotes all functions defined on $[-\pi, \pi]$.

In Fig. 4.2, we have a vector approximation problem that we already know how to work from calculus. Here Π is a plane through the origin and we are searching for a point y^* in Π that is closer to the given point b than any other point y in Π. Using **b**, **y**, and **y*** as the position vectors for the points b, y, and y^*, respectively, we know that **y*** is characterized by the fact that the remainder vector, **b** − **y***, is perpendicular to every position vector **y** in Π. That is, we can find **y*** by setting $(\mathbf{b} - \mathbf{y}^*)^T\mathbf{u}_1 = 0$ and $(\mathbf{b} - \mathbf{y}^*)^T\mathbf{u}_2 = 0$, where $\{\mathbf{u}_1, \mathbf{u}_2\}$ is any basis for Π.

Figure 4.3 gives another way of visualizing this problem. We see a striking similarity between Fig. 4.1 and 4.3. It gives us the inspiration to ask if we can find $s_n^*(x)$ in Fig. 4.1 by choosing its coefficients, a_k and b_k, so that the remainder function, $f(x) - s_n^*(x)$, is in some way "perpendicular" to every $s_n(x)$ in $\mathscr{S}_n$.

As we will show in Section 4.6, this is precisely the approach we use to compute $s_n^*(x)$. Thus the geometric character of the vector-space setting provides our intuition with a possible procedure for solution. We must then use the

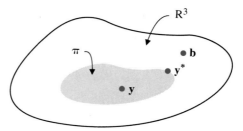

Figure 4.3
An abstract representation of the
problem of finding the closest vector,
$\mathbf{y}^*$, in a subspace Π to a vector $\mathbf{b}$.

algebraic character to:

a) argue that our intuition is valid, and

b) perform the actual calculation of the coefficients a_k and b_k for $s_n^*(x)$.

4.2   Vector Spaces

We begin our study of vector spaces by recalling the basic properties of R^n.
First recall that there are two algebraic operations in R^n:

1. Vectors in R^n can be added.

2. Any vector in R^n can be multiplied by a scalar.

Furthermore, these two operations satisfy the 10 basic properties given in Theo-
rem 1 of Section 2.2. For example, if $\mathbf{u}$ and $\mathbf{v}$ are in R^n, then $\mathbf{u} + \mathbf{v}$ is also in R^n.
Moreover, $\mathbf{u} + \mathbf{v} = \mathbf{v} + \mathbf{u}$ and $(\mathbf{u} + \mathbf{v}) + \mathbf{w} = \mathbf{u} + (\mathbf{v} + \mathbf{w})$ for all $\mathbf{u}, \mathbf{v}$, and $\mathbf{w}$ in R^n.
There are numerous sets other than R^n on which there are defined algebraic
operations of addition and scalar multiplication. Moreover, in many cases these
operations will satisfy the same 10 rules listed in Theorem 1 of Section 2.2. For
example, we have already defined matrix addition and scalar multiplication on
the set of all $(m \times n)$ matrices. Furthermore, it follows from Theorems 7, 8, and
9 of Section 1.7 that these operations satisfy the properties given in Theorem 1
of Section 2.2 (see Example 2 below). Thus, with R^n as a model, we could just
as easily study the set of all $(m \times n)$ matrices and derive most of the properties
and concepts given in Chapter 2, but in the context of matrices. Rather than
study each such set individually, however, it is more efficient to define a "vector
space" in the abstract as any set of objects that has algebraic operations that
satisfy a given list of basic properties. Using only these assumed properties, we
can prove other properties and develop further concepts. The results obtained
in this manner then apply to any specific vector space. For example, later in this
chapter the term *linearly independent* will be applied to a set of matrices, a set
of polynomials, or a set of continuous functions.

Drawing on this discussion, we see that a general vector space should consist of a set of elements (or vectors), V, and a set of scalars, S, together with two algebraic operations:

1. An "addition," which is defined between any two elements of V and which produces a sum that is in V;

2. A "scalar multiplication," which defines how to multiply any element of V by a scalar from S.

In practice the set V can consist of any collection of objects for which meaningful operations of addition and scalar multiplication can be defined. For example, V might be the set of all (2×3) matrices, the set R^4 of all four-dimensional vectors, a set of functions, a set of polynomials, or the set of all solutions to a linear homogeneous differential equation. We will take the set S of scalars to be the set of real numbers, although for added flexibility other sets of scalars may be used (for example, S could be the set of complex numbers). Throughout this chapter the term *scalar* will always denote a real number.

Using R^n as a model and the properties of R^n listed in Theorem 1 of Section 2.2 as a guide, we now define a general vector space. Note that the definition says nothing about the set V but rather specifies rules that the algebraic operations must satisfy.

DEFINITION 1   A set of elements V is said to be a ***vector space*** over a scalar field S if an addition operation is defined between any two elements of V and a scalar multiplication operation is defined between any element of S and any vector in V. Moreover, if $\mathbf{u}$, $\mathbf{v}$, and $\mathbf{w}$ are vectors in V, and if a and b are any two scalars, then these 10 properties must hold.

Closure properties:
 c1) $\mathbf{u} + \mathbf{v}$ is a vector in V.
 c2) $a\mathbf{v}$ is a vector in V.

Properties of addition:
 a1) $\mathbf{u} + \mathbf{v} = \mathbf{v} + \mathbf{u}$.
 a2) $\mathbf{u} + (\mathbf{v} + \mathbf{w}) = (\mathbf{u} + \mathbf{v}) + \mathbf{w}$.
 a3) There is a vector θ in V such that $\mathbf{v} + \theta = \mathbf{v}$ for all $\mathbf{v}$ in V.
 a4) Given a vector $\mathbf{v}$ in V, there is a vector $-\mathbf{v}$ in V such that $\mathbf{v} + (-\mathbf{v}) = \boldsymbol{\theta}$.

Properties of scalar multiplication:
 m1) $a(b\mathbf{v}) = (ab)\mathbf{v}$.
 m2) $a(\mathbf{u} + \mathbf{v}) = a\mathbf{u} + a\mathbf{v}$.
 m3) $(a + b)\mathbf{v} = a\mathbf{v} + b\mathbf{v}$.
 m4) $1\mathbf{v} = \mathbf{v}$ for all $\mathbf{v}$ in V.

The first two conditions, (c1) and (c2), in Definition 1, called closure properties, ensure that the sum of any two vectors in V remains in V and that any scalar multiple of a vector in V remains in V. In condition (a3), θ is naturally called the *zero vector* (or the *additive identity*). In (a4), the vector $-\mathbf{v}$ is called the *additive inverse* of $\mathbf{v}$, and (a4) asserts that the equation $\mathbf{v} + \mathbf{x} = \boldsymbol{\theta}$ has a solution in V. When the set of scalars S is the set of real numbers, V is called a *real vector space;* and as we have said, we will consider only real vector spaces.

Examples of Vector Spaces

We already have two familiar examples of vector spaces, namely, R^n and the set of all $(m \times n)$ matrices. It is easy to verify that these are vector spaces, and the verification is sketched in the first two examples below.

EXAMPLE 1   For any positive integer n, verify that R^n is a real vector space.

SOLUTION   Theorem 1 of Section 2.2 shows that R^n satisfies the properties listed in Definition 1, so R^n is a real vector space. $\square$

Example 2 may strike the reader as being a little unusual since we are considering matrices as elements in a vector space. The example, however, illustrates the flexibility of the vector-space concept; any set of entities that has addition and scalar multiplication operations can be a vector space, provided that the addition and the scalar multiplication satisfy the requirements of Definition 1.

EXAMPLE 2   Verify that the set of all (2×3) matrices with real entries is a real vector space.

SOLUTION   Let A and B be any (2×3) matrices, and let addition and scalar multiplication be defined as in Definitions 6 and 7 of Section 1.6. Therefore, $A + B$ and aA are defined by

$$A + B = \begin{bmatrix} a_{11} & a_{12} & a_{13} \\ a_{21} & a_{22} & a_{23} \end{bmatrix} + \begin{bmatrix} b_{11} & b_{12} & b_{13} \\ b_{21} & b_{22} & b_{23} \end{bmatrix}$$

$$= \begin{bmatrix} a_{11} + b_{11} & a_{12} + b_{12} & a_{13} + b_{13} \\ a_{21} + b_{21} & a_{22} + b_{22} & a_{23} + b_{23} \end{bmatrix}$$

$$aA = a\begin{bmatrix} a_{11} & a_{12} & a_{13} \\ a_{21} & a_{22} & a_{23} \end{bmatrix} = \begin{bmatrix} aa_{11} & aa_{12} & aa_{13} \\ aa_{21} & aa_{22} & aa_{23} \end{bmatrix}.$$

From these definitions it is obvious that both the sum $A + B$ and the scalar multiple aA are again (2×3) matrices; so (c1) and (c2) of Definition 1 hold. Properties (a1), (a2), (a3), and (a4) follow from Theorem 7 of Section 1.7; and (m1), (m2), and (m3) are proved in Theorems 8 and 9 of Section 1.7. Property (m4) is immediate from the definition of scalar multiplication [clearly $1A = A$ for any (2×3) matrix A]. For emphasis we recall that the zero element in this vector space is the matrix

$$\mathscr{O} = \begin{bmatrix} 0 & 0 & 0 \\ 0 & 0 & 0 \end{bmatrix},$$

and clearly $A + \mathcal{O} = A$ for any (2×3) matrix A. We further observe that $(-1)A$ is the additive inverse for A because

$$A + (-1)A = \mathcal{O}.$$

[That is, $(-1)A$ is a matrix we can add to A to produce the zero element $\mathcal{O}$.] A duplication of these arguments shows that for any m and n the set of all $(m \times n)$ matrices with real entries is a real vector space.

The next three examples show that certain sets of functions have a natural vector-space structure.

EXAMPLE 3 Let $\mathcal{P}_2$ denote the set of all real polynomials of degree 2 or less. Verify that $\mathcal{P}_2$ is a real vector space.

SOLUTION Note that a natural addition is associated with polynomials. For example, let $p(x)$ and $q(x)$ be the polynomials

$$p(x) = 2x^2 - x + 3 \quad \text{and} \quad q(x) = x^2 + 2x - 1.$$

Then the sum $r(x) = p(x) + q(x)$ is the polynomial $r(x) = 3x^2 + x + 2$. Scalar multiplication is defined similarly; so if $s(x) = 2q(x)$, then

$$s(x) = 2x^2 + 4x - 2.$$

Given this natural addition and scalar multiplication associated with the set $\mathcal{P}_2$, it seems reasonable to expect that $\mathcal{P}_2$ is a real vector space.

To establish this conclusion rigorously, we must be a bit more careful. To begin, we define $\mathcal{P}_2$ to be the set of all expressions (or functions) of the form

$$p(x) = a_2x^2 + a_1x + a_0, \tag{1}$$

where a_2, a_1, and a_0 are any real constants. Thus the following polynomials are vectors in $\mathcal{P}_2$:

$$p_1(x) = x^2 - x + 3, \qquad p_2(x) = x^2 + 1, \qquad p_3(x) = x - 2$$
$$p_4(x) = 2x, \qquad p_5(x) = 7, \qquad p_6(x) = 0.$$

For instance, we see that $p_2(x)$ has the form of Eq. (1), with $a_2 = 1$, $a_1 = 0$, and $a_0 = 1$. Similarly, $p_4(x)$ is in $\mathcal{P}_2$ because $p_4(x)$ is a function of the form (1), where $a_2 = 0$, $a_1 = 2$, and $a_0 = 0$. Finally, $p_6(x)$ has the form (1) with $a_2 = 0$, $a_1 = 0$, and $a_0 = 0$. To define addition precisely, let

$$p(x) = a_2x^2 + a_1x + a_0 \quad \text{and} \quad q(x) = b_2x^2 + b_1x + b_0$$

be two vectors in $\mathcal{P}_2$. We define the sum $r(x) = p(x) + q(x)$ to be the polynomial

$$r(x) = (a_2 + b_2)x^2 + (a_1 + b_1)x + (a_0 + b_0);$$

and we define the scalar multiple $s(x) = cp(x)$ to be the polynomial

$$s(x) = (ca_2)x^2 + (ca_1)x + (ca_0).$$

We leave it to the reader to verify that these algebraic operations meet the requirements of Definition 1; we note only that we choose the zero vector to be the polynomial that is identically zero. That is, the zero element in $\mathscr{P}_2$ is the polynomial $\theta(x)$, where $\theta(x) \equiv 0$; or in terms of Eq. (1), $\theta(x)$ is defined by

$$\theta(x) = 0x^2 + 0x + 0.$$ ☐

EXAMPLE 4 In this example we take $\mathscr{P}_n$ to be the set of all real polynomials of degree n or less. That is, $\mathscr{P}_n$ consists of all functions $p(x)$ of the form

$$p(x) = a_n x^n + a_{n-1} x^{n-1} + \cdots + a_2 x^2 + a_1 x + a_0,$$

where $a_n, a_{n-1}, \ldots, a_2, a_1, a_0$ are any real constants. With addition and scalar multiplication defined as in Example 3, it is easy to show that $\mathscr{P}_n$ is a real vector space. ☐

The next example presents one of the more important vector spaces in applications.

EXAMPLE 5 Let $C[a, b]$ be the set of functions defined by

$$C[a, b] = \{ f(x): f(x) \text{ is a real-valued continuous function, } a \le x \le b \}.$$

Verify that $C[a, b]$ is a real vector space.

SOLUTION $C[a, b]$ has a natural addition, just as $\mathscr{P}_n$. If f and g are vectors in $C[a, b]$, then we define the sum $h = f + g$ to be the function h given by

$$h(x) = f(x) + g(x), \qquad a \le x \le b.$$

Similarly, if c is a scalar, then the scalar multiple $q = cf$ is the function

$$q(x) = cf(x), \qquad a \le x \le b.$$

As a concrete example, if $f(x) = e^x$ and $g(x) = \sin x$, then $3f + g$ is the function r, where the action of r is defined by $r(x) = 3e^x + \sin x$. Note that the closure properties, (c1) and (c2), follow from elementary results of calculus—sums and

Function Spaces

The giant step of expanding vector spaces from R^n to spaces of functions was a combined effort of many mathematicians. Probably foremost among them, however, was David Hilbert (1862–1943), for whom Hilbert spaces are named. Hilbert had great success in solving several important contemporary problems by emphasizing abstraction and an axiomatic approach. His ideas on abstract spaces came largely from his work on important integral equations in physics. Hilbert related integral equations to problems of infinitely-many equations in infinitely-many unknowns, a natural extension of a fundamental problem in the setting of R^n. Great credit for expansion of vector-space ideas is also given to the work of Riesz, Fischer, Fréchet, and Weyl. In particular, Hermann Weyl (1885–1955) was known for his stress on the rigorous application of axiomatic logic rather than visual plausibility, which was all too often accepted as proof.

scalar multiples of continuous functions are again continuous functions. The remaining eight properties of Definition 1 are easily seen to hold in $C[a,b]$; the verification proceeds exactly as in $\mathscr{P}_n$. □

Note that any polynomial can be regarded as a continuous function on any interval $[a,b]$. Thus for any given positive integer n, $\mathscr{P}_n$ is not only a subset of $C[a,b]$ but also a vector space contained in the vector space $C[a,b]$. This concept of a vector space that contains a smaller vector space (or a "vector subspace") is quite important and is one topic of the next section.

Further Vector-Space Properties

The algebraic operations in a vector space have additional properties that can be derived from the 10 fundamental properties listed in Definition 1. The first of these, the cancellation laws for vector addition, are straightforward to prove and will be left as exercises.

Cancellation Laws for Vector Addition

Let V be a vector space, and let $\mathbf{u}$, $\mathbf{v}$, and $\mathbf{w}$ be vectors in V.

1. If $\mathbf{u} + \mathbf{v} = \mathbf{u} + \mathbf{w}$, then $\mathbf{v} = \mathbf{w}$.
2. If $\mathbf{v} + \mathbf{u} = \mathbf{w} + \mathbf{u}$, then $\mathbf{v} = \mathbf{w}$.

Some additional properties of vector spaces are summarized in Theorem 1.

THEOREM 1

If V is a vector space, then:

1. The zero vector $\boldsymbol{\theta}$ is unique.
2. For each $\mathbf{v}$, the additive inverse $-\mathbf{v}$ is unique.
3. $0\mathbf{v} = \boldsymbol{\theta}$ for every $\mathbf{v}$ in V, where 0 is the zero scalar.
4. $a\boldsymbol{\theta} = \boldsymbol{\theta}$ for every scalar a.
5. If $a\mathbf{v} = \boldsymbol{\theta}$, then $a = 0$ or $\mathbf{v} = \boldsymbol{\theta}$.
6. $(-1)\mathbf{v} = -\mathbf{v}$.

PROOF [We prove properties 1, 4, and 6 and leave the remaining properties as exercises.] We first prove property 1. Suppose that ξ is a vector in V such that $\mathbf{v} + \xi = \mathbf{v}$ for all $\mathbf{v}$ in V. Then setting $\mathbf{v} = \boldsymbol{\theta}$, we have

$$\boldsymbol{\theta} + \xi = \boldsymbol{\theta}. \tag{2}$$

By property (a3) of Definition 1, we know also that

$$\xi + \boldsymbol{\theta} = \xi. \tag{3}$$

But from property (a1) of Definition 1, we know that $\xi + \theta = \theta + \xi$; so using Eq. (2), property (a1), and Eq. (3), we conclude that

$$\theta = \theta + \xi = \xi + \theta = \xi,$$

or $\theta = \xi$.

We next prove property 4 of Theorem 1. We do so by observing that $\theta + \theta = \theta$, from property (a3) of Definition 1. Therefore if a is any scalar, we have from property (m2) of Definition 1 that

$$a\theta = a(\theta + \theta) = a\theta + a\theta. \tag{4}$$

Since $a\theta = a\theta + \theta$ by property (a3) of Definition 1, Eq. (4) becomes

$$a\theta + \theta = a\theta + a\theta.$$

The cancellation laws now yield $\theta = a\theta$.

Finally, we outline a proof for property 6 of Theorem 1 by displaying a sequence of equalities (the last equality is based on property 3, which is an exercise):

$$\mathbf{v} + (-1)\mathbf{v} = (1)\mathbf{v} + (-1)\mathbf{v} = [1 + (-1)]\mathbf{v} = 0\mathbf{v} = \boldsymbol{\theta}.$$

Thus $(-1)\mathbf{v}$ is a solution to the equation $\mathbf{v} + \mathbf{x} = \boldsymbol{\theta}$. But from property 2 of Theorem 1, the additive inverse $-\mathbf{v}$ is the only solution of $\mathbf{v} + \mathbf{x} = \boldsymbol{\theta}$; so we must have $(-1)\mathbf{v} = -\mathbf{v}$. Thus property 6 constitutes a formula for the additive inverse. This formula is not totally unexpected, but neither is it so obvious as it might seem, since a number of vector-space properties were required to prove it.

EXAMPLE 6 We conclude this section by introducing the *zero vector space*. The zero vector space contains only one vector, θ; and the arithmetic operations are defined by

$$\theta + \theta = \theta$$
$$k\theta = \theta.$$

It is easy to verify that the set $\{\boldsymbol{\theta}\}$ with the operations defined above is a vector space.

Exercises 4.2

For **u**, **v**, and **w** given in Exercises 1–3, calculate $\mathbf{u} - 2\mathbf{v}$, $\mathbf{u} - (2\mathbf{v} - 3\mathbf{w})$, and $-2\mathbf{u} - \mathbf{v} + 3\mathbf{w}$.

1. In the vector space of (2×3) matrices

$$\mathbf{u} = \begin{bmatrix} 2 & 1 & 3 \\ -1 & 1 & 2 \end{bmatrix}, \qquad \mathbf{v} = \begin{bmatrix} 1 & 4 & -1 \\ 5 & 2 & 7 \end{bmatrix},$$

$$\mathbf{w} = \begin{bmatrix} 4 & -5 & 11 \\ -13 & -1 & -1 \end{bmatrix}$$

2. In the vector space $\mathscr{P}_2$

$$\mathbf{u} = x^2 - 2, \qquad \mathbf{v} = x^2 + 2x - 1, \qquad \mathbf{w} = 2x + 1$$

3. In the vector space $C[0,1]$

$$\mathbf{u} = e^x, \quad \mathbf{v} = \sin x, \quad \mathbf{w} = \sqrt{x^2 + 1}$$

4. For $\mathbf{u}$, $\mathbf{v}$, and $\mathbf{w}$ in Exercise 2, find nonzero scalars c_1, c_2, c_3 such that $c_1\mathbf{u} + c_2\mathbf{v} + c_3\mathbf{w} = \boldsymbol{\theta}$. Are there nonzero scalars c_1, c_2, c_3 such that $c_1\mathbf{u} + c_2\mathbf{v} + c_3\mathbf{w} = \boldsymbol{\theta}$ for $\mathbf{u}$, $\mathbf{v}$, and $\mathbf{w}$ in Exercise 1?

5. For $\mathbf{u}$, $\mathbf{v}$, and $\mathbf{w}$ in Exercise 2, find scalars c_1, c_2, c_3 such that $c_1\mathbf{u} + c_2\mathbf{v} + c_3\mathbf{w} = x^2 + 6x + 1$. Show that there are no scalars c_1, c_2, c_3 such that $c_1\mathbf{u} + c_2\mathbf{v} + c_3\mathbf{w} = x^2$.

In Exercises 6–11, the given set is a subset of a vector space. Which of these subsets are also vector spaces in their own right? To answer this question, determine whether the subset satisfies the 10 properties of Definition 1. (*Note:* Because these sets are subsets of a vector space, properties (a1), (a2), (m1), (m2), (m3), and (m4) are automatically satisfied.)

6. $S = \{\mathbf{v}$ in $R^4: v_1 + v_4 = 0\}$
7. $S = \{\mathbf{v}$ in $R^4: v_1 + v_4 = 1\}$
8. $P = \{p(x)$ in $\mathscr{P}_2: p(0) = 0\}$
9. $P = \{p(x)$ in $\mathscr{P}_2: p''(0) \neq 0\}$
10. $P = \{p(x)$ in $\mathscr{P}_2: p(x) = p(-x)$ for all $x\}$
11. $P = \{p(x)$ in $\mathscr{P}_2: p(x)$ has degree 2$\}$.

In Exercises 12–16, V is the vector space of all real (3×4) matrices. Which of the given subsets of V are also vector spaces?

12. $S = \{A$ in $V: a_{11} = 0\}$
13. $S = \{A$ in $V: a_{11} + a_{23} = 0\}$
14. $S = \{A$ in $V: |a_{11}| + |a_{21}| = 1\}$
15. $S = \{A$ in $V: a_{32} \neq 0\}$
16. $S = \{A$ in $V:$ each a_{ij} is an integer$\}$

17. Let Q denote the set of all (2×2) nonsingular matrices with the usual matrix addition and scalar multiplication. Show that Q is not a vector space by exhibiting specific matrices in Q that violate property (c1) of Definition 1. Also show that properties (c2) and (a3) are not met.

18. Let Q denote the set of all (2×2) singular matrices with the usual matrix addition and scalar multiplication. Determine whether Q is a vector space.

19. Let Q denote the set of all (2×2) symmetric matrices with the usual matrix addition and scalar multiplication. Verify that Q is a vector space.

20. Prove the cancellation laws for vector addition.

21. Prove property 2 of Theorem 1. (*Hint:* Cf. the proof of Theorem 16 in Section 1.10.)

22. Prove property 3 of Theorem 1. (*Hint:* Note that $0\mathbf{v} = (0 + 0)\mathbf{v}$. Now mimic the proof given for property 4.)

23. Prove property 5 of Theorem 1. (If $a \neq 0$ then multiply both sides of $a\mathbf{v} = \boldsymbol{\theta}$ by a^{-1}. Use properties (m1) and (m4) of Definition 1 and use property 4 of Theorem 1.)

24. Prove that the zero vector space, defined in Example 6, is indeed a vector space.

In Exercises 25–29, the given set is a subset of $C[-1,1]$. Which of these are also vector spaces?

25. $F = \{f(x)$ in $C[-1,1]: f(-1) = f(1)\}$
26. $F = \{f(x)$ in $C[-1,1]: f(x)=0$ for $-1/2 \le x \le 1/2\}$
27. $F = \{f(x)$ in $C[-1,1]: f(1) = 1\}$
28. $F = \{f(x)$ in $C[-1,1]: f(1) = 0\}$
29. $F = \left\{f(x)$ in $C[-1,1]: \int_{-1}^{1} f(x)\,dx = 0\right\}$

30. The set $C^2[a,b]$ is defined to be the set of all real-valued functions $f(x)$ defined on $[a,b]$, where $f(x)$, $f'(x)$, and $f''(x)$ are continuous on $[a,b]$. Verify that $C^2[a,b]$ is a vector space by citing the appropriate theorems on continuity and differentiability from calculus.

31. The following are subsets of the vector space $C^2[-1,1]$. Which of these are vector spaces?
    a) $F = \{f(x)$ in $C^2[-1,1]: f''(x) + f(x) = 0, -1 \le x \le 1\}$
    b) $F = \{f(x)$ in $C^2[-1,1]: f''(x) + f(x) = x^2, -1 \le x \le 1\}$

32. Show that the set $\mathscr{P}$ of all real polynomials is a vector space.

33. Let $\mathscr{F}(R)$ denote the set of all real-valued functions defined on the reals. Thus

    $$\mathscr{F}(R) = \{f: f \text{ is a function}, f: R \to R\}.$$

    With addition of functions and scalar multiplication defined as in Example 5, show that $\mathscr{F}(R)$ is a vector space.

34. Let

$$V = \left\{\mathbf{x}: \mathbf{x} = \begin{bmatrix} x_1 \\ x_2 \end{bmatrix}, \text{ where } x_1 \text{ and } x_2 \text{ are in } R\right\}.$$

For **u** and **v** in V and c in R, define the operations of addition and scalar multiplication on V by

$$\mathbf{u} + \mathbf{v} = \begin{bmatrix} u_1 \\ u_2 \end{bmatrix} + \begin{bmatrix} v_1 \\ v_2 \end{bmatrix} = \begin{bmatrix} u_1 + v_1 + 1 \\ u_2 + v_2 - 1 \end{bmatrix} \quad \text{and}$$

$$c\mathbf{u} = \begin{bmatrix} cu_1 \\ cu_2 \end{bmatrix}. \tag{5}$$

a) Show that the operations defined in (5) satisfy properties (c1), (c2), (a1)–(a4), (m1), and (m4) of Definition 1.
b) Give examples to illustrate that properties (m2) and (m3) are not satisfied by the operations defined in (5).

35. Let

$$V = \{\mathbf{x}: \mathbf{x} = \begin{bmatrix} x_1 \\ x_2 \end{bmatrix}, \text{ where } x_1 \text{ and } x_2 \text{ are in } R\}.$$

For **u** and **v** in V and c in R, define the operations of addition and scalar multiplication on V by

$$\mathbf{u} + \mathbf{v} = \begin{bmatrix} u_1 \\ u_2 \end{bmatrix} + \begin{bmatrix} v_1 \\ v_2 \end{bmatrix} = \begin{bmatrix} u_1 + v_1 \\ u_2 + v_2 \end{bmatrix} \quad \text{and}$$

$$c\mathbf{u} = \begin{bmatrix} 0 \\ 0 \end{bmatrix}. \tag{6}$$

Show that the operations defined in (6) satisfy all the properties of Definition 1 except (m4). [Note that the addition given in (6) is the usual addition of R^2. Since R^2 is a vector space, all the additive properties of Definition 1 are satisfied.]

36. Let

$$V = \{\mathbf{x}: \mathbf{x} = \begin{bmatrix} x_1 \\ x_2 \end{bmatrix}, \text{ where } x_2 > 0\}.$$

For **u** and **v** in V and c in R, define addition and scalar multiplication by

$$\mathbf{u} + \mathbf{v} = \begin{bmatrix} u_1 \\ u_2 \end{bmatrix} + \begin{bmatrix} v_1 \\ v_2 \end{bmatrix} = \begin{bmatrix} u_1 + v_1 \\ u_2 v_2 \end{bmatrix} \quad \text{and}$$

$$c\mathbf{u} = \begin{bmatrix} cu_1 \\ u_2^c \end{bmatrix}. \tag{7}$$

With the operations defined in (7), show that V is a vector space.

4.3   Subspaces

Chapter 2 demonstrated that whenever W is a p-dimensional subspace of R^n, then W behaves essentially like R^p (for instance, any set of $p + 1$ vectors in W is linearly dependent). The situation is much the same in a general vector space V. In this setting, certain subsets of V inherit the vector-space structure of V and are vector spaces in their own right.

DEFINITION 2   If V and W are real vector spaces, and if W is a nonempty subset of V, then W is called a *subspace* of V.

Subspaces have considerable practical importance and are useful in problems involving approximation, optimization, differential equations, and so on. The vector-space/subspace framework allows us to pose and rigorously answer questions such as, How can we find good polynomial approximations to complicated functions? and How can we generate good approximate solutions to differential equations? Questions such as these are at the heart of many technical problems; and vector-space techniques, together with the computational power of the digital computer, are useful in helping to answer them.

As was the case in R^n, it is fairly easy to recognize when a subset of a vector space V is actually a subspace. Specifically, the following restatement of Theorem 2 of Section 2.2 holds in any vector space.

THEOREM 2

Let W be a subset of a vector space V. Then W is a subspace of V if and only if the following conditions are met:

s1) The zero vector, θ, of V is in W.
s2) $\mathbf{u} + \mathbf{v}$ is in W whenever $\mathbf{u}$ and $\mathbf{v}$ are in W.
s3) $a\mathbf{u}$ is in W whenever $\mathbf{u}$ is in W and a is any scalar.

The proof of Theorem 2 coincides with the proof given in Section 2.2 with one minor exception. In R^n it is easily seen that $-\mathbf{v} = (-1)\mathbf{v}$ for any vector $\mathbf{v}$. In a general vector space V, this is a consequence of Theorem 1 of Section 4.2.

Examples of Subspaces

If we are given that W is a subset of a known vector space V, Theorem 2 simplifies the task of determining whether or not W is itself a vector space. Instead of testing all 10 properties of Definition 1, Theorem 2 states that we need only verify that properties (s1)–(s3) hold. Furthermore, just as in Chapter 2, a subset W of V will be specified by certain defining relationships that tell whether a vector $\mathbf{u}$ is in W. Thus to verify that (s1) holds, it must be shown that the zero vector, θ, of V satisfies the specification given for W. To check (s2) and (s3), we select two arbitrary vectors, say $\mathbf{u}$ and $\mathbf{v}$, that satisfy the defining relationships of W (that is, $\mathbf{u}$ and $\mathbf{v}$ are in W). We then test $\mathbf{u} + \mathbf{v}$ and $a\mathbf{u}$ to see whether they also satisfy the defining relationships of W. (That is, do $\mathbf{u} + \mathbf{v}$ and $a\mathbf{u}$ belong to W?) The next three examples illustrate the use of Theorem 2.

EXAMPLE 1 Let V be the vector space of all real (2×2) matrices, and let W be the subset of V specified by

$$W = \{A: A = \begin{bmatrix} 0 & a_{12} \\ a_{21} & 0 \end{bmatrix}, \quad a_{12} \text{ and } a_{21} \text{ any real scalars}\}.$$

Verify that W is a subspace of V.

SOLUTION The zero vector for V is the (2×2) zero matrix $\mathcal{O}$, and $\mathcal{O}$ is in W since it satisfies the defining relationships of W. If A and B are any two vectors in W, then A and B have the form

$$A = \begin{bmatrix} 0 & a_{12} \\ a_{21} & 0 \end{bmatrix}, \quad B = \begin{bmatrix} 0 & b_{12} \\ b_{21} & 0 \end{bmatrix}.$$

Thus $A + B$ and aA have the form

$$A + B = \begin{bmatrix} 0 & a_{12} + b_{12} \\ a_{21} + b_{21} & 0 \end{bmatrix}, \quad aA = \begin{bmatrix} 0 & aa_{12} \\ aa_{21} & 0 \end{bmatrix}.$$

Therefore, $A + B$ and aA are in W, and we conclude that W is a subspace of the set of all real (2×2) matrices. ☐

EXAMPLE 2 Let W be the subset of $C[a, b]$ (cf. Example 5 of Section 4.2) defined by

$$W = \{f(x) \text{ in } C[a, b]: f(a) = f(b)\}.$$

Verify that W is a subspace of $C[a, b]$.

SOLUTION The zero vector in $C[a, b]$ is the zero function, $\theta(x)$, defined by $\theta(x) = 0$ for all x in the interval $[a, b]$. In particular, $\theta(a) = \theta(b)$ since $\theta(a) = 0$ and $\theta(b) = 0$. Therefore, $\theta(x)$ is in W. Now let $g(x)$ and $h(x)$ be any two functions that are in W, that is,

$$g(a) = g(b) \quad \text{and} \quad h(a) = h(b). \tag{1}$$

The sum of $g(x)$ and $h(x)$ is the function $s(x)$ defined by $s(x) = g(x) + h(x)$. To see that $s(x)$ is in W, note that property (1) gives

$$s(a) = g(a) + h(a) = g(b) + h(b) = s(b).$$

Similarly, if c is a scalar, then it is immediate from property (1) that $cg(a) = cg(b)$. It follows that $cg(x)$ is in W. Theorem 2 now implies that W is a subspace of $C[a, b]$. ☐

The next example illustrates how to use Theorem 2 to show that a subset of a vector space is not a vector space. Recall from Chapter 2 that if a subset fails to satisfy any one of the properties (s1), (s2), or (s3), then it is not a subspace.

EXAMPLE 3 Let V be the vector space of all (2×2) matrices, and let W be the subspace of V specified by

$$W = \{A: A = \begin{bmatrix} a & b \\ c & d \end{bmatrix}, \quad ad = 0 \quad \text{and} \quad bc = 0\}.$$

Show that W is not a subspace of V.

SOLUTION It is straightforward to show that W satisfies properties (s1) and (s3) of Theorem 2. Thus to demonstrate that W is not a subspace of V, we must show that (s2) fails. It suffices to give a specific example that illustrates the failure of (s2). For example, define A and B by

$$A = \begin{bmatrix} 1 & 0 \\ 0 & 0 \end{bmatrix} \quad \text{and} \quad B = \begin{bmatrix} 0 & 0 \\ 0 & 1 \end{bmatrix}.$$

Then A and B are in W, but $A + B$ is not, since

$$A + B = \begin{bmatrix} 1 & 0 \\ 0 & 1 \end{bmatrix}.$$

In particular, $ad = (1)(1) = 1$, so $ad \neq 0$. $\qquad\square$

If $n \leq m$, then $\mathscr{P}_n$ is a subspace of $\mathscr{P}_m$. We can verify this assertion directly from Definition 2 since we have already shown that $\mathscr{P}_n$ and $\mathscr{P}_m$ are each real vector spaces, and $\mathscr{P}_n$ is a subset of $\mathscr{P}_m$. Similarly, for any n, $\mathscr{P}_n$ is a subspace of $C[a, b]$. Again this assertion follows directly from Definition 2 since any polynomial is continuous on any interval $[a, b]$. Therefore, $\mathscr{P}_n$ can be considered a subspace of $C[a, b]$, as well as a vector space in its own right.

Spanning Sets

The vector-space structure as given in Definition 1 guarantees that the notion of a linear combination makes sense in a general vector space. Specifically, the vector $\mathbf{v}$ is a ***linear combination*** of the vectors $\mathbf{v}_1, \mathbf{v}_2, \ldots, \mathbf{v}_m$ provided that there exist scalars $a_1, a_2, \ldots, a_m$ such that

$$\mathbf{v} = a_1\mathbf{v}_1 + a_2\mathbf{v}_2 + \cdots + a_m\mathbf{v}_m.$$

The next example illustrates this concept in the vector space $\mathscr{P}_2$.

EXAMPLE 4 In $\mathscr{P}_2$ let $p(x)$, $p_1(x)$, $p_2(x)$, and $p_3(x)$ be defined by $p(x) = -1 + 2x^2$, $p_1(x) = 1 + 2x - 2x^2$, $p_2(x) = -1 - x$, and $p_3(x) = -3 - 4x + 4x^2$. Express $p(x)$ as a linear combination of $p_1(x)$, $p_2(x)$, and $p_3(x)$.

SOLUTION Setting $p(x) = a_1p_1(x) + a_2p_2(x) + a_3p_3(x)$ yields

$$-1 + 2x^2 = a_1(1 + 2x - 2x^2) + a_2(-1 - x) + a_3(-3 - 4x + 4x^2).$$

Equating coefficients yields the system of equations

$$\begin{aligned} a_1 - a_2 - 3a_3 &= -1 \\ 2a_1 - a_2 - 4a_3 &= 0 \\ -2a_1 \qquad + 4a_3 &= 2. \end{aligned}$$

This system has the unique solution $a_1 = 3$, $a_2 = -2$, and $a_3 = 2$. We can easily check that

$$p(x) = 3p_1(x) - 2p_2(x) + 2p_3(x). \qquad\square$$

The very useful concept of a "spanning set" is suggested by the discussion above.

DEFINITION 3   Let V be a vector space, and let $Q = \{\mathbf{v}_1, \mathbf{v}_2, \ldots, \mathbf{v}_m\}$ be a set of vectors in V. If every vector $\mathbf{v}$ in V is a linear combination of vectors in Q,

$$\mathbf{v} = a_1\mathbf{v}_1 + a_2\mathbf{v}_2 + \cdots + a_m\mathbf{v}_m,$$

then we say that Q is a **spanning set** for V.

For many vector spaces V, it is relatively easy to find a "natural" spanning set. For example, it is easily seen that $\{1, x, x^2\}$ is a spanning set for $\mathcal{P}_2$ and, in general, $\{1, x, \ldots, x^n\}$ is a spanning set for $\mathcal{P}_n$. The vector space of all (2×2) matrices is spanned by the set $\{E_{11}, E_{12}, E_{21}, E_{22}\}$, where

$$E_{11} = \begin{bmatrix} 1 & 0 \\ 0 & 0 \end{bmatrix}, \quad E_{12} = \begin{bmatrix} 0 & 1 \\ 0 & 0 \end{bmatrix}, \quad E_{21} = \begin{bmatrix} 0 & 0 \\ 1 & 0 \end{bmatrix}, \quad \text{and} \quad E_{22} = \begin{bmatrix} 0 & 0 \\ 0 & 1 \end{bmatrix}.$$

More generally, if the $(m \times n)$ matrix E_{ij} is the matrix with 1 as the ijth entry and zeros elsewhere, then $\{E_{ij}: 1 \le i \le m, 1 \le j \le n\}$ is a spanning set for the vector space of $(m \times n)$ real matrices.

If $Q = \{\mathbf{v}_1, \mathbf{v}_2, \ldots, \mathbf{v}_k\}$ is a set of vectors in a vector space V, then, as in Section 2.3, the **span** of Q, denoted $\text{Sp}(Q)$, is the set of all linear combinations of $\mathbf{v}_1, \mathbf{v}_2, \ldots, \mathbf{v}_k$:

$$\text{Sp}(Q) = \{\mathbf{v}: \mathbf{v} = a_1\mathbf{v}_1 + a_2\mathbf{v}_2 + \cdots + a_k\mathbf{v}_k\}.$$

From closure properties (c1) and (c2) of Definition 1, it is obvious that $\text{Sp}(Q)$ is a subset of V. In fact, the proof of Theorem 3 in Section 2.3 is valid in a general vector space, so we have the following theorem.

THEOREM 3   If V is a vector space and $Q = \{\mathbf{v}_1, \mathbf{v}_2, \ldots, \mathbf{v}_k\}$ is a set of vectors in V, then $\text{Sp}(Q)$ is a subspace of V.

The connection between spanning sets and the span of a set is fairly obvious. If W is a subspace of V and $Q \subseteq W$, then Q is a spanning set for W if and only if $W = \text{Sp}(Q)$. As the next three examples illustrate, it is often easy to obtain a spanning set for a subspace W when an algebraic specification for W is given.

EXAMPLE 5   Let V be the vector space of all real (2×2) matrices, and let W be the subspace given in Example 1:

$$W = \{A: A = \begin{bmatrix} 0 & a_{12} \\ a_{21} & 0 \end{bmatrix}, \quad a_{12} \text{ and } a_{21} \text{ any real scalars}\}.$$

Find a spanning set for W.

SOLUTION One obvious spanning set for W is seen to be the set of vectors $Q = \{A_1, A_2\}$, where

$$A_1 = \begin{bmatrix} 0 & 1 \\ 0 & 0 \end{bmatrix} \quad \text{and} \quad A_2 = \begin{bmatrix} 0 & 0 \\ 1 & 0 \end{bmatrix}.$$

To verify this assertion, suppose A is in W, where

$$A = \begin{bmatrix} 0 & a_{12} \\ a_{21} & 0 \end{bmatrix}.$$

Then clearly $A = a_{12}A_1 + a_{21}A_2$, and therefore Q is a spanning set for W.

EXAMPLE 6 Let W be the subspace of $\mathscr{P}_2$ defined by

$$W = \{p(x): p(x) = a_0 + a_1x + a_2x^2, \quad \text{where} \quad a_2 = -a_1 + 2a_0\}.$$

Exhibit a spanning set for W.

SOLUTION Let $p(x) = a_0 + a_1x + a_2x^2$ be a vector in W. From the specification of W, we know that $a_2 = -a_1 + 2a_0$. That is,

$$\begin{aligned} p(x) &= a_0 + a_1x + a_2x^2 \\ &= a_0 + a_1x + (-a_1 + 2a_0)x^2 \\ &= a_0(1 + 2x^2) + a_1(x - x^2). \end{aligned}$$

Since every vector p in W is a linear combination of $p_1(x) = 1 + 2x^2$ and $p_2(x) = x - x^2$, we see that $\{p_1(x), p_2(x)\}$ is a spanning set for W.

A square matrix, $A = (a_{ij})$, is called **skew symmetric** if $A^T = -A$. Recall that the ijth entry of A^T is a_{ji}, the jith entry of A. Thus the entries of A must satisfy $a_{ji} = -a_{ij}$ in order for A to be skew symmetric. In particular, each entry, a_{ii}, on the main diagonal must be zero since $a_{ii} = -a_{ii}$.

EXAMPLE 7 Let W be the set of all (3×3) skew-symmetric matrices. Show that W is a subspace of the vector space V of all (3×3) matrices, and exhibit a spanning set for W.

SOLUTION Let $\mathcal{O}$ denote the (3×3) zero matrix. Clearly $\mathcal{O}^T = \mathcal{O} = -\mathcal{O}$, so $\mathcal{O}$ is in W. If A and B are in W, then $A^T = -A$ and $B^T = -B$. Therefore,

$$(A + B)^T = A^T + B^T = -A - B = -(A + B).$$

It follows that $A + B$ is skew symmetric; that is, $A + B$ is in W. Likewise, if c is a scalar, then

$$(cA)^T = cA^T = c(-A) = -(cA),$$

so cA is in W. By Theorem 2, W is a subspace of V. Moreover, the remarks preceding the example imply that W can be described by

$$W = \{A: A = \begin{bmatrix} 0 & a & b \\ -a & 0 & c \\ -b & -c & 0 \end{bmatrix}, \quad a, b, c \text{ any real numbers}\}.$$

From this description it is easily seen that a natural spanning set for W is the set $Q = \{A_1, A_2, A_3\}$, where

$$A_1 = \begin{bmatrix} 0 & 1 & 0 \\ -1 & 0 & 0 \\ 0 & 0 & 0 \end{bmatrix}, \quad A_2 = \begin{bmatrix} 0 & 0 & 1 \\ 0 & 0 & 0 \\ -1 & 0 & 0 \end{bmatrix}, \quad \text{and} \quad A_3 = \begin{bmatrix} 0 & 0 & 0 \\ 0 & 0 & 1 \\ 0 & -1 & 0 \end{bmatrix}. \quad \square$$

Finally, note that in Definition 3 we have implicitly assumed that spanning sets are finite. This is not a required assumption, and frequently $\mathrm{Sp}(Q)$ is defined as the set of all finite linear combinations of vectors from Q, where Q may be either an infinite set or a finite set. We do not need this full generality, and we will explore this idea no further other than to note later that one contrast between the vector space R^n and a general vector space V is that V might not possess a finite spanning set. An example of a vector space where the most natural spanning set is infinite is the vector space $\mathscr{P}$, consisting of all polynomials (we place no upper limit on the degree). Then, for instance, $\mathscr{P}_n$ is a subspace of $\mathscr{P}$ for each n, $n = 1, 2, 3, \dots$. A natural spanning set for $\mathscr{P}$ (in the generalized sense described above) is the infinite set

$$Q = \{1, x, x^2, \dots, x^k, \dots\}.$$

Exercises 4.3

Let V be the vector space of all (2×3) matrices. Which of the subsets in Exercises 1–4 are subspaces of V?

1. $W = \{A \text{ in } V: a_{11} + a_{13} = 1\}$
2. $W = \{A \text{ in } V: a_{11} - a_{12} + 2a_{13} = 0\}$
3. $W = \{A \text{ in } V: a_{11} - a_{12} = 0, a_{12} + a_{13} = 0,$ and $a_{23} = 0\}$
4. $W = \{A \text{ in } V: a_{11}a_{12}a_{13} = 0\}$

In Exercises 5–8, which of the given subsets of $\mathscr{P}_2$ are subspaces of $\mathscr{P}_2$?

5. $W = \{p(x) \text{ in } \mathscr{P}_2: p(0) + p(2) = 0\}$
6. $W = \{p(x) \text{ in } \mathscr{P}_2: p(1) = p(3)\}$
7. $W = \{p(x) \text{ in } \mathscr{P}_2: p(1)p(3) = 0\}$

8. $W = \{p(x) \text{ in } \mathscr{P}_2: p(1) = -p(-1)\}$

In Exercises 9–12, which of the given subsets of $C[-1, 1]$ are subspaces of $C[-1, 1]$?

9. $F = \{f(x) \text{ in } C[-1, 1]: f(-1) = -f(1)\}$
10. $F = \{f(x) \text{ in } C[-1, 1]: f(x) \geq 0 \text{ for all } x \text{ in } [-1, 1]\}$
11. $F = \{f(x) \text{ in } C[-1, 1]: f(-1) = -2 \text{ and } f(1) = 2\}$
12. $F = \{f(x) \text{ in } C[-1, 1]: f(1/2) = 0\}$

In Exercises 13–16, which of the given subsets of $C^2[-1, 1]$ (cf. Exercise 30 of Section 4.2) are subspaces of $C^2[-1, 1]$?

13. $F = \{f(x) \text{ in } C^2[-1, 1]: f''(0) = 0\}$

14. $F = \{f(x) \text{ in } C^2[-1, 1]: f''(x) - e^x f'(x) + xf(x) = 0,$
    $-1 \le x \le 1\}$

15. $F = \{f(x) \text{ in } C^2[-1, 1]: f''(x) + f(x) = \sin x,$
    $-1 \le x \le 1\}$

16. $F = \{f(x) \text{ in } C^2[-1, 1]: f''(x) = 0, -1 \le x \le 1\}$

In Exercises 17–21, express the given vector as a linear combination of the vectors in the given set Q.

17. $p(x) = -1 - 3x + 3x^2$ and $Q = \{p_1(x), p_2(x), p_3(x)\}$, where $p_1(x) = 1 + 2x + x^2$, $p_2(x) = 2 + 5x$, and $p_3(x) = 3 + 8x - 2x^2$

18. $p(x) = -2 - 4x + x^2$ and

    $Q = \{p_1(x), p_2(x), p_3(x), p_4(x)\}$,

    and where $p_1(x) = 1 + 2x^2 + x^3$, $p_2(x) = 1 + x + 2x^3$, $p_3(x) = -1 - 3x + 4x^2 - 4x^3$, and $p_4(x) = 1 + 2x - x^2 + x^3$

19. $A = \begin{bmatrix} -2 & -4 \\ 1 & 0 \end{bmatrix}$ and $Q = \{B_1, B_2, B_3, B_4\}$, where

    $B_1 = \begin{bmatrix} 1 & 0 \\ 2 & 1 \end{bmatrix}$, $B_2 = \begin{bmatrix} 1 & 1 \\ 0 & 2 \end{bmatrix}$, $B_3 = \begin{bmatrix} -1 & -3 \\ 4 & -4 \end{bmatrix}$, and

    $B_4 = \begin{bmatrix} 1 & 2 \\ -1 & 1 \end{bmatrix}$

20. $f(x) = e^x$ and $Q = \{\sinh x, \cosh x\}$

21. $f(x) = \cos 2x$ and $Q = \{\sin^2 x, \cos^2 x\}$

22. Let V be the vector space of all (2×2) matrices. The subset W of V defined by

    $W = \{A \text{ in } V: a_{11} - a_{12} = 0, \quad a_{12} + a_{22} = 0\}$

    is a subspace of V. Find a spanning set for W. (*Hint:* Observe that A is in W if and only if A has the form

    $A = \begin{bmatrix} a_{11} & a_{11} \\ a_{21} & -a_{11} \end{bmatrix}$,

    where a_{11} and a_{21} are arbitrary.)

23. Let W be the subset of $\mathscr{P}_3$ defined by

    $W = \{p(x) \text{ in } \mathscr{P}_3: p(1) = p(-1) \text{ and } p(2) = p(-2)\}$.

    Show that W is a subspace of $\mathscr{P}_3$ and find a spanning set for W.

24. Let W be the subset of $\mathscr{P}_3$ defined by

    $W = \{p(x) \text{ in } \mathscr{P}_3: p(1) = 0 \text{ and } p'(-1) = 0\}$.

    Show that W is a subspace of $\mathscr{P}_3$ and find a spanning set for W.

25. Find a spanning set for each of the subsets that is a subspace in Exercises 1–8.

26. Show that the set W of all symmetric (3×3) matrices is a subspace of the vector space of all (3×3) matrices. Find a spanning set for W.

27. The trace of an $(n \times n)$ matrix $A = (a_{ij})$, denoted $\text{tr}(A)$, is defined to be the sum of the diagonal elements of A; that is, $\text{tr}(A) = a_{11} + a_{22} + \cdots + a_{nn}$. Let V be the vector space of all (3×3) matrices, and let W be defined by

    $W = \{A \text{ in } V: \text{tr}(A) = 0\}$.

    Show that W is a subspace of V and exhibit a spanning set for W.

28. Let A be an $(n \times n)$ matrix. Show that $B = (A + A^T)/2$ is symmetric and that $C = (A - A^T)/2$ is skew symmetric.

29. Use Exercise 28 to show that every $(n \times n)$ matrix can be expressed as the sum of a symmetric matrix and a skew-symmetric matrix.

30. Use Exercises 26 and 29 and Example 7 to construct a spanning set for the vector space of all (3×3) matrices where the spanning set consists entirely of symmetric and skew-symmetric matrices. Specify how a (3×3) matrix $A = (a_{ij})$ can be expressed by using this spanning set.

31. Let V be the set of all (3×3) upper-triangular matrices and note that V is a vector space. Each of the subsets W is a subspace of V. Find a spanning set for W.
    a) $W = \{A \text{ in } V: a_{11} = 0, a_{22} = 0, a_{33} = 0\}$
    b) $W = \{A \text{ in } V: a_{11} + a_{22} + a_{33} = 0, a_{12} + a_{23} = 0\}$
    c) $W = \{A \text{ in } V: a_{11} = a_{12}, a_{13} = a_{23}, a_{22} = a_{33}\}$
    d) $W = \{A \text{ in } V: a_{11} = a_{22}, a_{22} - a_{33} = 0, a_{12} + a_{23} = 0\}$

32. Let $p(x) = a_0 + a_1 x + a_2 x^2$ be a vector in $\mathscr{P}_2$. Find b_0, b_1, and b_2 in terms of a_0, a_1, and a_2 so that $p(x) = b_0 + b_1(x + 1) + b_2(x + 1)^2$. (*Hint:* Equate the coefficients of like powers of x.) Represent $q(x) = 1 - x + 2x^2$ and $r(x) = 2 - 3x + x^2$ in terms of the spanning set $\{1, x + 1, (x + 1)^2\}$.

33. Let A be an arbitrary matrix in the vector space of all (2×2) matrices:

    $A = \begin{bmatrix} a & b \\ c & d \end{bmatrix}$.

Find scalars x_1, x_2, x_3, x_4 in terms of a, b, c, and d such that $A = x_1 B_1 + x_2 B_2 + x_3 B_3 + x_4 B_4$, where

$$B_1 = \begin{bmatrix} 1 & 0 \\ 1 & -2 \end{bmatrix}, \quad B_2 = \begin{bmatrix} 2 & 1 \\ 1 & -2 \end{bmatrix},$$

$$B_3 = \begin{bmatrix} -1 & 3 \\ -3 & 6 \end{bmatrix}, \quad \text{and} \quad B_4 = \begin{bmatrix} 1 & 1 \\ -2 & 5 \end{bmatrix}.$$

Represent the matrices

$$C = \begin{bmatrix} 0 & 2 \\ -1 & 1 \end{bmatrix} \quad \text{and} \quad D = \begin{bmatrix} 2 & 1 \\ 0 & 1 \end{bmatrix}$$

in terms of the spanning set $\{B_1, B_2, B_3, B_4\}$

4.4   Linear Independence, Bases, and Coordinates

One of the central ideas of Chapters 1 and 2 was linear independence. As we will presently see, this concept generalizes directly to vector spaces. With the concepts of linear independence and spanning sets, it is then easy to extend the idea of a basis to our vector-space setting. The notion of a basis is one of the most fundamental concepts in the study of vector spaces. For example, in certain vector spaces a basis can be used to produce a "coordinate system" for the space. As a consequence, a real vector space with a basis of n vectors behaves "essentially" like R^n. Moreover, this coordinate system sometimes permits a geometric perspective in an otherwise nongeometric setting.

Linear Independence

We begin by restating Definition 11 of Section 1.8 in a general vector-space setting.

DEFINITION 4   Let V be a vector space, and let $\{v_1, v_2, \ldots, v_p\}$ be a set of vectors in V. This set is **linearly dependent** if there are scalars $a_1, a_2, \ldots, a_p$, not all of which are zero, such that

$$a_1 v_1 + a_2 v_2 + \cdots + a_p v_p = \theta. \tag{1}$$

The set $\{v_1, v_2, \ldots, v_p\}$ is **linearly independent** if it is not linearly dependent; that is, the only scalars for which Eq. (1) holds are the scalars $a_1 = a_2 = \cdots = a_p = 0$.

Note that as a consequence of property 3 of Theorem 1 in Section 4.2, the vector equation (1) in Definition 4 always has the trivial solution $a_1 = a_2 = \cdots = a_p = 0$. Thus the set $\{v_1, v_2, \ldots, v_p\}$ is linearly independent if the trivial solution is the only solution to Eq. (1). If another solution exists, then the set is linearly dependent.

As before, it is easy to prove that a set $\{v_1, v_2, \ldots, v_p\}$ is linearly dependent if and only if some v_i is a linear combination of the other $p - 1$ vectors in the set. The only real distinction between linear independence/dependence in R^n and in a general vector space is that we cannot always test for dependence by solving a homogeneous system of equations. That is, in a general vector space we may have to go directly to the defining equation

$$a_1\mathbf{v}_1 + a_2\mathbf{v}_2 + \cdots + a_p\mathbf{v}_p = \boldsymbol{\theta}$$

and attempt to determine whether there are nontrivial solutions. Examples 2 and 3 illustrate this point.

EXAMPLE 1 Let V be the vector space of (2×2) matrices, and let W be the subspace

$$W = \{A: A = \begin{bmatrix} 0 & a_{12} \\ a_{21} & 0 \end{bmatrix}, \quad a_{12} \text{ and } a_{21} \text{ any real scalars}\}.$$

Define matrices B_1, B_2, and B_3 in W by

$$B_1 = \begin{bmatrix} 0 & 2 \\ 1 & 0 \end{bmatrix}, \quad B_2 = \begin{bmatrix} 0 & 1 \\ 0 & 0 \end{bmatrix}, \quad \text{and} \quad B_3 = \begin{bmatrix} 0 & 2 \\ 3 & 0 \end{bmatrix}.$$

Show that the set $\{B_1, B_2, B_3\}$ is linearly dependent and express B_3 as a linear combination of B_1 and B_2. Show that $\{B_1, B_2\}$ is a linearly independent set.

SOLUTION According to Definition 4, the set $\{B_1, B_2, B_3\}$ is linearly dependent provided that there exist nontrivial solutions to the equation

$$a_1 B_1 + a_2 B_2 + a_3 B_3 = \mathcal{O}, \tag{2}$$

where $\mathcal{O}$ is the zero element in V [that is, $\mathcal{O}$ is the (2×2) zero matrix]. Writing Eq. (2) in detail, we see that a_1, a_2, a_3 are solutions of (2) if

$$\begin{bmatrix} 0 & 2a_1 \\ a_1 & 0 \end{bmatrix} + \begin{bmatrix} 0 & a_2 \\ 0 & 0 \end{bmatrix} + \begin{bmatrix} 0 & 2a_3 \\ 3a_3 & 0 \end{bmatrix} = \begin{bmatrix} 0 & 0 \\ 0 & 0 \end{bmatrix}.$$

With corresponding entries equated, a_1, a_2, a_3 must satisfy

$$2a_1 + a_2 + 2a_3 = 0 \quad \text{and} \quad a_1 + 3a_3 = 0.$$

This (2×3) homogeneous system has nontrivial solutions by Theorem 4 of Section 1.4, and one such solution is $a_1 = -3$, $a_2 = 4$, $a_3 = 1$. In particular,

$$-3B_1 + 4B_2 + B_3 = \mathcal{O}; \tag{3}$$

so the set $\{B_1, B_2, B_3\}$ is a linearly dependent set of vectors in W. It is an immediate consequence of Eq. (3) that

$$B_3 = 3B_1 - 4B_2.$$

To see that the set $\{B_1, B_2\}$ is linearly independent, let a_1 and a_2 be scalars such that $a_1 B_1 + a_2 B_2 = \mathcal{O}$. Then we must have

$$2a_1 + a_2 = 0 \quad \text{and} \quad a_1 = 0.$$

Hence $a_1 = 0$ and $a_2 = 0$; so if $a_1 B_1 + a_2 B_2 = \mathcal{O}$, then $a_1 = a_2 = 0$. Thus $\{B_1, B_2\}$ is a linearly independent set of vectors in W.

Establishing linear independence/dependence in a vector space of functions such as $\mathscr{P}_n$ or $C[a, b]$ may sometimes require techniques from calculus. We illustrate one such technique in the following example.

<u>EXAMPLE 2</u>   Show that $\{1, x, x^2\}$ is a linearly independent set in $\mathscr{P}_2$.

SOLUTION   Suppose that a_0, a_1, a_2 are any scalars that satisfy the defining equation

$$a_0 + a_1 x + a_2 x^2 = \theta(x), \tag{4}$$

where $\theta(x)$ is the zero polynomial. If Eq. (4) is to be an identity holding for all values of x, then [since $\theta'(x) = \theta(x)$] we can differentiate both sides of (4) to obtain

$$a_1 + 2a_2 x = \theta(x). \tag{5}$$

Similarly, differentiating both sides of Eq. (5), we obtain

$$2a_2 = \theta(x). \tag{6}$$

From Eq. (6) we must have $a_2 = 0$. If $a_2 = 0$, then Eq. (5) requires $a_1 = 0$; hence in Eq. (4), $a_0 = 0$ as well. Therefore, the only scalars that satisfy Eq. (4) are $a_0 = a_1 = a_2 = 0$, and thus $\{1, x, x^2\}$ is linearly independent in $\mathscr{P}_2$. (Also see the material on Wronskians in Section 5.5.)   ☐

The following example illustrates another procedure for showing that a set of functions is linearly independent:

<u>EXAMPLE 3</u>   Show that $\{\sqrt{x}, 1/x, x^2\}$ is a linearly independent subset of $C[1, 10]$.

SOLUTION   If the equation

$$a_1\sqrt{x} + a_2(1/x) + a_3 x^2 = 0 \tag{7}$$

holds for all x, $1 \le x \le 10$, then it must hold for any three values of x in the interval. Successively letting $x = 1$, $x = 4$, and $x = 9$ in Eq. (7) yields the system of equations

$$\begin{aligned} a_1 + \quad a_2 + \quad a_3 &= 0 \\ 2a_1 + (1/4)a_2 + 16a_3 &= 0 \\ 3a_1 + (1/9)a_2 + 81a_3 &= 0. \end{aligned} \tag{8}$$

It is easily shown that the trivial solution $a_1 = a_2 = a_3 = 0$ is the unique solution for system (8). It follows that the set $\{\sqrt{x}, 1/x, x^2\}$ is linearly independent.

Note that a nontrivial solution for system (8) would have yielded no information regarding the linear independence/dependence of the given set of functions. We could have concluded only that Eq. (7) holds when $x = 1$, $x = 4$, or $x = 9$.   ☐

Vector-Space Bases

It is now straightforward to combine the concepts of linear independence and spanning sets to define a basis for a vector space.

DEFINITION 5   Let V be a vector space, and let $B = \{v_1, v_2, \ldots, v_p\}$ be a spanning set for V. If B is linearly independent, then B is a **basis** for V.

Thus as before, a basis for V is a linearly independent spanning set for V. (Again we note the implicit assumption that a basis contains only a finite number of vectors.)

There is often a "natural" basis for a vector space. We have seen in Chapter 2 that the set of unit vectors $\{e_1, e_2, \ldots, e_n\}$ in R^n is a basis for R^n. In the preceding section we noted that the set $\{1, x, x^2\}$ is a spanning set for $\mathcal{P}_2$. Example 2 showed further that $\{1, x, x^2\}$ is linearly independent and hence is a basis for $\mathcal{P}_2$. More generally, the set $\{1, x, \ldots, x^n\}$ is a natural basis for $\mathcal{P}_n$. Similarly, the matrices

$$E_{11} = \begin{bmatrix} 1 & 0 \\ 0 & 0 \end{bmatrix}, \quad E_{12} = \begin{bmatrix} 0 & 1 \\ 0 & 0 \end{bmatrix}, \quad E_{21} = \begin{bmatrix} 0 & 0 \\ 1 & 0 \end{bmatrix}, \quad \text{and} \quad E_{22} = \begin{bmatrix} 0 & 0 \\ 0 & 1 \end{bmatrix}$$

constitute a basis for the vector space of all (2×2) real matrices; see Exercise 11. In general, the set of $(m \times n)$ matrices $\{E_{ij} \colon 1 \le i \le m,\ 1 \le j \le n\}$ defined in Section 4.3 is a natural basis for the vector space of all $(m \times n)$ real matrices.

Examples 5, 6, and 7 in Section 4.3 demonstrated a procedure for obtaining a natural spanning set for a subspace W when an algebraic specification for W is given. The spanning set obtained in this manner is often a basis for W. The following example provides another illustration.

EXAMPLE 4   Let V be the vector space of all (2×2) real matrices, and let W be the subspace defined by

$$W = \left\{ A \colon A = \begin{bmatrix} a & a + b \\ a - b & b \end{bmatrix}, \quad a \text{ and } b \text{ any real numbers} \right\}.$$

Exhibit a basis for W.

SOLUTION   In the specification for W, a and b are unconstrained variables. Assigning values $a = 1, b = 0$ and then $a = 0, b = 1$ yields the matrices

$$B_1 = \begin{bmatrix} 1 & 1 \\ 1 & 0 \end{bmatrix} \quad \text{and} \quad B_2 = \begin{bmatrix} 0 & 1 \\ -1 & 1 \end{bmatrix}$$

in W. Since

$$\begin{bmatrix} a & a + b \\ a - b & b \end{bmatrix} = a \begin{bmatrix} 1 & 1 \\ 1 & 0 \end{bmatrix} + b \begin{bmatrix} 0 & 1 \\ -1 & 1 \end{bmatrix},$$

the set $\{B_1, B_2\}$ is clearly a spanning set for W. The equation

$$c_1 B_1 + c_2 B_2 = \mathcal{O}$$

(where $\mathcal{O}$ is the (2×2) zero matrix) is equivalent to

$$\begin{bmatrix} c_1 & c_1 + c_2 \\ c_1 - c_2 & c_2 \end{bmatrix} = \begin{bmatrix} 0 & 0 \\ 0 & 0 \end{bmatrix}.$$

Equating entries immediately yields $c_1 = c_2 = 0$; so the set $\{B_1, B_2\}$ is linearly independent and hence is a basis for W.

Coordinate Vectors

As we noted in Chapter 2, a basis is a minimal spanning set; as such, a basis contains no redundant information. This lack of redundance is an important feature of a basis in the general vector-space setting and allows every vector to be represented uniquely in terms of the basis (see Theorem 4 below). We cannot make such an assertion of unique representation about a spanning set that is linearly dependent.

THEOREM 4   Let V be a vector space, and let $B = \{\mathbf{v}_1, \mathbf{v}_2, \ldots, \mathbf{v}_p\}$ be a basis for V. For each vector $\mathbf{w}$ in V, there exists a unique set of scalars $w_1, w_2, \ldots, w_p$ such that

$$\mathbf{w} = w_1 \mathbf{v}_1 + w_2 \mathbf{v}_2 + \cdots + w_p \mathbf{v}_p.$$

PROOF   Let $\mathbf{w}$ be a vector in V and suppose that $\mathbf{w}$ is represented in two ways as

$$\mathbf{w} = w_1 \mathbf{v}_1 + w_2 \mathbf{v}_2 + \cdots + w_p \mathbf{v}_p$$
$$\mathbf{w} = u_1 \mathbf{v}_1 + u_2 \mathbf{v}_2 + \cdots + u_p \mathbf{v}_p.$$

Subtracting, we obtain

$$\theta = (w_1 - u_1)\mathbf{v}_1 + (w_2 - u_2)\mathbf{v}_2 + \cdots + (w_p - u_p)\mathbf{v}_p.$$

Therefore, since $\{\mathbf{v}_1, \mathbf{v}_2, \ldots, \mathbf{v}_p\}$ is a linearly independent set, it follows that $w_1 - u_1 = 0, w_2 - u_2 = 0, \ldots, w_p - u_p = 0$. That is, a vector $\mathbf{w}$ cannot be represented in two different ways in terms of a basis B.

Now, let V be a vector space with a basis $B = \{\mathbf{v}_1, \mathbf{v}_2, \ldots, \mathbf{v}_p\}$. Given that each vector $\mathbf{w}$ in V has a unique representation in terms of B as

$$\mathbf{w} = w_1 \mathbf{v}_1 + w_2 \mathbf{v}_2 + \cdots + w_p \mathbf{v}_p, \tag{9}$$

it follows that the scalars $w_1, w_2, \ldots, w_p$ serve to characterize $\mathbf{w}$ completely in terms of the basis B. In particular, we can identify $\mathbf{w}$ unambiguously with the vector $[\mathbf{w}]_B$ in R^p, where

$$[\mathbf{w}]_B = \begin{bmatrix} w_1 \\ w_2 \\ \vdots \\ w_p \end{bmatrix}.$$

We will call the unique scalars $w_1, w_2, \ldots, w_p$ in Eq. (9) the **coordinates of w** with respect to the basis B, and we will call the vector $[\mathbf{w}]_B$ in R^p the **coordinate vector of w** with respect to B. This idea is a useful one; for example, we will show that a set of vectors $\{\mathbf{u}_1, \mathbf{u}_2, \ldots, \mathbf{u}_r\}$ in V is linearly independent if and only if the coordinate vectors $[\mathbf{u}_1]_B, [\mathbf{u}_2]_B, \ldots, [\mathbf{u}_r]_B$ are linearly independent in R^n. Since we know how to determine whether vectors in R^p are linearly independent or not, we can use the idea of coordinates to reduce a problem of linear independence/dependence in a general vector space to an equivalent problem in R^p, which we can work. Finally, we note that the subscript B is necessary when we write $[\mathbf{w}]_B$, since the coordinate vector for $\mathbf{w}$ changes when we change the basis.

EXAMPLE 5 Let V be the vector space of all real (2×2) matrices. Let $B = \{E_{11}, E_{12}, E_{21}, E_{22}\}$ and $Q = \{E_{11}, E_{21}, E_{12}, E_{22}\}$, where

$$E_{11} = \begin{bmatrix} 1 & 0 \\ 0 & 0 \end{bmatrix}, \quad E_{12} = \begin{bmatrix} 0 & 1 \\ 0 & 0 \end{bmatrix}, \quad E_{21} = \begin{bmatrix} 0 & 0 \\ 1 & 0 \end{bmatrix}, \quad \text{and} \quad E_{22} = \begin{bmatrix} 0 & 0 \\ 0 & 1 \end{bmatrix}.$$

Let the matrix A be defined by

$$A = \begin{bmatrix} 2 & -1 \\ -3 & 4 \end{bmatrix}.$$

Find $[A]_B$ and $[A]_Q$.

SOLUTION We have already noted that B is the natural basis for V. Since Q contains the same vectors as B, but in a different order, Q is also a basis for V. It is easy to see that

$$A = 2E_{11} - E_{12} - 3E_{21} + 4E_{22},$$

so

$$[A]_B = \begin{bmatrix} 2 \\ -1 \\ -3 \\ 4 \end{bmatrix}.$$

Similarly,

$$A = 2E_{11} - 3E_{21} - E_{12} + 4E_{22},$$

so

$$[A]_Q = \begin{bmatrix} 2 \\ -3 \\ -1 \\ 4 \end{bmatrix}.$$

It is apparent in the preceding example that the ordering of the basis vectors determined the ordering of the components of the coordinate vectors. A basis with such an implicitly fixed ordering is usually called an **ordered basis.** Although we do not intend to dwell on this point, we do have to be careful to work with a fixed ordering in a basis.

If V is a vector space with (ordered) basis $B = \{v_1, v_2, \ldots, v_p\}$, then the correspondence

$$v \to [v]_B$$

provides an identification between vectors in V and elements of R^p. For instance, the preceding example identified a (2×2) matrix with a vector in R^4. The following lemma lists some of the properties of this correspondence. (The lemma hints at the idea of an isomorphism that will be developed in detail later.)

LEMMA

Let V be a vector space that has a basis $B = \{v_1, v_2, \ldots, v_p\}$. If u and v are vectors in V and if c is a scalar, then the following hold:

$$[u + v]_B = [u]_B + [v]_B$$

and

$$[cu]_B = c[u]_B.$$

PROOF   Suppose that u and v are expressed in terms of the basis vectors in B as

$$u = a_1 v_1 + a_2 v_2 + \cdots + a_p v_p$$

and

$$v = b_1 v_1 + b_2 v_2 + \cdots + b_p v_p.$$

Then clearly

$$u + v = (a_1 + b_1)v_1 + (a_2 + b_2)v_2 + \cdots + (a_p + b_p)v_p$$

and

$$cu = (ca_1)v_1 + (ca_2)v_2 + \cdots + (ca_p)v_p.$$

Therefore,

$$[u]_B = \begin{bmatrix} a_1 \\ a_2 \\ \vdots \\ a_p \end{bmatrix}, \qquad [v]_B = \begin{bmatrix} b_1 \\ b_2 \\ \vdots \\ b_p \end{bmatrix},$$

$$[\mathbf{u} + \mathbf{v}]_B = \begin{bmatrix} a_1 + b_1 \\ a_2 + b_2 \\ \vdots \\ a_p + b_p \end{bmatrix}, \quad \text{and} \quad [c\mathbf{u}] = \begin{bmatrix} ca_1 \\ ca_2 \\ \vdots \\ ca_p \end{bmatrix}.$$

We can now easily see that $[\mathbf{u} + \mathbf{v}]_B = [\mathbf{u}]_B + [\mathbf{v}]_B$ and $[c\mathbf{u}]_B = c[\mathbf{u}]_B$.

The following example illustrates the preceding lemma.

EXAMPLE 6 In $\mathscr{P}_2$, let $p(x) = 3 - 2x + x^2$ and $q(x) = -2 + 3x - 4x^2$. Show that

$$[p(x) + q(x)]_B = [p(x)]_B + [q(x)]_B \quad \text{and} \quad [2p(x)]_B = 2[p(x)]_B,$$

where B is the natural basis for $\mathscr{P}_2$: $B = \{1, x, x^2\}$.

SOLUTION The coordinate vectors for $p(x)$ and $q(x)$ are

$$[p(x)]_B = \begin{bmatrix} 3 \\ -2 \\ 1 \end{bmatrix} \quad \text{and} \quad [q(x)]_B = \begin{bmatrix} -2 \\ 3 \\ -4 \end{bmatrix}.$$

Furthermore, $p(x) + q(x) = 1 + x - 3x^2$ and $2p(x) = 6 - 4x + 2x^2$. Thus

$$[p(x) + q(x)]_B = \begin{bmatrix} 1 \\ 1 \\ -3 \end{bmatrix} \quad \text{and} \quad [2p(x)]_B = \begin{bmatrix} 6 \\ -4 \\ 2 \end{bmatrix}.$$

Therefore, $[p(x) + q(x)]_B = [p(x)]_B + [q(x)]_B$ and $[2p(x)]_B = 2[p(x)]_B$.

Suppose that the vector space V has basis $B = \{\mathbf{v}_1, \mathbf{v}_2, \ldots, \mathbf{v}_p\}$, and let $\{\mathbf{u}_1, \mathbf{u}_2, \ldots, \mathbf{u}_m\}$ be a subset of V. The two properties in the preceding lemma can easily be combined and extended to give

$$[c_1\mathbf{u}_1 + c_2\mathbf{u}_2 + \cdots + c_m\mathbf{u}_m]_B = c_1[\mathbf{u}_1]_B + c_2[\mathbf{u}_2]_B + \cdots + c_m[\mathbf{u}_m]_B. \quad (10)$$

This observation will be useful in proving the next theorem.

THEOREM 5

Suppose that V is a vector space with a basis $B = \{\mathbf{v}_1, \mathbf{v}_2, \ldots, \mathbf{v}_p\}$. Let $S = \{\mathbf{u}_1, \mathbf{u}_2, \ldots, \mathbf{u}_m\}$ be a subset of V, let $T = \{[\mathbf{u}_1]_B, [\mathbf{u}_2]_B, \ldots, [\mathbf{u}_m]_B\}$.

1. A vector $\mathbf{u}$ in V is in $\text{Sp}(S)$ if and only if $[\mathbf{u}]_B$ is in $\text{Sp}(T)$.
2. The set S is linearly independent in V if and only if the set T is linearly independent in R^p.

PROOF The vector equation

$$\mathbf{u} = x_1\mathbf{u}_1 + x_2\mathbf{u}_2 + \cdots + x_m\mathbf{u}_m \tag{11}$$

in V is equivalent to the equation

$$[\mathbf{u}]_B = [x_1\mathbf{u}_1 + x_2\mathbf{u}_2 + \cdots + x_m\mathbf{u}_m]_B \tag{12}$$

in R^p. It follows from Eq. (10) that Eq. (12) is equivalent to

$$[\mathbf{u}]_B = x_1[\mathbf{u}_1]_B + x_2[\mathbf{u}_2]_B + \cdots + x_m[\mathbf{u}_m]_B. \tag{13}$$

Therefore, the vector equation (11) in V is equivalent to the vector equation (13) in R^p. In particular, Eq. (11) has a solution $x_1 = c_1, x_2 = c_2, \ldots, x_m = c_m$ if and only if Eq. (13) has the same solution. Thus $\mathbf{u}$ is in $Sp(S)$ if and only if $[\mathbf{u}]_B$ is in $Sp(T)$.

To avoid confusion in the proof of property 2, let θ_V denote the zero vector for V and let θ_p denote the p-dimensional zero vector in R^p. Then $[\theta_V]_B = \theta_p$. Thus setting $\mathbf{u} = \theta_V$ in Eqs. (11) and (13) implies that the vector equations

$$\theta_V = x_1\mathbf{u}_1 + x_2\mathbf{u}_2 + \cdots + x_m\mathbf{u}_m \tag{14}$$

and

$$\theta_p = x_1[\mathbf{u}_1]_B + x_2[\mathbf{u}_2]_B + \cdots + x_m[\mathbf{u}_m]_B \tag{15}$$

have the same solutions. In particular, Eq. (14) has only the trivial solution if and only if Eq. (15) has only the trivial solution; that is, S is a linearly independent set in V if and only if T is linearly independent in R^p. ☐

An immediate corollary to Theorem 5 is as follows.

COROLLARY

Let V be a vector space with a basis $B = \{\mathbf{v}_1, \mathbf{v}_2, \ldots, \mathbf{v}_p\}$. Let $S = \{\mathbf{u}_1, \mathbf{u}_2, \ldots, \mathbf{u}_m\}$ be a subset of V, and let $T = \{[\mathbf{u}_1]_B, [\mathbf{u}_2]_B, \ldots, [\mathbf{u}_m]_B\}$. Then S is a basis for V if and only if T is a basis for R^p.

PROOF By Theorem 5, S is both linearly independent and a spanning set for V if and only if T is both linearly independent and a spanning set for R^p. ☐

Theorem 5 and its corollary allow us to use the techniques developed in Chapter 2 to solve analogous problems in vector spaces other than R^p. The next two examples will provide illustrations.

EXAMPLE 7 Use the corollary to Theorem 5 to show that the set $\{1, 1 + x, 1 + 2x + x^2\}$ is a basis for $\mathcal{P}_2$.

SOLUTION   Let B be the standard basis for $\mathscr{P}_2$: $B = \{1, x, x^2\}$. The coordinate vectors of 1, $1 + x$, and $1 + 2x + x^2$ are

$$[1]_B = \begin{bmatrix} 1 \\ 0 \\ 0 \end{bmatrix}, \quad [1 + x]_B = \begin{bmatrix} 1 \\ 1 \\ 0 \end{bmatrix}, \quad \text{and} \quad [1 + 2x + x^2]_B = \begin{bmatrix} 1 \\ 2 \\ 1 \end{bmatrix}.$$

Clearly the coordinate vectors $[1]_B$, $[1 + x]_B$, and $[1 + 2x + x^2]_B$ are linearly independent in R^3. Since R^3 has dimension 3, the coordinate vectors constitute a basis for R^3. It now follows that $\{1, 1 + x, 1 + 2x + x^2\}$ is a basis for $\mathscr{P}_2$.

EXAMPLE 8 Let V be the vector space of all (2×2) matrices, and let the subset S of V be defined by $S = \{A_1, A_2, A_3, A_4\}$, where

$$A_1 = \begin{bmatrix} 1 & 2 \\ -1 & 3 \end{bmatrix}, \quad A_2 = \begin{bmatrix} 0 & -1 \\ 1 & 4 \end{bmatrix},$$

$$A_3 = \begin{bmatrix} -1 & 0 \\ 1 & -10 \end{bmatrix}, \quad \text{and} \quad A_4 = \begin{bmatrix} 3 & 7 \\ -2 & 6 \end{bmatrix}.$$

Use the corollary to Theorem 5 and the techniques of Section 2.4 to obtain a basis for $\text{Sp}(S)$.

SOLUTION   If B is the natural basis for V, $B = \{E_{11}, E_{12}, E_{21}, E_{22}\}$, then

$$[A_1]_B = \begin{bmatrix} 1 \\ 2 \\ -1 \\ 3 \end{bmatrix}, \quad [A_2]_B = \begin{bmatrix} 0 \\ -1 \\ 1 \\ 4 \end{bmatrix},$$

$$[A_3]_B = \begin{bmatrix} -1 \\ 0 \\ 1 \\ -10 \end{bmatrix}, \quad \text{and} \quad [A_4]_B = \begin{bmatrix} 3 \\ 7 \\ -2 \\ 6 \end{bmatrix}.$$

Let $T = \{[A_1]_B, [A_2]_B, [A_3]_B, [A_4]_B\}$. Several techniques for obtaining a basis for $\text{Sp}(T)$ were illustrated in Section 2.4. For example, using the method demonstrated in Example 7 of Section 2.4, we form the matrix

$$C = \begin{bmatrix} 1 & 0 & -1 & 3 \\ 2 & -1 & 0 & 7 \\ -1 & 1 & 1 & -2 \\ 3 & 4 & -10 & 6 \end{bmatrix}.$$

The matrix C^T can be reduced to the matrix

$$D^T = \begin{bmatrix} 1 & 2 & -1 & 3 \\ 0 & -1 & 1 & 4 \\ 0 & 0 & 2 & 1 \\ 0 & 0 & 0 & 0 \end{bmatrix},$$

which is in the echelon form. Thus

$$D = \begin{bmatrix} 1 & 0 & 0 & 0 \\ 2 & -1 & 0 & 0 \\ -1 & 1 & 2 & 0 \\ 3 & 4 & 1 & 0 \end{bmatrix},$$

and the nonzero columns of D constitute a basis for $\mathrm{Sp}(T)$. Denote the nonzero columns of D by $\mathbf{w}_1, \mathbf{w}_2,$ and $\mathbf{w}_3$, respectively. Thus

$$\mathbf{w}_1 = \begin{bmatrix} 1 \\ 2 \\ -1 \\ 3 \end{bmatrix}, \quad \mathbf{w}_2 = \begin{bmatrix} 0 \\ -1 \\ 1 \\ 4 \end{bmatrix}, \quad \text{and} \quad \mathbf{w}_3 = \begin{bmatrix} 0 \\ 0 \\ 2 \\ 1 \end{bmatrix},$$

and $\{\mathbf{w}_1, \mathbf{w}_2, \mathbf{w}_3\}$ is a basis for $\mathrm{Sp}(T)$. If $B_1, B_2,$ and B_3 are (2×2) matrices such that $[B_1]_B = \mathbf{w}_1$, $[B_2]_B = \mathbf{w}_2$, and $[B_3]_B = \mathbf{w}_3$, then it follows from Theorem 5 that $\{B_1, B_2, B_3\}$ is a basis for $\mathrm{Sp}(S)$. If

$$B_1 = E_{11} + 2E_{12} - E_{21} + 3E_{22},$$

then clearly $[B_1]_B = \mathbf{w}_1$. B_2 and B_3 are obtained in the same fashion, and

$$B_1 = \begin{bmatrix} 1 & 2 \\ -1 & 3 \end{bmatrix}, \quad B_2 = \begin{bmatrix} 0 & -1 \\ 1 & 4 \end{bmatrix}, \quad \text{and} \quad B_3 = \begin{bmatrix} 0 & 0 \\ 2 & 1 \end{bmatrix}. \qquad \square$$

Examples 7 and 8 illustrate an important point. Although Theorem 5 shows that questions regarding the span or the linear dependence/independence of a subset of V can be translated to an equivalent problem in R^p, we do need one basis for V as a point of reference. For example, in $\mathscr{P}_2$, once we know that $B = \{1, x, x^2\}$ is a basis, we can use Theorem 5 to pass from a problem in $\mathscr{P}_2$ to an analogous problem in R^3. In order to obtain the first basis B, however, we cannot use Theorem 5.

EXAMPLE 9   In $\mathscr{P}_4$, consider the set of vectors $S = \{p_1, p_2, p_3, p_4, p_5\}$, where $p_1(x) = x^4 + 3x^3 + 2x + 4$, $p_2(x) = x^3 - x^2 + 5x + 1$, $p_3(x) = x^4 + x + 3$, $p_4(x) = x^4 + x^3 - x + 2$, and $p_5(x) = x^4 + x^2$. Is S a basis for $\mathscr{P}_4$?

SOLUTION   Let B denote the standard basis for $\mathscr{P}_4$, $B = \{1, x, x^2, x^3, x^4\}$. By the corollary to Theorem 5, S is a basis for $\mathscr{P}_4$ if and only if T is a basis for

R^5, where $T = \{[p_1]_B, [p_2]_B, [p_3]_B, [p_4]_B, [p_5]_B\}$. In particular, the coordinate vectors in T are

$$[p_1]_B = \begin{bmatrix} 4 \\ 2 \\ 0 \\ 3 \\ 1 \end{bmatrix}, \quad [p_2]_B = \begin{bmatrix} 1 \\ 5 \\ -1 \\ 1 \\ 0 \end{bmatrix}, \quad [p_3]_B = \begin{bmatrix} 3 \\ 1 \\ 0 \\ 0 \\ 1 \end{bmatrix},$$

$$[p_4]_B = \begin{bmatrix} 2 \\ -1 \\ 0 \\ 1 \\ 1 \end{bmatrix}, \quad \text{and} \quad [p_5]_B = \begin{bmatrix} 0 \\ 0 \\ 1 \\ 0 \\ 1 \end{bmatrix}.$$

Since R^5 has dimension 5 and T contains 5 vectors, T will be a basis for R^5 if T is a linearly independent set. To check whether T is linearly independent, we form the matrix A whose columns are the vectors in T and use MATLAB to reduce A to echelon form. As can be seen from the results in Fig. 4.4 the columns of A are linearly independent. Hence, T is a basis for R^5. Therefore, S is a basis for $\mathscr{P}_4$.

```
A =

     4     1     3     2     0
     2     5     1    -1     0
     0    -1     0     0     1
     3     1     0     1     0
     1     0     1     1     1

>> rref ( A )

ans =

     1     0     0     0     0
     0     1     0     0     0
     0     0     1     0     0
     0     0     0     1     0
     0     0     0     0     1
```

Figure 4.4
MATLAB was used for Example 9 to determine whether the columns of A are linearly independent. Since A is row equivalent to the identity, its columns are linearly independent.

In Exercises 1–4, W is a subspace of the vector space V of all (2×2) matrices. A matrix A in W is written as

$$A = \begin{bmatrix} a & b \\ c & d \end{bmatrix}.$$

In each case exhibit a basis for W.

1. $W = \{A: a + b + c + d = 0\}$
2. $W = \{A: a = -d, \quad b = 2d, \quad c = -3d\}$
3. $W = \{A: a = 0\}$
4. $W = \{A: b = a - c, \quad d = 2a + c\}$

In Exercises 5–8, W is a subspace of $\mathcal{P}_2$. In each case exhibit a basis for W.

5. $W = \{p(x) = a_0 + a_1 x + a_2 x^2: a_2 = a_0 - 2a_1\}$
6. $W = \{p(x) = a_0 + a_1 x + a_2 x^2: a_0 = 3a_2, a_1 = -a_2\}$
7. $W = \{p(x) = a_0 + a_1 x + a_2 x^2: p(0) = 0\}$
8. $W = \{p(x) = a_0 + a_1 x + a_2 x^2: p(1) = p'(1) = 0\}$
9. Find a basis for the subspace V of $\mathcal{P}_4$, where $V = \{p(x) \text{ in } \mathcal{P}_4: p(0) = 0, \ p'(1) = 0, \ p''(-1) = 0\}$.
10. Prove that the set of all real (2×2) symmetric matrices is a subspace of the vector space of all real (2×2) matrices. Find a basis for this subspace (cf. Exercise 26 of Section 4.3).
11. Let V be the vector space of all (2×2) real matrices. Show that $B = \{E_{11}, E_{12}, E_{21}, E_{22}\}$ (cf. Example 5) is a basis for V.
12. With respect to the basis $B = \{1, x, x^2\}$ for $\mathcal{P}_2$, find the coordinate vector for each of the following.
   a) $p(x) = x^2 - x + 1$
   b) $p(x) = x^2 + 4x - 1$
   c) $p(x) = 2x + 5$
13. With respect to the basis $B = \{E_{11}, E_{12}, E_{21}, E_{22}\}$ for the vector space V of all (2×2) matrices, find the coordinate vector for each of the following.

   a) $A_1 = \begin{bmatrix} 2 & -1 \\ 3 & 2 \end{bmatrix}$ b) $A_2 = \begin{bmatrix} 1 & 0 \\ -1 & 1 \end{bmatrix}$

   c) $A_3 = \begin{bmatrix} 2 & 3 \\ 0 & 0 \end{bmatrix}$

14. Prove that $\{1, x, x^2, \ldots, x^n\}$ is a linearly independent set in $\mathcal{P}_n$ by supposing that $p(x) = \theta(x)$, where

$p(x) = a_0 + a_1 x_1 + \cdots + a_n x^n$. Next, take successive derivatives as in Example 2.

In Exercises 15–17, use the basis B of Exercise 11 and property 2 of Theorem 5 to test for linear independence in the vector space of (2×2) matrices.

15. $A_1 = \begin{bmatrix} 2 & 1 \\ 2 & 1 \end{bmatrix}$, $A_2 = \begin{bmatrix} 3 & 0 \\ 0 & 2 \end{bmatrix}$, $A_3 = \begin{bmatrix} 1 & 1 \\ 2 & 1 \end{bmatrix}$

16. $A_1 = \begin{bmatrix} 1 & 3 \\ 2 & 1 \end{bmatrix}$, $A_2 = \begin{bmatrix} 4 & -2 \\ 0 & 6 \end{bmatrix}$, $A_3 = \begin{bmatrix} 6 & 4 \\ 4 & 8 \end{bmatrix}$

17. $A_1 = \begin{bmatrix} 2 & 2 \\ 1 & 3 \end{bmatrix}$, $A_2 = \begin{bmatrix} 1 & 4 \\ 0 & 5 \end{bmatrix}$,

   $A_3 = \begin{bmatrix} 4 & 10 \\ 1 & 13 \end{bmatrix}$

In Exercises 18–21, use Exercise 14 and property 2 of Theorem 5 to test for linear independence in $\mathcal{P}_3$.

18. $\{x^3 - x, x^2 - 1, x + 4\}$
19. $\{x^2 + 2x - 1, x^2 - 5x + 2, 3x^2 - x\}$
20. $\{x^3 - x^2, x^2 - x, x - 1, x^3 - 1\}$
21. $\{x^3 + 1, x^2 + 1, x + 1, 1\}$
22. In $\mathcal{P}_2$, let $S = \{p_1(x), p_2(x), p_3(x), p_4(x)\}$, where $p_1(x) = 1 + 2x + x^2$, $p_2(x) = 2 + 5x$, $p_3(x) = 3 + 7x + x^2$, and $p_4(x) = 1 + x + 3x^2$. Use the method illustrated in Example 8 to obtain a basis for $\mathrm{Sp}(S)$. (*Hint:* Use the basis $B = \{1, x, x^2\}$ to obtain coordinate vectors for $p_1(x)$, $p_2(x)$, $p_3(x)$, and $p_4(x)$. Now use the method illustrated in Example 7 of Section 2.4.)
23. Let S be the subset of $\mathcal{P}_2$ given in Exercise 22. Find a subset of S that is a basis for $\mathrm{Sp}(S)$. (*Hint:* Proceed as in Exercise 22, but use the technique illustrated in Example 6 of Section 2.4.)
24. Let V be the vector space of all (2×2) matrices and let $S = \{A_1, A_2, A_3, A_4\}$, where

   $A_1 = \begin{bmatrix} 1 & 2 \\ -1 & 3 \end{bmatrix}$, $A_2 = \begin{bmatrix} -2 & 1 \\ 2 & -1 \end{bmatrix}$,

   $A_3 = \begin{bmatrix} -1 & -1 \\ 1 & -3 \end{bmatrix}$, and $A_4 = \begin{bmatrix} -2 & 2 \\ 2 & 0 \end{bmatrix}$.

   As in Example 8, find a basis for $\mathrm{Sp}(S)$.

25. Let V and S be as in Exercise 24. Find a subset of S that is a basis for $\text{Sp}(S)$. (*Hint:* Use Theorem 5 and the technique illustrated in Example 6 of Section 2.4.)

26. In $\mathscr{P}_2$, let $Q = \{p_1(x), p_2(x), p_3(x)\}$, where $p_1(x) = -1 + x + 2x^2$, $p_2(x) = x + 3x^2$, and $p_3(x) = 1 + 2x + 8x^2$. Use the basis $B = \{1, x, x^2\}$ to show that Q is a basis for $\mathscr{P}_2$.

27. Let Q be the basis for $\mathscr{P}_2$ given in Exercise 26. Find $[p(x)]_Q$ for $p(x) = 1 + x + x^2$.

28. Let Q be the basis for $\mathscr{P}_2$ given in Exercise 26. Find $[p(x)]_Q$ for $p(x) = a_0 + a_1 x + a_2 x^2$.

29. In the vector space V of (2×2) matrices, let $Q = \{A_1, A_2', A_3, A_4\}$ where

$$A_1 = \begin{bmatrix} 1 & 0 \\ 0 & 0 \end{bmatrix}, \quad A_2 = \begin{bmatrix} 1 & -1 \\ 0 & 0 \end{bmatrix},$$

$$A_3 = \begin{bmatrix} 0 & 2 \\ 0 & 0 \end{bmatrix}, \quad \text{and} \quad A_4 = \begin{bmatrix} -3 & 0 \\ 2 & 1 \end{bmatrix}.$$

Use the corollary to Theorem 5 and the natural basis for V to show that Q is a basis for V.

30. With V and Q as in Exercise 29, find $[A]_Q$ for

$$A = \begin{bmatrix} 7 & 3 \\ -3 & -1 \end{bmatrix}.$$

31. With V and Q as in Exercise 29, find $[A]_Q$ for

$$A = \begin{bmatrix} a & b \\ c & d \end{bmatrix}.$$

32. Give an alternative proof that $\{1, x, x^2\}$ is a linearly independent set in $\mathscr{P}_2$ as follows: Let $p(x) = a_0 + a_1 x + a_2 x^2$ and suppose that $p(x) = \theta(x)$. Then $p(-1) = 0$, $p(0) = 0$, and $p(1) = 0$. These three equations can be used to show that $a_0 = a_1 = a_2 = 0$.

33. The set $\{\sin x, \cos x\}$ is a subset of the vector space $C[-\pi, \pi]$. Prove that the set is linearly independent. (*Hint:* Set $f(x) = c_1 \sin x + c_2 \cos x$ and assume that $f(x) = \theta(x)$. Then $f(0) = 0$ and $f(\pi/2) = 0$.)

In Exercises 34 and 35, V is the set of functions

$$V = \{f(x): f(x) = ae^x + be^{2x} + ce^{3x} + de^{4x}$$

for real numbers $a, b, c, d\}$

It can be shown that V is a vector space.

34. Show that $B = \{e^x, e^{2x}, e^{3x}, e^{4x}\}$ is a basis for V. (*Hint:* To see that B is a linearly independent set, let $h(x) = c_1 e^x + c_2 e^{2x} + c_3 e^{3x} + c_4 e^{4x}$ and assume that $h(x) = \theta(x)$. Then $h'(x) = \theta(x)$, $h''(x) = \theta(x)$, and $h'''(x) = \theta(x)$. Therefore, $h(0) = 0$, $h'(0) = 0$, $h''(0) = 0$, and $h'''(0) = 0$.)

35. Let $S = \{g_1(x), g_2(x), g_3(x)\}$ be the subset of V, where $g_1(x) = e^x - e^{4x}$, $g_2(x) = e^{2x} + e^{3x}$, and $g_3(x) = -e^x + e^{3x} + e^{4x}$. Use Theorem 5 and basis B of Exercise 34 to show that S is a linearly independent set.

36. Prove that if $Q = \{v_1, v_2, \dots, v_m\}$ is a linearly independent subset of a vector space V, and if w is a vector in V such that w is not in $\text{Sp}(Q)$, then $\{v_1, v_2, \dots, v_m, w\}$ is also a linearly independent set in V. [*Note:* θ is always in $\text{Sp}(Q)$.]

37. Let $S = \{v_1, v_2, \dots, v_n\}$ be a subset of a vector space V, where $n \geq 2$. Prove that set S is linearly dependent if and only if at least one of the vectors, v_j, can be expressed as a linear combination of the remaining vectors.

38. Use Exercise 37 to obtain necessary and sufficient conditions for a set $\{u, v\}$ of two vectors to be linearly dependent. Determine by inspection whether each of the following sets is linearly dependent or linearly independent.
a) $\{1 + x, x^2\}$
b) $\{x, e^x\}$
c) $\{x, 3x\}$
d) $\left\{ \begin{bmatrix} -1 & 2 \\ 1 & 3 \end{bmatrix}, \begin{bmatrix} 2 & -4 \\ -2 & -6 \end{bmatrix} \right\}$
e) $\left\{ \begin{bmatrix} 0 & 0 \\ 0 & 0 \end{bmatrix}, \begin{bmatrix} 1 & 0 \\ 0 & 1 \end{bmatrix} \right\}$

4.5   Dimension

We now use Theorem 5 to generalize the idea of dimension to the general vector-space setting. We begin with two theorems that will be needed to show that dimension is a well-defined concept. These theorems are direct applications of the corollary to Theorem 5, and the proofs are left to the exercises because they are essentially the same as the proofs of the analogous theorems from Section 2.5.

> **THEOREM 6**
>
> If V is a vector space and if $B = \{v_1, v_2, \ldots, v_p\}$ is a basis of V, then any set of $p + 1$ vectors in V is linearly dependent.

> **THEOREM 7**
>
> Let V be a vector space, and let $B = \{v_1, v_2, \ldots, v_p\}$ be a basis for V. If $Q = \{u_1, u_2, \ldots, u_m\}$ is also a basis for V, then $m = p$.

If V is a vector space that has a basis of p vectors, then no ambiguity can arise if we define the dimension of V to be p (since the number of vectors in a basis for V is an invariant property of V by Theorem 7). There is, however, one extreme case, which is also included below in Definition 6. That is, there may not be a finite set of vectors that spans V; and in this case we call V an infinite-dimensional vector space.

> **DEFINITION 6**   Let V be a vector space.
>
> 1. If V has a basis $B = \{v_1, v_2, \ldots, v_n\}$ of n vectors, then V has **dimension** n, and we write $\dim(V) = n$. [If $V = \{\theta\}$, then $\dim(V) = 0$.]
> 2. If V is nontrivial and does not have a basis containing a finite number of vectors, then V is an **infinite-dimensional** vector space.

We already know from Chapter 2 that R^n has dimension n. In the preceding section it was shown that $\{1, x, x^2\}$ is a basis for $\mathscr{P}_2$, so $\dim(\mathscr{P}_2) = 3$. Similarly, the set $\{1, x, \ldots, x^n\}$ is a basis for $\mathscr{P}_n$, so $\dim(\mathscr{P}_n) = n + 1$. The vector space V consisting of all (2×2) real matrices has a basis with four vectors, namely, $B = \{E_{11}, E_{12}, E_{21}, E_{22}\}$. Therefore, $\dim(V) = 4$. More generally, the space of all $(m \times n)$ real matrices has dimension mn because the $(m \times n)$ matrices E_{ij}, $1 \leq i \leq m$, $1 \leq j \leq n$, constitute a basis for the space.

EXAMPLE 1   Let W be the subspace of the set of all (2×2) matrices defined by

$$W = \{A = \begin{bmatrix} a & b \\ c & d \end{bmatrix} : 2a - b + 3c + d = 0\}.$$

Determine the dimension of W.

SOLUTION   The algebraic specification for W can be rewritten as $d = -2a + b - 3c$. Thus an element of W is completely determined by the three independent variables a, b, and c. In succession, let $a = 1$, $b = 0$, $c = 0$; $a = 0$, $b = 1$, $c = 0$;

and $a = 0, b = 0, c = 1$. This yields three matrices

$$A_1 = \begin{bmatrix} 1 & 0 \\ 0 & -2 \end{bmatrix}, \quad A_2 = \begin{bmatrix} 0 & 1 \\ 0 & 1 \end{bmatrix}, \quad \text{and} \quad A_3 = \begin{bmatrix} 0 & 0 \\ 1 & -3 \end{bmatrix}$$

in W. The matrix A is in W if and only if $A = aA_1 + bA_2 + cA_3$, so $\{A_1, A_2, A_3\}$ is a spanning set for W. It is easy to show that the set $\{A_1, A_2, A_3\}$ is linearly independent, so it is a basis for W. It follows that $\dim(W) = 3$. ▫

An example of an infinite-dimensional vector space is given below in Example 2. As Example 2 illustrates, we can show that a vector space V is infinite dimensional if we can show that V contains subspaces of dimension k for $k = 1, 2, 3, \ldots$.

If W is a subspace of a vector space V, and if $\dim(W) = k$, then it is almost obvious that $\dim(V) \geq \dim(W) = k$ (we leave the proof of this as an exercise). This observation can be used to show that $C[a, b]$ is an infinite-dimensional vector space.

EXAMPLE 2 Show that $C[a, b]$ is an infinite-dimensional vector space.

SOLUTION To show that $C[a, b]$ is not a finite-dimensional vector space, we merely note that $\mathscr{P}_n$ is a subspace of $C[a, b]$ for every n. But $\dim(\mathscr{P}_n) = n + 1$; and so $C[a, b]$ contains subspaces of arbitrarily large dimension. Thus $C[a, b]$ must be an infinite-dimensional vector space. ▫

Properties of a p-Dimensional Vector Space

The next two theorems summarize some of the properties of a p-dimensional vector space V and show how properties of R^p carry over into V.

THEOREM 8

Let V be a finite-dimensional vector space with $\dim(V) = p$.

1. Any set of $p + 1$ or more vectors in V is linearly dependent.
2. Any set of p linearly independent vectors in V is a basis for V.

This theorem is a direct generalization from R^p (Exercise 20). To complete our discussion of finite-dimensional vector spaces, we state the following lemma.

LEMMA

Let V be a vector space, and let $Q = \{\mathbf{u}_1, \mathbf{u}_2, \ldots, \mathbf{u}_p\}$ be a spanning set for V. Then there is a subset Q' of Q that is a basis for V.

PROOF   (We only sketch the proof of this lemma because the proof follows familiar lines.) If Q is linearly independent, then Q itself is a basis for V. If Q is linearly dependent, we can express some vector from Q in terms of the other $p - 1$ vectors in Q. Without loss of generality, let us suppose we can express $\mathbf{u}_1$ in terms of $\mathbf{u}_2, \mathbf{u}_3, \ldots, \mathbf{u}_p$. In that event we have

$$\text{Sp}\{\mathbf{u}_2, \mathbf{u}_3, \ldots, \mathbf{u}_p\} = \text{Sp}\{\mathbf{u}_1, \mathbf{u}_2, \mathbf{u}_3, \ldots, \mathbf{u}_p\} = V;$$

and if $\{\mathbf{u}_2, \mathbf{u}_3, \ldots, \mathbf{u}_p\}$ is linearly independent, it is a basis for V. If $\{\mathbf{u}_2, \mathbf{u}_3, \ldots, \mathbf{u}_p\}$ is linearly dependent, we continue discarding redundant vectors until we obtain a linearly independent spanning set, Q'.

The following theorem is a companion to Theorem 8.

THEOREM 9

Let V be a finite-dimensional vector space with $\dim(V) = p$.

1. Any spanning set for V must contain at least p vectors.
2. Any set of p vectors that spans V is a basis for V.

PROOF   Property 1 follows immediately from the lemma above, for if there were a spanning set Q for V that contained fewer than p vectors, then we could find a subset Q' of Q that is a basis for V containing fewer than p vectors. This finding would contradict Theorem 7, so property 1 must be valid.

Property 2 also follows from the lemma, for we know there is a subset Q' of Q such that Q' is a basis for V. Since $\dim(V) = p$, Q' must have p vectors, and since $Q' \subseteq Q$, where Q has p vectors, we must have that $Q' = Q$.

EXAMPLE 3   Let V be the vector space of all (2×2) real matrices. In V, set

$$A_1 = \begin{bmatrix} 1 & 0 \\ -1 & 0 \end{bmatrix}, \qquad A_2 = \begin{bmatrix} 0 & 1 \\ 2 & 0 \end{bmatrix}, \qquad A_3 = \begin{bmatrix} 0 & 0 \\ -1 & 3 \end{bmatrix},$$

$$A_4 = \begin{bmatrix} 1 & 0 \\ -1 & 1 \end{bmatrix}, \quad \text{and} \quad A_5 = \begin{bmatrix} 2 & 1 \\ 3 & 1 \end{bmatrix}.$$

For each of the sets $\{A_1, A_2, A_3\}$, $\{A_1, A_2, A_3, A_4\}$, and $\{A_1, A_2, A_3, A_4, A_5\}$, determine whether the set is a basis for V.

SOLUTION   We have already noted that $\dim(V) = 4$ and that $B = \{E_{11}, E_{12}, E_{21}, E_{22}\}$ is a basis for V. It follows from property 1 of Theorem 9 that the set $\{A_1, A_2, A_3\}$ does not span V. Likewise, property 1 of Theorem 8, implies that $\{A_1, A_2, A_3, A_4, A_5\}$ is a linearly dependent set. By property 2 of Theorem 8, the set $\{A_1, A_2, A_3, A_4\}$ is a basis for V if and only if it is a linearly independent set. It is straightforward to see that the set of coordinate vectors

$\{[A_1]_B, [A_2]_B, [A_3]_B, [A_4]_B\}$ is a linearly independent set. By Theorem 5 of Section 4.4, the set $\{A_1, A_2, A_3, A_4\}$ is also linearly independent; thus the set is a basis for V.

Exercises 4.5

1. Let V be the set of all real (3×3) matrices, and let V_1 and V_2 be subsets of V, where V_1 consists of all the (3×3) lower-triangular matrices and V_2 consists of all the (3×3) upper-triangular matrices.

   a) Show that V_1 and V_2 are subspaces of V.
   b) Find bases for V_1 and V_2.
   c) Calculate dim(V), dim(V_1), and dim(V_2).

2. Suppose that V_1 and V_2 are subspaces of a vector space V. Show that $V_1 \cap V_2$ is also a subspace of V. It is not necessarily true that $V_1 \cap V_2$ is a subspace of V. Let $V = R^2$ and find two subspaces of R^2 whose union is not a subspace of R^2.

3. Let V, V_1, and V_2 be as in Exercise 1. By Exercise 2, $V_1 \cap V_2$ is a subspace of V. Describe $V_1 \cap V_2$ and calculate its dimension.

4. Let V be as in Exercise 1, and let W be the subset of all the (3×3) symmetric matrices in V. Clearly W is a subspace of V. What is dim(W)?

5. Recall that a square matrix A is called skew symmetric if $A^T = -A$. Let V be as in Exercise 1 and let W be the subset of all the (3×3) skew-symmetric matrices in V. Calculate dim(W).

6. Let W be the subspace of $\mathscr{P}_2$ consisting of polynomials $p(x) = a_0 + a_1 x + a_2 x^2$ such that $2a_0 - a_1 + 3a_2 = 0$. Determine dim(W).

7. Let W be the subspace of $\mathscr{P}_4$ defined thus: $p(x)$ is in W if and only if $p(1) + p(-1) = 0$ and $p(2) + p(-2) = 0$. What is dim(W)?

In Exercises 8–13, a subset S of a vector space V is given. In each case choose one of the statements (i), (ii), or (iii) that holds for S and verify that this is the case.
  i) S is a basis for V.
 ii) S does not span V.
iii) S is linearly dependent.

8. $S = \{1 + x - x^2, x + x^3, -x^2 + x^3\}$; $V = \mathscr{P}_3$

9. $S = \{1 + x^2, x - x^2, 1 + x, 2 - x + x^2\}$; $V = \mathscr{P}_2$

10. $S = \{1 + x + x^2, x + x^2, x^2\}$; $V = \mathscr{P}_2$

11. $S = \left\{ \begin{bmatrix} 0 & 1 \\ 1 & 0 \end{bmatrix}, \begin{bmatrix} 1 & 0 \\ 0 & 1 \end{bmatrix} \right\}$;
V is the set of all (2×2) real matrices

12. $S = \left\{ \begin{bmatrix} 0 & 0 \\ 0 & 1 \end{bmatrix}, \begin{bmatrix} 0 & 1 \\ 0 & 1 \end{bmatrix}, \begin{bmatrix} 1 & 1 \\ 1 & 1 \end{bmatrix}, \begin{bmatrix} 0 & 1 \\ 1 & 1 \end{bmatrix} \right\}$;
V is the set of all (2×2) real matrices

13. $S = \left\{ \begin{bmatrix} 1 & 0 \\ -1 & 0 \end{bmatrix}, \begin{bmatrix} 1 & 2 \\ 1 & -2 \end{bmatrix}, \begin{bmatrix} 1 & -1 \\ 1 & 4 \end{bmatrix}, \right.$
$\left. \begin{bmatrix} 3 & 4 \\ 0 & 4 \end{bmatrix}, \begin{bmatrix} 0 & 1 \\ -1 & 3 \end{bmatrix} \right\}$;
V is the set of all (2×2) real matrices

14. Let W be the subspace of $C[-\pi, \pi]$ consisting of functions of the form $f(x) = a \sin x + b \cos x$. Determine the dimension of W.

15. Let V denote the set of all infinite sequences of real numbers:
$$V = \{\mathbf{x}: \mathbf{x} = \{x_i\}_{i=1}^{\infty}, \quad x_i \text{ in } R\}.$$
If $\mathbf{x} = \{x_i\}_{i=1}^{\infty}$ and $\mathbf{y} = \{y_i\}_{i=1}^{\infty}$ are in V, then $\mathbf{x} + \mathbf{y}$ is the sequence $\{x_i + y_i\}_{i=1}^{\infty}$. If c is a real number, then $c\mathbf{x}$ is the sequence $\{cx_i\}_{i=1}^{\infty}$.
   a) Prove that V is a vector space.
   b) Show that V has infinite dimension. (*Hint:* For each positive integer, k, let $\mathbf{s}_k$ denote the sequence $\mathbf{s}_k = \{e_{ki}\}_{i=1}^{\infty}$, where $e_{kk} = 1$, but $e_{ki} = 0$ for $i \neq k$. For each positive integer n, show that $\{\mathbf{s}_1, \mathbf{s}_2, \ldots, \mathbf{s}_n\}$ is a linearly independent subset of V.)

16. Let V be a vector space, and let W be a subspace of V, where dim$(W) = k$. Prove that if V is finite dimensional, then dim$(V) \geq k$. (*Hint:* W must contain a set of k linearly independent vectors.)

17. Let W be a subspace of a finite-dimensional vector space V, where W contains at least one nonzero vector. Prove that W has a basis and that dim$(W) \leq$ dim(V). (*Hint:* Use Exercise 36 of Section 4.4 to show that W has a basis.)

18. Prove Theorem 6. (*Hint:* Let $\{\mathbf{u}_1, \mathbf{u}_2, \ldots, \mathbf{u}_k\}$ be a subset of V, where $k \geq p + 1$. Consider the vectors $[\mathbf{u}_1]_B, [\mathbf{u}_2]_B, \ldots, [\mathbf{u}_k]_B$ in R^p and apply Theorem 5 of Section 4.4.)

19. Prove Theorem 7.

20. Prove Theorem 8.

21. (*Change of basis;* see also Section 4.10). Let V be a vector space, where $\dim(V) = n$, and let $B = \{\mathbf{v}_1, \mathbf{v}_2, \ldots, \mathbf{v}_n\}$ and $C = \{\mathbf{u}_1, \mathbf{u}_2, \ldots, \mathbf{u}_n\}$ be two bases for V. Let $\mathbf{w}$ be any vector in V and suppose that $\mathbf{w}$ has these representations in terms of the bases B and C:

    $$\mathbf{w} = d_1\mathbf{v}_1 + d_2\mathbf{v}_2 + \cdots + d_n\mathbf{v}_n$$

    $$\mathbf{w} = c_1\mathbf{u}_1 + c_2\mathbf{u}_2 + \cdots + c_n\mathbf{u}_n.$$

    By considering Eq. (10) of Section 4.4, convince yourself that the coordinate vectors for $\mathbf{w}$ satisfy

    $$[\mathbf{w}]_B = A[\mathbf{w}]_C,$$

    where A is the $(n \times n)$ matrix whose ith column is equal to $[\mathbf{u}_i]_B$, $1 \leq i \leq n$. As an application, consider the two bases for $\mathscr{P}_2$: $C = \{1, x, x^2\}$ and $B = \{1, x + 1, (x + 1)^2\}$.
    a) Calculate the (3×3) matrix A described above.
    b) Using the identity $[p]_B = A[p]_C$, calculate the coordinate vector of $p(x) = x^2 + 4x + 8$ with respect to B.

22. The matrix A in Exercise 21 is called a transition matrix and shows how to transform a representa-

tion with respect to one basis into a representation with respect to another. Use the matrix in part (a) of Exercise 21 to convert $p(x) = c_0 + c_1x + c_2x^2$ to the form $p(x) = a_0 + a_1(x + 1) + a_2(x + 1)^2$, where:
a) $p(x) = x^2 + 3x - 2$;
b) $p(x) = 2x^2 - 5x + 8$;
c) $p(x) = -x^2 - 2x + 3$;
d) $p(x) = x - 9$.

23. By Theorem 5 of Section 4.4, an $(n \times n)$ transition matrix (see Exercises 21 and 22) is always nonsingular. Thus if $[\mathbf{w}]_B = A[\mathbf{w}]_C$, then $[\mathbf{w}]_C = A^{-1}[\mathbf{w}]_B$. Calculate A^{-1} for the matrix in part (a) of Exercise 21 and use the result to transform each of the polynomials below to the form $a_0 + a_1x + a_2x^2$.
a) $p(x) = 2 - 3(x + 1) + 7(x + 1)^2$
b) $p(x) = 1 + 4(x + 1) - (x + 1)^2$
c) $p(x) = 4 + (x + 1)$
d) $p(x) = 9 - (x + 1)^2$

24. Find a matrix A such that $[p]_B = A[p]_C$ for all $p(x)$ in $\mathscr{P}_3$, where $C = \{1, x, x^2, x^3\}$ and $B = \{1, x, x(x - 1), x(x - 1)(x - 2)\}$. Use A to convert each of the following to the form $p(x) = a_0 + a_1x + a_2x(x - 1) + a_3x(x - 1)(x - 2)$.
a) $p(x) = x^3 - 2x^2 + 5x - 9$
b) $p(x) = x^2 + 7x - 2$
c) $p(x) = x^3 + 1$
d) $p(x) = x^3 + 2x^2 + 2x + 3$

4.6   Inner-Product Spaces, Orthogonal Bases, and Projections (Optional)

Up to now we have considered a vector space solely as an entity with an algebraic structure. We know, however, that R^n possesses more than just an algebraic structure; in particular, we know that we can measure the "size" or "length" of a vector $\mathbf{x}$ in R^n by the quantity $\|\mathbf{x}\| = \sqrt{\mathbf{x}^T\mathbf{x}}$. Similarly, we can define the distance from $\mathbf{x}$ to $\mathbf{y}$ as $\|\mathbf{x} - \mathbf{y}\|$. The ability to measure distances means that R^n has a geometric structure, which supplements the algebraic structure. The geometric structure can be employed to study problems of convergence, continuity, and the like. In this section we briefly describe how a suitable measure of distance might be imposed on a general vector space. Our development will be sketchy, and we will leave most of the details to the reader; but the ideas parallel those in Sections 2.6 and 2.8.

Inner-Product Spaces

To begin, we observe that the geometrical structure for R^n is based on the scalar product $\mathbf{x}^T\mathbf{y}$. Essentially the scalar product is a real-valued function of two vector variables: Given $\mathbf{x}$ and $\mathbf{y}$ in R^n, the scalar product produces a number $\mathbf{x}^T\mathbf{y}$. Thus to derive a geometric structure for a vector space V, we should look for a generalization of the scalar-product function. A consideration of the properties of the scalar-product function leads to the definition of an "inner-product" function for a vector space. (With reference to Definition 7 below, we note that the expression $\mathbf{u}^T\mathbf{v}$ does not make sense in a general vector space V. Thus not only does the nomenclature change—scalar product becomes inner product— but also the notation changes as well, with $\langle \mathbf{u}, \mathbf{v} \rangle$ denoting the inner product of $\mathbf{u}$ and $\mathbf{v}$.)

DEFINITION 7 An *inner product* on a real vector space V is a function that assigns a real number, $\langle \mathbf{u}, \mathbf{v} \rangle$, to each pair of vectors $\mathbf{u}$ and $\mathbf{v}$ in V, and that satisfies these properties:

1. $\langle \mathbf{u}, \mathbf{u} \rangle \geq 0$ and $\langle \mathbf{u}, \mathbf{u} \rangle = 0$ if and only if $\mathbf{u} = \boldsymbol{\theta}$.
2. $\langle \mathbf{u}, \mathbf{v} \rangle = \langle \mathbf{v}, \mathbf{u} \rangle$.
3. $\langle a\mathbf{u}, \mathbf{v} \rangle = a\langle \mathbf{u}, \mathbf{v} \rangle$.
4. $\langle \mathbf{u}, \mathbf{v} + \mathbf{w} \rangle = \langle \mathbf{u}, \mathbf{v} \rangle + \langle \mathbf{u}, \mathbf{w} \rangle$.

The usual scalar product in R^n is an inner product in the sense of Definition 7, where $\langle \mathbf{x}, \mathbf{y} \rangle = \mathbf{x}^T\mathbf{y}$. To illustrate the flexibility of Definition 7, we also note that there are other sorts of inner products for R^n. The following example gives another inner product for R^2.

<u>EXAMPLE 1</u> Let V be the vector space R^2, and let A be the (2×2) matrix

$$A = \begin{bmatrix} 3 & 2 \\ 2 & 4 \end{bmatrix}.$$

Verify that the function $\langle \mathbf{u}, \mathbf{v} \rangle = \mathbf{u}^T A\mathbf{v}$ is an inner product for R^2.

SOLUTION Let $\mathbf{u}$ be a vector in R^2:

$$\mathbf{u} = \begin{bmatrix} u_1 \\ u_2 \end{bmatrix}.$$

Then

$$\langle \mathbf{u}, \mathbf{u} \rangle = \mathbf{u}^T A\mathbf{u} = [u_1, u_2] \begin{bmatrix} 3 & 2 \\ 2 & 4 \end{bmatrix} \begin{bmatrix} u_1 \\ u_2 \end{bmatrix},$$

so $\langle \mathbf{u}, \mathbf{u} \rangle = 3u_1^2 + 4u_1 u_2 + 4u_2^2 = 2u_1^2 + (u_1 + 2u_2)^2$. Thus $\langle \mathbf{u}, \mathbf{u} \rangle \geq 0$ and $\langle \mathbf{u}, \mathbf{u} \rangle = 0$ if and only if $u_1 = u_2 = 0$. This shows that property 1 of Definition 7 is satisfied.

To see that property 2 of Definition 7 holds, note that A is symmetric; that is, $A^T = A$. Also observe that if $\mathbf{u}$ and $\mathbf{v}$ are in R^2, then $\mathbf{u}^T A \mathbf{v}$ is a (1×1) matrix, so $(\mathbf{u}^T A \mathbf{v})^T = \mathbf{u}^T A \mathbf{v}$. It now follows that $\langle \mathbf{u}, \mathbf{v} \rangle = \mathbf{u}^T A \mathbf{v} = (\mathbf{u}^T A \mathbf{v})^T = \mathbf{v}^T A^T (\mathbf{u}^T)^T = \mathbf{v}^T A^T \mathbf{u} = \langle \mathbf{v}, \mathbf{u} \rangle$.

Properties 3 and 4 of Definition 7 follow easily from the properties of matrix multiplication, so $\langle \mathbf{u}, \mathbf{v} \rangle$ is an inner product for R^2.

In Example 1, an inner product for R^2 was defined in terms of a matrix A:

$$\langle \mathbf{u}, \mathbf{v} \rangle = \mathbf{u}^T A \mathbf{v}.$$

In general, we might ask the following question:

"For what $(n \times n)$ matrices A does the operation $\mathbf{u}^T A \mathbf{v}$ define an inner product on R^n?"

The answer to this question is suggested by the solution to Example 1. In particular (see Exercises 3 and 32), the operation $\langle \mathbf{u}, \mathbf{v} \rangle = \mathbf{u}^T A \mathbf{v}$ is an inner product for R^n if and only if A is a symmetric positive-definite matrix.

There are a number of ways in which inner products can be defined on spaces of functions. For example, Exercise 6 will show that

$$\langle p, q \rangle = p(0)q(0) + p(1)q(1) + p(2)q(2)$$

defines one inner product for $\mathscr{P}_2$. The following example gives yet another inner product for $\mathscr{P}_2$.

EXAMPLE 2 For $p(t)$ and $q(t)$ in $\mathscr{P}_2$, verify that

$$\langle p, q \rangle = \int_0^1 p(t)q(t)\, dt$$

is an inner product.

SOLUTION To check property 1 of Definition 7, note that

$$\langle p, p \rangle = \int_0^1 p(t)^2\, dt,$$

and $p(t)^2 \geq 0$ for $0 \leq t \leq 1$. Thus $\langle p, p \rangle$ is the area under the curve $p(t)^2$, $0 \leq t \leq 1$. Hence $\langle p, p \rangle \geq 0$, and equality holds if and only if $p(t) = 0$, $0 \leq t \leq 1$ (see Fig. 4.5).

Properties 2, 3, and 4 of Definition 7 are straightforward to verify, and we include here only the verification of property 4. If $p(t)$, $q(t)$, and $r(t)$ are in $\mathscr{P}_2$, then

$$\langle p, q + r \rangle = \int_0^1 p(t)[q(t) + r(t)]\, dt = \int_0^1 [p(t)q(t) + p(t)r(t)]\, dt$$

$$= \int_0^1 p(t)q(t)\, dt + \int_0^1 p(t)r(t)\, dt = \langle p, q \rangle + \langle p, r \rangle,$$

as required by property 4.

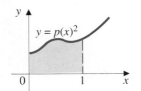

Figure 4.5
The value $\langle p, p \rangle$ is equal to the area under the graph of $y = p(x)^2$.

After the key step of defining a vector-space analog of the scalar product, the rest is routine. For purposes of reference we call a vector space with an inner product an *inner-product space.* As in R^n, we can use the inner product as a measure of size: If V is an inner-product space, then for each $\mathbf{v}$ in V we define $\|\mathbf{v}\|$ (the *norm* of $\mathbf{v}$) as

$$\|\mathbf{v}\| = \sqrt{\langle \mathbf{v}, \mathbf{v} \rangle}.$$

Note that $\langle \mathbf{v}, \mathbf{v} \rangle \geq 0$ for all $\mathbf{v}$ in V, so the norm function is always defined.

EXAMPLE 3 Use the inner product for $\mathscr{P}_2$ defined in Example 2 to determine $\|t^2\|$.

SOLUTION By definition, $\|t^2\| = \sqrt{\langle t^2, t^2 \rangle}$. But $\langle t^2, t^2 \rangle = \int_0^1 t^2 t^2\, dt = \int_0^1 t^4\, dt = 1/5$, so $\|t^2\| = 1/\sqrt{5}$. ☐

Before continuing, we pause to illustrate one way in which the inner-product space framework is used in practice. One of the many inner products for the vector space $C[0, 1]$ is

$$\langle f, g \rangle = \int_0^1 f(x)g(x)\, dx.$$

If f is a relatively complicated function in $C[0, 1]$, we might wish to approximate f by a simpler function, say a polynomial. For definiteness suppose we want to find a polynomial p in $\mathscr{P}_2$ that is a good approximation to f. The phrase "good approximation" is too vague to be used in any calculation, but the inner-product space framework allows us to measure size and thus to pose some meaningful problems. In particular, we can ask for a polynomial p^* in $\mathscr{P}_2$ such that

$$\|f - p^*\| \leq \|f - p\|$$

for all p in $\mathscr{P}_2$. Finding such a polynomial p^* in this setting is equivalent to minimizing

$$\int_0^1 [f(x) - p(x)]^2\, dx$$

among all p in $\mathscr{P}_2$. We will present a procedure for doing this shortly.

Orthogonal Bases

If $\mathbf{u}$ and $\mathbf{v}$ are vectors in an inner-product space V, we say that $\mathbf{u}$ and $\mathbf{v}$ are *orthogonal* if $\langle \mathbf{u}, \mathbf{v} \rangle = 0$. Similarly, $B = \{\mathbf{v}_1, \mathbf{v}_2, \ldots, \mathbf{v}_p\}$ is an *orthogonal set* in V if $\langle \mathbf{v}_i, \mathbf{v}_j \rangle = 0$ when $i \neq j$. In addition, if an orthogonal set of vectors B is a basis for V, we call B an *orthogonal basis.* The next two theorems correspond to their analogs in R^n, and we leave the proofs to the exercises. [See (5a), (5b), and Theorem 14 of Section 2.6.]

> **THEOREM 10**
>
> Let $B = \{\mathbf{v}_1, \mathbf{v}_2, \ldots, \mathbf{v}_n\}$ be an orthogonal basis for an inner-product space V. If $\mathbf{u}$ is any vector in V, then
>
> $$\mathbf{u} = \frac{\langle \mathbf{v}_1, \mathbf{u} \rangle}{\langle \mathbf{v}_1, \mathbf{v}_1 \rangle} \mathbf{v}_1 + \frac{\langle \mathbf{v}_2, \mathbf{u} \rangle}{\langle \mathbf{v}_2, \mathbf{v}_2 \rangle} \mathbf{v}_2 + \cdots + \frac{\langle \mathbf{v}_n, \mathbf{u} \rangle}{\langle \mathbf{v}_n, \mathbf{v}_n \rangle} \mathbf{v}_n.$$

> **THEOREM 11**
>
> **Gram–Schmidt Orthogonalization** Let V be an inner-product space, and let $\{\mathbf{u}_1, \mathbf{u}_2, \ldots, \mathbf{u}_n\}$ be a basis for V. Let $\mathbf{v}_1 = \mathbf{u}_1$, and for $2 \leq k \leq n$ define $\mathbf{v}_k$ by
>
> $$\mathbf{v}_k = \mathbf{u}_k - \sum_{j=1}^{k-1} \frac{\langle \mathbf{u}_k, \mathbf{v}_j \rangle}{\langle \mathbf{v}_j, \mathbf{v}_j \rangle} \mathbf{v}_j.$$
>
> Then $\{\mathbf{v}_1, \mathbf{v}_2, \ldots, \mathbf{v}_n\}$ is an orthogonal basis for V.

EXAMPLE 4 Let the inner product on $\mathscr{P}_2$ be the one given in Example 2. Starting with the natural basis $\{1, x, x^2\}$, use Gram–Schmidt orthogonalization to obtain an orthogonal basis for $\mathscr{P}_2$.

SOLUTION If we let $\{p_0, p_1, p_2\}$ denote the orthogonal basis, we have $p_0(x) = 1$ and find $p_1(x)$ from

$$p_1(x) = x - \frac{\langle p_0, x \rangle}{\langle p_0, p_0 \rangle} p_0(x).$$

We calculate

$$\langle p_0, x \rangle = \int_0^1 x\, dx = 1/2 \quad \text{and} \quad \langle p_0, p_0 \rangle = \int_0^1 dx = 1;$$

so $p_1(x) = x - 1/2$. The next step of the Gram–Schmidt orthogonalization process is to form

$$p_2(x) = x^2 - \frac{\langle p_1, x^2 \rangle}{\langle p_1, p_1 \rangle} p_1(x) - \frac{\langle p_0, x^2 \rangle}{\langle p_0, p_0 \rangle} p_0(x).$$

The required constants are

$$\langle p_1, x^2 \rangle = \int_0^1 (x^3 - x^2/2)\, dx = 1/12$$

$$\langle p_1, p_1 \rangle = \int_0^1 (x^2 - x + 1/4)\, dx = 1/12$$

$$\langle p_0, x^2 \rangle = \int_0^1 x^2\, dx = 1/3$$

$$\langle p_0, p_0 \rangle = \int_0^1 dx = 1.$$

Therefore, $p_2(x) = x^2 - p_1(x) - p_0(x)/3 = x^2 - x + 1/6$, and $\{p_0, p_1, p_2\}$ is an orthogonal basis for $\mathscr{P}_2$ with respect to the inner product. □

EXAMPLE 5 Let $B = \{p_0, p_1, p_2\}$ be the orthogonal basis for $\mathscr{P}_2$ obtained in Example 4. Find the coordinates of x^2 relative to B.

SOLUTION By Theorem 10, $x^2 = a_0 p_0(x) + a_1 p_1(x) + a_2 p_2(x)$, where

$$a_0 = \langle p_0, x^2 \rangle / \langle p_0, p_0 \rangle$$
$$a_1 = \langle p_1, x^2 \rangle / \langle p_1, p_1 \rangle$$
$$a_2 = \langle p_2, x^2 \rangle / \langle p_2, p_2 \rangle.$$

The necessary calculations are

$$\langle p_0, x^2 \rangle = \int_0^1 x^2 \, dx = 1/3$$

$$\langle p_1, x^2 \rangle = \int_0^1 [x^3 - (1/2)x^2] \, dx = 1/12$$

$$\langle p_2, x^2 \rangle = \int_0^1 [x^4 - x^3 + (1/6)x^2] \, dx = 1/180$$

$$\langle p_0, p_0 \rangle = \int_0^1 dx = 1$$

$$\langle p_1, p_1 \rangle = \int_0^1 [x^2 - x + 1/4] \, dx = 1/12$$

$$\langle p_2, p_2 \rangle = \int_0^1 [x^2 - x + 1/6]^2 \, dx = 1/180.$$

Thus $a_0 = 1/3$, $a_1 = 1$, and $a_2 = 1$. We can easily check that $x^2 = (1/3)p_0(x) + p_1(x) + p_2(x)$. □

Orthogonal Projections

We return now to the previously discussed problem of finding a polynomial p^* in $\mathscr{P}_2$ that is the best approximation of a function f in $C[0, 1]$. Note that the problem amounts to determining a vector p^* in a subspace of an inner-product space, where p^* is "closer" to f than any other vector in the subspace. The essential aspects of this problem can be stated formally as the following general problem:

Let V be an inner-product space and let W be a subspace of V. Given a vector $\mathbf{v}$ in V, find a vector $\mathbf{w}^*$ in W such that

$$\|\mathbf{v} - \mathbf{w}^*\| \le \|\mathbf{v} - \mathbf{w}\| \quad \text{for all } \mathbf{w} \text{ in } W. \tag{1}$$

A vector $\mathbf{w}^*$ in W satisfying inequality (1) is called the ***projection of v onto W***, or (frequently) the ***best least-squares approximation*** to $\mathbf{v}$. Intuitively $\mathbf{w}^*$ is the nearest vector in W to $\mathbf{v}$.

The solution process for this problem is almost exactly the same as that for the least-squares problem in R^n. One distinction in our general setting is that the subspace W might not be finite dimensional. If W is an infinite-dimensional subspace of V, then there may or may not be a projection of $\mathbf{v}$ onto W. If W is finite dimensional, then a projection always exists, is unique, and can be found explicitly. The next two theorems outline this concept, and again we leave the proofs to the reader since they parallel the proof of Theorem 17 of Section 2.8.

THEOREM 12

Let V be an inner-product space, and let W be a subspace of V. Let $\mathbf{v}$ be a vector in V, and suppose $\mathbf{w}^*$ is a vector in W such that

$$\langle \mathbf{v} - \mathbf{w}^*, \mathbf{w} \rangle = 0 \quad \text{for all } \mathbf{w} \text{ in } W.$$

Then $\|\mathbf{v} - \mathbf{w}^*\| \leq \|\mathbf{v} - \mathbf{w}\|$ for all $\mathbf{w}$ in W with equality holding only for $\mathbf{w} = \mathbf{w}^*$.

THEOREM 13

Let V be an inner-product space, and let $\mathbf{v}$ be a vector in V. Let W be an n-dimensional subspace of V, and let $\{\mathbf{u}_1, \mathbf{u}_2, \ldots, \mathbf{u}_n\}$ be an orthogonal basis for W. Then

$$\|\mathbf{v} - \mathbf{w}^*\| \leq \|\mathbf{v} - \mathbf{w}\| \quad \text{for all } \mathbf{w} \text{ in } W$$

if and only if

$$\mathbf{w}^* = \frac{\langle \mathbf{v}, \mathbf{u}_1 \rangle}{\langle \mathbf{u}_1, \mathbf{u}_1 \rangle}\mathbf{u}_1 + \frac{\langle \mathbf{v}, \mathbf{u}_2 \rangle}{\langle \mathbf{u}_2, \mathbf{u}_2 \rangle}\mathbf{u}_2 + \cdots + \frac{\langle \mathbf{v}, \mathbf{u}_n \rangle}{\langle \mathbf{u}_n, \mathbf{u}_n \rangle}\mathbf{u}_n. \tag{2}$$

In view of Theorem 13, it follows that when W is a finite-dimensional subspace of an inner-product space V, we can always find projections by first finding an orthogonal basis for W (by using Theorem 11) and then calculating the projection $\mathbf{w}^*$ from Eq. (2).

To illustrate the process of finding a projection, we return to the inner-product space $C[0, 1]$ with the subspace $\mathscr{P}_2$. As a specific but rather unrealistic function, f, we choose $f(x) = \cos x$, x in radians. The inner product is

$$\langle f, g \rangle = \int_0^1 f(x)g(x)\,dx.$$

EXAMPLE 6   In the vector space $C[0, 1]$, let $f(x) = \cos x$. Find the projection of f onto the subspace $\mathscr{P}_2$.

SOLUTION   Let $\{p_0, p_1, p_2\}$ be the orthogonal basis for $\mathscr{P}_2$ found in Example 4. (Note that the inner product used in Example 4 coincides with the

Table 4.1

x	$p^*(x)$	$\cos x$	$p^*(x) - \cos x$
0.0	1.0034	1.0000	.0034
0.2	.9789	.9801	−.0012
0.4	.9198	.9211	−.0013
0.6	.8263	.8253	.0010
0.8	.6983	.6967	.0016
1.0	.5359	.5403	−.0044

present inner product on $C[0, 1]$.) By Theorem 13, the projection of f onto $\mathscr{P}_2$ is the polynomial p^* defined by

$$p^*(x) = \frac{\langle f, p_0 \rangle}{\langle p_0, p_0 \rangle} p_0(x) + \frac{\langle f, p_1 \rangle}{\langle p_1, p_1 \rangle} p_1(x) + \frac{\langle f, p_2 \rangle}{\langle p_2, p_2 \rangle} p_2(x),$$

where

$$\langle f, p_0 \rangle = \int_0^1 \cos(x)\, dx \simeq .841471$$

$$\langle f, p_1 \rangle = \int_0^1 (x - 1/2)\cos(x)\, dx \simeq -.038962$$

$$\langle f, p_2 \rangle = \int_0^1 (x^2 - x + 1/6)\cos(x)\, dx \simeq -.002394.$$

From Example 5, we have $\langle p_0, p_0 \rangle = 1$, $\langle p_1, p_1 \rangle = 1/12$, and $\langle p_2, p_2 \rangle = 1/180$. Therefore, $p^*(x)$ is given by

$$p^*(x) = \langle f, p_0 \rangle p_0(x) + 12\langle f, p_1 \rangle p_1(x) + 180\langle f, p_2 \rangle p_2(x)$$
$$\simeq .841471 p_0(x) - .467544 p_1(x) - .430920 p_2(x).$$

In order to assess how well $p^*(x)$ approximates $\cos x$ in the interval $[0, 1]$, we can tabulate $p^*(x)$ and $\cos x$ at various values of x (see Table 4.1).

EXAMPLE 7   The function $\mathrm{Si}(x)$ (important in applications such as optics) is defined as follows:

$$\mathrm{Si}(x) = \int_0^x \frac{\sin u}{u}\, du, \quad \text{for } x \neq 0. \tag{3}$$

The integral in (3) is not an elementary one and so, for a given value of x, $\mathrm{Si}(x)$ must be evaluated using a numerical integration procedure. In this example, we approximate $\mathrm{Si}(x)$ by a cubic polynomial for $0 \leq x \leq 1$. In particular, it can be shown that if we define $\mathrm{Si}(0) = 0$, then $\mathrm{Si}(x)$ is continuous for all x. Thus we can ask: "What is the projection of $\mathrm{Si}(x)$ onto the subspace $\mathscr{P}_3$ of $C[0, 1]$?" This projection will serve as an approximation to $\mathrm{Si}(x)$ for $0 \leq x \leq 1$.

6: $\displaystyle\int_0^1 x^3\ P1\ (x)\ dx$

7: $\dfrac{3}{40}$

8: $\displaystyle\int_0^1 x^3\ P2\ (x)\ dx$

9: $\dfrac{1}{120}$

15: $P3\ (x)\ :=x^3\ -\ \dfrac{1}{4}\ P0\ (x)\ -\ \dfrac{9}{10}\ P1\ (x)\ -\ \dfrac{3}{2}\ P2\ (x)$

16: $P3\ (x)\ :=x^3\ -\ \dfrac{3\,x^2}{2}\ +\ \dfrac{3\,x}{5}\ -\ \dfrac{1}{20}$

17: $\displaystyle\int_0^1 P3\ (x)\ P3\ (x)\ dx$

18: $\dfrac{1}{2800}$

49: $\displaystyle\int_0^1 180\ P2\ (x)\ \int_0^x \dfrac{SIN\ (u)}{u}\ du\ dx$

50: $-\ 0.0804033$

51: $\displaystyle\int_0^1 2800\ P3\ (x)\ \int_0^x \dfrac{SIN\ (u)}{u}\ du\ dx$

52: $-\ 0.0510442$

Figure 4.6
Some of the steps used by Derive to generate the projection of
Si(x) onto $\mathscr{P}_3$ in Example 7.

SOLUTION We used the computer algebra system Derive to carry out the calculations. Some of the steps are shown in Fig. 4.6. To begin, let $\{p_0, p_1, p_2, p_3\}$ be the orthogonal basis for $\mathscr{P}_3$ found by the Gram–Schmidt process. From Example 4, we already have that $p_0(x) = 1$, $p_1(x) = x - 1/2$, and $p_2(x) = x^2 - x + 1/6$. To find p_3, we first calculate the inner products

$$\langle p_0, x^3 \rangle, \langle p_1, x^3 \rangle, \langle p_2, x^3 \rangle$$

(see steps 6–9 in Fig. 4.6 for $\langle p_1, x^3 \rangle$ and $\langle p_2, x^3 \rangle$).

Using Theorem 11, we find p_3 and, for later use, $\langle p_3, p_3 \rangle$:

$$p_3(x) = x^3 - (3/2)x^2 + (3/5)x - 1/20$$
$$\langle p_3, p_3 \rangle = 1/2800$$

(see steps 15–18 in Fig. 4.6). Finally, by Theorem 13, the projection of $\mathrm{Si}(x)$ onto $\mathscr{P}_3$ is the polynomial p^* defined by

$$p^*(x) = \frac{\langle \mathrm{Si}, p_0 \rangle}{\langle p_0, p_0 \rangle} p_0(x) + \frac{\langle \mathrm{Si}, p_1 \rangle}{\langle p_1, p_1 \rangle} p_1(x) + \frac{\langle \mathrm{Si}, p_2 \rangle}{\langle p_2, p_2 \rangle} p_2(x) + \frac{\langle \mathrm{Si}, p_3 \rangle}{\langle p_3, p_3 \rangle} p_3(x)$$

$$= \langle \mathrm{Si}, p_0 \rangle p_0(x) + 12 \langle \mathrm{Si}, p_1 \rangle p_1(x) + 180 \langle \mathrm{Si}, p_2 \rangle p_2(x) + 2800 \langle \mathrm{Si}, p_3 \rangle p_3(x).$$

In the expression above for p^*, the inner products $\langle \mathrm{Si}, p_k \rangle$ for $k = 0, 1, 2$, and 3 are given by

$$\langle \mathrm{Si}, p_k \rangle = \int_0^1 p_k(x) \, \mathrm{Si}(x) \, dx = \int_0^1 p_k(x) \left\{ \int_0^x \frac{\sin u}{u} \, du \right\} dx$$

(see steps 49–52 in Fig. 4.6 for 180 $\langle \mathrm{Si}, p_2 \rangle$ and 2800 $\langle \mathrm{Si}, p_3 \rangle$).

Now, since $\mathrm{Si}(x)$ must be estimated numerically, it follows that the inner products $\langle \mathrm{Si}, p_k \rangle$ must be estimated as well. Using Derive to approximate the inner products, we obtain the projection (or best least-squares approximation)

$$p^*(x) = .486385 p_0(x) + .951172 p_1(x) - .0804033 p_2(x) - .0510442 p_3(x).$$

To assess how well $p^*(x)$ approximates $\mathrm{Si}(x)$ in $[0, 1]$, we tabulate each function at a few selected points (see Table 4.2). As can be seen from Table 4.2, it appears that $p^*(x)$ is a very good approximation to $\mathrm{Si}(x)$.

Table 4.2

x	$p^*(x)$	$\mathrm{Si}(x)$	$p^*(x) - \mathrm{Si}(x)$
0.0	.000049	.000000	.000049
0.2	.199578	.199556	.000028
0.4	.396449	.396461	−.000012
0.6	.588113	.588128	−.000015
0.8	.772119	.772095	.000024
1.0	.946018	.946083	−.000065

1. Prove that $\langle \mathbf{x}, \mathbf{y} \rangle = 4x_1 y_1 + x_2 y_2$ is an inner product on R^2, where

$$\mathbf{x} = \begin{bmatrix} x_1 \\ x_2 \end{bmatrix} \quad \text{and} \quad \mathbf{y} = \begin{bmatrix} y_1 \\ y_2 \end{bmatrix}.$$

2. Prove that $\langle \mathbf{x}, \mathbf{y} \rangle = a_1 x_1 y_1 + a_2 x_2 y_2 + \cdots + a_n x_n y_n$ is an inner product on R^n, where $a_1, a_2, \ldots, a_n$ are positive real numbers and where

$$\mathbf{x} = [x_1, x_2, \ldots, x_n]^T \text{ and } \mathbf{y} = [y_1, y_2, \ldots, y_n]^T.$$

3. A real $(n \times n)$ symmetric matrix A is called *positive definite* if $\mathbf{x}^T A \mathbf{x} > 0$ for all $\mathbf{x}$ in R^n, $\mathbf{x} \neq \theta$. Let A be a symmetric positive-definite matrix and verify that

$$\langle \mathbf{x}, \mathbf{y} \rangle = \mathbf{x}^T A \mathbf{y}$$

defines an inner product on R^n; that is, verify that the four properties of Definition 7 are satisfied.

4. Prove that the symmetric matrix A given below is positive definite. Prove this by choosing an arbitrary vector $\mathbf{x}$ in R^2, $\mathbf{x} \neq \theta$, and calculating $\mathbf{x}^T A \mathbf{x}$.

$$A = \begin{bmatrix} 1 & 1 \\ 1 & 2 \end{bmatrix}$$

5. In $\mathscr{P}_2$ let $p(x) = a_0 + a_1 x + a_2 x^2$ and $q(x) = b_0 + b_1 x + b_2 x^2$. Prove that $\langle p, q \rangle = a_0 b_0 + a_1 b_1 + a_2 b_2$ is an inner product on $\mathscr{P}_2$.

6. Prove that $\langle p, q \rangle = p(0)q(0) + p(1)q(1) + p(2)q(2)$ is an inner product on $\mathscr{P}_2$.

7. Let $A = (a_{ij})$ and $B = (b_{ij})$ be (2×2) matrices. Show that $\langle A, B \rangle = a_{11}b_{11} + a_{12}b_{12} + a_{21}b_{21} + a_{22}b_{22}$ is an inner product for the vector space of all (2×2) matrices.

8. For $\mathbf{x} = [1, -2]^T$ and $\mathbf{y} = [0, 1]^T$ in R^2, find $\langle \mathbf{x}, \mathbf{y} \rangle$, $\|\mathbf{x}\|$, $\|\mathbf{y}\|$, and $\|\mathbf{x} - \mathbf{y}\|$ using the inner product given in Exercise 1.

9. Repeat Exercise 8 with the inner product defined in Exercise 3 and the matrix A given in Exercise 4.

10. In $\mathscr{P}_2$ let $p(x) = -1 + 2x + x^2$ and $q(x) = 1 - x + 2x^2$. Using the inner product given in Exercise 5, find $\langle p, q \rangle$, $\|p\|$, $\|q\|$, and $\|p - q\|$.

11. Repeat Exercise 10 using the inner product defined in Exercise 6.

12. Show that $\{1, x, x^2\}$ is an orthogonal basis for $\mathscr{P}_2$ with the inner product defined in Exercise 5 but not with the inner product in Exercise 6.

13. In R^2 let $S = \{\mathbf{x} : \|\mathbf{x}\| = 1\}$. Sketch a graph of S if $\langle \mathbf{x}, \mathbf{y} \rangle = \mathbf{x}^T \mathbf{y}$. Now graph S using the inner product given in Exercise 1.

14. Let A be the matrix given in Exercise 4, and for $\mathbf{x}$, $\mathbf{y}$ in R^2 define $\langle \mathbf{x}, \mathbf{y} \rangle = \mathbf{x}^T A \mathbf{y}$ (cf. Exercise 3). Starting with the natural basis $\{\mathbf{e}_1, \mathbf{e}_2\}$, use Theorem 11 to obtain an orthogonal basis $\{\mathbf{u}_1, \mathbf{u}_2\}$ for R^2.

15. Let $\{\mathbf{u}_1, \mathbf{u}_2\}$ be the orthogonal basis for R^2 obtained in Exercise 14 and let $\mathbf{v} = [3, 4]^T$. Use Theorem 10 to find scalars a_1, a_2 such that $\mathbf{v} = a_1 \mathbf{u}_1 + a_2 \mathbf{u}_2$.

16. Use Theorem 11 to calculate an orthogonal basis $\{p_0, p_1, p_2\}$ for $\mathscr{P}_2$ with respect to the inner product in Exercise 6. Start with the natural basis $\{1, x, x^2\}$ for $\mathscr{P}_2$.

17. Use Theorem 10 to write $q(x) = 2 + 3x - 4x^2$ in terms of the orthogonal basis $\{p_0, p_1, p_2\}$ obtained in Exercise 16.

18. Show that the function defined in Exercise 6 is not an inner product for $\mathscr{P}_3$. (*Hint:* Find $p(x)$ in $\mathscr{P}_3$ such that $\langle p, p \rangle = 0$, but $p \neq \theta$.)

19. Starting with the natural basis $\{1, x, x^2, x^3, x^4\}$, generate an orthogonal basis for $\mathscr{P}_4$ with respect to the inner product

$$\langle p, q \rangle = \sum_{i=-2}^{2} p(i)q(i).$$

20. If V is an inner-product space, show that $\langle \mathbf{v}, \theta \rangle = 0$ for each vector $\mathbf{v}$ in V.

21. Let V be an inner-product space, and let $\mathbf{u}$ be a vector in V such that $\langle \mathbf{u}, \mathbf{v} \rangle = 0$ for every vector $\mathbf{v}$ in V. Show that $\mathbf{u} = \theta$.

22. Let a be a scalar and $\mathbf{v}$ a vector in an inner-product space V. Prove that $\|a\mathbf{v}\| = |a| \|\mathbf{v}\|$.

23. Prove that if $\{\mathbf{v}_1, \mathbf{v}_2, \ldots, \mathbf{v}_k\}$ is an orthogonal set of nonzero vectors in an inner-product space, then this set is linearly independent.

24. Prove Theorem 10.

25. Approximate x^3 with a polynomial in $\mathscr{P}_2$. (*Hint:* Use the inner product

$$\langle p, q \rangle = \int_0^1 p(t)q(t)\, dt$$

and let $\{p_0, p_1, p_2\}$ be the orthogonal basis for $\mathscr{P}_2$ obtained in Example 4. Now apply Theorem 13.)

26. In Examples 4 and 7 we found $p_0(x), \ldots, p_3(x)$, which are orthogonal with respect to

$$\langle f, g \rangle = \int_0^1 f(x)g(x)\,dx.$$

Continue the process and find $p_4(x)$ so that $\{p_0, p_1, \ldots, p_4\}$ is an orthogonal basis for $\mathscr{P}_4$. (Clearly there is an infinite sequence of polynomials $p_0, p_1, \ldots, p_n, \ldots$ that satisfy

$$\int_0^1 p_i(x)p_j(x)\,dx = 0, \qquad i \neq j.$$

These are called the ***Legendre polynomials.***)

27. With the orthogonal basis for $\mathscr{P}_3$ obtained in Example 7, use Theorem 13 to find the projection of $f(x) = \cos x$ in $\mathscr{P}_3$. Construct a table similar to Table 4.1 and note the improvement.

28. An inner product on $C[-1, 1]$ is

$$\langle f, g \rangle = \frac{2}{\pi} \int_{-1}^1 \frac{f(x)g(x)}{\sqrt{1 - x^2}}\,dx.$$

Starting with the set $\{1, x, x^2, x^3, \ldots\}$, use the Gram–Schmidt process to find polynomials $T_0(x)$, $T_1(x)$, $T_2(x)$, and $T_3(x)$ such that $\langle T_i, T_j \rangle = 0$ when $i \neq j$. These polynomials are called the ***Chebyshev polynomials of the first kind.*** (*Hint:* Make a change of variables $x = \cos \theta$.)

29. A sequence of orthogonal polynomials usually satisfies a "three-term recurrence relation." For example, the Chebyshev polynomials are related by

$$T_{n+1}(x) = 2xT_n(x) - T_{n-1}(x), \qquad n = 1, 2, \ldots, \quad \textbf{(R)}$$

where $T_0(x) = 1$ and $T_1(x) = x$. Verify that the polynomials defined by the relation (R) above are indeed orthogonal in $C[-1, 1]$ with respect to the inner product in Exercise 28. Verify this as follows:

a) Make the change of variables $x = \cos \theta$ and use induction to show that $T_k(\cos \theta) = \cos k\theta$, $k = 0, 1, \ldots$, where $T_k(x)$ is defined by (R).

b) Using part (a), show that $\langle T_i, T_j \rangle = 0$ when $i \neq j$.

c) Use induction to show that $T_k(x)$ is a polynomial of degree, k, $k = 0, 1, \ldots$.

d) Use (R) to calculate T_2, T_3, T_4, and T_5.

30. Let $C[-1, 1]$ have the inner product of Exercise 28, and let f be in $C[-1, 1]$. Use Theorem 13 to prove that $\|f - p^*\| \leq \|f - p\|$ for all p in $\mathscr{P}_n$ if

$$p^*(x) = \frac{a_0}{2} + \sum_{j=1}^n a_j T_j(x),$$

where $a_j = \langle f, T_j \rangle$, $j = 0, 1, \ldots, n$

31. The iterated trapezoid rule provides a good estimate of $\int_a^b f(x)\,dx$ when $f(x)$ is periodic in $[a, b]$. In particular, let N be a positive integer and let $h = (b - a)/N$. Next, define x_i by $x_i = a + ih$, $i = 0, 1, \ldots, N$, and suppose $f(x)$ is in $C[a, b]$. If we define $A(f)$ by

$$A(f) = \frac{h}{2}f(x_0) + h \sum_{j=1}^{N-1} f(x_j) + \frac{h}{2}f(x_N),$$

then $A(f)$ is the iterated trapezoid rule applied to $f(x)$. Using the result in Exercise 30, write a computer program that generates a good approximation to $f(x)$ in $C[-1, 1]$. That is, for an input function $f(x)$ and a specified value of n, calculate estimates of $a_0, a_1, \ldots, a_n$, where

$$a_k = \langle f, T_k \rangle \simeq A(fT_k).$$

To do this calculation, make the usual change of variables $x = \cos \theta$ so that

$$a_k = \frac{2}{\pi} \int_0^\pi f(\cos \theta) \cos(k\theta)\,d\theta, \qquad k = 0, 1, \ldots, n.$$

Use the iterated trapezoid rule to estimate each a_k. Test your program on $f(x) = e^{2x}$ and note that (R) can be used to evaluate $p^*(x)$ at any point x in $[-1, 1]$.

32. Show that if A is a real $(n \times n)$ matrix and if the expression $\langle \mathbf{u}, \mathbf{v} \rangle = \mathbf{u}^T A \mathbf{v}$ defines an inner product on R^n, then A must be symmetric and positive definite (see Exercise 3 for the definition of positive definite). (*Hint:* Consider $\langle \mathbf{e}_i, \mathbf{e}_j \rangle$.)

4.7 | Linear Transformations

Linear transformations on subspaces of R^n were introduced in Section 2.7. The definition given there extends naturally to the general vector-space setting. In this section and the next, we develop the basic properties of linear transformations, and in Section 4.8 we will use linear transformations and the concept of coordinate vectors to show that an n-dimensional vector space is essentially just R^n.

If $T: R^n \to R^m$ is a linear transformation, there exists an $(m \times n)$ matrix A such that $T(\mathbf{x}) = A\mathbf{x}$. Although this is not the case in the general vector-space setting, we will show in Section 4.9 that there is still a close relationship between linear transformations and matrices, provided that the domain space is finite dimensional.

We begin with the definition of a linear transformation.

DEFINITION 8   Let U and V be vector spaces, and let T be a function from U to V, $T: U \to V$. We say that T is a *linear transformation* if for all $\mathbf{u}$ and $\mathbf{w}$ in U and all scalars a

$$T(\mathbf{u} + \mathbf{w}) = T(\mathbf{u}) + T(\mathbf{w})$$

and

$$T(a\mathbf{u}) = aT(\mathbf{u}).$$

Examples of Linear Transformations

To illustrate Definition 8, we now provide several examples of linear transformations.

EXAMPLE 1   Let $T: \mathscr{P}_2 \to R^1$ be defined by $T(p) = p(2)$. Verify that T is a linear transformation.

SOLUTION   First note that R^1 is just the set R of real numbers, but in this context R is regarded as a vector space. To illustrate the definition of T, if $p(x) = x^2 - 3x + 1$, then $T(p) = p(2) = -1$.

To verify that T is a linear transformation, let $p(x)$ and $q(x)$ be in $\mathscr{P}_2$ and let a be a scalar. Then $T(p + q) = (p + q)(2) = p(2) + q(2) = T(p) + T(q)$. Likewise, $T(ap) = (ap)(2) = ap(2) = aT(p)$. Thus T is a linear transformation.

In general, if W is any subspace of $C[a, b]$ and if x_0 is any number in $[a, b]$, then the function $T: W \to R^1$ defined by $T(f) = f(x_0)$ is a linear transformation.

EXAMPLE 2   Let V be a p-dimensional vector space with basis $B = \{\mathbf{v}_1, \mathbf{v}_2, \dots, \mathbf{v}_p\}$. Show that $T: V \to R^p$ defined by $T(\mathbf{v}) = [\mathbf{v}]_B$ is a linear transformation.

SOLUTION   That T is a linear transformation is a direct consequence of the lemma in Section 4.4. Specifically, if $\mathbf{u}$ and $\mathbf{v}$ are vectors in V, then $T(\mathbf{u} + \mathbf{v}) = [\mathbf{u} + \mathbf{v}]_B = [\mathbf{u}]_B + [\mathbf{v}]_B = T(\mathbf{u}) + T(\mathbf{v})$. Also, if a is a scalar, then $T(a\mathbf{u}) = [a\mathbf{u}]_B = a[\mathbf{u}]_B = aT(\mathbf{u})$.

EXAMPLE 3 Let $T: C[0, 1] \to R^1$ be defined by

$$T(f) = \int_0^1 f(t)\, dt.$$

Prove that T is a linear transformation.

SOLUTION If $f(x)$ and $g(x)$ are functions in $C[0, 1]$, then

$$T(f + g) = \int_0^1 [f(t) + g(t)]\, dt$$

$$= \int_0^1 f(t)\, dt + \int_0^1 g(t)\, dt$$

$$= T(f) + T(g).$$

Likewise, if a is a scalar, the properties of integration give

$$T(af) = \int_0^1 af(t)\, dt$$

$$= a \int_0^1 f(t)\, dt$$

$$= aT(f).$$

Therefore, T is a linear transformation.

EXAMPLE 4 Let $C^1[0, 1]$ denote the set of all functions that have a continuous first derivative in the interval $[0, 1]$. (Note that $C^1[0, 1]$ is a subspace of $C[0, 1]$.) Let $k(x)$ be a fixed function in $C[0, 1]$ and define $T: C^1[0, 1] \to C[0, 1]$ by

$$T(f) = f' + kf.$$

Verify that T is a linear transformation.

SOLUTION To illustrate the definition of T, suppose, for example, that $k(x) = x^2$. If $f(x) = \sin x$, then $T(f)$ is the function defined by $T(f)(x) = f'(x) + k(x)f(x) = \cos x + x^2 \sin x$.

To see that T is a linear transformation, let g and h be functions in $C^1[0, 1]$. Then

$$T(g + h) = (g + h)' + k(g + h)$$
$$= g' + h' + kg + kh$$
$$= (g' + kg) + (h' + kh)$$
$$= T(g) + T(h).$$

Also, for a scalar c, $T(cg) = (cg)' + k(cg) = c(g' + kg) = cT(g)$. Hence T is a linear transformation.

The linear transformation in Example 4 is an example of a differential operator. We will return to differential operators later and only mention here that the term *operator* is traditional in the study of differential equations. Operator is another term for function or transformation, and we could equally well speak of T as a differential transformation.

For any vector space V, the mapping $I: V \to V$ defined by $I(\mathbf{v}) = \mathbf{v}$ is a linear transformation called the *identity transformation.* Between any two vector spaces U and V, there is always at least one linear transformation, called the *zero transformation.* If $\boldsymbol{\theta}_V$ is the zero vector in V, then the zero transformation $T: U \to V$ is defined by $T(\mathbf{u}) = \boldsymbol{\theta}_V$ for all $\mathbf{u}$ in U.

Properties of Linear Transformations

One of the important features of the two linearity properties in Definition 8 is that if $T: U \to V$ is a linear transformation and if U is a finite-dimensional vector space, then the action of T on U is completely determined by the action of T on a basis for U. To see why this statement is true, suppose U has a basis $B = \{\mathbf{u}_1, \mathbf{u}_2, \dots, \mathbf{u}_p\}$. Then given any $\mathbf{u}$ in U, we know that $\mathbf{u}$ can be expressed uniquely as

$$\mathbf{u} = a_1\mathbf{u}_1 + a_2\mathbf{u}_2 + \cdots + a_p\mathbf{u}_p.$$

From this expression it follows that $T(\mathbf{u})$ is given by

$$\begin{aligned} T(\mathbf{u}) &= T(a_1\mathbf{u}_1 + a_2\mathbf{u}_2 + \cdots + a_p\mathbf{u}_p) \\ &= a_1 T(\mathbf{u}_1) + a_2 T(\mathbf{u}_2) + \cdots + a_p T(\mathbf{u}_p). \end{aligned} \tag{1}$$

Clearly Eq. (1) shows that if we know the vectors $T(\mathbf{u}_1), T(\mathbf{u}_2), \dots, T(\mathbf{u}_p)$, then we know $T(\mathbf{u})$ for any $\mathbf{u}$ in U; T is completely determined once T is defined on the basis. The next example illustrates this concept.

EXAMPLE 5 Let $T: \mathcal{P}_3 \to \mathcal{P}_2$ be a linear transformation such that $T(1) = 1 - x$, $T(x) = x + x^2$, $T(x^2) = 1 + 2x$, and $T(x^3) = 2 - x^2$. Find $T(2 - 3x + x^2 - 2x^3)$.

SOLUTION Applying Eq. (1) yields

$$\begin{aligned} T(2 - 3x + x^2 - 2x^3) &= 2T(1) - 3T(x) + T(x^2) - 2T(x^3) \\ &= 2(1 - x) - 3(x + x^2) + (1 + 2x) - 2(2 - x^2) \\ &= -1 - 3x - x^2. \end{aligned}$$

Similarly,

$$\begin{aligned} T(a_0 &+ a_1 x + a_2 x^2 + a_3 x^3) \\ &= a_0 T(1) + a_1 T(x) + a_2 T(x^2) + a_3 T(x^3) \\ &= a_0(1 - x) + a_1(x + x^2) + a_2(1 + 2x) + a_3(2 - x^2) \\ &= (a_0 + a_2 + 2a_3) + (-a_0 + a_1 + 2a_2)x + (a_1 - a_3)x^2. \end{aligned}$$

Before giving further properties of linear transformations, we require several definitions. Let $T: U \to V$ be a linear transformation, and for clarity let us denote the zero vectors in U and V as θ_U and θ_V, respectively. The **null space** (or **kernel**) of T, denoted by $\mathcal{N}(T)$, is the subset of U defined by

$$\mathcal{N}(T) = \{\mathbf{u} \text{ in } U: T(\mathbf{u}) = \theta_V\}.$$

The **range** of T, denoted by $\mathcal{R}(T)$, is the subset of V defined by

$$\mathcal{R}(T) = \{\mathbf{v} \text{ in } V: \mathbf{v} = T(\mathbf{u}) \text{ for some } \mathbf{u} \text{ in } U\}.$$

As before, the dimension of $\mathcal{N}(T)$ is called the **nullity** of T and is denoted by nullity(T). Likewise, the dimension of $\mathcal{R}(T)$ is called the **rank** of T and is denoted by rank(T). Finally, we say a linear transformation is **one to one** if $T(\mathbf{u}) = T(\mathbf{w})$ implies $\mathbf{u} = \mathbf{w}$ for all $\mathbf{u}$ and $\mathbf{w}$ in U. Some of the elementary properties of linear transformations are given in the theorem below.

THEOREM 14

Let $T: U \to V$ be a linear transformation. Then:

1. $T(\theta_U) = \theta_V$.
2. $\mathcal{N}(T)$ is a subspace of U.
3. $\mathcal{R}(T)$ is a subspace of V.
4. T is one to one if and only if $\mathcal{N}(T) = \{\theta_U\}$; that is, T is one to one if and only if nullity$(T) = 0$.

PROOF   To prove property 1, note that $0\theta_U = \theta_U$, so

$$T(\theta_U) = T(0\theta_U) = 0T(\theta_U) = \theta_V.$$

To prove property 2, we must verify that $\mathcal{N}(T)$ satisfies the three properties of Theorem 2 in Section 4.3. It follows from property 1 that θ_U is in $\mathcal{N}(T)$. Next, let $\mathbf{u}_1$ and $\mathbf{u}_2$ be in $\mathcal{N}(T)$ and let a be a scalar. Then $T(\mathbf{u}_1 + \mathbf{u}_2) = T(\mathbf{u}_1) + T(\mathbf{u}_2) = \theta_V + \theta_V = \theta_V$, so $\mathbf{u}_1 + \mathbf{u}_2$ is in $\mathcal{N}(T)$. Similarly, $T(a\mathbf{u}_1) = aT(\mathbf{u}_1) = a\theta_V = \theta_V$, so $a\mathbf{u}_1$ is in $\mathcal{N}(T)$. Therefore, $\mathcal{N}(T)$ is a subspace of U.

The proof of property 3 is left as an exercise. To prove property 4, suppose that $\mathcal{N}(T) = \{\theta_U\}$. In order to show that T is one to one, let $\mathbf{u}$ and $\mathbf{w}$ be vectors in U such that $T(\mathbf{u}) = T(\mathbf{w})$. Then $\theta_V = T(\mathbf{u}) - T(\mathbf{w}) = T(\mathbf{u}) + (-1)T(\mathbf{w}) = T[\mathbf{u} + (-1)\mathbf{w}] = T(\mathbf{u} - \mathbf{w})$. It follows that $\mathbf{u} - \mathbf{w}$ is in $\mathcal{N}(T)$. But $\mathcal{N}(T) = \{\theta_U\}$, so $\mathbf{u} - \mathbf{w} = \theta_U$. Therefore, $\mathbf{u} = \mathbf{w}$ and T is one to one. The converse is Exercise 24.

When $T: R^n \to R^m$ is given by $T(\mathbf{x}) = A\mathbf{x}$, with A an $(m \times n)$ matrix, then $\mathcal{N}(T)$ is the null space of A and $\mathcal{R}(T)$ is the range of A. In this setting, property 4 of Theorem 14 states that a consistent system of equations $A\mathbf{x} = \mathbf{b}$ has a unique solution if and only if the trivial solution is the unique solution for the homogeneous system $A\mathbf{x} = \theta$.

The following theorem gives additional properties of a linear transformation $T: U \to V$, where U is a finite-dimensional vector space.

THEOREM 15

Let $T: U \to V$ be a linear transformation and let U be p-dimensional, where $B = \{\mathbf{u}_1, \mathbf{u}_2, \ldots, \mathbf{u}_p\}$ is a basis for U.

1. $\mathcal{R}(T) = \mathrm{Sp}\{T(\mathbf{u}_1), T(\mathbf{u}_2), \ldots, T(\mathbf{u}_p)\}$.
2. T is one to one if and only if $\{T(\mathbf{u}_1), T(\mathbf{u}_2), \ldots, T(\mathbf{u}_p)\}$ is linearly independent in V.
3. $\mathrm{rank}(T) + \mathrm{nullity}(T) = p$.

PROOF Property 1 is immediate from Eq. (1). That is, if $\mathbf{v}$ is in $\mathcal{R}(T)$, then $\mathbf{v} = T(\mathbf{u})$ for some $\mathbf{u}$ in U. But B is a basis for U; so $\mathbf{u}$ is of the form $\mathbf{u} = a_1\mathbf{u}_1 + a_2\mathbf{u}_2 + \cdots + a_p\mathbf{u}_p$; and hence $T(\mathbf{u}) = \mathbf{v} = a_1 T(\mathbf{u}_1) + a_2 T(\mathbf{u}_2) + \cdots + a_p T(\mathbf{u}_p)$. Therefore, $\mathbf{v}$ is in $\mathrm{Sp}\{T(\mathbf{u}_1), T(\mathbf{u}_2), \ldots, T(\mathbf{u}_p)\}$.

To prove property 2, we can use property 4 of Theorem 14; T is one to one if and only if $\boldsymbol{\theta}_U$ is the only vector in $\mathcal{N}(T)$. In particular, let us suppose that $\mathbf{u}$ is some vector in $\mathcal{N}(T)$, where $\mathbf{u} = b_1\mathbf{u}_1 + b_2\mathbf{u}_2 + \cdots + b_p\mathbf{u}_p$. Then $T(\mathbf{u}) = \boldsymbol{\theta}_V$, or

$$b_1 T(\mathbf{u}_1) + b_2 T(\mathbf{u}_2) + \cdots + b_p T(\mathbf{u}_p) = \boldsymbol{\theta}_V. \tag{2}$$

If $\{T(\mathbf{u}_1), T(\mathbf{u}_2), \ldots, T(\mathbf{u}_p)\}$ is a linearly independent set in V, then the only scalars satisfying Eq. (2) are $b_1 = b_2 = \cdots = b_p = 0$. Therefore, $\mathbf{u}$ must be $\boldsymbol{\theta}_U$; so T is one to one. On the other hand, if T is one to one, then there cannot be a nontrivial solution to Eq. (2); for if there were, $\mathcal{N}(T)$ would contain the nonzero vector $\mathbf{u}$.

To prove property 3, we first note that $0 \le \mathrm{rank}(T) \le p$ by property 1. We leave the two extreme cases, $\mathrm{rank}(T) = p$ and $\mathrm{rank}(T) = 0$, to the exercises and consider only $0 < \mathrm{rank}(T) < p$. [Note that $\mathrm{rank}(T) < p$ implies that $\mathrm{nullity}(T) \ge 1$, so T is not one to one. We mention this point because we will need to choose a basis for $\mathcal{N}(T)$ below.]

It is conventional to let r denote $\mathrm{rank}(T)$, so let us suppose $\mathcal{R}(T)$ has a basis of r vectors, $\{\mathbf{v}_1, \mathbf{v}_2, \ldots, \mathbf{v}_r\}$. From the definition of $\mathcal{R}(T)$, we know there are vectors $\mathbf{w}_1, \mathbf{w}_2, \ldots, \mathbf{w}_r$, in U such that

$$T(\mathbf{w}_i) = \mathbf{v}_i, \qquad 1 \le i \le r. \tag{3}$$

Now suppose that $\mathrm{nullity}(T) = k$ and let $\{\mathbf{x}_1, \mathbf{x}_2, \ldots, \mathbf{x}_k\}$ be a basis for $\mathcal{N}(T)$. We now show that the set

$$Q = \{\mathbf{x}_1, \mathbf{x}_2, \ldots, \mathbf{x}_k, \mathbf{w}_1, \mathbf{w}_2, \ldots, \mathbf{w}_r\}$$

is a basis for U (therefore, $k + r = p$, which proves property 3).

We first establish that Q is linearly independent set in U by considering

$$c_1\mathbf{x}_1 + c_2\mathbf{x}_2 + \cdots + c_k\mathbf{x}_k + a_1\mathbf{w}_1 + a_2\mathbf{w}_2 + \cdots + a_r\mathbf{w}_r = \boldsymbol{\theta}_U. \tag{4}$$

Applying T to both sides of Eq. (4) yields

$$T(c_1\mathbf{x}_1 + \cdots + c_k\mathbf{x}_k + a_1\mathbf{w}_1 + \cdots + a_r\mathbf{w}_r) = T(\boldsymbol{\theta}_U). \tag{5a}$$

Using Eq. (1) and property 1 of Theorem 14, Eq. (5a) becomes

$$c_1 T(\mathbf{x}_1) + \cdots + c_k T(\mathbf{x}_k) + a_1 T(\mathbf{w}_1) + \cdots + a_r T(\mathbf{w}_r) = \boldsymbol{\theta}_V. \tag{5b}$$

Since each $\mathbf{x}_i$ is in $\mathcal{N}(T)$ and $T(\mathbf{w}_i) = \mathbf{v}_i$, Eq. (5b) becomes

$$a_1\mathbf{v}_1 + a_2\mathbf{v}_2 + \cdots + a_r\mathbf{v}_r = \boldsymbol{\theta}_V. \tag{5c}$$

Since the set $\{\mathbf{v}_1, \mathbf{v}_2, \ldots, \mathbf{v}_r\}$ is linearly independent, $a_1 = a_2 = \cdots = a_r = 0$. The vector equation (4) now becomes

$$c_1\mathbf{x}_1 + c_2\mathbf{x}_2 + \cdots + c_k\mathbf{x}_k = \boldsymbol{\theta}_U. \tag{6}$$

But $\{\mathbf{x}_1, \mathbf{x}_2, \ldots, \mathbf{x}_k\}$ is a linearly independent set in U, so we must have $c_1 = c_2 = \cdots = c_k = 0$. Therefore, Q is a linearly independent set.

To complete the argument, we need to show that Q is a spanning set for U. So let $\mathbf{u}$ be any vector in U. Then $\mathbf{v} = T(\mathbf{u})$ is a vector in $\mathcal{R}(T)$; so

$$T(\mathbf{u}) = b_1\mathbf{v}_1 + b_2\mathbf{v}_2 + \cdots + b_r\mathbf{v}_r.$$

Consider an associated vector $\mathbf{x}$ in U, where $\mathbf{x}$ is defined by

$$\mathbf{x} = b_1\mathbf{w}_1 + b_2\mathbf{w}_2 + \cdots + b_r\mathbf{w}_r. \tag{7}$$

We observe that $T(\mathbf{u} - \mathbf{x}) = \boldsymbol{\theta}_V$; so obviously $\mathbf{u} - \mathbf{x}$ is in $\mathcal{N}(T)$ and can be written as

$$\mathbf{u} - \mathbf{x} = d_1\mathbf{x}_1 + d_2\mathbf{x}_2 + \cdots + d_k\mathbf{x}_k. \tag{8}$$

Placing $\mathbf{x}$ on the right-hand side of Eq. (8) and using Eq. (7), we have shown that $\mathbf{u}$ is a linear combination of vectors in Q. Thus Q is a basis for U, and property 3 is proved since $k + r$ must equal p.

As the following example illustrates, property 1 of Theorem 15 and the techniques of Section 4.4 give a method for obtaining a basis for $\mathcal{R}(T)$.

EXAMPLE 6 Let V be the vector space of all (2×2) matrices, and let $T: \mathscr{P}_3 \to V$ be the linear transformation defined by

$$T(a_0 + a_1x + a_2x^2 + a_3x^3) = \begin{bmatrix} a_0 + a_2 & a_0 + a_3 \\ a_1 + a_2 & a_1 + a_3 \end{bmatrix}.$$

Emmy Noether

Emmy Noether (1882–1935) is the most heralded female mathematician of the early twentieth century. Overcoming great obstacles for women in mathematics at the time, she received her doctorate from Göttingen and went on to work with David Hilbert and Felix Klein on the general theory of relativity. Among her most highly regarded results are the representation of noncommutative algebras as linear transformations and "Noether's Theorem," which is used to explain the correspondences between certain invariants and physical conservation laws. She fled from Germany in 1933 and spent the last two years of her life on the faculty at Bryn Mawr College in Philadelphia.

Find a basis for $\mathcal{R}(T)$ and determine rank(T) and nullity(T). Finally, show that T is not one to one.

SOLUTION   By property 1 of Theorem 15, $\mathcal{R}(T) = \text{Sp}\{T(1), T(x), T(x^2), T(x^3)\}$. Thus $\mathcal{R}(T) = \text{Sp}(S)$, where $S = \{A_1, A_2, A_3, A_4\}$ and

$$A_1 = \begin{bmatrix} 1 & 1 \\ 0 & 0 \end{bmatrix}, \quad A_2 = \begin{bmatrix} 0 & 0 \\ 1 & 1 \end{bmatrix}, \quad A_3 = \begin{bmatrix} 1 & 0 \\ 1 & 0 \end{bmatrix}, \quad \text{and} \quad A_4 = \begin{bmatrix} 0 & 1 \\ 0 & 1 \end{bmatrix}.$$

Let B be the natural basis for V: $B = \{E_{11}, E_{12}, E_{21}, E_{22}\}$. Form the (4×4) matrix C with column vectors $[A_1]_B, [A_2]_B, [A_3]_B, [A_4]_B$; thus

$$C = \begin{bmatrix} 1 & 0 & 1 & 0 \\ 1 & 0 & 0 & 1 \\ 0 & 1 & 1 & 0 \\ 0 & 1 & 0 & 1 \end{bmatrix}.$$

The matrix C^T reduces to the matrix

$$D^T = \begin{bmatrix} 1 & 1 & 0 & 0 \\ 0 & 1 & 0 & 1 \\ 0 & 0 & 1 & 1 \\ 0 & 0 & 0 & 0 \end{bmatrix}$$

in echelon form. Therefore,

$$D = \begin{bmatrix} 1 & 0 & 0 & 0 \\ 1 & 1 & 0 & 0 \\ 0 & 0 & 1 & 0 \\ 0 & 1 & 1 & 0 \end{bmatrix},$$

and the nonzero columns of D constitute a basis for the subspace $\text{Sp}\{[A_1]_B, [A_2]_B, [A_3]_B, [A_4]_B\}$ of R^4. If the matrices B_1, B_2, and B_3 are defined by

$$B_1 = \begin{bmatrix} 1 & 1 \\ 0 & 0 \end{bmatrix}, \quad B_2 = \begin{bmatrix} 0 & 1 \\ 0 & 1 \end{bmatrix}, \quad \text{and} \quad B_3 = \begin{bmatrix} 0 & 0 \\ 1 & 1 \end{bmatrix},$$

then $[B_1]_B$, $[B_2]_B$, and $[B_3]_B$ are the nonzero columns of D. It now follows from Theorem 5 of Section 4.4 that $\{B_1, B_2, B_3\}$ is a basis for $\mathcal{R}(T)$.

By property 3 of Theorem 15,

$$\dim(\mathcal{P}_3) = \text{rank}(T) + \text{nullity}(T).$$

We have just shown that rank$(T) = 3$. Since $\dim(\mathcal{P}_3) = 4$, it follows that nullity$(T) = 1$. In particular, T is not one to one by property 4 of Theorem 14.

□

<u>EXAMPLE 7</u>   Let $T: \mathcal{P}_2 \to R^1$ be defined by $T(p(x)) = p(2)$. Exhibit a basis for $\mathcal{N}(T)$ and determine the rank and nullity of T.

SOLUTION   By definition, $T(a_0 + a_1 x + a_2 x^2) = a_0 + 2a_1 + 4a_2$. Thus

$$\mathcal{N}(T) = \{p(x): p(x) \text{ is in } \mathcal{P}_2 \quad \text{and} \quad a_0 + 2a_1 + 4a_2 = 0\}.$$

In the algebraic specification for $\mathcal{N}(T)$, a_1 and a_2 can be designated as unconstrained variables, and $a_0 = -2a_1 - 4a_2$. Thus $p(x)$ in $\mathcal{N}(T)$ can be decomposed as

$$p(x) = (-2a_1 - 4a_2) + a_1 x + a_2 x^2 = a_1(-2 + x) + a_2(-4 + x^2).$$

It follows that $\{-2 + x, -4 + x^2\}$ is a basis for $\mathcal{N}(T)$. In particular, nullity$(T) = 2$. Then rank$(T) = \dim(\mathcal{P}_2) - $ nullity$(T) = 3 - 2 = 1$.

We have already noted that if A is an $(m \times n)$ matrix and $T: R^n \to R^m$ is defined by $T(\mathbf{x}) = A\mathbf{x}$, then $\mathcal{R}(T) = \mathcal{R}(A)$ and $\mathcal{N}(T) = \mathcal{N}(A)$. The following corollary, given as a remark in Section 2.5, is now an immediate consequence of these observations and property 3 of Theorem 15.

COROLLARY

If A is an $(m \times n)$ matrix, then

$$n = \text{rank}(A) + \text{nullity}(A).$$

Exercises 4.7

In Exercises 1–4, V is the vector space of all (2×2) matrices and A has the form

$$A = \begin{bmatrix} a & b \\ c & d \end{bmatrix}.$$

Determine whether the function $T: V \to R^1$ is a linear transformation.

1. $T(A) = \det(A)$

2. $T(A) = a + 2b - c + d$

3. $T(A) = \text{tr}(A)$, where $\text{tr}(A)$ denotes the trace of A and is defined by $\text{tr}(A) = a + d$.

4. $T(A) = (a - d)(b - c)$

In Exercises 5–8, determine whether T is a linear transformation.

5. $T: C^1[-1, 1] \to R^1$ defined by $T(f) = f'(0)$

6. $T: C[0, 1] \to C[0, 1]$ defined by $T(f) = g$, where $g(x) = e^x f(x)$

7. $T: \mathcal{P}_2 \to \mathcal{P}_2$ defined by $T(a_0 + a_1 x + a_2 x^2) = (a_0 + 1) + (a_1 + 1)x + (a_2 + 1)x^2$

8. $T: \mathcal{P}_2 \to \mathcal{P}_2$ defined by $T(p(x)) = p(0) + xp'(x)$

9. Suppose that $T: \mathcal{P}_2 \to \mathcal{P}_3$ is a linear transformation, where $T(1) = 1 + x^2$, $T(x) = x^2 - x^3$, and $T(x^2) = 2 + x^3$.
   a) Find $T(p)$, where $p(x) = 3 - 2x + 4x^2$.
   b) Give a formula for T; that is, find $T(a_0 + a_1 x + a_2 x^2)$.

10. Suppose that $T: \mathcal{P}_2 \to \mathcal{P}_4$ is a linear transformation, where $T(1) = x^4$, $T(x + 1) = x^3 - 2x$, and $T(x^2 + 2x + 1) = x$. Find $T(p)$ and $T(q)$, where $p(x) = x^2 + 5x - 1$ and $q(x) = x^2 + 9x + 5$.

11. Let V be the set of all (2×2) matrices and suppose that $T: V \to \mathcal{P}_2$ is a linear transformation such that

$T(E_{11}) = 1 - x$, $T(E_{12}) = 1 + x + x^2$, $T(E_{21}) = 2x - x^2$, and $T(E_{22}) = 2 + x - 2x^2$.

a) Find $T(A)$, where

$$A = \begin{bmatrix} -2 & 2 \\ 3 & 4 \end{bmatrix}.$$

b) Give a formula for T; that is, find

$$T\left(\begin{bmatrix} a & b \\ c & d \end{bmatrix}\right).$$

12. With V as in Exercise 11, define $T: V \to R^2$ by

$$T\left(\begin{bmatrix} a & b \\ c & d \end{bmatrix}\right) = \begin{bmatrix} a + 2d \\ b - c \end{bmatrix}.$$

a) Prove that T is a linear transformation.
b) Give an algebraic specification for $\mathcal{N}(T)$.
c) Exhibit a basis for $\mathcal{N}(T)$.
d) Determine the nullity and the rank of T.
e) Without doing any calculations, argue that $\mathcal{R}(T) = R^2$.
f) Prove $\mathcal{R}(T) = R^2$ as follows: Let $\mathbf{v}$ be in R^2,

$$\mathbf{v} = \begin{bmatrix} x \\ y \end{bmatrix}.$$

Exhibit a (2×2) matrix A in V such that $T(A) = \mathbf{v}$.

13. Let $T: \mathscr{P}_4 \to \mathscr{P}_2$ be the linear transformation defined by $T(p) = p''(x)$.

a) Exhibit a basis for $\mathcal{R}(T)$ and conclude that $\mathcal{R}(T) = \mathscr{P}_2$.
b) Determine the nullity of T and conclude that T is not one to one.
c) Give a direct proof that $\mathcal{R}(T) = \mathscr{P}_2$; that is, for $p(x) = a_0 + a_1 x + a_2 x^2$ in $\mathscr{P}_2$, exhibit a polynomial $q(x)$ in $\mathscr{P}_4$ such that $T(q) = p$.

14. Define $T: \mathscr{P}_4 \to \mathscr{P}_3$ by

$$T(a_0 + a_1 x + a_2 x^2 + a_3 x^3 + a_4 x^4)$$
$$= (a_0 - a_1 + 2a_2 - a_3 + a_4)$$
$$\quad + (-a_0 + 3a_1 - 2a_2 + 3a_3 - a_4)x$$
$$\quad + (2a_0 - 3a_1 + 5a_2 - a_3 + a_4)x^2$$
$$\quad + (3a_0 - a_1 + 7a_2 + 2a_3 + 2a_4)x^3.$$

Find a basis for $\mathcal{R}(T)$ (cf. Example 6) and show that T is not one to one.

15. Identify $\mathcal{N}(T)$ and $\mathcal{R}(T)$ for the linear transformation T given in Example 1.

16. Identify $\mathcal{N}(T)$ and $\mathcal{R}(T)$ for the linear transformation T given in Example 3.

17. Let $I: V \to V$ be defined by $I(\mathbf{v}) = \mathbf{v}$ for each $\mathbf{v}$ in V.
a) Prove that I is a linear transformation.
b) Determine $\mathcal{N}(I)$ and $\mathcal{R}(I)$.

18. Let U and V be vector spaces and define $T: U \to V$ by $T(\mathbf{u}) = \boldsymbol{\theta}_V$ for each $\mathbf{u}$ in U.
a) Prove that T is a linear transformation.
b) Determine $\mathcal{N}(T)$ and $\mathcal{R}(T)$.

19. Suppose that $T: \mathscr{P}_4 \to \mathscr{P}_2$ is a linear transformation. Enumerate the various possibilities for rank(T) and nullity(T). Can T possibly be one to one?

20. Let $T: U \to V$ be a linear transformation and let U be finite dimensional. Prove that if $\dim(U) > \dim(V)$, then T cannot be one to one.

21. Suppose that $T: R^3 \to \mathscr{P}_3$ is a linear transformation. Enumerate the various possibilities for rank(T) and nullity(T). Is $\mathcal{R}(T) = \mathscr{P}_3$ a possibility?

22. Let $T: U \to V$ be a linear transformation and let U be finite dimensional. Prove that if $\dim(U) < \dim(V)$, then $\mathcal{R}(T) = V$ is not possible.

23. Prove property 3 of Theorem 14.

24. Complete the proof of property 4 of Theorem 14 by showing that if T is one to one, then $\mathcal{N}(T) = \{\boldsymbol{\theta}_U\}$.

25. Complete the proof of property 3 of Theorem 15 as follows:
a) If rank$(T) = p$, prove that nullity$(T) = 0$.
b) If rank$(T) = 0$, show that nullity$(T) = p$.

26. Let $T: R^n \to R^n$ be defined by $T(\mathbf{x}) = A\mathbf{x}$, where A is an $(n \times n)$ matrix. Use property 4 of Theorem 14 to show that T is one to one if and only if A is nonsingular.

27. Let V be the vector space of all (2×2) matrices and define $T: V \to V$ by $T(A) = A^T$.
a) Show that T is a linear transformation.
b) Determine the nullity and rank of T. Conclude that T is one to one and $\mathcal{R}(T) = V$.
c) Show directly that $\mathcal{R}(T) = V$; that is, for B in V exhibit a matrix C in V such that $T(C) = B$.

4.8 Operations with Linear Transformations

We know that a useful arithmetic structure is associated with matrices: Matrices can be added and multiplied, nonsingular matrices have inverses, and so on. Much of this structure is available also for linear transformations. For our explanation we will need some definitions. Let U and V be vector spaces and let T_1 and T_2 be linear transformations, where $T_1: U \rightarrow V$ and $T_2: U \rightarrow V$. By the *sum* $T_3 = T_1 + T_2$, we mean the function $T_3: U \rightarrow V$, where $T_3(\mathbf{u}) = T_1(\mathbf{u}) + T_2(\mathbf{u})$ for all $\mathbf{u}$ in U. The following example illustrates this concept.

EXAMPLE 1   Let $T_1: \mathscr{P}_4 \rightarrow \mathscr{P}_2$ be given by $T_1(p) = p''(x)$, and suppose that $T_2: \mathscr{P}_4 \rightarrow \mathscr{P}_2$ is defined by $T_2(p) = xp(1)$. If $S = T_1 + T_2$, give the formula for S.

SOLUTION   By definition, the sum $T_1 + T_2$ is the linear transformation $S: \mathscr{P}_4 \rightarrow \mathscr{P}_2$ defined by $S(p) = T_1(p) + T_2(p) = p''(x) + xp(1)$.

$\square$

If $T: U \rightarrow V$ is a linear transformation and a is a scalar, then aT denotes the function $aT: U \rightarrow V$ defined by $aT(\mathbf{u}) = a(T(\mathbf{u}))$ for all $\mathbf{u}$ in U. Again, we illustrate with an example.

EXAMPLE 2   Let V be the vector space of all (2×2) matrices and define $T: V \rightarrow R^1$ by $T(A) = 2a - b + 3c + 4d$, where

$$A = \begin{bmatrix} a & b \\ c & d \end{bmatrix}.$$

Give the formula for $3T$.

$$U \xrightarrow{S} V \xrightarrow{T} W$$
$$\mathbf{u} \longrightarrow S(\mathbf{u}) \longrightarrow T(S(\mathbf{u}))$$
$$\underline{\qquad T \circ S \qquad}\uparrow$$

Figure 4.7
The composition of linear transformations is a linear transformation (see Example 3).

SOLUTION   By definition, $3T(A) = 3(T(A)) = 3(2a - b + 3c + 4d) = 6a - 3b + 9c + 12d$.

$\square$

It is straightforward to show that the functions $T_1 + T_2$ and aT defined above are linear transformations (cf. Exercises 13 and 14).
Now let U, V, and W be vector spaces and let S and T be linear transformations, where $S: U \rightarrow V$ and $T: V \rightarrow W$. The *composition,* $L = T \circ S$, of S and T is defined to be function $L: U \rightarrow W$ given by $L(\mathbf{u}) = T(S(\mathbf{u}))$ for all $\mathbf{u}$ in U (see Fig. 4.7).

EXAMPLE 3   Let $S: U \rightarrow V$ and $T: V \rightarrow W$ be linear transformations. Verify that the composition $L = T \circ S$ is also a linear transformation.

SOLUTION   Let $\mathbf{u}_1, \mathbf{u}_2$ be vectors in U. Then $L(\mathbf{u}_1 + \mathbf{u}_2) = T(S(\mathbf{u}_1 + \mathbf{u}_2))$. Since S is a linear transformation, $S(\mathbf{u}_1 + \mathbf{u}_2) = S(\mathbf{u}_1) + S(\mathbf{u}_2)$. But T is also a linear transformation, so $L(\mathbf{u}_1 + \mathbf{u}_2) = T(S(\mathbf{u}_1) + S(\mathbf{u}_2)) = T(S(\mathbf{u}_1)) + T(S(\mathbf{u}_2)) = L(\mathbf{u}_1) + L(\mathbf{u}_2)$. Similarly, if $\mathbf{u}$ is in U and a is a scalar, $L(a\mathbf{u}) = T(S(a\mathbf{u})) = T(aS(\mathbf{u})) = aT(S(\mathbf{u})) = aL(\mathbf{u})$. This shows that $L = T \circ S$ is a linear transformation. ☐

The next two examples provide specific illustrations of the composition of two linear transformations.

EXAMPLE 4 Let U be the vector space of all (2×2) matrices. Define $S: U \rightarrow \mathscr{P}_2$ by $S(A) = (a - c) + (b + 2c)x + (3c - d)x^2$, where

$$A = \begin{bmatrix} a & b \\ c & d \end{bmatrix}.$$

Define $T: \mathscr{P}_2 \rightarrow R^2$ by

$$T(a_0 + a_1 x + a_2 x^2) = \begin{bmatrix} a_0 - a_1 \\ 2a_1 + a_2 \end{bmatrix}.$$

Give the formula for $T \circ S$ and show that $S \circ T$ is not defined.

SOLUTION The composition $T \circ S: U \rightarrow R^2$ is defined by $(T \circ S)(A) = T(S(A)) = T[(a - c) + (b + 2c)x + (3c - d)x^2]$. Thus

$$(T \circ S)\left(\begin{bmatrix} a & b \\ c & d \end{bmatrix}\right) = \begin{bmatrix} a - b - 3c \\ 2b + 7c - d \end{bmatrix}.$$

If $p(x)$ is in $\mathscr{P}_2$, then $T(p(x)) = \mathbf{v}$, where $\mathbf{v}$ is in R^2. Thus $(S \circ T)(p(x)) = S(T(p(x))) = S(\mathbf{v})$. But $\mathbf{v}$ is not in the domain of S, so $S(\mathbf{v})$ is not defined. Therefore, $S \circ T$ is undefined. ☐

Example 4 illustrates that, as with matrix multiplication, $T \circ S$ may be defined, whereas $S \circ T$ is not defined. The next example illustrates that even when both are defined, they may be different transformations.

EXAMPLE 5 Let $S: \mathscr{P}_4 \rightarrow \mathscr{P}_4$ be given by $S(p) = p''(x)$ and define $T: \mathscr{P}_4 \rightarrow \mathscr{P}_4$ by $T(q) = xq(1)$. Give the formulas for $T \circ S$ and $S \circ T$.

SOLUTION The linear transformation $T \circ S: \mathscr{P}_4 \rightarrow \mathscr{P}_4$ is defined by

$$(T \circ S)(p) = T(S(p)) = T(p''(x)) = xp''(1),$$

and $S \circ T: \mathscr{P}_4 \rightarrow \mathscr{P}_4$ is given by

$$(S \circ T)(p) = S(T(p)) = S(xp(1)) = [xp(1)]'' = \theta(x).$$

In particular, $S \circ T \neq T \circ S$. ☐

Invertible Transformations

As we have previously noted, linear transformations can be viewed as an extension of the notion of a matrix to general vector spaces. In this subsection we introduce those linear transformations that correspond to nonsingular (or invertible) matrices. First, suppose X and Y are any sets and $f: X \rightarrow Y$ is a function; and suppose $\mathcal{R}(f)$ denotes the range of f where $\mathcal{R}(f) \subseteq Y$. Recall that f is **onto** provided that $\mathcal{R}(f) = Y$; that is, f is onto if for each element y in Y there exists an element x in X such that $f(x) = y$.

In order to show that a linear transformation $T: U \rightarrow V$ is onto, it is frequently convenient to use the results of Section 4.7 and a dimension argument to determine whether $\mathcal{R}(T) = V$. To be more specific, suppose V has finite dimension. If $\mathcal{R}(T)$ has the same dimension, then, since $\mathcal{R}(T)$ is a subspace of V, it must be the case that $\mathcal{R}(T) = V$. Thus in order to show that T is onto when the dimension of V is finite, it suffices to demonstrate that $\text{rank}(T) = \dim(V)$. Alternatively, an elementwise argument can be used to show that T is onto. The next two examples illustrate both procedures.

EXAMPLE 6 Let U be the subspace of (2×2) matrices defined by

$$U = \left\{ A: A = \begin{bmatrix} a & -b \\ b & a \end{bmatrix}, \quad \text{where } a \text{ and } b \text{ are in } R \right\},$$

and let $V = \{ f(x) \text{ in } C[0, 1]: f(x) = ce^x + de^{-x}, \text{ where } c \text{ and } d \text{ are in } R \}$. Define $T: U \rightarrow V$ by

$$T\left(\begin{bmatrix} a & -b \\ b & a \end{bmatrix} \right) = (a + b)e^x + (a - b)e^{-x}.$$

Show that $\mathcal{R}(T) = V$.

SOLUTION Note that U has basis $\{A_1, A_2\}$, where

$$A_1 = \begin{bmatrix} 1 & 0 \\ 0 & 1 \end{bmatrix} \quad \text{and} \quad A_2 = \begin{bmatrix} 0 & -1 \\ 1 & 0 \end{bmatrix}.$$

It follows from Theorem 15, property 1, of Section 4.7 that $\mathcal{R}(T) = \text{Sp}\{T(A_1), T(A_2)\} = \text{Sp}\{e^x + e^{-x}, e^x - e^{-x}\}$. It is easily shown that the set $\{e^x + e^{-x}, e^x - e^{-x}\}$ is linearly independent. It follows that $\text{rank}(T) = 2$. Since $\{e^x, e^{-x}\}$ is a linearly independent set and $V = \text{Sp}\{e^x, e^{-x}\}$, the set is a basis for V. In particular, $\dim(V) = 2$. Since $\mathcal{R}(T) \subseteq V$ and $\text{rank}(T) = \dim(V)$, it follows that $\mathcal{R}(T) = V$. □

EXAMPLE 7 Let $T: \mathcal{P} \rightarrow \mathcal{P}$ be defined by $T(p) = p''(x)$. Show that $\mathcal{R}(T) = \mathcal{P}$.

SOLUTION Recall that $\mathcal{P}$ is the vector space of all polynomials, with no bound on the degree. We have previously seen that $\mathcal{P}$ does not have a finite basis, so the techniques of Example 6 do not apply. To show that $\mathscr{R}(T) = \mathscr{P}$, let $q(x) = a_0 + a_1 x + \cdots + a_n x^n$ be an arbitrary polynomial in $\mathscr{P}$. We must exhibit a polynomial $p(x)$ in $\mathscr{P}$ such that $T(p) = p''(x) = q(x)$. It is easy to see that $p(x) = (1/2)a_0 x^2 + (1/6)a_1 x^3 + \cdots + [1/(n+1)(n+2)]a_n x^{n+2}$ is one choice for $p(x)$. Thus T is onto. □

Let $f: X \to Y$ be a function. If f is both one to one and onto, then the *inverse* of f, $f^{-1}: Y \to X$, is the function defined by

$$f^{-1}(y) = x \quad \text{if and only if} \quad f(x) = y. \tag{1}$$

Therefore, if $T: U \to V$ is a linear transformation that is both one to one and onto, then the inverse function $T^{-1}: V \to U$ is defined. The next two examples illustrate this concept.

EXAMPLE 8   Let $T: \mathscr{P}_4 \to \mathscr{P}_3$ be defined by $T(p) = p''(x)$. Show that T^{-1} is not defined.

SOLUTION   It is easy to see that $\mathscr{N}(T) = \mathscr{P}_1$. In particular, by property 4 of Theorem 14 (Section 4.7), T is not one to one. Thus T^{-1} is not defined. To illustrate specifically, note that $T(x) = T(x+1) = \theta(x)$. Thus by formula (1) above, we have both $T^{-1}(\theta(x)) = x$ and $T^{-1}(\theta(x)) = x + 1$. Since $T^{-1}(\theta(x))$ is not uniquely determined, T^{-1} does not exist.

Since $\mathscr{N}(T) = \mathscr{P}_1$, it follows that nullity$(T) = 2$. By property 3 of Theorem 15 (Section 4.7), rank$(T) = \dim(\mathscr{P}_4) - \text{nullity}(T) = 5 - 2 = 3$. But $\dim(\mathscr{P}_3) = 4$, so T is not onto. In particular, x^3 is in $\mathscr{P}_3$, and it is easy to see that there is no polynomial $p(x)$ in $\mathscr{P}_4$ such that $T(p(x)) = p''(x) = x^3$. Thus $T^{-1}(x^3)$ remains undefined by formula (1). □

Example 8 illustrates the following: If $T: U \to V$ is not one to one, then there exists $\mathbf{v}$ in V such that $T^{-1}(\mathbf{v})$ is not uniquely determined by formula (1), since there exists $\mathbf{u}_1$ and $\mathbf{u}_2$ in U such that $\mathbf{u}_1 \neq \mathbf{u}_2$ but $T(\mathbf{u}_1) = \mathbf{v} = T(\mathbf{u}_2)$. On the other hand, if T is not onto, there exists $\mathbf{v}$ in V such that $T^{-1}(\mathbf{v})$ is not defined by formula (1), since there exists no vector $\mathbf{u}$ in U such that $T(\mathbf{u}) = \mathbf{v}$.

EXAMPLE 9   Let $T: U \to V$ be the linear transformation defined in Example 6. Show that T is both one to one and onto, and give the formula for T^{-1}.

SOLUTION   We showed in Example 6 that T is onto. In order to show that T is one to one, it suffices, by Theorem 14, property 4 of Section 4.7, to show that if $A \in \mathscr{N}(T)$, then $A = \mathcal{O}$ [where $\mathcal{O}$ is the (2×2) zero matrix]. Thus suppose that

$$A = \begin{bmatrix} a & -b \\ b & a \end{bmatrix}$$

and $T(A) = \theta(x)$; that is, $(a + b)e^x + (a - b)e^{-x} = \theta(x)$. Since the set $\{e^x, e^{-x}\}$ is linearly independent, it follows that $a + b = 0$ and $a - b = 0$. Therefore, $a = b = 0$ and $A = \mathcal{O}$.

To determine the formula for T^{-1}, let $f(x) = ce^x + de^{-x}$ be in V. By formula (1), $T^{-1}(f) = A$, where A is a matrix such that $T(A) = f(x)$; that is,

$$T(A) = (a + b)e^x + (a - b)e^{-x} = ce^x + de^{-x}. \tag{2}$$

Since $\{e^x, e^{-x}\}$ is a linearly independent set, Eq. (2) requires that $a + b = c$ and $a - b = d$. This yields $a = (1/2)c + (1/2)d$ and $b = (1/2)c - (1/2)d$. Therefore, the formula for T^{-1} is given by

$$T^{-1}(ce^x + de^{-x}) = (1/2)\begin{bmatrix} c + d & -c + d \\ c - d & c + d \end{bmatrix}. \qquad \square$$

A linear transformation $T: U \to V$ that is both one to one and onto is called an ***invertible*** linear transformation. Thus if T is invertible, then the mapping $T^{-1}: V \to U$ exists and is defined by formula (1). The next theorem lists some of the properties of T^{-1}.

THEOREM 16

Let U and V be vector spaces, and let $T: U \to V$ be an invertible linear transformation. Then:

1. $T^{-1}: V \to U$ is a linear transformation.
2. T^{-1} is invertible and $(T^{-1})^{-1} = T$.
3. $T^{-1} \circ T = I_U$ and $T \circ T^{-1} = I_V$, where I_U and I_V are the identity transformations on U and V, respectively.

PROOF For property 1, we need to show that $T^{-1}: V \to U$ satisfies Definition 8. Suppose that $\mathbf{v}_1$ and $\mathbf{v}_2$ are vectors in V. Since T is onto, there are vectors $\mathbf{u}_1$ and $\mathbf{u}_2$ in U such that $T(\mathbf{u}_1) = \mathbf{v}_1$ and $T(\mathbf{u}_2) = \mathbf{v}_2$. By formula (1),

$$T^{-1}(\mathbf{v}_1) = \mathbf{u}_1 \quad \text{and} \quad T^{-1}(\mathbf{v}_2) = \mathbf{u}_2. \tag{3}$$

Furthermore, $\mathbf{v}_1 + \mathbf{v}_2 = T(\mathbf{u}_1) + T(\mathbf{u}_2) = T(\mathbf{u}_1 + \mathbf{u}_2)$, so by formula (1),

$$T^{-1}(\mathbf{v}_1 + \mathbf{v}_2) = \mathbf{u}_1 + \mathbf{u}_2 = T^{-1}(\mathbf{v}_1) + T^{-1}(\mathbf{v}_2).$$

It is equally easy to see that $T^{-1}(c\mathbf{v}) = cT^{-1}(\mathbf{v})$ for all $\mathbf{v}$ in V and for any scalar c (cf. Exercise 15).

The proof of property 2 requires showing that T^{-1} is both one to one and onto. To see that T^{-1} is one to one, let $\mathbf{v}$ be in $\mathcal{N}(T^{-1})$. Then $T^{-1}(\mathbf{v}) = \boldsymbol{\theta}_U$, so by formula (1), $T(\boldsymbol{\theta}_U) = \mathbf{v}$. By Theorem 14, property 1, of Section 4.7, $\mathbf{v} = \boldsymbol{\theta}_V$ so Theorem 14, property 4, implies that T is one to one. To see that T^{-1} is onto, let $\mathbf{u}$ be an arbitrary vector in U. If $\mathbf{v} = T(\mathbf{u})$, then $\mathbf{v}$ is in V and, by formula (1), $T^{-1}(\mathbf{v}) = \mathbf{u}$. Therefore, T^{-1} is onto, and it follows that T is invertible.

That $(T^{-1})^{-1} = T$ is an easy consequence of formula (1), as are the equalities given in property 3, and the proofs are left as exercises.

As might be guessed from the corresponding theorems for nonsingular matrices, other properties of invertible transformations can be established. For example, if $T: U \rightarrow V$ is an invertible transformation, then for each vector **b** in V, $\mathbf{x} = T^{-1}(\mathbf{b})$ is the unique solution of $T(\mathbf{x}) = \mathbf{b}$. Also, if S and T are invertible and $S \circ T$ is defined, then $S \circ T$ is invertible and $(S \circ T)^{-1} = T^{-1} \circ S^{-1}$.

Isomorphic Vector Spaces

Suppose that a linear transformation $T: U \rightarrow V$ is invertible. Since T is both one to one and onto, T establishes an exact pairing between elements of U and V. Moreover, because T is a linear transformation, this pairing preserves algebraic properties. Therefore, although U and V may be different sets, they may be regarded as indistinguishable (or equivalent) algebraically. Stated another way, U and V both represent just one underlying vector space but perhaps with different "labels" for the elements. The invertible linear transformation T acts as a translation from one set of labels to another.

If U and V are vector spaces and if $T: U \rightarrow V$ is an invertible linear transformation, then U and V are said to be *isomorphic* vector spaces. Also, an invertible transformation T is called an *isomorphism.* For instance, the vector spaces U and V given in Example 6 are isomorphic, as shown in Example 9. The next example provides another illustration.

EXAMPLE 10 Let U be the subspace of $\mathcal{P}_3$ defined by

$$U = \{p(x) = a_0 + a_1 x + a_2 x^2 + a_3 x^3 : a_3 = -2a_0 + 3a_1 + a_2\}.$$

Show that U is isomorphic to R^3.

SOLUTION Note that $\dim(U) = 3$ and the set $\{1 - 2x^3, x + 3x^3, x^2 + x^3\}$ is a basis for U. Moreover, each polynomial $p(x)$ in U can be decomposed as

$$p(x) = a_0 + a_1 x + a_2 x^2 + a_3 x^3$$
$$= a_0(1 - 2x^3) + a_1(x + 3x^3) + a_2(x^2 + x^3). \tag{4}$$

It is reasonable to expect that an isomorphism $T: U \rightarrow R^3$ will map a basis of U to a basis of R^3. Since $\{\mathbf{e}_1, \mathbf{e}_2, \mathbf{e}_3\}$ is a basis for R^3, we seek a linear transformation T such that

$$T(1 - 2x^3) = \mathbf{e}_1, \quad T(x + 3x^3) = \mathbf{e}_2, \quad \text{and} \quad T(x^2 + x^3) = \mathbf{e}_3. \tag{5}$$

It follows from Eq. (4) above and from Eq. (1) of Section 4.7 that if such a linear transformation exists, then it is defined by

$$T(a_0 + a_1 x + a_2 x^2 + a_3 x^3) = a_0 T(1 - 2x^3) + a_1 T(x + 3x^3) + a_2 T(x^2 + x^3)$$
$$= a_0 \mathbf{e}_1 + a_1 \mathbf{e}_2 + a_2 \mathbf{e}_3.$$

That is,

$$T(a_0 + a_1 x + a_2 x^2 + a_3 x^3) = \begin{bmatrix} a_0 \\ a_1 \\ a_2 \end{bmatrix}. \tag{6}$$

It is straightforward to show that the function T defined by Eq. (6) is a linear transformation. Moreover, the constraints placed on T by (5) imply, by Theorem 15, property 1, of Section 4.7, that $\mathscr{R}(T) = R^3$. Likewise, by Theorem 15, property 2, T is one to one. Therefore, T is an isomorphism and U and R^3 are isomorphic vector spaces. $\quad\square$

The previous example is actually just a special case of the following theorem, which states that every real n-dimensional vector space is isomorphic to R^n.

THEOREM 17

If U is a real n-dimensional vector space, then U and R^n are isomorphic.

PROOF To prove this theorem, we need only exhibit the isomorphism, and a coordinate system on U will provide the means. Let $B = \{\mathbf{u}_1, \mathbf{u}_2, \ldots, \mathbf{u}_n\}$ be a basis for U, and let $T: U \to R^n$ be the linear transformation defined by

$$T(\mathbf{u}) = [\mathbf{u}]_B.$$

Since B is a basis, $\boldsymbol{\theta}_U$ is the only vector in $\mathscr{N}(T)$; and therefore T is one to one. Furthermore, $T(\mathbf{u}_i)$ is the vector $\mathbf{e}_i$ in R^n; so

$$\mathscr{R}(T) = \mathrm{Sp}\{T(\mathbf{u}_1), T(\mathbf{u}_2), \ldots, T(\mathbf{u}_n)\} = \mathrm{Sp}\{\mathbf{e}_1, \mathbf{e}_2, \ldots, \mathbf{e}_n\} = R^n.$$

Hence T is one to one and onto. $\quad\square$

As an illustration of Theorem 17, note that $\dim(\mathscr{P}_2) = 3$, so $\mathscr{P}_2$ and R^3 are isomorphic. Moreover, if $B = \{1, x, x^2\}$ is the natural basis for $\mathscr{P}_2$, then the linear transformation $T: \mathscr{P}_2 \to R^3$ defined by $T(p) = [p]_B$ is an isomorphism; thus

$$T(a_0 + a_1 x + a_2 x^2) = \begin{bmatrix} a_0 \\ a_1 \\ a_2 \end{bmatrix}.$$

The isomorphism T "pairs" the elements of $\mathscr{P}_2$ with elements of R^3, $p(x) \leftrightarrow [p(x)]_B$. Furthermore, under this correspondence the sum of two polynomials, $p(x) + q(x)$, is paired with the sum of the corresponding coordinate vectors:

$$p(x) + q(x) \leftrightarrow [p(x)]_B + [q(x)]_B.$$

Similarly, a scalar multiple, $ap(x)$, of a polynomial $p(x)$ is paired with the corresponding scalar multiple of $[p(x)]_B$:

$$ap(x) \leftrightarrow a[p(x)]_B.$$

In this sense, $\mathscr{P}_2$ and R^3 have the same algebraic character.

It is easy to show that if U is isomorphic to V and V is isomorphic to W, then U and W are also isomorphic (cf. Exercise 19). Using this fact, we obtain the following corollary of Theorem 17.

COROLLARY

If U and V are real n-dimensional vector spaces, then U and V are isomorphic.

Exercises 4.8

In Exercises 1–6, the linear transformations S, T, and H are defined as follows:

$S: \mathcal{P}_3 \to \mathcal{P}_4$ is defined by $S(p) = p'(0)$.

$T: \mathcal{P}_3 \to \mathcal{P}_4$ is defined by $T(p) = (x + 2)p(x)$.

$H: \mathcal{P}_4 \to \mathcal{P}_3$ is defined by $H(p) = p'(x) + p(0)$.

1. Give the formula for $S + T$. Calculate $(S + T)(x)$ and $(S + T)(x^2)$.

2. Give the formula for $2T$. Calculate $(2T)(x)$.

3. Give the formula for $H \circ T$. What is the domain for $H \circ T$? Calculate $(H \circ T)(x)$.

4. Give the formula for $T \circ H$. What is the domain for $T \circ H$? Calculate $(T \circ H)(x)$.

5. a) Prove that T is one to one but not onto.
   b) Attempt to define $T^{-1}: \mathcal{P}_4 \to \mathcal{P}_3$ as in formula (1) by setting $T^{-1}(q) = p$ if and only if $T(p) = q$. What is $T^{-1}(x)$?

6. a) Prove that H is onto but not one to one.
   b) Attempt to define $H^{-1}: \mathcal{P}_3 \to \mathcal{P}_4$ as in formula (1) by setting $H^{-1}(q) = p$ if and only if $H(p) = q$. Show that $H^{-1}(x)$ is not uniquely determined.

7. The functions e^x, e^{2x}, and e^{3x} are linearly independent in $C[0, 1]$. Let V be the subspace of $C[0, 1]$ defined by $V = \mathrm{Sp}\{e^x, e^{2x}, e^{3x}\}$, and let $T: V \to V$ be given by $T(p) = p'(x)$. Show that T is invertible and calculate $T^{-1}(e^x)$, $T^{-1}(e^{2x})$, and $T^{-1}(e^{3x})$. What is $T^{-1}(ae^x + be^{2x} + ce^{3x})$?

8. Let V be the subspace of $C[0, 1]$ defined by $V = \mathrm{Sp}\{\sin x, \cos x, e^{-x}\}$, and let $T: V \to V$ be given by $T(f) = f'(x)$. Given that the set $\{\sin x, \cos x, e^{-x}\}$ is linearly independent, show that T is invertible. Calculate $T^{-1}(\sin x)$, $T^{-1}(\cos x)$, and $T^{-1}(e^{-x})$ and give the formula for T^{-1}; that is, determine $T^{-1}(a \sin x + b \cos x + ce^{-x})$.

9. Let V be the vector space of all (2×2) matrices and define $T: V \to V$ by $T(A) = A^T$. Show that T is invertible and give the formula for T^{-1}.

10. Let V be the vector space of all (2×2) matrices, and let Q be a given nonsingular (2×2) matrix. If $T: V \to V$ is defined by $T(A) = Q^{-1}AQ$, prove that T is invertible and give the formula for T^{-1}.

11. Let V be the vector space of all (2×2) matrices.
    a) Use Theorem 17 to show that V is isomorphic to R^4.
    b) Use the corollary to Theorem 17 to show that V is isomorphic to $\mathcal{P}_3$.
    c) Exhibit an isomorphism $T: V \to \mathcal{P}_3$. (Hint: Cf. Example 10.)

12. Let U be the vector space of all (2×2) symmetric matrices.
    a) Use Theorem 17 to show that U is isomorphic to R^3.
    b) Use the corollary to Theorem 17 to show that U is isomorphic to $\mathcal{P}_2$.
    c) Exhibit an isomorphism $T: U \to \mathcal{P}_2$.

13. Let $T_1: U \to V$ and $T_2: U \to V$ be linear transformations. Prove that $S: U \to V$, where $S = T_1 + T_2$, is a linear transformation.

14. If $T: U \to V$ is a linear transformation and a is a scalar, show that $aT: U \to V$ is a linear transformation.

15. Complete the proof of property 1 of Theorem 16 by showing that $T^{-1}(c\mathbf{v}) = cT^{-1}(\mathbf{v})$ for all $\mathbf{v}$ in V and for an arbitrary scalar c.

16. Complete the proof of property 2 of Theorem 16 by showing that $(T^{-1})^{-1} = T$. (Hint: Use formula 1.)

17. Prove property 3 of Theorem 16.

18. Let $S: U \to V$ and $T: V \to W$ be linear transformations.

a) Prove that if S and T are both one to one, then $T \circ S$ is one to one.

b) Prove that if S and T are both onto, then $T \circ S$ is onto.

c) Prove that if S and T are both invertible, then $T \circ S$ is invertible and $(T \circ S)^{-1} = S^{-1} \circ T^{-1}$.

19. Let U, V, and W be vector spaces such that U and V are isomorphic and V and W are isomorphic. Use Exercise 18 to show that U and W are isomorphic.

20. Let U and V both be n-dimensional vector spaces, and suppose that $T: U \rightarrow V$ is a linear transformation.

a) If T is one to one, prove that T is invertible. (*Hint:* Use property 3 of Theorem 15 to prove that $\mathcal{R}(T) = V$.)

b) If T is onto, prove that T is invertible. (*Hint:* Use property 3 of Theorem 15 and property 4 of Theorem 14 to prove that T is one to one.)

21. Define $T: \mathcal{P} \rightarrow \mathcal{P}$ by $T(a_0 + a_1 x + \cdots + a_n x^n) = a_0 x + (1/2)a_1 x^2 + \cdots + (1/(n + 1))a_n x^{n+1}$. Prove that T is one to one but not onto. Why is this example not a contradiction of part (a) of Exercise 20?

22. Define $S: \mathcal{P} \rightarrow \mathcal{P}$ by $S(p) = p'(x)$. Prove that S is onto but not one to one. Why is this example not a contradiction of part (b) of Exercise 20?

In Exercises 23–25, $S: U \rightarrow V$ and $T: V \rightarrow W$ are linear transformations.

23. Show that $\mathcal{N}(S) \subseteq \mathcal{N}(T \circ S)$. Conclude that if $T \circ S$ is one to one, then S is one to one.

24. Show that $\mathcal{R}(T \circ S) \subseteq \mathcal{R}(T)$. Conclude that if $T \circ S$ is onto, then T is onto.

25. Assume that U, V, and W all have dimension n. Prove that if $T \circ S$ is invertible, then both T and S are invertible. (*Hint:* Use Exercises 23, 24, and 20.)

26. Let A be an $(m \times p)$ matrix and B a $(p \times n)$ matrix. Use Exercises 23 and 24 to show that nullity$(B) \leq$ nullity(AB) and rank$(AB) \leq$ rank(A).

27. Let A be an $(n \times n)$ matrix, and suppose that $T: R^n \rightarrow R^n$ is defined by $T(\mathbf{x}) = A\mathbf{x}$. Show that T is invertible if and only if A is nonsingular. If T is invertible, give a formula for T^{-1}.

28. Let A and B be $(n \times n)$ matrices such that AB is nonsingular. Use Exercises 25 and 27 to show that each of the matrices A and B is nonsingular.

29. Let U and V be vector spaces, and let $L(U, V) = \{T: T \text{ is a linear transformation from } U \text{ to } V\}$. With the operations of addition and scalar multiplication defined in this section, show that $L(U, V)$ is a vector space.

4.9 ■ Matrix Representations for Linear Transformations

In Section 2.7 we showed that a linear transformation $T: R^n \rightarrow R^m$ can be represented as multiplication by an $(m \times n)$ matrix A; that is, $T(\mathbf{x}) = A\mathbf{x}$ for all $\mathbf{x}$ in R^n. In the general vector-space setting, we have viewed a linear transformation $T: U \rightarrow V$ as an extension of this notion. Now suppose that U and V both have finite dimension, say dim$(U) = n$ and dim$(V) = m$. By Theorem 17 of Section 4.8, U is isomorphic to R^n and V is isomorphic to R^m. To be specific, let B be a basis for U and let C be a basis for V. Then each vector $\mathbf{u}$ in U can be represented by the vector $[\mathbf{u}]_B$ in R^n, and similarly each vector $\mathbf{v}$ in V can be represented by the vector $[\mathbf{v}]_C$ in R^m. In this section we show that T can be represented by an $(m \times n)$ matrix Q in the sense that if $T(\mathbf{u}) = \mathbf{v}$, then $Q[\mathbf{u}]_B = [\mathbf{v}]_B$.

The Matrix of a Transformation

We begin by defining the matrix of a linear transformation. Thus let $T: U \rightarrow V$ be a linear transformation, where dim$(U) = n$ and dim$(V) = m$. Let $B = \{\mathbf{u}_1, \mathbf{u}_2, \ldots, \mathbf{u}_n\}$ be a basis for U and let $C = \{\mathbf{v}_1, \mathbf{v}_2, \ldots, \mathbf{v}_m\}$ be a basis for V. The *matrix representation* for T with respect to the bases B and C is the $(m \times n)$

matrix Q defined by

$$Q = [\mathbf{Q}_1, \mathbf{Q}_2, \ldots, \mathbf{Q}_n],$$

where

$$\mathbf{Q}_j = [T(\mathbf{u}_j)]_C.$$

Thus to determine Q, we first represent each of the vectors $T(\mathbf{u}_1), T(\mathbf{u}_2), \ldots, T(\mathbf{u}_n)$ in terms of the basis C for V:

$$
\begin{aligned}
T(\mathbf{u}_1) &= q_{11}\mathbf{v}_1 + q_{21}\mathbf{v}_2 + \cdots + q_{m1}\mathbf{v}_m \\
T(\mathbf{u}_2) &= q_{12}\mathbf{v}_1 + q_{22}\mathbf{v}_2 + \cdots + q_{m2}\mathbf{v}_m \\
&\;\;\vdots \qquad\quad \vdots \qquad\qquad\quad \vdots \\
T(\mathbf{u}_n) &= q_{1n}\mathbf{v}_1 + q_{2n}\mathbf{v}_2 + \cdots + q_{mn}\mathbf{v}_m.
\end{aligned}
\tag{1}
$$

It follows from system (1) that

$$
\mathbf{Q}_1 = [T(\mathbf{u}_1)]_C = \begin{bmatrix} q_{11} \\ q_{21} \\ \vdots \\ q_{m1} \end{bmatrix}, \ldots, \mathbf{Q}_n = [T(\mathbf{u}_n)]_C = \begin{bmatrix} q_{1n} \\ q_{2n} \\ \vdots \\ q_{mn} \end{bmatrix}.
\tag{2}
$$

The following example provides a specific illustration.

EXAMPLE 1 Let U be the vector space of all (2×2) matrices and define $T: U \to \mathscr{P}_2$ by

$$
T\left(\begin{bmatrix} a & b \\ c & d \end{bmatrix}\right) = (a - d) + (a + 2b)x + (b - 3c)x^2.
$$

Find the matrix of T relative to the natural bases for U and $\mathscr{P}_2$.

SOLUTION   Let $B = \{E_{11}, E_{12}, E_{21}, E_{22}\}$ be the natural basis for U and let $C = \{1, x, x^2\}$ be the natural basis for $\mathscr{P}_2$. Then $T(E_{11}) = 1 + x$, $T(E_{12}) = 2x + x^2$, $T(E_{21}) = -3x^2$, and $T(E_{22}) = -1$. In this example system (1) above becomes

$$
\begin{aligned}
T(E_{11}) &= \;\; 1 + 1x + 0x^2 \\
T(E_{12}) &= \;\; 0 + 2x + 1x^2 \\
T(E_{21}) &= \;\; 0 + 0x - 3x^2 \\
T(E_{22}) &= -1 + 0x + 0x^2.
\end{aligned}
$$

Therefore, the matrix of T is the (3×4) matrix Q given by

$$
Q = \begin{bmatrix} 1 & 0 & 0 & -1 \\ 1 & 2 & 0 & 0 \\ 0 & 1 & -3 & 0 \end{bmatrix}.
$$

The Representation Theorem

The next theorem shows that if we translate from general vector spaces to coordinate vectors, the action of a linear transformation T translates to multiplication by its matrix representative.

$$u \xrightarrow{} T(\mathbf{u}) = \mathbf{v}$$

$$[\mathbf{u}]_B \xrightarrow{} Q[\mathbf{u}]_B = [\mathbf{v}]_C$$

Figure 4.8
The matrix of T

THEOREM 18

Let $T: U \to V$ be a linear transformation, where $\dim(U) = n$ and $\dim(V) = m$. Let B and C be bases for U and V, respectively, and let Q be the matrix of T relative to B and C. If $\mathbf{u}$ is a vector in U and if $T(\mathbf{u}) = \mathbf{v}$, then

$$Q[\mathbf{u}]_B = [\mathbf{v}]_C. \qquad (3)$$

Moreover, Q is the unique matrix that satisfies (3).

The representation of T by Q is illustrated in Fig. 4.8. Before giving the proof of Theorem 18, we illustrate the result with an example.

EXAMPLE 2   Let $T: U \to \mathscr{P}_2$ be the linear transformation defined in Example 1, and let Q be the matrix representation determined in that example. Show by direct calculation that if $T(A) = p(x)$, then

$$Q[A]_B = [p(x)]_C. \qquad (4)$$

SOLUTION   Recall that $B = \{E_{11}, E_{12}, E_{21}, E_{22}\}$ and $C = \{1, x, x^2\}$. Equation (4) is, of course, an immediate consequence of Eq. (3). To verify Eq. (4) directly, note that if

$$A = \begin{bmatrix} a & b \\ c & d \end{bmatrix},$$

then

$$[A]_B = \begin{bmatrix} a \\ b \\ c \\ d \end{bmatrix}.$$

Further, if $p(x) = T(A)$, then $p(x) = (a - d) + (a + 2b)x + (b - 3c)x^2$, so

$$[p(x)]_C = \begin{bmatrix} a - d \\ a + 2b \\ b - 3c \end{bmatrix}.$$

Therefore,

$$Q[A]_B = \begin{bmatrix} 1 & 0 & 0 & -1 \\ 1 & 2 & 0 & 0 \\ 0 & 1 & -3 & 0 \end{bmatrix} \begin{bmatrix} a \\ b \\ c \\ d \end{bmatrix} = \begin{bmatrix} a - d \\ a + 2b \\ b - 3c \end{bmatrix} = [p(x)]_C. \qquad \square$$

PROOF OF THEOREM 18   Let $B = \{u_1, u_2, \ldots, u_n\}$ be the given basis for U, let u be in U, and set $T(u) = v$. First write u in terms of the basis vectors:

$$u = a_1u_1 + a_2u_2 + \cdots + a_nu_n. \tag{5}$$

It follows from Eq. (5) that the coordinate vector for u is

$$[u]_B = \begin{bmatrix} a_1 \\ a_2 \\ \vdots \\ a_n \end{bmatrix}.$$

Furthermore, the action of T is completely determined by its action on a basis for U (cf. Eq. 1 of Section 4.7), so Eq. (5) implies that

$$T(u) = a_1 T(u_1) + a_2 T(u_2) + \cdots + a_n T(u_n) = v. \tag{6}$$

The vectors in Eq. (6) are in V, and passing to coordinate vectors relative to the basis C yields, by Eq. (10) of Section 4.4,

$$a_1[T(u_1)]_C + a_2[T(u_2)]_C + \cdots + a_n[T(u_n)]_C = [v]_C. \tag{7}$$

Recall that the matrix Q of T is the $(m \times n)$ matrix $Q = [Q_1, Q_2, \ldots, Q_n]$, where

$$Q_j = [T(u_j)]_C.$$

Thus Eq. (7) can be rewritten as

$$a_1Q_1 + a_2Q_2 + \cdots + a_nQ_n = [v]_C. \tag{8}$$

Since Eq. (8) is equivalent to the matrix equation

$$Q[u]_B = [v]_C,$$

this shows that Equation 3 of Theorem 18 holds. The uniqueness of Q is left as an exercise. □

EXAMPLE 3   Let $S: \mathscr{P}_2 \to \mathscr{P}_3$ be the differential operator defined by $S(f) = x^2f'' - 2f' + xf$. Find the (4×3) matrix P that represents S with respect to the natural bases $C = \{1, x, x^2\}$ and $D = \{1, x, x^2, x^3\}$ for $\mathscr{P}_2$ and $\mathscr{P}_3$, respectively. Also, illustrate that P satisfies Equation 3 of Theorem 18.

SOLUTION   To construct the (4×3) matrix P that represents S, we need to find the coordinate vectors of $S(1)$, $S(x)$, and $S(x^2)$ with respect to D. We calculate that $S(1) = x$, $S(x) = x^2 - 2$, and $S(x^2) = x^3 + 2x^2 - 4x$; so the coordinate vectors of $S(1)$, $S(x)$, and $S(x^2)$ are

$$[S(1)]_D = \begin{bmatrix} 0 \\ 1 \\ 0 \\ 0 \end{bmatrix}, \quad [S(x)]_D = \begin{bmatrix} -2 \\ 0 \\ 1 \\ 0 \end{bmatrix}, \quad \text{and} \quad [S(x^2)]_D = \begin{bmatrix} 0 \\ -4 \\ 2 \\ 1 \end{bmatrix}.$$

Thus the matrix representation for S is the (4×3) matrix

$$P = \begin{bmatrix} 0 & -2 & 0 \\ 1 & 0 & -4 \\ 0 & 1 & 2 \\ 0 & 0 & 1 \end{bmatrix}.$$

To see that Eq. (3) of Theorem 18 holds, let $p(x) = a_0 + a_1 x + a_2 x^2$ be in $\mathscr{P}_2$. Then $S(p) = -2a_1 + (a_0 - 4a_2)x + (a_1 + 2a_2)x^2 + a_2 x^3$. In this case

$$[p(x)]_C = \begin{bmatrix} a_0 \\ a_1 \\ a_2 \end{bmatrix},$$

and if $S(p) = q(x)$, then

$$[q(x)]_D = \begin{bmatrix} -2a_1 \\ a_0 - 4a_2 \\ a_1 + 2a_2 \\ a_2 \end{bmatrix}.$$

A straightforward calculation shows that $P[p(x)]_C = [q(x)]_D$. ☐

EXAMPLE 4 Let A be an $(m \times n)$ matrix and consider the linear transformation $T: R^n \to R^m$ defined by $T(\mathbf{x}) = A\mathbf{x}$. Show that relative to the natural bases for R^n and R^m, the matrix for T is A.

SOLUTION Let $B = \{\mathbf{e}_1, \mathbf{e}_2, \ldots, \mathbf{e}_n\}$ be the natural basis for R^n, and let C denote the natural basis for R^m. First note that for each vector $\mathbf{y}$ in R^m, $\mathbf{y} = [\mathbf{y}]_C$. Now let Q denote the matrix of T relative to B and C, $Q = [\mathbf{Q}_1, \mathbf{Q}_2, \ldots, \mathbf{Q}_n]$, and write $A = [\mathbf{A}_1, \mathbf{A}_2, \ldots, \mathbf{A}_n]$. The definition of Q gives

$$\mathbf{Q}_j = [T(\mathbf{e}_j)]_C = T(\mathbf{e}_j) = A\mathbf{e}_j = \mathbf{A}_j.$$

It now follows that $Q = A$. ☐

If V is a vector space, then linear transformations of the form $T: V \to V$ are of considerable interest and importance. In this case, the same basis B is normally chosen for both the domain and the range of T, and we refer to the representation as the matrix of T with respect to B. In this case, if Q is the matrix of T and if $\mathbf{v}$ is in V, then Eq. (3) of Theorem 18 becomes

$$Q[\mathbf{v}]_B = [T(\mathbf{v})]_B.$$

The next example illustrates this special case.

EXAMPLE 5 Let $T: \mathscr{P}_2 \to \mathscr{P}_2$ be the linear transformation defined by $T(p) = xp'(x)$. Find the matrix, Q, of T relative to the natural basis for $\mathscr{P}_2$.

SOLUTION Let $B = \{1, x, x^2\}$. Then $T(1) = 0$, $T(x) = x$, and $T(x^2) = 2x^2$. The coordinate vectors for $T(1)$, $T(x)$, $T(x^2)$ relative to B are

$$[T(1)]_B = \begin{bmatrix} 0 \\ 0 \\ 0 \end{bmatrix}, \quad [T(x)]_B = \begin{bmatrix} 0 \\ 1 \\ 0 \end{bmatrix}, \quad \text{and} \quad [T(x^2)]_B = \begin{bmatrix} 0 \\ 0 \\ 2 \end{bmatrix}.$$

It follows that the matrix of T with respect to B is the (3×3) matrix

$$Q = \begin{bmatrix} 0 & 0 & 0 \\ 0 & 1 & 0 \\ 0 & 0 & 2 \end{bmatrix}.$$

If $p(x) = a_0 + a_1 x + a_2 x^2$, then $T[p(x)] = x(a_1 + 2a_2 x) = a_1 x + 2a_2 x^2$. Thus

$$[p(x)]_B = \begin{bmatrix} a_0 \\ a_1 \\ a_2 \end{bmatrix} \quad \text{and} \quad [T(p(x))]_B = \begin{bmatrix} 0 \\ a_1 \\ 2a_2 \end{bmatrix}.$$

A direct calculation verifies that $Q[p(x)]_B = [T(p(x))]_B$. ☐

Algebraic Properties

In Section 4.8, we defined the algebraic operations of addition, scalar multiplication, and composition for linear transformations. We now examine the matrix representations of the resulting transformations. We begin with the following theorem.

THEOREM 19

Let U and V be vector spaces, with $\dim(U) = n$ and $\dim(V) = m$, and suppose that B and C are bases for U and V, respectively. Let T_1, T_2, and T be transformations from U to V and let Q_1, Q_2, and Q be the matrix representations with respect to B and C for T_1, T_2, and T, respectively. Then:

1. $Q_1 + Q_2$ is the matrix representation for $T_1 + T_2$ with respect to B and C.
2. For a scalar a, aQ is the matrix representation for aT with respect to B and C.

PROOF We include here only the proof of property 1. The proof of property 2 is left for Exercises 26 and 27.

To prove property 1, set $T_3 = T_1 + T_2$ and let Q_3 be the matrix of T_3. By Eq. (3) of Theorem 18, Q_3 satisfies the equation

$$Q_3[\mathbf{u}]_B = [T_3(\mathbf{u})]_C \tag{9}$$

for every vector $\mathbf{u}$ in U; moreover, any other matrix that satisfies Eq. (9) is equal to Q_3. We also know from Theorem 18 that

$$Q_1[\mathbf{u}]_B = [T_1(\mathbf{u})]_C \quad \text{and} \quad Q_2[\mathbf{u}]_B = [T_2(\mathbf{u})]_C \tag{10}$$

for every vector **u** in U. Using Eq. (10) in Section 4.4 gives

$$[T_1(\mathbf{u}) + T_2(\mathbf{u})]_C = [T_1(\mathbf{u})]_C + [T_2(\mathbf{u})]_C. \tag{11}$$

It follows from Eqs. (10) and (11) that

$$(Q_1 + Q_2)[\mathbf{u}]_B = [T_1(\mathbf{u}) + T_2(\mathbf{u})]_C = [T_3(\mathbf{u})]_C;$$

therefore, $Q_3 = Q_1 + Q_2$.

The following example illustrates the preceding theorem.

EXAMPLE 6   Let T_1 and T_2 be the linear transformations from $\mathscr{P}_2$ to R^2 defined by

$$T_1(p) = \begin{bmatrix} p(0) \\ p(1) \end{bmatrix} \quad \text{and} \quad T_2(p) = \begin{bmatrix} p'(0) \\ p(-1) \end{bmatrix}.$$

Set $T_3 = T_1 + T_2$ and $T_4 = 3T_1$ and let $B = \{1, x, x^2\}$ and $C = \{\mathbf{e}_1, \mathbf{e}_2\}$. Use the definition to calculate the matrices Q_1, Q_2, Q_3 and Q_4 for T_1, T_2, T_3, and T_4, respectively, relative to the bases B and C. Note that $Q_3 = Q_1 + Q_2$ and $Q_4 = 3Q_1$.

SOLUTION   Since

$$T_1(1) = \begin{bmatrix} 1 \\ 1 \end{bmatrix}, \qquad T_1(x) = \begin{bmatrix} 0 \\ 1 \end{bmatrix}, \quad \text{and} \quad T_1(x^2) = \begin{bmatrix} 0 \\ 1 \end{bmatrix},$$

it follows that Q_1 is the (2×3) matrix given by

$$Q_1 = \begin{bmatrix} 1 & 0 & 0 \\ 1 & 1 & 1 \end{bmatrix}.$$

Similarly,

$$T_2(1) = \begin{bmatrix} 0 \\ 1 \end{bmatrix}, \qquad T_2(x) = \begin{bmatrix} 1 \\ -1 \end{bmatrix}, \quad \text{and} \quad T_2(x^2) = \begin{bmatrix} 0 \\ 1 \end{bmatrix},$$

so Q_2 is given by

$$Q_2 = \begin{bmatrix} 0 & 1 & 0 \\ 1 & -1 & 1 \end{bmatrix}.$$

Now $T_3(p) = T_1(p) + T_2(p)$, so

$$T_3(p) = \begin{bmatrix} p(0) + p'(0) \\ p(1) + p(-1) \end{bmatrix}.$$

Proceeding as before, we obtain

$$T_3(1) = \begin{bmatrix} 1 \\ 2 \end{bmatrix}, \qquad T_3(x) = \begin{bmatrix} 1 \\ 0 \end{bmatrix}, \quad \text{and} \quad T_3(x^2) = \begin{bmatrix} 0 \\ 2 \end{bmatrix}.$$

Thus

$$Q_3 = \begin{bmatrix} 1 & 1 & 0 \\ 2 & 0 & 2 \end{bmatrix},$$

and clearly $Q_3 = Q_1 + Q_2$.

The formula for T_4 is

$$T_4(p) = 3T_1(p) = 3\begin{bmatrix} p(0) \\ p(1) \end{bmatrix} = \begin{bmatrix} 3p(0) \\ 3p(1) \end{bmatrix}.$$

The matrix, Q_4, for T_4 is easily obtained and is given by

$$Q_4 = \begin{bmatrix} 3 & 0 & 0 \\ 3 & 3 & 3 \end{bmatrix}.$$

In particular, $Q_4 = 3Q_1$.

The following theorem shows that the composition of two linear transformations corresponds to the product of the matrix representations.

THEOREM 20

Let $T: U \to V$ and $S: V \to W$ be linear transformations, and suppose $\dim(U) = n$, $\dim(V) = m$, and $\dim(W) = k$. Let B, C, and D be bases for U, V, and W, respectively. If the matrix for T relative to B and C is Q [Q is $(m \times n)$] and the matrix for S relative to C and D is P [P is $(k \times m)$], then the matrix representation for $S \circ T$ is PQ.

PROOF   The composition of T and S is illustrated in Fig. 4.9(a), and the matrix representation is illustrated in 4.9(b).

To prove Theorem 20, let N be the matrix of $S \circ T$ with respect to the bases B and D. Then N is the unique matrix with the property that

$$N[\mathbf{u}]_B = [(S \circ T)(\mathbf{u})]_D \qquad (12)$$

for every vector $\mathbf{u}$ in U. Similarly, Q and P are characterized by

$$Q[\mathbf{u}]_B = [T(\mathbf{u})]_C \quad \text{and} \quad P[\mathbf{v}]_C = [S(\mathbf{v})]_D \qquad (13)$$

for all $\mathbf{u}$ in U and $\mathbf{v}$ in V. It follows from Eq. (13) that

$$PQ[\mathbf{u}]_B = P[T(\mathbf{u})]_C = [S(T(\mathbf{u}))]_D = [(S \circ T)(\mathbf{u})]_D.$$

The uniqueness of N in Eq. (12) now implies that $PQ = N$.

$$U \xrightarrow{\;T\;} V \qquad \xrightarrow{\;S\;} W$$

(a) $\mathbf{u} \longrightarrow T(\mathbf{u}) \longrightarrow S[T(\mathbf{u})]$

$$R^n \xrightarrow{\;Q\;} R^m \qquad \xrightarrow{\;P\;} R^k$$

(b) $[\mathbf{u}]_B \longrightarrow Q[\mathbf{u}]_B \longrightarrow PQ[\mathbf{u}]_B$

Figure 4.9
The matrix for $S \circ T$

EXAMPLE 7 Let U be the vector space of (2×2) matrices. If $T: U \to \mathcal{P}_2$ is the transformation given in Example 1 and $S: \mathcal{P}_2 \to \mathcal{P}_3$ is the transformation described in Example 3, give the formula for $S \circ T$. By direct calculation, find the matrix of $S \circ T$ with respect to the bases $B = \{E_{11}, E_{12}, E_{21}, E_{22}\}$ and $D = \{1, x, x^2, x^3\}$ for U and $\mathcal{P}_3$, respectively. Finally, use Theorem 20 and the matrices found in Examples 1 and 3 to calculate the matrix for $S \circ T$.

SOLUTION Recall that $T: U \to \mathcal{P}_2$ is given by

$$T\left(\begin{bmatrix} a & b \\ c & d \end{bmatrix}\right) = (a - d) + (a + 2b)x + (b - 3c)x^2,$$

and $S: \mathcal{P}_2 \to \mathcal{P}_3$ is defined by

$$S(p) = x^2 p'' - 2p' + xp.$$

Therefore, $S \circ T: U \to \mathcal{P}_3$ is defined by

$$
\begin{aligned}
(S \circ T)(A) = S(T(A)) &= S((a - d) + (a + 2b)x + (b - 3c)x^2) \\
&= (-2a - 4b) + (a - 4b + 12c - d)x + (a + 4b - 6c)x^2 \\
&\quad + (b - 3c)x^3.
\end{aligned}
$$

The matrix, N, of $S \circ T$ relative to the given bases B and D is easily determined to be the (4×4) matrix

$$
N = \begin{bmatrix}
-2 & -4 & 0 & 0 \\
1 & -4 & 12 & -1 \\
1 & 4 & -6 & 0 \\
0 & 1 & -3 & 0
\end{bmatrix}.
$$

Moreover, $N = PQ$, where Q is the matrix for T found in Example 1 and P is the matrix for S determined in Example 3. ☐

A particularly useful case of Theorem 20 is the one in which S and T both map V to V, $\dim(V) = n$, and the same basis B is used for both the domain and the range. In this case, the composition $S \circ T$ is always defined, and the matrices P and Q for S and T, respectively, are both $(n \times n)$ matrices. Using Theorem 20, we can easily show that if T is invertible, then Q is nonsingular, and furthermore the matrix representation for T^{-1} is Q^{-1}. The matrix representation for the identity transformation on V, I_V, is the $(n \times n)$ identity matrix I. The matrix representation for the zero transformation on V is the $(n \times n)$ zero matrix. (Observe that the identity and the zero transformations always have the same matrix representations, regardless of what basis we choose for V. Thus changing the basis for V may change the matrix representation for T or may leave the representation unchanged.)

The Vector Space $L(U, V)$ (Optional)

If U and V are vector spaces, then $L(U, V)$ denotes the set of all linear transformations from U to V:

$$L(U, V) = \{T: T \text{ is a linear transformation}; \; T: U \to V\}.$$

If T, T_1, and T_2 are in $L(U, V)$ and a is a scalar, we have seen in Section 4.8 that $T_1 + T_2$ and aT are again in $L(U, V)$. In fact, with these operations of addition and scalar multiplication, we have the following.

| REMARK   The set $L(U, V)$ is a vector space.

The proof of this remark is Exercise 29 of Section 4.8. We note here only that the zero of $L(U, V)$ is the zero transformation $T_0: U \to V$ defined by $T_0(\mathbf{u}) = \boldsymbol{\theta}_V$ for all $\mathbf{u}$ in U. To see this, let T be in $L(U, V)$. Then $(T + T_0)(\mathbf{u}) = T(\mathbf{u}) + T_0(\mathbf{u}) = T(\mathbf{u}) + \boldsymbol{\theta}_V = T(\mathbf{u})$. This shows that $T + T_0 = T$, so T_0 is the zero vector in $L(U, V)$.

Now let R_{mn} denote the vector space of $(m \times n)$ real matrices. If $\dim(U) = n$ and $\dim(V) = m$, then we can define a function $\psi: L(U, V) \to R_{mn}$ as follows: Let B and C be bases for U and V, respectively. For a transformation T in $L(U, V)$, set $\psi(T) = Q$, where Q is the matrix of T with respect to B and C. We will now show that ψ is an isomorphism. In particular, the following theorem holds.

THEOREM 21   If U and V are vector spaces such that $\dim(U) = n$ and $\dim(V) = m$, then $L(U, V)$ is isomorphic to R_{mn}.

PROOF   It is an immediate consequence of Theorem 19 that the function ψ defined above is a linear transformation; that is, if S and T are in $L(U, V)$ and a is a scalar, then $\psi(S + T) = \psi(S) + \psi(T)$ and $\psi(aT) = a\psi(T)$.

To show that ψ maps $L(U, V)$ onto R_{mn}, let $Q = [q_{ij}]$ be an $(m \times n)$ matrix. Assume that $B = \{\mathbf{u}_1, \mathbf{u}_2, \dots, \mathbf{u}_n\}$ and $C = \{\mathbf{v}_1, \mathbf{v}_2, \dots, \mathbf{v}_m\}$ are the given bases for U and V, respectively. Define a subset $\{\mathbf{w}_1, \mathbf{w}_2, \dots, \mathbf{w}_n\}$ of V as follows:

$$
\begin{aligned}
\mathbf{w}_1 &= q_{11}\mathbf{v}_1 + q_{21}\mathbf{v}_2 + \cdots + q_{m1}\mathbf{v}_m \\
\mathbf{w}_2 &= q_{12}\mathbf{v}_1 + q_{22}\mathbf{v}_2 + \cdots + q_{m2}\mathbf{v}_m \\
&\;\;\vdots \qquad \vdots \qquad\qquad\quad \vdots \\
\mathbf{w}_n &= q_{1n}\mathbf{v}_1 + q_{2n}\mathbf{v}_2 + \cdots + q_{mn}\mathbf{v}_m.
\end{aligned}
\tag{14}
$$

Each vector $\mathbf{u}$ in U can be expressed uniquely in the form

$$\mathbf{u} = a_1\mathbf{u}_1 + a_2\mathbf{u}_2 + \cdots + a_n\mathbf{u}_n.$$

If $T: U \to V$ is the function defined by

$$T(\mathbf{u}) = a_1\mathbf{w}_1 + a_2\mathbf{w}_2 + \cdots + a_n\mathbf{w}_n,$$

then T is a linear transformation and $T(\mathbf{u}_j) = \mathbf{w}_j$ for each j, $1 \le j \le n$. By comparing systems (14) and (1), it becomes clear that the matrix of T with respect to B and C is Q; that is, $\psi(T) = Q$.

The proof that ψ is one to one is Exercise 33.   □

The following example illustrates the method, described in the proof above, for obtaining the transformation when its matrix representation is given.

EXAMPLE 8 Let Q be the (3×4) matrix

$$Q = \begin{bmatrix} 1 & 0 & -1 & 0 \\ 0 & 1 & 1 & 0 \\ 2 & 0 & 3 & 1 \end{bmatrix}.$$

Give the formula for a linear transformation $T: \mathscr{P}_3 \to \mathscr{P}_2$ such that the matrix of T relative to the natural bases for $\mathscr{P}_3$ and $\mathscr{P}_2$ is Q.

SOLUTION Let $B = \{1, x, x^2, x^3\}$ and let $C = \{1, x, x^2\}$. Following the proof of Theorem 21, we form a subset $\{q_0(x), q_1(x), q_2(x), q_3(x)\}$ of $\mathscr{P}_2$ by using the columns of Q. Thus

$$\begin{aligned} q_0(x) &= \quad (1)1 + 0x + 2x^2 = 1 + 2x^2 \\ q_1(x) &= \quad (0)1 + 1x + 0x^2 = x \\ q_2(x) &= (-1)1 + 1x + 3x^2 = -1 + x + 3x^2 \\ q_3(x) &= \quad (0)1 + 0x + 1x^2 = x^2. \end{aligned}$$

If $p(x) = a_0 + a_1x + a_2x^2 + a_3x^3$ is an arbitrary polynomial in $\mathscr{P}_3$, then define $T: \mathscr{P}_3 \to \mathscr{P}_2$ by $T(p(x)) = a_0q_0(x) + a_1q_1(x) + a_2q_2(x) + a_3q_3(x)$. Thus

$$T(p(x)) = (a_0 - a_2) + (a_1 + a_2)x + (2a_0 + 3a_2 + a_3)x^2.$$

It is straightforward to show that T is a linear transformation. Moreover, $T(1) = q_0(x)$, $T(x) = q_1(x)$, $T(x^2) = q_2(x)$, and $T(x^3) = q_3(x)$. It follows that Q is the matrix of T with respect to B and C.   □

Let U and V be vector spaces such that $\dim(U) = n$ and $\dim(V) = m$. Theorem 21 implies that $L(U, V)$ and R_{mn} are essentially the same vector space. Thus, for example, we can now conclude that $L(U, V)$ has dimension mn. Furthermore, if T is a linear transformation in $L(U, V)$ and Q is the corresponding matrix in R_{mn}, then properties of T can be ascertained by studying Q. For example, a vector $\mathbf{u}$ in U is in $\mathscr{N}(T)$ if and only if $[\mathbf{u}]_B$ is in $\mathscr{N}(Q)$ and a vector $\mathbf{v}$ in V is in $\mathscr{R}(T)$ if and only if $[\mathbf{v}]_C$ is in $\mathscr{R}(Q)$. It follows that nullity$(T) =$ nullity(Q) and rank$(T) = $ rank(Q). In summary, the correspondence between $L(U, V)$ and R_{mn} allows both the computational and the theoretical aspects of linear transformations to be studied in the more familiar context of matrices.

Exercises 4.9

In Exercises 1–10, the linear transformations S, T, H are defined as follows:

$S: \mathscr{P}_3 \to \mathscr{P}_4$ is defined by $S(p) = p'(0)$.

$T: \mathscr{P}_3 \to \mathscr{P}_4$ is defined by $T(p) = (x + 2)p(x)$.

$H: \mathscr{P}_4 \to \mathscr{P}_3$ is defined by $H(p) = p'(x) + p(0)$.

Also, $B = \{1, x, x^2, x^3\}$ is the natural basis for $\mathscr{P}_3$, and $C = \{1, x, x^2, x^3, x^4\}$ is the natural basis for $\mathscr{P}_4$.

1. Find the matrix for S with respect to B and C.

2. Find the matrix for T with respect to B and C.

3. a) Use the formula for $S + T$ (cf. Exercise 1 of Section 4.8) to calculate the matrix of $S + T$ relative to B and C.
   b) Use Theorem 19 and the matrices found in Exercises 1 and 2 to obtain the matrix representation of $S + T$.

4. a) Use the formula for $2T$ (cf. Exercise 2 of Section 4.8) to calculate the matrix for $2T$ with respect to B and C.
   b) Use Theorem 19 and the matrix found in Exercise 2 to find the matrix for $2T$.

5. Find the matrix for H with respect to C and B.

6. a) Use the formula for $H \circ T$ (cf. Exercise 3 of Section 4.8) to determine the matrix of $H \circ T$ with respect to B.
   b) Use Theorem 20 and the matrices obtained in Exercises 2 and 5 to obtain the matrix representation for $H \circ T$.

7. a) Use the formula for $T \circ H$ (cf. Exercise 4 of Section 4.8) to determine the matrix of $T \circ H$ with respect to C.
   b) Use Theorem 20 and the matrices obtained in Exercises 2 and 5 to obtain the matrix representation for $T \circ H$.

8. Let $p(x) = a_0 + a_1 x + a_2 x^2 + a_3 x^3$ be an arbitrary polynomial in $\mathscr{P}_3$.
   a) Exhibit the coordinate vectors $[p]_B$ and $[S(p)]_C$.
   b) If P is the matrix for S obtained in Exercise 1, demonstrate that $P[p]_B = [S(p)]_C$.

9. Let $p(x) = a_0 + a_1 x + a_2 x^2 + a_3 x^3$ be an arbitrary polynomial in $\mathscr{P}_3$.
   a) Exhibit the coordinate vectors $[p]_B$ and $[T(p)]_C$.
   b) If Q is the matrix for T obtained in Exercise 2, demonstrate that $Q[p]_B = [T(p)]_C$.

10. Let N be the matrix representation obtained for H in Exercise 5. Demonstrate that $N[q]_C = [H(q)]_B$ for $q(x) = a_0 + a_1 x + a_2 x^2 + a_3 x^3 + a_4 x^4$ in $\mathscr{P}_4$.

11. Let $T: V \to V$ be the linear transformation defined in Exercise 7 of Section 4.8, and let $B = \{e^x, e^{2x}, e^{3x}\}$.
    a) Find the matrix, Q, of T with respect to B.
    b) Find the matrix, P, of T^{-1} with respect to B.
    c) Show that $P = Q^{-1}$.

12. Let $T: V \to V$ be the linear transformation defined in Exercise 8 of Section 4.8, and let $B = \{\sin x, \cos x, e^{-x}\}$. Repeat Exercise 11.

13. Let V be the vector space of (2×2) matrices and define $T: V \to V$ by $T(A) = A^T$ (cf. Exercise 9 of Section 4.8). Let $B = \{E_{11}, E_{12}, E_{21}, E_{22}\}$ be the natural basis for V.
    a) Find the matrix, Q, of T with respect to B.
    b) For arbitrary A in V, show that $Q[A]_B = [A^T]_B$.

14. Let $S: \mathscr{P}_2 \to \mathscr{P}_3$ be given by $S(p) = x^3 p'' - x^2 p' + 3p$. Find the matrix representation of S with respect to the natural bases $B = \{1, x, x^2\}$ for $\mathscr{P}_2$ and $C = \{1, x, x^2, x^3\}$ for $\mathscr{P}_3$.

15. Let S be the transformation in Exercise 14, let the basis for $\mathscr{P}_2$ be $B = \{x + 1, x + 2, x^2\}$, and let the basis for $\mathscr{P}_3$ be $C = \{1, x, x^2, x^3\}$. Find the matrix representation for S.

16. Let S be the transformation in Exercise 14, let the basis for $\mathscr{P}_2$ be $B = \{1, x, x^2\}$, and let the basis for $\mathscr{P}_3$ be $D = \{3, 3x - x^2, 3x^2, x^3\}$. Find the matrix for S.

17. Let $T: \mathscr{P}_2 \to R^3$ be given by

$$T(p) = \begin{bmatrix} p(0) \\ 3p'(1) \\ p'(1) + p''(0) \end{bmatrix}.$$

Find the representation of T with respect to the natural bases for $\mathscr{P}_2$ and R^3.

18. Find the representation for the transformation in Exercise 17 with respect to the natural basis for $\mathscr{P}_2$ and the basis $\{\mathbf{u}_1, \mathbf{u}_2, \mathbf{u}_3\}$ for R^3, where

$$\mathbf{u}_1 = \begin{bmatrix} 1 \\ 0 \\ 1 \end{bmatrix}, \quad \mathbf{u}_2 = \begin{bmatrix} 0 \\ 1 \\ 1 \end{bmatrix}, \quad \text{and} \quad \mathbf{u}_3 = \begin{bmatrix} 1 \\ 1 \\ 1 \end{bmatrix}.$$

19. Let $T: V \to V$ be a linear transformation, where $B = \{\mathbf{v}_1, \mathbf{v}_2, \mathbf{v}_3, \mathbf{v}_4\}$ is a basis for V. Find the matrix

representation of T with respect to B if $T(\mathbf{v}_1) = \mathbf{v}_2$, $T(\mathbf{v}_2) = \mathbf{v}_3$, $T(\mathbf{v}_3) = \mathbf{v}_1 + \mathbf{v}_2$, and $T(\mathbf{v}_4) = \mathbf{v}_1 + 3\mathbf{v}_4$.

20. Let $T: R^3 \to R^2$ be given by $T(\mathbf{x}) = A\mathbf{x}$, where

$$A = \begin{bmatrix} 1 & 2 & 1 \\ 3 & 0 & 4 \end{bmatrix}.$$

Find the representation of T with respect to the natural bases for R^2 and R^3.

21. Let $T: \mathcal{P}_2 \to \mathcal{P}_2$ be defined by $T(a_0 + a_1 x + a_2 x^2) = (-4a_0 - 2a_1) + (3a_0 + 3a_1)x + (-a_0 + 2a_1 + 3a_2)x^2$. Determine the matrix of T relative to the natural basis B for $\mathcal{P}_2$.

22. Let T be the linear transformation defined in Exercise 21. If Q is the matrix representation found in Exercise 21, show that $Q[p]_B = [T(p)]_B$ for $p(x) = a_0 + a_1 x + a_2 x^2$.

23. Let T be the linear transformation defined in Exercise 21. Find the matrix of T with respect to the basis $C = \{1 - 3x + 7x^2, 6 - 3x + 2x^2, x^2\}$.

24. Complete the proof of Theorem 18 by showing that Q is the unique matrix that satisfies Equation 3. (*Hint:* Suppose $P = [\mathbf{P}_1, \mathbf{P}_2, \ldots, \mathbf{P}_n]$ is an $(m \times n)$ matrix such that $P[\mathbf{u}]_B = [T(\mathbf{u})]_C$ for each $\mathbf{u}$ in U. By taking $\mathbf{u}$ in B, show that $\mathbf{P}_j = \mathbf{Q}_j$ for $1 \leq j \leq n$.)

25. Give another proof of property 1 of Theorem 19 by constructing matrix representations for T_1, T_2, and $T_1 + T_2$.

26. Give a proof of property 2 of Theorem 19 by constructing matrix representations for T and aT.

27. Give a proof of property 2 of Theorem 19 that uses the uniqueness assertion in Theorem 18.

28. Let V be an n-dimensional vector space, and let $I_V: V \to V$ be the identity transformation on V. [Recall that $I_V(\mathbf{v}) = \mathbf{v}$ for all $\mathbf{v}$ in V.] Show that the matrix representation for I_V with respect to any basis for V is the $(n \times n)$ identity matrix I.

29. Let V be an n-dimensional vector space, and let $T_0: V \to V$ be the zero transformation in V; that is, $T_0(\mathbf{v}) = \boldsymbol{\theta}_V$ for all $\mathbf{v}$ in V. Show that the matrix representation for T_0 with respect to any basis for V is the $(n \times n)$ zero matrix.

30. Let V be an n-dimensional vector space with basis B, and let $T: V \to V$ be an invertible linear transformation. Let Q be the matrix of T with respect to B, and let P be the matrix of T^{-1} with respect to B. Prove that $P = Q^{-1}$. (*Hint:* Note that $T^{-1} \circ T = I_V$. Now apply Theorem 20 and Exercise 28.)

In Exercises 31 and 32, Q is the (3×4) matrix given by

$$Q = \begin{bmatrix} 1 & 0 & 2 & 0 \\ 0 & 1 & 0 & 1 \\ -1 & 1 & 0 & -1 \end{bmatrix}.$$

31.* Give the formula for a linear transformation $T: \mathcal{P}_3 \to \mathcal{P}_2$ such that Q is the matrix of T with respect to the natural bases for $\mathcal{P}_3$ and $\mathcal{P}_2$.

32.* Let V be the vector space of all (2×2) matrices. Give the formula for a linear transformation $S: \mathcal{P}_2 \to V$ such that Q^T is the matrix of S with respect to the natural bases for $\mathcal{P}_2$ and V.

33.* Complete the proof of Theorem 21 by showing that the mapping described in the proof of the theorem is one to one.

4.10 Change of Basis and Diagonalization

In Section 4.9, we saw that a linear transformation from U to V could be represented as an $(m \times n)$ matrix when $\dim(U) = n$ and $\dim(V) = m$. A consequence of this representation is that properties of transformations can be studied by examining their corresponding matrix representations. Moreover, we have a great deal of machinery in place for matrix theory; so matrices will provide a suitable analytical and computational framework for studying a linear transformation. To simplify matters somewhat, we consider only transformations from V to V, where $\dim(V) = n$. So let $T: V \to V$ be a linear transformation and

---

* Based on optional material.

suppose that Q is the matrix representation for T with respect to a basis B; that is,

$$\text{if } \mathbf{w} = T(\mathbf{u}), \quad \text{then} \quad [\mathbf{w}]_B = Q[\mathbf{u}]_B. \tag{1}$$

As we know, when we change the basis B for V, we may change the matrix representation for T. If we are interested in the properties of T, then it is reasonable to search for a basis for V that makes the matrix representation for T as simple as possible. Finding such a basis is the subject of this section.

Diagonalizable Transformations

A particularly nice matrix to deal with computationally is a diagonal matrix. If $T: V \to V$ is a linear transformation whose matrix representation with respect to B is a diagonal matrix,

$$D = \begin{bmatrix} d_1 & 0 & 0 & \cdots & 0 \\ 0 & d_2 & 0 & \cdots & 0 \\ \vdots & & & & \vdots \\ 0 & 0 & 0 & \cdots & d_n \end{bmatrix}, \tag{2}$$

then it is easy to analyze the action of T on V, as the following example illustrates.

EXAMPLE 1 Let V be a three-dimensional vector space with basis $B = \{\mathbf{v}_1, \mathbf{v}_2, \mathbf{v}_3\}$, and suppose that $T: V \to V$ is a linear transformation with matrix

$$D = \begin{bmatrix} 2 & 0 & 0 \\ 0 & 3 & 0 \\ 0 & 0 & 0 \end{bmatrix}$$

with respect to B. Describe the action of T in terms of the basis vectors and determine bases for $\mathcal{N}(T)$ and $\mathcal{R}(T)$.

SOLUTION It follows from the definition of D that $T(\mathbf{v}_1) = 2\mathbf{v}_1$, $T(\mathbf{v}_2) = 3\mathbf{v}_2$, and $T(\mathbf{v}_3) = \boldsymbol{\theta}$. If $\mathbf{u}$ is any vector in V and

$$\mathbf{u} = a\mathbf{v}_1 + b\mathbf{v}_2 + c\mathbf{v}_3,$$

then

$$T(\mathbf{u}) = aT(\mathbf{v}_1) + bT(\mathbf{v}_2) + cT(\mathbf{v}_3).$$

Therefore, the action of T on $\mathbf{u}$ is given by

$$T(\mathbf{u}) = 2a\mathbf{v}_1 + 3b\mathbf{v}_2.$$

It follows that $\mathbf{u}$ is in $\mathcal{N}(T)$ if and only if $a = b = 0$; that is,

$$\mathcal{N}(T) = \text{Sp}\{\mathbf{v}_3\}. \tag{3}$$

Further, $\mathscr{R}(T) = \mathrm{Sp}\{T(\mathbf{v}_1), T(\mathbf{v}_2), T(\mathbf{v}_3)\}$, and since $T(\mathbf{v}_3) = \boldsymbol{\theta}$, it follows that

$$\mathscr{R}(T) = \mathrm{Sp}\{T(\mathbf{v}_1), T(\mathbf{v}_2)\} = \mathrm{Sp}\{2\mathbf{v}_1, 3\mathbf{v}_2\}. \tag{4}$$

One can easily see that the spanning sets given in Eqs. (3) and (4) are linearly independent, so they are bases for $\mathscr{N}(T)$ and $\mathscr{R}(T)$, respectively.

If T is a linear transformation with a matrix representation that is diagonal, then T is called ***diagonalizable.*** Before characterizing diagonalizable linear transformations, we need to extend the concepts of eigenvalues and eigenvectors to the general vector-space setting. Specifically, a scalar λ is called an ***eigenvalue*** for a linear transformation $T: V \to V$ provided that there is a nonzero vector $\mathbf{v}$ in V such that $T(\mathbf{v}) = \lambda\mathbf{v}$. The vector $\mathbf{v}$ is called an ***eigenvector*** for T corresponding to λ. The following example illustrates these concepts.

EXAMPLE 2 Let $T: \mathscr{P}_2 \to \mathscr{P}_2$ be defined by

$$T(a_0 + a_1 x + a_2 x^2) = (2a_1 - 2a_2) + (2a_0 + 3a_2)x + 3a_2 x^2.$$

Show that $C = \{1 + x, 1 - x, x + x^2\}$ is a basis of V consisting of eigenvectors for T, and exhibit the matrix of T with respect to C.

SOLUTION It is straightforward to show that C is a basis for $\mathscr{P}_2$. Also,

$$\begin{aligned}
T(x + 1) &= & 2 &+ 2x &= & 2(1 + x) \\
T(1 - x) &= & -2 &+ 2x &= & -2(1 - x) \\
T(x + x^2) &= & 3x &+ 3x^2 &= & 3(x + x^2).
\end{aligned} \tag{5}$$

Thus T has eigenvalues 2, -2, and 3 with corresponding eigenvectors $1 + x$, $1 - x$, and $x + x^2$, respectively. Moreover, it follows from (5) that the matrix of T with respect to C is the (3×3) diagonal matrix

$$Q = \begin{bmatrix} 2 & 0 & 0 \\ 0 & -2 & 0 \\ 0 & 0 & 3 \end{bmatrix}.$$

In particular, T is a diagonalizable linear transformation.

The linear transformation in Example 2 provides an illustration of the following general result.

THEOREM 22

Let V be an n-dimensional vector space. A linear transformation $T: V \to V$ is diagonalizable if and only if there exists a basis for V consisting of eigenvectors for T.

PROOF   First, suppose that $B = \{\mathbf{v}_1, \mathbf{v}_2, \ldots, \mathbf{v}_n\}$ is a basis for V consisting entirely of eigenvectors—say, $T(\mathbf{v}_1) = d_1\mathbf{v}_1$, $T(\mathbf{v}_2) = d_2\mathbf{v}_2, \ldots, T(\mathbf{v}_n) = d_n\mathbf{v}_n$. It follows that the coordinate vectors for $T(\mathbf{v}_1), T(\mathbf{v}_2), \ldots, T(\mathbf{v}_n)$ are the n-dimensional vectors

$$[T(\mathbf{v}_1)]_B = \begin{bmatrix} d_1 \\ 0 \\ \vdots \\ 0 \end{bmatrix}, \qquad [T(\mathbf{v}_2)]_B = \begin{bmatrix} 0 \\ d_2 \\ \vdots \\ 0 \end{bmatrix}, \ldots, \qquad [T(\mathbf{v}_n)]_B = \begin{bmatrix} 0 \\ 0 \\ \vdots \\ d_n \end{bmatrix}. \qquad (6)$$

Therefore, the matrix representation for T with respect to B is the $(n \times n)$ diagonal matrix D given in (2). In particular, T is diagonalizable.

Conversely, assume that T is diagonalizable and that the matrix for T with respect to the basis $B = \{\mathbf{v}_1, \mathbf{v}_2, \ldots, \mathbf{v}_n\}$ is the diagonal matrix D given in (2). Then the coordinate vectors for $T(\mathbf{v}_1), T(\mathbf{v}_2), \ldots, T(\mathbf{v}_n)$ are given by (6), so it follows that

$$\begin{aligned} T(\mathbf{v}_1) &= d_1\mathbf{v}_1 + 0\mathbf{v}_2 + \cdots + 0\mathbf{v}_n = d_1\mathbf{v}_1 \\ T(\mathbf{v}_2) &= 0\mathbf{v}_1 + d_2\mathbf{v}_2 + \cdots + 0\mathbf{v}_n = d_2\mathbf{v}_2 \\ &\vdots \qquad\quad \vdots \qquad\qquad\quad \vdots \qquad\quad \vdots \\ T(\mathbf{v}_n) &= 0\mathbf{v}_1 + 0\mathbf{v}_2 + \cdots + d_n\mathbf{v}_n = d_n\mathbf{v}_n. \end{aligned}$$

Thus B consists of eigenvectors for T.   □

As with matrices, not every linear transformation is diagonalizable. Equivalently, if $T: V \to V$ is a linear transformation, it may be that no matter what basis we choose for V, we never obtain a matrix representation for T that is diagonal. Moreover, even if T is diagonalizable, Theorem 22 provides no procedure for calculating a basis for V consisting of eigenvectors for T. Before providing such a procedure, we will examine the relationship between matrix representations of a single transformation with respect to different bases. First we need to facilitate the process of changing bases.

The Transition Matrix

Let B and C be bases for an n-dimensional vector space V. Theorem 23, below, relates the coordinate vectors $[\mathbf{v}]_B$ and $[\mathbf{v}]_C$ for an arbitrary vector $\mathbf{v}$ in V. Using this theorem, we will be able to show later that if Q is the matrix of a linear transformation T with respect to B and if P is the matrix of T relative to C, then Q and P are similar. Since we know how to determine whether a matrix is similar to a diagonal matrix, we will be able to determine when T is diagonalizable.

Change of Basis Let B and C be bases for the vector space V, with $B = \{u_1, u_2, \ldots, u_n\}$, and let P be the $(n \times n)$ matrix given by $P = [P_1, P_2, \ldots, P_n]$, where the ith column of P is

$$P_i = [u_i]_C. \tag{7}$$

Then P is a nonsingular matrix and

$$[v]_C = P[v]_B \tag{8}$$

for each vector v in V.

PROOF Let I_V denote the identity transformation on V; that is, $I_V(v) = v$ for all v in V. Recall that the ith column of the matrix of I_V with respect to B and C is the coordinate vector $[I_V(u_i)]_C$. But $I_V(u_i) = u_i$, so it follows that the matrix P described above is just the matrix representation of I_V with respect to B and C. It now follows from Eq. (3) of Theorem 18 that

$$P[v]_B = [I_V(v)]_C = [v]_C$$

for each v in V; in particular, Eq. (8) is proved. The proof that P is nonsingular is left as Exercise 17. ☐

The matrix P given in Theorem 23 is called the **transition matrix.** Since P is nonsingular, we have, in addition to Eq. (8), the relationship

$$[v]_B = P^{-1}[v]_C \tag{9}$$

for each vector v in V. The following example illustrates the use of the transition matrix.

EXAMPLE 3 Let B and C be the bases for $\mathscr{P}_2$ given by $B = \{1, x, x^2\}$ and $C = \{1, x+1, (x+1)^2\}$. Find the transition matrix P such that

$$P[q]_B = [q]_C$$

for each polynomial $q(x)$ in $\mathscr{P}_2$.

SOLUTION Following Theorem 23, we determine the coordinates of 1, x, and x^2 in terms of 1, $x + 1$, and $(x + 1)^2$. This determination is easy, and we find

$$1 = 1$$
$$x = (x + 1) - 1$$
$$x^2 = (x + 1)^2 - 2(x + 1) + 1.$$

Thus with respect to C the coordinate vectors of B are

$$[1]_C = \begin{bmatrix} 1 \\ 0 \\ 0 \end{bmatrix}, \qquad [x]_C = \begin{bmatrix} -1 \\ 1 \\ 0 \end{bmatrix}, \quad \text{and} \quad [x^2]_C = \begin{bmatrix} 1 \\ -2 \\ 1 \end{bmatrix}.$$

The transition matrix P is therefore

$$P = \begin{bmatrix} 1 & -1 & 1 \\ 0 & 1 & -2 \\ 0 & 0 & 1 \end{bmatrix}.$$

In particular, any polynomial $q(x) = a_0 + a_1 x + a_2 x^2$ can be expressed in terms of 1, $x + 1$, and $(x + 1)^2$ by forming $[q]_C = P[q]_B$. Forming this, we find

$$[q]_C = \begin{bmatrix} a_0 - a_1 + a_2 \\ a_1 - 2a_2 \\ a_2 \end{bmatrix}.$$

So with respect to C, we can write $q(x)$ as $q(x) = (a_0 - a_1 + a_2) + (a_1 - 2a_2) \times (x + 1) + a_2(x + 1)^2$ [a result that we can verify directly by multiplying out the new expression for $q(x)$].   □

Matrix Representation and Change of Basis

In terms of the transition matrix, we can now state precisely the relationship between the matrix representations of a linear transformation with respect to two different bases B and C. Moreover, given a basis B, the relationship suggests how to determine a basis C such that the matrix relative to C is a "simpler" matrix.

THEOREM 24

Let B and C be bases for the n-dimensional vector space V, and let $T: V \to V$ be a linear transformation. If Q_1 is the matrix of T with respect to B and if Q_2 is the matrix of T with respect to C, then

$$Q_2 = P^{-1}Q_1 P, \tag{10}$$

where P is the transition matrix from C to B.

PROOF   First note that the notation is reversed from Theorem 23; P is the transition matrix from C to B, so

$$[\mathbf{v}]_B = P[\mathbf{v}]_C \tag{11}$$

for all $\mathbf{v}$ in V. Also,

$$P^{-1}[\mathbf{w}]_B = [\mathbf{w}]_C \tag{12}$$

for each $\mathbf{w}$ in V. If $\mathbf{v}$ is in V and if $T(\mathbf{v}) = \mathbf{w}$, then (1) implies that

$$Q_1[\mathbf{v}]_B = [\mathbf{w}]_B \quad \text{and} \quad Q_2[\mathbf{v}]_C = [\mathbf{w}]_C. \tag{13}$$

From the equations given in (11), (12), and (13), we obtain

$$P^{-1}Q_1P[\mathbf{v}]_C = P^{-1}Q_1[\mathbf{v}]_B = P^{-1}[\mathbf{w}]_B = [\mathbf{w}]_C;$$

that is, the matrix $P^{-1}Q_1P$ satisfies the same property as Q_2 in (13). By the uniqueness of Q_2, we conclude that $Q_2 = P^{-1}Q_1P$.

The following example provides an illustration of Theorem 24.

EXAMPLE 4 Let $T: \mathscr{P}_2 \to \mathscr{P}_2$ be the linear transformation given in Example 2, and let B and C be the bases for $\mathscr{P}_2$ given by $B = \{1, x, x^2\}$ and $C = \{1 + x, 1 - x, x + x^2\}$. Calculate the matrix of T with respect to B and use Theorem 24 to find the matrix of T with respect to C.

SOLUTION Recall that T is defined by

$$T(a_0 + a_1x + a_2x^2) = (2a_1 - 2a_2) + (2a_0 + 3a_2)x + 3a_2x^2.$$

In particular, $T(1) = 2x$, $T(x) = 2$, and $T(x^2) = -2 + 3x + 3x^2$. Thus

$$[T(1)]_B = \begin{bmatrix} 0 \\ 2 \\ 0 \end{bmatrix}, \qquad [T(x)]_B = \begin{bmatrix} 2 \\ 0 \\ 0 \end{bmatrix}, \quad \text{and} \quad [T(x^2)]_B = \begin{bmatrix} -2 \\ 3 \\ 3 \end{bmatrix}.$$

It follows that the matrix of T with respect to B is the matrix Q_1 given by

$$Q_1 = \begin{bmatrix} 0 & 2 & -2 \\ 2 & 0 & 3 \\ 0 & 0 & 3 \end{bmatrix}.$$

Now let P be the transition matrix from C to B; that is, $P[\mathbf{v}]_C = [\mathbf{v}]_B$ for each vector $\mathbf{v}$ in V (note that the roles of B and C are reversed from Theorem 23). By Theorem 23, P is the (3×3) matrix $P = [\mathbf{P}_1, \mathbf{P}_2, \mathbf{P}_3]$, where

$$\mathbf{P}_1 = [1 + x]_B, \qquad \mathbf{P}_2 = [1 - x]_B, \quad \text{and} \quad \mathbf{P}_3 = [x + x^2]_B.$$

Thus P is given by

$$P = \begin{bmatrix} 1 & 1 & 0 \\ 1 & -1 & 1 \\ 0 & 0 & 1 \end{bmatrix}.$$

The inverse of P can easily be calculated and is given by

$$P^{-1} = (1/2)\begin{bmatrix} 1 & 1 & -1 \\ 1 & -1 & 1 \\ 0 & 0 & 2 \end{bmatrix}.$$

By Theorem 24, the matrix of T with respect to C is the matrix Q_2 determined by $Q_2 = P^{-1}Q_1P$. This yields

$$Q_2 = \begin{bmatrix} 2 & 0 & 0 \\ 0 & -2 & 0 \\ 0 & 0 & 3 \end{bmatrix}. \qquad \square$$

Although the preceding example serves to illustrate the statement of Theorem 2, a comparison of Examples 2 and 4 makes it clear that when the basis C is given, it may be easier to calculate the matrix of T with respect to C directly from the definition. Theorem 24, however, suggests the following idea: If we are given a linear transformation $T: V \to V$ and the matrix representation, Q, for T with respect to a given basis B, then we should look for a "simple" matrix R (diagonal if possible) that is similar to Q, $R = S^{-1}QS$. In this case we can use S^{-1} as a transition matrix to obtain a new basis C for V, where $[\mathbf{u}]_C = S^{-1}[\mathbf{u}]_B$. With respect to the basis C, T will have the matrix representation R, where $R = S^{-1}QS$.

Given the transition matrix S^{-1}, it is an easy matter to find the actual basis vectors of C. In particular, suppose that $B = \{\mathbf{u}_1, \mathbf{u}_2, \ldots, \mathbf{u}_n\}$ is the given basis for V, and we wish to find vectors in $C = \{\mathbf{v}_1, \mathbf{v}_2, \ldots, \mathbf{v}_n\}$. Since $[\mathbf{u}]_C = S^{-1}[\mathbf{u}]_B$ for all $\mathbf{u}$ in V, we know that $S[\mathbf{u}]_C = [\mathbf{u}]_B$. Moreover, with respect to C, $[\mathbf{v}_i]_C = \mathbf{e}_i$. So from $S[\mathbf{v}_i]_C = [\mathbf{v}_i]_B$ we obtain

$$S\mathbf{e}_i = [\mathbf{v}_i]_B, \qquad 1 \le i \le n. \qquad (14)$$

But if $S = [\mathbf{S}_1, \mathbf{S}_2, \ldots, \mathbf{S}_n]$, then $S\mathbf{e}_i = \mathbf{S}_i$, and Eq. (14) tells us that the coordinate vector of $\mathbf{v}_i$ with respect to the known basis B is the ith column of S. The procedure just described can be summarized as follows:

SUMMARY

Let $T: V \to V$ be a linear transformation and let $B = \{\mathbf{u}_1, \mathbf{u}_2, \ldots, \mathbf{u}_n\}$ be a given basis for V.

Step 1. Calculate the matrix, Q, for T with respect to the basis B.

Step 2. Use matrix techniques to find a "simple" matrix R and a nonsingular matrix S such that $R = S^{-1}QS$.

Step 3. Determine vectors $\mathbf{v}_1, \mathbf{v}_2, \ldots, \mathbf{v}_n$ in V so that $[\mathbf{v}_i]_B = \mathbf{S}_i$, $1 \le i \le n$, where $\mathbf{S}_i$ is the ith column of S.

Then $C = \{\mathbf{v}_1, \mathbf{v}_2, \ldots, \mathbf{v}_n\}$ is a basis for V and R is the matrix of T with respect to C.

The case of particular interest is the one in which Q is similar to a diagonal matrix R. In this case, if we choose $\{\mathbf{S}_1, \mathbf{S}_2, \ldots, \mathbf{S}_n\}$ to be a basis of R^n consisting of eigenvectors for Q, then

$$R = S^{-1}QS = \begin{bmatrix} d_1 & 0 & \cdots & 0 \\ 0 & d_2 & \cdots & 0 \\ \vdots & & & \vdots \\ 0 & 0 & \cdots & d_n \end{bmatrix},$$

where $d_1, d_2, \ldots, d_n$ are the (not necessarily distinct) eigenvalues for Q and where $QS_i = d_iS_i$. Since R is the matrix of T with respect to C, $C = \{v_1, v_2, \ldots, v_n\}$ is a basis of V consisting of eigenvectors for T; specifically, $T(v_i) = d_iv_i$ for $1 \le i \le n$. The following example provides an illustration.

EXAMPLE 5 Show that the differential operator $T: \mathscr{P}_2 \to \mathscr{P}_2$ defined by $T(p) = x^2p'' + (2x - 1)p' + 3p$ is diagonalizable.

SOLUTION With respect to the basis $B = \{1, x, x^2\}$, T has the matrix representation

$$Q = \begin{bmatrix} 3 & -1 & 0 \\ 0 & 5 & -2 \\ 0 & 0 & 9 \end{bmatrix}.$$

Since Q is triangular, we see that the eigenvalues are 3, 5, and 9; and since Q has distinct eigenvalues, Q can be diagonalized where the matrix S of eigenvectors will diagonalize Q.

We calculate the eigenvectors u_1, u_2, u_3 for Q and form $S = [u_1, u_2, u_3]$, which yields

$$S = \begin{bmatrix} 1 & 1 & 1 \\ 0 & -2 & -6 \\ 0 & 0 & 12 \end{bmatrix}.$$

In this case it follows that

$$S^{-1}QS = \begin{bmatrix} 3 & 0 & 0 \\ 0 & 5 & 0 \\ 0 & 0 & 9 \end{bmatrix} = R.$$

In view of our remarks above, R is the matrix representation for T with respect to the basis $C = \{v_1, v_2, v_3\}$, where $[v_i]_B = S_i$, or

$$[v_1]_B = \begin{bmatrix} 1 \\ 0 \\ 0 \end{bmatrix}, \quad [v_2]_B = \begin{bmatrix} 1 \\ -2 \\ 0 \end{bmatrix}, \quad \text{and} \quad [v_3]_B = \begin{bmatrix} 1 \\ -6 \\ 12 \end{bmatrix}.$$

Therefore, the basis C is given precisely as $C = \{1, 1 - 2x, 1 - 6x + 12x^2\}$. Moreover, it is easy to see that $T(v_1) = 3v_1$, $T(v_2) = 5v_2$, and $T(v_3) = 9v_3$, where $v_1 = 1$, $v_2 = 1 - 2x$, and $v_3 = 1 - 6x + 12x^2$.

Exercises 4.10

1. Let $T: R^2 \to R^2$ be defined by

$$T\left(\begin{bmatrix} x_1 \\ x_2 \end{bmatrix}\right) = \begin{bmatrix} 2x_1 + x_2 \\ x_1 + 2x_2 \end{bmatrix}.$$

Define $\mathbf{u}_1, \mathbf{u}_2$ in R^2 by

$$\mathbf{u}_1 = \begin{bmatrix} -1 \\ 1 \end{bmatrix} \quad \text{and} \quad \mathbf{u}_2 = \begin{bmatrix} 1 \\ 1 \end{bmatrix}.$$

Show that $C = \{\mathbf{u}_1, \mathbf{u}_2\}$ is a basis of R^2 consisting of eigenvectors for T. Calculate the matrix of T with respect to C.

2. Let $T: \mathscr{P}_2 \to \mathscr{P}_2$ be defined by

$$T(a_0 + a_1 x + a_2 x^2) = (2a_0 - a_1 - a_2) + (a_0 - a_2)x$$
$$+ (-a_0 + a_1 + 2a_2)x^2.$$

Show that $C = \{1 + x - x^2, 1 + x^2, 1 + x\}$ is a basis of $\mathscr{P}_2$ consisting of eigenvectors for T, and find the matrix of T with respect to C.

3. Let V be the vector space of (2×2) matrices, and let $T: V \to V$ be defined by

$$T\left(\begin{bmatrix} a & b \\ c & d \end{bmatrix}\right) = \begin{bmatrix} -3a + 5d & 3b - 5c \\ -2c & 2d \end{bmatrix}.$$

If $C = \{A_1, A_2, A_3, A_4\}$, where

$$A_1 = \begin{bmatrix} 1 & 0 \\ 0 & 1 \end{bmatrix}, \quad A_2 = \begin{bmatrix} 0 & 1 \\ 1 & 0 \end{bmatrix},$$

$$A_3 = \begin{bmatrix} 0 & 1 \\ 0 & 0 \end{bmatrix}, \quad \text{and} \quad A_4 = \begin{bmatrix} 1 & 0 \\ 0 & 0 \end{bmatrix},$$

then show that C is a basis of V consisting of eigenvectors for T. Find the matrix of T with respect to C.

4. Let C be the basis for R^2 given in Exercise 1, and let B be the natural basis for R^2. Find the transition matrix and represent the following vectors in terms of C:

$$\mathbf{a} = \begin{bmatrix} 4 \\ 2 \end{bmatrix}, \quad \mathbf{b} = \begin{bmatrix} -2 \\ 0 \end{bmatrix},$$

$$\mathbf{c} = \begin{bmatrix} 9 \\ 5 \end{bmatrix}, \quad \text{and} \quad \mathbf{d} = \begin{bmatrix} a \\ b \end{bmatrix}$$

5. Let C be the basis for $\mathscr{P}_2$ given in Exercise 2 and let $B = \{1, x, x^2\}$. Find the transition matrix and

represent the following polynomials in terms of C:

$$p(x) = 2 + x, \quad q(x) = -1 + 2x + 2x^2,$$

$$s(x) = -1 + x^2, \quad \text{and} \quad r(x) = a_0 + a_1 x + a_2 x^2$$

6. Let V be the vector space of (2×2) matrices, and let C be the basis given in Exercise 3. If B is the natural basis for V, $B = \{E_{11}, E_{12}, E_{21}, E_{22}\}$, then find the transition matrix and express the following matrices in terms of the vectors in C:

$$A = \begin{bmatrix} 1 & 2 \\ 3 & 4 \end{bmatrix}, \quad B = \begin{bmatrix} -1 & 1 \\ 0 & 3 \end{bmatrix}, \quad \text{and}$$

$$C = \begin{bmatrix} a & b \\ c & d \end{bmatrix}.$$

7. Find the transition matrix for R^2 when $B = \{\mathbf{u}_1, \mathbf{u}_2\}$ and $C = \{\mathbf{w}_1, \mathbf{w}_2\}$:

$$\mathbf{w}_1 = \begin{bmatrix} 2 \\ 1 \end{bmatrix}, \quad \mathbf{w}_2 = \begin{bmatrix} 1 \\ 2 \end{bmatrix},$$

$$\mathbf{u}_1 = \begin{bmatrix} 1 \\ 1 \end{bmatrix}, \quad \text{and} \quad \mathbf{u}_2 = \begin{bmatrix} 3 \\ 1 \end{bmatrix}.$$

8. Repeat Exercise 7 for the basis vectors

$$\mathbf{w}_1 = \begin{bmatrix} 4 \\ 3 \end{bmatrix}, \quad \mathbf{w}_2 = \begin{bmatrix} 2 \\ 3 \end{bmatrix},$$

$$\mathbf{u}_1 = \begin{bmatrix} 4 \\ 1 \end{bmatrix}, \quad \text{and} \quad \mathbf{u}_2 = \begin{bmatrix} 2 \\ 1 \end{bmatrix}.$$

9. Let $B = \{1, x, x^2, x^3\}$ and $C = \{x, x + 1, x^2 - 2x, x^3 + 3\}$ be bases for $\mathscr{P}_3$. Find the transition matrix and use it to represent the following in terms of C:

$$p(x) = x^2 - 7x + 2, \quad q(x) = x^3 + 9x - 1, \quad \text{and}$$
$$r(x) = x^3 - 2x^2 + 6$$

10. Represent the following quadratic polynomials in the form $a_0 + a_1 x + a_2 x(x - 1)$ by constructing the appropriate transition matrix:

$$p(x) = x^2 + 5x - 3, \quad q(x) = 2x^2 - 6x + 8, \quad \text{and}$$
$$r(x) = x^2 - 5$$

11. Let $T: R^2 \to R^2$ be the linear transformation defined in Exercise 1. Find the matrix of T with respect to the natural basis $B = \{\mathbf{e}_1, \mathbf{e}_2\}$. If C is the basis for

R^2 given in Exercise 1, use Theorem 24 to calculate the matrix of T with respect to C.

12. Let $T: \mathscr{P}_2 \to \mathscr{P}_2$ be the linear transformation given in Exercise 2. Find the matrix representation of T with respect to the natural basis $B = \{1, x, x^2\}$ and then use Theorem 24 to calculate the matrix of T relative to the basis C given in Exercise 2.

13. Let V and T be as in Exercise 3. Find the matrix representation of T with respect to the natural basis $B = \{E_{11}, E_{12}, E_{21}, E_{22}\}$. If C is the basis for V given in Exercise 3, use Theorem 24 to determine the matrix of T with respect to C.

In Exercises 14–16, proceed through the following steps:
a) Find the matrix, Q, of T with respect to the natural basis B for V.
b) Show that Q is similar to a diagonal matrix; that is, find a nonsingular matrix S and a diagonal matrix R such that $R = S^{-1}QS$.
c) Exhibit a basis C of V such that R is the matrix representation of T with respect to C.
d) Calculate the transition matrix, P, from B to C.
e) Use the transition matrix P and the formula $R[\mathbf{v}]_C = [T(\mathbf{v})]_C$ to calculate $T(\mathbf{w}_1)$, $T(\mathbf{w}_2)$, and $T(\mathbf{w}_3)$.

14. $V = \mathscr{P}_1$ and $T: V \to V$ is defined by $T(a_0 + a_1 x) = (4a_0 + 3a_1) + (-2a_0 - 3a_1)x$. Also,
$$\mathbf{w}_1 = 2 + 3x, \qquad \mathbf{w}_2 = -1 + x, \quad \text{and} \quad \mathbf{w}_3 = x.$$

15. $V = \mathscr{P}_2$ and $T: V \to V$ is defined by $T(p) = xp'' + (x + 1)p' + p$. Also,
$$\mathbf{w}_1 = -8 + 7x + x^2, \qquad \mathbf{w}_2 = 5 + x^2, \quad \text{and}$$
$$\mathbf{w}_3 = 4 - 3x + 2x^2.$$

16. V is the vector space of (2×2) matrices and $T: V \to V$ is given by
$$T\left(\begin{bmatrix} a & b \\ c & d \end{bmatrix}\right) = \begin{bmatrix} a - b & 2b - 2c \\ 5c - 3d & 10d \end{bmatrix}.$$

Also,
$$\mathbf{w}_1 = \begin{bmatrix} 0 & 3 \\ 0 & 1 \end{bmatrix}, \qquad \mathbf{w}_2 = \begin{bmatrix} 2 & -3 \\ 1 & 0 \end{bmatrix}, \quad \text{and}$$
$$\mathbf{w}_3 = \begin{bmatrix} 8 & -7 \\ 0 & 2 \end{bmatrix}.$$

17. Complete the proof of Theorem 23 by showing that the transition matrix P is nonsingular. (*Hint:* We have already noted in the proof of Theorem 23 that P is the matrix representation of I_V with respect to the bases B and C. Let Q be the matrix representation of I_V with respect to C and B. Now apply Theorem 20 with $T = S = I_V$.)

18. Let V be an n-dimensional vector space with basis B, and assume that $T: V \to V$ is a linear transformation with matrix representation Q relative to B.
a) If $\mathbf{v}$ is an eigenvector for T associated with the eigenvalue λ, then prove that $[\mathbf{v}]_B$ is an eigenvector for Q associated with λ.
b) If the vector $\mathbf{x}$ in R^n is an eigenvector for Q corresponding to the eigenvalue λ and if $\mathbf{v}$ in V is a vector such that $[\mathbf{v}]_B = \mathbf{x}$, prove that $\mathbf{v}$ is an eigenvector for T corresponding to the eigenvalue λ. (*Hint:* Make use of Eq. (1).)

19. Let $T: V \to V$ be a linear transformation, and let λ be an eigenvalue for T. Show that λ^2 is an eigenvalue for $T^2 = T \circ T$.

20. Prove that a linear transformation $T: V \to V$ is one to one if and only if zero is not an eigenvalue for T. (*Hint:* Use Theorem 14, property 4, of Section 4.7.)

21. Let $T: V \to V$ be an invertible linear transformation. If λ is an eigenvalue for T, prove that λ^{-1} is an eigenvalue for T^{-1}. (Note that $\lambda \neq 0$ by Exercise 20.)

Supplementary Computational Exercises

1. Let V be the set of all (2×2) matrices with real entries and with the usual operation of addition. Suppose, however, that scalar multiplication in V is defined by
$$k\begin{bmatrix} a & b \\ c & d \end{bmatrix} = \begin{bmatrix} ka & 0 \\ 0 & kd \end{bmatrix}.$$
Determine whether V is a real vector space.

2. Recall that $\mathscr{F}(R)$ denotes the set of all functions from R to R; that is, $\mathscr{F}(R) = \{f: R \rightarrow R\}$. A function g in $\mathscr{F}(R)$ is called an even function if $g(-x) = g(x)$ for every x in R. Prove that the set of all even functions in $\mathscr{F}(R)$ is a subspace of $\mathscr{F}(R)$.

3. In each of parts (a)–(c), show that the set S is linearly dependent, and write one of the vectors in S as a linear combination of the remaining vectors.

a) $S = \{A_1, A_2, A_3, A_4\}$, where

$$A_1 = \begin{bmatrix} 1 & 0 \\ -1 & 1 \end{bmatrix}, \quad A_2 = \begin{bmatrix} -1 & 1 \\ 0 & 1 \end{bmatrix},$$

$$A_3 = \begin{bmatrix} -1 & 3 \\ -2 & 5 \end{bmatrix}, \quad \text{and} \quad A_4 = \begin{bmatrix} -3 & 2 \\ 2 & 0 \end{bmatrix}.$$

b) $S = \{p_1(x), p_2(x), p_3(x), p_4(x)\}$, where $p_1(x) = 1 - x^2 + x^3$, $p_2(x) = -1 + x + x^3$, $p_3(x) = -1 + 3x - 2x^2 + 5x^3$, and $p_4(x) = -3 + 2x + 2x^2$

c) $S = \{v_1, v_2, v_3, v_4\}$, where

$$v_1 = \begin{bmatrix} 1 \\ 0 \\ -1 \\ 1 \end{bmatrix}, \quad v_2 = \begin{bmatrix} -1 \\ 1 \\ 0 \\ 1 \end{bmatrix},$$

$$v_3 = \begin{bmatrix} -1 \\ 3 \\ -2 \\ 5 \end{bmatrix}, \quad \text{and} \quad v_4 = \begin{bmatrix} -3 \\ 2 \\ 2 \\ 0 \end{bmatrix}.$$

4. Let W be the subspace of the set of (2×2) real matrices defined by

$$W = \left\{ A = \begin{bmatrix} a & b \\ c & d \end{bmatrix} : a - 2b + 3c + d = 0 \right\}.$$

a) Exhibit a basis B for W.

b) Find a matrix A in W such that $[A]_B = [2, 1, -2]^T$.

5. In $\mathscr{P}_2$, let $S = \{p_1(x), p_2(x), p_3(x)\}$, where $p_1(x) = 1 - x + 2x^2$, $p_2(x) = 2 + 3x + x^2$, and $p_3(x) = 1 - 6x + 5x^2$.

a) Obtain an algebraic specification for Sp(S).

b) Determine which of the following polynomials are in Sp(S):

$$q_1(x) = 5 + 5x + 4x^2, \qquad q_2(x) = 5 - 5x + 8x^2,$$

$$q_3(x) = -5x + 3x^2, \quad \text{and} \quad q_4(x) = 5 + 7x^2.$$

c) Use the algebraic specification obtained in part (a) to determine a basis, B, of Sp(S).

d) For each polynomial $q_i(x)$, $i = 1, 2, 3, 4$, given in part (b), if $q_i(x)$ is in Sp(S), then find $[q_i(x)]_B$.

6. In parts (a)–(c), find a subset of S that is a basis for Sp(S). Express each element of S that does not appear in the basis as a linear combination of the basis vectors.

a) $S = \{A_1, A_2, A_3, A_4, A_5\}$, where

$$A_1 = \begin{bmatrix} 1 & -2 \\ 1 & -1 \end{bmatrix}, \quad A_2 = \begin{bmatrix} 2 & -3 \\ 4 & -3 \end{bmatrix},$$

$$A_3 = \begin{bmatrix} -1 & 1 \\ -3 & 2 \end{bmatrix}, \quad A_4 = \begin{bmatrix} 1 & -1 \\ 4 & 0 \end{bmatrix}, \quad \text{and}$$

$$A_5 = \begin{bmatrix} 12 & -17 \\ 30 & -11 \end{bmatrix}$$

b) $S = \{p_1(x), p_2(x), p_3(x), p_4(x), p_5(x)\}$, where

$$p_1(x) = 1 - 2x + x^2 - x^3,$$

$$p_2(x) = 2 - 3x + 4x^2 - 3x^3,$$

$$p_3(x) = -1 + x - 3x^2 + 2x^3,$$

$$p_4(x) = 1 - x + 4x^2, \quad \text{and}$$

$$p_5(x) = 12 - 17x + 30x^2 - 11x^3$$

c) $S = \{f_1(x), f_2(x), f_3(x), f_4(x), f_5(x)\}$, where

$$f_1(x) = e^x - 2e^{2x} + e^{3x} - e^{4x},$$

$$f_2(x) = 2e^x - 3e^{2x} + 4e^{3x} - 3e^{4x},$$

$$f_3(x) = -e^x + e^{2x} - 3e^{3x} + 2e^{4x},$$

$$f_4(x) = e^x - e^{2x} + 4e^{3x}, \quad \text{and}$$

$$f_5(x) = 12e^x - 17e^{2x} + 30e^{3x} - 11e^{4x}$$

In Exercises 7–11, use the fact that the matrix

$$[A \,|\, b] = \begin{bmatrix} 1 & -1 & 3 & 1 & 3 & 2 & | & a \\ 1 & 0 & 2 & 3 & 2 & 3 & | & b \\ 0 & -2 & 2 & -4 & 3 & 0 & | & c \\ 2 & -1 & 5 & 4 & 6 & 7 & | & d \end{bmatrix}$$

is row equivalent to

$$\begin{bmatrix} 1 & 0 & 2 & 3 & 0 & -1 & | & 4a - 3b - 2c \\ 0 & 1 & -1 & 2 & 0 & 3 & | & -3a + 3b + c \\ 0 & 0 & 0 & 0 & 1 & 2 & | & -2a + 2b + c \\ 0 & 0 & 0 & 0 & 0 & 0 & | & a - 3b - c + d \end{bmatrix}.$$

7. Find a basis for Sp$\{A_1, A_2, A_3, A_4\}$, where

$$A_1 = \begin{bmatrix} 1 & -1 & 3 \\ 1 & 3 & 2 \end{bmatrix}, \quad A_2 = \begin{bmatrix} 1 & 0 & 2 \\ 3 & 2 & 3 \end{bmatrix},$$

$$A_3 = \begin{bmatrix} 0 & -2 & 2 \\ -4 & 3 & 0 \end{bmatrix}, \quad \text{and} \quad A_4 = \begin{bmatrix} 2 & -1 & 5 \\ 4 & 6 & 7 \end{bmatrix}.$$

8. Let $S = \{p_1(x), p_2(x), p_3(x), p_4(x), p_5(x), p_6(x)\}$, where

$p_1(x) = 1 + x + 2x^3$,

$p_2(x) = -1 - 2x^2 - x^3$,

$p_3(x) = 3 + 2x + 2x^2 + 5x^3$,

$p_4(x) = 1 + 3x - 4x^2 + 4x^3$,

$p_5(x) = 3 + 2x + 3x^2 + 6x^3$,   and

$p_6(x) = 2 + 3x + 7x^3$.

Find a subset of S that is a basis for $\text{Sp}(S)$.

9. Let S be the set of polynomials given in Exercise 8. Show that $q(x) = 1 + 2x - x^2 + 4x^3$ is in $\text{Sp}(S)$, and express $q(x)$ as a linear combination of the basis vectors found in Exercise 8.

10. If

$$S = \left\{ \begin{bmatrix} 1 & 1 \\ 0 & 2 \end{bmatrix}, \begin{bmatrix} -1 & 0 \\ -2 & -1 \end{bmatrix}, \begin{bmatrix} 3 & 2 \\ 2 & 5 \end{bmatrix}, \right.$$

$$\left. \begin{bmatrix} 1 & 3 \\ -4 & 4 \end{bmatrix}, \begin{bmatrix} 3 & 2 \\ 3 & 6 \end{bmatrix}, \begin{bmatrix} 2 & 3 \\ 0 & 7 \end{bmatrix} \right\},$$

then give an algebraic specification for $\text{Sp}(S)$ and use the specification to determine a basis for $\text{Sp}(S)$.

11. Let V be the vector space for all (2×3) matrices, and suppose that $T: V \rightarrow \mathscr{P}_3$ is the linear transformation defined by

$$T\left(\begin{bmatrix} a_{11} & a_{12} & a_{13} \\ a_{21} & a_{22} & a_{23} \end{bmatrix} \right)$$

$$= (a_{11} - a_{12} + 3a_{13} + a_{21} + 3a_{22} + 2a_{23})$$
$$+ (a_{11} + 2a_{13} + 3a_{21} + 2a_{22} + 3a_{23})x$$
$$+ (-2a_{12} + 2a_{13} - 4a_{21} + 3a_{22})x^2$$
$$+ (2a_{11} - a_{12} + 5a_{13} + 4a_{21} + 6a_{22} + 7a_{23})x^3.$$

a) Calculate the matrix of T relative to the natural bases B and C for V and $\mathscr{P}_3$, respectively.
b) Determine the rank and the nullity of T.

c) Give an algebraic specification for $\mathscr{R}(T)$ and use the specification to determine a basis for $\mathscr{R}(T)$.
d) Show that $q(x) = 1 + 2x - x^2 + 4x^3$ is in $\mathscr{R}(T)$ and find a matrix A in V such that $T(A) = q(x)$.
e) Find a basis for $\mathscr{N}(T)$.

12. Show that there is a linear transformation $T: R^2 \rightarrow \mathscr{P}_2$ such that

$$T\left(\begin{bmatrix} 0 \\ 1 \end{bmatrix} \right) = 1 + 2x + x^2 \quad \text{and} \quad T\left(\begin{bmatrix} -1 \\ 1 \end{bmatrix} \right) = 2 - x.$$

Give a formula for

$$T\left(\begin{bmatrix} a \\ b \end{bmatrix} \right).$$

13. Show that there are infinitely many linear transformations $T: \mathscr{P}_2 \rightarrow R^2$ such that

$$T(x) = \begin{bmatrix} 1 \\ 0 \end{bmatrix} \quad \text{and} \quad T(x^2) = \begin{bmatrix} 0 \\ 1 \end{bmatrix}.$$

Give a formula for $T(a + bx + cx^2)$ for one such linear transformation.

14. Let V be the vector space of (2×2) matrices, and let $T: V \rightarrow \mathscr{P}_2$ be the linear transformation defined by

$$T\left(\begin{bmatrix} a & b \\ c & d \end{bmatrix} \right) = (a - b + c - 4d) + (b + c + 3d)x$$

$$+ (a + 2c - d)x^2.$$

a) Find the matrix of T relative to the natural bases, B and C, for V and $\mathscr{P}_2$, respectively.
b) Give an algebraic specification for $\mathscr{R}(T)$ and use the specification to obtain a basis S for $\mathscr{R}(T)$.
c) For each polynomial $q(x)$ in S, find a matrix A in V such that $T(A) = q(x)$. Let B_1 denote the set of matrices found.
d) Find a basis, B_2, for $\mathscr{N}(T)$.
e) Show that $B_1 \cup B_2$ is a basis for V.
(*Note:* This exercise illustrates the proof that $\text{rank}(T) + \text{nullity}(T) = \dim(V)$.)

Supplementary Conceptual Exercises

In Exercises 1–10, answer true or false. Justify your answer by providing a counterexample if the statement is false or an outline of a proof if the statement is true.

1. If a is a nonzero scalar and $\mathbf{u}$ and $\mathbf{v}$ are vectors in a vector space V such that $a\mathbf{u} = a\mathbf{v}$, then $\mathbf{u} = \mathbf{v}$.

2. If $\mathbf{v}$ is a nonzero vector in a vector space V and a and b are scalars such that $a\mathbf{v} = b\mathbf{v}$, then $a = b$.

3. Every vector space V contains a unique vector called the additive inverse of V.

4. If V consists of all real polynomials of degree

exactly n together with the zero polynomial, then V is a vector space.

5. If W is a subspace of the vector space V and $\dim(W) = \dim(V) = n$, then $W = V$.

6. If $\dim(V) = n$ and W is a subspace of V, then $\dim(W) \leq n$.

7. The subset $\{\theta\}$ of a vector space is linearly dependent.

8. Let S_1 and S_2 be subsets of a vector space V such that $S_1 \subseteq S_2$. If S_1 is linearly dependent, then so is S_2.

9. Let S_1 and S_2 be subsets of a vector space V such that $S_1 \subseteq S_2$. If S_1 is linearly independent, then so is S_2.

10. Suppose that $S_1 = \{\mathbf{v}_1,\ldots,\mathbf{v}_k\}$ and $S_2 = \{\mathbf{w}_1,\ldots,\mathbf{w}_l\}$ are subsets of a vector space V. If $V = \text{Sp}(S_1)$ and S_2 is linearly independent, then $l \leq k$.

In Exercises 11–19, give a brief answer.

11. Let W be a subspace of the vector space V. If $\mathbf{u}$ and $\mathbf{v}$ are elements of V such that $\mathbf{u} + \mathbf{v}$ and $\mathbf{u} - \mathbf{v}$ are in W, show that $\mathbf{u}$ and $\mathbf{v}$ are in W.

12. Let W be a subset of a vector space V that satisfies the following properties:
i) θ is in W.
ii) If $\mathbf{x}$ and $\mathbf{y}$ are in W and a is a scalar, then $a\mathbf{x} + \mathbf{y}$ is in W. Prove that W is a subspace of V.

13. If W is a subspace of a vector space V, show that $\text{Sp}(W) = W$.

14. Give examples of subsets of S_1 and S_2 of a vector space V such that $\text{Sp}(S_1) \cap \text{Sp}(S_2) \neq \text{Sp}(S_1 \cap S_2)$.

15. If U and W are subspaces of a vector space V, then $U + W = \{\mathbf{u} + \mathbf{w}: \mathbf{u} \text{ is in } U \text{ and } \mathbf{w} \text{ is in } W\}$.
a) Prove that $U + W$ is a subspace of V.
b) Let $S_1 = \{\mathbf{x}_1,\ldots,\mathbf{x}_m\}$ and $S_2 = \{\mathbf{y}_1,\ldots,\mathbf{y}_n\}$ be subsets of V. Show that $\text{Sp}(S_1 \cup S_2) = \text{Sp}(S_1) + \text{Sp}(S_2)$.

16. Let $B = \{\mathbf{v}_1,\ldots,\mathbf{v}_n\}$ be a basis for a vector space V, and let $\mathbf{v}$ be a nonzero vector in V. Prove that there exists a vector $\mathbf{v}_j$ in B, $1 \leq j \leq n$, such that $\mathbf{v}_j$ can be replaced by $\mathbf{v}$ and the resulting set, B', is still a basis for V.

17. Let $B = \{\mathbf{v}_1,\ldots,\mathbf{v}_n\}$ be a basis for a vector space V, and let $S: V \to W$ and $T: V \to W$ be linear transformations such that $S(\mathbf{v}_i) = T(\mathbf{v}_i)$ for $i = 1, 2,\ldots,n$. Show that $S = T$.

18. Let $T: V \to W$ be a linear transformation.
a) If T is one to one, then show that T carries linearly independent subsets of V to linearly independent subsets of W.
b) If T carries linearly independent subsets of V to linearly independent subsets of W, then prove that T is one to one.

19. Give an example of a linear transformation $T: R^2 \to R^2$ such that $\mathcal{N}(T) = \mathcal{R}(T)$.

5

Determinants

Introduction

Determinants have played a major role in the historical development of matrix theory, and they possess a number of properties that are theoretically pleasing. For example, in terms of linear algebra, determinants can be used to characterize nonsingular matrices, to express solutions of nonsingular systems $A\mathbf{x} = \mathbf{b}$, and to calculate the dimension of subspaces. In analysis, determinants are used to express vector cross products, to express the conversion factor (the Jacobian) when a change of variables is needed to evaluate a multiple integral, to serve as a convenient test (the Wronskian) for linear independence of sets of functions, and so on. We explore the theory and some of the applications of determinants in this chapter.

The material in Sections 5.2 and 5.3 duplicates the material in Sections 3.2 and 3.3 in order to present a contiguous coverage of determinants. The treatment is slightly different because the material in Chapter 5 is self-contained, whereas Chapter 3 uses a result (Theorem 5.13) that is stated in Chapter 3 but actually proved in Chapter 5. Hence the reader who has seen the results of Sections 3.2 and 3.3 may wish to proceed directly to Section 5.4.

5.2 Cofactor Expansions of Determinants

If A is an $(n \times n)$ matrix, the determinant of A, denoted $\det(A)$, is a number that we will associate with A. Determinants are usually defined either in terms of "cofactors" or in terms of "permutations," and we elect to use the cofactor definition here. We begin with the definition of det(A) when A is a (2×2) matrix.

DEFINITION 1 Let $A = (a_{ij})$ be a (2×2) matrix. The **determinant** of A is given by

$$\det(A) = a_{11}a_{22} - a_{12}a_{21}.$$

For notational purposes the determinant is often expressed by using vertical bars:

$$\det(A) = \begin{vmatrix} a_{11} & a_{12} \\ a_{21} & a_{22} \end{vmatrix}.$$

EXAMPLE 1 Find the determinants of the following matrices:

$$A = \begin{bmatrix} 1 & 2 \\ -1 & 3 \end{bmatrix}, \quad B = \begin{bmatrix} 4 & 1 \\ 2 & 1 \end{bmatrix}, \quad \text{and} \quad C = \begin{bmatrix} 3 & 4 \\ 6 & 8 \end{bmatrix}.$$

SOLUTION

$$\det(A) = \begin{vmatrix} 1 & 2 \\ -1 & 3 \end{vmatrix} = 1 \cdot 3 - 2(-1) = 5;$$

$$\det(B) = \begin{vmatrix} 4 & 1 \\ 2 & 1 \end{vmatrix} = 4 \cdot 1 - 1 \cdot 2 = 2;$$

$$\det(C) = \begin{vmatrix} 3 & 4 \\ 6 & 8 \end{vmatrix} = 3 \cdot 8 - 4 \cdot 6 = 0$$

We now define the determinant of an $(n \times n)$ matrix as a "weighted sum" of $[(n-1) \times (n-1)]$ determinants. It is convenient to make a preliminary definition.

DEFINITION 2 Let $A = (a_{ij})$ be an $(n \times n)$ matrix, and let M_{rs} denote the $[(n-1) \times (n-1)]$ matrix obtained by deleting the rth row and sth column from A. Then M_{rs} is called a **minor matrix** of A, and the number det(M_{rs}) is the **minor of the (r, s)th entry**, a_{rs}. In addition, the numbers

$$A_{ij} = (-1)^{i+j} \det(M_{ij})$$

are called **cofactors** (or **signed minors**).

EXAMPLE 2 Determine the minor matrices M_{11}, M_{23}, and M_{32} for the matrix A given by

$$A = \begin{bmatrix} 1 & -1 & 2 \\ 2 & 3 & -3 \\ 4 & 5 & 1 \end{bmatrix}.$$

Also, calculate the cofactors A_{11}, A_{23}, and A_{32}.

SOLUTION   Deleting row 1 and column 1 from A, we obtain M_{11}:

$$M_{11} = \begin{bmatrix} 3 & -3 \\ 5 & 1 \end{bmatrix}.$$

Similarly, the minor matrices M_{23} and M_{32} are

$$M_{23} = \begin{bmatrix} 1 & -1 \\ 4 & 5 \end{bmatrix} \quad \text{and} \quad M_{32} = \begin{bmatrix} 1 & 2 \\ 2 & -3 \end{bmatrix}.$$

The associated cofactors, $A_{ij} = (-1)^{i+j} \det(M_{ij})$, are given by

$$A_{11} = (-1)^{1+1} \begin{vmatrix} 3 & -3 \\ 5 & 1 \end{vmatrix} = 3 + 15 = 18;$$

$$A_{23} = (-1)^{2+3} \begin{vmatrix} 1 & -1 \\ 4 & 5 \end{vmatrix} = -(5 + 4) = -9;$$

$$A_{32} = (-1)^{3+2} \begin{vmatrix} 1 & 2 \\ 2 & -3 \end{vmatrix} = -(-3 - 4) = 7.$$

We use cofactors in our definition of the determinant.

DEFINITION 3   Let $A = (a_{ij})$ be an $(n \times n)$ matrix. Then the ***determinant*** of A is

$$\det(A) = a_{11}A_{11} + a_{12}A_{12} + \cdots + a_{1n}A_{1n},$$

where A_{1j} is the cofactor of a_{1j}, $1 \le j \le n$.

Determinants are defined only for square matrices. Note also the inductive nature of the definition. For example, if A is (3×3), then $\det(A) = a_{11}A_{11} + a_{12}A_{12} + a_{13}A_{13}$, and the cofactors A_{11}, A_{12}, and A_{13} can be evaluated from Definition 1. Similarly, the determinant of a (4×4) matrix is the sum of four (3×3) determinants, where each (3×3) determinant is in turn the sum of three (2×2) determinants.

EXAMPLE 3 Compute $\det(A)$, where

$$A = \begin{bmatrix} 3 & 2 & 1 \\ 2 & 1 & -3 \\ 4 & 0 & 1 \end{bmatrix}.$$

SOLUTION   The matrix A is (3×3). Using $n = 3$ in Definition 3, we have

$$\det(A) = a_{11}A_{11} + a_{12}A_{12} + a_{13}A_{13}$$

$$= 3 \begin{vmatrix} 1 & -3 \\ 0 & 1 \end{vmatrix} - 2 \begin{vmatrix} 2 & -3 \\ 4 & 1 \end{vmatrix} + 1 \begin{vmatrix} 2 & 1 \\ 4 & 0 \end{vmatrix}$$

$$= 3(1) - 2(14) + 1(-4) = -29.$$

*D*eterminants by Permutations

The determinant of an $(n \times n)$ matrix A can be defined in terms of permutations rather than cofactors. Specifically, let $S = \{1, 2,..., n\}$ denote the set consisting of the first n positive integers. A **permutation** $(j_1, j_2,..., j_n)$ of the set $S = \{1, 2,..., n\}$ is just a rearrangement of the numbers in S. An **inversion** of this permutation occurs whenever a number j_r is followed by a smaller number j_s. For example, the permutation $(1, 3, 2)$ has one inversion, but $(2, 3, 1)$ has two inversions. A permutation of S is called **odd (even)** if it has an odd (even) number of inversions.

It can be shown that $\det(A)$ is the sum of all possible terms of the form $\pm a_{1j_1} a_{2j_2}...a_{nj_n}$, where $(j_1, j_2,..., j_n)$ is a permutation of S and the sign is taken as + or −, depending on whether the permutation is even or odd. For instance,

$$\begin{vmatrix} a_{11} & a_{12} \\ a_{21} & a_{22} \end{vmatrix} = +a_{11}a_{22} - a_{12}a_{21};$$

$$\begin{vmatrix} a_{11} & a_{12} & a_{13} \\ a_{21} & a_{22} & a_{23} \\ a_{31} & a_{32} & a_{33} \end{vmatrix} = +a_{11}a_{22}a_{33} - a_{11}a_{23}a_{32} - a_{12}a_{21}a_{33} + a_{12}a_{23}a_{31} + a_{13}a_{21}a_{32} - a_{13}a_{22}a_{31}.$$

Since there are $n!$ different permutations when $S = \{1, 2,..., n\}$, you can see why this definition is not suitable for calculation. For example, calculating the determinant of a (10×10) matrix requires us to evaluate $10! = 3{,}628{,}800$ different terms of the form $\pm a_{1j_1} a_{2j_2}...a_{10j_{10}}$. The permutation definition is useful for theoretical purposes, however. For instance, the permutation definition gives immediately that $\det(A) = 0$ when A has a row of zeros.

EXAMPLE 4   Compute $\det(A)$, where

$$A = \begin{bmatrix} 1 & 2 & 0 & 2 \\ -1 & 2 & 3 & 1 \\ -3 & 2 & -1 & 0 \\ 2 & -3 & -2 & 1 \end{bmatrix}.$$

SOLUTION   The matrix A is (4×4). Using $n = 4$ in Definition 3, we have

$$\det(A) = a_{11}A_{11} + a_{12}A_{12} + a_{13}A_{13} + a_{14}A_{14} = A_{11} + 2A_{12} + 2A_{14}.$$

The required cofactors, A_{11}, A_{12}, and A_{14}, are calculated as in Example 3 (note that the cofactor A_{13} is not needed, since $a_{13} = 0$).
In detail,

$$A_{11} = \begin{vmatrix} 2 & 3 & 1 \\ 2 & -1 & 0 \\ -3 & -2 & 1 \end{vmatrix}$$

$$= 2\begin{vmatrix} -1 & 0 \\ -2 & 1 \end{vmatrix} - 3\begin{vmatrix} 2 & 0 \\ -3 & 1 \end{vmatrix} + 1\begin{vmatrix} 2 & -1 \\ -3 & -2 \end{vmatrix} = -15;$$

$$A_{12} = - \begin{vmatrix} -1 & 3 & 1 \\ -3 & -1 & 0 \\ 2 & -2 & 1 \end{vmatrix}$$

$$= - \left(-1 \begin{vmatrix} -1 & 0 \\ -2 & 1 \end{vmatrix} - 3 \begin{vmatrix} -3 & 0 \\ 2 & 1 \end{vmatrix} + 1 \begin{vmatrix} -3 & -1 \\ 2 & -2 \end{vmatrix} \right) = -18;$$

$$A_{14} = - \begin{vmatrix} -1 & 2 & 3 \\ -3 & 2 & -1 \\ 2 & -3 & -2 \end{vmatrix}$$

$$= - \left(-1 \begin{vmatrix} 2 & -1 \\ -3 & -2 \end{vmatrix} - 2 \begin{vmatrix} -3 & -1 \\ 2 & -2 \end{vmatrix} + 3 \begin{vmatrix} -3 & 2 \\ 2 & -3 \end{vmatrix} \right) = -6.$$

Thus it follows that

$$\det(A) = A_{11} + 2A_{12} + 2A_{14} = -15 - 36 - 12 = -63. \qquad \square$$

The definition of $\det(A)$ given in Definition 3 and used in Examples 3 and 4 is based on a cofactor expansion along the first row of A. In Section 5.5 (see Theorem 13), we prove that the value $\det(A)$ can be calculated from a cofactor expansion along any row or any column.

Also, note in Example 4 that the calculation of the (4×4) determinant was simplified because of the zero entry in the $(1,3)$ position. Clearly, if we had some procedure for creating zero entries, we could simplify the computation of determinants since the cofactor of a zero entry need not be calculated. We will develop such simplifications in the next section.

EXAMPLE 5 Compute the determinant of the lower-triangular matrix T, where

$$T = \begin{bmatrix} 3 & 0 & 0 & 0 \\ 1 & 2 & 0 & 0 \\ 2 & 3 & 2 & 0 \\ 1 & 4 & 5 & 1 \end{bmatrix}.$$

SOLUTION We have $\det(T) = t_{11}T_{11} + t_{12}T_{12} + t_{13}T_{13} + t_{14}T_{14}$. Since $t_{12} = 0$, $t_{13} = 0$, and $t_{14} = 0$, the calculation simplifies to

$$\det(T) = t_{11}T_{11} = 3 \begin{vmatrix} 2 & 0 & 0 \\ 3 & 2 & 0 \\ 4 & 5 & 1 \end{vmatrix}$$

$$= 3 \cdot 2 \begin{vmatrix} 2 & 0 \\ 5 & 1 \end{vmatrix}$$

$$= 3 \cdot 2 \cdot 2 \cdot 1 = 12. \qquad \square$$

In Example 5, we saw that the determinant of the lower-triangular matrix T was the product of the diagonal entries, $\det(T) = t_{11}t_{22}t_{33}t_{44}$. This simple relationship is valid for any lower-triangular matrix.

THEOREM 1	

Let $T = (t_{ij})$ be an $(n \times n)$ lower-triangular matrix. Then

$$\det(T) = t_{11} \cdot t_{22} \cdot \cdots \cdot t_{nn}.$$

PROOF   If T is a (2×2) lower-triangular matrix, then

$$\det(T) = \begin{vmatrix} t_{11} & 0 \\ t_{21} & t_{22} \end{vmatrix} = t_{11}t_{22}.$$

Proceeding inductively, suppose that the theorem is true for any $(k \times k)$ lower-triangular matrix, where $2 \leq k \leq n - 1$. If T is an $(n \times n)$ lower-triangular matrix, then

$$\det(T) = \begin{vmatrix} t_{11} & 0 & 0 & \cdots & 0 \\ t_{21} & t_{22} & 0 & \cdots & 0 \\ \vdots & & & & \vdots \\ t_{n1} & t_{n2} & t_{n3} & \cdots & t_{nn} \end{vmatrix} = t_{11}, T_{11}, \quad \text{where } T_{11} = \begin{vmatrix} t_{22} & 0 & \cdots & 0 \\ t_{32} & t_{33} & \cdots & 0 \\ \vdots & & & \vdots \\ t_{n2} & t_{n3} & \cdots & t_{nn} \end{vmatrix}.$$

Clearly, T_{11} is the determinant of an $[(n - 1) \times (n - 1)]$ lower-triangular matrix, so $T_{11} = t_{22}t_{33} \cdots t_{nn}$. Thus $\det(T) = t_{11}t_{22} \cdots t_{nn}$, and the theorem is proved.

EXAMPLE 6   Let I denote the $(n \times n)$ identity matrix. Calculate $\det(I)$.

SOLUTION   Since I is a lower-triangular matrix with diagonal entries equal to 1, we see from Theorem 1 that

$$\det(I) = \underbrace{1 \cdot 1 \cdot \cdots \cdot 1}_{n \text{ factors}} = 1.$$

Exercises 5.2

In Exercises 1–8, evaluate the determinant of the given matrix. If the determinant is zero, find a nonzero vector **x** such that $A\mathbf{x} = \boldsymbol{\theta}$. (We will see later that $\det(A) = 0$ if and only if A is singular.)

1. $\begin{bmatrix} 1 & 3 \\ 2 & 1 \end{bmatrix}$

2. $\begin{bmatrix} 6 & 7 \\ 7 & 3 \end{bmatrix}$

3. $\begin{bmatrix} 2 & 4 \\ 4 & 8 \end{bmatrix}$

4. $\begin{bmatrix} 1 & 3 \\ 0 & 2 \end{bmatrix}$

5. $\begin{bmatrix} 4 & 3 \\ 1 & 7 \end{bmatrix}$

6. $\begin{bmatrix} 2 & -1 \\ 1 & 1 \end{bmatrix}$

7. $\begin{bmatrix} 4 & 1 \\ -2 & 1 \end{bmatrix}$

8. $\begin{bmatrix} 1 & 3 \\ 2 & 6 \end{bmatrix}$

In Exercises 9–14, calculate the cofactors A_{11}, A_{12}, A_{13}, and A_{33} for the given matrix A.

9. $A = \begin{bmatrix} 1 & 2 & 1 \\ 0 & 1 & 3 \\ 2 & 1 & 1 \end{bmatrix}$

10. $A = \begin{bmatrix} 1 & 4 & 0 \\ 1 & 0 & 2 \\ 3 & 1 & 2 \end{bmatrix}$

11. $A = \begin{bmatrix} 2 & -1 & 3 \\ -1 & 2 & 2 \\ 3 & 2 & 1 \end{bmatrix}$

12. $A = \begin{bmatrix} 1 & 1 & 1 \\ 1 & 1 & 2 \\ 2 & 1 & 1 \end{bmatrix}$

13. $A = \begin{bmatrix} -1 & 1 & -1 \\ 2 & 1 & 0 \\ 0 & 1 & 3 \end{bmatrix}$ **14.** $A = \begin{bmatrix} 4 & 2 & 1 \\ 4 & 3 & 1 \\ 0 & 0 & 2 \end{bmatrix}$

In Exercises 15–20, use the results of Exercises 9–14 to find $\det(A)$, where:

15. A is in Exercise 9.

16. A is in Exercise 10.

17. A is in Exercise 11.

18. A is in Exercise 12.

19. A is in Exercise 13.

20. A is in Exercise 14.

In Exercises 21–24, calculate $\det(A)$.

21. $A = \begin{bmatrix} 2 & 1 & -1 & 2 \\ 3 & 0 & 0 & 1 \\ 2 & 1 & 2 & 0 \\ 3 & 1 & 1 & 2 \end{bmatrix}$

22. $A = \begin{bmatrix} 1 & -1 & 1 & 2 \\ 1 & 0 & 1 & 3 \\ 0 & 0 & 2 & 4 \\ 1 & 1 & -1 & 1 \end{bmatrix}$

23. $A = \begin{bmatrix} 2 & 0 & 2 & 0 \\ 1 & 3 & 1 & 2 \\ 0 & 1 & 2 & 1 \\ 0 & 3 & 1 & 4 \end{bmatrix}$

24. $A = \begin{bmatrix} 1 & 2 & 1 & 1 \\ 0 & 2 & 0 & 3 \\ 1 & 4 & 1 & 2 \\ 0 & 2 & 1 & 3 \end{bmatrix}$

In Exercises 25 and 26, show that the quantities $\det(A)$, $a_{21}A_{21} + a_{22}A_{22} + a_{23}A_{23}$, and $a_{31}A_{31} + a_{32}A_{32} + a_{33}A_{33}$ are all equal. (This is a special case of a general result given later in Theorem 13.)

25. $A = \begin{bmatrix} 1 & 3 & 2 \\ -1 & 4 & 1 \\ 2 & 2 & 3 \end{bmatrix}$ **26.** $A = \begin{bmatrix} 2 & 4 & 1 \\ 3 & 1 & 3 \\ 2 & 3 & 2 \end{bmatrix}$

In Exercises 27 and 28, show that $a_{11}A_{21} + a_{12}A_{22} + a_{13}A_{23} = 0$ and $a_{11}A_{31} + a_{12}A_{32} + a_{13}A_{33} = 0$. (This is a special case of a general result given later in the lemma to Theorem 14.)

27. A as in Exercise 25   **28.** A as in Exercise 26

In Exercises 29 and 30, form the (3×3) matrix of cofactors C where $c_{ij} = A_{ij}$ and then calculate BA where $B = C^T$. How can you use this result to find A^{-1}?

29. A as in Exercise 25   **30.** A as in Exercise 26

31. Verify that $\det(A) = 0$ when
$$A = \begin{bmatrix} 0 & a_{12} & a_{13} \\ 0 & a_{22} & a_{23} \\ 0 & a_{32} & a_{33} \end{bmatrix}.$$

32. Use the result of Exercise 31 to prove that if $U = (u_{ij})$ is a (4×4) upper-triangular matrix, then $\det(U) = u_{11}u_{22}u_{33}u_{44}$.

33. Let $A = (a_{ij})$ be a (2×2) matrix. Show that $\det(A^T) = \det(A)$.

34. An $(n \times n)$ symmetric matrix A is called positive definite if $\mathbf{x}^T A \mathbf{x} > 0$ for all $\mathbf{x}$ in R^n, $\mathbf{x} \neq \boldsymbol{\theta}$. Let A be a (2×2) symmetric matrix. Prove the following:
a) If A is positive definite, then $a_{11} > 0$ and $\det(A) > 0$.
b) If $a_{11} > 0$ and $\det(A) > 0$, then A is positive definite. (*Hint:* For part (a), consider $\mathbf{x} = \mathbf{e}_1$. Then consider $\mathbf{x} = [u, v]^T$ and use the fact that A is symmetric.)

35. a) Let A be an $(n \times n)$ matrix. If $n = 3$, $\det(A)$ can be found by evaluating three (2×2) determinants. If $n = 4$, $\det(A)$ can be found by evaluating twelve (2×2) determinants. Give a formula, $H(n)$, for the number of (2×2) determinants necessary to find $\det(A)$ for an arbitrary n.
b) Suppose you can perform additions, subtractions, multiplications, and divisions each at a rate of one per second. How long does it take to evaluate $H(n)$ determinants of order (2×2) when $n = 2$, $n = 5$, and $n = 10$?

5.3 Elementary Operations and Determinants

In this section we show how certain column operations simplify the calculation of determinants. In addition, the properties we develop will be used later to demonstrate some of the connections between determinant theory and linear algebra. We will use three elementary column operations, which are analogous to the elementary row operations defined in Chapter 1. For a matrix A, the elementary column operations are as follows:

1. Interchange two columns of A.
2. Multiply a column of A by a scalar c, $c \neq 0$.
3. Add a scalar multiple of one column of A to another column of A.

From Chapter 1, we know that row operations can be used to reduce a square matrix A to an upper-triangular matrix (that is, we know A can be reduced to echelon form, and a square matrix in echelon form is upper triangular). Similarly, it is easy to show that column operations can be used to reduce a square matrix to lower-triangular form. One reason for reducing an $(n \times n)$ matrix A to a lower-triangular matrix T is that $\det(T)$ is trivial to evaluate (see Theorem 1). Thus if we can calculate the effect that column operations have on the determinant, we can relate $\det(A)$ to $\det(T)$.

Before proceeding, we wish to make the following statement about elementary row and column operations. We will prove a succession of results dealing only with column operations. These results lead to a proof in Section 5.5 of the following theorem (see Theorem 12):

THEOREM

If A is an $(n \times n)$ matrix, then

$$\det(A^T) = \det(A). \tag{1}$$

Once Eq. (1) is formally established, we will immediately know that the theorems for column operations are also valid for row operations. (Row operations on A are precisely mirrored by column operations on A^T.) Therefore the following theorems are stated in terms of elementary row operations, as well as elementary column operations, although the row results will not be truly established until Theorem 12 is proved.

Elementary Operations

Our purpose is to describe how the determinant of a matrix A changes when an elementary column operation is applied to A. The description will take the form of a series of theorems. Because of the technical nature of the first three theorems, we defer their proofs to the end of the section.

Our first result relating to elementary operations is given in the theorem below. This theorem asserts that a column interchange (or a row interchange) will change the sign of the determinant.

THEOREM 2

Let $A = [\mathbf{A}_1, \mathbf{A}_2, \ldots, \mathbf{A}_n]$ be an $(n \times n)$ matrix. If B is obtained from A by interchanging two columns (or rows) of A, then $\det(B) = -\det(A)$.

The proof of Theorem 2 is at the end of this section.

__EXAMPLE 1__ Verify Theorem 2 for the (2×2) matrix

$$A = \begin{bmatrix} a_{11} & a_{12} \\ a_{21} & a_{22} \end{bmatrix}.$$

SOLUTION Let B denote the matrix obtained by interchanging the first and second columns of A. Thus B is given by

$$B = \begin{bmatrix} a_{12} & a_{11} \\ a_{22} & a_{21} \end{bmatrix}.$$

Now $\det(B) = a_{12}a_{21} - a_{11}a_{22}$, and $\det(A) = a_{11}a_{22} - a_{12}a_{21}$. Thus $\det(B) = -\det(A)$. ☐

__EXAMPLE 2__ Let A be the (3×3) matrix

$$A = \begin{bmatrix} 1 & 3 & 1 \\ 2 & 0 & 4 \\ 1 & 2 & 3 \end{bmatrix}.$$

The determinant of A is -10. Use the fact that $\det(A) = -10$ to find the determinants of B, C, and F, where

$$B = \begin{bmatrix} 3 & 1 & 1 \\ 0 & 2 & 4 \\ 2 & 1 & 3 \end{bmatrix}, \quad C = \begin{bmatrix} 1 & 1 & 3 \\ 2 & 4 & 0 \\ 1 & 3 & 2 \end{bmatrix}, \quad \text{and} \quad F = \begin{bmatrix} 1 & 1 & 3 \\ 4 & 2 & 0 \\ 3 & 1 & 2 \end{bmatrix}.$$

SOLUTION If A is given in column form as $A = [\mathbf{A}_1, \mathbf{A}_2, \mathbf{A}_3]$, then $B = [\mathbf{A}_2, \mathbf{A}_1, \mathbf{A}_3]$, $C = [\mathbf{A}_1, \mathbf{A}_3, \mathbf{A}_2]$, and $F = [\mathbf{A}_3, \mathbf{A}_1, \mathbf{A}_2]$. Since both B and C are obtained from A by a single column interchange, it follows from Theorem 2 that

$$\det(B) = \det(C) = -\det(A) = 10.$$

We can obtain F from A by two column interchanges as follows:

$$A \rightarrow G = [\mathbf{A}_2, \mathbf{A}_1, \mathbf{A}_3] \rightarrow F = [\mathbf{A}_3, \mathbf{A}_1, \mathbf{A}_2].$$

From Theorem 2, $\det(G) = -\det(A)$ and $\det(F) = -\det(G)$. Therefore $\det(F) = -\det(G) = -[-\det(A)] = \det(A) = -10$. ☐

By performing a sequence of column interchanges, we can produce any rearrangement of columns that we wish; and Theorem 2 can be used to find the determinant of the end result. For example, if $A = [\mathbf{A}_1, \mathbf{A}_2, \mathbf{A}_3, \mathbf{A}_4]$ is a (4×4) matrix and $B = [\mathbf{A}_4, \mathbf{A}_3, \mathbf{A}_1, \mathbf{A}_2]$ then we can relate $\det(B)$ to $\det(A)$ as follows: Form $B_1 = [\mathbf{A}_4, \mathbf{A}_2, \mathbf{A}_3, \mathbf{A}_1]$; then form $B_2 = [\mathbf{A}_4, \mathbf{A}_3, \mathbf{A}_2, \mathbf{A}_1]$; and then form B by interchanging the last two columns of B_2. In this sequence, $\det(B_1) = -\det(A)$ and $\det(B_2) = -\det(B_1)$, so $\det(B) = -\det(B_2) = \det(B_1) = -\det(A)$.

Our next theorem shows that multiplying all entries in a column of A by a scalar c has the effect of multiplying the determinant by c.

<table>
<tr><td>THEOREM 3</td><td>If A is an $(n \times n)$ matrix, and if B is the $(n \times n)$ matrix resulting from multiplying the kth column (or row) of A by a scalar c, then $\det(B) = c \det(A)$.</td></tr>
</table>

Again, the proof of Theorem 3 is rather technical, so we defer it to the end of this section. The next example, however, verifies Theorem 3 for a (2×2) matrix A.

EXAMPLE 3 Verify Theorem 3 for the (2×2) matrix

$$A = \begin{bmatrix} a_{11} & a_{12} \\ a_{21} & a_{22} \end{bmatrix}.$$

SOLUTION   Consider the matrices A' and A'' given by

$$A' = \begin{bmatrix} ca_{11} & a_{12} \\ ca_{21} & a_{22} \end{bmatrix} \quad \text{and} \quad A'' = \begin{bmatrix} a_{11} & ca_{12} \\ a_{21} & ca_{22} \end{bmatrix}.$$

Clearly, $\det(A') = ca_{11}a_{22} - ca_{21}a_{12} = c(a_{11}a_{22} - a_{21}a_{12}) = c\det(A)$. Similarly,

$$\det(A'') = ca_{11}a_{22} - ca_{21}a_{12} = c(a_{11}a_{22} - a_{21}a_{12}) = c\det(A).$$

The calculations above prove Theorem 3 for a (2×2) matrix A.

We emphasize that Theorem 3 is valid when $c = 0$. That is, if A has a column of zeros, then $\det(A) = 0$.

EXAMPLE 4 Let A be the (3×3) matrix

$$A = \begin{bmatrix} 1 & 3 & 1 \\ 2 & 0 & 4 \\ 1 & 2 & 3 \end{bmatrix}.$$

The determinant of A is -10. Use the fact that $\det(A) = -10$ to find the determinants of G, H, and J, where

$$G = \begin{bmatrix} 2 & 3 & 1 \\ 4 & 0 & 4 \\ 2 & 2 & 3 \end{bmatrix}, \quad H = \begin{bmatrix} 2 & -3 & 1 \\ 4 & 0 & 4 \\ 2 & -2 & 3 \end{bmatrix}, \quad \text{and} \quad J = \begin{bmatrix} 2 & -3 & 2 \\ 4 & 0 & 8 \\ 2 & -2 & 6 \end{bmatrix}.$$

SOLUTION   Let $A = [\mathbf{A}_1, \mathbf{A}_2, \mathbf{A}_3]$. Then

$$G = [2\mathbf{A}_1, \mathbf{A}_2, \mathbf{A}_3], H = [2\mathbf{A}_1, -\mathbf{A}_2, \mathbf{A}_3], \quad \text{and} \quad J = [2\mathbf{A}_1, -\mathbf{A}_2, 2\mathbf{A}_3].$$

By Theorem 3, $\det(G) = 2\det(A) = -20$.

Next, H is obtained from G by multiplying the second column of G by -1. Therefore, $\det(H) = -\det(G) = 20$. Finally, J is obtained from H by multiplying the third column of H by 2. Thus, $\det(J) = 2\det(H) = 40$.

The following result is a corollary of Theorem 3:

COROLLARY

Let A be an $(n \times n)$ matrix and let c be a scalar. Then

$$\det(cA) = c^n \det(A).$$

We leave the proof of the corollary as Exercise 32.

EXAMPLE 5 Find $\det(3A)$, where

$$A = \begin{bmatrix} 1 & 2 \\ 4 & 1 \end{bmatrix}.$$

SOLUTION Clearly, $\det(A) = -7$. Therefore, by the corollary, $\det(3A) = 3^2 \det(A) = -63$. As a check, note that the matrix $3A$ is given by

$$3A = \begin{bmatrix} 3 & 6 \\ 12 & 3 \end{bmatrix}.$$

Thus, $\det(3A) = 9 - 72 = -63$, confirming the calculation above. ☐

So far we have considered the effect of two elementary column operations: column interchanges and multiplication of a column by a scalar. We now wish to show that the addition of a constant multiple of one column to another column does not change the determinant. We need several preliminary results to prove this.

THEOREM 4

If A, B, and C are $(n \times n)$ matrices that are equal except that the sth column (or row) of A is equal to the sum of the sth columns (or rows) of B and C, then $\det(A) = \det(B) + \det(C)$.

As before, the proof of Theorem 4 is somewhat technical and is deferred to the end of this section.

EXAMPLE 6 Verify Theorem 4 where A, B, and C are (2×2) matrices.

SOLUTION Suppose that A, B, and C are (2×2) matrices such that the first column of A is equal to the sum of the first columns of B and C. Thus,

$$B = \begin{bmatrix} b_1 & \alpha \\ b_2 & \beta \end{bmatrix}, \qquad C = \begin{bmatrix} c_1 & \alpha \\ c_2 & \beta \end{bmatrix}, \quad \text{and} \quad A = \begin{bmatrix} b_1 + c_1 & \alpha \\ b_2 + c_2 & \beta \end{bmatrix}.$$

Calculating det(A), we have

$$\det(A) = (b_1 + c_1)\beta - \alpha(b_2 + c_2)$$
$$= (b_1\beta - \alpha b_2) + (c_1\beta - \alpha c_2)$$
$$= \det(B) + \det(C).$$

The case in which A, B, and C have the same first column is left as an exercise.

EXAMPLE 7 Given that det(B) = 22 and det(C) = 29, find det(A), where

$$A = \begin{bmatrix} 1 & 3 & 2 \\ 0 & 4 & 7 \\ 2 & 1 & 8 \end{bmatrix}, \quad B = \begin{bmatrix} 1 & 1 & 2 \\ 0 & 2 & 7 \\ 2 & 0 & 8 \end{bmatrix}, \quad \text{and} \quad C = \begin{bmatrix} 1 & 2 & 2 \\ 0 & 2 & 7 \\ 2 & 1 & 8 \end{bmatrix}.$$

SOLUTION In terms of column vectors, $\mathbf{A}_1 = \mathbf{B}_1 = \mathbf{C}_1$, $\mathbf{A}_3 = \mathbf{B}_3 = \mathbf{C}_3$, and $\mathbf{A}_2 = \mathbf{B}_2 + \mathbf{C}_2$. Thus,

$$\det(A) = \det(B) + \det(C) = 22 + 29 = 51.$$

THEOREM 5

Let A be an $(n \times n)$ matrix. If the jth column (or row) of A is a multiple of the kth column (or row) of A, then det(A) = 0.

PROOF Let $A = [\mathbf{A}_1, \mathbf{A}_2, \ldots, \mathbf{A}_j, \ldots, \mathbf{A}_k, \ldots, \mathbf{A}_n]$ and suppose that $\mathbf{A}_j = c\mathbf{A}_k$. Define B to be the matrix $B = [\mathbf{A}_1, \mathbf{A}_2, \ldots, \mathbf{A}_k, \ldots, \mathbf{A}_k, \ldots, \mathbf{A}_n]$ and observe that $\det(A) = c\det(B)$. Now if we interchange the jth and kth columns of B, the matrix B remains the same, but the determinant changes sign (Theorem 2). This $[\det(B) = -\det(B)]$ can happen only if det(B) = 0; and since $\det(A) = c\det(B)$, then det(A) = 0.

Two special cases of Theorem 5 are particularly interesting. If A has two identical columns ($c = 1$ in the proof above), or if A has a zero column ($c = 0$ in the proof), then det(A) = 0.

Theorems 4 and 5 can be used to analyze the effect of the last elementary column operation.

THEOREM 6

If A is an $(n \times n)$ matrix, and if a multiple of the kth column (or row) is added to the jth column (or row), then the determinant is not changed.

PROOF Let $A = [\mathbf{A}_1, \mathbf{A}_2, \ldots, \mathbf{A}_j, \ldots, \mathbf{A}_k, \ldots, \mathbf{A}_n]$ and let $B = [\mathbf{A}_1, \mathbf{A}_2, \ldots, \mathbf{A}_j + c\mathbf{A}_k, \ldots, \mathbf{A}_k, \ldots, \mathbf{A}_n]$. By Theorem 4, $\det(B) = \det(A) + \det(Q)$, where

$Q = [\mathbf{A}_1, \mathbf{A}_2, \ldots, c\mathbf{A}_k, \ldots, \mathbf{A}_k, \ldots, \mathbf{A}_n]$. By Theorem 5, $\det(Q) = 0$; so $\det(B) = \det(A)$, and the theorem is proved. □

As shown in the examples that follow, we can use elementary column operations to introduce zero entries into the first row of a matrix A. The analysis of how these operations affect the determinant will allow us to relate this effect back to $\det(A)$.

EXAMPLE 8 Use elementary column operations to simplify finding the determinant of the (4×4) matrix A:

$$A = \begin{bmatrix} 1 & 2 & 0 & 2 \\ -1 & 2 & 3 & 1 \\ -3 & 2 & -1 & 0 \\ 2 & -3 & -2 & 1 \end{bmatrix}.$$

SOLUTION In Example 4 of Section 5.2, a laborious cofactor expansion showed that $\det(A) = -63$. In column form, $A = [\mathbf{A}_1, \mathbf{A}_2, \mathbf{A}_3, \mathbf{A}_4]$, and clearly we can introduce a zero into the $(1,2)$ position by replacing $\mathbf{A}_2$ by $\mathbf{A}_2 - 2\mathbf{A}_1$. Similarly, replacing $\mathbf{A}_4$ by $\mathbf{A}_4 - 2\mathbf{A}_1$ creates a zero in the $(1,4)$ entry. Moreover, by Theorem 6, the determinant is unchanged. The details are

$$\det(A) = \begin{vmatrix} 1 & 2 & 0 & 2 \\ -1 & 2 & 3 & 1 \\ -3 & 2 & -1 & 0 \\ 2 & -3 & -2 & 1 \end{vmatrix} = \begin{vmatrix} 1 & 0 & 0 & 2 \\ -1 & 4 & 3 & 1 \\ -3 & 8 & -1 & 0 \\ 2 & -7 & -2 & 1 \end{vmatrix}$$

$$= \begin{vmatrix} 1 & 0 & 0 & 0 \\ -1 & 4 & 3 & 3 \\ -3 & 8 & -1 & 6 \\ 2 & -7 & -2 & -3 \end{vmatrix}.$$

Thus it follows that $\det(A)$ is given by

$$\det(A) = \begin{vmatrix} 4 & 3 & 3 \\ 8 & -1 & 6 \\ -7 & -2 & -3 \end{vmatrix}.$$

We now wish to create zeros in the $(1,2)$ and $(1,3)$ positions of this (3×3) determinant. To avoid using fractions, we multiply the second and third columns by 4 (using Theorem 3), and then add a multiple of -3 times column 1 to columns 2 and 3:

$$\det(A) = \begin{vmatrix} 4 & 3 & 3 \\ 8 & -1 & 6 \\ -7 & -2 & -3 \end{vmatrix} = \frac{1}{16} \begin{vmatrix} 4 & 12 & 12 \\ 8 & -4 & 24 \\ -7 & -8 & -12 \end{vmatrix} = \frac{1}{16} \begin{vmatrix} 4 & 0 & 0 \\ 8 & -28 & 0 \\ -7 & 13 & 9 \end{vmatrix}.$$

Thus we again find $\det(A) = -63$. □

EXAMPLE 9 Use column operations to find $\det(A)$, where

$$A = \begin{bmatrix} 0 & 1 & 3 & 1 \\ 1 & -2 & -2 & 2 \\ 3 & 4 & 2 & -2 \\ 4 & 3 & -1 & 1 \end{bmatrix}.$$

SOLUTION   As in Gaussian elimination, column interchanges are sometimes desirable and serve to keep order in the computation. Consider

$$\det(A) = \begin{vmatrix} 0 & 1 & 3 & 1 \\ 1 & -2 & -2 & 2 \\ 3 & 4 & 2 & -2 \\ 4 & 3 & -1 & 1 \end{vmatrix} = - \begin{vmatrix} 1 & 0 & 3 & 1 \\ -2 & 1 & -2 & 2 \\ 4 & 3 & 2 & -2 \\ 3 & 4 & -1 & 1 \end{vmatrix}.$$

Use column 1 to introduce zeros along the first row:

$$\det(A) = - \begin{vmatrix} 1 & 0 & 0 & 0 \\ -2 & 1 & 4 & 4 \\ 4 & 3 & -10 & -6 \\ 3 & 4 & -10 & -2 \end{vmatrix} = - \begin{vmatrix} 1 & 4 & 4 \\ 3 & -10 & -6 \\ 4 & -10 & -2 \end{vmatrix}.$$

Again column 1 can be used to introduce zeros:

$$\det(A) = - \begin{vmatrix} 1 & 0 & 0 \\ 3 & -22 & -18 \\ 4 & -26 & -18 \end{vmatrix} = - \begin{vmatrix} -22 & -18 \\ -26 & -18 \end{vmatrix} = 18 \begin{vmatrix} -22 & 1 \\ -26 & 1 \end{vmatrix},$$

and we calculate the (2×2) determinant to find $\det(A) = 72$. ☐

Proof of Theorems 2, 3, and 4 (Optional)

We conclude this section with the proofs of Theorems 2, 3, and 4. Note that these proofs are very similar and fairly straightforward.

PROOF OF THEOREM 2   The proof is by induction. The initial case ($k = 2$) was proved in Example 1.

Assuming the result is valid for any ($k \times k$) matrix, $2 \leq k \leq n - 1$, let B be obtained from A by interchanging the ith and jth columns. For $1 \leq s \leq n$, let M_{1s} and N_{1s} denote minor matrices of A and B, respectively.

If $s \neq i$ or j, then N_{1s} is the same as M_{1s} except for a single column interchange. Hence, by the induction hypotheses,

$$\det(N_{1s}) = -\det(M_{1s}), \qquad s \neq i \quad \text{or} \quad j.$$

For definiteness let us suppose that $i > j$. Note that N_{1i} contains no entries from the original jth column. Furthermore, the columns of N_{1i} can be rearranged to be the same as the columns of M_{1j} by $i - j - 1$ successive interchanges of adjacent columns. By the induction hypotheses, each such interchange causes a sign change, and so

$$\det(N_{1i}) = (-1)^{(i-j-1)} \det(M_{1j}).$$

Therefore,

$$\det(B) = \left(\sum_{\substack{s=1 \\ s \ne i \,or\, j}}^{n} a_{1s}(-1)^{1+s} \det(N_{1s}) \right) + a_{1j}(-1)^{1+i} \det(N_{1i})$$

$$+ a_{1i}(-1)^{1+j} \det(N_{1j})$$

$$= \left(\sum_{\substack{s=1 \\ s \ne i \,or\, j}}^{n} a_{1s}(-1)^{1+s} [-\det(M_{1s})] \right) + a_{1j}(-1)^{1+i}(-1)^{i-j-1} \det(M_{1j})$$

$$+ a_{1i}(-1)^{1+j}(-1)^{i-j-1} \det(M_{1i})$$

$$= \sum_{s=1}^{n} a_{1s}(-1)^{2+s} \det(M_{1s}) = -\det(A). \qquad \square$$

PROOF OF THEOREM 3   Again, the proof is by induction. The case $k = 2$ was proved in Example 3.

Assuming the result is valid for $(k \times k)$ matrices, $2 \le k \le n - 1$, let B be the $(n \times n)$ matrix, where

$$B = [\mathbf{A}_1, \ldots, \mathbf{A}_{s-1}, c\mathbf{A}_s, \mathbf{A}_{s+1}, \ldots, \mathbf{A}_n].$$

Let M_{1j} and N_{1j} be minor matrices of A and B, respectively, for $1 \le j \le n$. If $j \ne s$, then $N_{1j} = M_{1j}$ except that one column of N_{1j} is multiplied by c. By the induction hypothesis,

$$\det(N_{1j}) = c \det(M_{1j}), \qquad 1 \le j \le n, \qquad j \ne s.$$

Moreover, $N_{1s} = M_{1s}$. Hence

$$\det(B) = \left(\sum_{\substack{j=1 \\ j \ne s}}^{n} a_{1j}(-1)^{1+j} \det(N_{1j}) \right) + ca_{1s}(-1)^{1+s} \det(N_{1s})$$

$$= \left(\sum_{\substack{j=1 \\ j \ne s}}^{n} a_{1j}(-1)^{1+j} c \det(M_{1j}) \right) + ca_{1s}(-1)^{1+s} \det(M_{1s})$$

$$= c \sum_{j=1}^{n} a_{1j}(-1)^{1+j} \det(M_{1j}) = c \det(A). \qquad \square$$

PROOF OF THEOREM 4   We use induction where the case $k = 2$ is done in Example 6. Assuming the result is true for $(k \times k)$ matrices for $2 \le k \le n - 1$, let

$$A = [\mathbf{A}_1, \mathbf{A}_2, \ldots, \mathbf{A}_n], \, B = [\mathbf{A}_1, \ldots, \mathbf{A}_{s-1}, \mathbf{B}_s, \mathbf{A}_{s+1}, \ldots, \mathbf{A}_n], \quad \text{and}$$
$$C = [\mathbf{A}_1, \ldots, \mathbf{A}_{s-1}, \mathbf{C}_s, \mathbf{A}_{s+1}, \ldots, \mathbf{A}_n],$$

where $\mathbf{A}_s = \mathbf{B}_s + \mathbf{C}_s$, or

$$a_{is} = b_{is} + c_{is}, \qquad \text{for } 1 \le i \le n.$$

Let M_{1j}, N_{1j}, and P_{1j} be minor matrices of A, B, and C, respectively, for $1 \le j \le n$. If $j \ne s$, then M_{1j}, N_{1j}, and P_{1j} are equal except in one column, which we designate as the rth column. Now the rth columns of N_{1j} and P_{1j} sum to the rth column of M_{1j}. Hence, by the induction hypothesis,

$$\det(M_{1j}) = \det(N_{1j}) + \det(P_{1j}), \qquad 1 \le j \le n, \qquad j \ne s.$$

Clearly, if $j = s$, then $M_{1s} = N_{1s} = P_{1s}$. Hence

$$\det(B) + \det(C) = \left(\sum_{\substack{j=1 \\ j \ne s}}^{n} a_{1j}(-1)^{1+j}\det(N_{1j}) \right) + b_{1s}(-1)^{1+s}\det(N_{1s})$$

$$+ \left(\sum_{\substack{j=1 \\ j \ne s}}^{n} a_{1j}(-1)^{1+j}\det(P_{1j}) \right) + c_{1s}(-1)^{1+s}\det(P_{1s})$$

$$= \left(\sum_{\substack{j=1 \\ j \ne s}}^{n} a_{1j}(-1)^{1+j}[\det(N_{1j}) + \det(P_{1j})] \right)$$

$$+ (b_{1s} + c_{1s})(-1)^{1+s}\det(M_{1s})$$

$$= \sum_{j=1}^{n} a_{1j}(-1)^{1+j}\det(M_{1j}) = \det(A). \qquad \square$$

Exercises 5.3

In Exercises 1–6, use elementary column operations to create zeros in the last two entries in the first row and then calculate the determinant of the original matrix.

1. $\begin{bmatrix} 1 & 2 & 1 \\ 2 & 0 & 1 \\ 1 & -1 & 1 \end{bmatrix}$
2. $\begin{bmatrix} 2 & 4 & -2 \\ 0 & 2 & 3 \\ 1 & 1 & 2 \end{bmatrix}$

3. $\begin{bmatrix} 0 & 1 & 2 \\ 3 & 1 & 2 \\ 2 & 0 & 3 \end{bmatrix}$
4. $\begin{bmatrix} 2 & 2 & 4 \\ 1 & 0 & 1 \\ 2 & 1 & 2 \end{bmatrix}$

5. $\begin{bmatrix} 0 & 1 & 3 \\ 2 & 1 & 2 \\ 1 & 1 & 2 \end{bmatrix}$
6. $\begin{bmatrix} 1 & 1 & 1 \\ 2 & 1 & 2 \\ 3 & 0 & 2 \end{bmatrix}$

Suppose that $A = [\mathbf{A}_1, \mathbf{A}_2, \mathbf{A}_3, \mathbf{A}_4]$ is a (4×4) matrix, where $\det(A) = 3$. In Exercises 7–12, find $\det(B)$.

7. $B = [2\mathbf{A}_1, \mathbf{A}_2, \mathbf{A}_4, \mathbf{A}_3]$
8. $B = [\mathbf{A}_2, 3\mathbf{A}_3, \mathbf{A}_1, -2\mathbf{A}_4]$
9. $B = [\mathbf{A}_1 + 2\mathbf{A}_2, \mathbf{A}_2, \mathbf{A}_3, \mathbf{A}_4]$
10. $B = [\mathbf{A}_1, \mathbf{A}_1 + 2\mathbf{A}_2, \mathbf{A}_3, \mathbf{A}_4]$
11. $B = [\mathbf{A}_1 + 2\mathbf{A}_2, \mathbf{A}_2 + 3\mathbf{A}_3, \mathbf{A}_3, \mathbf{A}_4]$
12. $B = [2\mathbf{A}_1 - \mathbf{A}_2, 2\mathbf{A}_2 - \mathbf{A}_3, \mathbf{A}_3, \mathbf{A}_4]$

In Exercises 13–15, use only column interchanges to produce a triangular matrix and then give the determinant of the original matrix.

13. $\begin{bmatrix} 1 & 0 & 0 & 0 \\ 2 & 0 & 0 & 3 \\ 1 & 1 & 0 & 1 \\ 1 & 4 & 2 & 2 \end{bmatrix}$
14. $\begin{bmatrix} 0 & 0 & 2 & 0 \\ 0 & 0 & 1 & 3 \\ 0 & 4 & 1 & 3 \\ 2 & 1 & 5 & 6 \end{bmatrix}$

15. $\begin{bmatrix} 0 & 1 & 0 & 0 \\ 0 & 2 & 0 & 3 \\ 2 & 1 & 0 & 6 \\ 3 & 2 & 2 & 4 \end{bmatrix}$

In Exercises 16–18, use elementary column operations to create zeros in the $(1,2)$, $(1,3)$, $(1,4)$, $(2,3)$, and $(2,4)$ positions. Then evaluate the original determinant.

16. $\begin{vmatrix} 1 & 2 & 0 & 3 \\ 2 & 5 & 1 & 1 \\ 2 & 0 & 4 & 3 \\ 0 & 1 & 6 & 2 \end{vmatrix}$

17. $\begin{vmatrix} 2 & 4 & -2 & -2 \\ 1 & 3 & 1 & 2 \\ 1 & 3 & 1 & 3 \\ -1 & 2 & 1 & 2 \end{vmatrix}$

18. $\begin{vmatrix} 1 & 1 & 2 & 1 \\ 0 & 1 & 4 & 1 \\ 2 & 1 & 3 & 0 \\ 2 & 2 & 1 & 2 \end{vmatrix}$

19. Use elementary row operations on the determinant in Exercise 16 to create zeros in the $(2, 1), (3, 1), (4, 1)$, $(3, 2)$, and $(4, 2)$ positions. Assuming the column results in this section also hold for rows, give the value of the original determinant and verify that it is the same as in Exercise 16.

20. Repeat Exercise 19, using the determinant in Exercise 17.

21. Repeat Exercise 19, using the determinant in Exercise 18.

22. Find a (2×2) matrix A and a (2×2) matrix B, where $\det(A + B)$ is not equal to $\det(A) + \det(B)$. Find a different A and B, both nonzero, such that $\det(A + B) = \det(A) + \det(B)$.

23. For any real number a, $a \neq 0$, show that

$$\begin{vmatrix} a + 1 & a + 4 & a + 7 \\ a + 2 & a + 5 & a + 8 \\ a + 3 & a + 6 & a + 9 \end{vmatrix} = 0, \quad \begin{vmatrix} a & 4a & 7a \\ 2a & 5a & 8a \\ 3a & 6a & 9a \end{vmatrix} = 0,$$

and $\begin{vmatrix} a & a^4 & a^7 \\ a^2 & a^5 & a^8 \\ a^3 & a^6 & a^9 \end{vmatrix} = 0.$

24. Let $A = [\mathbf{A}_1, \mathbf{A}_2, \mathbf{A}_3]$ be a (3×3) matrix and set

$$B = \begin{bmatrix} 2 & 0 & 0 \\ 3 & -1 & 0 \\ 1 & 3 & 4 \end{bmatrix}.$$

a) Show that
$$AB = [2\mathbf{A}_1 + 3\mathbf{A}_2 + \mathbf{A}_3, -\mathbf{A}_2 + 3\mathbf{A}_3, 4\mathbf{A}_3].$$

b) Use column operations to show that $\det(AB) = -8\det(A)$.

c) Conclude that $\det(AB) = \det(A)\det(B)$.

25. Let U be an $(n \times n)$ upper-triangular matrix and consider the cofactors U_{1j}, $2 \le j \le n$. Show that $U_{1j} = 0$, $2 \le j \le n$. (*Hint:* Some column in U_{1j} is always the zero column.)

26. Use the result of Exercise 25 to prove inductively that $\det(U) = u_{11}u_{22}\ldots u_{nn}$, where $U = (u_{ij})$ is an $(n \times n)$ upper-triangular matrix.

27. Let $y = mx + b$ be the equation of the line through the points (x_1, y_1) and (x_2, y_2) in the plane. Show that the equation is given also by

$$\begin{vmatrix} x & y & 1 \\ x_1 & y_1 & 1 \\ x_2 & y_2 & 1 \end{vmatrix} = 0.$$

28. Let (x_1, y_1), (x_2, y_2), and (x_3, y_3) be the vertices of a triangle in the plane where these vertices are numbered counterclockwise. Prove that the area of the triangle is given by

$$\text{Area} = \frac{1}{2}\begin{vmatrix} x_1 & y_1 & 1 \\ x_2 & y_2 & 1 \\ x_3 & y_3 & 1 \end{vmatrix}.$$

29. Let $\mathbf{x}$ and $\mathbf{y}$ be vectors in R^3 and let $A = I + \mathbf{xy}^T$. Show that $\det(A) = 1 + \mathbf{y}^T\mathbf{x}$. (*Hint:* If $B = \mathbf{xy}^T$, $B = [\mathbf{B}_1, \mathbf{B}_2, \mathbf{B}_3]$, then $A = [\mathbf{B}_1 + \mathbf{e}_1, \mathbf{B}_2 + \mathbf{e}_2, \mathbf{B}_3 + \mathbf{e}_3]$. Therefore, $\det(A) = \det[\mathbf{B}_1, \mathbf{B}_2 + \mathbf{e}_2, \mathbf{B}_3 + \mathbf{e}_3] + \det[\mathbf{e}_1, \mathbf{B}_2 + \mathbf{e}_2, \mathbf{B}_3 + \mathbf{e}_3]$. Use Theorems 4 and 5 to show that the first determinant is equal to $\det[\mathbf{B}_1, \mathbf{e}_2, \mathbf{B}_3 + \mathbf{e}_3]$, and so on.)

30. Use column operations to prove that

$$\begin{vmatrix} 1 & a & a^2 \\ 1 & b & b^2 \\ 1 & c & c^2 \end{vmatrix} = (b - a)(c - a)(c - b).$$

31. Evaluate the (4×4) determinant

$$\begin{vmatrix} 1 & a & a^2 & a^3 \\ 1 & b & b^2 & b^3 \\ 1 & c & c^2 & c^3 \\ 1 & d & d^2 & d^3 \end{vmatrix}.$$

(*Hint:* Proceed as in Exercise 30.)

32. Prove the corollary to Theorem 3.

5.4 Cramer's Rule

In Section 5.3, we saw how to calculate the effect that a column operation or a row operation has on a determinant. In this section, we use that information to analyze the relationships between determinants, nonsingular matrices, and solutions of systems $A\mathbf{x} = \mathbf{b}$. We begin with the following lemma, which will be helpful in the proof of the subsequent theorems.

LEMMA 1

Let $A = [\mathbf{A}_1, \mathbf{A}_2, \ldots, \mathbf{A}_n]$ be an $(n \times n)$ matrix, and let $\mathbf{b}$ be any vector in R^n. For each i, $1 \leq i \leq n$, let B_i be the $(n \times n)$ matrix:

$$B_i = [\mathbf{A}_1, \ldots, \mathbf{A}_{i-1}, \mathbf{b}, \mathbf{A}_{i+1}, \ldots, \mathbf{A}_n].$$

If the system of equations $A\mathbf{x} = \mathbf{b}$ is consistent and x_i is the ith component of a solution, then

$$x_i \det(A) = \det(B_i). \tag{1}$$

PROOF To keep the notation simple, we give the proof of Eq. (1) only for $i = 1$. Since the system $A\mathbf{x} = \mathbf{b}$ is assumed to be consistent, there are values $x_1, x_2, \ldots, x_n$ such that

$$x_1 \mathbf{A}_1 + x_2 \mathbf{A}_2 + \cdots + x_n \mathbf{A}_n = \mathbf{b}.$$

Using the properties of determinants, we have

$$\begin{aligned}
x_1 \det(A) &= \det[x_1 \mathbf{A}_1, \mathbf{A}_2, \ldots, \mathbf{A}_n] \\
&= \det[\mathbf{b} - x_2 \mathbf{A}_2 - \cdots - x_n \mathbf{A}_n, \mathbf{A}_2, \ldots, \mathbf{A}_n] \\
&= \det[\mathbf{b}, \mathbf{A}_2, \ldots, \mathbf{A}_n] - x_2 \det[\mathbf{A}_2, \mathbf{A}_2, \ldots, \mathbf{A}_n] \\
&\quad - \cdots - x_n \det[\mathbf{A}_n, \mathbf{A}_2, \ldots, \mathbf{A}_n].
\end{aligned}$$

By Theorem 5, the last $n - 1$ determinants are zero, so we have

$$x_1 \det(A) = \det[\mathbf{b}, \mathbf{A}_2, \ldots, \mathbf{A}_n];$$

and this equality verifies Eq. (1) for $i = 1$. Clearly, the same argument is valid for any i.

As the following theorem shows, one consequence of Lemma 1 is that a singular matrix has determinant zero.

THEOREM 7

If A is an $(n \times n)$ singular matrix, then $\det(A) = 0$.

PROOF Since A is singular, $A\mathbf{x} = \boldsymbol{\theta}$ has a nontrivial solution. Let x_i be the ith component of a nontrivial solution and choose i so that $x_i \neq 0$. By Lemma 1, $x_i \det(A) = \det(B_i)$, where $B_i = [\mathbf{A}_1, \ldots, \mathbf{A}_{i-1}, \boldsymbol{\theta}, \mathbf{A}_{i+1}, \ldots, \mathbf{A}_n]$. It follows from Theorem 3 that $\det(B_i) = 0$. Thus, $x_i \det(A) = 0$, and since $x_i \neq 0$, then $\det(A) = 0$.

Theorem 9, stated below, establishes the converse of Theorem 7: If $\det(A) = 0$, then A is a singular matrix. Theorem 9 will be an easy consequence of the product rule for determinants.

The Determinant of a Product

Theorem 8, below, states that if A and B are $(n \times n)$ matrices, then $\det(AB) = \det(A)\det(B)$. This result is somewhat surprising in view of the complexity of matrix multiplication. We also know, in general, that $\det(A + B)$ is distinct from $\det(A) + \det(B)$.

THEOREM 8

If A and B are $(n \times n)$ matrices, then

$$\det(AB) = \det(A)\det(B).$$

Before sketching a proof of Theorem 8, note that if A is an $(n \times n)$ matrix and if B is obtained from A by a sequence of elementary column operations, then, by the properties of determinants given in Theorems 2, 3, and 6, $\det(A) = k\det(B)$, where the scalar k is completely determined by the elementary column operations. To illustrate, suppose that B is obtained by the following sequence of elementary column operations:

1. Interchange the first and third columns.
2. Multiply the second column by 3.
3. Add 2 times the second column to the first column.

It now follows from Theorems 2, 3, and 6 that $\det(B) = -3\det(A)$ or, equivalently, $\det(A) = (-1/3)\det(B)$. Moreover, the scalar $-1/3$ is completely determined by the operations; that is, the scalar is independent of the matrices involved.

The proof of Theorem 8 is based on the observation above and on the following lemma.

LEMMA 2

Let A and B be $(n \times n)$ matrices and let $C = AB$. Let $\hat{C}$ denote the result of applying an elementary column operation to C and let $\hat{B}$ denote the result of applying the same column operation to B. Then $\hat{C} = A\hat{B}$.

The proof of Lemma 2 is left to the exercises. The intent of the lemma is given schematically in Fig. 5.1.

Lemma 2 tells us that the same result is produced whether we apply a column operation to the product AB or whether we apply the operation to B first (producing $\hat{B}$) and then form the product $A\hat{B}$. For example, suppose that A and B are (3×3) matrices. Consider the operation of interchanging column 1 and column 3:

$$B = [\mathbf{B}_1, \mathbf{B}_2, \mathbf{B}_3] \rightarrow \hat{B} = [\mathbf{B}_3, \mathbf{B}_2, \mathbf{B}_1]; \qquad A\hat{B} = [A\mathbf{B}_3, A\mathbf{B}_2, A\mathbf{B}_1]$$
$$AB = [A\mathbf{B}_1, A\mathbf{B}_2, A\mathbf{B}_3] \rightarrow \widehat{AB} = [A\mathbf{B}_3, A\mathbf{B}_2, A\mathbf{B}_1]; \qquad \widehat{AB} = A\hat{B}.$$

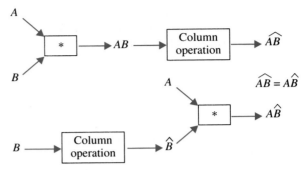

Figure 5.1
Schematic diagram of Lemma 2

PROOF OF THEOREM 8　Suppose that A and B are $(n \times n)$ matrices. If B is singular, then Theorem 8 is immediate, for in this case AB is also singular. Thus, by Theorem 7, $\det(B) = 0$ and $\det(AB) = 0$. Consequently, $\det(AB) = \det(A)\det(B)$.

Next, suppose that B is nonsingular. In this case, B can be transformed to the $(n \times n)$ identity matrix I by a sequence of elementary column operations. (To see this, note that B^T is nonsingular by Theorem 17, property 4, of Section 1.10. It now follows from the results of Section 1.11 that B^T can be reduced to I by a sequence of elementary row operations. But performing row operations on B^T is equivalent to performing column operations on B.) Therefore, $\det(B) = k\det(I) = k$, where k is determined entirely by the sequence of elementary column operations. By Lemma 2, the same sequence of operations reduces the matrix AB to the matrix $AI = A$. Thus, $\det(AB) = k\det(A) = \det(B)\det(A) = \det(A)\det(B)$. $\qquad\square$

EXAMPLE 1　Show by direct calculation that $\det(AB) = \det(A)\det(B)$ for the matrices

$$A = \begin{bmatrix} 2 & 1 \\ 1 & 3 \end{bmatrix} \quad \text{and} \quad B = \begin{bmatrix} -1 & 3 \\ 2 & -2 \end{bmatrix}.$$

SOLUTION　We have $\det(A) = 5$ and $\det(B) = -4$. Since AB is given by

$$AB = \begin{bmatrix} 0 & 4 \\ 5 & -3 \end{bmatrix},$$

it follows that $\det(AB) = -20 = (5)(-4) = \det(A)\det(B)$. $\qquad\square$

The following theorem is now an easy consequence of Theorem 8.

THEOREM 9

If the $(n \times n)$ matrix A is nonsingular, then $\det(A) \neq 0$. Moreover, $\det(A^{-1}) = 1/\det(A)$.

PROOF Since A is nonsingular, A^{-1} exists and $AA^{-1} = I$. By Theorem 8, $1 = \det(I) = \det(AA^{-1}) = \det(A)\det(A^{-1})$. In particular, $\det(A) \neq 0$ and $\det(A^{-1}) = 1/\det(A)$. ☐

Theorems 7 and 9 show that an $(n \times n)$ matrix A is singular if and only if $\det(A) = 0$. This characterization of singular matrices is especially useful when we want to examine matrices that depend on a parameter. The next example illustrates one such application.

EXAMPLE 2 Find all values λ such that the matrix $B(\lambda)$ is singular, where

$$B(\lambda) = \begin{bmatrix} 2 - \lambda & 0 & 0 \\ 2 & 3 - \lambda & 4 \\ 1 & 2 & 1 - \lambda \end{bmatrix}.$$

SOLUTION By Theorems 7 and 9, $B(\lambda)$ is singular if and only if $\det[B(\lambda)] = 0$. The equation $\det[B(\lambda)] = 0$ is determined by

$$\begin{aligned} 0 &= \det[B(\lambda)] \\ &= (2 - \lambda)[(3 - \lambda)(1 - \lambda) - 8] \\ &= (2 - \lambda)[\lambda^2 - 4\lambda - 5] \\ &= (2 - \lambda)(\lambda - 5)(\lambda + 1). \end{aligned}$$

Thus, $B(\lambda)$ is singular if and only if λ is one of the values $\lambda = 2$, $\lambda = 5$, or $\lambda = -1$.

The three matrices discovered by solving $\det[B(\lambda)] = 0$ are listed below. As we can see, each of these matrices is singular:

$$B(2) = \begin{bmatrix} 0 & 0 & 0 \\ 2 & 1 & 4 \\ 1 & 2 & -1 \end{bmatrix}, \qquad B(5) = \begin{bmatrix} -3 & 0 & 0 \\ 2 & -2 & 4 \\ 1 & 2 & -4 \end{bmatrix}, \qquad B(-1) = \begin{bmatrix} 3 & 0 & 0 \\ 2 & 4 & 4 \\ 1 & 2 & 2 \end{bmatrix}. \quad ☐$$

Solving $A\mathbf{x} = \mathbf{b}$ with Cramer's Rule

A major result in determinant theory is Cramer's rule, which gives a formula for the solution of any system $A\mathbf{x} = \mathbf{b}$ when A is nonsingular.

THEOREM 10

Cramer's Rule Let $A = [A_1, A_2, \ldots, A_n]$ be a nonsingular $(n \times n)$ matrix, and let $\mathbf{b}$ be any vector in R^n. For each i, $1 \leq i \leq n$, let B_i be the matrix $B_i = [A_1, \ldots, A_{i-1}, \mathbf{b}, A_{i+1}, \ldots, A_n]$. Then the ith component, x_i, of the solution of $A\mathbf{x} = \mathbf{b}$ is given by

$$x_i = \frac{\det(B_i)}{\det(A)}. \tag{2}$$

PROOF   Since A is nonsingular, $\det(A) \neq 0$. Formula (2) is now an immediate consequence of (1) in Lemma 1. ☐

EXAMPLE 3   Use Cramer's rule to solve the system

$$3x_1 + 2x_2 = 4$$
$$5x_1 + 4x_2 = 6.$$

SOLUTION   To solve this system by Cramer's rule, we write the system as $A\mathbf{x} = \mathbf{b}$ and we form $B_1 = [\mathbf{b}, A_2]$ and $B_2 = [A_1, \mathbf{b}]$:

$$A = \begin{bmatrix} 3 & 2 \\ 5 & 4 \end{bmatrix}, \qquad B_1 = \begin{bmatrix} 4 & 2 \\ 6 & 4 \end{bmatrix}, \qquad B_2 = \begin{bmatrix} 3 & 4 \\ 5 & 6 \end{bmatrix}.$$

Note that $\det(A) = 2$, $\det(B_1) = 4$, and $\det(B_2) = -2$. Thus from Eq. (2) the solution is

$$x_1 = \frac{4}{2} = 2 \quad \text{and} \quad x_2 = \frac{-2}{2} = -1.$$ ☐

EXAMPLE 4   Use Cramer's rule to solve the system

$$x_1 - x_2 + x_3 = 0$$
$$x_1 + x_2 - 2x_3 = 1$$
$$x_1 + 2x_2 + x_3 = 6.$$

SOLUTION   Writing the system as $A\mathbf{x} = \mathbf{b}$, we have

$$A = \begin{bmatrix} 1 & -1 & 1 \\ 1 & 1 & -2 \\ 1 & 2 & 1 \end{bmatrix}, \qquad B_1 = \begin{bmatrix} 0 & -1 & 1 \\ 1 & 1 & -2 \\ 6 & 2 & 1 \end{bmatrix},$$

$$B_2 = \begin{bmatrix} 1 & 0 & 1 \\ 1 & 1 & -2 \\ 1 & 6 & 1 \end{bmatrix}, \qquad B_3 = \begin{bmatrix} 1 & -1 & 0 \\ 1 & 1 & 1 \\ 1 & 2 & 6 \end{bmatrix}.$$

A calculation shows that $\det(A) = 9$, $\det(B_1) = 9$, $\det(B_2) = 18$, and $\det(B_3) = 9$. Thus by Eq. (2) the solution is

$$x_1 = \frac{9}{9} = 1, \qquad x_2 = \frac{18}{9} = 2, \quad \text{and} \quad x_3 = \frac{9}{9} = 1.$$ ☐

As a computational tool, Cramer's rule is rarely competitive with Gaussian elimination. It is, however, a valuable theoretical tool. Three specific examples illustrating the use of Cramer's rule in theoretical applications are as follows.

Cramer's Rule

In 1750, Gabriel Cramer (1704–1752) published a work in which, in the appendix, he stated the determinant procedure named after him for solving n linear equations in n unknowns. The first discoverer of this rule, however, was almost surely the Scottish mathematician Colin Maclaurin (1698–1746). It appeared in a paper of Maclaurin's in 1748, published two years after his death. This perhaps compensates for the fact that the famous series named after Maclaurin was not first discovered by him. (Ironically, the Maclaurin series is a special case of a Taylor series, named after the English mathematician Brook Taylor. However, as with the Maclaurin series, Taylor was not the first discoverer of the Taylor series!)

1. The method of variation of parameters (see W. E. Boyce and R. C. DiPrima, *Elementary Differential Equations and Boundary Value Problems*, p. 277. New York: John Wiley and Sons, 1986).

2. The theory of continued fractions (see Peter Henrici, *Applied and Computational Complex Analysis*, Volume 2, pp. 520–521. New York: John Wiley and Sons, 1977).

3. Characterization of best approximations (see E. W. Cheney, *Introduction to Approximation Theory*, p. 74. New York: McGraw-Hill, 1966).

Exercises 5.4

In Exercises 1–3, use column operations to reduce the given matrix A to lower-triangular form. Find the determinant of A.

1. $A = \begin{bmatrix} 0 & 1 & 3 \\ 1 & 2 & 1 \\ 3 & 4 & 1 \end{bmatrix}$

2. $A = \begin{bmatrix} 1 & 2 & 1 \\ 2 & 4 & 3 \\ 2 & 1 & 3 \end{bmatrix}$

3. $A = \begin{bmatrix} 2 & 2 & 4 \\ 1 & 3 & 4 \\ -1 & 2 & 1 \end{bmatrix}$

In Exercises 4–6, use column operations to reduce the given matrix A to the identity matrix. Find the determinant of A.

4. $A = \begin{bmatrix} 1 & 0 & 1 \\ 2 & 1 & 1 \\ 1 & 2 & 1 \end{bmatrix}$

5. $A = \begin{bmatrix} 1 & 0 & -2 \\ 3 & 1 & 3 \\ 0 & 1 & 2 \end{bmatrix}$

6. $A = \begin{bmatrix} 2 & 2 & 2 \\ 4 & 3 & 4 \\ 2 & 1 & 2 \end{bmatrix}$

7. Let A and B be (3×3) matrices such that $\det(A) = 2$ and $\det(B) = 3$. Find the value of each of the following.
   a) $\det(AB)$ b) $\det(AB^2)$
   c) $\det(A^{-1}B)$ d) $\det(2A^{-1})$
   e) $\det(2A)^{-1}$

8. Show that the matrices
   $$\begin{bmatrix} \sin\theta & -\cos\theta \\ \cos\theta & \sin\theta \end{bmatrix} \quad \text{and} \quad \begin{bmatrix} \sin\theta & -\cos\theta & 2 \\ \cos\theta & \sin\theta & 3 \\ 0 & 0 & 1 \end{bmatrix}$$
   are nonsingular for all values of θ.

In Exercises 9–14, find all values λ such that the given matrix $B(\lambda)$ is singular.

9. $B(\lambda) = \begin{bmatrix} \lambda & 0 \\ 3 & 2 - \lambda \end{bmatrix}$

10. $B(\lambda) = \begin{bmatrix} \lambda & 1 \\ 1 & \lambda \end{bmatrix}$

11. $B(\lambda) = \begin{bmatrix} 2 & \lambda \\ \lambda & 2 \end{bmatrix}$

12. $B(\lambda) = \begin{bmatrix} 1 & \lambda & \lambda^2 \\ 1 & 1 & 1 \\ 1 & 3 & 9 \end{bmatrix}$

13. $B(\lambda) = \begin{bmatrix} \lambda & 1 & 1 \\ 1 & \lambda & 1 \\ 1 & 1 & \lambda \end{bmatrix}$

14. $B(\lambda) = \begin{bmatrix} 2-\lambda & 0 & 3 \\ 2 & \lambda & 1 \\ 1 & 0 & -\lambda \end{bmatrix}$

In Exercises 15–21, use Cramer's rule to solve the given system.

15. $x_1 + x_2 = 3$
$x_1 - x_2 = -1$

16. $x_1 + 3x_2 = 4$
$x_1 - x_2 = 0$

17. $x_1 - 2x_2 + x_3 = -1$
$x_1 + x_3 = 3$
$x_1 - 2x_2 = 0$

18. $x_1 + x_2 + x_3 = 2$
$x_1 + 2x_2 + x_3 = 2$
$x_1 + 3x_2 - x_3 = -4$

19. $x_1 + x_2 + x_3 - x_4 = 2$
$x_2 - x_3 + x_4 = 1$
$x_3 - x_4 = 0$
$x_3 + 2x_4 = 3$

20. $2x_1 - x_2 + x_3 = 3$
$x_1 + x_2 = 3$
$x_2 - x_3 = 1$

21. $x_1 + x_2 + x_3 = a$
$x_2 + x_3 = b$
$x_3 = c$

22. Suppose that A is an $(n \times n)$ matrix such that $A^2 = I$. Show that $|\det(A)| = 1$.

23. Prove Lemma 2. (*Hint:* Let

$B = [\mathbf{B}_1, \mathbf{B}_2, \ldots, \mathbf{B}_i, \ldots, \mathbf{B}_j, \ldots, \mathbf{B}_n]$

and consider the matrix $\hat{B}$ produced by interchanging column i and column j. Also consider the matrix $\hat{B}$ produced by replacing $\mathbf{B}_i$ by $\mathbf{B}_i + a\mathbf{B}_j$.)

24. We know that AB and BA are not usually equal. However, show that if A and B are $(n \times n)$, then $\det(AB) = \det(BA)$.

25. Suppose that S is a nonsingular $(n \times n)$ matrix, and suppose that A and B are $(n \times n)$ matrices such that $SAS^{-1} = B$. Prove that $\det(A) = \det(B)$.

26. Suppose that A is $(n \times n)$ and $A^2 = A$. What is $\det(A)$?

27. If $\det(A) = 3$, what is $\det(A^5)$?

28. Let A be a nonsingular matrix and suppose that all the entries of both A and A^{-1} are integers. Prove that $\det(A) = \pm 1$. (*Hint:* Use Theorem 9.)

29. Let A and C be square matrices, and let Q be a matrix of the form

$$Q = \begin{bmatrix} A & \mathcal{O} \\ B & C \end{bmatrix}.$$

Convince yourself that $\det(Q) = \det(A)\det(C)$. (*Hint:* Reduce C to lower-triangular form with column operations; then reduce A.)

30. Verify the result in Exercise 29 for the matrix

$$Q = \begin{bmatrix} 1 & 2 & 0 & 0 & 0 \\ 2 & 1 & 0 & 0 & 0 \\ 3 & 5 & 1 & 2 & 2 \\ 7 & 2 & 3 & 5 & 1 \\ 1 & 8 & 1 & 4 & 1 \end{bmatrix}.$$

5.5 Applications of Determinants: Inverses and Wronskians

Now that we have $\det(AB) = \det(A)\det(B)$, we are ready to prove that $\det(A^T) = \det(A)$ and to establish some other useful properties of determinants. First, however, we need the preliminary result stated in Theorem 11.

THEOREM 11

Let A be an $(n \times n)$ matrix. Then there is a nonsingular $(n \times n)$ matrix Q such that $AQ = L$, where L is lower triangular. Moreover, $\det(Q^T) = \det(Q)$.

The proof of Theorem 11 is based on the following fact: The result of any elementary column operation applied to A can be represented in matrix terms as AQ_i, where Q_i is an "elementary matrix." We discuss this fact and give the proof of Theorem 11 at the end of this section.

Theorem 11 can be used to prove the following important result.

THEOREM 12

If A is an $(n \times n)$ matrix, then $\det(A^T) = \det(A)$.

PROOF   By Theorem 11, there is an $(n \times n)$ matrix Q such that $AQ = L$, where L is a lower-triangular matrix. Moreover, Q is nonsingular and $\det(Q^T) = \det(Q)$. Now, given $AQ = L$, it follows that

$$Q^T A^T = L^T.$$

Applying Theorem 8 to $AQ = L$ and to $Q^T A^T = L^T$, we obtain

$$\det(A)\det(Q) = \det(L)$$
$$\det(Q^T)\det(A^T) = \det(L^T).$$

Since L and L^T are triangular matrices with the same diagonal entries, it follows (see Theorem 1 of Section 5.2 and Exercise 26 of Section 5.3) that $\det(L) = \det(L^T)$. Hence, from the two equalities above, we have

$$\det(A)\det(Q) = \det(Q^T)\det(A^T).$$

Finally, since $\det(Q) = \det(Q^T)$ and $\det(Q) \neq 0$, we see that $\det(A) = \det(A^T)$.

◻

At this point we know that Theorems 2–6 of Section 5.3 are valid for rows as well as for columns. In particular, we can use row operations to reduce a matrix A to a triangular matrix T and conclude that $\det(A) = \pm\det(T)$.

EXAMPLE 1   We return to the (4×4) matrix A in Example 8 of Section 5.3, where $\det(A) = -63$:

$$\det(A) = \begin{vmatrix} 1 & 2 & 0 & 2 \\ -1 & 2 & 3 & 1 \\ -3 & 2 & -1 & 0 \\ 2 & -3 & -2 & 1 \end{vmatrix}.$$

By using row operations, we can reduce $\det(A)$ to

$$\det(A) = \begin{vmatrix} 1 & 2 & 0 & 2 \\ 0 & 4 & 3 & 3 \\ 0 & 8 & -1 & 6 \\ 0 & -7 & -2 & -3 \end{vmatrix}.$$

Now we switch rows 2 and 3 and then switch columns 2 and 3 in order to get the number -1 into the pivot position. Following this switch, we create zeros in the $(2,3)$ and $(2,4)$ positions with row operations; and we find

$$\det(A) = \begin{vmatrix} 1 & 0 & 2 & 2 \\ 0 & -1 & 8 & 6 \\ 0 & 3 & 4 & 3 \\ 0 & -2 & -7 & -3 \end{vmatrix}$$

$$= \begin{vmatrix} 1 & 0 & 2 & 2 \\ 0 & -1 & 8 & 6 \\ 0 & 0 & 28 & 21 \\ 0 & 0 & -23 & -15 \end{vmatrix}.$$

(The sign of the first determinant above is the same as $\det(A)$ because the first determinant is the result of two interchanges.) A quick calculation shows that the last determinant has the value -63.

$\square$

The next theorem shows that we can evaluate $\det(A)$ by using an expansion along any row or any column we choose. Computationally, this ability is useful when some row or column contains a number of zero entries.

THEOREM 13

Let $A = (a_{ij})$ be an $(n \times n)$ matrix. Then:

$$\det(A) = a_{i1}A_{i1} + a_{i2}A_{i2} + \cdots + a_{in}A_{in} \tag{1}$$

$$\det(A) = a_{1j}A_{1j} + a_{2j}A_{2j} + \cdots + a_{nj}A_{nj}. \tag{2}$$

PROOF   We establish only Eq. (1), which is an expansion of $\det(A)$ along the ith row. Expansion of $\det(A)$ along the jth column in Eq. (2) is proved the same way.

Form a matrix B from A in the following manner: Interchange row i first with row $i - 1$ and then with row $i - 2$; continue until row i is the top row of B. In other words, bring row i to the top and push the other rows down so that they retain their same relative ordering. This procedure requires $i - 1$ interchanges; so $\det(A) = (-1)^{i-1}\det(B)$. An inspection shows that the cofactors $B_{11}, B_{12}, \ldots, B_{1n}$ are also related to the cofactors $A_{i1}, A_{i2}, \ldots, A_{in}$ by $B_{1k} = (-1)^{i-1}A_{ik}$. To see this relationship, one need only observe that if M is the minor of the $(1, k)$ entry of B, then M is the minor of the (i, k) entry of A. Therefore, $B_{1k} = (-1)^{k+1}M$ and $A_{ik} = (-1)^{i+k}M$, which shows that $B_{1k} = (-1)^{i-1}A_{ik}$. With this equality and Definition 2 of Section 5.2,

$$\det(B) = b_{11}B_{11} + b_{12}B_{12} + \cdots + b_{1n}B_{1n}$$
$$= a_{i1}B_{11} + a_{i2}B_{12} + \cdots + a_{in}B_{1n}$$
$$= (-1)^{i-1}(a_{i1}A_{i1} + a_{i2}A_{i2} + \cdots + a_{in}A_{in}).$$

Since $\det(A) = (-1)^{i-1}\det(B)$, formula (1) is proved.

$\square$

The Adjoint Matrix and the Inverse

We next show how determinants can be used to obtain a formula for the inverse of a nonsingular matrix. We first prove a lemma, which is similar in appearance to Theorem 13. In words, the lemma states that the sum of the products of entries from the ith row with cofactors from the kth row is zero when $i \neq k$ (and by Theorem 13 this sum is the determinant when $i = k$).

LEMMA

If A is an $(n \times n)$ matrix and if $i \neq k$, then $a_{i1} A_{k1} + a_{i2} A_{k2} + \cdots + a_{in} A_{kn} = 0$.

PROOF   For i and k, given $i \neq k$, let B be the $(n \times n)$ matrix obtained from A by deleting the kth row of A and replacing it by the ith row of A; that is, B has two equal rows, the ith and kth, and B is the same as A for all rows but the kth.

In this event it is clear that $\det(B) = 0$, that the cofactor B_{kj} is equal to A_{kj}, and that the entry b_{kj} is equal to a_{ij}. Putting these together gives

$$0 = \det(B) = b_{k1} B_{k1} + b_{k2} B_{k2} + \cdots + b_{kn} B_{kn}$$
$$= a_{i1} A_{k1} + a_{i2} A_{k2} + \cdots + a_{in} A_{kn};$$

thus the lemma is proved.   ☐

This lemma can be used to derive a formula for A^{-1}. In particular, let A be an $(n \times n)$ matrix, and let C denote the "matrix of cofactors"; $C = (c_{ij})$ is $(n \times n)$, and $c_{ij} = A_{ij}$. The **adjoint** matrix of A, denoted Adj(A), is equal to C^T. With these preliminaries, we prove Theorem 14.

THEOREM 14

If A is an $(n \times n)$ nonsingular matrix, then

$$A^{-1} = \frac{1}{\det(A)} \text{Adj}(A).$$

PROOF   Let $B = (b_{ij})$ be the matrix product of A and Adj(A). Then the ijth entry of B is

$$b_{ij} = a_{i1} A_{j1} + a_{i2} A_{j2} + \cdots + a_{in} A_{jn},$$

and by the lemma and Theorem 13, $b_{ij} = 0$ when $i \neq j$, while $b_{ii} = \det(A)$. Therefore, B is equal to a multiple of $\det(A)$ times I, and the theorem is proved.   ☐

EXAMPLE 2   Let A be the matrix

$$A = \begin{bmatrix} 1 & -1 & 2 \\ 2 & 1 & -3 \\ 4 & 1 & 1 \end{bmatrix}.$$

We calculate the nine required cofactors and find

$$
\begin{aligned}
A_{11} &= 4 & A_{12} &= -14 & A_{13} &= -2 \\
A_{21} &= 3 & A_{22} &= -7 & A_{23} &= -5 \\
A_{31} &= 1 & A_{32} &= 7 & A_{33} &= 3.
\end{aligned}
$$

The adjoint matrix (the transpose of the cofactor matrix) is

$$
\text{Adj}(A) = \begin{bmatrix} 4 & 3 & 1 \\ -14 & -7 & 7 \\ -2 & -5 & 3 \end{bmatrix}.
$$

A multiplication shows that the product of A and $\text{Adj}(A)$ is

$$
\begin{bmatrix} 14 & 0 & 0 \\ 0 & 14 & 0 \\ 0 & 0 & 14 \end{bmatrix};
$$

so $A^{-1} = (1/14)\,\text{Adj}(A)$, where of course $\det(A) = 14$.

Theorem 14 is especially useful when we need to calculate the inverse of a matrix that contains variables. For instance, consider the (3×3) matrix

$$
A = \begin{bmatrix} a & 1 & b \\ 1 & 1 & 1 \\ b & 1 & a \end{bmatrix}. \tag{3}
$$

Although A has some variable entries, we can calculate $\det(A)$ and $\text{Adj}(A)$ and hence find A^{-1}.

EXAMPLE 3 Let A be the (3×3) matrix displayed in (3). Find A^{-1}.

SOLUTION Although we can do this calculation by hand, it is more convenient to use a computer algebra system. We used Derive and found A^{-1} as shown in Fig. 5.2.

The Wronskian

As a final application of determinant theory, we develop a simple test for the linear independence of a set of functions. Suppose that $f_0(x), f_1(x), \ldots, f_n(x)$ are real-valued functions defined on an interval $[a, b]$. If there exist scalars $a_0, a_1, \ldots, a_n$ (not all of which are zero) such that

$$
a_0 f_0(x) + a_1 f_1(x) + \cdots + a_n f_n(x) = 0 \tag{4}
$$

for all x in $[a, b]$, then $\{f_0(x), f_1(x), \ldots, f_n(x)\}$ is a linearly dependent set of functions (see Section 4.4). If the only scalars for which Eq. (4) holds for all x in $[a, b]$ are $a_0 = a_1 = \cdots = a_n = 0$, then the set is linearly independent.

A test for linear independence can be formulated from Eq. (4) as follows: If $a_0, a_1, \ldots, a_n$ are scalars satisfying Eq. (4) and if the functions $f_i(x)$ are sufficiently differentiable, then we can differentiate both sides of the identity (4) and have $a_0 f_0^{(i)}(x) + a_1 f_1^{(i)}(x) + \cdots + a_n f_n^{(i)}(x) = 0$, $1 \le i \le n$. In matrix terms, these

$$
7: \quad
\begin{bmatrix}
a & 1 & b \\
1 & 1 & 1 \\
b & 1 & a
\end{bmatrix}^{-1}
$$

$$
8: \quad
\begin{bmatrix}
\dfrac{a-1}{(a+b-2)(a-b)} & -\dfrac{1}{a+b-2} & \dfrac{1-b}{(a+b-2)(a-b)} \\[2ex]
-\dfrac{1}{a+b-2} & \dfrac{a+b}{a+b-2} & -\dfrac{1}{a+b-2} \\[2ex]
\dfrac{1-b}{(a+b-2)(a-b)} & -\dfrac{1}{a+b-2} & \dfrac{a-1}{(a+b-2)(a-b)}
\end{bmatrix}
$$

Figure 5.2
Using Derive to find the inverse of a matrix with variable
entries, as in Example 3

equations are

$$
\begin{bmatrix}
f_0(x) & f_1(x) & \cdots & f_n(x) \\
f'_0(x) & f'_1(x) & \cdots & f'_n(x) \\
\vdots & & & \vdots \\
f_0^{(n)}(x) & f_1^{(n)}(x) & \cdots & f_n^{(n)}(x)
\end{bmatrix}
\begin{bmatrix}
a_0 \\ a_1 \\ \vdots \\ a_n
\end{bmatrix}
=
\begin{bmatrix}
0 \\ 0 \\ \vdots \\ 0
\end{bmatrix}.
$$

If we denote the coefficient matrix above as $W(x)$, then $\det[W(x)]$ is called the
Wronskian for $\{f_0(x), f_1(x), \ldots, f_n(x)\}$. If there is a point x_0 in $[a,b]$ such that
$\det[W(x_0)] \neq 0$, then the matrix $W(x)$ is nonsingular at $x = x_0$, and the impli-
cation is that $a_0 = a_1 = \cdots = a_n = 0$. In summary, if the Wronskian is nonzero
at any point in $[a,b]$, then $\{f_0(x), f_1(x), \ldots, f_n(x)\}$ is a linearly independent set
of functions. Note, however, that $\det[W(x)] = 0$ for all x in $[a,b]$ does not imply
linear dependence (see Example 4).

EXAMPLE 4 Let $F_1 = \{x, \cos x, \sin x\}$ and $F_2 = \{\sin^2 x, |\sin x| \sin x\}$ for $-1 \leq x \leq 1$. The
respective Wronskians are

$$
w_1(x) =
\begin{vmatrix}
x & \cos x & \sin x \\
1 & -\sin x & \cos x \\
0 & -\cos x & -\sin x
\end{vmatrix}
= x
$$

and

$$
w_2(x) =
\begin{vmatrix}
\sin^2 x & |\sin x| \sin x \\
\sin 2x & |\sin 2x|
\end{vmatrix}
= 0.
$$

Since $w_1(x) \neq 0$ for $x \neq 0$, F_1 is linearly independent. Even though $w_2(x) = 0$ for
all x in $[-1, 1]$, F_2 is also linearly independent, for if $a_1 \sin^2 x + a_2 |\sin x| \sin x =
0$, then at $x = 1$, $a_1 + a_2 = 0$; and at $x = -1$, $a_1 - a_2 = 0$; so $a_1 = a_2 = 0$.

Wronskians

Wronskians are named after the Polish mathematician Josef Maria Hoëné-Wroński (1778–1853). Unfortunately, the violent character of his personal life often detracted from the respect he was due from his mathematical work. The Wronskian provides a partial test for linear independence. If the Wronskian is nonzero for some x_0 in $[a, b]$, then $f_0(x), f_1(x),..., f_n(x)$ are linearly independent (see the first part of Example 4). If the Wronskian is zero for all x in $[a, b]$, then the test gives no information (see the second part of Example 4).

The Wronskian does provide a complete test for linear independence, however, when $f_0(x), f_1(x), ..., f_n(x)$ are solutions of an $(n+1)$st-order linear differential equation of the form

$$y^{(n+1)} + g_n(x)y^{(n)} + \cdots + g_1(x)y' + g_0(x)y = 0,$$

where $g_0(x), g_1(x),..., g_n(x)$ are all continuous on (a, b). In this case, $f_0(x), f_1(x),..., f_n(x)$ are linearly independent if and only if the Wronskian is never zero for any x in (a, b).

Elementary Matrices (Optional)

In this subsection, we observe that the result of applying a sequence of elementary column operations to a matrix A can be represented in matrix terms as multiplication of A by a sequence of "elementary matrices." In particular, let I denote the $(n \times n)$ identity matrix, and let E be the matrix that results when an elementary column operation is applied to I. Such a matrix E is called an *elementary matrix*.

For example, consider the (3×3) matrices

$$E_1 = \begin{bmatrix} 1 & 0 & 3 \\ 0 & 1 & 0 \\ 0 & 0 & 1 \end{bmatrix} \quad \text{and} \quad E_2 = \begin{bmatrix} 0 & 1 & 0 \\ 1 & 0 & 0 \\ 0 & 0 & 1 \end{bmatrix}.$$

As we can see, E_1 is obtained from I by adding 3 times the first column of I to the third column of I. Similarly, E_2 is obtained from I by interchanging the first and second columns of I. Thus E_1 and E_2 are specific examples of (3×3) elementary matrices.

The next theorem shows how elementary matrices can be used to represent elementary column operations as matrix products.

THEOREM 15

Let E be the $(n \times n)$ elementary matrix that results from performing a certain column operation on the $(n \times n)$ identity. If A is any $(n \times n)$ matrix, then AE is the matrix that results when this same column operation is performed on A.

PROOF We prove Theorem 15 only for the case in which the column operation is to add c times column i to column j. The rest of the proof is left to the exercises.

Let E denote the elementary matrix derived by adding c times the ith column of I to the jth column of I. Since I is given by $I = [\mathbf{e}_1, \mathbf{e}_2, \ldots, \mathbf{e}_i, \ldots, \mathbf{e}_j, \ldots, \mathbf{e}_n]$, we can represent the elementary matrix E in column form as

$$E = [\mathbf{e}_1, \mathbf{e}_2, \ldots, \mathbf{e}_i, \ldots, \mathbf{e}_j + c\mathbf{e}_i, \ldots, \mathbf{e}_n].$$

Consequently, in column form, AE is the matrix

$$AE = [A\mathbf{e}_1, A\mathbf{e}_2, \ldots, A\mathbf{e}_i, \ldots, A(\mathbf{e}_j + c\mathbf{e}_i), \ldots, A\mathbf{e}_n].$$

Next, if $A = [\mathbf{A}_1, \mathbf{A}_2, \ldots, \mathbf{A}_n]$, then $A\mathbf{e}_k = \mathbf{A}_k$, $1 \le k \le n$. Therefore, AE has the form

$$AE = [\mathbf{A}_1, \mathbf{A}_2, \ldots, \mathbf{A}_i, \ldots, \mathbf{A}_j + c\mathbf{A}_i, \ldots, \mathbf{A}_n].$$

From this column representation for AE, it follows that AE is the matrix that results when c times column i of A is added to column j.

We now use Theorem 15 to prove Theorem 11. Let A be an $(n \times n)$ matrix. Then A can be reduced to a lower-triangular matrix L by using a sequence of column operations. Equivalently, by Theorem 15, there is a sequence of elementary matrices $E_1, E_2, \ldots, E_r$ such that

$$AE_1 E_2 \cdots E_r = L. \tag{5}$$

In Eq. (5), an elementary matrix E_k represents either a column interchange or the addition of a multiple of one column to another. It can be shown that:

a) If E_k represents a column interchange, then E_k is symmetric.

b) If E_k represents the addition of a multiple of column i to column j, where $i < j$, then E_k is an upper-triangular matrix with all main diagonal entries equal to 1.

Now in Eq. (5), let Q denote the matrix $Q = E_1 E_2 \cdots E_r$ and observe that Q is nonsingular because each E_k is nonsingular. To complete the proof of Theorem 11, we need to verify that $\det(Q^T) = \det(Q)$.

From the remarks in (a) and (b) above, $\det(E_k^T) = \det(E_k)$, $1 \le k \le r$, since each matrix E_k is either symmetric or triangular. Thus

$$\begin{aligned}
\det(Q^T) &= \det(E_r^T \cdots E_2^T E_1^T) \\
&= \det(E_r^T) \cdots \det(E_2^T) \det(E_1^T) \\
&= \det(E_r) \cdots \det(E_2) \det(E_1) \\
&= \det(Q).
\end{aligned}$$

An illustration of the discussion above is provided by the next example.

EXAMPLE 5 Let A be the (3×3) matrix

$$A = \begin{bmatrix} 0 & 1 & 3 \\ 1 & 2 & 1 \\ 3 & 4 & 2 \end{bmatrix}.$$

Display elementary matrices E_1, E_2, and E_3 such that $AE_1E_2E_3 = L$, where L is lower triangular.

SOLUTION   Matrix A can be reduced to a lower-triangular matrix by the following sequence of column operations:

$$A = \begin{bmatrix} 0 & 1 & 3 \\ 1 & 2 & 1 \\ 3 & 4 & 2 \end{bmatrix} \xrightarrow{C_1 \leftrightarrow C_2} \begin{bmatrix} 1 & 0 & 3 \\ 2 & 1 & 1 \\ 4 & 3 & 2 \end{bmatrix}$$

$$\xrightarrow{C_3 - 3C_1} \begin{bmatrix} 1 & 0 & 0 \\ 2 & 1 & -5 \\ 4 & 3 & -10 \end{bmatrix} \xrightarrow{C_3 + 5C_2} \begin{bmatrix} 1 & 0 & 0 \\ 2 & 1 & 0 \\ 4 & 3 & 5 \end{bmatrix}.$$

Therefore, $AE_1E_2E_3 = L$, where

$$E_1 = \begin{bmatrix} 0 & 1 & 0 \\ 1 & 0 & 0 \\ 0 & 0 & 1 \end{bmatrix}, \quad E_2 = \begin{bmatrix} 1 & 0 & -3 \\ 0 & 1 & 0 \\ 0 & 0 & 1 \end{bmatrix}, \quad \text{and} \quad E_3 = \begin{bmatrix} 1 & 0 & 0 \\ 0 & 1 & 5 \\ 0 & 0 & 1 \end{bmatrix}.$$

Note that E_1 is symmetric and E_2 and E_3 are upper triangular. ▢

Exercises 5.5

In Exercises 1–4, use row operations to reduce the given determinant to upper-triangular form and determine the value of the original determinant.

1. $\begin{vmatrix} 1 & 2 & 1 \\ 2 & 3 & 2 \\ -1 & 4 & 1 \end{vmatrix}$

2. $\begin{vmatrix} 0 & 3 & 1 \\ 1 & 2 & 1 \\ 2 & -2 & 2 \end{vmatrix}$

3. $\begin{vmatrix} 0 & 1 & 3 \\ 1 & 2 & 2 \\ 3 & 1 & 0 \end{vmatrix}$

4. $\begin{vmatrix} 1 & 0 & 1 \\ 0 & 2 & 4 \\ 3 & 2 & 1 \end{vmatrix}$

In Exercises 5–10, find the adjoint matrix for the given matrix A. Next, use Theorem 14 to calculate the inverse of the given matrix.

5. $\begin{bmatrix} 1 & 2 \\ 3 & 4 \end{bmatrix}$

6. $\begin{bmatrix} a & b \\ c & d \end{bmatrix}$

7. $\begin{bmatrix} 1 & 0 & 1 \\ 2 & 1 & 2 \\ 1 & 1 & 2 \end{bmatrix}$

8. $\begin{bmatrix} 2 & 1 & 0 \\ 3 & 0 & 1 \\ 0 & 1 & 1 \end{bmatrix}$

9. $\begin{bmatrix} 1 & 1 & 1 \\ 1 & 2 & 2 \\ 1 & 3 & 1 \end{bmatrix}$

10. $\begin{bmatrix} 1 & 2 & 3 \\ 0 & 1 & 2 \\ 0 & 0 & 1 \end{bmatrix}$

In Exercises 11–16, calculate the Wronskian. Also, determine whether the given set of functions is linearly independent on the interval $[-1, 1]$.

11. $\{1, x, x^2\}$

12. $\{e^x, e^{2x}, e^{3x}\}$

13. $\{1, \cos^2 x, \sin^2 x\}$

14. $\{1, \cos x, \cos 2x\}$

15. $\{x^2, x|x|\}$

16. $\{x^2, 1 + x^2, 2 - x^2\}$

In Exercises 17–20, find elementary matrices E_1, E_2, and E_3 such that $AE_1E_2E_3 = L$, where L is lower triangular. Calculate the product $Q = E_1E_2E_3$ and verify that $AQ = L$ and $\det(Q) = \det(Q^T)$.

17. $A = \begin{bmatrix} 0 & 1 & 3 \\ 1 & 2 & 4 \\ 2 & 2 & 1 \end{bmatrix}$

18. $A = \begin{bmatrix} 0 & -1 & 2 \\ 1 & 3 & -1 \\ 1 & 2 & 1 \end{bmatrix}$

19. $A = \begin{bmatrix} 1 & 2 & -1 \\ 3 & 5 & 1 \\ 4 & 0 & 2 \end{bmatrix}$

20. $A = \begin{bmatrix} 2 & 4 & -6 \\ 1 & 1 & 1 \\ 3 & 2 & 1 \end{bmatrix}$

In Exercises 21–24, calculate $\det[A(x)]$ and show that the given matrix $A(x)$ is nonsingular for any real value of x. Use Theorem 14 to find an expression for the inverse of $A(x)$.

21. $A(x) = \begin{bmatrix} x & 1 \\ -1 & x \end{bmatrix}$ **22.** $A(x) = \begin{bmatrix} 1 & x \\ -x & 2 \end{bmatrix}$

23. $A(x) = \begin{bmatrix} 2 & x & 0 \\ -x & 2 & x \\ 0 & -x & 2 \end{bmatrix}$

24. $A(x) = \begin{bmatrix} \sin x & 0 & \cos x \\ 0 & 1 & 0 \\ -\cos x & 0 & \sin x \end{bmatrix}$

25. Let L and U be the (3×3) matrices

$$L = \begin{bmatrix} 1 & 0 & 0 \\ a & 1 & 0 \\ b & c & 1 \end{bmatrix} \quad \text{and} \quad U = \begin{bmatrix} 1 & a & b \\ 0 & 1 & c \\ 0 & 0 & 1 \end{bmatrix}.$$

Use Theorem 14 to show that L^{-1} is lower triangular and U^{-1} is upper triangular.

26. Let L be a nonsingular (4×4) lower-triangular matrix. Show that L^{-1} is also a lower-triangular matrix. (*Hint:* Consider a variation of Exercise 25.)

27. Let A be an $(n \times n)$ matrix, where $\det(A) = 1$ and A contains only integer entries. Show that A^{-1} contains only integer entries.

28. Let E denote the $(n \times n)$ elementary matrix corresponding to an interchange of the ith and jth columns of I. Let A be any $(n \times n)$ matrix.

a) Show that matrix AE is equal to the result of interchanging columns i and j of A.

b) Show that matrix E is symmetric.

29. An $(n \times n)$ matrix A is called **skew symmetric** if $A^T = -A$. Show that if A is skew symmetric, then $\det(A) = (-1)^n \det(A)$. If n is odd, show that A must be singular.

30. An $(n \times n)$ real matrix is **orthogonal** provided that $A^T = A^{-1}$. If A is an orthogonal matrix, prove that $\det(A) = \pm 1$.

31. Let A be an $(n \times n)$ nonsingular matrix. Prove that $\det[\text{Adj}(A)] = [\det(A)]^{n-1}$. (*Hint:* Use Theorem 14.)

32. Let A be an $(n \times n)$ nonsingular matrix.

a) Show that

$$[\text{Adj}(A)]^{-1} = \frac{1}{\det(A)} A.$$

(*Hint:* Use Theorem 14.)

b) Show that

$$\text{Adj}(A^{-1}) = \frac{1}{\det(A)} A.$$

[*Hint:* Use Theorem 14 to obtain a formula for $(A^{-1})^{-1}$.]

Supplementary Computational Exercises

1. Express

$$\begin{vmatrix} a_{11} + b_{11} & a_{12} + b_{12} \\ a_{21} + b_{21} & a_{22} + b_{22} \end{vmatrix}$$

as a sum of four determinants in which there are no sums in the entries.

2. Let $A = [\mathbf{A}_1, \mathbf{A}_2, \dots, \mathbf{A}_n]$ be an $(n \times n)$ matrix and let $B = [\mathbf{A}_n, \mathbf{A}_{n-1}, \dots, \mathbf{A}_1]$. How are $\det(A)$ and $\det(B)$ related when n is odd? when n is even?

3. If A is an $(n \times n)$ matrix such that $A^3 = A$, then list all possible values for $\det(A)$.

4. If A is a nonsingular (2×2) matrix and c is a scalar such that $A^T = cA$, what are the possible values for c? If A is a nonsingular (3×3) matrix, what are the possible values for c?

5. Let $A = (a_{ij})$ be a (3×3) matrix such that $\det(A) = 2$, and let A_{ij} denote the ijth cofactor of A. If

$$B = \begin{bmatrix} A_{31} & A_{21} & A_{11} \\ A_{32} & A_{22} & A_{12} \\ A_{33} & A_{23} & A_{13} \end{bmatrix},$$

then calculate AB.

6. Let $A = (a_{ij})$ be a (3×3) matrix with $a_{11} = 1$, $a_{12} = 2$, and $a_{13} = -1$. Let

$$C = \begin{bmatrix} -7 & 5 & 4 \\ -4 & 3 & 2 \\ 9 & -7 & -5 \end{bmatrix}$$

be the matrix of cofactors for A. (That is, $A_{11} = -7$, $A_{12} = 5$, and so on.) Find A.

7. Let $\mathbf{b} = [b_1, b_2, \ldots, b_n]^T$.
   a) For $1 \le i \le n$, let A_i be the $(n \times n)$ matrix $A_i = [\mathbf{e}_1, \ldots, \mathbf{e}_{i-1}, \mathbf{b}, \mathbf{e}_{i+1}, \ldots, \mathbf{e}_n]$. Apply Cramer's rule to the system $I_n\mathbf{x} = \mathbf{b}$ to show that $\det(A_i) = b_i$.
   b) If B is the $(n \times n)$ matrix $B = [\mathbf{b}, \ldots, \mathbf{b}]$, then use part (a) and Theorem 4 to determine a formula for $\det(B + I)$.

8. If the Wronskian for $\{f_0(x), f_1(x), f_2(x)\}$ is $(x^2 + 1)e^x$, then calculate the Wronskian for $\{xf_0(x), xf_1(x), xf_2(x)\}$.

Supplementary Conceptual Exercises

In Exercises 1–8, answer true or false. Justify your answer by providing a counterexample if the statement is false or an outline of a proof if the statement is true.

1. If A, B, and C are $(n \times n)$ matrices such that $AB = AC$ and $\det(A) \ne 0$, then $B = C$.

2. If A and B are $(n \times n)$ matrices, then $\det(AB) = \det(BA)$.

3. If A is an $(n \times n)$ matrix and c is a scalar, then $\det(cI_n - A) = c^n - \det(A)$.

4. If A is an $(n \times n)$ matrix and c is a scalar, then $\det(cA) = c \det(A)$.

5. If A is an $(n \times n)$ matrix such that $A^k = \mathcal{O}$ for some positive integer k, then $\det(A) = 0$.

6. If $A_1, A_2, \ldots, A_m$ are $(n \times n)$ matrices such that $B = A_1A_2 \ldots A_m$ is nonsingular, then each A_i is nonsingular.

7. If the matrix A is symmetric, then so is $\text{Adj}(A)$.

8. If A is an $(n \times n)$ matrix such that $\det(A) = 1$, then $\text{Adj}[\text{Adj}(A)] = A$.

In Exercises 9–15, give a brief answer.

9. Show that $A^2 + I = \mathcal{O}$ is not possible if A is an $(n \times n)$ matrix and n is odd.

10. Let A and B be $(n \times n)$ matrices such that $AB = I$. Prove that $BA = I$. [*Hint:* Show that $\det(A) \ne 0$ and conclude that A^{-1} exists.]

11. If A is an $(n \times n)$ matrix and c is a scalar, show that $\det(A^T - cI) = \det(A - cI)$.

12. Let A and B be $(n \times n)$ matrices such that B is nonsingular, and let c be a scalar.
    a) Show that $\det(A - cI) = \det(B^{-1}AB - cI)$.
    b) Show that $\det(AB - cI) = \det(BA - cI)$.

13. If A is a nonsingular $(n \times n)$ matrix, then prove that $\text{Adj}(A)$ is also nonsingular. (*Hint:* Consider the product $A[\text{Adj}(A)]$.)

14. a) If A and B are nonzero $(n \times n)$ matrices such that $AB = \mathcal{O}$, then prove that both A and B are singular. (*Hint:* What would you conclude if either A or B were nonsingular?)
    b) Use part (a) to prove that if A is a singular $(n \times n)$ matrix, then $\text{Adj}(A)$ is also a singular matrix. (*Hint:* Consider the product $A[\text{Adj}(A)]$.)

15. If $A = (a_{ij})$ is an $(n \times n)$ orthogonal matrix (that is, $A^T = A^{-1}$), then prove that $A_{ij} = a_{ij}\det(A)$, where A_{ij} is the ijth cofactor of A. [*Hint:* Express A^{-1} in terms of $\text{Adj}(A)$.]

6

Eigenvalues and

Applications

6.1 Quadratic Forms*

An expression of the sort

$$q(x, y) = ax^2 + bxy + cy^2$$

is called a **quadratic form in x and y.** Similarly, the expression

$$q(x, y, z) = ax^2 + by^2 + cz^2 + dxy + exz + fyz$$

is a quadratic form in the variables x, y, and z.

In general, a **quadratic form** in the variables $x_1, x_2, \ldots, x_n$ is an expression of the form

$$q(\mathbf{x}) = q(x_1, x_2, \ldots, x_n) = \sum_{i=1}^{n} \sum_{j=1}^{n} b_{ij} x_i x_j. \tag{1}$$

In Eq. (1), the coefficients b_{ij} are given constants and, for simplicity, we assume that b_{ij} are *real* constants.

The term *form* means homogeneous polynomial; that is, $q(a\mathbf{x}) = a^k q(\mathbf{x})$. The adjective *quadratic* implies that the form is homogeneous of degrees 2; that is, $q(a\mathbf{x}) = a^2 q(\mathbf{x})$. Quadratic forms occur naturally in applications such as mechanics, vibrations, geometry, optimization, and so on.

---

* The sections in this chapter need not be read in the order they are given; see the preface for details.

Matrix Representations for Quadratic Forms

As we see in Eq. (1), a quadratic form is nothing more than a polynomial in several variables, where each term of the polynomial has degree 2 exactly. It turns out that such polynomials can be represented in the form $q(\mathbf{x}) = \mathbf{x}^T A \mathbf{x}$, where A is a uniquely determined symmetric matrix. For example, consider the quadratic form

$$q(x, y) = 2x^2 + 4xy - 3y^2.$$

Using the properties of matrix multiplication, we can verify that

$$q(x, y) = [x, y] \begin{bmatrix} 2 & 2 \\ 2 & -3 \end{bmatrix} \begin{bmatrix} x \\ y \end{bmatrix}.$$

There is a simple procedure for finding the symmetric matrix $A = (a_{ij})$ such that $q(\mathbf{x}) = \mathbf{x}^T A \mathbf{x}$. In particular, consider the general quadratic form given in Eq. (1). The procedure, as applied to (1), is simply:

1. Define $a_{ii} = b_{ii}$, $1 \leq i \leq n$.
2. Define $a_{ij} = (b_{ij} + b_{ji})/2$, $1 \leq i, j \leq n, i \neq j$.

When these steps are followed, the $(n \times n)$ matrix $A = (a_{ij})$ will be symmetric and will satisfy the conditions $q(\mathbf{x}) = \mathbf{x}^T A \mathbf{x}$.

EXAMPLE 1 Represent the quadratic form in Eq. (2) as $q(\mathbf{x}) = \mathbf{x}^T A \mathbf{x}$:

$$q(\mathbf{x}) = x_1^2 + x_2^2 + 3x_3^2 + 6x_1x_2 + 4x_1x_3 - 10x_2x_3. \tag{2}$$

SOLUTION In the context of Eq. (1), the constant b_{ij} is the coefficient of the term $x_i x_j$. With respect to Eq. (2), $b_{11} = b_{22} = 1$, $b_{33} = 3$, $b_{12} = 6$, $b_{13} = 4$, $b_{23} = -10$, and the other coefficients b_{ij} are zero. Following the simple two-step procedure, we obtain

$$A = \begin{bmatrix} 1 & 3 & 2 \\ 3 & 1 & -5 \\ 2 & -5 & 3 \end{bmatrix}.$$

A quick check shows that $\mathbf{x}^T A \mathbf{x} = q(\mathbf{x})$:

$$[x_1, x_2, x_3] \begin{bmatrix} 1 & 3 & 2 \\ 3 & 1 & -5 \\ 2 & -5 & 3 \end{bmatrix} \begin{bmatrix} x_1 \\ x_2 \\ x_3 \end{bmatrix}$$

$$= x_1^2 + x_2^2 + 3x_3^2 + 6x_1x_2 + 4x_1x_3 - 10x_2x_3.$$

The procedure used in Example 1 is stated formally in the next theorem. Also, for brevity in Theorem 1, we have combined steps (1) and (2) using an equivalent formulation.

THEOREM 1

For $\mathbf{x}$ in R^n, let $q(\mathbf{x})$ denote the quadratic form

$$q(\mathbf{x}) = \sum_{i=1}^{n} \sum_{j=1}^{n} b_{ij} x_i x_j.$$

Let $A = (a_{ij})$ be the $(n \times n)$ matrix defined by

$$a_{ij} = (b_{ij} + b_{ji})/2, \quad 1 \le i, j \le n.$$

The matrix A is symmetric and, moreover, $q(\mathbf{x}) = \mathbf{x}^T A \mathbf{x}$. In addition, there is no other symmetric matrix B such that $\mathbf{x}^T B \mathbf{x} = q(\mathbf{x})$.

PROOF   The fact that A is symmetric comes from the expression that defines a_{ij}. In particular, observe that a_{ij} and a_{ji} are given, respectively, by

$$a_{ij} = \frac{b_{ij} + b_{ji}}{2} \quad \text{and} \quad a_{ji} = \frac{b_{ji} + b_{ij}}{2}.$$

Thus, $a_{ij} = a_{ji}$, which shows that A is symmetric. Also note that $a_{ii} = (b_{ii} + b_{ii})/2 = b_{ii}$, which agrees with step (1) of the previously given two-step procedure. The rest of the proof is left to the exercises.

In Theorem 1, if we relax the condition that A be symmetric, then we no longer have uniqueness. That is, there are many nonsymmetric matrices B such that $\mathbf{x}^T B \mathbf{x} = q(\mathbf{x})$.

Diagonalizing Quadratic Forms

Theorem 1 shows that a quadratic form $q(\mathbf{x})$ can be represented as $q(\mathbf{x}) = \mathbf{x}^T A \mathbf{x}$, where A is a symmetric matrix. For many applications, however, it is useful to have the even simpler representation described in this subsection.

Recall that a real symmetric matrix A can be diagonalized with an orthogonal matrix. That is, there is a square matrix Q such that:

1. $Q^T Q = I$.
2. $Q^T A Q = D$, where D is diagonal.
3. The diagonal entries of D are the eigenvalues of A.

Now consider the quadratic form $q(\mathbf{x}) = \mathbf{x}^T A \mathbf{x}$, where A is $(n \times n)$ and symmetric. If we make the substitution $\mathbf{x} = Q\mathbf{y}$, we obtain

$$
\begin{aligned}
q(\mathbf{x}) &= q(Q\mathbf{y}) \\
&= (Q\mathbf{y})^T A (Q\mathbf{y}) \\
&= \mathbf{y}^T Q^T A Q \mathbf{y} \\
&= \mathbf{y}^T D \mathbf{y} \\
&= \lambda_1 y_1^2 + \lambda_2 y_2^2 + \cdots + \lambda_n y_n^2. \quad\quad \textbf{(3)}
\end{aligned}
$$

The representation in Eq. (3) gives some qualitative information about $q(\mathbf{x})$. For instance, suppose that the matrix A has only positive eigenvalues. In this case, if $q(\mathbf{x})$ is evaluated at some specific vector $\mathbf{x}^*$ in R^n, $\mathbf{x}^* \neq \boldsymbol{\theta}$, then $q(\mathbf{x}^*)$ will always be a positive number.

EXAMPLE 2 Find the substitution $\mathbf{x} = Q\mathbf{y}$ that diagonalizes the quadratic form

$$q(\mathbf{x}) = q(r, s) = r^2 + 4rs - 2s^2.$$

SOLUTION Following Theorem 1, we first represent $q(\mathbf{x})$ as $q(\mathbf{x}) = \mathbf{x}^T A \mathbf{x}$:

$$q(\mathbf{x}) = [r, s]\begin{bmatrix} 1 & 2 \\ 2 & -2 \end{bmatrix}\begin{bmatrix} r \\ s \end{bmatrix}.$$

For the (2×2) matrix A above, the eigenvalues and eigenvectors are (for a and b nonzero),

$$\lambda = 2, \quad \mathbf{w}_1 = a\begin{bmatrix} 2 \\ 1 \end{bmatrix} \quad \text{and} \quad \lambda = -3, \quad \mathbf{w}_2 = b\begin{bmatrix} 1 \\ -2 \end{bmatrix}.$$

An orthogonal matrix Q that diagonalizes A can be formed from normalized eigenvectors:

$$Q = \frac{1}{\sqrt{5}}\begin{bmatrix} 1 & 2 \\ -2 & 1 \end{bmatrix}.$$

The substitution $\mathbf{x} = Q\mathbf{y}$ is given by

$$\begin{bmatrix} r \\ s \end{bmatrix} = \frac{1}{\sqrt{5}}\begin{bmatrix} 1 & 2 \\ -2 & 1 \end{bmatrix}\begin{bmatrix} u \\ v \end{bmatrix},$$

or

$$r = \frac{1}{\sqrt{5}}(u + 2v), \quad s = \frac{1}{\sqrt{5}}(-2u + v).$$

Using the substitution above in the quadratic form, we obtain

$$r^2 + 4rs - 2s^2 = \frac{1}{5}(u + 2v)^2 + \frac{4}{5}(u + 2v)(-2u + v) - \frac{2}{5}(-2u + v)^2$$

$$= \frac{1}{5}[-15u^2 + 10v^2] = 2v^2 - 3u^2.$$ ☐

Classifying Quadratic Forms

We can think of a quadratic form as defining a function from R^n to R. Specifically, if $q(\mathbf{x}) = \mathbf{x}^T A \mathbf{x}$, where A is a real $(n \times n)$ symmetric matrix, then we can define a real-valued function with domain R^n by the rule

$$y = q(\mathbf{x}) = \mathbf{x}^T A \mathbf{x}.$$

The quadratic form is classified as:

a) **Positive definite** if $q(\mathbf{x}) > 0$ for all $\mathbf{x}$ in R^n, $\mathbf{x} \neq \boldsymbol{\theta}$.

b) **Positive semidefinite** if $q(\mathbf{x}) \geq 0$ for all $\mathbf{x}$ in $\mathbf{R}^n$, $\mathbf{x} \neq \boldsymbol{\theta}$.

c) **Negative definite** if $q(\mathbf{x}) < 0$ for all $\mathbf{x}$ in R^n, $\mathbf{x} \neq \boldsymbol{\theta}$.

d) **Negative semidefinite** if $q(\mathbf{x}) \leq 0$ for all $\mathbf{x}$ in R^n, $\mathbf{x} \neq \boldsymbol{\theta}$.

e) **Indefinite** if $q(\mathbf{x})$ assumes both positive and negative values.

The diagonalization process shown in Eq. (3) allows us to classify any specific quadratic form $q(\mathbf{x}) = \mathbf{x}^T A \mathbf{x}$ in terms of the eigenvalues of A. The details are given in the next theorem.

THEOREM 2

Let $q(\mathbf{x})$ be a quadratic form with representation $q(\mathbf{x}) = \mathbf{x}^T A \mathbf{x}$, where A is a symmetric $(n \times n)$ matrix. Let the eigenvalues of A be $\lambda_1, \lambda_2, \ldots, \lambda_n$. The quadratic form is:

a) Positive definite if and only if $\lambda_i > 0$, for $1 \leq i \leq n$.

b) Positive semidefinite if and only if $\lambda_i \geq 0$, for $1 \leq i \leq n$.

c) Negative definite if and only if $\lambda_i < 0$, for $1 \leq i \leq n$.

d) Negative semidefinite if and only if $\lambda_i \leq 0$, for $1 \leq i \leq n$.

e) Indefinite if and only if A has both positive and negative eigenvalues.

The proof is based on the diagonalization shown in Eq. (3) and is left as an exercise.

EXAMPLE 3 Classify the quadratic form

$$q(r, s) = 3r^2 - 4rs + 3s^2.$$

SOLUTION We first find the matrix representation for $q(\mathbf{x}) = q(r, s)$:

$$q(\mathbf{x}) = q(r, s) = [r, s]\begin{bmatrix} 3 & -2 \\ -2 & 3 \end{bmatrix}\begin{bmatrix} r \\ s \end{bmatrix} = \mathbf{x}^T A \mathbf{x}.$$

By Theorem 2, the quadratic form can be classified once we know the eigenvalues of A.

The characteristic polynomial is

$$p(t) = \det(A - tI) = \begin{vmatrix} 3 - t & -2 \\ -2 & 3 - t \end{vmatrix}$$

$$= t^2 - 6t + 5$$

$$= (t - 5)(t - 1).$$

Thus, since all the eigenvalues of A are positive, $q(\mathbf{x})$ is a positive-definite quadratic form.

Because the quadratic form in Example 3 is so simple, it is possible to show directly that the form is positive definite. In particular,

$$q(r, s) = 3r^2 - 4rs + 3s^2$$
$$= 2(r^2 - 2rs + s^2) + r^2 + s^2$$
$$= 2(r - s)^2 + r^2 + s^2. \qquad (4)$$

From Eq. (4), it follows that $q(\mathbf{x}) > 0$ for every nonzero vector $\mathbf{x} = [r, s]^T$.

EXAMPLE 4   Verify that the quadratic form below is indefinite:

$$q(\mathbf{x}) = q(r, s) = r^2 + 4rs - 2s^2.$$

Also, find vectors $\mathbf{x}_1$ and $\mathbf{x}_2$ such that $q(\mathbf{x}_1) > 0$ and $q(\mathbf{x}_2) < 0$.

SOLUTION   Example 2 showed that $q(\mathbf{x}) = \mathbf{x}^T A \mathbf{x}$, where A has eigenvalues $\lambda_1 = 2$ and $\lambda_2 = -3$. By Theorem 2, $q(\mathbf{x})$ is indefinite.
   If $\mathbf{x}_1$ is an eigenvector corresponding to $\lambda_1 = 2$, then $q(\mathbf{x}_1)$ is given by

$$q(\mathbf{x}_1) = \mathbf{x}_1^T A \mathbf{x}_1 = \mathbf{x}_1^T (\lambda_1 \mathbf{x}_1) = \mathbf{x}_1^T (2\mathbf{x}_1) = 2\mathbf{x}_1^T \mathbf{x}_1.$$

Thus, $q(\mathbf{x}_1) > 0$. Similarly, if $A\mathbf{x}_2 = \lambda_2 \mathbf{x}_2$, where $\lambda_2 = -3$, then $q(\mathbf{x}_2)$ is negative since $q(\mathbf{x}_2) = -3\mathbf{x}_2^T \mathbf{x}_2$. ☐

Note that as in (4), the quadratic form in Example 4 can be seen to be indefinite by observing

$$q(r, s) = r^2 + 4rs - 2s^2$$
$$= r^2 + 4rs + 4s^2 - 6s^2$$
$$= (r + 2s)^2 - 6s^2. \qquad (5)$$

From Eq. (5), if r and s are numbers such that $r + 2s = 0$ with s nonzero, then $q(r, s)$ is negative. On the other hand, $q(r, 0)$ is positive for any nonzero value of r. Therefore, Eq. (5) confirms that $q(r, s)$ takes on both positive and negative values. As specific instances, note that $q(2, -1) = -6$ and $q(2, 0) = 4$.

Conic Sections and Quadric Surfaces

The ideas associated with quadratic forms are useful when we want to describe the solution set of a quadratic equation in several variables. For example, consider the general quadratic equation in x and y given below:

$$ax^2 + bxy + cy^2 + dx + ey + f = 0 \qquad (6)$$

(In Eq. 6 we assume that at least one of a, b, or c is nonzero.)
   As we will see, the theory associated with quadratic forms allows us to make a special change of variables that will eliminate the cross-product term in Eq. (6). In particular, there is always a change of variables of the form

$$x = a_1 u + a_2 v, \qquad y = b_1 u + b_2 v,$$

which, when these expressions are substituted into Eq. (6), produces a new equation of the form

$$a'u^2 + b'v^2 + c'u + d'v + e' = 0 \qquad (7)$$

If Eq. (7) has solutions, then the pairs (u, v) that satisfy (7) will define a curve in the uv-plane. Recall from analytic geometry that the solution set of Eq. (7) defines the following curves in the uv-plane:

a) An ellipse when $a'b' > 0$.

b) A hyperbola when $a'b' < 0$.

c) A parabola when $a'b' = 0$ and one of a' or b' is nonzero.

In terms of the original variables x and y, the solution set for Eq. (6) is a curve in the xy-plane. Because of the special nature of the change of variables, the solution set for Eq. (6) can be obtained simply by rotating the curve defined by Eq. (7).

To begin a study of the general quadratic equation in (6), we first rewrite (6) in the form

$$\mathbf{x}^T A \mathbf{x} + \mathbf{a}^T \mathbf{x} + f = 0, \qquad (8)$$

where

$$\mathbf{x} = \begin{bmatrix} x \\ y \end{bmatrix}, \qquad A = \begin{bmatrix} a & b/2 \\ b/2 & c \end{bmatrix}, \quad \text{and} \quad \mathbf{a} = \begin{bmatrix} d \\ e \end{bmatrix}.$$

Now if Q is an orthogonal matrix that diagonalizes A, then the substitution $\mathbf{x} = Q\mathbf{y}$ will remove the cross-product term from Eq. (6). Specifically, suppose that $Q^T A Q = D$, where D is a diagonal matrix. In Eq. (8), the substitution $\mathbf{x} = Q\mathbf{y}$ leads to

$$\mathbf{y}^T D \mathbf{y} + \mathbf{a}^T Q \mathbf{y} + f = 0. \qquad (9)$$

For $\mathbf{y} = [u, v]^T$, Eq. (9) has the simple form

$$\lambda_1 u^2 + \lambda_2 v^2 + c'u + d'v + f = 0. \qquad (10)$$

In Eq. (10), λ_1 and λ_2 are the $(1, 1)$ and $(2, 2)$ entries of D, respectively. (Note that λ_1 and λ_2 are the eigenvalues of A.)

As we noted above, if Eq. (10) has solutions, then the solution set will define an ellipse, a hyperbola, or a parabola in the uv-plane. Since the change of variables $\mathbf{x} = Q\mathbf{y}$ is defined by an orthogonal matrix Q, the pairs $\mathbf{x} = [x, y]^T$ that satisfy Eq. (6) are obtained simply by rotating pairs $\mathbf{y} = [u, v]^T$ that satisfy Eq. (7).

An example will illustrate the ideas.

EXAMPLE 5 Describe and graph the solution set of

$$x^2 + 4xy - 2y^2 + 2\sqrt{5}x + 4\sqrt{5}y - 1 = 0. \qquad (11)$$

SOLUTION The equation has the form $\mathbf{x}^T A \mathbf{x} + \mathbf{a}^T \mathbf{x} + f = 0$, where

$$\mathbf{x} = \begin{bmatrix} x \\ y \end{bmatrix}, \qquad A = \begin{bmatrix} 1 & 2 \\ 2 & -2 \end{bmatrix}, \qquad \mathbf{a} = \begin{bmatrix} 2\sqrt{5} \\ 4\sqrt{5} \end{bmatrix}, \quad \text{and} \quad f = -1.$$

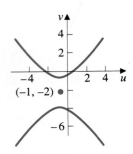

Figure 6.1
The graph of the hyperbola
$$\frac{(v+2)^2}{3} - \frac{(u+1)^2}{2} = 1$$

From Example 2, we know that $Q^T A Q = D$, where

$$Q = \frac{1}{\sqrt{5}} \begin{bmatrix} 1 & 2 \\ -2 & 1 \end{bmatrix} \quad \text{and} \quad D = \begin{bmatrix} -3 & 0 \\ 0 & 2 \end{bmatrix}.$$

For $\mathbf{y} = [u, v]^T$, we make the substitution $\mathbf{x} = Q\mathbf{y}$ in Eq. (11), obtaining

$$2v^2 - 3u^2 - 6u + 8v - 1 = 0.$$

Completing the square, we can express the equation above as

$$2(v^2 + 4v + 4) - 3(u^2 + 2u + 1) = 6,$$

or

$$\frac{(v+2)^2}{3} - \frac{(u+1)^2}{2} = 1. \tag{12}$$

From analytic geometry, Eq. (12) defines a hyperbola in the uv-plane, where the center of the hyperbola has coordinates $(-1, -2)$; see Fig. 6.1, which gives a sketch of the hyperbola. (For reference, the vertices of the hyperbola have coordinates $(-1, -2 \pm \sqrt{3})$ and the foci have coordinates $(-1, -2 \pm \sqrt{13})$.)

Finally, Fig. 6.2 shows the solution set of Eq. (11), sketched in the xy-plane. The hyperbola shown in Fig. 6.2 is a rotation of hyperbola (12) sketched in Fig. 6.1.

Note that, as Example 5 illustrated, if Eq. (6) has real solutions, then the solution set is easy to sketch when a change of variables is used to eliminate any cross-product terms. In some circumstances, however, the solution set of quadratic equation (6) may consist of a single point or there may be no real solutions at all. For instance, the solution set of $x^2 + y^2 = 0$ consists of the single pair $(0, 0)$, whereas the equation $x^2 + y^2 = -1$ has no real solutions.

For quadratic equations involving more than two variables, all the cross-product terms can be eliminated by using the same technique employed with

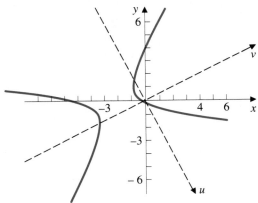

Figure 6.2
The graph of the hyperbola
$$x^2 + 4xy - 2y^2 + 2\sqrt{5}x + 4\sqrt{5}y - 1 = 0$$

Eq. (6). For instance, consider the general quadratic equation in the variables x, y, and z:

$$ax^2 + by^2 + cz^2 + dxy + exz + fyz + px + qy + rz + s = 0. \qquad (13)$$

As with Eq. (6), we can express Eq. (13) in matrix-vector terms as

$$\mathbf{x}^T A\mathbf{x} + \mathbf{a}^T\mathbf{x} + s = 0, \qquad (14)$$

where

$$\mathbf{x} = \begin{bmatrix} x \\ y \\ z \end{bmatrix}, \qquad A = \begin{bmatrix} a & d/2 & e/2 \\ d/2 & b & f/2 \\ e/2 & f/2 & c \end{bmatrix}, \quad \text{and} \quad \mathbf{a} = \begin{bmatrix} p \\ q \\ r \end{bmatrix}.$$

If Q is an orthogonal matrix such that $Q^T A Q = D$, where D is diagonal, then the substitution $\mathbf{x} = Q\mathbf{y}$ will reduce Eq. (14) to

$$\mathbf{y}^T D\mathbf{y} + \mathbf{a}^T(Q\mathbf{y}) + s = 0. \qquad (15)$$

For $\mathbf{y} = [u, v, w]^T$, Eq. (15) has no cross-product terms and will have the form

$$\lambda_1 u^2 + \lambda_2 v^2 + \lambda_3 w^2 + a'u + b'v + c'w + s = 0. \qquad (16)$$

(Again, the scalars λ_1, λ_2, and λ_3 are the diagonal entries of D or, equivalently, the eigenvalues of A.)

If Eq. (16) has real solutions, then the triples (u, v, w) that satisfy (16) will define a surface in three-space. Such surfaces are called quadric surfaces, and detailed descriptions (along with graphs) can be found in most calculus books.

The geometric nature of a quadric surface depends on the λ_i and the scalars a', b', c', and s in Eq. (16). As a simple example, consider the equation

$$\lambda_1 u^2 + \lambda_2 v^2 + \lambda_3 w^2 = d, \qquad d > 0. \qquad (17)$$

If the λ_i are all positive, then the surface defined by Eq. (17) is an *ellipsoid*. If one of the λ_i is negative and the other two are positive, then the surface is a *hyperboloid of one sheet*. If two of the λ_i are negative and the other is positive, the surface is a *hyperboloid of two sheets*. If the λ_i are all negative, then Eq. (17) has no real solutions. The various other surfaces associated with solution sets of Eq. (16) can be found in a calculus book.

The Principal Axis Theorem

The general quadratic equation in n variables has the form

$$\sum_{i=1}^{n} \sum_{j=1}^{n} b_{ij} x_i x_j + \sum_{i=1}^{n} c_i x_i + e = 0. \qquad (18)$$

As we know from the earlier discussions, Eq. (18) can be expressed in matrix-vector terms as

$$\mathbf{x}^T A\mathbf{x} + \mathbf{a}^T\mathbf{x} + e = 0,$$

where A is a real $(n \times n)$ symmetric matrix and $\mathbf{a} = [c_1, c_2, \ldots, c_n]^T$.

The following theorem tells us that it is always possible to make a change of variables that will eliminate the cross-product terms in Eq. (18).

THEOREM 3

The Principal Axis Theorem Let quadratic equation (18) be expressed as

$$\mathbf{x}^T A \mathbf{x} + \mathbf{a}^T \mathbf{x} + e = 0,$$

where A is real $(n \times n)$ symmetric matrix and $\mathbf{a} = [c_1, c_2, \ldots, c_n]^T$. Let Q be an orthogonal matrix such that $Q^T A Q = D$, where D is diagonal. For $\mathbf{y} = [y_1, y_2, \ldots, y_n]^T$, the substitution $\mathbf{x} = Q\mathbf{y}$ transforms Eq. (18) to an equation of the form

$$\lambda_1 y_1^2 + \lambda_2 y_2^2 + \cdots + \lambda_n y_n^2 + d_1 y_1 + d_2 y_2 + \cdots + d_n y_n + e = 0.$$

Exercises 6.1

In Exercises 1–6, find a symmetric matrix A such that $q(\mathbf{x}) = \mathbf{x}^T A \mathbf{x}$.

1. $q(\mathbf{x}) = 2x^2 + 4xy - 3y^2$

2. $q(\mathbf{x}) = -x^2 + 6xy + y^2$

3. $q(\mathbf{x}) = x^2 - 4y^2 + 3z^2 + 2xy - 6xz + 8yz$

4. $q(\mathbf{x}) = u^2 + 4w^2 - z^2 + 2uv + 10uw - 4uz + 4vw - 2vz + 6wz$

5. $q(\mathbf{x}) = [x, y] \begin{bmatrix} 2 & 0 \\ 4 & 1 \end{bmatrix} \begin{bmatrix} x \\ y \end{bmatrix}$

6. $q(\mathbf{x}) = [x, y, z] \begin{bmatrix} 1 & 3 & 1 \\ 5 & 2 & 4 \\ 3 & 2 & 1 \end{bmatrix} \begin{bmatrix} x \\ y \\ z \end{bmatrix}$

In Exercises 7–12, find a substitution $\mathbf{x} = Q\mathbf{y}$ that diagonalizes the given quadratic form, where Q is orthogonal. Also, use Theorem 2 to classify the form as positive definite, positive semidefinite, and so on.

7. $q(\mathbf{x}) = 2x^2 + 6xy + 2y^2$

8. $q(\mathbf{x}) = 5x^2 - 4xy + 5y^2$

9. $q(\mathbf{x}) = x^2 + y^2 + z^2 + 4(xy + xz + yz)$

10. $q(\mathbf{x}) = x^2 + y^2 + z^2 + 2(xy + xz + yz)$

11. $q(\mathbf{x}) = 3x^2 - 2xy + 3y^2$

12. $q(\mathbf{x}) = u^2 + v^2 + w^2 + z^2 - 2(uv + uw + uz + vw + vz + wz)$

In Exercises 13–20, find a substitution $\mathbf{x} = Q\mathbf{y}$ (where Q is orthogonal) that eliminates the cross-product term

in the given equation. Sketch a graph of the transformed equation, where $\mathbf{y} = [u, v]^T$.

13. $2x^2 + \sqrt{3}xy + y^2 = 10$

14. $3x^2 + 2xy + 3y^2 = 8$

15. $x^2 + 6xy - 7y^2 = 8$

16. $3x^2 + 4xy + 5y^2 = 4$

17. $xy = 4$

18. $3x^2 + 2\sqrt{3}xy + y^2 + 4x = 4$

19. $3x^2 - 2xy + 3y^2 = 16$

20. $x^2 + 2xy + y^2 = -1$

21. Consider the quadratic form given by $q(\mathbf{x}) = \mathbf{x}^T A \mathbf{x}$, where A is an $(n \times n)$ symmetric matrix. Suppose that C is any $(n \times n)$ symmetric matrix such that $\mathbf{x}^T A \mathbf{x} = \mathbf{x}^T C \mathbf{x}$, for all $\mathbf{x}$ in R^n. Show that $C = A$. (*Hint:* Let $\mathbf{x} = \mathbf{e}_i$ and verify that $c_{ii} = a_{ii}$, $1 \le i \le n$. Next, consider $\mathbf{x} = \mathbf{e}_r + \mathbf{e}_s$, where $1 \le r, s \le n$.)

22. Consider the quadratic form $q(\mathbf{x}) = \mathbf{x}^T A \mathbf{x}$, where A is an $(n \times n)$ symmetric matrix. Let A have eigenvalues $\lambda_1, \lambda_2, \ldots, \lambda_n$.
   a) Show that if the quadratic form is positive definite, then $\lambda_i > 0$ for $1 \le i \le n$. (*Hint:* Choose $\mathbf{x}$ to be an eigenvector of A.)
   b) Show that if $\lambda_i > 0$ for $1 \le i \le n$, then the quadratic form is positive definite. (*Hint:* Recall Eq. 3.)

(*Note:* Exercise 22 proves property (a) of Theorem 2.)

23. Prove property (b) of Theorem 2.

24. Prove properties (c) and (d) of Theorem 2.

25. Prove property (e) of Theorem 2. [*Note:* The proof of property (e) is somewhat different from the proof of properties (a)–(d).]

26. Let A be an $(n \times n)$ symmetric matrix and consider the function R defined on R^n by

$$R(\mathbf{x}) = \frac{\mathbf{x}^T A \mathbf{x}}{\mathbf{x}^T \mathbf{x}}, \qquad \mathbf{x} \neq \theta.$$

The number $R(\mathbf{x})$ is called a ***Rayleigh quotient***. Let A have eigenvalues $\lambda_1, \lambda_2, \ldots, \lambda_n$, where $\lambda_1 \leq \lambda_2 \leq \lambda_3 \leq \cdots \leq \lambda_n$. Prove that for every $\mathbf{x}$ in R^n, $\lambda_1 \leq R(\mathbf{x}) \leq \lambda_n$. (*Hint:* By the corollary to Theorem 23 in Section 3.7, R^n has an orthonormal basis $\{\mathbf{u}_1, \mathbf{u}_2, \ldots, \mathbf{u}_n\}$, where $A\mathbf{u}_i = \lambda_i \mathbf{u}_i$, $1 \leq i \leq n$. For a given $\mathbf{x}$, $\mathbf{x} \neq \theta$, we can express $\mathbf{x}$ as $\mathbf{x} = a_1 \mathbf{u}_1 +$ $a_2 \mathbf{u}_2 + \cdots + a_n \mathbf{u}_n$. Using this expansion, calculate $\mathbf{x}^T A \mathbf{x}$ and $\mathbf{x}^T \mathbf{x}$.)

27. Let A be an $(n \times n)$ symmetric matrix, as in Exercise 26. Let D denote the set of all vectors $\mathbf{x}$ in R^n such that $\|\mathbf{x}\| = 1$ and consider the quadratic form $q(\mathbf{x}) = \mathbf{x}^T A \mathbf{x}$. Show that the maximum value of $q(\mathbf{x})$, $\mathbf{x}$ in D, is λ_n and the minimum value of $q(\mathbf{x})$, $\mathbf{x}$ in D, is λ_1. (*Hint:* Use the results of Exercise 26. Be sure to verify that the maximum and minimum values are attained.)

28. Let A be an $(n \times n)$ symmetric matrix, and let S be an $(n \times n)$ nonsingular matrix. Define the matrix B by $B = S^T A S$.
   a) Verify that B is symmetric.
   b) Consider the quadratic forms $q_1(\mathbf{x}) = \mathbf{x}^T A \mathbf{x}$ and $q_2(\mathbf{x}) = \mathbf{x}^T B \mathbf{x}$. Show that $q_1(\mathbf{x})$ is positive definite if and only if $q_2(\mathbf{x})$ is positive definite.

6.2 Systems of Differential Equations

In Section 3.8, we provided a brief introduction to the problem of solving a system of differential equations:

$$\begin{aligned}
x_1'(t) &= a_{11}x_1(t) + a_{12}x_2(t) + \cdots + a_{1n}x_n(t) \\
x_2'(t) &= a_{21}x_1(t) + a_{22}x_2(t) + \cdots + a_{2n}x_n(t) \\
&\ \ \vdots \qquad\qquad \vdots \qquad\qquad \vdots \\
x_n'(t) &= a_{n1}x_1(t) + a_{n2}x_2(t) + \cdots + a_{nn}x_n(t).
\end{aligned} \tag{1}$$

A solution to system (1) is a set of functions $x_1(t), x_2(t), \ldots, x_n(t)$ that simultaneously satisfy these equations.

In order to express system (1) in matrix terms, let us define the vector-valued function $\mathbf{x}(t)$ by

$$\mathbf{x}(t) = \begin{bmatrix} x_1(t) \\ x_2(t) \\ \vdots \\ x_n(t) \end{bmatrix}.$$

With $\mathbf{x}(t)$ defined as above, we can write system (1) as

$$\mathbf{x}'(t) = A\mathbf{x}(t), \tag{2}$$

where the vector $\mathbf{x}'(t)$ and the $(n \times n)$ matrix A are given by

$$\mathbf{x}'(t) = \begin{bmatrix} x_1'(t) \\ x_2'(t) \\ \vdots \\ x_n'(t) \end{bmatrix} \quad \text{and} \quad A = \begin{bmatrix} a_{11} & a_{12} & \cdots & a_{1n} \\ a_{21} & a_{22} & \cdots & a_{2n} \\ \vdots & & & \vdots \\ a_{n1} & a_{n2} & \cdots & a_{nn} \end{bmatrix}.$$

The General Solution of x′ = Ax

As in Section 3.8, let us assume that $\mathbf{x}' = A\mathbf{x}$ has a solution of the form

$$\mathbf{x}(t) = e^{\lambda t}\mathbf{u}. \tag{3}$$

For $\mathbf{x}(t) = e^{\lambda t}\mathbf{u}$, we have $\mathbf{x}'(t) = \lambda e^{\lambda t}\mathbf{u}$. Therefore, inserting the trial form (3) into $\mathbf{x}' = A\mathbf{x}$ leads to the condition

$$\lambda e^{\lambda t}\mathbf{u} = A e^{\lambda t}\mathbf{u},$$

which can be rewritten as

$$e^{\lambda t}[A\mathbf{u} - \lambda\mathbf{u}] = \theta. \tag{4}$$

Since $e^{\lambda t}$ is never zero, we see from (4) that $\mathbf{x}(t) = e^{\lambda t}\mathbf{u}$ will be a nontrivial solution of $\mathbf{x}' = A\mathbf{x}$ if and only if λ is an eigenvalue of A and $\mathbf{u}$ is a corresponding eigenvector.

In general, suppose that the $(n \times n)$ matrix A has eigenvalues $\lambda_1, \lambda_2, \ldots, \lambda_n$ and corresponding eigenvectors $\mathbf{u}_1, \mathbf{u}_2, \ldots, \mathbf{u}_n$. Then the vector-valued functions $\mathbf{x}_1(t) = e^{\lambda_1 t}\mathbf{u}_1, \mathbf{x}_2(t) = e^{\lambda_2 t}\mathbf{u}_2, \ldots, \mathbf{x}_n(t) = e^{\lambda_n t}\mathbf{u}_n$ are all solutions of $\mathbf{x}' = A\mathbf{x}$. It is easy to verify, moreover, that any linear combination of $\mathbf{x}_1(t), \mathbf{x}_2(t), \ldots, \mathbf{x}_n(t)$ is also a solution. That is,

$$\mathbf{x}(t) = a_1 e^{\lambda_1 t}\mathbf{u}_1 + a_2 e^{\lambda_2 t}\mathbf{u}_2 + \cdots + a_n e^{\lambda_n t}\mathbf{u}_n \tag{5}$$

will solve $\mathbf{x}' = A\mathbf{x}$ for any choice of scalars $a_1, a_2, \ldots, a_n$.

The question then arises: "Are there solutions to $\mathbf{x}' = A\mathbf{x}$ other than the ones listed in (5)?" The answer is: "If the eigenvectors $\mathbf{u}_1, \mathbf{u}_2, \ldots, \mathbf{u}_n$ are linearly independent, then every solution of $\mathbf{x}' = A\mathbf{x}$ has the form (5)." A proof of this fact can be found in a differential equations text. Equivalently, we can summarize the discussion above as follows:

Let A be an $(n \times n)$ nondefective matrix with linearly independent eigenvectors $\mathbf{u}_1, \mathbf{u}_2, \ldots, \mathbf{u}_n$. Then $\mathbf{x}(t)$ solves $\mathbf{x}' = A\mathbf{x}$ if and only if $\mathbf{x}(t)$ has the form (5).

For A nondefective, the expression (5) is known as the **general solution** of $\mathbf{x}' = A\mathbf{x}$.

EXAMPLE 1   Write the following system of differential equations in the form $\mathbf{x}' = A\mathbf{x}$ and find the general solution:

$$
\begin{aligned}
u' &= 3u + v - w \\
v' &= 12u \quad\;\; - 5w \\
w' &= 4u + 2v - w.
\end{aligned}
$$

SOLUTION   This system has the form $\mathbf{x}' = A\mathbf{x}$, where

$$\mathbf{x}(t) = \begin{bmatrix} u(t) \\ v(t) \\ w(t) \end{bmatrix} \quad \text{and} \quad A = \begin{bmatrix} 3 & 1 & -1 \\ 12 & 0 & -5 \\ 4 & 2 & -1 \end{bmatrix}.$$

The eigenvalues are $\lambda_1 = -1$, $\lambda_2 = 1$, and $\lambda_3 = 2$. Corresponding eigenvectors are

$$\mathbf{u}_1 = \begin{bmatrix} 1 \\ -2 \\ 2 \end{bmatrix}, \quad \mathbf{u}_2 = \begin{bmatrix} 3 \\ 1 \\ 7 \end{bmatrix}, \quad \text{and} \quad \mathbf{u}_3 = \begin{bmatrix} 1 \\ 1 \\ 2 \end{bmatrix}.$$

Therefore, the general solution is

$$\mathbf{x}(t) = a_1 e^{-t}\mathbf{u}_1 + a_2 e^t\mathbf{u}_2 + a_3 e^{2t}\mathbf{u}_3$$

$$= a_1 e^{-t}\begin{bmatrix} 1 \\ -2 \\ 2 \end{bmatrix} + a_2 e^t\begin{bmatrix} 3 \\ 1 \\ 7 \end{bmatrix} + a_3 e^{2t}\begin{bmatrix} 1 \\ 1 \\ 2 \end{bmatrix}.$$

In terms of the original variables, the general solution is $u(t) = a_1 e^{-t} + 3a_2 e^t + a_3 e^{2t}$, $v(t) = -2a_1 e^{-t} + a_2 e^t + a_3 e^{2t}$, $w(t) = 2a_1 e^{-t} + 7a_2 e^t + 2a_3 e^{2t}$, where a_1, a_2, and a_3 are arbitrary. ☐

In practice, we are often presented with an initial condition as well as a differential equation. Such a problem,

$$\mathbf{x}'(t) = A\mathbf{x}(t), \qquad \mathbf{x}(0) = \mathbf{x}_0, \tag{6}$$

is called an **initial-value problem.** That is, let $\mathbf{x}_0$ be a given initial vector. Then, among all solutions of $\mathbf{x}' = A\mathbf{x}$, we want to identify that special solution that satisfies the initial condition $\mathbf{x}(0) = \mathbf{x}_0$.

When A is nondefective, it is easy to solve the initial-value problem (6). In particular, every solution of $\mathbf{x}' = A\mathbf{x}$ has the form (5), and for $\mathbf{x}(t)$ as in (5) we have

$$\mathbf{x}(0) = a_1\mathbf{u}_1 + a_2\mathbf{u}_2 + \cdots + a_n\mathbf{u}_n.$$

Since the eigenvectors $\mathbf{u}_1, \mathbf{u}_2, \ldots, \mathbf{u}_n$ are linearly independent, we can always choose scalars $\alpha_1, \alpha_2, \ldots, \alpha_n$ such that $\mathbf{x}_0 = \alpha_1\mathbf{u}_1 + \alpha_2\mathbf{u}_2 + \cdots + \alpha_n\mathbf{u}_n$; therefore, $\mathbf{x}(t) = \alpha_1 e^{\lambda_1 t}\mathbf{u}_1 + \alpha_2 e^{\lambda_2 t}\mathbf{u}_2 + \cdots + \alpha_n e^{\lambda_n t}\mathbf{u}_n$ is the unique solution of $\mathbf{x}' = A\mathbf{x}$, $\mathbf{x}(0) = \mathbf{x}_0$.

EXAMPLE 2 Solve the initial-value problem $\mathbf{x}' = A\mathbf{x}$, $\mathbf{x}(0) = \mathbf{x}_0$, where

$$A = \begin{bmatrix} 3 & 1 & -1 \\ 12 & 0 & -5 \\ 4 & 2 & -1 \end{bmatrix} \quad \text{and} \quad \mathbf{x}_0 = \begin{bmatrix} 7 \\ -3 \\ 16 \end{bmatrix}.$$

SOLUTION   From Example 1, the general solution of $\mathbf{x}' = A\mathbf{x}$ is $\mathbf{x}(t) = a_1 e^{-t}\mathbf{u}_1 + a_2 e^t\mathbf{u}_2 + a_3 e^{2t}\mathbf{u}_2$, where

$$\mathbf{u}_1 = \begin{bmatrix} 1 \\ -2 \\ 2 \end{bmatrix}, \quad \mathbf{u}_2 = \begin{bmatrix} 3 \\ 1 \\ 7 \end{bmatrix}, \quad \text{and} \quad \mathbf{u}_3 = \begin{bmatrix} 1 \\ 1 \\ 2 \end{bmatrix}.$$

Therefore, the condition $\mathbf{x}(0) = \mathbf{x}_0$ reduces to $a_1\mathbf{u}_1 + a_2\mathbf{u}_2 + a_3\mathbf{u}_3 = \mathbf{x}_0$, or

$$a_1\begin{bmatrix} 1 \\ -2 \\ 2 \end{bmatrix} + a_2\begin{bmatrix} 3 \\ 1 \\ 7 \end{bmatrix} + a_3\begin{bmatrix} 1 \\ 1 \\ 2 \end{bmatrix} = \begin{bmatrix} 7 \\ -3 \\ 16 \end{bmatrix}.$$

Solving, we find $a_1 = 2$, $a_2 = 2$, and $a_3 = -1$. Thus the solution of the initial-value problem is

$$\begin{aligned}
\mathbf{x}(t) &= 2e^{-t}\begin{bmatrix} 1 \\ -2 \\ 2 \end{bmatrix} + 2e^{t}\begin{bmatrix} 3 \\ 1 \\ 7 \end{bmatrix} - e^{2t}\begin{bmatrix} 1 \\ 1 \\ 2 \end{bmatrix} \\
&= \begin{bmatrix} 2e^{-t} + 6e^{t} - e^{2t} \\ -4e^{-t} + 2e^{t} - e^{2t} \\ 4e^{-t} + 14e^{t} - 2e^{2t} \end{bmatrix}.
\end{aligned}$$

The problem of solving $\mathbf{x}' = A\mathbf{x}$ when A is defective is discussed in Section 6.8. Also, see Exercises 9 and 10 at the end of this section.

Solution by Diagonalization

As noted in Eq. (5), if an $(n \times n)$ matrix A has n linearly independent eigenvectors, then the general solution of $\mathbf{x}'(t) = A\mathbf{x}(t)$ is given by

$$\begin{aligned}
\mathbf{x}(t) &= b_1\mathbf{x}_1(t) + b_2\mathbf{x}_2(t) + \cdots + b_n\mathbf{x}_n(t) \\
&= b_1 e^{\lambda_1 t}\mathbf{u}_1 + b_2 e^{\lambda_2 t}\mathbf{u}_2 + \cdots + b_n e^{\lambda_n t}\mathbf{u}_n.
\end{aligned}$$

Now, given that A has a set of n linearly independent eigenvectors, the solution of $\mathbf{x}'(t) = A\mathbf{x}(t)$ can also be described in terms of diagonalization. This alternative solution process has some advantages, especially for nonhomogeneous systems of the form $\mathbf{x}'(t) = A\mathbf{x}(t) + \mathbf{f}(t)$.

Suppose that A is an $(n \times n)$ matrix with n linearly independent eigenvectors. As we know, A is then diagonalizable. In particular, suppose that

$$S^{-1}AS = D, \qquad D \text{ diagonal}.$$

Next, consider the equation $\mathbf{x}'(t) = A\mathbf{x}(t)$. Let us make the substitution

$$\mathbf{x}(t) = S\mathbf{y}(t).$$

With this substitution, the equation $\mathbf{x}'(t) = A\mathbf{x}(t)$ becomes

$$S\mathbf{y}'(t) = AS\mathbf{y}(t),$$

or

$$\mathbf{y}'(t) = S^{-1}AS\mathbf{y}(t),$$

or

$$\mathbf{y}'(t) = D\mathbf{y}(t). \tag{7}$$

Since D is diagonal, system (7) has the form

$$\begin{bmatrix} y'_1(t) \\ y'_2(t) \\ \vdots \\ y'_n(t) \end{bmatrix} = \begin{bmatrix} \lambda_1 & 0 & 0 & \cdots & 0 \\ 0 & \lambda_2 & 0 & \cdots & 0 \\ & \vdots & & & \\ 0 & 0 & 0 & \cdots & \lambda_n \end{bmatrix} \begin{bmatrix} y_1(t) \\ y_2(t) \\ \vdots \\ y_n(t) \end{bmatrix}.$$

Because D is diagonal, the equation above implies that the component functions, $y_i(t)$, are related by

$$y'_i(t) = \lambda_i y_i(t), \qquad 1 \le i \le n.$$

Then, since the general solution of the scalar equation $w' = \lambda w$ is given by $w(t) = ce^{\lambda t}$, it follows that

$$y_i(t) = c_i e^{\lambda_i t}, \qquad 1 \le i \le n.$$

Therefore, the general solution of $\mathbf{y}'(t) = D\mathbf{y}(t)$ in system (7) is given by

$$\mathbf{y}(t) = \begin{bmatrix} c_1 e^{\lambda_1 t} \\ c_2 e^{\lambda_2 t} \\ \vdots \\ c_n e^{\lambda_n t} \end{bmatrix}, \tag{8}$$

where $c_1, c_2, \ldots, c_n$ are arbitrary constants. In terms of $\mathbf{x}(t)$, we have $\mathbf{x}(t) = S\mathbf{y}(t)$ and $\mathbf{x}(0) = S\mathbf{y}(0)$, where $\mathbf{y}(0) = [c_1, c_2, \ldots, c_n]^T$. For an initial-value problem $\mathbf{x}'(t) = A\mathbf{x}(t)$, $\mathbf{x}(0) = \mathbf{x}_0$, we would choose $c_1, c_2, \ldots, c_n$ so that $S\mathbf{y}(0) = \mathbf{x}_0$ or $\mathbf{y}(0) = S^{-1}\mathbf{x}_0$.

EXAMPLE 3 Use the diagonalization procedure to solve the initial-value problem

$$\begin{aligned} u'(t) &= -2u(t) + v(t) + w(t), & u(0) &= 1 \\ v'(t) &= u(t) - 2v(t) + w(t), & v(0) &= 3 \\ w'(t) &= u(t) + v(t) - 2w(t), & w(0) &= -1. \end{aligned}$$

SOLUTION First, we write the problem as $\mathbf{x}'(t) = A\mathbf{x}(t)$, $\mathbf{x}(0) = \mathbf{x}_0$, where

$$\mathbf{x}(t) = \begin{bmatrix} u(t) \\ v(t) \\ w(t) \end{bmatrix}, \qquad A = \begin{bmatrix} -2 & 1 & 1 \\ 1 & -2 & 1 \\ 1 & 1 & -2 \end{bmatrix}, \quad \text{and} \quad \mathbf{x}_0 = \begin{bmatrix} 1 \\ 3 \\ -1 \end{bmatrix}.$$

The eigenvalues and eigenvectors of A are

$$\lambda_1 = 0, \quad \mathbf{u}_1 = \begin{bmatrix} 1 \\ 1 \\ 1 \end{bmatrix}; \qquad \lambda_2 = -3, \quad \mathbf{u}_2 = \begin{bmatrix} 1 \\ 0 \\ -1 \end{bmatrix}; \qquad \lambda_3 = -3, \quad \mathbf{u}_3 = \begin{bmatrix} 1 \\ -1 \\ 0 \end{bmatrix}.$$

Thus we can construct a diagonalizing matrix S such that $S^{-1}AS = D$ by choosing $S = [\mathbf{u}_1, \mathbf{u}_2, \mathbf{u}_3]$:

$$S = \begin{bmatrix} 1 & 1 & 1 \\ 1 & 0 & -1 \\ 1 & -1 & 0 \end{bmatrix}, \qquad D = \begin{bmatrix} 0 & 0 & 0 \\ 0 & -3 & 0 \\ 0 & 0 & -3 \end{bmatrix}.$$

Next, solving $\mathbf{y}'(t) = D\mathbf{y}(t)$, we obtain

$$\mathbf{y}(t) = \begin{bmatrix} c_1 \\ c_2 e^{-3t} \\ c_3 e^{-3t} \end{bmatrix}.$$

From this, $\mathbf{x}(t) = S\mathbf{y}(t)$, or

$$\mathbf{x}(t) = \begin{bmatrix} c_1 + c_2 e^{-3t} + c_3 e^{-3t} \\ c_1 \qquad\qquad - c_3 e^{-3t} \\ c_1 - c_2 e^{-3t} \end{bmatrix}.$$

To satisfy the initial condition $\mathbf{x}(0) = \mathbf{x}_0 = [1, 3, -1]^T$, we choose $c_1 = 1$, $c_2 = 2$, and $c_3 = -2$. Thus, $\mathbf{x}(t)$ is given by

$$\mathbf{x}(t) = \begin{bmatrix} u(t) \\ v(t) \\ w(t) \end{bmatrix} = \begin{bmatrix} 1 \\ 1 + 2e^{-3t} \\ 1 - 2e^{-3t} \end{bmatrix}.$$

(*Note:* For large t, $\mathbf{x}(t) \approx [1, 1, 1]^T$.)

Complex Solutions

As we have seen, solutions to $\mathbf{x}'(t) = A\mathbf{x}(t)$ are built up from functions of the form

$$\mathbf{x}_i(t) = e^{\lambda_i t} \mathbf{u}_i. \tag{9}$$

In many applications, the function $\mathbf{x}_i(t)$ represents a particular state of the physical system modeled by $\mathbf{x}'(t) = A\mathbf{x}(t)$. Furthermore, for many applications, the state vector $\mathbf{x}_i(t)$ has component functions that are oscillatory in nature (for instance, see Example 5). Now, in general, a function of the form $y(t) = e^{\lambda t}$ has an oscillatory nature if and only if λ is a complex scalar. To explain this fact, we need to give the definition of $e^{\lambda t}$ when λ is complex. In advanced texts it is shown, for $\lambda = a + ib$, that

$$e^{\lambda t} = e^{(a + ib)t} = e^{at}(\cos bt + i \sin bt). \tag{10}$$

An example is presented below of a system $\mathbf{x}'(t) = A\mathbf{x}(t)$, where A has complex eigenvalues.

EXAMPLE 4 Solve the initial-value problem

$$\begin{aligned} u'(t) &= 3u(t) + v(t), & u(0) &= 2 \\ v'(t) &= -2u(t) + v(t), & v(0) &= 8. \end{aligned}$$

SOLUTION The system can be written as $\mathbf{x}'(t) = A\mathbf{x}(t)$, $\mathbf{x}(0) = \mathbf{x}_0$, where

$$\mathbf{x}(t) = \begin{bmatrix} u(t) \\ v(t) \end{bmatrix}, \qquad A = \begin{bmatrix} 3 & 1 \\ -2 & 1 \end{bmatrix}, \quad \text{and} \quad \mathbf{x}_0 = \begin{bmatrix} 2 \\ 8 \end{bmatrix}.$$

The eigenvalues are $\lambda_1 = 2 + i$ and $\lambda_2 = 2 - i$ with corresponding eigenvectors

$$\mathbf{u}_1 = \begin{bmatrix} 1 + i \\ -2 \end{bmatrix} \quad \text{and} \quad \mathbf{u}_2 = \begin{bmatrix} 1 - i \\ -2 \end{bmatrix}.$$

The general solution of $\mathbf{x}'(t) = A\mathbf{x}(t)$ is given by

$$\mathbf{x}(t) = a_1 e^{\lambda_1 t}\mathbf{u}_1 + a_2 e^{\lambda_2 t}\mathbf{u}_2. \tag{11}$$

From Eq. (10) it is clear that $e^{\lambda t}$ has the value 1 when $t = 0$, whether λ is complex or real. Thus to satisfy the condition $\mathbf{x}(0) = \mathbf{x}_0 = [2, 8]^T$, we need to choose a_1 and a_2 in Eq. (11) so that $a_1\mathbf{u}_1 + a_2\mathbf{u}_2 = \mathbf{x}_0$:

$$a_1(1 + i) + a_2(1 - i) = 2$$
$$a_1(-2) \quad + a_2(-2) \quad = 8.$$

Solving the system above by using Gaussian elimination, we obtain $a_1 = -2 - 3i$ and $a_2 = -2 + 3i$.

Having the coefficients a_1 and a_2 in Eq. (11), some complex arithmetic calculations will give the functions u and v that satisfy the given initial-value problem. In particular, since $\lambda_1 = 2 + i$, it follows that

$$e^{\lambda_1 t} = e^{(2 + i)t} = e^{2t}(\cos t + i \sin t).$$

Similarly, from the fact that $\cos(-t) = \cos t$ and $\sin(-t) = -\sin t$,

$$e^{\lambda_2 t} = e^{(2 - i)t} = e^{2t}(\cos t - i \sin t).$$

Thus, $\mathbf{x}(t)$ is given by

$$\mathbf{x}(t) = e^{\lambda_1 t}(a_1 \mathbf{u}_1) + e^{\lambda_2 t}(a_2 \mathbf{u}_2)$$

$$= e^{2t}(\cos t + i \sin t)\begin{bmatrix} 1 - 5i \\ 4 + 6i \end{bmatrix} + e^{2t}(\cos t - i \sin t)\begin{bmatrix} 1 + 5i \\ 4 - 6i \end{bmatrix}$$

$$= \begin{bmatrix} e^{2t}(2\cos t + 10\sin t) \\ e^{2t}(8\cos t - 12\sin t) \end{bmatrix}.$$

That is,

$$u(t) = 2e^{2t}(\cos t + 5\sin t)$$
$$v(t) = 4e^{2t}(2\cos t - 3\sin t).$$

An example of a physical system that leads to a system of differential equations is illustrated in Fig. 6.3. This figure shows a spring–mass system, where $y_1 = 0$ and $y_2 = 0$ indicate the equilibrium position of the masses, and $y_1(t)$ and $y_2(t)$ denote the displacements at time t. For a single spring and mass as in

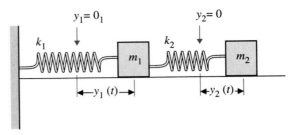

Figure 6.3
A coupled spring–mass system

Fig. 6.4, we can use Hooke's law and $F = ma$ to deduce $my''(t) = -ky(t)$; that is, the restoring force of the spring is proportional to the displacement, $y(t)$, of the mass from the equilibrium position, $y = 0$. The constant of proportionality is the spring constant k, and the minus sign indicates that the force is directed toward equilibrium.

In Fig. 6.3, the spring attached to m_2 is "stretched" (or "compressed") by the amount $y_2(t) - y_1(t)$, so we can write $m_2 y_2''(t) = -k_2[y_2(t) - y_1(t)]$. The mass m_1 is being "pulled" by two springs, so we have $m_1 y_1''(t) = -k_1 y_1(t) + k_2[y_2(t) - y_1(t)]$. Thus the motion of the physical system is governed by

$$y_1''(t) = -\frac{k_1 + k_2}{m_1} y_1(t) + \frac{k_2}{m_1} y_2(t)$$

$$y_2''(t) = \frac{k_2}{m_2} y_1(t) - \frac{k_2}{m_2} y_2(t). \tag{12}$$

To solve these equations, we write them in matrix form as $\mathbf{y}''(t) = A\mathbf{y}(t)$, and we use a trial solution of the form $\mathbf{y}(t) = e^{\omega t}\mathbf{u}$, where $\mathbf{u}$ is a constant vector. Since $\mathbf{y}''(t) = \omega^2 e^{\omega t}\mathbf{u}$, we will have a solution if

$$\omega^2 e^{\omega t}\mathbf{u} - e^{\omega t}A\mathbf{u} = \mathbf{0} \tag{13}$$

or if $(A - \omega^2 I)\mathbf{u} = \mathbf{0}$. Thus to solve $\mathbf{y}''(t) = A\mathbf{y}(t)$, we solve $(A - \lambda I)\mathbf{u} = \mathbf{0}$ and then choose ω so that $\omega^2 = \lambda$. (It can be shown that λ will be negative and real, so ω must be a complex number.)

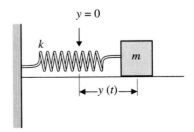

Figure 6.4
A single spring–mass system

<u>EXAMPLE 5</u> Consider the spring–mass system illustrated in Fig. 6.3 and described mathematically in system (12). Suppose that $m_1 = m_2 = 1$, $k_1 = 3$, and $k_2 = 2$. Find $y_1(t)$ and $y_2(t)$ if the initial conditions are

$$y_1(0) = 0, \qquad y_2(0) = 10$$
$$y_1'(0) = 0, \qquad y_2'(0) = 0.$$

SOLUTION System (12) has the form $\mathbf{y}''(t) = A\mathbf{y}(t)$, where

$$\mathbf{y}(t) = \begin{bmatrix} y_1(t) \\ y_2(t) \end{bmatrix} \quad \text{and} \quad A = \begin{bmatrix} -5 & 2 \\ 2 & -2 \end{bmatrix}.$$

By Eq. (13), a function $\mathbf{x}$ of the form

$$\mathbf{x}(t) = e^{\beta t}\mathbf{u}$$

will satisfy $\mathbf{y}''(t) = A\mathbf{y}(t)$ if β^2 is an eigenvalue of A.

The eigenvalues of A are $\lambda_1 = -1$ and $\lambda_2 = -6$, with corresponding eigenvectors

$$\mathbf{u}_1 = \begin{bmatrix} 1 \\ 2 \end{bmatrix} \quad \text{and} \quad \mathbf{u}_2 = \begin{bmatrix} -2 \\ 1 \end{bmatrix}.$$

Thus four solutions of $\mathbf{y}''(t) = A\mathbf{y}(t)$ are

$$\mathbf{x}_1(t) = e^{it}\mathbf{u}_1, \qquad \mathbf{x}_2(t) = e^{-it}\mathbf{u}_1,$$
$$\mathbf{x}_3(t) = e^{\sqrt{6}it}\mathbf{u}_2, \qquad \mathbf{x}_4(t) = e^{-\sqrt{6}it}\mathbf{u}_2.$$

The general solution is

$$\mathbf{y}(t) = a_1\mathbf{x}_1(t) + a_2\mathbf{x}_2(t) + a_3\mathbf{x}_3(t) + a_4\mathbf{x}_4(t). \tag{14}$$

To satisfy the initial conditions, we need to choose the a_i above so that

$$\mathbf{y}(0) = \begin{bmatrix} 0 \\ 10 \end{bmatrix} \quad \text{and} \quad \mathbf{y}'(0) = \begin{bmatrix} 0 \\ 0 \end{bmatrix}.$$

An evaluation of Eq. (14) shows that

$$\mathbf{y}(0) = (a_1 + a_2)\mathbf{u}_1 + (a_3 + a_4)\mathbf{u}_2$$
$$\mathbf{y}'(0) = i(a_1 - a_2)\mathbf{u}_1 + i\sqrt{6}(a_3 - a_4)\mathbf{u}_2.$$

Since $\mathbf{y}'(0) = \mathbf{0}$, we see that $a_1 = a_2$ and $a_3 = a_4$. With this information in the condition $\mathbf{y}(0) = [0, 10]^T$, it follows that $a_1 = 2$ and $a_3 = 1$.

Finally, by Eq. (14), we obtain

$$\mathbf{y}(t) = 2[\mathbf{x}_1(t) + \mathbf{x}_2(t)] + [\mathbf{x}_3(t) + \mathbf{x}_4(t)]$$
$$= \begin{bmatrix} 4\cos t - 4\cos(\sqrt{6}\,t) \\ 8\cos t + 2\cos(\sqrt{6}\,t) \end{bmatrix}.$$

□

Exercises 6.2

In Exercises 1–8, write the given system of differential equations in the form $\mathbf{x}'(t) = A\mathbf{x}(t)$. Express the general solution in the form (5) and determine the particular solution that satisfies the given initial condition. (*Note:* Exercises 5 and 6 involve complex eigenvalues.)

1. $u'(t) = 5u(t) - 2v(t)$
$v'(t) = 6u(t) - 2v(t)$, $\quad \mathbf{x}_0 = \begin{bmatrix} 5 \\ 8 \end{bmatrix}$

2. $u'(t) = 2u(t) - v(t)$
$v'(t) = -u(t) + 2v(t)$, $\quad \mathbf{x}_0 = \begin{bmatrix} 2 \\ -1 \end{bmatrix}$

3. $u'(t) = u(t) + v(t)$
$v'(t) = 2u(t) + 2v(t)$, $\quad \mathbf{x}_0 = \begin{bmatrix} 5 \\ 1 \end{bmatrix}$

4. $u'(t) = 5u(t) - 6v(t)$
$v'(t) = 3u(t) - 4v(t)$, $\quad \mathbf{x}_0 = \begin{bmatrix} 3 \\ 2 \end{bmatrix}$

5. $u'(t) = .5u(t) + .5v(t)$
$v'(t) = -.5u(t) + .5v(t)$, $\quad \mathbf{x}_0 = \begin{bmatrix} 4 \\ 4 \end{bmatrix}$

6. $u'(t) = 6u(t) + 8v(t)$
$v'(t) = -u(t) + 2v(t)$, $\quad \mathbf{x}_0 = \begin{bmatrix} 8 \\ 0 \end{bmatrix}$

7. $u'(t) = 4u(t) + w(t)$
$v'(t) = -2u(t) + v(t)$
$w'(t) = -2u(t) + w(t)$, $\quad \mathbf{x}_0 = \begin{bmatrix} -1 \\ 1 \\ 0 \end{bmatrix}$

8. $u'(t) = 3u(t) + v(t) - 2w(t)$
$v'(t) = -u(t) + 2v(t) + w(t)$, $\quad \mathbf{x}_0 = \begin{bmatrix} -2 \\ 4 \\ -8 \end{bmatrix}$
$w'(t) = 4u(t) + v(t) - 3w(t)$

9. Consider the system

$u'(t) = u(t) - v(t)$
$v'(t) = u(t) + 3v(t).$

a) Write this system in the form $\mathbf{x}'(t) = A\mathbf{x}(t)$ and observe that there is only one solution of the form $\mathbf{x}_1(t) = e^{\lambda t}\mathbf{u}$. What is the solution?

b) Having λ and $\mathbf{u}$, find a vector $\mathbf{y}_0$ for which $\mathbf{x}_2(t) = te^{\lambda t}\mathbf{u} + e^{\lambda t}\mathbf{y}_0$ is a solution. (*Hint:* Substitute $\mathbf{x}_2(t)$ into $\mathbf{x}'(t) = A\mathbf{x}(t)$ to determine $\mathbf{y}_0$. The vector $\mathbf{y}_0$ is called a generalized eigenvector. See Section 6.8.)

c) Show that we can always choose constants c_1 and c_2 such that

$\mathbf{y}(t) = c_1\mathbf{x}_1(t) + c_2\mathbf{x}_2(t)$

satisfies $\mathbf{y}(0) = \mathbf{x}_0$ for any $\mathbf{x}_0$ in R^2.

10. Repeat Exercise 9 for the system

$u'(t) = 2u(t) - v(t)$
$v'(t) = 4u(t) + 6v(t)$

and find the solution that satisfies $u(0) = 1$, $v(0) = 1$.

6.3 Transformation to Hessenberg Form

In order to find the eigenvalues of an $(n \times n)$ matrix A, we would like to find a matrix H that has the same eigenvalues as A but in which the eigenvalues of H are relatively easy to determine. We already know from Section 3.7 that similar matrices have the same eigenvalues, so we shall look for a matrix H such that

$$H = S^{-1}AS$$

and such that H has some special sort of form that facilitates finding the characteristic polynomial for H.

We might hope that we could choose H to be a diagonal or triangular matrix since this choice would make the eigenvalue problem for H trivial. Unfortunately we cannot expect easily to reduce an arbitrary matrix A to a similar matrix H, where H is triangular or diagonal. To see why, recall that if $p(t)$ is any polynomial, then we can construct a matrix B for which $p(t)$ is the characteristic polynomial of B (see Exercise 27 of Section 3.4). If it were easy to reduce B

to a similar matrix H that was triangular or diagonal, then we would have an easy means of finding the roots of $p(t) = 0$. But as we have commented, Abel showed that finding the roots of a polynomial equation cannot be an "easy" problem. Since we cannot expect to find an efficient procedure to transform an $(n \times n)$ matrix A into a similar matrix H that is triangular, we ask for the next best thing—a way to transform A into an almost triangular or Hessenberg matrix. In this section, we establish the details of reduction to Hessenberg form, and in the next section, we state an algorithm that can be used to find the characteristic polynomial of a Hessenberg matrix.

We also prove that this algorithm is mathematically sound, and in the process we develop more of the theoretical foundation for the eigenvalue problem.

To begin, we say that an $(n \times n)$ matrix $H = (h_{ij})$ is a **Hessenberg** matrix if $h_{ij} = 0$ whenever $i > j + 1$. Thus H is a Hessenberg matrix if all the entries below the subdiagonal of H are zero, where the **subdiagonal** of H means the entries $h_{21}, h_{32}, h_{43}, \ldots, h_{n,n-1}$. For example, a (6×6) Hessenberg matrix has the form

$$H = \begin{bmatrix} \times & \times & \times & \times & \times & \times \\ \times & \times & \times & \times & \times & \times \\ 0 & \times & \times & \times & \times & \times \\ 0 & 0 & \times & \times & \times & \times \\ 0 & 0 & 0 & \times & \times & \times \\ 0 & 0 & 0 & 0 & \times & \times \end{bmatrix}.$$

Note that the definition of a Hessenberg matrix insists only that the entries below the subdiagonal are zero; it is irrelevant whether the other entries are zero. Thus, for example, diagonal and upper-triangular matrices are in Hessenberg form; and as an extreme example, the $(n \times n)$ zero matrix is a Hessenberg matrix. Every (2×2) matrix is (trivially) a Hessenberg matrix since there are no entries below the subdiagonal. We will see shortly that Hessenberg form plays the same role for the eigenvalue problem as echelon form does for the problem of solving $A\mathbf{x} = \mathbf{b}$.

EXAMPLE 1 The following matrices are in Hessenberg form:

$$H_1 = \begin{bmatrix} 1 & 2 \\ 3 & 1 \end{bmatrix}, \qquad H_2 = \begin{bmatrix} 1 & 2 & 1 \\ 2 & 3 & 1 \\ 0 & 4 & 2 \end{bmatrix}, \qquad H_3 = \begin{bmatrix} 1 & 2 & 0 & 3 \\ 2 & 0 & 1 & 4 \\ 0 & 1 & 3 & 2 \\ 0 & 0 & 0 & 5 \end{bmatrix}. \qquad \square$$

Our approach to finding the eigenvalues of A has two parts:

1. Find a Hessenberg matrix H that is similar to A.
2. Calculate the characteristic polynomial for H.

As we show below, both of these steps are (relatively) easy. Transforming A to Hessenberg form is accomplished by simple row and column operations that resemble the operations used previously to reduce a matrix to echelon form. Next, the characteristic polynomial for a Hessenberg matrix can be found simply by solving a triangular system of equations. The main theoretical result of this section is Theorem 4, which asserts that every $(n \times n)$ matrix is similar to a Hessenberg matrix. The proof is constructive and shows how the similarity transformation is made.

In order to make the $(n \times n)$ case easier to understand, we begin by showing how a (4×4) matrix can be reduced to Hessenberg form. Let A be the (4×4) matrix

$$A = \begin{bmatrix} a_{11} & a_{12} & a_{13} & a_{14} \\ a_{21} & a_{22} & a_{23} & a_{24} \\ a_{31} & a_{32} & a_{33} & a_{34} \\ a_{41} & a_{42} & a_{43} & a_{44} \end{bmatrix}, \tag{1}$$

and suppose for the moment that $a_{21} \neq 0$. Define a matrix Q_1 by

$$Q_1 = \begin{bmatrix} 1 & 0 & 0 & 0 \\ 0 & 1 & 0 & 0 \\ 0 & \dfrac{-a_{31}}{a_{21}} & 1 & 0 \\ 0 & \dfrac{-a_{41}}{a_{21}} & 0 & 1 \end{bmatrix}, \tag{2a}$$

and observe that Q_1^{-1} is given by

$$Q_1^{-1} = \begin{bmatrix} 1 & 0 & 0 & 0 \\ 0 & 1 & 0 & 0 \\ 0 & \dfrac{a_{31}}{a_{21}} & 1 & 0 \\ 0 & \dfrac{a_{41}}{a_{21}} & 0 & 1 \end{bmatrix}, \tag{2b}$$

(That is, Q_1^{-1} is obtained from Q_1 by changing the sign of the off-diagonal entries of Q_1; equivalently, $Q_1 + Q_1^{-1} = 2I$.)

It is easy to see that forming the product $Q_1 A$ has the effect of adding a multiple of $-a_{31}/a_{21}$ times row 2 of A to row 3 and adding a multiple of $-a_{41}/a_{21}$ times row 2 of A to row 4 of A. Thus $Q_1 A$ has zeros in the $(3,1)$ and $(4,1)$ positions. The matrix $Q_1 A Q_1^{-1}$ is similar to A, and we note that the zeros in the $(3,1)$ and $(4,1)$ positions are not disturbed when the product $(Q_1 A) Q_1^{-1}$ is formed. (This fact is easy to see since Q_1^{-1} has the form $Q_1^{-1} = [\mathbf{e}_1, \mathbf{q}, \mathbf{e}_3, \mathbf{e}_4]$; so the first, third, and fourth columns of $Q_1 A$ are not disturbed when $(Q_1 A) Q_1^{-1}$ is formed.) In summary, when Q_1 and Q_1^{-1} are defined by (2), then $A_1 = Q_1 A Q_1^{-1}$ has the form

$$A_1 = \begin{bmatrix} b_{11} & b_{12} & b_{13} & b_{14} \\ b_{21} & b_{22} & b_{23} & b_{24} \\ 0 & b_{32} & b_{33} & b_{34} \\ 0 & b_{42} & b_{43} & b_{44} \end{bmatrix}. \tag{3}$$

Matrix A_1 is similar to A and represents the first step in Hessenberg reduction. As a point of interest, we note that there is an easy way to see how to construct Q_1. That is, if we wished to create zeros in the (3, 1) and (4, 1) entries of A by using elementary row operations, we could multiply row 2 by $-a_{31}/a_{21}$ and add the result to row 3, and next multiply row 2 by $-a_{41}/a_{21}$ and add the result to row 4. The matrix Q_1 is formed from the (4 × 4) identity I by performing these same row operations on I. [It is not usually possible to use row 1 to create zeros in the (2, 1), (3, 1), and (4, 1) positions and still produce a similar matrix.]

The next step in Hessenberg reduction is analogous to the first. We can introduce a zero into the (4, 2) position of A_1 if we multiply row 3 of A_1 by $-b_{42}/b_{32}$ and add the result to row 4. Following the discussion above, we define Q_2 to be the matrix

$$Q_2 = \begin{bmatrix} 1 & 0 & 0 & 0 \\ 0 & 1 & 0 & 0 \\ 0 & 0 & 1 & 0 \\ 0 & 0 & \dfrac{-b_{42}}{b_{32}} & 1 \end{bmatrix}, \tag{4a}$$

and we note as before that Q_2^{-1} is obtained from Q_2 by changing the sign of the off-diagonal entries:

$$Q_2^{-1} = \begin{bmatrix} 1 & 0 & 0 & 0 \\ 0 & 1 & 0 & 0 \\ 0 & 0 & 1 & 0 \\ 0 & 0 & \dfrac{b_{42}}{b_{32}} & 1 \end{bmatrix}. \tag{4b}$$

By a direct multiplication, it is easy to see that $H = Q_2 A_1 Q_2^{-1}$ is a Hessenberg matrix. Since H is similar to A_1 and A_1 is similar to A, we see that H is similar to A. In fact, $H = Q_2 A_1 Q_2^{-1} = Q_2 (Q_1 A Q_1^{-1}) Q_2^{-1} = (Q_2 Q_1) A (Q_2 Q_1)^{-1}$.

Except for the possibility that $a_{21} = 0$ and/or $b_{32} = 0$, this discussion shows how to reduce an arbitrary (4 × 4) matrix to Hessenberg form. We will describe how to handle zero pivot elements after an example.

EXAMPLE 2   Reduce the (4 × 4) matrix A to Hessenberg form, where

$$A = \begin{bmatrix} 1 & -2 & 4 & 1 \\ 2 & 0 & 5 & 2 \\ 2 & -2 & 9 & 3 \\ -6 & -1 & -16 & -6 \end{bmatrix}.$$

SOLUTION   Following (2a) and (2b), we define Q_1 and Q_1^{-1} to be

$$Q_1 = \begin{bmatrix} 1 & 0 & 0 & 0 \\ 0 & 1 & 0 & 0 \\ 0 & -1 & 1 & 0 \\ 0 & 3 & 0 & 1 \end{bmatrix} \quad \text{and} \quad Q_1^{-1} = \begin{bmatrix} 1 & 0 & 0 & 0 \\ 0 & 1 & 0 & 0 \\ 0 & 1 & 1 & 0 \\ 0 & -3 & 0 & 1 \end{bmatrix}.$$

Given this definition,

$$Q_1 A = \begin{bmatrix} 1 & -2 & 4 & 1 \\ 2 & 0 & 5 & 2 \\ 0 & -2 & 4 & 1 \\ 0 & -1 & -1 & 0 \end{bmatrix},$$

and

$$A_1 = Q_1 A Q_1^{-1} = \begin{bmatrix} 1 & -1 & 4 & 1 \\ 2 & -1 & 5 & 2 \\ 0 & -1 & 4 & 1 \\ 0 & -2 & -1 & 0 \end{bmatrix}.$$

The final step of Hessenberg reduction is to use (4a) and (4b) to define Q_2 and Q_2^{-1}:

$$Q_2 = \begin{bmatrix} 1 & 0 & 0 & 0 \\ 0 & 1 & 0 & 0 \\ 0 & 0 & 1 & 0 \\ 0 & 0 & -2 & 1 \end{bmatrix}, \qquad Q_2^{-1} = \begin{bmatrix} 1 & 0 & 0 & 0 \\ 0 & 1 & 0 & 0 \\ 0 & 0 & 1 & 0 \\ 0 & 0 & 2 & 1 \end{bmatrix}.$$

We obtain $H = Q_2 A_1 Q_2^{-1}$,

$$H = \begin{bmatrix} 1 & -1 & 6 & 1 \\ 2 & -1 & 9 & 2 \\ 0 & -1 & 6 & 1 \\ 0 & 0 & -13 & -2 \end{bmatrix};$$

and H is a Hessenberg matrix that is similar to A.

To complete our discussion of how to reduce a (4×4) matrix to Hessenberg form, we must show how to proceed when $a_{21} = 0$ in (1) or when $b_{32} = 0$ in (3). This situation is easily handled by using one of the permutation matrices (see Exercises 15–22 at the end of the section):

$$P_1 = \begin{bmatrix} 1 & 0 & 0 & 0 \\ 0 & 0 & 1 & 0 \\ 0 & 1 & 0 & 0 \\ 0 & 0 & 0 & 1 \end{bmatrix}, \qquad P_2 = \begin{bmatrix} 1 & 0 & 0 & 0 \\ 0 & 0 & 0 & 1 \\ 0 & 0 & 1 & 0 \\ 0 & 1 & 0 & 0 \end{bmatrix}, \qquad P_3 = \begin{bmatrix} 1 & 0 & 0 & 0 \\ 0 & 1 & 0 & 0 \\ 0 & 0 & 0 & 1 \\ 0 & 0 & 1 & 0 \end{bmatrix}.$$

Each of these matrices is its own inverse: $P_1 P_1 = I$, $P_2 P_2 = I$, $P_3 P_3 = I$. Thus, $P_1 A P_1$ is similar to A, as are $P_2 A P_2$ and $P_3 A P_3$. The action of these similarity transformations is easy to visualize; for example, forming $P_1 A$ has the effect of interchanging rows 2 and 3 of A, whereas forming $(P_1 A)P_1$ switches columns 2 and 3 of $P_1 A$. In detail, $P_1 A P_1$ is given by

$$P_1 A P_1 = \begin{bmatrix} a_{11} & a_{13} & a_{12} & a_{14} \\ a_{31} & a_{33} & a_{32} & a_{34} \\ a_{21} & a_{23} & a_{22} & a_{24} \\ a_{41} & a_{43} & a_{42} & a_{44} \end{bmatrix}.$$

If $a_{21} = 0$ in (1), but $a_{31} \neq 0$, then $P_1 A P_1$ is a matrix similar to A with a nonzero entry in the (2, 1) position. We can clearly carry out the first stage of Hessenberg reduction on $P_1 A P_1$. If $a_{21} = a_{31} = 0$ in (1), but $a_{41} \neq 0$, then $P_2 A P_2$ has a nonzero entry in the (2, 1) position; and we can now carry out the first stage of Hessenberg reduction. Finally, if $a_{21} = a_{31} = a_{41} = 0$, the first stage is not necessary. In A_1 in (3), if $b_{32} = 0$, but $b_{42} \neq 0$, then forming $P_3 A_1 P_3$ will produce a similar matrix with a nonzero entry in the (3, 2) position. Moreover, the first column of A_1 will be left unchanged, and so the second step of Hessenberg reduction can be executed. (In general, interchanging two rows of a matrix A and then interchanging the same two columns produces a matrix similar to A. Also note that the permutation matrices P_1, P_2, and P_3 are derived from the identity matrix I by performing the desired row-interchange operations on I.)

The discussion above proves that every (4×4) matrix is similar to a Hessenberg matrix H and also shows how to construct H. The situation with respect to $(n \times n)$ matrices is exactly analogous, and we can now state the main result of this section.

THEOREM 4

Let A be an $(n \times n)$ matrix. Then there is a nonsingular $(n \times n)$ matrix Q such that $QAQ^{-1} = H$, where H is a Hessenberg matrix.

A proof of Theorem 4 can be constructed along the lines of the discussion for the (4×4) case. Since no new ideas are involved, we omit the proof.

EXAMPLE 3 Reduce A to Hessenberg form, where

$$A = \begin{bmatrix} 1 & 1 & 8 & -2 \\ 0 & 3 & 5 & -1 \\ 1 & -1 & -3 & 2 \\ 3 & -1 & -4 & 9 \end{bmatrix}.$$

SOLUTION In A, the entry a_{21} is zero, so we want to interchange rows 2 and 3. We construct the appropriate permutation matrix P by interchanging rows 2 and 3 of I, obtaining

$$P = \begin{bmatrix} 1 & 0 & 0 & 0 \\ 0 & 0 & 1 & 0 \\ 0 & 1 & 0 & 0 \\ 0 & 0 & 0 & 1 \end{bmatrix}.$$

Clearly, $PP = I$, so that $P^{-1} = P$.

With this, B is similar to A, where $B = PAP$:

$$B = PAP = \begin{bmatrix} 1 & 8 & 1 & -2 \\ 1 & -3 & -1 & 2 \\ 0 & 5 & 3 & -1 \\ 3 & -4 & -1 & 9 \end{bmatrix}.$$

Next, we define a matrix Q_1 and form $A_1 = Q_1 B Q_1^{-1}$, where

$$Q_1 = \begin{bmatrix} 1 & 0 & 0 & 0 \\ 0 & 1 & 0 & 0 \\ 0 & 0 & 1 & 0 \\ 0 & -3 & 0 & 1 \end{bmatrix} \quad \text{and} \quad A_1 = Q_1 B Q_1^{-1} = \begin{bmatrix} 1 & 2 & 1 & -2 \\ 1 & 3 & -1 & 2 \\ 0 & 2 & 3 & -1 \\ 0 & 14 & 2 & 3 \end{bmatrix}.$$

Finally, we form a matrix Q_2 and calculate $H = Q_2 A_1 Q_2^{-1}$:

$$Q_2 = \begin{bmatrix} 1 & 0 & 0 & 0 \\ 0 & 1 & 0 & 0 \\ 0 & 0 & 1 & 0 \\ 0 & 0 & -7 & 1 \end{bmatrix} \quad \text{and} \quad H = Q_2 A_1 Q_2^{-1} = \begin{bmatrix} 1 & 2 & -13 & -2 \\ 1 & 3 & 13 & 2 \\ 0 & 2 & -4 & -1 \\ 0 & 0 & 51 & 10 \end{bmatrix}. \qquad \square$$

Computational Considerations

A variety of similarity transformations, besides the elementary ones we have described, have been developed to reduce a matrix to Hessenberg form. Particularly effective are Householder transformations, a sequence of explicitly defined transformations involving orthogonal matrices (see Chapter 7 and Section 6.5).

Although a reduction process like transformation to Hessenberg form may seem quite tedious, we show in the next section that it is easy to calculate the characteristic polynomial of a Hessenberg matrix. Also, however tedious Hessenberg reduction may seem, the alternative of calculating the characteristic polynomial from $p(t) = \det(A - tI)$ is worse. To illustrate this point, we note that in order to gauge the efficiency of an algorithm (particularly an algorithm that will be implemented on a computer), operations counts are frequently used as a first approximation. By an operations count, we mean a count of the number of multiplications and additions that must be performed in order to execute the algorithm. Given an $(n \times n)$ matrix A, it is not hard to see (see Chapter 7) that a total of approximately n^3 multiplications and n^3 additions are needed to reduce A to Hessenberg form and then to find the characteristic polynomial. By contrast, if A is $(n \times n)$, calculating $p(t)$ from

$$p(t) = \det(A - tI)$$

requires on the order of $n!$ multiplications and $n!$ additions. In the language of computer science, reduction to Hessenberg form is a polynomial-time algorithm, whereas computing $\det(A - tI)$ is an exponential-time algorithm. In a polynomial-time algorithm, execution time grows at a rate proportional to n^k as n grows (where k is a constant); in an exponential time algorithm, execution time grows at least as fast as b^n (where b is a constant larger than 1). The distinction is more than academic because exponential-time algorithms can be used on only the smallest problems, and the basic question is whether or not we can produce acceptable answers to practical problems in a reasonable amount of time. In fact, in some areas of applications, the only known algorithms are

Table 6.1

n	n^3	$n!$
3	27	6
4	64	24
5	125	120
6	216	720
7	343	5,040
8	512	40,320
9	729	362,880
10	1,000	3,628,800
11	1,331	39,916,800
12	1,728	479,001,600

exponential-time algorithms, and hence realistic problems cannot be solved except by an inspired guess. (An example of such a problem is the "traveling salesman's" problem, which arises in operations research.)

Table 6.1 should illustrate the difference between polynomial time and exponential time for the problem of calculating the characteristic polynomial. We can draw some rough conclusions from this table. For instance, if an algorithm requiring n^3 operations is used on a (12×12) matrix, and if the algorithm executes in 1 second, then we would expect any algorithm requiring $n!$ operations to take on the order of 77 hours to execute when applied to the same (12×12) matrix. For larger values of n, the comparison between polynomial-time and exponential-time algorithms borders on the absurd. For example, if an algorithm requiring n^3 operations executes in 1 second for $n = 20$, we would suspect that an algorithm requiring 20! operations would take something like 8×10^{10} hours, or approximately 9,000,000 years.

Exercises 6.3

In Exercises 1–10, reduce the given matrix to Hessenberg form by using similarity transformations. Display the matrices used in the similarity transformations.

1. $\begin{bmatrix} -7 & 4 & 3 \\ 8 & -3 & 3 \\ 32 & -15 & 13 \end{bmatrix}$

2. $\begin{bmatrix} -6 & 3 & -14 \\ -1 & 2 & -2 \\ 2 & 0 & 5 \end{bmatrix}$

3. $\begin{bmatrix} 1 & 3 & 1 \\ 0 & 2 & 4 \\ 1 & 1 & 3 \end{bmatrix}$

4. $\begin{bmatrix} 1 & 2 & -1 \\ 3 & 2 & 1 \\ -6 & 1 & 3 \end{bmatrix}$

5. $\begin{bmatrix} 3 & -1 & -1 \\ 4 & -1 & -2 \\ -12 & 5 & 0 \end{bmatrix}$

6. $\begin{bmatrix} 4 & 0 & 3 \\ 0 & 1 & 2 \\ 3 & 2 & 1 \end{bmatrix}$

7. $\begin{bmatrix} 1 & -1 & -1 & -1 \\ -1 & 1 & -1 & -1 \\ -1 & -1 & 1 & -1 \\ -1 & -1 & -1 & 1 \end{bmatrix}$

8. $\begin{bmatrix} 6 & 1 & 4 & 4 \\ 1 & 6 & 4 & 4 \\ 4 & 4 & 6 & 1 \\ 4 & 4 & 1 & 6 \end{bmatrix}$

9. $\begin{bmatrix} 1 & 2 & 1 & 3 \\ 0 & 1 & 1 & 2 \\ 0 & 3 & 1 & 1 \\ 1 & 2 & 0 & 2 \end{bmatrix}$

10. $\begin{bmatrix} 2 & -2 & 0 & -1 \\ -1 & -1 & -2 & 1 \\ 2 & 2 & 1 & 4 \\ 1 & 1 & -3 & 9 \end{bmatrix}$

11. Consider the general (4×4) Hessenberg matrix H, where a_2, b_3, and c_4 are nonzero:

$$H = \begin{bmatrix} a_1 & b_1 & c_1 & d_1 \\ a_2 & b_2 & c_2 & d_2 \\ 0 & b_3 & c_3 & d_3 \\ 0 & 0 & c_4 & d_4 \end{bmatrix}. \qquad (5)$$

Suppose that λ is any eigenvalue of H, for simplicity, we assume λ is real. Show that the geometric multiplicity of λ must be equal to 1. (*Hint:* Consider the columns of $H - \lambda I$.)

12. Let H be a (4×4) Hessenberg matrix as in (5), but where a_2, b_3, and c_4 are not necessarily nonzero. Suppose that H is similar to a symmetric matrix A. Let λ be an eigenvalue of A, where λ has an algebraic multiplicity greater than 1. Use Exercise 11 to conclude that at least one of a_2, b_3, or c_4 must be zero.

13. Let A be the matrix in Exercise 7 and let H be the Hessenberg matrix found in Exercise 7. Determine the characteristic equation for H and solve the equation to find the eigenvalues of A. (Exercise 12 explains why some subdiagonal entry of H must be zero.)

14. Repeat Exercise 13 for the matrix in Exercise 8.

Exercises 15–22 deal with permutation matrices. Recall (Section 3.7) that an $(n \times n)$ matrix P is a **permutation**

matrix if P is formed by rearranging the columns of the $(n \times n)$ identity matrix. For example, some (3×3) permutation matrices are $P = [\mathbf{e}_3, \mathbf{e}_2, \mathbf{e}_1]$, $P = [\mathbf{e}_2, \mathbf{e}_3, \mathbf{e}_1]$, and $P = [\mathbf{e}_1, \mathbf{e}_3, \mathbf{e}_2]$. By convention the identity matrix, $P = [\mathbf{e}_1, \mathbf{e}_2, \mathbf{e}_3]$, is also considered a permutation matrix.

15. List, in column form, all the possible (3×3) permutation matrices (there are six).

16. List, in column form, all possible (4×4) permutation matrices (there are 24).

17. How many different $(n \times n)$ permutation matrices are there? (*Hint:* How many positions can $\mathbf{e}_1$ occupy? Once $\mathbf{e}_1$ is fixed, how many rearrangements of the remaining $n - 1$ columns are there?)

18. Let P be an $(n \times n)$ permutation matrix. Verify that P is an orthogonal matrix. (*Hint:* Recall (5) in Section 3.7.)

19. Let A be an $(n \times n)$ matrix and P an $(n \times n)$ permutation matrix, $P = [\mathbf{e}_i, \mathbf{e}_j, \mathbf{e}_k, \dots, \mathbf{e}_r]$. Show that $AP = [\mathbf{A}_i, \mathbf{A}_j, \mathbf{A}_k, \dots, \mathbf{A}_r]$; that is, forming AP rearranges the columns of A through the same rearrangement that produced P from I.

20. As in Exercise 19, show that $P^T A$ rearranges the rows of A in the same pattern as the columns of P. (*Hint:* Consider $A^T P$.)

21. Let P and Q be two $(n \times n)$ permutation matrices. Show that PQ is also a permutation matrix.

22. Let P be an $(n \times n)$ permutation matrix. Show that there is a positive integer k such that $P^k = I$. (*Hint:* Consider the sequence, $P, P^2, P^3, \dots$. Can there be infinitely many different matrices in this sequence?)

6.4 Eigenvalues of Hessenberg Matrices

In Section 6.3, we saw that an $(n \times n)$ matrix A is similar to a Hessenberg matrix H. In Section 6.7, we will prove a rather important result, the Cayley–Hamilton theorem (see the corollary to Theorem 15 in Section 6.7). In this section, we see that the Cayley–Hamilton theorem can be used as a tool for calculating the characteristic polynomial of Hessenberg matrices.

As we have noted, Hessenberg form plays somewhat the same role for the eigenvalue problem as echelon form plays with respect to solving $A\mathbf{x} = \mathbf{b}$. For instance, a square matrix A, whether invertible or not, can always be reduced to echelon form by using a sequence of simple row operations. Likewise, a square matrix A, whether diagonalizable or not, can always be reduced to Hessenberg form by using a sequence of simple similarity transformations.

For the problem $A\mathbf{x} = \mathbf{b}$, echelon form is easily achieved and reveals much about the possible solutions of $A\mathbf{x} = \mathbf{b}$, even when A is not invertible. Similarly (see Section 6.8), Hessenberg form provides a convenient framework for discussing generalized eigenvectors. We will need this concept in the event that A is not diagonalizable.

Finally, once the system $A\mathbf{x} = \mathbf{b}$ is made row equivalent to $U\mathbf{x} = \mathbf{c}$, where U is upper triangular (in echelon form), then it is fairly easy to complete the solution process. Likewise, if A is similar to H, where H is in Hessenberg form, then it is relatively easy to find the characteristic polynomial of H (recall that the similar matrices A and H have the same characteristic polynomial).

The Characteristic Polynomial of a Hessenberg Matrix

For hand calculations, an efficient method for determining the characteristic polynomial of a matrix A is as follows:

1. Reduce A to a Hessenberg matrix H, as in Section 6.3.
2. Find the characteristic polynomial for H according to the algorithm described in this subsection.

The algorithm referred to in step (2) is known as ***Krylov's method.*** We outline the steps for Krylov's method in Eqs. (3)–(5) below. For a general $(n \times n)$ matrix A, we note that Krylov's method can fail. For an $(n \times n)$ Hessenberg matrix H, however, the procedure is always effective.

Let H be an $(n \times n)$ Hessenberg matrix. In Section 3.4, we defined the characteristic polynomial for H by $p(t) = \det(H - tI)$. In this section, it will be more convenient to define $p(t)$ by

$$p(t) = \det(tI - H). \tag{1}$$

Note that the properties of determinants show that

$$\det(tI - H) = \det[(-1)(H - tI)]$$
$$= (-1)^n \det(H - tI).$$

Thus the zeros of $p(t)$ in Eq. (1) are the eigenvalues of H. We will call $p(t)$ the characteristic polynomial for H, even though it differs by a factor of $(-1)^n$ from our previous definition.

The algorithm described below in Eqs. (3)–(5) is valid for Hessenberg matrices with *nonzero* subdiagonal elements. In this regard, a Hessenberg matrix, $H = (h_{ij})$, is said to be ***unreduced*** if

$$h_{k,k-1} \neq 0, \qquad k = 2, 3, \ldots, n. \tag{2}$$

If H has at least one subdiagonal entry that is zero, then H will be called ***reduced.***

For example, the Hessenberg matrix H_1 is unreduced, whereas H_2 is reduced:

$$H_1 = \begin{bmatrix} 2 & 1 & 3 & 4 \\ 1 & 3 & 2 & 1 \\ 0 & 4 & 2 & 5 \\ 0 & 0 & 1 & 3 \end{bmatrix}, \qquad H_2 = \begin{bmatrix} 4 & 2 & 5 & 1 \\ 3 & 6 & 4 & 2 \\ 0 & 0 & 1 & 3 \\ 0 & 0 & 4 & 7 \end{bmatrix}.$$

That is, H_2 is reduced since H_2 has a zero on the subdiagonal, in the $(3, 2)$ position.

With these preliminaries, we can state the following algorithm for determining $p(t) = \det(tI - H)$.

ALGORITHM 1   Let H be an unreduced Hessenberg matrix, and let $\mathbf{w}_0$ denote the $(n \times 1)$ unit vector $\mathbf{e}_1$.

a) Compute the vectors $\mathbf{w}_1, \mathbf{w}_2, \ldots, \mathbf{w}_n$ by

$$\mathbf{w}_k = H\mathbf{w}_{k-1}, \qquad \text{for } k = 1, 2, \ldots, n. \tag{3}$$

b) Solve the linear system

$$a_0\mathbf{w}_0 + a_1\mathbf{w}_1 + \cdots + a_{n-1}\mathbf{w}_{n-1} = -\mathbf{w}_n \tag{4}$$

for $a_0, a_1, \ldots, a_{n-1}$.

c) Use the values from (b) as coefficients in $p(t)$:

$$p(t) = t^n + a_{n-1}t^{n-1} + \cdots + a_1 t + a_0. \tag{5}$$

It will be shown in Section 6.7 that $p(t)$ in Eq. (5) is the same as $p(t)$ in Eq. (1). The theoretical basis for Algorithm 1 is discussed in Exercise 23.

EXAMPLE 1   Use Algorithm 1 to find the eigenvalues of

$$H = \begin{bmatrix} 5 & -2 \\ 6 & -2 \end{bmatrix}.$$

SOLUTION   Note that $h_{21} = 6 \neq 0$, so H is unreduced. With $\mathbf{w}_0 = \mathbf{e}_1$, Eq. (3) yields

$$\mathbf{w}_0 = \begin{bmatrix} 1 \\ 0 \end{bmatrix}, \qquad \mathbf{w}_1 = H\mathbf{w}_0 = \begin{bmatrix} 5 \\ 6 \end{bmatrix}, \quad \text{and} \quad \mathbf{w}_2 = H\mathbf{w}_1 = \begin{bmatrix} 13 \\ 18 \end{bmatrix}.$$

From Eq. (4), we have

$$a_0 \begin{bmatrix} 1 \\ 0 \end{bmatrix} + a_1 \begin{bmatrix} 5 \\ 6 \end{bmatrix} = -\begin{bmatrix} 13 \\ 18 \end{bmatrix},$$

or

$$a_0 + 5a_1 = -13$$
$$6a_1 = -18.$$

The solution is $a_1 = -3$ and $a_0 = 2$. Thus, by Eq. (5),

$$p(t) = t^2 - 3t + 2 = (t - 2)(t - 1).$$

Hence $\lambda_1 = 2$ and $\lambda_2 = 1$ are the eigenvalues of H. For the simple example above, the reader can easily check that

$$\det(tI - H) = \begin{vmatrix} t - 5 & 2 \\ -6 & t + 2 \end{vmatrix} = t^2 - 3t + 2. \qquad \square$$

EXAMPLE 2 Use Algorithm 1 to find the eigenvalues of

$$H = \begin{bmatrix} 2 & 2 & -1 \\ -1 & -1 & 1 \\ 0 & 2 & 1 \end{bmatrix}.$$

SOLUTION Note that h_{21} and h_{32} are nonzero, so H is unreduced. With $\mathbf{w}_0 = \mathbf{e}_1 = [1, 0, 0]^T$, Eq. (3) yields

$$\mathbf{w}_1 = H\mathbf{w}_0 = \begin{bmatrix} 2 \\ -1 \\ 0 \end{bmatrix}, \quad \mathbf{w}_2 = H\mathbf{w}_1 = \begin{bmatrix} 2 \\ -1 \\ -2 \end{bmatrix}, \quad \text{and} \quad \mathbf{w}_3 = H\mathbf{w}_2 = \begin{bmatrix} 4 \\ -3 \\ -4 \end{bmatrix}.$$

The system $a_0\mathbf{w}_0 + a_1\mathbf{w}_1 + a_2\mathbf{w}_2 = -\mathbf{w}_3$ is

$$\begin{aligned} a_0 + 2a_1 + 2a_2 &= -4 \\ -a_1 - a_2 &= 3 \\ -2a_2 &= 4. \end{aligned}$$

The solution is $a_2 = -2$, $a_1 = -1$, and $a_0 = 2$. So from Eq. (5),

$$p(t) = t^3 - 2t^2 - t + 2 = (t + 1)(t - 1)(t - 2).$$

Thus the eigenvalues of H are $\lambda_1 = 2$, $\lambda_2 = 1$, and $\lambda_3 = -1$. □

EXAMPLE 3 Use Algorithm 1 to find the eigenvalues of

$$H = \begin{bmatrix} 1 & 1 & 1 & 1 \\ 2 & 0 & 1 & 1 \\ 0 & -1 & -2 & -2 \\ 0 & 0 & 2 & 2 \end{bmatrix}.$$

SOLUTION Since h_{21}, h_{32}, and h_{43} are nonzero, H is unreduced. With $\mathbf{w}_0 = \mathbf{e}_1 = [1, 0, 0, 0]^T$, Eq. (3) yields

$$\mathbf{w}_1 = \begin{bmatrix} 1 \\ 2 \\ 0 \\ 0 \end{bmatrix}, \quad \mathbf{w}_2 = \begin{bmatrix} 3 \\ 2 \\ -2 \\ 0 \end{bmatrix}, \quad \mathbf{w}_3 = \begin{bmatrix} 3 \\ 4 \\ 2 \\ -4 \end{bmatrix}, \quad \text{and} \quad \mathbf{w}_4 = \begin{bmatrix} 5 \\ 4 \\ 0 \\ -4 \end{bmatrix}.$$

The system $a_0\mathbf{w}_0 + a_1\mathbf{w}_1 + a_2\mathbf{w}_2 + a_3\mathbf{w}_3 = -\mathbf{w}_4$ is

$$\begin{aligned} a_0 + a_1 + 3a_2 + 3a_3 &= -5 \\ 2a_1 + 2a_2 + 4a_3 &= -4 \\ -2a_2 + 2a_3 &= 0 \\ -4a_3 &= 4. \end{aligned}$$

Hence $a_3 = -1, a_2 = -1, a_1 = 1$, and $a_0 = 0$; so

$$p(t) = t^4 - t^3 - t^2 + t = t(t - 1)^2(t + 1).$$

The eigenvalues are $\lambda_1 = \lambda_2 = 1, \lambda_3 = 0$, and $\lambda_4 = -1$. This example illustrates that Algorithm 1 is effective even when H is singular ($\lambda_3 = 0$).

As Examples 1–3 indicate, the system (4) that is solved to obtain the coefficients of $p(t)$ is both triangular and nonsingular. This fact is proved in Theorem 5. Knowing that system (4) is nonsingular tells us that Algorithm 1 cannot fail.

Of course, the characteristic polynomial $p(t)$ can also be obtained by expanding the determinant $\det(tI - H)$. Algorithm 1, however, is more efficient (requires fewer arithmetic operations) than a determinant expansion. Besides increased efficiency, we introduce this version of Krylov's method because the technique provides insight into matrix polynomials, generalized eigenvectors, and other important aspects of the eigenvalue problem.

THEOREM 5

Let H be an unreduced ($n \times n$) Hessenberg matrix, and let $\mathbf{w}_0$ denote the ($n \times 1$) unit vector $\mathbf{e}_1$. Then the vectors $\mathbf{w}_0, \mathbf{w}_1, \ldots, \mathbf{w}_{n-1}$, defined by

$$\mathbf{w}_i = H\mathbf{w}_{i-1}, \qquad i = 1, 2, \ldots, n - 1.$$

form a basis for R^n.

PROOF Since any set of n linearly independent vectors in R^n is a basis for R^n, we can prove this theorem by showing that $\{\mathbf{w}_0, \mathbf{w}_1, \ldots, \mathbf{w}_{n-1}\}$ is a linearly independent set of vectors. To prove this, we observe first that $\mathbf{w}_0$ and $\mathbf{w}_1$ are given by

$$\mathbf{w}_0 = \begin{bmatrix} 1 \\ 0 \\ 0 \\ \vdots \\ 0 \end{bmatrix} \quad \text{and} \quad \mathbf{w}_1 = \begin{bmatrix} h_{11} \\ h_{21} \\ 0 \\ \vdots \\ 0 \end{bmatrix}.$$

Forming $\mathbf{w}_2 = H\mathbf{w}_1$, we find that

$$\mathbf{w}_2 = \begin{bmatrix} h_{11}h_{11} + h_{12}h_{21} \\ h_{21}h_{11} + h_{22}h_{21} \\ h_{32}h_{21} \\ 0 \\ \vdots \\ 0 \end{bmatrix}.$$

Since H was given as an unreduced Hessenberg matrix, the second component of $\mathbf{w}_1$ and the third component of $\mathbf{w}_2$ are nonzero.

In general (see Exercise 23) it can be shown that ith component of $\mathbf{w}_{i-1}$ is the product $h_{i,i-1}h_{i-1,i-2}\cdots h_{32}h_{21}$, and the kth component of $\mathbf{w}_{i-1}$ is zero for $k = i+1, i+2, \ldots, n$. Thus the $(n \times n)$ matrix

$$W = [\mathbf{w}_0, \mathbf{w}_1, \ldots, \mathbf{w}_{n-1}]$$

is upper triangular, and the diagonal entries of W are all nonzero. In light of this, we conclude that $\{\mathbf{w}_0, \mathbf{w}_1, \ldots, \mathbf{w}_{n-1}\}$ is a set of n linearly independent vectors in R^n and hence is a basis for R^n.

Reduced Hessenberg Matrices

We now consider a reduced Hessenberg matrix H and illustrate that Algorithm 1 *cannot* be used on H.

EXAMPLE 4 Demonstrate why Algorithm 1 fails on the reduced Hessenberg matrix

$$H = \begin{bmatrix} 1 & 2 & 1 & 3 \\ 2 & 1 & 1 & 1 \\ 0 & 0 & 2 & 1 \\ 0 & 0 & 1 & 1 \end{bmatrix}.$$

SOLUTION Since $h_{32} = 0$, H is reduced. From Eq. (3), with $\mathbf{w}_0 = \mathbf{e}_1$,

$$\mathbf{w}_1 = \begin{bmatrix} 1 \\ 2 \\ 0 \\ 0 \end{bmatrix}, \quad \mathbf{w}_2 = \begin{bmatrix} 5 \\ 4 \\ 0 \\ 0 \end{bmatrix}, \quad \mathbf{w}_3 = \begin{bmatrix} 13 \\ 14 \\ 0 \\ 0 \end{bmatrix}, \quad \text{and} \quad \mathbf{w}_4 = \begin{bmatrix} 41 \\ 40 \\ 0 \\ 0 \end{bmatrix}.$$

The vectors above are linearly dependent, and the solutions of $a_0\mathbf{w}_0 + a_1\mathbf{w}_1 + a_2\mathbf{w}_2 + a_3\mathbf{w}_3 = -\mathbf{w}_4$ are

$$\begin{aligned} a_0 &= -3a_2 - 6a_3 - 21 \\ a_1 &= -2a_2 - 7a_3 - 20, \end{aligned} \tag{6}$$

where a_2 and a_3 are arbitrary. The coefficients of the characteristic polynomial ($a_0 = -3, a_1 = 7, a_2 = 4, a_3 = -5$) are one of the solutions in (6), but in general it is impossible to discern this solution by inspection.

We now prove a result that shows how the eigenvalue problem for H uncouples into smaller problems when H is a reduced Hessenberg matrix. This theorem is based on the observation that a Hessenberg matrix that has a zero subdiagonal entry can be partitioned in a natural and useful way. To illustrate, we consider a (5×5) reduced Hessenberg matrix:

$$H = \begin{bmatrix} 2 & 1 & 3 & 5 & 7 \\ 6 & 2 & 1 & 3 & 8 \\ 0 & 1 & 2 & 1 & 3 \\ 0 & 0 & 0 & 4 & 1 \\ 0 & 0 & 0 & 1 & 6 \end{bmatrix}.$$

We can partition H into four submatrices, H_{11}, H_{12}, H_{22}, and $\mathcal{O}$, as indicated below:

$$H = \left[\begin{array}{ccc|cc} 2 & 1 & 3 & 5 & 7 \\ 6 & 2 & 1 & 3 & 8 \\ 0 & 1 & 2 & 1 & 3 \\ \hline 0 & 0 & 0 & 4 & 1 \\ 0 & 0 & 0 & 1 & 6 \end{array} \right] \qquad (7)$$

$$= \begin{bmatrix} H_{11} & H_{12} \\ \mathcal{O} & H_{22} \end{bmatrix}.$$

For a matrix H partitioned as in (7), we can show that $\det(H) = \det(H_{11}) \det(H_{22})$ and that $\det(tI - H) = \det(tI - H_{11}) \det(tI - H_{22})$. This fact leads to Theorem 6, stated below. We provide a different proof of Theorem 6 in order to demonstrate how to find eigenvectors for a block matrix.

A matrix written in partitioned form, such as

$$H = \begin{bmatrix} H_{11} & H_{12} \\ \mathcal{O} & H_{22} \end{bmatrix},$$

is usually called a block matrix—the entries in H are blocks, or submatrices, of H. In fact, H is called block upper triangular since the only block below the diagonal blocks is a zero block.

When some care is exercised to see that all the products are defined, the blocks in a block matrix can be treated as though they were scalars when forming the product of two block matrices. For example, suppose Q is a (5×5) matrix partitioned in the same fashion as H in (7):

$$Q = \begin{bmatrix} Q_{11} & Q_{12} \\ Q_{21} & Q_{22} \end{bmatrix},$$

so that Q_{11} is (3×3), Q_{12} is (3×2), Q_{21} is (2×3), and Q_{22} is (2×2). Then it is not hard to show that the product HQ is also given in block form as

$$HQ = \begin{bmatrix} H_{11}Q_{11} + H_{12}Q_{21} & H_{11}Q_{12} + H_{12}Q_{22} \\ H_{22}Q_{21} & H_{22}Q_{22} \end{bmatrix}.$$

(Note that all the products make sense in the block representation of HQ.)

With these preliminaries, we now state an important theorem.

THEOREM 6

Let B be an $(n \times n)$ matrix of the form

$$B = \begin{bmatrix} B_{11} & B_{12} \\ \mathcal{O} & B_{22} \end{bmatrix},$$

where B_{11} is $(k \times k)$, B_{12} is $[k \times (n - k)]$, $\mathcal{O}$ is the $[(n - k) \times k]$ zero matrix, and B_{22} is $[(n - k) \times (n - k)]$. Then λ is an eigenvalue of B if and only if λ is an eigenvalue either of B_{11} or of B_{22}.

PROOF   Let $\mathbf{x}$ be any $(n \times 1)$ vector and write $\mathbf{x}$ in partitioned form as

$$\mathbf{x} = \begin{bmatrix} \mathbf{u} \\ \mathbf{v} \end{bmatrix}, \tag{8}$$

where $\mathbf{u}$ is $(k \times 1)$ and $\mathbf{v}$ is $[(n - k) \times 1]$. It is easy to see that the equation $B\mathbf{x} = \lambda\mathbf{x}$ is equivalent to

$$\begin{aligned} B_{11}\mathbf{u} + B_{12}\mathbf{v} &= \lambda\mathbf{u} \\ B_{22}\mathbf{v} &= \lambda\mathbf{v}. \end{aligned} \tag{9}$$

Suppose first that λ is an eigenvalue of B. Then there is a vector $\mathbf{x}$, $\mathbf{x} \neq \boldsymbol{\theta}$, such that $B\mathbf{x} = \lambda\mathbf{x}$. If $\mathbf{v} \neq \boldsymbol{\theta}$ in Eq. (8), then we see from (9) that λ is an eigenvalue of B_{22}. On the other hand, if $\mathbf{v} = \boldsymbol{\theta}$ in (8), then we must have $\mathbf{u} \neq \boldsymbol{\theta}$; and (9) guarantees that λ is an eigenvalue of B_{11}.

Conversely, if λ is an eigenvalue of B_{11}, then there is a nonzero vector $\mathbf{u}_1$ such that $B_{11}\mathbf{u}_1 = \lambda\mathbf{u}_1$. In (8) we set $\mathbf{u} = \mathbf{u}_1$ and $\mathbf{v} = \boldsymbol{\theta}$ to produce a solution of (9), and this result shows that any eigenvalue of B_{11} is also an eigenvalue of B. Finally, suppose that λ is not an eigenvalue of B_{11} but is an eigenvalue of B_{22}. Then there is a nonzero vector $\mathbf{v}_1$ such that $B_{22}\mathbf{v}_1 = \lambda\mathbf{v}_1$; and so $\mathbf{v}_1$ satisfies the last equation in (9). To satisfy the first equation in (9), we must solve

$$(B_{11} - \lambda I)\mathbf{u} = -B_{12}\mathbf{v}_1.$$

But since λ is not an eigenvalue of B_{11}, we know that $B_{11} - \lambda I$ is nonsingular, and so we can solve (9). Thus every eigenvalue of B_{22} is also an eigenvalue of B. ☐

As another example, consider the (7×7) Hessenberg matrix

$$H = \begin{bmatrix} 2 & 3 & 1 & 6 & -1 & 3 & 8 \\ 5 & 7 & 2 & 8 & 2 & 2 & 1 \\ 0 & 0 & 4 & 1 & 3 & -5 & 2 \\ 0 & 0 & 6 & 1 & 2 & 4 & 3 \\ 0 & 0 & 0 & 4 & 1 & 2 & 1 \\ 0 & 0 & 0 & 0 & 0 & 6 & 5 \\ 0 & 0 & 0 & 0 & 0 & 7 & 3 \end{bmatrix}. \tag{10}$$

We first partition H as

$$H = \begin{bmatrix} H_{11} & H_{12} \\ \mathcal{O} & H_{22} \end{bmatrix},$$

where H_{11} is the upper (2×2) block

$$H_{11} = \begin{bmatrix} 2 & 3 \\ 5 & 7 \end{bmatrix}, \qquad H_{22} = \begin{bmatrix} 4 & 1 & 3 & -5 & 2 \\ 6 & 1 & 2 & 4 & 3 \\ 0 & 4 & 1 & 2 & 1 \\ 0 & 0 & 0 & 6 & 5 \\ 0 & 0 & 0 & 7 & 3 \end{bmatrix}.$$

Now the eigenvalues of H are precisely the eigenvalues of H_{11} and H_{22}. The block H_{11} is unreduced, so we can apply the algorithm to find the characteristic polynomial for H_{11}. H_{22} is reduced, however, so we partition H_{22} as

$$H_{22} = \begin{bmatrix} C_{11} & C_{12} \\ \mathcal{O} & C_{22} \end{bmatrix},$$

where C_{11} and C_{22} are

$$C_{11} = \begin{bmatrix} 4 & 1 & 3 \\ 6 & 1 & 2 \\ 0 & 4 & 1 \end{bmatrix} \quad \text{and} \quad C_{22} = \begin{bmatrix} 6 & 5 \\ 7 & 3 \end{bmatrix}.$$

The eigenvalues of H_{22} are precisely the eigenvalues of C_{11} and C_{22}, and we can apply the algorithm to find the characteristic polynomial for C_{11} and C_{22}. In summary, the eigenvalue problem for H has uncoupled into three eigenvalue problems for the unreduced Hessenberg matrices H_{11}, C_{11}, and C_{22}.

Exercises 6.4

In Exercises 1–8, use Algorithm 1 to find the characteristic polynomial for the given matrix.

1. $\begin{bmatrix} 2 & 0 \\ 1 & 1 \end{bmatrix}$

2. $\begin{bmatrix} 0 & 0 \\ 3 & 0 \end{bmatrix}$

3. $\begin{bmatrix} 1 & 0 & 1 \\ 2 & 1 & 0 \\ 0 & 1 & 2 \end{bmatrix}$

4. $\begin{bmatrix} 1 & 2 & 1 \\ 1 & 3 & -1 \\ 0 & 1 & 2 \end{bmatrix}$

5. $\begin{bmatrix} 2 & 4 & 1 \\ 1 & 1 & 3 \\ 0 & 1 & 5 \end{bmatrix}$

6. $\begin{bmatrix} 0 & 0 & 1 \\ 1 & 0 & 0 \\ 0 & 1 & 0 \end{bmatrix}$

7. $\begin{bmatrix} 0 & 1 & 0 & 1 \\ 1 & 2 & 1 & 1 \\ 0 & 1 & 0 & 1 \\ 0 & 0 & 2 & 1 \end{bmatrix}$

8. $\begin{bmatrix} 0 & 2 & 1 & 2 \\ 1 & 0 & 1 & -1 \\ 0 & 2 & 0 & 2 \\ 0 & 0 & 1 & 1 \end{bmatrix}$

In Exercises 9–12, partition the given matrix H into blocks, as in the proof of Theorem 6. Find the eigenvalues of the diagonal blocks and for each distinct eigenvalue, find an eigenvector, as in Eq. (9).

9. $H = \begin{bmatrix} 1 & -1 & 1 & 4 \\ 1 & 3 & -2 & 1 \\ 0 & 0 & 2 & -1 \\ 0 & 0 & -1 & 2 \end{bmatrix}$

10. $H = \begin{bmatrix} 1 & 1 & 2 & 1 \\ 1 & 1 & 1 & 3 \\ 0 & 0 & 3 & 0 \\ 0 & 0 & 1 & 4 \end{bmatrix}$

11. $H = \begin{bmatrix} -2 & 0 & -2 & 1 \\ -1 & 1 & -2 & 3 \\ 0 & 1 & -1 & -2 \\ 0 & 0 & 0 & 2 \end{bmatrix}$

12. $H = \begin{bmatrix} 2 & 3 & 1 & 4 \\ 3 & 2 & 0 & 1 \\ 0 & 0 & 3 & 0 \\ 0 & 0 & 1 & 3 \end{bmatrix}$

13. Consider the block matrix B given by

$$B = \begin{bmatrix} a & b & c & d \\ e & f & g & h \\ 0 & 0 & w & x \\ 0 & 0 & y & z \end{bmatrix} = \begin{bmatrix} B_{11} & B_{12} \\ O & B_{22} \end{bmatrix}.$$

Verify, by expanding $\det(B)$, that $\det(B) = \det(B_{11})\det(B_{22})$.

14. Use the result of Exercise 13 to calculate $\det(H)$, where H is the matrix in Exercise 9.

15. There is one (3×3) permutation matrix P such that P is an unreduced Hessenberg matrix. List this permutation matrix in column form.

16. As in Exercise 15, there is a unique (4×4) per-

mutation matrix P that is both unreduced and Hessenberg. List P in column form.

17. Give the column form for the unique $(n \times n)$ permutation matrix P that is unreduced and Hessenberg.

18. Apply Algorithm 1 to determine the characteristic polynomial for the $(n \times n)$ matrix P in Exercise 17. [*Hint:* Consider $n = 3$ and $n = 4$ to see the nature of system (4).]

19. Let H be an unreduced $(n \times n)$ Hessenberg matrix and let λ be an eigenvalue of H. Show that the geometric multiplicity of λ is equal to 1. (*Hint:* See Exercise 11 of Section 6.3.)

20. Let H be an unreduced $(n \times n)$ Hessenberg matrix and let λ be an eigenvalue of H. Use Exercise 19 to show that if H is symmetric, then H has n distinct eigenvalues.

21. Consider the (2×2) matrix H, where

$$H = \begin{bmatrix} a & b \\ b & c \end{bmatrix}.$$

Calculate the characteristic polynomial for H and use the quadratic formula to show that H has two distinct eigenvalues if H is an unreduced matrix.

22. Let H be an unreduced $(n \times n)$ Hessenberg matrix and suppose $H\mathbf{u} = \lambda\mathbf{u}$, $\mathbf{u} \neq \boldsymbol{\theta}$. Show that the nth component of $\mathbf{u}$ is nonzero.

23. Complete the proof of Theorem 5 by using induction to show that the ith component of $\mathbf{w}_{i-1}$ is nonzero.

6.5 Householder Transformations

In this section, we consider another method for reducing a matrix A to Hessenberg form, using Householder transformations. A Householder transformation (or Householder matrix) is a symmetric orthogonal matrix that has an especially simple form. As we will see, one reason for wanting to use Householder matrices in a similarity transformation is that symmetry is preserved.

> **DEFINITION 1** Let $\mathbf{u}$ be a nonzero vector in R^n and let I be the $(n \times n)$ identity matrix. The $(n \times n)$ matrix Q given by
>
> $$Q = I - \frac{2}{\mathbf{u}^T\mathbf{u}}\mathbf{u}\mathbf{u}^T \tag{1}$$
>
> is called a ***Householder transformation*** or a ***Householder matrix***.

Householder matrices are a basic tool for applied linear algebra and are widely used even in applications not directly involving eigenvalues. For instance, we will see in the next section that Householder matrices can be used to good effect in least-squares problems.

The following theorem shows that a Householder matrix is both symmetric and orthogonal.

THEOREM 7

Let Q be a Householder matrix as in Eq. (1). Then:

a) $Q^T = Q$.
b) $Q^T Q = I$.

PROOF We leave the proof of property (a) to the exercises. To prove property (b), it is sufficient to show that $QQ = I$, since $Q^T = Q$.

To simplify the notation, let b denote the scalar $2/(\mathbf{u}^T\mathbf{u})$ in Eq. (1). Thus

$$Q = I - b\mathbf{u}\mathbf{u}^T, \qquad b = 2/(\mathbf{u}^T\mathbf{u}).$$

Forming QQ, we have

$$
\begin{aligned}
QQ &= (I - b\mathbf{u}\mathbf{u}^T)(I - b\mathbf{u}\mathbf{u}^T) \\
&= I - 2b\mathbf{u}\mathbf{u}^T + b^2(\mathbf{u}\mathbf{u}^T)(\mathbf{u}\mathbf{u}^T) \\
&= I - 2b\mathbf{u}\mathbf{u}^T + b^2\mathbf{u}(\mathbf{u}^T\mathbf{u})\mathbf{u}^T \\
&= I - 2b\mathbf{u}\mathbf{u}^T + b^2(\mathbf{u}^T\mathbf{u})(\mathbf{u}\mathbf{u}^T).
\end{aligned}
\tag{2}
$$

[*Note:* We used the associativity of matrix multiplication to write $(\mathbf{u}\mathbf{u}^T) \times (\mathbf{u}\mathbf{u}^T) = \mathbf{u}(\mathbf{u}^T\mathbf{u})\mathbf{u}^T$.]

Next, observe that $\mathbf{u}^T\mathbf{u}$ is a scalar and that

$$b^2(\mathbf{u}^T\mathbf{u}) = \frac{4}{(\mathbf{u}^T\mathbf{u})^2}(\mathbf{u}^T\mathbf{u}) = \frac{4}{\mathbf{u}^T\mathbf{u}} = 2b.$$

Thus from Eq. (2) it follows that $QQ = I$.

$\square$

Operations with Householder Matrices

In practice it is neither necessary nor desirable to calculate explicitly the entries of a Householder matrix Q. In particular, if we need to form matrix products such as QA and AQ, or if we need to form a matrix-vector product $Q\mathbf{x}$, then the result can be found merely by exploiting the form of Q.

For instance, consider the problem of calculating $Q\mathbf{x}$, where Q is an $(n \times n)$ Householder matrix and $\mathbf{x}$ is in R^n. As in the proof of Theorem 7, we write Q as

$$Q = I - b\mathbf{u}\mathbf{u}^T, \qquad b = 2/(\mathbf{u}^T\mathbf{u}).$$

Now $Q\mathbf{x}$ is given by

$$Q\mathbf{x} = (I - b\mathbf{u}\mathbf{u}^T)\mathbf{x} = \mathbf{x} - b(\mathbf{u}\mathbf{u}^T)\mathbf{x}. \qquad (3)$$

In this expression, note that $b(\mathbf{u}\mathbf{u}^T)\mathbf{x} = b\mathbf{u}(\mathbf{u}^T\mathbf{x})$ and that $\mathbf{u}^T\mathbf{x}$ is a scalar. Thus, from Eq. (3), $Q\mathbf{x}$ has the form $\mathbf{x} - \gamma\mathbf{u}$, where γ is the scalar $b(\mathbf{u}^T\mathbf{x})$:

$$Q\mathbf{x} = \mathbf{x} - \gamma\mathbf{u}, \qquad \gamma = 2\mathbf{u}^T\mathbf{x}/(\mathbf{u}^T\mathbf{u}). \qquad (4)$$

Hence to form $Q\mathbf{x}$ we need only calculate the scalar γ and then perform the vector subtraction, $\mathbf{x} - \gamma\mathbf{u}$, as indicated by (4).

Similarly, if A is an $(n \times p)$ matrix, then we can form the product QA without actually having to calculate Q. Specifically, if $A = [\mathbf{A}_1, \mathbf{A}_2, \ldots, \mathbf{A}_p]$ is an $(n \times p)$ matrix, then

$$QA = [Q\mathbf{A}_1, Q\mathbf{A}_2, \ldots, Q\mathbf{A}_p].$$

As in (4), the columns of QA are found from

$$Q\mathbf{A}_k = \mathbf{A}_k - \gamma_k\mathbf{u}, \qquad \gamma_k = 2\mathbf{u}^T\mathbf{A}_k/(\mathbf{u}^T\mathbf{u}).$$

EXAMPLE 1 Let Q denote the Householder matrix of the form (1), where $\mathbf{u} = [1, 2, 0, 1]^T$. Calculate $Q\mathbf{x}$ and QA, where

$$\mathbf{x} = \begin{bmatrix} 1 \\ 1 \\ 4 \\ 3 \end{bmatrix} \quad \text{and} \quad A = \begin{bmatrix} 1 & 6 \\ 2 & 0 \\ 1 & 5 \\ -2 & 3 \end{bmatrix}.$$

SOLUTION By Eq. (1), Q is the (4×4) matrix $Q = I - b\mathbf{u}\mathbf{u}^T$, where $b = 2/(\mathbf{u}^T\mathbf{u}) = 2/6 = 1/3$. In detail, $Q\mathbf{x}$ is given by

$$Q\mathbf{x} = (I - (1/3)\mathbf{u}\mathbf{u}^T)\mathbf{x} = \mathbf{x} - \frac{\mathbf{u}^T\mathbf{x}}{3}\mathbf{u} = \mathbf{x} - \frac{6}{3}\mathbf{u} = \mathbf{x} - 2\mathbf{u}.$$

Thus $Q\mathbf{x}$ is the vector

$$Q\mathbf{x} = \mathbf{x} - 2\mathbf{u} = \begin{bmatrix} 1 \\ 1 \\ 4 \\ 3 \end{bmatrix} - 2\begin{bmatrix} 1 \\ 2 \\ 0 \\ 1 \end{bmatrix} = \begin{bmatrix} -1 \\ -3 \\ 4 \\ 1 \end{bmatrix}.$$

The matrix QA is found by forming $QA = [Q\mathbf{A}_1, Q\mathbf{A}_2]$. Briefly, the calculations are

$$Q\mathbf{A}_1 = \mathbf{A}_1 - \left(\frac{\mathbf{u}^T\mathbf{A}_1}{3}\right)\mathbf{u} = \mathbf{A}_1 - \mathbf{u}$$

$$Q\mathbf{A}_2 = \mathbf{A}_2 - \left(\frac{\mathbf{u}^T\mathbf{A}_2}{3}\right)\mathbf{u} = \mathbf{A}_2 - 3\mathbf{u}.$$

Thus QA is the (4 × 2) matrix given by

$$QA = \begin{bmatrix} 0 & 3 \\ 0 & -6 \\ 1 & 5 \\ -3 & 0 \end{bmatrix}.$$

Householder Reduction to Hessenberg Form

In Section 6.3, we saw how an $(n \times n)$ matrix A could be reduced to Hessenberg form by using a sequence of similarity transformations

$$Q_{n-2} \cdots Q_2 Q_1 A Q_1^{-1} Q_2^{-1} \cdots Q_{n-2}^{-1} = H. \tag{5}$$

We will see that the matrices Q_i above can be chosen to be Householder matrices.

In the next subsection, we will give the details of how these Householder matrices are constructed. First, however, we want to comment on the significance of using orthogonal matrices in a similarity transformation.

Let us define a matrix Q by

$$Q = Q_{n-2} \cdots Q_2 Q_1,$$

where the Q_i are as in Eq. (5). Next, recall that Q^{-1} is given by

$$Q^{-1} = Q_1^{-1} Q_2^{-1} \cdots Q_{n-2}^{-1}.$$

Thus Eq. (5) can be written compactly as

$$QAQ^{-1} = H,$$

where H is a Hessenberg matrix.

THEOREM 8

Let A be an $(n \times n)$ matrix and let $Q = Q_{n-2} \cdots Q_2 Q_1$, where the Q_i are $(n \times n)$ and nonsingular. Also, suppose that $QAQ^{-1} = H$. If the matrices Q_i are orthogonal, $1 \le i \le n - 2$, then:

a) The product matrix Q is also orthogonal, so that

$$QAQ^{-1} = QAQ^T = H.$$

b) If A is symmetric, then H is also symmetric.

We leave the proof of Theorem 8 to the exercises.

As Theorem 8 indicates, a sequence of similarity transformations of the form (5) will preserve symmetry when the matrices Q_i are orthogonal. So if the Q_i in (5) are orthogonal and A is symmetric, then the Hessenberg matrix H in (5) is also symmetric. A symmetric Hessenberg matrix has a special form—it is "tridiagonal." The form of a general (6 × 6) tridiagonal matrix T is given in Fig. 6.5.

$$T = \begin{bmatrix} \times & \times & 0 & 0 & 0 & 0 \\ \times & \times & \times & 0 & 0 & 0 \\ 0 & \times & \times & \times & 0 & 0 \\ 0 & 0 & \times & \times & \times & 0 \\ 0 & 0 & 0 & \times & \times & \times \\ 0 & 0 & 0 & 0 & \times & \times \end{bmatrix}$$

Figure 6.5
A tridiagonal matrix

As its name suggests, a tridiagonal matrix has three diagonals: a sub-diagonal, the main diagonal, and a superdiagonal. Of course, every tridiagonal matrix is a Hessenberg matrix. Moreover, every symmetric Hessenberg matrix is necessarily tridiagonal.

Once we see how to design the orthogonal matrices Q_i in (5), we will have the following result.

Let A be an $(n \times n)$ symmetric matrix. It is easy to construct an orthogonal matrix Q such that $QAQ^T = T$, where T is tridiagonal.

Although we cannot diagonalize a symmetric matrix A without knowing all the eigenvalues of A, we can always reduce A to tridiagonal form by using a sequence of Householder transformations. In this sense, tridiagonal form represents the closest we can get to diagonal form without actually finding the eigenvalues of A.

Constructing Householder Matrices

We now return to the main objective of this section. Given a general $(n \times n)$ matrix A, find orthogonal matrices $Q_1, Q_2, \ldots, Q_{n-2}$ such that

$$Q_{n-2} \cdots Q_2 Q_1 A Q_1^T Q_2^T \cdots Q_{n-2}^T = H,$$

where H is a Hessenberg matrix.

As with the procedure described in Section 6.3, the product $Q_1 A Q_1^T$ will have zeros in column 1, below the subdiagonal. Similarly, the product $Q_2(Q_1 A Q_1^T)Q_2^T$ will have zeros in columns 1 and 2, and so on. That is, we will be able to design Householder matrices $Q_1, Q_2, \ldots$ such that

$$Q_1 A Q_1^T = \begin{bmatrix} \times & \times & \times & \times & \cdots & \times \\ \times & \times & \times & \times & \cdots & \times \\ 0 & \times & \times & \times & \cdots & \times \\ 0 & \times & \times & \times & \cdots & \times \\ \vdots & & & & & \\ 0 & \times & \times & \times & \cdots & \times \end{bmatrix}, \quad Q_2 Q_1 A Q_1^T Q_2^T = \begin{bmatrix} \times & \times & \times & \times & \cdots & \times \\ \times & \times & \times & \times & \cdots & \times \\ 0 & \times & \times & \times & \cdots & \times \\ 0 & 0 & \times & \times & \cdots & \times \\ \vdots & & & & & \\ 0 & 0 & \times & \times & \cdots & \times \end{bmatrix}.$$

To accomplish each of the individual steps of the reduction process described above, we want to be able to design a Householder matrix that solves the following problem.

Problem

Let $\mathbf{v}$ be an $(n \times 1)$ vector,

$$\mathbf{v} = \begin{bmatrix} v_1 \\ v_2 \\ v_3 \\ \vdots \\ v_n \end{bmatrix}.$$

Given an integer k, $1 \le k \le n$, find a Householder matrix Q such that $Q\mathbf{v} = \mathbf{w}$ and that $\mathbf{w}$ has the form

$$\mathbf{w} = \begin{bmatrix} v_1 \\ v_2 \\ \vdots \\ v_{k-1} \\ s \\ 0 \\ 0 \\ \vdots \\ 0 \end{bmatrix}. \tag{6}$$

In words, the problem posed above is as follows: Given a vector $\mathbf{v}$ in R^n, find a Householder matrix Q so that forming the product $Q\mathbf{v}$ results in a vector $\mathbf{w} = Q\mathbf{v}$, where $\mathbf{w}$ has zeros in the $k+1, k+2, \ldots, n$ components. Furthermore, $\mathbf{w}$ and $\mathbf{v}$ should agree in the first $k-1$ components.

It is easy to form a Householder matrix Q such that the vector $\mathbf{w} = Q\mathbf{v}$ has the form (6). Specifically, suppose that $\mathbf{u}$ is a vector and $Q = I - b\mathbf{u}\mathbf{u}^T$, $b = 2/(\mathbf{u}^T\mathbf{u})$. Since $Q\mathbf{v}$ is given by

$$Q\mathbf{v} = \mathbf{v} - \gamma\mathbf{u},$$

we see that the form (6) can be achieved if $\mathbf{u}$ satisfies these conditions:

a) $\gamma = 1$.

b) $u_{k+1} = v_{k+1}, u_{k+2} = v_{k+2}, \ldots, u_n = v_n$.

c) $u_1 = 0, u_2 = 0, \ldots, u_{k-1} = 0$.

The following algorithm will solve the problem posed above.

ALGORITHM 2 Given an integer k, $1 \le k \le n$, and a vector $\mathbf{v} = [v_1, v_2, \ldots, v_n]^T$, construct $\mathbf{u} = [u_1, u_2, \ldots, u_n]^T$ as follows:

1. $u_1 = u_2 = \cdots = u_{k-1} = 0$.
2. $u_k = v_k - s$, where

$$s = \pm\sqrt{v_k^2 + v_{k+1}^2 + \cdots + v_n^2}.$$

3. $u_i = v_i$ for $i = k+1, k+2, \ldots, n$.

In step (2), choose the sign of s so that $v_k s \le 0$.

For the vector $\mathbf{u}$ defined by Algorithm 2, the Householder matrix $Q = I - b\mathbf{u}\mathbf{u}^T$, $b = 2/(\mathbf{u}^T\mathbf{u})$, has the property that the product $Q\mathbf{v}$ is of the desired form (6). In detail, $Q\mathbf{v}$ is given by

$$Q\mathbf{v} = \mathbf{v} - \mathbf{u} = \begin{bmatrix} v_1 \\ v_2 \\ \vdots \\ v_{k-1} \\ v_k \\ v_{k+1} \\ \vdots \\ v_n \end{bmatrix} - \begin{bmatrix} 0 \\ 0 \\ \vdots \\ 0 \\ v_k - s \\ v_{k+1} \\ \vdots \\ v_n \end{bmatrix} = \begin{bmatrix} v_1 \\ v_2 \\ \vdots \\ v_{k-1} \\ s \\ 0 \\ \vdots \\ 0 \end{bmatrix}. \tag{7}$$

To verify (7), it is necessary only to calculate $Q\mathbf{v}$ according to (4). Thus

$$Q\mathbf{v} = \mathbf{v} - \gamma\mathbf{u}, \qquad \gamma = 2\mathbf{u}^T\mathbf{v}/(\mathbf{u}^T\mathbf{u}). \tag{8}$$

From the definition of $\mathbf{u}$, it is clear that

$$\begin{aligned}
\mathbf{u}^T\mathbf{u} &= (v_k - s)^2 + (v_{k+1})^2 + \cdots + (v_n)^2 \\
&= v_k^2 - 2sv_k + s^2 + v_{k+1}^2 + \cdots + v_n^2 \\
&= 2s^2 - 2sv_k.
\end{aligned}$$

Similarly,

$$\begin{aligned}
\mathbf{u}^T\mathbf{v} &= (v_k - s)v_k + (v_{k+1})^2 + \cdots + (v_n)^2 \\
&= s^2 - sv_k.
\end{aligned}$$

Therefore, from (8),

$$\gamma = \frac{2\mathbf{u}^T\mathbf{v}}{\mathbf{u}^T\mathbf{u}} = \frac{2(s^2 - sv_k)}{2s^2 - 2sv_k} = 1.$$

So, since $\gamma = 1$, the calculation in (7) follows from (8).

EXAMPLE 2 Let $\mathbf{v} = [1, 12, 3, 4]^T$. Use Algorithm 2 to determine Householder matrices Q_1 and Q_2, where:

a) $Q_1\mathbf{v} = \begin{bmatrix} 1 \\ s_1 \\ 0 \\ 0 \end{bmatrix}$;

b) $Q_2 \mathbf{v} = \begin{bmatrix} 1 \\ 12 \\ s_2 \\ 0 \end{bmatrix}$.

SOLUTION

a) Q is defined by selecting a vector $\mathbf{u}$ according to Algorithm 2. Since $k = 2$, we calculate

$$s = \pm\sqrt{v_2^2 + v_3^2 + v_4^2} = \pm\sqrt{144 + 9 + 16} = \pm 13.$$

Choosing the sign of s so that $sv_2 \le 0$, we have $s = -13$. Thus Q_1 is defined by the vector $\mathbf{u} = [0, 25, 3, 4]^T$.

b) $k = 3$, and we find

$$s = \pm\sqrt{v_3^2 + v_4^2} = \pm\sqrt{9 + 16} = \pm 5.$$

Choosing $s = -5$, the vector $\mathbf{u}$ that defines Q_2 is given by $\mathbf{u} = [0, 0, 8, 4]^T$.

☐

EXAMPLE 3 Let $\mathbf{x} = [4, 2, 5, 5]^T$. Find the product $Q_2\mathbf{x}$, where Q_2 is the Householder matrix in Example 2.

SOLUTION   As we know from Eq. (4), the product $Q\mathbf{x}$ is given by

$$Q\mathbf{x} = \mathbf{x} - \gamma\mathbf{u}, \qquad \gamma = 2\mathbf{u}^T\mathbf{x}/(\mathbf{u}^T\mathbf{u}).$$

From Example 2, $\mathbf{u} = [0, 0, 8, 4]^T$. Thus

$$\gamma = \frac{2\mathbf{u}^T\mathbf{x}}{\mathbf{u}^T\mathbf{u}} = \frac{120}{80} = \frac{3}{2}.$$

Therefore,

$$Q\mathbf{x} = \mathbf{x} - \left(\frac{3}{2}\right)\mathbf{u} = \begin{bmatrix} 4 \\ 2 \\ 5 \\ 5 \end{bmatrix} - \begin{bmatrix} 0 \\ 0 \\ 12 \\ 6 \end{bmatrix} = \begin{bmatrix} 4 \\ 2 \\ -7 \\ -1 \end{bmatrix}.$$

☐

For a given vector $\mathbf{v}$ in R^n, Algorithm 2 tells us how to construct a Householder matrix Q so that $Q\mathbf{v}$ has zeros in its last $n - k$ components. We now indicate how the algorithm can be applied to reduce a matrix A to Hessenberg form. As in Section 6.3, we illustrate the process for a (4×4) matrix and merely note that the process extends to $(n \times n)$ matrices in an obvious fashion.

Let A be the (4×4) matrix $A = [\mathbf{A}_1, \mathbf{A}_2, \mathbf{A}_3, \mathbf{A}_4]$. Construct a Householder matrix Q so that the vector $Q\mathbf{A}_1$ has zeros in its last two components. Thus, forming $QA = [Q\mathbf{A}_1, Q\mathbf{A}_2, Q\mathbf{A}_3, Q\mathbf{A}_4]$ will produce a matrix of the form

$$QA = \begin{bmatrix} c_{11} & c_{12} & c_{13} & c_{14} \\ c_{21} & c_{22} & c_{23} & c_{24} \\ 0 & c_{32} & c_{33} & c_{34} \\ 0 & c_{42} & c_{43} & c_{44} \end{bmatrix}.$$

Next, form $B = QAQ$ and note that B is similar to A since Q is a Householder matrix. Also (see Exercise 25), it can be shown that forming $B = (QA)Q$ will not disturb the zero entries in the $(3, 1)$ and $(4, 1)$ positions. Thus, $B = QAQ$ has the form

$$B = \begin{bmatrix} b_{11} & b_{12} & b_{13} & b_{14} \\ b_{21} & b_{22} & b_{23} & b_{24} \\ 0 & b_{32} & b_{33} & b_{34} \\ 0 & b_{42} & b_{43} & b_{44} \end{bmatrix}. \tag{9}$$

For the matrix above, $B = [\mathbf{B}_1, \mathbf{B}_2, \mathbf{B}_3, \mathbf{B}_4]$, choose a Householder matrix S such that the vector $S\mathbf{B}_2$ has a zero in its last component. It can be shown (see Exercise 25) that $S\mathbf{B}_1 = \mathbf{B}_1$. Thus, $SB = [S\mathbf{B}_1, S\mathbf{B}_2, S\mathbf{B}_3, S\mathbf{B}_4]$ has the form

$$SB = \begin{bmatrix} b_{11} & d_{12} & d_{13} & d_{14} \\ b_{21} & d_{22} & d_{23} & d_{24} \\ 0 & d_{32} & d_{33} & d_{34} \\ 0 & 0 & d_{43} & d_{44} \end{bmatrix}. \tag{10}$$

Finally, the matrix SBS is similar to B and hence to A. Moreover (see Exercise 25), forming $(SB)S$ does not disturb the zero entries in (10). Therefore, $SBS = S(QAQ)S$ has the desired Hessenberg form

$$SBS = SQAQS = \begin{bmatrix} h_{11} & h_{12} & h_{13} & h_{14} \\ h_{21} & h_{22} & h_{23} & h_{24} \\ 0 & h_{32} & h_{33} & h_{34} \\ 0 & 0 & h_{43} & h_{44} \end{bmatrix}. \tag{11}$$

The next example illustrates the final stage of reduction to Hessenberg form for a (4×4) matrix, the process of going from (9) to (11) above.

EXAMPLE 4   Find a Householder matrix S such that $SBS = H$, where H is a Hessenberg matrix and where B is given by

$$B = \begin{bmatrix} 1 & 2 & 4 & 2 \\ 3 & 3 & -4 & 2 \\ 0 & 3 & 9 & -1 \\ 0 & -4 & -2 & 8 \end{bmatrix}.$$

Also, calculate the matrix SB.

SOLUTION   We seek a Householder matrix S such that the vector $S\mathbf{B}_2$ has a zero in the fourth component, where $\mathbf{B}_2 = [2, 3, 3, -4]^T$.

Using $k = 3$ in Algorithm 2, we define a vector $\mathbf{u}$, where

$$\mathbf{u} = \begin{bmatrix} 0 \\ 0 \\ 8 \\ -4 \end{bmatrix}.$$

The appropriate Householder matrix is $S = I - b\mathbf{u}\mathbf{u}^T$, $b = 2/(\mathbf{u}^T\mathbf{u}) = 1/40$.

Next, we calculate the matrix SB by using $SB = [S\mathbf{B_1}, S\mathbf{B_2}, S\mathbf{B_3}, S\mathbf{B_4}]$, where

$$S\mathbf{B_i} = \mathbf{B_i} - \gamma_i\mathbf{u}; \qquad \gamma_i = 2\mathbf{u}^T\mathbf{B_i}/(\mathbf{u}^T\mathbf{u}) = \mathbf{u}^T\mathbf{B_i}/40.$$

The details are:

a) $\gamma_1 = \mathbf{u}^T\mathbf{B_1}/40 = 0$, so $S\mathbf{B_1} = \mathbf{B_1} = [1, 3, 0, 0]^T$.

b) $\gamma_2 = \mathbf{u}^T\mathbf{B_2}/40 = 1$, so $S\mathbf{B_2} = \mathbf{B_2} - \mathbf{u} = [2, 3, -5, 0]^T$.

c) $\gamma_3 = \mathbf{u}^T\mathbf{B_3}/40 = 2$, so $S\mathbf{B_3} = \mathbf{B_3} - 2\mathbf{u} = [4, -4, -7, 6]^T$.

d) $\gamma_4 = \mathbf{u}^T\mathbf{B_4}/40 = -1$, so $S\mathbf{B_4} = \mathbf{B_4} + \mathbf{u} = [2, 2, 7, 4]^T$.

Thus the matrix SB is given by

$$SB = \begin{bmatrix} 1 & 2 & 4 & 2 \\ 3 & 3 & -4 & 2 \\ 0 & -5 & -7 & 7 \\ 0 & 0 & 6 & 4 \end{bmatrix}.$$

□

(*Note:* The Householder matrix S does not disturb the first column of B, since $\mathbf{u}^T\mathbf{B_1} = 0$ and hence $\gamma_1 = 0$; see Exercise 25.)

In order to complete the similarity transformation begun in Example 4, we need to calculate the matrix $(SB)S$. Although we know how to form QA and $Q\mathbf{x}$ when Q is a Householder matrix, we have not yet discussed how to form the product AQ.

The easiest way to form AQ is to proceed as follows:

1. Calculate $M = QA^T$.

2. Form $M^T = (QA^T)^T = AQ^T = AQ$.

(*Note:* In step 2, $AQ^T = AQ$ since a Householder matrix Q is symmetric.)

EXAMPLE 5 For the matrix SB in Example 4, calculate H, where $H = SBS$.

SOLUTION Following the two-step procedure above, we first calculate $S(SB)^T$. From Example 4 we have

$$(SB)^T = \begin{bmatrix} 1 & 3 & 0 & 0 \\ 2 & 3 & -5 & 0 \\ 4 & -4 & -7 & 6 \\ 2 & 2 & 7 & 4 \end{bmatrix}.$$

For notation, let the columns of $(SB)^T$ be denoted as $\mathbf{R_i}$, so $(SB)^T = [\mathbf{R_1}, \mathbf{R_2}, \mathbf{R_3}, \mathbf{R_4}]$. As in Example 4, the matrix $S(SB)^T$ has column vectors $S\mathbf{R_i}$, where

$$S\mathbf{R_i} = \mathbf{R_i} - \gamma_i\mathbf{u}, \qquad \gamma_i = \mathbf{u}^T\mathbf{R_i}/40, \qquad 1 \le i \le 4.$$

With $\mathbf{u}$ from Example 4, the scalars are

$$\gamma_1 = 3/5, \qquad \gamma_2 = -1, \qquad \gamma_3 = -21/10, \quad \text{and} \quad \gamma_4 = 4/5.$$

Therefore, the matrix $S(SB)^T$ is given by

$$S(SB)^T = \begin{bmatrix} 1.0 & 3.0 & 0.0 & 0.0 \\ 2.0 & 3.0 & -5.0 & 0.0 \\ -.8 & 4.0 & 9.8 & -.4 \\ 4.4 & -2.0 & -1.4 & 7.2 \end{bmatrix}.$$

The transpose of the matrix above is the Hessenberg matrix H, where $H = SBS$.

Exercises 6.5

Let $Q = I - b\mathbf{u}\mathbf{u}^T$ be the Householder matrix defined by (1), where $\mathbf{u} = [1, -1, 1, -1]^T$. In Exercises 1–8, calculate the indicated product.

1. $Q\mathbf{x}$,   for $\mathbf{x} = \begin{bmatrix} 3 \\ 2 \\ 5 \\ 8 \end{bmatrix}$

2. $Q\mathbf{x}$,   for $\mathbf{x} = \begin{bmatrix} 0 \\ 1 \\ 1 \\ 8 \end{bmatrix}$

3. QA,   for $A = \begin{bmatrix} 2 & 1 \\ 6 & 3 \\ 4 & 2 \\ 2 & 4 \end{bmatrix}$

4. QA,   for $A = \begin{bmatrix} 0 & 1 & 2 \\ 2 & 2 & 1 \\ 1 & 4 & 3 \\ 3 & 7 & 2 \end{bmatrix}$

5. $\mathbf{x}^TQ$,   for $\mathbf{x} = \begin{bmatrix} 3 \\ 2 \\ 2 \\ 5 \end{bmatrix}$

6. $\mathbf{x}^TQ$,   for $\mathbf{x} = \begin{bmatrix} 1 \\ 3 \\ 2 \\ 2 \end{bmatrix}$

7. AQ,   for $A = \begin{bmatrix} 2 & 1 & 2 & 1 \\ 1 & 0 & 1 & 4 \end{bmatrix}$

8. BQ, where B is the (4×4) matrix in Example 4.

For the given vectors $\mathbf{v}$ and $\mathbf{w}$ in Exercises 9–14, determine a vector $\mathbf{u}$ such that $(I - b\mathbf{u}\mathbf{u}^T)\mathbf{v} = \mathbf{w}$.

9. $\mathbf{v} = \begin{bmatrix} 1 \\ 2 \\ 2 \\ 1 \end{bmatrix}$, $\mathbf{w} = \begin{bmatrix} 1 \\ a \\ 0 \\ 0 \end{bmatrix}$ **10.** $\mathbf{v} = \begin{bmatrix} 1 \\ 1 \\ 1 \\ 1 \end{bmatrix}$, $\mathbf{w} = \begin{bmatrix} a \\ 0 \\ 0 \\ 0 \end{bmatrix}$

11. $\mathbf{v} = \begin{bmatrix} 2 \\ 1 \\ 4 \\ 3 \end{bmatrix}$, $\mathbf{w} = \begin{bmatrix} 2 \\ 1 \\ a \\ 0 \end{bmatrix}$ **12.** $\mathbf{v} = \begin{bmatrix} 2 \\ 0 \\ -2 \\ 2 \\ 1 \end{bmatrix}$, $\mathbf{w} = \begin{bmatrix} 2 \\ 0 \\ a \\ 0 \\ 0 \end{bmatrix}$

13. $\mathbf{v} = \begin{bmatrix} 0 \\ 0 \\ 0 \\ -3 \\ 4 \end{bmatrix}$, $\mathbf{w} = \begin{bmatrix} 0 \\ 0 \\ 0 \\ a \\ 0 \end{bmatrix}$ **14.** $\mathbf{v} = \begin{bmatrix} 1 \\ 1 \\ 4 \\ 0 \\ 0 \end{bmatrix}$, $\mathbf{w} = \begin{bmatrix} 1 \\ 1 \\ a \\ 0 \\ 0 \end{bmatrix}$

In Exercises 15–20, find a Householder matrix Q such that $QAQ = H$, with H a Hessenberg matrix. List the vector $\mathbf{u}$ that defines Q and gives the matrix H.

15. $A = \begin{bmatrix} 1 & 3 & 4 \\ 3 & 1 & 1 \\ 4 & 1 & 1 \end{bmatrix}$ **16.** $A = \begin{bmatrix} 1 & 0 & 5 \\ 0 & 2 & 1 \\ 5 & 1 & 2 \end{bmatrix}$

17. $A = \begin{bmatrix} 0 & -4 & 3 \\ -4 & 0 & 1 \\ 3 & 1 & 2 \end{bmatrix}$ **18.** $A = \begin{bmatrix} 1 & 2 & 0 & 0 \\ 2 & 1 & 3 & 4 \\ 0 & 3 & 1 & 1 \\ 0 & 4 & 1 & 1 \end{bmatrix}$

19. $A = \begin{bmatrix} 2 & 1 & 1 & 2 \\ 3 & 4 & 0 & 1 \\ 0 & -3 & 1 & 1 \\ 0 & 4 & 2 & 3 \end{bmatrix}$ **20.** $A = \begin{bmatrix} 1 & 2 & 3 & 0 \\ 4 & 1 & 2 & 3 \\ 0 & 0 & 2 & 1 \\ 0 & 1 & 3 & 2 \end{bmatrix}$

21. Let Q denote the Householder matrix defined by (1). Verify that Q is symmetric.

22. Let Q be the Householder matrix defined by (1) and calculate the product $Q\mathbf{u}$. If $\mathbf{v}$ is any vector orthogonal to $\mathbf{u}$, what is the result of forming $Q\mathbf{v}$?

23. Consider the $(n \times n)$ Householder matrix $Q = I - b\mathbf{u}\mathbf{u}^T$, $b = 2/(\mathbf{u}^T\mathbf{u})$. Show that Q has eigenvalues $\lambda = -1$ and $\lambda = 1$. (*Hint:* Use the Gram–Schmidt process to argue that R^n has an orthogonal basis $\{\mathbf{u}, \mathbf{w}_2, \mathbf{w}_3, \ldots, \mathbf{w}_n\}$. Also, recall Exercise 22.)

24. Prove Theorem 8.

25. Consider a (4×4) matrix B of the form shown in (9), where $b_{31} = 0$ and $b_{41} = 0$. Let $\mathbf{u}$ be a vector of the form $\mathbf{u} = [0, a, b, c]^T$, and let $Q = I - b\mathbf{u}\mathbf{u}^T$ be the associated Householder matrix.

a) Show that forming the product BQ does not change the first column of B. (*Hint:* Form BQ by using the two-step procedure illustrated in Example 5.)

b) Let $\mathbf{u}$ be a vector of the form $\mathbf{u} = [0, 0, a, b]^T$, and let $Q = I - b\mathbf{u}\mathbf{u}^T$ be the associated Householder matrix. Show that forming QBQ does not alter the first column of B.

6.6   The QR Factorization and Least-Squares Solutions

The Householder transformations of Section 6.5 can be used very effectively to construct an algorithm to find the least-squares solution of an overdetermined linear system, $A\mathbf{x} = \mathbf{b}$. This construction also yields a useful way of expressing A as a product of two other matrices, called the QR factorization. The QR factorization is a principal instrument in many of the software packages for numerical linear algebra.

Reduction to Trapezoidal Form

The following theorem is proved by construction, and hence its proof serves as an algorithm for the desired factorization.

THEOREM 9

Let A be an $(m \times n)$ matrix with $m \geq n$. There exists an $(m \times m)$ orthogonal matrix S such that

$$SA = \begin{bmatrix} R \\ \mathcal{O} \end{bmatrix},$$

where R is an $(n \times n)$ upper-triangular matrix and $\mathcal{O}$ is the $[(m-n) \times n]$ zero matrix. (If $m = n$, $SA = R$.)

PROOF   Let $A = [\mathbf{A}_1, \mathbf{A}_2, \ldots, \mathbf{A}_n]$, where the column vectors $\mathbf{A}_i$ are in R^m. Let S_1 be the $(m \times m)$ Householder matrix such that

$$S_1\mathbf{A}_1 = [s_1, 0, 0, \ldots, 0]^T.$$

Thus the product $S_1 A = [S_1 \mathbf{A}_1, S_1 \mathbf{A}_2, \ldots, S_1 \mathbf{A}_n]$ has the form

$$S_1 A = \begin{bmatrix} s_1 & c_{12} & \cdots & c_{1n} \\ 0 & c_{22} & \cdots & c_{2n} \\ 0 & c_{32} & \cdots & c_{3n} \\ \vdots & \vdots & & \\ 0 & c_{m2} & \cdots & c_{mn} \end{bmatrix}.$$

For notation, let $B = S_1 A$ and write B as $B = [\mathbf{B}_1, \mathbf{B}_2, \ldots, \mathbf{B}_n]$. Next, choose the Householder S_2 such that

$$S_2 \mathbf{B}_2 = [c_{12}, s_2, 0, 0, \ldots, 0]^T.$$

As in reduction to Hessenberg form, notice that $S_2 \mathbf{B}_1 = \mathbf{B}_1$. Thus the product $S_2 B = S_2 S_1 A$ has the form

$$S_2 S_1 A = \begin{bmatrix} s_1 & c_{12} & d_{13} & \cdots & d_{1n} \\ 0 & s_2 & d_{23} & & d_{2n} \\ 0 & 0 & d_{33} & & d_{3n} \\ \vdots & \vdots & \vdots & & \vdots \\ 0 & 0 & d_{m3} & \cdots & d_{mn} \end{bmatrix}.$$

Continuing in this fashion, we ultimately find Householder matrices $S_1, S_2, \ldots, S_n$ such that the product $S_n \cdots S_2 S_1 A$ has the form

$$S_n \cdots S_2 S_1 A = \begin{bmatrix} \times & \times & \times & \times & \cdots & \times \\ 0 & \times & \times & \times & \cdots & \times \\ 0 & 0 & \times & \times & \cdots & \times \\ 0 & 0 & 0 & \times & \cdots & \times \\ \vdots & & & & & \\ 0 & 0 & 0 & 0 & \cdots & \times \\ 0 & 0 & 0 & 0 & \cdots & 0 \\ \vdots & & & & & \\ 0 & 0 & 0 & 0 & \cdots & 0 \end{bmatrix} \begin{matrix} \\ \\ \\ \\ \\ \leftarrow \text{row } n \\ \\ \\ \leftarrow \text{row } m \end{matrix}$$
$$\uparrow$$
$$\text{column } n$$

Equivalently, with $S = S_n \cdots S_2 S_1$, we find that S is orthogonal and

$$SA = \begin{bmatrix} R \\ \mathcal{O} \end{bmatrix}, \qquad \text{where } \begin{cases} R \text{ is } (n \times n) \text{ upper triangular} \\ \mathcal{O} \text{ is the } [(m-n) \times n] \text{ zero matrix.} \end{cases}$$

(*Note:* The block matrix $\begin{bmatrix} R \\ \mathcal{O} \end{bmatrix}$ in Theorem 9 is called an ***upper-trapezoidal*** matrix. Also note that we are not interested in preserving similarity in Theorem 9. Thus we do not form $S_1 A S_1$ or $S_2 S_1 A S_1 S_2$ in the construction described in the proof of Theorem 9.)

EXAMPLE 1 Following the proof of Theorem 9, find Householder matrices S_1 and S_2 such that $S_2 S_1 A$ is in trapezoidal form, where

$$A = \begin{bmatrix} 1 & -2/3 \\ -1 & 3 \\ 0 & -2 \\ -1 & 1 \\ 1 & 0 \end{bmatrix}.$$

SOLUTION   Following Algorithm 2 in Section 6.5, we define a vector $\mathbf{u}$ by

$$\mathbf{u} = \begin{bmatrix} 3 \\ -1 \\ 0 \\ -1 \\ 1 \end{bmatrix}.$$

The first Householder matrix S_1 is then $S_1 = I - b\mathbf{u}\mathbf{u}^T$, where $b = 2/(\mathbf{u}^T\mathbf{u}) = 1/6$.

We next calculate $S_1 A = [S_1\mathbf{A}_1, S_1\mathbf{A}_2]$, where $S_1\mathbf{A}_i = \mathbf{A}_i - \gamma_i\mathbf{u}$, $\gamma_i = \mathbf{u}^T\mathbf{A}_i/6$. The scalars γ_i are $\gamma_1 = 1$ and $\gamma_2 = -1$. Thus the matrix $S_1 A$ has columns $\mathbf{A}_1 - \mathbf{u}$ and $\mathbf{A}_2 + \mathbf{u}$:

$$S_1 A = \begin{bmatrix} -2 & 7/3 \\ 0 & 2 \\ 0 & -2 \\ 0 & 0 \\ 0 & 1 \end{bmatrix}.$$

We now define the second Householder matrix S_2, where S_2 is designed so that $S_2(S_1 A)$ has zeros in positions $(3, 2)$, $(4, 2)$, and $(5, 2)$.

Following Algorithm 2, define a vector $\mathbf{v}$ by

$$\mathbf{v} = \begin{bmatrix} 0 \\ 5 \\ -2 \\ 0 \\ 1 \end{bmatrix}$$

and set $S_2 = I - b\mathbf{v}\mathbf{v}^T$, $b = 2/(\mathbf{v}^T\mathbf{v}) = 1/15$. Forming $S_2(S_1 A)$, we obtain

$$S_2 S_1 A = \begin{bmatrix} -2 & 7/3 \\ 0 & -3 \\ 0 & 0 \\ 0 & 0 \\ 0 & 0 \end{bmatrix} = \begin{bmatrix} R \\ \mathcal{O} \end{bmatrix}; \qquad R = \begin{bmatrix} -2 & 7/3 \\ 0 & -3 \end{bmatrix}. \qquad \square$$

Least-Squares Solutions

Suppose that $A\mathbf{x} = \mathbf{b}$ represents a system of m linear equations in n unknowns, where $m \geq n$. If the system is inconsistent, it is often necessary to find a "least-squares solution" to $A\mathbf{x} = \mathbf{b}$. By a least-squares solution (recall Section 2.8), we mean a vector $\mathbf{x}^*$ in R^n such that

$$\|A\mathbf{x}^* - \mathbf{b}\| \leq \|A\mathbf{x} - \mathbf{b}\|, \qquad \text{for all } \mathbf{x} \text{ in } R^n. \tag{1}$$

In Section 2.8, we saw a simple procedure for solving (1). That is, $\mathbf{x}^*$ can be obtained by solving the normal equations:

$$A^T A \mathbf{x} = A^T \mathbf{b}.$$

In this subsection, we consider an alternative procedure. The alternative is not so efficient for hand calculations, but it is the preferred procedure for machine calculations. The reason is based on the observation that the matrix $A^T A$ is often "ill conditioned." Thus it is frequently difficult to compute numerically an accurate solution to $A^T A \mathbf{x} = A^T \mathbf{b}$.

Recall from Section 3.7 that orthogonal matrices preserve the length of a vector under multiplication. That is, if $\mathbf{y}$ is any vector in R^m and Q is an $(m \times m)$ orthogonal matrix, then

$$\|Q\mathbf{y}\| = \|\mathbf{y}\|.$$

In the context of (1), let Q be an $(m \times m)$ orthogonal matrix. Also, suppose that $\mathbf{x}^*$ is a vector in R^n such that

$$\|A\mathbf{x}^* - \mathbf{b}\| \leq \|A\mathbf{x} - \mathbf{b}\|, \qquad \text{for all } \mathbf{x} \text{ in } R^n.$$

Then, since $\|A\mathbf{x}^* - \mathbf{b}\| = \|Q(A\mathbf{x}^* - \mathbf{b})\| = \|QA\mathbf{x}^* - Q\mathbf{b}\|$ and $\|A\mathbf{x} - \mathbf{b}\| = \|Q(A\mathbf{x} - \mathbf{b})\| = \|QA\mathbf{x} - Q\mathbf{b}\|$, we have

$$\|QA\mathbf{x}^* - Q\mathbf{b}\| \leq \|QA\mathbf{x} - Q\mathbf{b}\|, \qquad \text{for all } \mathbf{x} \text{ in } R^n. \tag{2}$$

Similarly, if a vector $\mathbf{x}^*$ satisfies (2), then that same vector also satisfies (1).

In other words:

> If Q is an orthogonal matrix, then finding the least-squares solution of $QA\mathbf{x} = Q\mathbf{b}$ is equivalent to finding the least-squares solution of $A\mathbf{x} = \mathbf{b}$.

Now, using the construction in Theorem 9, we can form an orthogonal matrix Q so that the least-squares solution of $QA\mathbf{x} = Q\mathbf{b}$ is easy to find.

In particular, for an $(m \times n)$ matrix A, let S be an orthogonal matrix such that SA is in trapezoid form. Consider the problem of finding the least-squares solution of $SA\mathbf{x} = S\mathbf{b}$, where

$$SA = \begin{bmatrix} R \\ O \end{bmatrix}, \qquad S\mathbf{b} = \begin{bmatrix} \mathbf{c} \\ \mathbf{d} \end{bmatrix}, \qquad \text{where } \mathbf{c} \text{ is in } R^n \text{ and } \mathbf{d} \text{ is in } R^{m-n}.$$

For any vector $\mathbf{x}$ in R^n, we have

$$SA\mathbf{x} - S\mathbf{b} = \begin{bmatrix} R\mathbf{x} \\ \theta \end{bmatrix} - \begin{bmatrix} \mathbf{c} \\ \mathbf{d} \end{bmatrix} = \begin{bmatrix} R\mathbf{x} - \mathbf{c} \\ -\mathbf{d} \end{bmatrix}.$$

Thus, $\|SA\mathbf{x} - S\mathbf{b}\|$ can be found from the relationship

$$\|SA\mathbf{x} - S\mathbf{b}\|^2 = \|R\mathbf{x} - \mathbf{c}\|^2 + \|\mathbf{d}\|^2. \tag{3}$$

By (3), a vector $\mathbf{x}^*$ in R^n minimizes $\|SA\mathbf{x} - S\mathbf{b}\|$ if and only if $\mathbf{x}^*$ minimizes $\|R\mathbf{x} - \mathbf{c}\|$.

As an example to illustrate these ideas, consider the (5×3) trapezoidal matrix SA, where

$$SA = \begin{bmatrix} 1 & 2 & 1 \\ 0 & 2 & 4 \\ 0 & 0 & 3 \\ 0 & 0 & 0 \\ 0 & 0 & 0 \end{bmatrix} = \begin{bmatrix} R \\ \mathcal{O} \end{bmatrix}. \tag{4}$$

In (4), R is the upper (3×3) block of SA and $\mathcal{O}$ is the (2×3) zero matrix. Now for $\mathbf{x}$ in R^3 and SA given by (4), note that $SA\mathbf{x}$ has the form

$$SA\mathbf{x} = \begin{bmatrix} 1 & 2 & 1 \\ 0 & 2 & 4 \\ 0 & 0 & 3 \\ 0 & 0 & 0 \\ 0 & 0 & 0 \end{bmatrix} \begin{bmatrix} x_1 \\ x_2 \\ x_3 \end{bmatrix} = \begin{bmatrix} x_1 + 2x_2 + x_3 \\ 2x_2 + 4x_3 \\ 3x_3 \\ 0 \\ 0 \end{bmatrix} = \begin{bmatrix} R\mathbf{x} \\ \theta \end{bmatrix}. \tag{5}$$

In (5) the vector $R\mathbf{x}$ is three-dimensional and θ is the two-dimensional zero vector.

In general, as noted above, a vector $\mathbf{x}^*$ in R^n minimizes $\|SA\mathbf{x} - S\mathbf{b}\|$ if and only if $\mathbf{x}^*$ minimizes $\|R\mathbf{x} - \mathbf{c}\|$ in (3). Since R is upper triangular, the problem of minimizing $\|R\mathbf{x} - \mathbf{c}\|$ is fairly easy. In particular, if R is nonsingular, then there is a unique minimizer, $\mathbf{x}^*$, and $\mathbf{x}^*$ is the solution of $R\mathbf{x} = \mathbf{c}$. The nonsingular case is summarized in the next theorem.

THEOREM 10

Let A be an $(m \times n)$ matrix and suppose that the column vectors of A are linearly independent. Let S be an orthogonal matrix such that SA is upper trapezoidal. Given a vector $\mathbf{b}$ in R^m, let SA and $S\mathbf{b}$ be denoted as

$$SA = \begin{bmatrix} R \\ \mathcal{O} \end{bmatrix} \quad \text{and} \quad S\mathbf{b} = \begin{bmatrix} \mathbf{c} \\ \mathbf{d} \end{bmatrix}.$$

Then:

a) R is nonsingular.
b) There is a unique least-squares solution of $A\mathbf{x} = \mathbf{b}$, $\mathbf{x}^*$.
c) The vector $\mathbf{x}^*$ satisfies $R\mathbf{x}^* = \mathbf{c}$.

PROOF   In Exercise 19, the reader is asked to show that R is nonsingular when the columns of A are linearly independent. Then, since there is a unique vector $\mathbf{x}^*$ such that $R\mathbf{x}^* = \mathbf{c}$, the rest of the conclusions in Theorem 10 follow from (3). $\qquad\square$

EXAMPLE 2   Use Theorem 10 to find the least-squares solution of $A\mathbf{x} = \mathbf{b}$, where

$$A = \begin{bmatrix} 1 & -2/3 \\ -1 & 3 \\ 0 & -2 \\ -1 & 1 \\ 1 & 0 \end{bmatrix} \quad \text{and} \quad \mathbf{b} = \begin{bmatrix} 1 \\ 3 \\ -4 \\ -3 \\ 3 \end{bmatrix}.$$

SOLUTION   In Example 1, we found Householder matrices S_1 and S_2 such that $S_2 S_1 A$ is in trapezoidal form. The matrices S_1 and S_2 were defined by vectors $\mathbf{u}$ and $\mathbf{v}$ (respectively), where

$$\mathbf{u} = \begin{bmatrix} 3 \\ -1 \\ 0 \\ -1 \\ 1 \end{bmatrix} \quad \text{and} \quad \mathbf{v} = \begin{bmatrix} 0 \\ 5 \\ -2 \\ 0 \\ 1 \end{bmatrix}.$$

The vector $S\mathbf{b} = S_2 S_1 \mathbf{b}$ is found from

$$S_1\mathbf{b} = \mathbf{b} - \mathbf{u} = \begin{bmatrix} -2 \\ 4 \\ -4 \\ -2 \\ 2 \end{bmatrix} \quad \text{and} \quad S_2(S_1\mathbf{b}) = S_1\mathbf{b} - 2\mathbf{v} = \begin{bmatrix} -2 \\ -6 \\ 0 \\ -2 \\ 0 \end{bmatrix} = \begin{bmatrix} \mathbf{c} \\ \mathbf{d} \end{bmatrix}.$$

By Example 1, the matrix SA is given by

$$SA = \begin{bmatrix} -2 & 7/3 \\ 0 & -3 \\ 0 & 0 \\ 0 & 0 \\ 0 & 0 \end{bmatrix} = \begin{bmatrix} R \\ \mathcal{O} \end{bmatrix}.$$

Thus the least-squares solution, $\mathbf{x}^*$, is found by solving $R\mathbf{x} = \mathbf{c}$, where $\mathbf{c} = [-2, -6]^T$:

$$\begin{aligned} -2x_1 + (7/3)x_2 &= -2 \\ -3x_2 &= -6. \end{aligned}$$

The solution of $R\mathbf{x} = \mathbf{c}$ is $\mathbf{x}^* = [10/3, 2]^T$. $\qquad\square$

The *QR* Factorization

The main result of this subsection is the following theorem.

THEOREM 11

Let A be an $(m \times n)$ matrix with $m \geq n$, where A has rank n. There is an $(m \times n)$ matrix Q with orthonormal column vectors such that

$$A = QR.$$

Moreover, in the factorization $A = QR$, the matrix R is upper triangular and nonsingular.

PROOF   From Theorem 9, we know there is an orthogonal matrix S such that

$$SA = \begin{bmatrix} R \\ \mathcal{O} \end{bmatrix},$$

where R is an $(n \times n)$ upper-triangular matrix and R is nonsingular.

Since S is orthogonal, the reduction displayed above is equivalent to

$$A = S^T \begin{bmatrix} R \\ \mathcal{O} \end{bmatrix}. \tag{6}$$

To simplify the notation, we let B denote the $(m \times m)$ matrix S^T so that (6) becomes

$$A = B \begin{bmatrix} R \\ \mathcal{O} \end{bmatrix} = [\mathbf{B}_1, \mathbf{B}_2, \ldots, \mathbf{B}_n, \ldots, \mathbf{B}_m] \begin{bmatrix} R \\ \mathcal{O} \end{bmatrix}.$$

Examination of the product on the right-hand side above shows that

$$A = [\mathbf{B}_1, \mathbf{B}_2, \ldots, \mathbf{B}_n] R. \tag{7}$$

That is, the column vectors $\mathbf{B}_{n+1}, \ldots, \mathbf{B}_m$ are multiplied by the zero entries of $\mathcal{O}$. Hence (7) yields a factorization of A that is different from (6) but is still valid.

The proof of Theorem 11 is complete once we define the $(m \times n)$ matrix Q to be given by

$$Q = [\mathbf{B}_1, \mathbf{B}_2, \ldots, \mathbf{B}_n].$$

That is, Q consists of the first n columns of the matrix S^T, where S is the $(m \times m)$ orthogonal matrix defined in Theorem 9. $\square$

Note that if $m = n$ so that A is a square matrix, then the factors Q and R are also square and Q is an orthogonal matrix. This feature is illustrated in the next example.

EXAMPLE 3 Find a QR factorization for the matrix A:

$$A = \begin{bmatrix} 3 & 1 & 2 \\ 0 & 3 & -1 \\ 4 & 8 & 6 \end{bmatrix}.$$

SOLUTION   Following the construction shown in the proof of Theorem 9, we use Householder matrices to reduce A to upper-triangular form.

First, define a Householder matrix S_1 from $S_1 = I - b\mathbf{u}\mathbf{u}^T$, where $\mathbf{u} = [8, 0, 4]^T$. Then we have

$$S_1 A = \begin{bmatrix} -5 & -7 & -6 \\ 0 & 3 & -1 \\ 0 & 4 & 2 \end{bmatrix}.$$

Next, define $S_2 = I - b\mathbf{u}\mathbf{u}^T$, where $\mathbf{u} = [0, 8, 4]^T$. Forming $S_2(S_1 A)$, we obtain

$$S_2 S_1 A = \begin{bmatrix} -5 & -7 & -6 \\ 0 & -5 & -1 \\ 0 & 0 & 2 \end{bmatrix} = R.$$

With the above, the desired QR factorization is given by

$$A = S_1 S_2 R = QR, \qquad Q = S_1 S_2.$$

If we wished to do so, we could form the product $S_1 S_2$ and list the matrix Q explicitly. For most applications, however, there is no need to know the individual entries of Q.

EXAMPLE 4 Use the QR factorization found in Example 3 to solve $A\mathbf{x} = \mathbf{b}_1$ and $A\mathbf{x} = \mathbf{b}_2$, where

$$\mathbf{b}_1 = \begin{bmatrix} 1 \\ 8 \\ 8 \end{bmatrix} \quad \text{and} \quad \mathbf{b}_2 = \begin{bmatrix} 10 \\ -4 \\ 10 \end{bmatrix}.$$

SOLUTION   The factorization found in Example 3 states that $A\mathbf{x} = \mathbf{b}_k$ can be written as $(QR)\mathbf{x} = \mathbf{b}_k$, $k = 1, 2$. Equivalently,

$$(S_1 S_2 R)\mathbf{x} = \mathbf{b}_k \quad \text{or} \quad R\mathbf{x} = S_2 S_1 \mathbf{b}_k, \qquad k = 1, 2.$$

Since S_1 and S_2 are Householder matrices, it is easy to form the vectors $S_2 S_1 \mathbf{b}_k$, $k = 1, 2$. We find that

$$S_2 S_1 \mathbf{b}_1 = \begin{bmatrix} -7 \\ -8 \\ -4 \end{bmatrix} \quad \text{and} \quad S_2 S_1 \mathbf{b}_2 = \begin{bmatrix} -14 \\ 4 \\ 2 \end{bmatrix}.$$

Solving $R\mathbf{x} = S_2 S_1 \mathbf{b}_k$, we obtain the solutions to $A\mathbf{x} = \mathbf{b}_k$:

$$\mathbf{x} = \begin{bmatrix} 1 \\ 2 \\ -2 \end{bmatrix} \quad \text{and} \quad \mathbf{x} = \begin{bmatrix} 3 \\ -1 \\ 1 \end{bmatrix}.$$

The QR Algorithm

In practice, the eigenvalues of an $(n \times n)$ matrix A are usually found by transforming A to Hessenberg form H and then applying some version of the QR algorithm to H. The simplest and most basic version is given below.

The QR Algorithm

Given an $(n \times n)$ matrix B, let $B^{(1)} = B$. For each positive integer k, find the QR factorization of $B^{(k)}$. That is, $B^{(k)} = Q^{(k)}R^{(k)}$, where $Q^{(k)}$ is orthogonal and $R^{(k)}$ is upper triangular. Then set $B^{(k+1)} = R^{(k)}Q^{(k)}$ and repeat the process.

Since $R^{(k)} = [Q^{(k)}]^T B^{(k)}$, it follows that

$$B^{(k+1)} = [Q^{(k)}]^T B^{(k)} Q^{(k)}.$$

Hence each $B^{(k)}$ is similar to B. If all the eigenvalues of B have distinct absolute values, the QR iterates, $B^{(k)}$, converge to an upper-triangular matrix T with the eigenvalues of B on its diagonal. Under other conditions, the iterates converge to other forms whose eigenvalues are discernible.

EXAMPLE 5   Perform one step of the QR algorithm on matrix A in Example 3.

SOLUTION   Let $A^{(1)} = A$ and let $Q^{(1)}$ and $R^{(1)}$ be the orthogonal and upper-triangular matrices, respectively, that were computed in Example 3.

If we form the product $A^{(2)} = R^{(1)}Q^{(1)}$ by using the two-step method illustrated in Example 5 of Section 6.5, we find

$$A^{(2)} = R^{(1)}Q^{(1)}$$

$$= \begin{bmatrix} 7.8 & 3.88 & 5.84 \\ .8 & 3.48 & 3.64 \\ -1.6 & -.96 & .72 \end{bmatrix}.$$

(*Note:* We can draw no conclusions from just one iteration, but already in $A^{(2)}$ we can see that the size of the $(2, 1)$, $(3, 1)$, and $(3, 2)$ entries begins to diminish.)

Exercises 6.6

In Exercises 1–4, use Theorem 10 to find a vector $\mathbf{x}^*$ such that $\|A\mathbf{x}^* - \mathbf{b}\| \le \|A\mathbf{x} - \mathbf{b}\|$ for all $\mathbf{x}$.

1. $A = \begin{bmatrix} 1 & 2 \\ 0 & 1 \\ 0 & 0 \end{bmatrix}$, $\quad \mathbf{b} = \begin{bmatrix} 3 \\ 1 \\ 3 \end{bmatrix}$

2. $A = \begin{bmatrix} 2 & 3 \\ 0 & 1 \\ 0 & 0 \end{bmatrix}$, $\quad \mathbf{b} = \begin{bmatrix} 1 \\ -1 \\ 2 \end{bmatrix}$

3. $A = \begin{bmatrix} 1 & 2 & 1 \\ 0 & 1 & 3 \\ 0 & 0 & 2 \\ 0 & 0 & 0 \end{bmatrix}$, $\quad \mathbf{b} = \begin{bmatrix} 6 \\ 7 \\ 4 \\ -1 \end{bmatrix}$

4. $A = \begin{bmatrix} 2 & 0 & 3 \\ 0 & 1 & 2 \\ 0 & 0 & 3 \\ 0 & 0 & 0 \end{bmatrix}$, $\quad \mathbf{b} = \begin{bmatrix} 5 \\ 4 \\ 3 \\ 5 \end{bmatrix}$

In Exercises 5–10, find a Householder matrix S such that $SA = R$, with R upper triangular. List R and the vector $\mathbf{u}$ that defines $S = I - b\mathbf{u}\mathbf{u}^T$.

5. $A = \begin{bmatrix} 3 & 5 \\ 4 & 10 \end{bmatrix}$

6. $A = \begin{bmatrix} 0 & 3 \\ 1 & 5 \end{bmatrix}$

7. $A = \begin{bmatrix} 0 & 2 \\ 4 & 6 \end{bmatrix}$

8. $A = \begin{bmatrix} -4 & 20 \\ 3 & -10 \end{bmatrix}$

9. $A = \begin{bmatrix} 1 & 2 & 1 \\ 0 & 0 & 6 \\ 0 & 1 & 8 \end{bmatrix}$

10. $A = \begin{bmatrix} 3 & 1 & 2 \\ 0 & 3 & 5 \\ 0 & 4 & 10 \end{bmatrix}$

In Exercises 11–14, use Householder matrices to reduce the given matrix A to upper-trapezoidal form.

11. $A = \begin{bmatrix} 1 & -5/3 \\ 2 & 12 \\ 2 & 8 \\ 4 & 15 \end{bmatrix}$

12. $A = \begin{bmatrix} 1 & 2 \\ 1 & 3 \\ 1 & 3 \\ 1 & 6 \end{bmatrix}$

13. $A = \begin{bmatrix} 2 & 4 \\ 0 & 3 \\ 0 & 0 \\ 0 & 4 \end{bmatrix}$

14. $A = \begin{bmatrix} 3 & 5 \\ 0 & 2 \\ 0 & 1 \\ 0 & 2 \end{bmatrix}$

In Exercises 15–18, use Theorem 10 to find the least-squares solution to the problem $A\mathbf{x} = \mathbf{b}$.

15. A in Exercise 11, $\quad \mathbf{b} = \begin{bmatrix} 1 \\ 10 \\ 0 \\ 1 \end{bmatrix}$

16. A in Exercise 12, $\quad \mathbf{b} = \begin{bmatrix} 5 \\ 0 \\ 2 \\ 1 \end{bmatrix}$

17. A in Exercise 13, $\quad \mathbf{b} = \begin{bmatrix} 2 \\ 8 \\ 16 \\ 8 \end{bmatrix}$

18. A in Exercise 14, $\quad \mathbf{b} = \begin{bmatrix} 5 \\ 3 \\ 0 \\ 2 \end{bmatrix}$

19. Prove property (a) of Theorem 10.

6.7 Matrix Polynomials and the Cayley–Hamilton Theorem

The objective of this section is twofold. First, we wish to give a partial justification of the algorithm (presented in Section 6.4) for finding the characteristic polynomial of a Hessenberg matrix. Second, we want to lay some of the necessary foundation for the material in Section 6.8, which describes how a basis for

R^n can be constructed by using eigenvectors and generalized eigenvectors. These ideas are indispensable if we want to solve a difference equation $\mathbf{x}_k = A\mathbf{x}_{k-1}$ or a differential equation $\mathbf{x}'(t) = A\mathbf{x}(t)$, where A is defective.

Matrix Polynomials

To complete our discussion of the algorithm presented in Section 6.4, it is convenient to introduce the idea of a matrix polynomial. By way of example, consider the polynomial

$$q(t) = t^2 + 3t - 2.$$

If A is an $(n \times n)$ matrix, then we can define a matrix expression corresponding to $q(t)$:

$$q(A) = A^2 + 3A - 2I,$$

where I is the $(n \times n)$ identity. In effect, we have inserted A for t in $q(t) = t^2 + 3t - 2$ and defined A^0 by $A^0 = I$. In general, if $q(t)$ is the kth-degree polynomial

$$q(t) = b_k t^k + \cdots + b_2 t^2 + b_1 t + b_0,$$

and if A is an $(n \times n)$ matrix, we define $q(A)$ by

$$q(A) = b_k A^k + \cdots + b_2 A^2 + b_1 A + b_0 I,$$

where I is the $(n \times n)$ identity matrix. Since $q(A)$ is obviously an $(n \times n)$ matrix, we might ask for the eigenvalues and eigenvectors of $q(A)$. It is easy to show that if λ is an eigenvalue of A, then $q(\lambda)$ is an eigenvalue of $q(A)$. [Note that $q(\lambda)$ is the scalar obtained by substituting the value $t = \lambda$ into $q(t)$.]

THEOREM 12

Suppose that $q(t)$ is a kth-degree polynomial and that A is an $(n \times n)$ matrix such that $A\mathbf{x} = \lambda\mathbf{x}$, where $\mathbf{x} \neq \boldsymbol{0}$. Then $q(A)\mathbf{x} = q(\lambda)\mathbf{x}$.

PROOF Suppose that $A\mathbf{x} = \lambda\mathbf{x}$, where $\mathbf{x} \neq \boldsymbol{0}$. As we know, a consequence is that $A^2\mathbf{x} = \lambda^2\mathbf{x}$, and in general

$$A^i\mathbf{x} = \lambda^i\mathbf{x}, \qquad i = 2, 3, \ldots.$$

Therefore, if $q(t) = b_k t^k + \cdots + b_2 t^2 + b_1 t + b_0$, then

$$
\begin{aligned}
q(A)\mathbf{x} &= (b_k A^k + \cdots + b_2 A^2 + b_1 A + b_0 I)\mathbf{x} \\
&= b_k A^k \mathbf{x} + \cdots + b_2 A^2 \mathbf{x} + b_1 A\mathbf{x} + b_0 \mathbf{x} \\
&= b_k \lambda^k \mathbf{x} + \cdots + b_2 \lambda^2 \mathbf{x} + b_1 \lambda\mathbf{x} + b_0 \mathbf{x} \\
&= q(\lambda)\mathbf{x}.
\end{aligned}
$$

Thus if λ is an eigenvalue of A, then $q(\lambda)$ is an eigenvalue of $q(A)$. ▫

The next example provides an illustration of Theorem 12.

EXAMPLE 1 Let $q(t)$ denote the polynomial $q(t) = t^2 + 5t + 4$. Find the eigenvalues and eigenvectors for $q(A)$, where A is the matrix given by

$$A = \begin{bmatrix} 2 & 0 \\ 3 & 1 \end{bmatrix}.$$

SOLUTION The eigenvalues of A are $\lambda = 2$ and $\lambda = 1$. Therefore, by Theorem 12, the eigenvalues of $q(A)$ are given by

$$q(2) = 18 \quad \text{and} \quad q(1) = 10.$$

By way of verification, we calculate $q(A)$:

$$q(A) = A + 5A + 4I = \begin{bmatrix} 4 & 0 \\ 9 & 1 \end{bmatrix} + \begin{bmatrix} 10 & 0 \\ 15 & 5 \end{bmatrix} + \begin{bmatrix} 4 & 0 \\ 0 & 4 \end{bmatrix} = \begin{bmatrix} 18 & 0 \\ 24 & 10 \end{bmatrix}.$$

As the calculation above confirms, $q(A)$ has eigenvalues $\lambda = q(2) = 18$ and $\lambda = q(1) = 10$. ☐

An interesting special case of Theorem 12 is provided when $q(t)$ is the characteristic polynomial for A. In particular, suppose that λ is an eigenvalue of A and that $p(t)$ is the characteristic polynomial for A so that $p(\lambda) = 0$. Since $p(\lambda)$ is an eigenvalue of $p(A)$ and $p(\lambda) = 0$, we conclude that zero is an eigenvalue for $p(A)$; that is, $p(A)$ is a singular matrix. In fact, we will be able to prove more than this; we will show that $p(A)$ is the zero matrix $[p(A) = \mathcal{O}$ is the conclusion of the Cayley–Hamilton theorem].

EXAMPLE 2 Calculate the matrix $p(A)$, where

$$A = \begin{bmatrix} 1 & -2 \\ 2 & 3 \end{bmatrix}$$

and where $p(t)$ is the characteristic polynomial for A.

SOLUTION The characteristic polynomial for A is given by $p(t) = \det(A - tI) = t^2 - 4t + 7$. Therefore, the matrix $p(A)$ is given by

$$p(A) = A^2 - 4A + 7I = \begin{bmatrix} -3 & -8 \\ 8 & 5 \end{bmatrix} - \begin{bmatrix} 4 & -8 \\ 8 & 12 \end{bmatrix} + \begin{bmatrix} 7 & 0 \\ 0 & 7 \end{bmatrix} = \begin{bmatrix} 0 & 0 \\ 0 & 0 \end{bmatrix}.$$ ☐

Thus Example 4 provides a particular instance of the Cayley–Hamilton theorem: If $p(t) = \det(A - tI)$, then $p(A) = \mathcal{O}$.

The theorems that follow show that the algorithm given in Section 6.4 leads to a polynomial $p(t)$ whose zeros are the eigenvalues of H. In the process of verifying this, we will prove an interesting version of the Cayley–Hamilton theorem that is applicable to an unreduced Hessenberg matrix. Before beginning, we make an observation about the sequence of vectors $\mathbf{w}_0, \mathbf{w}_1, \mathbf{w}_2, \ldots$ defined in Algorithm 1:

$$\mathbf{w}_i = H\mathbf{w}_{i-1}, \quad i = 1, 2, \ldots.$$

Since $\mathbf{w}_0 = \mathbf{e}_1$, then $\mathbf{w}_1 = H\mathbf{w}_0 = H\mathbf{e}_1$. Given that $\mathbf{w}_1 = H\mathbf{e}_1$, we see that

$$\mathbf{w}_2 = H\mathbf{w}_1 = H(H\mathbf{e}_1) = H^2\mathbf{e}_1;$$

and in general $\mathbf{w}_k = H^k\mathbf{e}_1$. Thus we can interpret the sequence $\mathbf{w}_0, \mathbf{w}_1, \mathbf{w}_2, \ldots, \mathbf{w}_n$ as being given by $\mathbf{e}_1, H\mathbf{e}_1, H^2\mathbf{e}_1, \ldots, H^n\mathbf{e}_1$.

With this interpretation, we rewrite the equation $a_0\mathbf{w}_0 + a_1\mathbf{w}_1 + \cdots + a_{n-1}\mathbf{w}_{n-1} + \mathbf{w}_n = \boldsymbol{\theta}$ given in Algorithm 1 as

$$a_0\mathbf{e}_1 + a_1 H\mathbf{e}_1 + a_2 H^2\mathbf{e}_1 + \cdots + a_{n-1}H^{n-1}\mathbf{e}_1 + H^n\mathbf{e}_1 = \boldsymbol{\theta}; \tag{1}$$

or by regrouping, (1) is the same as

$$(a_0 I + a_1 H + a_2 H^2 + \cdots + a_{n-1}H^{n-1} + H^n)\mathbf{e}_1 = \boldsymbol{\theta}. \tag{2}$$

Now Theorem 5 asserts that if H is an unreduced $(n \times n)$ Hessenberg matrix, then the vectors $\mathbf{e}_1, H\mathbf{e}_1, H^2\mathbf{e}_1, \ldots, H^{n-1}\mathbf{e}_1$ are linearly independent, and that there is a unique set of scalars $a_0, a_1, \ldots, a_{n-1}$ that satisfy (1). Defining $p(t)$ from (1) as

$$p(t) = a_0 + a_1 t + a_2 t^2 + \cdots + a_{n-1}t^{n-1} + t^n,$$

we see from (2) that $p(H)\mathbf{e}_1 = \boldsymbol{\theta}$. With these preliminaries, we prove the following result.

THEOREM 13

Let H be an $(n \times n)$ unreduced Hessenberg matrix; let $a_0, a_1, \ldots, a_{n-1}$ be the unique scalars satisfying

$$a_0\mathbf{e}_1 + a_1 H\mathbf{e}_1 + a_2 H^2\mathbf{e}_1 + \cdots + a_{n-1}H^{n-1}\mathbf{e}_1 + H^n\mathbf{e}_1 = \boldsymbol{\theta};$$

and let $p(t) = a_0 + a_1 t + a_2 t^2 + \cdots + a_{n-1}t^{n-1} + t^n$. Then:

a) $p(H)$ is the zero matrix.
b) If $q(t) = b_0 + b_1 t + b_2 t^2 + \cdots + b_{k-1}t^{k-1} + t^k$ is any monic kth-degree polynomial, and if $q(H)$ is the zero matrix, then $k \geq n$. Moreover, if $k = n$, then $q(t) \equiv p(t)$.

PROOF   For property (a), since $\{\mathbf{e}_1, H\mathbf{e}_1, H^2\mathbf{e}_1, \ldots, H^{n-1}\mathbf{e}_1\}$ is a basis for R^n, we can express any vector $\mathbf{y}$ in R^n as a linear combination:

$$\mathbf{y} = c_0\mathbf{e}_1 + c_1 H\mathbf{e}_1 + c_2 H^2\mathbf{e}_1 + \cdots + c_{n-1}H^{n-1}\mathbf{e}_1.$$

Therefore, $p(H)\mathbf{y}$ is the vector

$$p(H)\mathbf{y} = c_0 p(H)\mathbf{e}_1 + c_1 p(H)H\mathbf{e}_1 + \cdots + c_{n-1}p(H)H^{n-1}\mathbf{e}_1. \tag{3}$$

Now although matrix products do not normally commute, it is easy to see that $p(H)H^i = H^i p(H)$. Therefore, from (3), we can represent $p(H)\mathbf{y}$ as

$$p(H)\mathbf{y} = c_0 p(H)\mathbf{e}_1 + c_1 H p(H)\mathbf{e}_1 + \cdots + c_{n-1}H^{n-1}p(H)\mathbf{e}_1;$$

and since $p(H)\mathbf{e}_1 = \boldsymbol{\theta}$ [see (1) and (2)], then $p(H)\mathbf{y} = \boldsymbol{\theta}$ for *any* $\mathbf{y}$ in R^n. In particular, $p(H)\mathbf{e}_j = \mathbf{0}$ for $j = 1, 2, \ldots, n$; and since $p(H)\mathbf{e}_j$ is the jth column of $p(H)$, it follows that $p(H) = \mathcal{O}$.

For the proof of property (b), suppose that $q(H)$ is the zero matrix, where

$$q(H) = b_0 + b_1 H + b_2 H^2 + \cdots + b_{k-1}H^{k-1} + H^k.$$

Then $q(H)\mathbf{y} = \mathbf{0}$ for every $\mathbf{y}$ in R^n, and in particular for $\mathbf{y} = \mathbf{e}_1$ we have $q(H)\mathbf{e}_1 = \mathbf{0}$, or

$$b_0\mathbf{e}_1 + b_1 H\mathbf{e}_1 + b_2 H^2\mathbf{e}_1 + \cdots + b_{k-1}H^{k-1}\mathbf{e}_1 + H^k\mathbf{e}_1 = \mathbf{0}. \tag{4}$$

However, the vectors $\mathbf{e}_1, H\mathbf{e}_1, \ldots, H^k\mathbf{e}_1$ are linearly independent when $k \leq n - 1$; so (4) can hold only if $k \geq n$ [recall that the leading coefficient of $q(t)$ is 1; so we are excluding the possibility that $q(t)$ is the zero polynomial]. Moreover, if $k = n$, we can satisfy (4) only with the choice $b_0 = a_0, b_1 = a_1, \ldots, b_{n-1} = a_{n-1}$ by Theorem 5; so if $k = n$, then $q(t) \equiv p(t)$. □

Since it was shown above that $p(H)$ is the zero matrix whenever $p(t)$ is the polynomial defined by the algorithm of Section 6.4, it is now an easy matter to show that the zeros of $p(t)$ are precisely the eigenvalues of H.

THEOREM 14

Let H be an $(n \times n)$ unreduced Hessenberg matrix, and let $p(t)$ be the polynomial defined by Algorithm 1. Then λ is a root of $p(t) = 0$ if and only if λ is an eigenvalue of H.

PROOF We show first that every eigenvalue of H is a zero of $p(t)$. Thus we suppose that $H\mathbf{x} = \lambda\mathbf{x}$, where $\mathbf{x} \neq \mathbf{0}$. By Theorem 12, we know that

$$p(H)\mathbf{x} = p(\lambda)\mathbf{x};$$

and since $p(H)$ is the zero matrix of Theorem 13, we must also conclude that

$$\mathbf{0} = p(\lambda)\mathbf{x}.$$

But since $\mathbf{x} \neq \mathbf{0}$, the equality $\mathbf{0} = p(\lambda)\mathbf{x}$ implies that $p(\lambda) = 0$. Thus every eigenvalue of H is a zero of $p(t)$.

Conversely, suppose that λ is a zero of $p(t)$. Then we can write $p(t)$ in the form

$$p(t) = (t - \lambda)q(t), \tag{5}$$

where $q(t)$ is a monic polynomial of degree $n - 1$. Now equating coefficients of like powers shows that if $u(t) = r(t)s(t)$, where u, r, and s are polynomials, then we also have a corresponding matrix identity

$$u(A) = r(A)s(A)$$

for any square matrix A. Thus from (5) we can assert that

$$p(H) = (H - \lambda I)q(H). \tag{6}$$

If $H - \lambda I$ were nonsingular, we could rewrite (6) as

$$(H - \lambda I)^{-1} p(H) = q(H);$$

and since $p(H)$ is the zero matrix by property (a) of Theorem 13, then $q(H)$ would be the zero matrix also. However, $q(t)$ is a monic polynomial of degree $n - 1$, so property (b) of Theorem 13 assures us that $q(H)$ is *not* the zero matrix. Thus if λ is a root of $p(t) = 0$, we know that $H - \lambda I$ must be singular, and hence λ is an eigenvalue of H.   □

We conclude this section by outlining a proof of the Cayley–Hamilton theorem for an arbitrary matrix. If H is a Hessenberg matrix of the form

$$H = \begin{bmatrix} H_{11} & H_{12} & \cdots & H_{1r} \\ \mathcal{O} & H_{22} & \cdots & H_{2r} \\ \vdots & & & \\ \mathcal{O} & \mathcal{O} & \cdots & H_{rr} \end{bmatrix}, \tag{7}$$

where $H_{11}, H_{22}, \ldots, H_{rr}$ are unreduced Hessenberg blocks, then we define $p(t)$ to be the characteristic polynomial for H, where

$$p(t) = p_1(t) p_2(t) \cdots p_r(t)$$

and where $p_i(t)$ is the characteristic polynomial for H_{ii}, $1 \le i \le r$.

THEOREM 15

If $p(t)$ is the characteristic polynomial for a Hessenberg matrix H, then $p(H)$ is the zero matrix.

PROOF   We sketch the proof for the case $r = 2$. If H has the form

$$H = \begin{bmatrix} H_{11} & H_{12} \\ \mathcal{O} & H_{22} \end{bmatrix},$$

where H_{11} and H_{22} are square blocks, then it can be shown that H^k is a block matrix of the form

$$H^k = \begin{bmatrix} H_{11}^k & V_k \\ \mathcal{O} & H_{22}^k \end{bmatrix}.$$

Given this, it follows that if $q(t)$ is any polynomial, then $q(H)$ is a block matrix of the form

$$q(H) = \begin{bmatrix} q(H_{11}) & W \\ \mathcal{O} & q(H_{22}) \end{bmatrix}.$$

From these preliminaries, if H_{11} and H_{22} are unreduced blocks, then

$$p(H) = p_1(H) p_2(H) = \begin{bmatrix} p_1(H_{11}) & R \\ \mathcal{O} & p_1(H_{22}) \end{bmatrix} \begin{bmatrix} p_2(H_{11}) & S \\ \mathcal{O} & p_2(H_{22}) \end{bmatrix};$$

and since $p_1(H_{11})$ and $p_2(H_{22})$ are zero blocks, it is easy to see that $p(H)$ is the zero matrix. This argument can be repeated inductively to show that $p(H)$ is the zero matrix when H has the form (7) for $r > 2$. ◻

Finally, we note that the essential features of polynomial expressions are preserved by similarity transformations. For example, if $H = SAS^{-1}$ and if $q(t)$ is any polynomial, then (Exercise 5, Section 6.7)

$$q(H) = Sq(A)S^{-1}.$$

Thus if A is similar to H and if $p(H)$ is the zero matrix, then $p(A)$ is the zero matrix as well.

The remarks made above allow us to state the Cayley–Hamilton theorem as a corollary of Theorem 15.

COROLLARY

The Cayley–Hamilton Theorem If $p(t)$ is the characteristic polynomial for an $(n \times n)$ matrix A, then $p(A)$ is the zero matrix.

Exercises 6.7

1. Let $q(t) = t^2 - 4t + 3$. Calculate the matrices $q(A)$, $q(B)$, and $q(C)$.

$$A = \begin{bmatrix} 1 & -1 \\ 1 & 3 \end{bmatrix}, \quad B = \begin{bmatrix} 2 & -1 \\ -1 & 2 \end{bmatrix},$$

$$C = \begin{bmatrix} -2 & 0 & -2 \\ -1 & 1 & -2 \\ 0 & 1 & -1 \end{bmatrix}$$

2. The polynomial $p(t) = (t - 1)^3 = t^3 - 3t^2 + 3t - 1$ is the characteristic polynomial for A, B, C, and I.

$$A = \begin{bmatrix} 1 & 0 & 0 \\ 1 & 1 & 0 \\ 0 & 0 & 1 \end{bmatrix}, \quad B = \begin{bmatrix} 1 & 0 & 0 \\ 0 & 1 & 0 \\ 0 & 1 & 1 \end{bmatrix},$$

$$C = \begin{bmatrix} 1 & 0 & 0 \\ 1 & 1 & 0 \\ 0 & 1 & 1 \end{bmatrix}, \quad I = \begin{bmatrix} 1 & 0 & 0 \\ 0 & 1 & 0 \\ 0 & 0 & 1 \end{bmatrix}.$$

a) Verify that $p(A)$, $p(B)$, $p(C)$, and $p(I)$ are each the zero matrix.

b) For A and B, find a quadratic polynomial $q(t)$ such that $q(A) = q(B) = \mathcal{O}$.

3. Suppose that $q(t)$ is any polynomial and $p(t)$ is the characteristic polynomial for a matrix A. If we divide $p(t)$ into $q(t)$, we obtain an identity

$$q(t) = s(t)p(t) + r(t),$$

where the degree of $r(t)$ is less than the degree of $p(t)$. From this result, it can be shown that $q(A) = s(A)p(A) + r(A)$; and since $p(A) = \mathcal{O}$, $q(A) = r(A)$.

a) Let $p(t) = t^2 - 4t + 3$ and $q(t) = t^5 - 4t^4 + 4t^3 - 5t^2 + 8t - 1$. Find $s(t)$ and $r(t)$ so that $q(t) = s(t)p(t) + r(t)$.

b) Observe that $p(t)$ is the characteristic polynomial for the matrix B in Exercise 1. Calculate the matrix $q(B)$ without forming the powers B^5, B^4, and so on.

4. Consider the (7×7) Hessenberg matrix H given in (10) of Section 6.4, where H is partitioned with three unreduced diagonal blocks, H_{11}, H_{22}, and H_{33}. Verify that $\det(H - tI) = -p_1(t)p_2(t)p_3(t)$, where $p_1(t)$, $p_2(t)$, and $p_3(t)$ are the characteristic polynomials for H_{11}, H_{22}, and H_{33} as given by Algorithm 1.

5. Suppose that $H = SAS^{-1}$ and $q(t)$ is any polynomial. Show that $q(H) = Sq(A)S^{-1}$. (*Hint:* Show that $H^k = SA^kS^{-1}$ by direct multiplication.)

Exercises 6–8 give another proof that a symmetric matrix is diagonalizable.

6. Let A be an $(n \times n)$ symmetric matrix. Let λ_1 and λ_2 be distinct eigenvalues of A with corresponding eigenvectors $\mathbf{u}_1$ and $\mathbf{u}_2$. Prove that $\mathbf{u}_1^T\mathbf{u}_2 = 0$. (*Hint:* Given that $A\mathbf{u}_1 = \lambda_1\mathbf{u}_1$ and $A\mathbf{u}_2 = \lambda_2\mathbf{u}_2$, show that $\mathbf{u}_1^T A\mathbf{u}_2 = \mathbf{u}_2^T A\mathbf{u}_1$.)

7. Let W be a subspace of R^n, where $\dim(W) = d, d \geq 1$. Let A be an $(n \times n)$ matrix, and suppose that $A\mathbf{x}$ is in W whenever $\mathbf{x}$ is in W.
   a) Let $\mathbf{x}_0$ be any fixed vector in W. Prove that $A^j\mathbf{x}_0$ is in W for $j = 1, 2, \ldots$. There is a smallest value k for which the set of vectors $\{\mathbf{x}_0, A\mathbf{x}_0, A^2\mathbf{x}_0, \ldots, A^k\mathbf{x}_0\}$ is linearly dependent; and thus there are unique scalars $a_0, a_1, \ldots, a_{k-1}$ such that
   $$a_0\mathbf{x}_0 + a_1 A\mathbf{x}_0 + \cdots + a_{k-1}A^{k-1}\mathbf{x}_0 + A^k\mathbf{x}_0 = \mathbf{0}.$$
   Use these scalars to define the polynomial $m(t)$, where $m(t) = t^k + a_{k-1}t^{k-1} + \cdots + a_1 t + a_0$. Observe that $m(A)\mathbf{x}_0 = \mathbf{0}$; $m(t)$ is called the minimal annihilating polynomial for $\mathbf{x}_0$. By construction there is no monic polynomial $q(t)$, where $q(t)$ has degree less than k and $q(A)\mathbf{x}_0 = \mathbf{0}$.
   b) Let r be a root of $m(t) = 0$ so that $m(t) = (t - r)s(t)$. Prove that r is an eigenvalue of A. (*Hint:* Is the vector $s(A)\mathbf{x}_0$ nonzero?) Note that part (b) shows that every root of $m(t) = 0$ is an eigenvalue of A. If A is symmetric, then $m(t) = 0$ has only real roots, so $s(A)\mathbf{x}_0$ is in W.

8. Exercise 6 shows that eigenvectors of a symmetric matrix belonging to distinct eigenvalues are orthogonal. We now show that if A is a symmetric $(n \times n)$ matrix, then A has a set of n orthogonal eigenvectors. Let $\{\mathbf{u}_1, \mathbf{u}_2, \ldots, \mathbf{u}_k\}$ be a set of k eigenvectors for A, $1 \leq k < n$, where $\mathbf{u}_i^T\mathbf{u}_j = 0, i \neq j$. Let W be the subset of R^n defined by
$$W = \{\mathbf{x} : \mathbf{x}^T\mathbf{u}_i = 0, \qquad i = 1, 2, \ldots, k\}$$
From the Gram–Schmidt theorem, the subset W contains nonzero vectors.
   a) Prove that W is a subspace of R^n.
   b) Suppose that A is $(n \times n)$ and symmetric. Prove that $A\mathbf{x}$ is in W whenever $\mathbf{x}$ is in W. From Exercise 7, A has an eigenvector, $\mathbf{u}$, in W. If we label $\mathbf{u}$ as $\mathbf{u}_{k+1}$, then by construction $\{\mathbf{u}_1, \mathbf{u}_2, \ldots, \mathbf{u}_k, \mathbf{u}_{k+1}\}$ is a set of orthogonal eigenvectors for A. It follows that A has a set of n orthogonal eigenvectors, $\mathbf{u}_1, \mathbf{u}_2, \ldots, \mathbf{u}_n$. Using these, we can form Q so that $Q^TAQ = D$, where D is diagonal and $Q^TQ = I$.

6.8 Generalized Eigenvectors and Solutions of Systems of Differential Equations

In this section, we develop the idea of a generalized eigenvector in order to give the complete solution to the system of differential equations $\mathbf{x}' = A\mathbf{x}$. When an $(n \times n)$ matrix A has real eigenvalues, the eigenvectors and generalized eigenvectors of A form a basis for R^n. We show how to construct the complete solution of $\mathbf{x}' = A\mathbf{x}$ from this special basis. (When some of the eigenvalues of A are complex, a few modifications are necessary to obtain the complete solution of $\mathbf{x}' = A\mathbf{x}$ in a real form. In any event, the eigenvectors and generalized eigenvectors of A form a basis for C^n, where C^n denotes the set of all n-dimensional vectors with real or complex components.)

To begin, let A be an $(n \times n)$ matrix. The problem we wish to solve is called an initial-value problem and is formulated as follows: Given a vector $\mathbf{x}_0$ in R^n, find a function $\mathbf{x}(t)$ such that

$$\mathbf{x}(0) = \mathbf{x}_0$$
$$\mathbf{x}'(t) = A\mathbf{x}(t) \quad \text{for all } t. \tag{1}$$

If we can find n functions $\mathbf{x}_1(t), \mathbf{x}_2(t), \ldots, \mathbf{x}_n(t)$ that satisfy

$$\mathbf{x}_1'(t) = A\mathbf{x}_1(t), \qquad \mathbf{x}_2'(t) = A\mathbf{x}_2(t), \ldots, \mathbf{x}_n'(t) = A\mathbf{x}_n(t)$$

and such that $\{\mathbf{x}_1(0), \mathbf{x}_2(0), \ldots, \mathbf{x}_n(0)\}$ is linearly independent, then we can always solve (1). To show why, we merely note that there must be constants $c_1, c_2, \ldots, c_n$ such that

$$\mathbf{x}_0 = c_1\mathbf{x}_1(0) + c_2\mathbf{x}_2(0) + \cdots + c_n\mathbf{x}_n(0)$$

and then note that the function

$$\mathbf{y}(t) = c_1\mathbf{x}_1(t) + c_2\mathbf{x}_2(t) + \cdots + c_n\mathbf{x}_n(t)$$

satisfies the requirements of (1). Thus to solve $\mathbf{x}' = A\mathbf{x}$, $\mathbf{x}(0) = \mathbf{x}_0$, we are led to search for n solutions $\mathbf{x}_1(t), \mathbf{x}_2(t), \ldots, \mathbf{x}_n(t)$ of $\mathbf{x}' = A\mathbf{x}$ for which $\{\mathbf{x}_1(0), \mathbf{x}_2(0), \ldots, \mathbf{x}_n(0)\}$ is linearly independent.

If A has a set of k linearly independent eigenvectors $\{\mathbf{u}_1, \mathbf{u}_2, \ldots, \mathbf{u}_k\}$, where

$$A\mathbf{u}_i = \lambda_i\mathbf{u}_i, \qquad i = 1, 2, \ldots, k,$$

then, as in Section 6.2, we can immediately construct k solutions to $\mathbf{x}' = A\mathbf{x}$, namely,

$$\mathbf{x}_1(t) = e^{\lambda_1 t}\mathbf{u}_1, \qquad \mathbf{x}_2(t) = e^{\lambda_2 t}\mathbf{u}_2, \ldots, \mathbf{x}_k(t) = e^{\lambda_k t}\mathbf{u}_k.$$

Also, since $\mathbf{x}_i(0) = \mathbf{u}_i$, it follows that $\{\mathbf{x}_1(0), \mathbf{x}_2(0), \ldots, \mathbf{x}_k(0)\}$ is a linearly independent set. The difficulty arises when $k < n$, for then we must produce an additional set of $n - k$ solutions of $\mathbf{x}' = A\mathbf{x}$. In this connection, recall that an $(n \times n)$ matrix A is called *defective* if A has fewer than n linearly independent eigenvectors. (*Note:* Distinct eigenvalues give rise to linearly independent eigenvectors; so A can be defective only if the characteristic equation $p(t) = 0$ has fewer than n distinct roots.)

Generalized Eigenvectors

A complete analysis of the initial-value problem is simplified considerably if we assume A is a Hessenberg matrix. If A is not a Hessenberg matrix, then a simple change of variables can be used to convert $\mathbf{x}' = A\mathbf{x}$ to an equivalent problem $\mathbf{y}' = H\mathbf{y}$, where H is a Hessenberg matrix. In particular, suppose that $QAQ^{-1} = H$ and let $\mathbf{y}(t) = Q\mathbf{x}(t)$. Therefore, we see that $\mathbf{x}(t) = Q^{-1}\mathbf{y}(t)$ and $\mathbf{x}'(t) = Q^{-1}\mathbf{y}'(t)$. Thus, $\mathbf{x}'(t) = A\mathbf{x}(t)$ is the same as

$$Q^{-1}\mathbf{y}'(t) = AQ^{-1}\mathbf{y}(t).$$

Multiplying both sides by Q, we obtain the related equation $\mathbf{y}' = H\mathbf{y}$, where $H = QAQ^{-1}$ is a Hessenberg matrix. Given that we can always make this change of variables, we will focus for the remainder of this section on the problem of solving

$$\mathbf{x}'(t) = H\mathbf{x}(t), \qquad \mathbf{x}(0) = \mathbf{x}_0. \tag{2}$$

As we know, if H is $(n \times n)$ and has n linearly independent eigenvectors, we can always solve (2). To see how to solve (2) when H is defective, let us suppose that $p(t)$ is the characteristic polynomial for H. If we write $p(t)$ in factored form as

$$p(t) = (t - \lambda_1)^{m_1}(t - \lambda_2)^{m_2} \cdots (t - \lambda_k)^{m_k},$$

where $m_1 + m_2 + \cdots + m_k = n$, then we say that the eigenvalue λ_i has *algebraic multiplicity* m_i. Given λ_i, we want to construct m_i solutions of $\mathbf{x}' = H\mathbf{x}$ that are associated with λ_i. For example, suppose that λ is an eigenvalue of H of algebraic multiplicity 2. We have one solution of $\mathbf{x}' = H\mathbf{x}$, namely, $\mathbf{x}(t) = e^{\lambda t}\mathbf{u}$, where $H\mathbf{u} = \lambda\mathbf{u}$; and we would like another solution. To find this additional solution, we note that the theory from elementary differential equations suggests that we look for another solution to $\mathbf{x}' = H\mathbf{x}$ that is of the form $\mathbf{x}(t) = te^{\lambda t}\mathbf{a} + e^{\lambda t}\mathbf{b}$, where $\mathbf{a} \neq \boldsymbol{\theta}$, $\mathbf{b} \neq \boldsymbol{\theta}$. To see what conditions $\mathbf{a}$ and $\mathbf{b}$ must satisfy, we calculate

$$\mathbf{x}'(t) = t\lambda e^{\lambda t}\mathbf{a} + e^{\lambda t}\mathbf{a} + \lambda e^{\lambda t}\mathbf{b}$$
$$H\mathbf{x}(t) = te^{\lambda t}H\mathbf{a} + e^{\lambda t}H\mathbf{b}.$$

After we equate $\mathbf{x}'(t)$ with $H\mathbf{x}(t)$ and group like powers of t, our guess leads to the conditions

$$t\lambda e^{\lambda t}\mathbf{a} = te^{\lambda t}H\mathbf{a}$$
$$e^{\lambda t}(\mathbf{a} + \lambda\mathbf{b}) = e^{\lambda t}H\mathbf{b}. \tag{3}$$

If (3) is to hold for all t, we will need

$$\lambda\mathbf{a} = H\mathbf{a}$$
$$\mathbf{a} + \lambda\mathbf{b} = H\mathbf{b},$$

or equivalently,

$$(H - \lambda I)\mathbf{a} = \boldsymbol{\theta}$$
$$(H - \lambda I)\mathbf{b} = \mathbf{a}, \tag{4}$$

where $\mathbf{a}$ and $\mathbf{b}$ are nonzero vectors. From (4) we see that $\mathbf{a}$ is an eigenvector and that $(H - \lambda I)^2\mathbf{b} = \boldsymbol{\theta}$, but $(H - \lambda I)\mathbf{b} \neq \boldsymbol{\theta}$. We will call $\mathbf{b}$ a generalized eigenvector of order 2. If we can find vectors $\mathbf{a}$ and $\mathbf{b}$ that satisfy (4), then we have two solutions of $\mathbf{x}' = H\mathbf{x}$ associated with λ, namely,

$$\mathbf{x}_1(t) = e^{\lambda t}\mathbf{a}$$
$$\mathbf{x}_2(t) = te^{\lambda t}\mathbf{a} + e^{\lambda t}\mathbf{b}.$$

Moreover, $\mathbf{x}_1(0) = \mathbf{a}$, $\mathbf{x}_2(0) = \mathbf{b}$, and it is easy to see that $\mathbf{x}_1(0)$ and $\mathbf{x}_2(0)$ are linearly independent. [If $c_1\mathbf{a} + c_2\mathbf{b} = \boldsymbol{\theta}$, then $(H - \lambda I)(c_1\mathbf{a} + c_2\mathbf{b}) = \boldsymbol{\theta}$. Since $(H - \lambda I)\mathbf{a} = \boldsymbol{\theta}$, it follows that $c_2(H - \lambda I)\mathbf{b} = c_2\mathbf{a} = \boldsymbol{\theta}$, which shows that $c_2 = 0$. Finally if $c_2 = 0$, then $c_1\mathbf{a} = \boldsymbol{\theta}$, which means that $c_1 = 0$.]

<u>EXAMPLE 1</u>   Solve the initial-value problem $\mathbf{x}'(t) = A\mathbf{x}(t)$, $\mathbf{x}(0) = \mathbf{x}_0$, where

$$A = \begin{bmatrix} 1 & -1 \\ 1 & 3 \end{bmatrix} \quad \text{and} \quad \mathbf{x}_0 = \begin{bmatrix} 5 \\ -7 \end{bmatrix}.$$

SOLUTION   For matrix A, the characteristic polynomial $p(t) = \det(A - tI)$ is given by

$$p(t) = (t - 2)^2.$$

Thus the only eigenvalue of A is $\lambda = 2$. The only eigenvectors for $\lambda = 2$ are those vectors $\mathbf{u}$ of the form

$$\mathbf{u} = a\begin{bmatrix} 1 \\ -1 \end{bmatrix}, \qquad a \neq 0.$$

Since A is defective, we look for a generalized eigenvector associated with $\lambda = 2$. That is, as in (4) we look for a vector $\mathbf{v}$ such that

$$(A - 2I)\mathbf{v} = \mathbf{u}, \qquad \mathbf{u} = \begin{bmatrix} 1 \\ -1 \end{bmatrix}.$$

In detail, the equation $(A - 2I)\mathbf{v} = \mathbf{u}$ is given by

$$\begin{bmatrix} -1 & -1 \\ 1 & 1 \end{bmatrix}\begin{bmatrix} v_1 \\ v_2 \end{bmatrix} = \begin{bmatrix} 1 \\ -1 \end{bmatrix}.$$

Now, although matrix $A - 2I$ is singular, the equation above does have a solution, namely, $v_1 = 1, v_2 = -2$.

Thus we have found two solutions to $\mathbf{x}'(t) = A\mathbf{x}(t)$, $\mathbf{x}_1(t)$ and $\mathbf{x}_2(t)$, where

$$\mathbf{x}_1(t) = e^{2t}\mathbf{u} \quad \text{and} \quad \mathbf{x}_2(t) = te^{2t}\mathbf{u} + e^{2t}\mathbf{v}.$$

The general solution of $\mathbf{x}'(t) = A\mathbf{x}(t)$ is

$$\mathbf{x}(t) = a_1 e^{2t}\mathbf{u} + a_2(te^{2t}\mathbf{u} + e^{2t}\mathbf{v}).$$

To satisfy the initial condition $\mathbf{x}(0) = \mathbf{x}_0 = [5, -7]^T$, we need a_1 and a_2 so that

$$\mathbf{x}(0) = a_1\mathbf{u} + a_2\mathbf{v} = \mathbf{x}_0.$$

Solving for a_1 and a_2, we find $a_1 = 3$ and $a_2 = 2$.

Therefore, the solution is

$$\mathbf{x}(t) = 3e^{2t}\mathbf{u} + 2(te^{2t}\mathbf{u} + e^{2t}\mathbf{v})$$

$$= \begin{bmatrix} 5e^{2t} + 2te^{2t} \\ -7e^{2t} - 2te^{2t} \end{bmatrix}. \qquad \square$$

In this section, we will see that the solution procedure illustrated in Example 1 can be applied to an unreduced Hessenberg matrix. To formalize the procedure, we need a definition.

DEFINITION 2   Let A be an $(n \times n)$ matrix. A nonzero vector $\mathbf{v}$ such that

$$(A - \lambda I)^j\mathbf{v} = \boldsymbol{\theta}$$

$$(A - \lambda I)^{j-1}\mathbf{v} \neq \boldsymbol{\theta}$$

is called a ***generalized eigenvector of order*** j corresponding to λ.

Note that an eigenvector can be regarded as a generalized eigenvector of order 1.

If a matrix H has a generalized eigenvector $\mathbf{v}_m$ of order m corresponding to λ, then the following sequence of vectors can be defined:

$$
\begin{aligned}
(H - \lambda I)\mathbf{v}_m &= \mathbf{v}_{m-1} \\
(H - \lambda I)\mathbf{v}_{m-1} &= \mathbf{v}_{m-2} \\
&\;\;\vdots \qquad\qquad \vdots \\
(H - \lambda I)\mathbf{v}_2 &= \mathbf{v}_1.
\end{aligned}
\tag{5}
$$

It is easy to show that each vector $\mathbf{v}_r$ in (5) is a generalized eigenvector of order r and that $\{\mathbf{v}_1, \mathbf{v}_2, \ldots, \mathbf{v}_m\}$ is a linearly independent set (see Exercise 6). In addition, each generalized eigenvector $\mathbf{v}_r$ leads to a solution $\mathbf{x}_r(t)$ of $\mathbf{x}' = H\mathbf{x}$, where

$$
\mathbf{x}_r(t) = e^{\lambda t}\left(\mathbf{v}_r + t\mathbf{v}_{r-1} + \cdots + \frac{t^{r-1}}{(r-1)!}\mathbf{v}_1 \right)
\tag{6}
$$

(see Exercise 7).

We begin the analysis by proving two theorems that show that an $(n \times n)$ unreduced Hessenberg matrix H has a set of n linearly independent eigenvectors and generalized eigenvectors. Then, following several examples, we comment on the general case.

THEOREM 16

Let H be an $(n \times n)$ unreduced Hessenberg matrix, and let λ be an eigenvalue of H, where λ has algebraic multiplicity m. Then H has a generalized eigenvector of order m corresponding to λ.

PROOF   Let $p(t) = (t - \lambda)^m q(t)$ be the characteristic polynomial for H, where $q(\lambda) \neq 0$. Let $\mathbf{v}_m$ be the vector $\mathbf{v}_m = q(H)\mathbf{e}_1$. By Theorem 13, $(H - \lambda I)^{m-1}q(H)\mathbf{e}_1 \neq \boldsymbol{\theta}$, so $(H - \lambda I)^{m-1}\mathbf{v}_m \neq \boldsymbol{\theta}$. Also by Theorem 13, $(H - \lambda I)^m \mathbf{v}_m = (H - \lambda I)^m q(H)\mathbf{e}_1 = p(H)\mathbf{e}_1 = \boldsymbol{\theta}$, so we see that $\mathbf{v}_m$ is a generalized eigenvector of order m.

Theorem 16 is an existence result that is quite valuable. If we know that an unreduced Hessenberg matrix H has an eigenvalue of multiplicity m, then we know that the sequence of vectors in (5) is defined. Therefore, we can start with an eigenvector $\mathbf{v}_1$, then find $\mathbf{v}_2$, then find $\mathbf{v}_3$, and so on. (If H is a reduced Hessenberg matrix, the sequence (5) may not exist.)

EXAMPLE 2   Consider the unreduced Hessenberg matrix H, where

$$
H = \begin{bmatrix} 1 & 0 & 0 \\ 1 & 1 & 0 \\ 0 & 1 & 1 \end{bmatrix}.
$$

Note that the eigenvalue $\lambda = 1$ has algebraic multiplicity 3. Find generalized eigenvectors of orders 2 and 3.

SOLUTION   We work backward up the chain of vectors in (5), starting with an eigenvector $\mathbf{v}_1$. Now all eigenvectors corresponding to $\lambda = 1$ have the form $\mathbf{u} = a[0, 0, 1]^T$, $a \neq 0$. If we choose $\mathbf{v}_1 = [0, 0, 1]^T$, the equation $(H - I)\mathbf{v}_2 = \mathbf{v}_1$ has the form

$$\begin{bmatrix} 0 & 0 & 0 \\ 1 & 0 & 0 \\ 0 & 1 & 0 \end{bmatrix} \begin{bmatrix} x_1 \\ x_2 \\ x_3 \end{bmatrix} = \begin{bmatrix} 0 \\ 0 \\ 1 \end{bmatrix}.$$

The solution to the equation above is $\mathbf{v}_2 = [0, 1, a]^T$, where a is arbitrary. For simplicity we choose $a = 0$ and obtain $\mathbf{v}_2 = [0, 1, 0]^T$.

Next, we need to solve $(H - I)\mathbf{v}_3 = \mathbf{v}_2$:

$$\begin{bmatrix} 0 & 0 & 0 \\ 1 & 0 & 0 \\ 0 & 1 & 0 \end{bmatrix} \begin{bmatrix} x_1 \\ x_2 \\ x_3 \end{bmatrix} = \begin{bmatrix} 0 \\ 1 \\ 0 \end{bmatrix}.$$

The solution to the equation above is $\mathbf{v}_3 = [1, 0, a]^T$, where a is arbitrary. One solution is $\mathbf{v}_3 = [1, 0, 0]^T$.

To summarize, an eigenvector and two generalized eigenvectors for H are

$$\mathbf{v}_1 = \begin{bmatrix} 0 \\ 0 \\ 1 \end{bmatrix}, \qquad \mathbf{v}_2 = \begin{bmatrix} 0 \\ 1 \\ 0 \end{bmatrix}, \qquad \mathbf{v}_3 = \begin{bmatrix} 1 \\ 0 \\ 0 \end{bmatrix}.$$

(*Note:* For $i = 1, 2, 3$, $\mathbf{v}_i$ is a generalized eigenvector of order i.) ☐

Example 2 illustrates a situation in which a (3×3) unreduced Hessenberg matrix has a set of eigenvalues and generalized eigenvectors that form a basis for R^3. As the next theorem demonstrates, Example 2 is not atypical.

THEOREM 17

Let H be an $(n \times n)$ unreduced Hessenberg matrix. There is a set $\{\mathbf{u}_1, \mathbf{u}_2, \ldots, \mathbf{u}_n\}$ of linearly independent vectors in which each $\mathbf{u}_i$ is an eigenvector or a generalized eigenvector of H.

PROOF   Suppose that H has eigenvalues $\lambda_1, \lambda_2, \ldots, \lambda_k$, where λ_i has multiplicity m_i. Thus the characteristic polynomial has the form

$$p(t) = (t - \lambda_1)^{m_1}(t - \lambda_2)^{m_2} \cdots (t - \lambda_k)^{m_k},$$

where $m_1 + m_2 + \cdots + m_k = n$. By Theorem 16, each eigenvalue λ_i has an associated generalized eigenvector of order m_i. For each eigenvalue λ_i, we can use (5) to generate a set of m_i generalized eigenvectors having order $1, 2, \ldots, m_i$. Let us denote this collection of n generalized eigenvectors as

$$\mathbf{v}_1, \mathbf{v}_2, \ldots, \mathbf{v}_{m_1}, \mathbf{w}_1, \mathbf{w}_2, \ldots, \mathbf{w}_r, \tag{7}$$

where $m_1 + r = n$. In (7), $\mathbf{v}_j$ is a generalized eigenvector of order j corresponding to the eigenvalue λ_1, whereas each of the vectors $\mathbf{w}_j$ is a generalized eigenvector for one of $\lambda_2, \lambda_3, \ldots, \lambda_k$.

To show that the vectors in (7) are linearly independent, consider

$$a_1 \mathbf{v}_1 + a_2 \mathbf{v}_2 + \cdots + a_{m_1} \mathbf{v}_{m_1} + b_1 \mathbf{w}_1 + b_2 \mathbf{w}_2 + \cdots + b_r \mathbf{w}_r = \mathbf{0}. \tag{8}$$

Now for $q(t) = (t - \lambda_2)^{m_2} \ldots (t - \lambda_k)^{m_k}$ and for $1 \leq j \leq r$, we note that

$$q(H)\mathbf{w}_j = \mathbf{0},$$

since $\mathbf{w}_j$ is a generalized eigenvector of order m_i or less corresponding to some λ_i. [That is, $(H - \lambda_i I)^{m_i}\mathbf{w}_j = \mathbf{0}$ for some λ_i and $(H - \lambda_i I)^{m_i}$ is one of the factors of $q(H)$. Thus, $q(H)\mathbf{w}_j = \mathbf{0}$ for any j, $1 \leq j \leq r$.]

Now multiplying both sides of (8) by $q(H)$, we obtain

$$a_1 q(H)\mathbf{v}_1 + a_2 q(H)\mathbf{v}_2 + \cdots + a_{m_1} q(H)\mathbf{v}_{m_1} = \mathbf{0}. \tag{9}$$

Finally, we can use (5) to show that $a_1, a_2, \ldots, a_{m_1}$ are all zero in (9) (see Exercise 8). Since we could have made this argument for any of the eigenvalues $\lambda_2, \lambda_3, \ldots, \lambda_k$, it follows that all the coefficients b_j in (8) are also zero.

□

EXAMPLE 3   Find the general solution of $\mathbf{x}' = H\mathbf{x}$, where

$$H = \begin{bmatrix} 1 & 0 & 0 \\ 1 & 3 & 0 \\ 0 & 1 & 1 \end{bmatrix}.$$

SOLUTION   The characteristic polynomial is $p(t) = (t - 1)^2(t - 3)$; so $\lambda = 1$ is an eigenvalue of multiplicity 2, whereas $\lambda = 3$ is an eigenvalue of multiplicity 1. Eigenvectors corresponding to $\lambda = 1$ and $\lambda = 3$ are (respectively)

$$\mathbf{v}_1 = \begin{bmatrix} 0 \\ 0 \\ 1 \end{bmatrix} \quad \text{and} \quad \mathbf{w}_1 = \begin{bmatrix} 0 \\ 2 \\ 1 \end{bmatrix}.$$

Thus we have two solutions of $\mathbf{x}' = H\mathbf{x}$, namely, $\mathbf{x}(t) = e^t\mathbf{v}_1$ and $\mathbf{x}(t) = e^{3t}\mathbf{w}_1$. We need one more solution to solve the initial-value problem for any $\mathbf{x}_0$ in R^3. To find a third solution, we need a vector $\mathbf{v}_2$ that is a generalized eigenvector of order 2 corresponding to $\lambda = 1$. According to the remarks above, we solve $(H - I)\mathbf{x} = \mathbf{v}_1$ and obtain

$$\mathbf{v}_2 = \begin{bmatrix} -2 \\ 1 \\ 0 \end{bmatrix}.$$

By (6), a third solution to $x' = Hx$ is given by $x(t) = e^t(v_2 + tv_1)$. Clearly $\{v_1, v_2, w_1\}$ is a basis for R^3; so if $x_0 = c_1 v_1 + c_2 v_2 + c_3 w_1$, then

$$x(t) = c_1 e^t v_1 + c_2 e^t (v_2 + tv_1) + c_3 e^{3t} w_1$$

will satisfy $x' = Hx$, $x(0) = x_0$.   □

EXAMPLE 4   Find the general solution of $x' = Ax$, where

$$A = \begin{bmatrix} -1 & -8 & 1 \\ -1 & -3 & 2 \\ -4 & -16 & 7 \end{bmatrix}.$$

SOLUTION   Reducing A to Hessenberg form, we have $H = QAQ^{-1}$, where H, Q, and Q^{-1} are

$$H = \begin{bmatrix} -1 & -4 & 1 \\ -1 & 5 & 2 \\ 0 & -8 & -1 \end{bmatrix}, \quad Q = \begin{bmatrix} 1 & 0 & 0 \\ 0 & 1 & 0 \\ 0 & -4 & 1 \end{bmatrix}, \quad \text{and} \quad Q^{-1} = \begin{bmatrix} 1 & 0 & 0 \\ 0 & 1 & 0 \\ 0 & 4 & 1 \end{bmatrix}.$$

The change of variables $y(t) = Qx(t)$ converts $x' = Ax$, $x(0) = x_0$ to the problem $y' = Hy$, $y(0) = Qx_0$.

The characteristic polynomial for H is $p(t) = (t - 1)^3$, so $\lambda = 1$ is an eigenvalue of multiplicity 3 of H. Up to a scalar multiple, the only eigenvector of H is

$$v_1 = \begin{bmatrix} 4 \\ -1 \\ 4 \end{bmatrix}.$$

We obtain two generalized eigenvectors for H by solving $(H - I)x = v_1$ to get v_2, and $(H - I)x = v_2$ to get v_3. These generalized eigenvectors are

$$v_2 = \begin{bmatrix} 1 \\ -1 \\ 2 \end{bmatrix} \quad \text{and} \quad v_3 = \begin{bmatrix} 3 \\ -1 \\ 3 \end{bmatrix}.$$

Thus the general solution of $y' = Hy$ is

$$y(t) = e^t \left[c_1 v_1 + c_2 (v_2 + tv_1) + c_3 \left(v_3 + tv_2 + \frac{t^2}{2} v_1 \right) \right],$$

and we can recover $x(t)$ from $x(t) = Q^{-1}y(t)$.

If H is an $(n \times n)$ *reduced* Hessenberg matrix, it can be shown that R^n has a basis consisting of eigenvectors and generalized eigenvectors of H. This general result is fairly difficult to establish, however, and we do not do so here.

□

1. Find a full set of eigenvectors and generalized eigenvectors for each of the following.

a) $\begin{bmatrix} 1 & -1 \\ 1 & 3 \end{bmatrix}$ b) $\begin{bmatrix} -2 & 0 & -2 \\ -1 & 1 & -2 \\ 0 & 1 & -1 \end{bmatrix}$

c) $\begin{bmatrix} -6 & 31 & -14 \\ -1 & 6 & -2 \\ 0 & 2 & 1 \end{bmatrix}$

2. Find a full set of eigenvectors and generalized eigenvectors for the following. (*Note:* $\lambda = 2$ is the only eigenvalue of B.)

$$A = \begin{bmatrix} 1 & 0 & 0 & 0 \\ 1 & 1 & 0 & 0 \\ 0 & 1 & 1 & 0 \\ 0 & 0 & 1 & 1 \end{bmatrix}, \qquad B = \begin{bmatrix} 2 & 3 & -21 & -3 \\ 2 & 7 & -41 & -5 \\ 0 & 1 & -5 & -1 \\ 0 & 0 & 4 & 4 \end{bmatrix}$$

3. Solve $\mathbf{x}' = A\mathbf{x}$, $\mathbf{x}(0) = \mathbf{x}_0$ by transforming A to Hessenberg form, where

$$\mathbf{x}_0 = \begin{bmatrix} -1 \\ -1 \\ 1 \end{bmatrix}, \quad \text{and}$$

a) $A = \begin{bmatrix} 8 & -6 & 21 \\ 1 & -1 & 3 \\ -3 & 2 & -8 \end{bmatrix}$,

b) $A = \begin{bmatrix} 2 & 1 & -1 \\ -3 & -1 & 1 \\ 9 & 3 & -4 \end{bmatrix}$,

c) $A = \begin{bmatrix} 1 & 1 & -1 \\ -3 & -2 & 1 \\ 9 & 3 & -5 \end{bmatrix}$.

4. Give the general solution of $\mathbf{x}' = A\mathbf{x}$, where A is from Exercise 2.

5. Repeat Exercise 4, where A is in part (c) of Exercise 1.

6. Prove that each vector $\mathbf{v}_r$ in (5) is a generalized eigenvector of order r and that $\{\mathbf{v}_1, \mathbf{v}_2, \dots, \mathbf{v}_m\}$ is linearly independent.

7. Prove that the functions $\mathbf{x}_r(t)$ defined in (6) are solutions of $\mathbf{x}' = H\mathbf{x}$.

8. Prove that the coefficients $a_1, a_2, \dots$ in (9), are all zero. [*Hint:* Multiply (9) by $(H - \lambda_1 I)^{m_1 - 1}$.]

1. Consider the quadratic form $q(\mathbf{x}) = x_1^2 + 3x_1 x_2 + x_2^2$. Describe all possible real (2×2) matrices A such that $q(\mathbf{x}) = \mathbf{x}^T A \mathbf{x}$.

2. Let

$$A = \begin{bmatrix} 2 & 6 + a \\ a & 2 \end{bmatrix},$$

where a is a real constant. For what values a is A defective?

3. Let

$$B = \begin{bmatrix} 2 & 6 \\ 0 & 2 \end{bmatrix}$$

and consider the quadratic form defined by $q(\mathbf{x}) = \mathbf{x}^T B \mathbf{x}$.

a) Find a vector $\mathbf{x}$ such that $q(\mathbf{x}) < 0$.

b) Note that B has only positive eigenvalues. Why does this fact not contradict Theorem 2 in Section 6.1?

4. Let $q(t) = t^2 + 3t + 2$ and let A be a nonsingular $(n \times n)$ matrix. Show that $q(A)$ and $q(A^{-1})$ commute in the sense that $q(A)q(A^{-1}) = q(A^{-1})q(A)$.

5. A positive definite matrix A can be factored as $A = LL^T$, where L is a nonsingular lower-triangular matrix. (Such a factorization is called the **Cholesky decomposition**.) Find the Cholesky decomposition for each of the following.

a) $A = \begin{bmatrix} 4 & 6 \\ 6 & 10 \end{bmatrix}$ b) $A = \begin{bmatrix} 1 & 3 & 1 \\ 3 & 13 & 7 \\ 1 & 7 & 6 \end{bmatrix}$

Supplementary Conceptual Exercises

1. Let A be a (3×3) nonsingular matrix. Use the Cayley–Hamilton theorem to show that A^{-1} can be represented as $A^{-1} = aI + bA + cA^2$.

2. Let A and B be similar $(n \times n)$ matrices and let $p(t)$ denote a kth-degree polynomial. Show that $p(A)$ and $p(B)$ are also similar.

3. Let A be an $(n \times n)$ symmetric matrix and suppose that the quadratic form $q(\mathbf{x}) = \mathbf{x}^T A \mathbf{x}$ is positive definite. Show that the diagonal entries of A, a_{ii} for $1 \le i \le n$, are all positive.

4. Let A be a (3×3) matrix.
   a) Use the Cayley–Hamilton theorem to show that A^4 can be represented as $A^4 = aI + bA + cA^2$.
   b) Make an informal argument that A^k can be represented as a quadratic polynomial in A for $k = 5, 6, \ldots$.

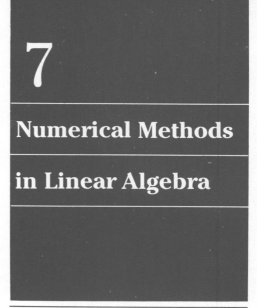

7

Numerical Methods
in Linear Algebra

Computer Arithmetic and Roundoff (Optional)*

The advent of the computer has greatly increased the range of problems in which matrix theory and linear algebra are applicable. Problems that are too large and complex to consider solving by hand are now solved routinely in seconds when the proper techniques are coded and executed on the computer. As computer technology advances, we can expect that problems too large for today's machines will become routine tomorrow. However, every computer has computational limitations that must be considered, and two of these important limitations form the subject of this chapter:

1. Efficient algorithms are necessary; even the largest computer may be unable to perform the number of calculations required by a poorly conceived algorithm.

2. Computer arithmetic is inherently inaccurate; every arithmetic operation in the computer is a potential source of error.

In essence, our goal is to design computer programs that solve mathematical problems in a reasonable amount of time with a reasonable standard of accuracy.

In this section, we consider some of the limitations to accuracy. In particular, we describe ***roundoff error,*** which loosely means any error introduced into a computation when performing arithmetic operations restricted to a finite

---

* This section contains background material on computer arithmetic. A casual reading of Section 7.1 is all that is required for the later sections, for only indirect reference will be made to Section 7.1.

number of decimal places. We will explain roundoff error in more detail later; for now we want to observe only that the presence of roundoff error means that arithmetic operations in the machine obey hardly any of the rules for arithmetic in the real-number system. For example, in the real-number system we expect that

$$a + b = b + a$$
$$a + (b + c) = (a + b) + c$$
$$ab = ba \tag{1}$$
$$a(bc) = (ab)c$$
$$a(b + c) = ab + ac.$$

In any digital device, from the smallest hand calculator to the largest computer, the associative and distributive laws in (1) do not hold. For example, if we form $a + (b + c)$ in a machine and then form $(a + b) + c$, we should not necessarily expect the same result.

EXAMPLE 1   Different machines produce different sorts of roundoff errors. As an example, consider the three numbers

$$a = \frac{3.}{7.}, \qquad b = \frac{5.}{9.}, \qquad c = \frac{7.}{11.}.$$

On an HP calculator, the products $a(bc)$ and $(ab)c$ are different, whereas $a(b + c)$ is equal to $ab + ac$. By contrast, when the same operations were coded in FORTRAN and run on an IBM mainframe (using one of the several FORTRAN compilers), the opposite result was produced. That is, $(ab)c$ and $a(bc)$ were equal, whereas $a(b + c)$ was not equal to $ab + ac$. With regard to hand calculators, we note that some of them display fewer digits than the calculators use internally. So, for example, to see the difference between $a(bc)$ and $(ab)c$ for various numbers a, b, and c, it may be necessary to calculate $r = (ab)c$, store r in memory, then calculate $s = a(bc)$, and form

$$q = \frac{1}{r - s}. \qquad \square$$

An example should serve to illustrate how roundoff errors occur when number representations are limited to a finite number of places. For instance, suppose the restriction is that every number a must be represented in the form

$$a = \pm .a_1 a_2 a_3 a_4 a_5 \times 10^c, \tag{2}$$

where a_i is an integer with $0 \le a_i < 10$ for $1 \le i \le 5$, and where $a_1 \ne 0$ unless $a = 0$. For example, if $a = 2147.6$ and $f = 16.211$, then we will represent a and f as

$$a = .21476 \times 10^4 \quad \text{and} \quad f = .16211 \times 10^2.$$

(These restrictions are exactly like those under which a computer or hand calculator must operate.)

In order to add a and f, we must first align the decimal places, so we express f in the form $f = .0016211 \times 10^4$. We can now form the sum $s = a + f$ and obtain

$$s = a + f = .2163811 \times 10^4.$$

Since we are restricted to representing s in the form (2), we will have to discard two of the seven digits in s and represent $s = a + f$ as

$$a + f = .21638 \times 10^4.$$

This representation is obviously an error (a roundoff error), and any further operations we carry out using s will be contaminated by this error. Multiplying a times f produces an error also since

$$af = .348147436 \times 10^5,$$

and we will have to store af as $.34814 \times 10^5$ or as $.34815 \times 10^5$ (depending on the rounding rule we choose to use). Again, a roundoff error has been committed and will contaminate all further calculations.

We next illustrate a particularly disastrous kind of roundoff error. Suppose that $a = .21476 \times 10^4$ and $q = .13129 \times 10^{-2}$. To form $a + q$, we must again align the decimal places, and we write q as

$$q = .00000013129 \times 10^4.$$

Therefore, $a + q$ is given by

$$a + q = .21476013129 \times 10^4,$$

and the sum $a + q$ is stored as

$$a + q = .21476 \times 10^4.$$

In effect, we have said that $a + q = a$; and although this statement is not a bad approximation, it is an error. On philosophical grounds, it is a very bad error— we have lost all the information contained in q. In general, we should be careful when there is a possibility of adding numbers of widely varying magnitudes, since accuracy may be degraded. There are many other potential (some quite subtle) sources of computational error, but it would be inappropriate to discuss them in this brief introduction. We will mention some of the principal strategies of error control for numerical procedures in linear algebra in subsequent sections.

The principles illustrated above carry over to the computer; only the details are different. At the machine level, a number can be thought of as being represented by a finite string of 0's and 1's. The arithmetic operations are actually carried out (most usually) in base 2, base 8, base 10, or base 16; and the base dictates the actual arrangement of the 0's and 1's in the string. To be specific, suppose that the base of the arithmetic system is b. Then any real number a is represented in the form

$$a = \pm .a_1 a_2 \cdots a_m \times b^c, \tag{3}$$

	h_1	h_2	h_3	h_4	h_5	h_6
Byte 1	Byte 2		Byte 3		Byte 4	

Figure 7.1

where each a_i is an integer, $0 \le a_i < b$, and where $a_1 \ne 0$ unless $a = 0$. In (3), m is fixed and determined by the "word length" of the computer. [Many hand calculators use $b = 10$ and a value of m on the order of 10 to 12. The IBM mainframe computers use $b = 16$ (hexadecimal arithmetic) and $m = 6$.]

The IBM computers serve as convenient examples to illustrate some of the details of (3). In these computers a "real" number is represented as a string of thirty-two 0's and 1's, as in Fig. 7.1. In Fig. 7.1, the compartments labeled Byte 1, Byte 2, Byte 3, and Byte 4 each represent a string of eight 0's and 1's. (The 0's and 1's represent binary digits or "bits," and there are eight bits to a byte.) Bytes 2, 3, and 4 each consist of two strings of four 0's and 1's—these are $h_1, h_2, \ldots, h_6$ in Fig. 7.1. A real number a is then represented internally in the form

$$a = \pm .h_1 h_2 h_3 h_4 h_5 h_6 \times 16^c, \tag{4}$$

where each h_i is a hexadecimal digit, an integer between 0 and 15. These hexadecimal digits are represented in base 2 (as 0's and 1's) according to Table 7.1. In this scheme A stands for 10, B for 11, and so on. For example, the number $d = .B3C \times 16^2$ is the same as

$$d = \left(\frac{11}{16} + \frac{3}{16^2} + \frac{12}{16^3} \right) \times 16^2.$$

Thus, $d = 11 \times 16 + 3 + 12/16 = 176 + 3 + .75$, or $d = 179.75$ in base 10. In the context of Fig. 7.1, the number d has the form $d = .B3C000 \times 16^2$; so Byte 2 would contain $h_1 h_2$ or 10110011, Byte 3 would have $h_3 h_4$ or 11000000, and Byte 4 is 00000000.

In the hexadecimal representation (4), if $a \ge 0$, then 0 occupies the first location of Byte 1, whereas 1 is present if $a < 0$. The remaining seven locations of Byte 1 store the exponent c. One of the vagaries of hexadecimal arithmetic

Table 7.1

0	0000	8	1000
1	0001	9	1001
2	0010	A	1010
3	0011	B	1011
4	0100	C	1100
5	0101	D	1101
6	0110	E	1110
7	0111	F	1111

is that certain nice decimal numbers, such as .1, have infinite hexadecimal expansions. That is, if $(.h_1 h_2 h_3 \ldots)_{16}$ represents

$$\frac{h_1}{16} + \frac{h_2}{16^2} + \frac{h_3}{16^3} + \cdots,$$

then $1/10$ has the form $(.1999999\ldots)_{16}$.

The representation (3) is for "real" variables; "integer" variables are treated somewhat differently. In addition, different software systems on the same computer may use different rules of arithmetic.

Available Numerical Software

Most computer centers have libraries of subroutines, such as the ISML library and the NAG library, that are very useful for scientific computing. These general-purpose libraries contain subroutines for numerical linear algebra as well as subroutines for rootfinding, optimization, differential equations, and so forth. There are also software packages devoted primarily to numerical linear algebra, such as LINPACK and EISPACK. Finally, there are more specialized packages aimed at particular types of problems. For example, ITPACK and the Yale Sparse Matrix Package are designed for sparse matrix problems. A good bibliography of software for linear algebra can be found in W. W. Hager, *Applied Numerical Linear Algebra* (Englewood Cliffs: Prentice Hall, 1988).

Exercises 7.1

1. The following binary (base-2) numbers are expressed as in (3). Convert them to their decimal (base-10) equivalents.

   $a = .101 \times 2^3,\qquad b = .11011 \times 2^3,\qquad c = .10011 \times 2^6$

2. Convert the following octal (base-8) numbers to their decimal equivalents.

   $a = .1314 \times 8^3,\qquad b = .7113 \times 8^2,$

   $c = .624 \times 8^4$

3. Convert the following hexadecimal numbers to their decimal equivalents.

   $a = .B29 \times 16^3,\qquad b = .91C \times 16^2,$

   $c = .333 \times 16^2$

4. Does every hexadecimal number with a finite expansion (such as those in Exercise 3) have a finite decimal expansion?

5. Convert the following decimal numbers to base 2 in the form of (3).

   $a = 14,\qquad b = 29.5,\qquad c = 6.75,$

   $d = 134.875$

6. Convert the decimal numbers in Exercise 5 to base 8.

7. Convert the decimal numbers in Exercise 5 to base 16.

8. Give an example of a decimal number that has a finite decimal expansion but an infinite expansion in base 8.

9. Prove that the hexadecimal expansion of the decimal number $1/10$ is $.19999\ldots$. That is, show that

   $$1/10 = 1/16 + \sum_{n=2}^{\infty} 9/16^n.$$

   (*Hint:* The infinite series is a geometric series.)

10. Verify that $(ab)c \neq a(bc)$ on a hand calculator, where $a = 1./7.$, $b = 1./9.$, and $c = 1./11.$

11. Suppose we are using a machine that does base-10 arithmetic and carries five places [in (3), $b = 10$ and $m = 5$]. Suppose that the result of any arithmetic operation is truncated to five places. (For example, if $a + b = .314263 \times 10^2$, then $a + b$ is stored as $.31426 \times 10^2$. If $a + b = .314269 \times 10^2$, then $a + b$ is also stored as $.31426 \times 10^2$.) Given the machine

numbers

$$a = .20211 \times 10^2, \qquad b = .31323 \times 10^{-1},$$

$$c = .50000 \times 10^1, \quad \text{and} \quad d = .60000 \times 10^{-4},$$

calculate what the machine will store as the result of performing $a + b$, ab, $a(b + 3c)$, $c + (d + d)$, and $(c + d) + d$.

7.2 Gaussian Elimination

Besides being easy to understand and serving as a valuable theoretical tool, Gaussian elimination is one of the most popular computational procedures for solving a linear system $A\mathbf{x} = \mathbf{b}$. In this section, we discuss some of the aspects that are related to implementing Gaussian elimination on the computer. In order not to obscure the main points, we restrict our discussion to systems $A\mathbf{x} = \mathbf{b}$, where A is square. The basic operations in Gaussian elimination or reduction to echelon form are row interchanges and the addition of a multiple of one row to another. Clearly the only source of computational error is that of adding one row to another, so we analyze this basic operation first.

To begin, consider the $(n \times n)$ system

$$
\begin{aligned}
a_{11}x_1 + a_{12}x_2 + \cdots + a_{1n}x_n &= b_1 \\
a_{21}x_1 + a_{22}x_2 + \cdots + a_{2n}x_n &= b_2 \\
&\vdots \\
a_{n1}x_1 + a_{n2}x_2 + \cdots + a_{nn}x_n &= b_n.
\end{aligned}
\tag{1}
$$

As we know, if $a_{11} \neq 0$, we can eliminate x_1 from the kth equation by multiplying row 1 by $-a_{k1}/a_{11}$ and adding the result to row k for $2 \leq k \leq n$. But as we saw in Section 7.1, if we add a large number to a small one, we essentially erase much of the information contained in the small number. Similarly, if $-a_{k1}/a_{11}$ is very large, then much of the information in equation k is lost when we add a multiple of $-a_{k1}/a_{11}$ times the first row to row k. On the other hand, if $-a_{k1}/a_{11}$ is quite small (relative to the entries in the kth row), then we will disturb the entries in the kth row only slightly; that is, the information contained in the kth equation will be left very nearly intact.

In summary, roundoff-error considerations suggest that if we replace row k by a "large" multiple of row 1, then we have in effect replaced row k by a multiple of row 1. Alternatively, if we replace row k by a "small" multiple of row 1, then we have left row k reasonably intact. Since the multiplier for row 1 that is used to eliminate row k is $-a_{k1}/a_{11}$, it follows that we would like a_{k1}/a_{11} to be as small as possible in absolute value. This observation leads to a "pivoting strategy":

1. Search column 1 for the largest entry, say a_{r1}; that is,

$$|a_{r1}| \geq |a_{i1}|, \qquad 1 \leq i \leq n.$$

2. Interchange row 1 and row r in (1).

By interchanging row 1 and row r, we obtain an equivalent system:

$$a_{r1}x_1 + a_{r2}x_2 + \cdots + a_{rn}x_n = b_r$$
$$a_{21}x_1 + a_{22}x_2 + \cdots + a_{2n}x_n = b_2$$
$$\vdots \qquad\qquad \vdots \quad \vdots \qquad\qquad (2)$$
$$a_{n1}x_1 + a_{n2}x_2 + \cdots + a_{nn}x_n = b_n,$$

and we are assured that when we eliminate x_1 from row k by multiplying the first row of (2) by $-a_{k1}/a_{r1}$, the multiplier $-a_{k1}/a_{r1}$ is as small as possible.

EXAMPLE 1 As an illustration of the observations above, consider the (2×2) system

$$-.001x_1 + x_2 = 1$$
$$x_1 + x_2 = 2. \qquad\qquad (3)$$

To eliminate x_1 from row 2, we multiply row 1 by 1000, add the result to row 2, and obtain an equivalent system:

$$-.001x_1 + x_2 = 1$$
$$1001x_2 = 1002.$$

If we were working with a machine that carried only three places, the coefficients in the last equation would be rounded to three places:

$$-.001x_1 + x_2 = 1$$
$$1000x_2 = 1000. \qquad\qquad (4)$$

Solving (4), we would have $x_2 = 1$ and $x_1 = 0$; and this result is certainly not a solution of (3), either mathematically or even in our hypothetical three-place machine. In effect, we have erased the information in the second equation of (3) by adding a multiple of 1000 times row 1 to row 2—in our three-place machine, $x_1 = 0$ and $x_2 = 1$ satisfy the first equation but not the second equation of (3); the information in the second equation has been lost in (4).

Suppose now that we pivot, interchanging the first and second equations and obtain

$$x_1 + x_2 = 2$$
$$-.001x_1 + x_2 = 1. \qquad\qquad (5)$$

If we eliminate x_1 in the second equation by multiplying the first by .001 and adding the result to the second, we obtain

$$x_1 + x_2 = 2$$
$$1.001x_2 = 1.002,$$

which becomes in our three-place machine

$$x_1 + x_2 = 2$$
$$x_2 = 1. \qquad\qquad (6)$$

Solving (6), we get $x_2 = 1$ and $x_1 = 1$; and either mathematically or in our machine $x_1 = 1$ and $x_2 = 1$ is a better solution of (5) than our first solution was. In fact, in the machine, $x_1 = 1$ and $x_2 = 1$ solve (5) "exactly." The mathematical solution is, of course,

$$x_1 = \frac{1000}{1001} = 1 - \frac{1}{1001}, \qquad x_2 = \frac{1002}{1001} = 1 + \frac{1}{1001},$$

so our second "machine solution" is not too far off. □

The pivoting strategy described above is called *partial pivoting*. In detail, we interchange equation r and the first equation in (1), where $|a_{r1}| \geq |a_{i1}|$, $1 \leq i \leq n$; and then we eliminate x_1 from the remaining equations. The result is a system of the form

$$a'_{11}x_1 + a'_{12}x_2 + \cdots + a'_{1n}x_n = b'_1$$
$$a'_{22}x_2 + \cdots + a'_{2n}x_n = b'_2$$
$$\vdots \qquad \qquad \vdots \qquad \vdots$$
$$a'_{n2}x_2 + \cdots + a'_{nn}x_n = b'_n.$$

Given the system above, we search column 2, rows $2, 3, \ldots, n$ for an entry a'_{j2}, where $|a'_{j2}| \geq |a'_{i2}|$, $2 \leq i \leq n$. We then interchange rows j and 2; and we eliminate x_2 in rows $3, 4, \ldots, n$ by using the multiplier $-a'_{k2}/a'_{j2}$ to eliminate x_2 in row k. Having done this procedure, we move to column 3 and search for the largest entry in rows $3, 4, \ldots, n$, and so on.

Instead of partial pivoting, we might consider *total pivoting*, which requires more of a search. That is, in (1) we could ask for the largest entry a_{rs}, where $|a_{rs}| \geq |a_{ij}|$, $1 \leq i \leq n$, $1 \leq j \leq n$. Having a_{rs}, we could interchange row 1 and row r, and then interchange column 1 and column s in order to move a_{rs} to the $(1, 1)$ position. While this strategy is desirable, it is expensive and requires n^2 comparisons to find the largest entry, a_{rs}. In many Gaussian elimination programs, partial pivoting is used rather than total pivoting.

The other computational aspect of Gaussian elimination that bears mentioning is that the procedure requires on the order of $n^3/3$ multiplications and $n^3/3$ additions to solve an $(n \times n)$ system. To elaborate, we note that one crude measure of the "efficiency" of an algorithm is an *operations count,* a count of the number of arithmetic operations that must be executed when the algorithm is used. In the exercises, the reader is asked to show that given an $(n \times n)$ system $A\mathbf{x} = \mathbf{b}$, then about $n^3/3$ multiplications and $n^3/3$ additions are necessary to reduce the augmented matrix $[A \,|\, \mathbf{b}]$ to echelon form. Given the echelon form, it then takes about n^2 multiplications and n^2 additions to solve the system. The number n^2 tells us why it is not efficient to solve $A\mathbf{x} = \mathbf{b}$ by calculating A^{-1} and then forming $A^{-1}\mathbf{b}$. To see why, we need only note that forming the product $A^{-1}\mathbf{b}$ when A^{-1} is $(n \times n)$ takes n^2 multiplications and n^2 additions, the same number of operations we perform when we solve a system that is in echelon form. Thus the only way that forming $A^{-1}\mathbf{b}$ could be competitive with simply solving

$A\mathbf{x} = \mathbf{b}$ directly is if fewer operations were required to obtain A^{-1} then to reduce $[A\,|\,\mathbf{b}]$ to echelon form. Since finding A^{-1} is most easily done by reducing the $(n \times 2n)$ matrix $[A\,|\,I]$ to echelon form, it is almost obvious that we need to do more work to find A^{-1} than to reduce $[A\,|\,\mathbf{b}]$ (a count shows that more than twice as much work is required to find A^{-1} than to solve $A\mathbf{x} = \mathbf{b}$). In addition, more than efficiency is at stake—a procedure that uses twice as many arithmetic operations has the potential of making twice as many errors. We can expect that $A^{-1}\mathbf{b}$ is not so accurate a solution to $A\mathbf{x} = \mathbf{b}$ as is the solution we calculate directly.

In Fig. 7.2, we have listed a FORTRAN program, Subroutine GAUSS, which implements Gaussian elimination with partial pivoting. Our purpose in this listing is to provide the interested reader with an easy-to-understand and easy-to-use computational tool. This program, as well as others we include later, was written with an emphasis on simplicity and readability with a minimum of user options. In the interest of clarity, we included only enough comment statements to highlight the main computational segments of the program. The following is a description of the parameters that are not explained in the comments.

Subroutine GAUSS uses Gaussian elimination with partial pivoting to solve $A\mathbf{x} = \mathbf{b}$. The required inputs are the following:

A	An $(N \times N)$ matrix.
B	An $(N \times 1)$ vector.
N	The size of A.
MAINDM	The declared dimension of the array containing A in the calling program.

The outputs from Subroutine GAUSS are the following:

X	The machine solution of $A\mathbf{x} = \mathbf{b}$.
IERROR	A flag set equal to 2 if Gaussian elimination cannot proceed because of a zero pivot. IERROR is set equal to 1 if Gaussian elimination is successful.
RNORM	This number is set equal to the norm of the residual vector, $\mathbf{b} - A\mathbf{x}$, where $\mathbf{x}$ is the machine solution.

In brief, GAUSS sets up the augmented matrix $[A\,|\,\mathbf{b}]$ and stores this in array AUG. AUG is reduced to echelon form; and if no zero diagonal entries are found, the solution X is calculated. As a crude error test, RNORM $= \|\mathbf{b} - A\mathbf{x}\|$ is calculated and returned to the calling program.

An example of a simple program that uses Subroutine GAUSS to solve a linear system is given in Fig. 7.3. As a guide to interpreting the output of a linear-equation solver such as GAUSS, we first observe that the machine may not recognize a coefficient matrix A as being singular when it is indeed singular. For example, with the inputs

$$N = 2, \quad A = \begin{bmatrix} 1 & 4 \\ 3 & 12 \end{bmatrix},$$
$$B = \begin{bmatrix} 4 \\ 5 \end{bmatrix},$$

```
        SUBROUTINE GAUSS(A,B,X,N,MAINDM,IERROR,RNORM)
        DIMENSION A(MAINDM,MAINDM),B(MAINDM),X(MAINDM)
        DIMENSION AUG(50,51)
        NM1=N-1
        NP1=N+1
C
C   SET UP THE AUGMENTED MATRIX FOR AX=B.
C
        DO 2 I=1,N
          DO 1 J=1,N
          AUG(I,J)=A(I,J)
    1     CONTINUE
        AUG(I,NP1)=B(I)
    2   CONTINUE
C
C   THE OUTER LOOP USES ELEMENTARY ROW OPERATIONS TO TRANSFORM
C   THE AUGMENTED MATRIX TO ECHELON FORM.
C
        DO 8 I=1,NM1
C
C   SEARCH FOR THE LARGEST ENTRY IN COLUMN I, ROWS I THROUGH N.
C   IPIVOT IS THE ROW INDEX OF THE LARGEST ENTRY.
C
        PIVOT=0.
          DO 3 J=I,N
          TEMP=ABS(AUG(J,I))
          IF(PIVOT.GE.TEMP)  GO TO 3
          PIVOT=TEMP
          IPIVOT=J
    3     CONTINUE
        IF(PIVOT.EQ.0.)  GO TO 13
        IF(IPIVOT.EQ.I)  GO TO  5
C
C   INTERCHANGE ROW I AND ROW IPIVOT.
C
        DO 4 K=I,NP1
        TEMP=AUG(I,K)
        AUG(I,K)=AUG(IPIVOT,K)
        AUG(IPIVOT,K)=TEMP
    4     CONTINUE
C
C   ZERO ENTRIES (I+1,I), (I+2,I),...,(N,I) IN THE AUGMENTED MATRIX.
C
    5   IP1=I+1
        DO 7 K=IP1,N
        Q=-AUG(K,I)/AUG(I,I)
        AUG(K,I)=0.
          DO 6 J=IP1,NP1
          AUG(K,J)=Q*AUG(I,J)+AUG(K,J)
    6     CONTINUE
    7     CONTINUE
    8   CONTINUE
      IF(AUG(N,N).EQ.0.)  GO TO 13
C
C   BACKSOLVE TO OBTAIN A SOLUTION TO AX=B.
C
      X(N)=AUG(N,NP1)/AUG(N,N)
        DO 10 K=1,NM1
        Q=0.
          DO 9 J=1,K
          Q=Q+AUG(N-K,NP1-J)*X(NP1-J)
    9     CONTINUE
        X(N-K)=(AUG(N-K,NP1)-Q)/AUG(N-K,N-K)
   10     CONTINUE
C
C   CALCULATE THE NORM OF THE RESIDUAL VECTOR, B-AX.
C   SET IERROR=1 AND RETURN.
C
      RSQ=0.
        DO 12 I=1,N
        Q=0.
          DO 11 J=1,N
          Q=Q+A(I,J)*X(J)
   11     CONTINUE
        RESI=B(I)-Q
        RMAG=ABS(RESI)
        RSQ=RSQ+RMAG**2
   12     CONTINUE
      RNORM=SQRT(RSQ)
      IERROR=1
      RETURN
C
C   ABNORMAL RETURN --- REDUCTION TO ECHELON FORM PRODUCES A ZERO
C   ENTRY ON THE DIAGONAL.  THE MATRIX A MAY BE SINGULAR.
C
   13 IERROR=2
      RETURN
      END
```

Figure 7.2

```
        DIMENSION A(20,20),B(20),X(20)
        MAINDM=20
      1 READ 100,N
        IF(N.LE.1)  STOP
        DO 2 I=1,N
        READ 101,(A(I,J),J=1,N),B(I)
      2 CONTINUE
        CALL GAUSS(A,B,X,N,MAINDM,IERROR,RNORM)
        PRINT 102,IERROR
        IF(IERROR.EQ.2)  GO TO 1
        PRINT 103,RNORM
        PRINT 104,(X(I),I=1,N)
    100 FORMAT(I2)
    101 FORMAT(21F4.0)
    102 FORMAT(8H IERROR=,I3)
    103 FORMAT(7H RNORM=,E20.6)
    104 FORMAT(1H ,6E16.6)
        GO TO 1
        END
```

Figure 7.3

the following was output*:

$$\text{IERROR} = 1, \qquad \text{RNORM} = .117047\text{E}02, \qquad X = \begin{bmatrix} -.420336\text{E}17 \\ .105084\text{E}17 \end{bmatrix}.$$

Here the size of $\|\mathbf{b} - A\mathbf{x}\|$ gives a clue that the machine solution is suspect.

Runs with nonsingular matrices are much more satisfactory. For example, with an input of

$$N = 4, \qquad A = \begin{bmatrix} 5 & 7 & 6 & 5 \\ 7 & 10 & 8 & 7 \\ 6 & 8 & 10 & 9 \\ 5 & 7 & 9 & 10 \end{bmatrix}, \qquad B = \begin{bmatrix} 23 \\ 32 \\ 33 \\ 31 \end{bmatrix}, \qquad (7)$$

the output was

$$\text{IERROR} = 1, \qquad \text{RNORM} = .502430\text{E} - 14, \qquad X = \begin{bmatrix} .100000\text{E}01 \\ .100000\text{E}01 \\ .100000\text{E}01 \\ .100000\text{E}01 \end{bmatrix};$$

and the machine solution is correct to as many places as listed.

As a final note of caution, suppose that $\mathbf{x}_m$ denotes the machine solution to $A\mathbf{x} = \mathbf{b}$ and $\mathbf{x}_t$ denotes the actual (mathematical) solution. We would like some reasonable estimate of how far $\mathbf{x}_m$ is from $\mathbf{x}_t$, but unfortunately this estimate is hard to obtain when we do not know $\mathbf{x}_t$. The number RNORM serves as a rough guide; and if $\text{RNORM} = \|\mathbf{b} - A\mathbf{x}_m\|$ is not small, then we suspect that the matrix A is badly behaved in some fashion. But even if RNORM is small, we may have to be careful. To see why, suppose that $\mathbf{r} = \mathbf{b} - A\mathbf{x}_m$ and that the true solution is $\mathbf{x}_t$ so that $\mathbf{b} = A\mathbf{x}_t$. In this case,

$$\mathbf{r} = \mathbf{b} - A\mathbf{x}_m = A\mathbf{x}_t - A\mathbf{x}_m,$$

---

* This output was from an IBM mainframe, using double-precision arithmetic.

and if A is nonsingular, we can write $\mathbf{r} = A\mathbf{x}_t - A\mathbf{x}_m = A(\mathbf{x}_t - \mathbf{x}_m)$ or

$$A^{-1}\mathbf{r} = \mathbf{x}_t - \mathbf{x}_m.$$

From this expression we see that even if $\mathbf{r}$ is "small," $\mathbf{x}_t - \mathbf{x}_m$ may be "large" if A^{-1} is "large."

It can be shown that the size of A^{-1} is reflected by the size of α, where

$$\alpha = \max_{\mathbf{x} \text{ in } R^n} \frac{\|\mathbf{x}\|}{\|A\mathbf{x}\|}. \tag{8}$$

Estimating α is usually preferable to the actual calculation of A^{-1}, but we will not discuss the estimation of α here.

An example of a matrix A with a relatively large inverse is the (5×5) Hilbert matrix A, where

$$A = \begin{bmatrix} 1 & 1/2 & 1/3 & 1/4 & 1/5 \\ 1/2 & 1/3 & 1/4 & 1/5 & 1/6 \\ 1/3 & 1/4 & 1/5 & 1/6 & 1/7 \\ 1/4 & 1/5 & 1/6 & 1/7 & 1/8 \\ 1/5 & 1/6 & 1/7 & 1/8 & 1/9 \end{bmatrix},$$

$$A^{-1} = \begin{bmatrix} 25 & -300 & 1050 & -1400 & 630 \\ -300 & 4800 & -18900 & 26880 & -12600 \\ 1050 & -18900 & 79380 & -117600 & 56700 \\ -1400 & 26880 & -117600 & 179200 & -88200 \\ 630 & -12600 & 56700 & -88200 & 44100 \end{bmatrix}. \tag{9}$$

We invite the reader to test Subroutine GAUSS or other available system solvers on systems with Hilbert coefficient matrices.

Solving Several Systems Simultaneously

We now consider the problem of solving several different linear systems with the same $(n \times n)$ coefficient matrix A. That is, for each integer k, $1 \leq k \leq m$, find the various solutions $\mathbf{x} = \mathbf{x}_k$ to

$$A\mathbf{x} = \mathbf{b}_k \qquad \text{for } k = 1, 2, \ldots, m. \tag{10}$$

Calling Subroutine Gauss m times would clearly be inefficient, since there is no need to triangularize A more than once. Instead we recommend a combination of two subroutines called FACTOR and SOLVE. Subroutine FACTOR would use elementary row operations to reduce A to an upper-triangular matrix T. The output from FACTOR would be T, the necessary Gaussian multiples m_{ij}, and a record of the necessary row interchanges. For each k, $1 \leq k \leq m$, Subroutine SOLVE would accept $\mathbf{b}_k$ and the output from FACTOR. The Gaussian multiples and a record of the row interchanges would be used to perform the same elementary operations on $\mathbf{b}_k$ that were performed on A to form T. Denoting the resultant vector as $\mathbf{c}_k$, the SOLVE routine would then back-solve the system $T\mathbf{x} = \mathbf{c}_k$.

It can be shown that each call to SOLVE requires n^2 multiplications, and so the total calls to SOLVE require mn^2 multiplications. Each product $A^{-1}\mathbf{b}_k$ requires n^2 multiplications or mn^2 multiplications altogether for all systems. Hence solving (8) by computing A^{-1} and forming $A^{-1}\mathbf{b}_k$ can be more efficient than the above FACTOR/SOLVE approach only if forming A^{-1} is cheaper than triangularizing A.

The FACTOR/SOLVE approach with partial pivoting is illustrated by the following computations on $A\mathbf{x} = \mathbf{b}_k$ with

$$A = \begin{bmatrix} 1 & -1 & 0 \\ 2 & -1 & 1 \\ 2 & -2 & -1 \end{bmatrix} \quad \text{and} \quad \mathbf{b}_k = \begin{bmatrix} 2 \\ 4 \\ 3 \end{bmatrix}.$$

FACTOR:

$$A \xrightarrow{R_1 \leftrightarrow R_2} \begin{bmatrix} 2 & -1 & 1 \\ 1 & -1 & 0 \\ 2 & -2 & -1 \end{bmatrix} \xrightarrow[m_{31} = -1]{m_{21} = -1/2} \begin{bmatrix} 2 & -1 & 1 \\ 0 & -1/2 & -1/2 \\ 0 & -1 & -2 \end{bmatrix}$$

$$\xrightarrow{R_2 \leftrightarrow R_3} \begin{bmatrix} 2 & -1 & 1 \\ 0 & -1 & -2 \\ 0 & -1/2 & -1/2 \end{bmatrix} \xrightarrow{m_{32} = -1/2} \begin{bmatrix} 2 & -1 & 1 \\ 0 & -1 & -2 \\ 0 & 0 & 1/2 \end{bmatrix} = T$$

SOLVE:

$$\mathbf{b}_k \xrightarrow{R_1 \leftrightarrow R_2} \begin{bmatrix} 4 \\ 2 \\ 3 \end{bmatrix} \xrightarrow[m_{31} = -1]{m_{21} = -1/2} \begin{bmatrix} 4 \\ 0 \\ -1 \end{bmatrix} \xrightarrow{R_2 \leftrightarrow R_3} \begin{bmatrix} 4 \\ -1 \\ 0 \end{bmatrix} \xrightarrow{m_{32} = -1/2} \begin{bmatrix} 4 \\ -1 \\ 1/2 \end{bmatrix} = \mathbf{c}_k$$

Backsolving $T\mathbf{x} = \mathbf{c}_k$ yields $\mathbf{x}_k = [1, -1, 1]^T$.

Exercises 7.2

1. Write a program that uses Subroutine GAUSS to calculate A^{-1}, where A is $(n \times n)$. To do this, set up an $(n \times n)$ array AINV(I, J) to hold A^{-1}, where the jth column of AINV is the result of solving $A\mathbf{x} = \mathbf{e}_j$ by GAUSS. Test your program on matrix A in (7); it is known that

$$A^{-1} = \begin{bmatrix} 68 & -41 & -17 & 10 \\ -41 & 25 & 10 & -6 \\ -17 & 10 & 5 & -3 \\ 10 & -6 & -3 & 2 \end{bmatrix}.$$

2. Use Subroutine GAUSS to solve the system $A\mathbf{x} = \mathbf{b}$ given in (7). Use your *machine* version of A^{-1}, found by the program in Exercise 1, to calculate $A^{-1}\mathbf{b}$; and

compare the two answers by calculating the norm of the residual vector.

3. Use GAUSS to solve $A\mathbf{x} = \mathbf{b}$, where A, $\mathbf{b}$, and the solution $\mathbf{x}_t$ are

$$A = \begin{bmatrix} 1 & -2 & 3 & 1 \\ -2 & 1 & -2 & -1 \\ 3 & -2 & 1 & 5 \\ 1 & -1 & 5 & 3 \end{bmatrix}, \quad \mathbf{b} = \begin{bmatrix} 3 \\ -4 \\ 7 \\ 8 \end{bmatrix},$$

$$\mathbf{x}_t = \begin{bmatrix} 1 \\ 1 \\ 1 \\ 1 \end{bmatrix}.$$

4. Small changes in the right-hand side of $A\mathbf{x} = \mathbf{b}$ may lead to relatively large changes in the solution. As an illustration, solve $A\mathbf{x} = \mathbf{b}$, where A is the matrix in (7) and $\mathbf{b}$ is each of the following.

a) $\mathbf{b} = \begin{bmatrix} 23.01 \\ 31.99 \\ 32.99 \\ 31.01 \end{bmatrix}$

b) $\mathbf{b} = \begin{bmatrix} 23.1 \\ 31.9 \\ 32.9 \\ 31.1 \end{bmatrix}$.

5. Let A_1 and A_2 be the matrices obtained from A in (7) by replacing the $(1, 1)$ entry of A by 5.01 and 4.99, respectively. Using the program in Exercise 1, calculate A_1^{-1} and A_2^{-1}. Compare your answers with A^{-1}.

6. FORTRAN has provisions for complex arithmetic, and it is easy to modify Subroutine GAUSS to solve systems with complex coefficients. The required changes are these (see the listing in Fig. 7.2):
a) Replace the two dimension declarations by

COMPLEX A(MAINDM, MAINDM),

B(MAINDM), X(MAINDM)

COMPLEX AUG(50, 51)

COMPLEX CTEMP, Q, RESI, CABS.

b) In the row-interchange loop, change TEMP to CTEMP so that the loop reads

DO 4 K = I, NP1

CTEMP = AUG(I, K)

AUG(I, K) = AUG(IPIVOT, K)

AUG(IPIVOT, K) = CTEMP

CONTINUE.

c) Change the divide check statement just before the backsolving segment to

IF(CABS(AUG(N, N)).EQ.0.) GO TO 13.

In any program that calls GAUSS, the arrays A, B, and X must also be declared complex. There are two ways to make a complex assignment. For example, if S is declared to be complex and we want to assign $3 + 4i$ to S, we can write $S = (3., 4.)$ or we can write $S = \mathrm{CMPLX}(3., 4.)$. To read in a complex matrix A, we can read the real and imaginary parts of A into two real arrays U and V.

In particular, if $a_{jk} = u_{jk} + iv_{jk}$, then the statement

A(J, K) = CMPLX(U(J, K), V(J, K))

will set up the matrix A. Finally, all complex-valued functions and variables must be declared complex. Thus the functions CABS, CSQRT, CMPLX, CSIN, and so on, must be declared to be of complex type in the beginning.
    Modify GAUSS to do complex arithmetic, and use it to solve the systems in Exercises 25 and 26 of Section 3.6.

7. If Q is an $(n \times n)$ matrix and $\mathbf{b}$ is in R^n, verify that forming the product $Q\mathbf{b}$ requires n^2 multiplications.

8. If T is an $(n \times n)$ upper-triangular matrix and $\mathbf{b}$ is a vector in R^n, verify that it requires $1 + 2 + 3 + \cdots + n = n(n + 1)/2$ multiplications and/or divisions to solve $T\mathbf{x} = \mathbf{b}$. (*Hint:* Write out the system $T\mathbf{x} = \mathbf{b}$, and note that it takes one division to find x_n, two multiplications and/or divisions to obtain x_{n-1}, and so on. Observe that Exercises 7 and 8 show that calculating $A^{-1}\mathbf{b}$ requires more effort than solving $T\mathbf{x} = \mathbf{b}$.)

9. Let $[A \,|\, \mathbf{b}]$ be the augmented matrix for the $(n \times n)$ system $A\mathbf{x} = \mathbf{b}$. To create zeros in the $(r, 1)$ entry of $[A \,|\, \mathbf{b}]$ with a row operation requires the formation of a multiple $m = a_{r1}/a_{11}$ (one division) and the replacement of the (r, j) entry of $[A \,|\, \mathbf{b}]$ by $a_{rj} - ma_{1j}$ for $j = 2, 3, \ldots, n + 1$. Thus $n + 1$ multiplications and/or divisions are required for each row in the first step of reducing $[A \,|\, \mathbf{b}]$ to echelon form; a total of $(n - 1)(n + 1)$ multiplications and/or divisions is needed.
a) Verify that $(n - 2)n$ multiplications and/or divisions are necessary to create zeros in the $(3, 2)$, $(4, 2), \ldots, (n, 2)$ entries.
b) Verify that $(n - i)(n - i + 2)$ multiplications and/or divisions are required at the ith stage of Gaussian elimination (creating zeros in the entries below the main diagonal in column i).
c) The process of Gaussian elimination stops after column $n - 1$, so a total of

$$S = (n - 1)(n + 1)$$

$$+ (n - 2)n + \cdots + (n - i)(n - i + 2) + \cdots + 1 \cdot 3$$

multiplications and/or divisions is required. Evaluate this sum by expressing it as

$$S = \sum_{i=1}^{n-1} (n - i)(n - i + 2).$$

7.3   The Power Method for Eigenvalues

Finding the eigenvalues of an $(n \times n)$ matrix A is a harder computational problem than is solving $A\mathbf{x} = \mathbf{b}$. In particular, as we commented in Chapter 3, we cannot usually expect to find the eigenvalues of an $(n \times n)$ matrix in a finite number of steps when $n \geq 5$ (since this process amounts to solving a polynomial equation). Given the development of Chapter 6, one might feel that the following is a reasonable computational scheme for finding the eigenvalues of A:

1. Reduce A to Hessenberg form, H.
2. Find the characteristic polynomial, $p(t)$, for H.
3. Find the roots of $p(t) = 0$.

Unfortunately, this approach may well lead to severe errors, and it is relatively easy to see why. In particular, when A is reduced to Hessenberg form, roundoff errors will occur, and the Hessenberg matrix found by the machine will not be quite what it should be (if exact arithmetic were used). As the example below shows, even though two matrices are almost the same, it is possible for their eigenvalues to be substantially different.

EXAMPLE 1   This example is due to Forsythe and is an extreme instance of two nearly equal matrices with different eigenvalues. Let H and $H + E$ be $(n \times n)$ Hessenberg matrices, where

$$H = \begin{bmatrix} 1 & 0 & 0 & \cdots & 0 & 0 \\ 1 & 1 & 0 & \cdots & 0 & 0 \\ 0 & 1 & 1 & \cdots & 0 & 0 \\ \vdots & & & & & \vdots \\ 0 & 0 & 0 & \cdots & 1 & 0 \\ 0 & 0 & 0 & \cdots & 1 & 1 \end{bmatrix} \quad \text{and} \quad H + E = \begin{bmatrix} 1 & 0 & 0 & \cdots & 0 & \varepsilon \\ 1 & 1 & 0 & \cdots & 0 & 0 \\ 0 & 1 & 1 & \cdots & 0 & 0 \\ \vdots & & & & & \vdots \\ 0 & 0 & 0 & \cdots & 1 & 0 \\ 0 & 0 & 0 & \cdots & 1 & 1 \end{bmatrix}.$$

Thus H consists entirely of 0's except for the main diagonal and the subdiagonal which contain 1's. The matrix $H + E$ is equal to H except for the entry ε in the $(1, n)$ position.

It can be shown (Exercise 3) that the characteristic polynomials of H and $H + E$ are

$$p(t) = (1 - t)^n \quad \text{and} \quad q(t) = (1 - t)^n + (-1)^{n+1}\varepsilon, \quad \text{respectively.}$$

To see how the eigenvalues might differ, suppose that $n = 10$ and $\varepsilon = 2^{-10}$. We see that $q(t) = 0$ means $(1 - t)^{10} = 2^{-10}$; and one solution to this equation is $1 - t = 2^{-1}$, or $t = .5$ (a change in H of amount 2^{-10} produces a 50 percent change in one eigenvalue of H). Not every perturbation of entries in H will lead to such a dramatic change in the eigenvalues. For example, if the $(1, 1)$ entry of H is changed to $1 + \varepsilon$, then an eigenvalue of H has been perturbed only by amount ε.

Computational experience, as well as examples such as the one above, suggests that no matter how carefully we carry out Hessenberg reduction on A, the inevitable roundoff errors will produce a Hessenberg matrix in the machine whose eigenvalues are not the same as those of A (and may differ substantially from the eigenvalues of A). To overcome this difficulty, we employ a device that is relatively common in numerical methods, that of "refinement." Briefly, our approach will be to reduce A to Hessenberg form H, find the eigenvalues of H, and then regard these eigenvalues of H as *estimates* to the eigenvalues of A. We next apply the inverse power method (as described in Section 7.4) to the original matrix A and refine (or correct) these estimates by iteration. Before the inverse power method is described, it is convenient to consider a closely related algorithm, the power method.

The power method is an iterative procedure that can be used to estimate the "dominant" eigenvalue of a matrix. To begin, suppose that A is an $(n \times n)$ matrix and that A has eigenvalues $\lambda_1, \lambda_2, \ldots, \lambda_n$ with corresponding eigenvectors $\mathbf{u}_1, \mathbf{u}_2, \ldots, \mathbf{u}_n$, so $A\mathbf{u}_j = \lambda_j \mathbf{u}_j$, $1 \le j \le n$. (Of course, we do not know the eigenvalues or the eigenvectors; the objective of the power method is to estimate an eigenvalue accurately.) For simplicity, we will put a rather severe restriction on the eigenvectors; we will assume that $\{\mathbf{u}_1, \mathbf{u}_2, \ldots, \mathbf{u}_n\}$ is linearly independent. (Recall that the eigenvectors may have complex components, so linear independence means that the only scalars, either real or complex, that satisfy $b_1\mathbf{u}_1 + b_2\mathbf{u}_2 + \cdots + b_n\mathbf{u}_n = \boldsymbol{\theta}$ are the scalars $b_1 = b_2 = \cdots = b_n = 0$. Moreover, linear independence means that any $(n \times 1)$ vector $\mathbf{v}$, where $\mathbf{v}$ may have real or complex components, can be expressed as a linear combination of $\mathbf{u}_1, \mathbf{u}_2, \ldots, \mathbf{u}_n$.) Given this assumption, let us choose some initial vector $\mathbf{v}_0$, where $\mathbf{v}_0 \ne \boldsymbol{\theta}$. By our linear independence assumption, we know that $\mathbf{v}_0$ can be expressed in the form

$$\mathbf{v}_0 = a_1\mathbf{u}_1 + a_2\mathbf{u}_2 + \cdots + a_n\mathbf{u}_n. \tag{1}$$

(Again, we do not know the eigenvectors, so we do not know the coefficients $a_1, a_2, \ldots, a_n$ in (1); all we know is that $\mathbf{v}_0$ can be expressed in the form above. A typical initial vector $\mathbf{v}_0$ is one that has each component equal to 1.) If we form the sequence $\mathbf{v}_k = A\mathbf{v}_{k-1}$ for $k = 1, 2, \ldots$, then as we saw in Chapter 3, each vector $\mathbf{v}_k$ in the calculated sequence has the form

$$\mathbf{v}_k = a_1\lambda_1^k\mathbf{u}_1 + a_2\lambda_2^k\mathbf{u}_2 + \cdots + a_n\lambda_n^k\mathbf{u}_n. \tag{2}$$

Now suppose the eigenvalues are ordered so that $|\lambda_1| \ge |\lambda_2| \ge |\lambda_3| \ge \cdots \ge |\lambda_n|$, and let us write (2) as

$$\mathbf{v}_k = \lambda_1^k\left(a_1\mathbf{u}_1 + a_2\left(\frac{\lambda_2}{\lambda_1}\right)^k\mathbf{u}_2 + \cdots + a_n\left(\frac{\lambda_n}{\lambda_1}\right)^k\mathbf{u}_n\right). \tag{3}$$

If $|\lambda_1| > |\lambda_2|$, then the terms $(\lambda_i/\lambda_1)^k$ are small for large k, where $2 \le i \le n$. Thus from (3) we expect (if $a_1 \ne 0$) that

$$\mathbf{v}_k \simeq \lambda_1^k a_1\mathbf{u}_1. \tag{4}$$

(That is, we expect that $\mathbf{v}_k$ is nearly a multiple of $\mathbf{u}_1$. Even though we do not know a_1, λ_1, or $\mathbf{u}_1$, we do have the calculated vector $\mathbf{v}_k$. Under the assumptions that $a_1 \ne 0$ and $|\lambda_1| > |\lambda_2|$, we do know that the vectors $\mathbf{v}_0, \mathbf{v}_1, \mathbf{v}_2, \ldots$ are aligning themselves along $\mathbf{u}_1$.) To obtain an estimate to λ_1, we utilize two successive calculated vectors $\mathbf{v}_k$ and $\mathbf{v}_{k+1}$, where, as above, we expect that

$$\mathbf{v}_{k+1} \simeq \lambda_1^{k+1} a_1 \mathbf{u}_1 \quad \text{or} \quad \mathbf{v}_{k+1} \simeq \lambda_1 \mathbf{v}_k.$$

Now if we form the quotient $\beta_k = \mathbf{w}^T \mathbf{v}_{k+1} / \mathbf{w}^T \mathbf{v}_k$, where $\mathbf{w}$ is any vector such that $\mathbf{w}^T \mathbf{u}_1 \neq 0$, we have

$$\beta_k = \frac{\mathbf{w}^T \mathbf{v}_{k+1}}{\mathbf{w}^T \mathbf{v}_k} \simeq \frac{\lambda_1^{k+1} a_1 \mathbf{w}^T \mathbf{u}_1}{\lambda_1^k a_1 \mathbf{w}^T \mathbf{u}_1} = \lambda_1. \tag{5}$$

The approximation in (5) is the essence of the power method. With respect to (5), we note that to the extent that (4) is valid, a reasonable choice for $\mathbf{w}$ is the vector $\mathbf{v}_k$ itself. This choice leads to the approximation

$$\beta_k = \frac{\mathbf{v}_k^T \mathbf{v}_{k+1}}{\mathbf{v}_k^T \mathbf{v}_k} \simeq \lambda_1. \tag{6}$$

The left-hand side of (6) is called the **Rayleigh quotient;** and it can be shown that if $a_1 \neq 0$ in (1) and if $|\lambda_1| > |\lambda_2|$, then

$$\lim_{k \to \infty} \frac{\mathbf{v}_k^T \mathbf{v}_{k+1}}{\mathbf{v}_k^T \mathbf{v}_k} = \lambda_1.$$

In summary, the power method proceeds as follows:

1. Guess an initial vector $\mathbf{v}_0$.
2. Form the sequence $\mathbf{v}_k = A\mathbf{v}_{k-1}, k = 1, 2, \ldots$.
3. For each k, calculate the Rayleigh quotient in (6).

The output of the power method is in part the sequence of numbers $\beta_0, \beta_1, \beta_2, \ldots$ in (6), which, one hopes, converges to the dominant eigenvalue of A. (By dominant, we mean, of course, the eigenvalue that is largest in absolute value.)

EXAMPLE 2 Let A be the matrix

$$\begin{bmatrix} 1 & -1 & 2 \\ -2 & 0 & 5 \\ 6 & -3 & 6 \end{bmatrix}.$$

It is easy to verify that the eigenvalues of A are $\lambda_1 = 5$, $\lambda_2 = 3$, and $\lambda_3 = -1$; and corresponding eigenvectors are

$$\mathbf{u}_1 = \begin{bmatrix} 5 \\ 16 \\ 18 \end{bmatrix}, \quad \mathbf{u}_2 = \begin{bmatrix} 1 \\ 6 \\ 4 \end{bmatrix}, \quad \text{and} \quad \mathbf{u}_3 = \begin{bmatrix} -1 \\ -2 \\ 0 \end{bmatrix}.$$

With the initial vector $\mathbf{v}_0$ given below, we generate

$$\mathbf{v}_0 = \begin{bmatrix} 0 \\ 1 \\ 3 \end{bmatrix}, \quad \mathbf{v}_1 = \begin{bmatrix} 5 \\ 15 \\ 15 \end{bmatrix}, \quad \mathbf{v}_2 = \begin{bmatrix} 20 \\ 65 \\ 75 \end{bmatrix}, \quad \text{and} \quad \mathbf{v}_3 = \begin{bmatrix} 105 \\ 335 \\ 375 \end{bmatrix}.$$

Table 7.2

k	β_k		k	β_k	
1	0.4666666E	01	14	0.5002243E	01
2	0.7127658E	01	15	0.5001345E	01
3	0.5770836E	01	16	0.5000809E	01
4	0.5460129E	01	17	0.5000484E	01
5	0.5245014E	01	18	0.5000289E	01
6	0.5142666E	01	19	0.5000171E	01
7	0.5083027E	01	20	0.5000103E	01
8	0.5049140E	01	21	0.5000067E	01
9	0.5029204E	01	22	0.5000038E	01
10	0.5017425E	01	23	0.5000021E	01
11	0.5010424E	01	24	0.5000011E	01
12	0.5006239E	01	25	0.5000006E	01
13	0.5003741E	01	26	0.5000005E	01

The corresponding Rayleigh quotients are

$$\frac{\mathbf{v}_0^T\mathbf{v}_1}{\mathbf{v}_0^T\mathbf{v}_0} = \frac{60}{10} = 6, \qquad \frac{\mathbf{v}_1^T\mathbf{v}_2}{\mathbf{v}_1^T\mathbf{v}_1} = \frac{2200}{475} \simeq 4.63, \quad \text{and} \quad \frac{\mathbf{v}_2^T\mathbf{v}_3}{\mathbf{v}_2^T\mathbf{v}_2} = \frac{52000}{10250} \simeq 5.07,$$

and they appear to be converging to the dominant eigenvalue $\lambda_1 = 5$. To illustrate the point that the vectors $\mathbf{v}_k$ are aligning themselves in the direction of $\mathbf{u}_1$, we observe that

$$\mathbf{v}_1 = \begin{bmatrix} 5 \\ 15 \\ 15 \end{bmatrix}, \qquad (1/4)\mathbf{v}_2 = \begin{bmatrix} 5 \\ 16.25 \\ 18.75 \end{bmatrix}, \quad \text{and}$$

$$(1/21)\mathbf{v}_3 = \begin{bmatrix} 5 \\ 15.95 \\ 17.86 \end{bmatrix} \quad \text{(approximately)}.$$

The vector $\mathbf{v}_3$, for example, is close to being a scalar multiple of $\mathbf{u}_1$.

As a more realistic illustration, we used a computer program that implements the power method and produced the estimates listed in Table 7.2. The program used "scaling," which we define shortly, and Rayleigh quotients to produce the sequence of estimates to λ_1.   □

Several issues related to the power method have not been resolved. First, if the power method finds just the dominant eigenvalue, how do we find the other $n - 1$ eigenvalues of A and how do we find the eigenvectors? The answer is that the power method is not suited to finding the other eigenvalues, and we will use the inverse power method to find them if we have reasonably good estimates. The eigenvectors are easily found when we use "scaling," as described later. Second, if we choose an initial vector $\mathbf{v}_0$ at random, what if $a_1 = 0$ in (1)? This is more of a problem in theory than it is in practice. However, a thorough discussion would lead us too far afield, and the interested reader can consult any good numerical analysis book. Third, what if A does not have a single dominant eigenvalue; what if $|\lambda_1| = |\lambda_2|$? This situation, it develops, is a real flaw in the power method. To see the dimensions of this problem, let us consider (3) in a bit more detail. First of all, if $\lambda_1 = \lambda_2$ and if $|\lambda_2| > |\lambda_3| \geq \cdots \geq |\lambda_n|$, then (3) leads to a slightly different approximation in (4):

$$\mathbf{v}_k \simeq \lambda_1^k (a_1 \mathbf{u}_1 + a_2 \mathbf{u}_2).$$

However, for large k we still expect that [as in (5)]

$$\beta_k = \frac{\mathbf{w}^T \mathbf{v}_{k+1}}{\mathbf{w}^T \mathbf{v}_k} \simeq \frac{\lambda_1^{k+1} \mathbf{w}^T (a_1 \mathbf{u}_1 + a_2 \mathbf{u}_2)}{\lambda_1^k \mathbf{w}^T (a_1 \mathbf{u}_1 + a_2 \mathbf{u}_2)} = \lambda_1.$$

In general, it is not hard to show that a multiple dominant eigenvalue will not affect the convergence of the power method. However, if $\lambda_1 = -\lambda_2$, then we do have problems. In this case, (3) shows that

$$\mathbf{v}_k \simeq \lambda_1^k [a_1 \mathbf{u}_1 + (-1)^k a_2 \mathbf{u}_2],$$

and the quotients $\beta_k = \mathbf{w}^T \mathbf{v}_{k+1} / \mathbf{w}^T \mathbf{v}_k$ will exhibit an oscillatory behavior and will not converge. Furthermore, if A is a real matrix and if λ_1 is complex, then we know that $\bar{\lambda}_1$ is also an eigenvalue of A. In this event, $|\lambda_1| = |\bar{\lambda}_1|$, so it can be expected that the power method will have difficulty whenever the dominant eigenvalue of A is complex. This difficulty can be overcome, but overcoming it requires quite an effort.

A thoughtful reader with some computational experience may have noted another potential problem. If $|\lambda_1| > 1$, then the components of $\mathbf{v}_k$ are growing rapidly with k, and overflow threatens. Similarly, if $|\lambda_1| < 1$, we have the potential for underflow. (Every digital device has a "largest" and "smallest" number that it can represent. The IBM mainframe computers cannot represent positive numbers larger than about $.73 \times 10^{76}$ or positive numbers smaller than $.53 \times 10^{-78}$. Attempting to calculate a number whose magnitude is outside this range may cause a program to cease execution.) This problem is easily remedied by *scaling* the vectors $\mathbf{v}_k$. In particular, suppose at the kth stage that $\|\mathbf{v}_k\| = 1$. We then calculate

$$\mathbf{z}_{k+1} = A\mathbf{v}_k$$

and form the Rayleigh quotient

$$\beta_k = \mathbf{v}_k^T \mathbf{z}_{k+1} \simeq \lambda_1.$$

(Note that $\mathbf{v}_k^T \mathbf{v}_k = 1$ since $\|\mathbf{v}_k\| = 1$.) Next, we define $\mathbf{v}_{k+1}$ by

$$\mathbf{v}_{k+1} = \mathbf{z}_{k+1} / \|\mathbf{z}_{k+1}\|.$$

Since $\|\mathbf{v}_{k+1}\| = 1$, the $(k + 1)$st stage proceeds just as the kth stage above. It is not hard to prove that this scaling process does not affect the convergence of the power method [this statement is nearly obvious from (2) since scaling $\mathbf{v}_k$ is the same as adjusting the coefficients $a_1, a_2, \ldots, a_n$]. Furthermore, as we expect from (4), it can be shown that the vectors $\mathbf{v}_k$ converge to an eigenvector of A corresponding to λ_1 when scaling is used and when the power method itself is converging.

We want to make one final observation about the power method. If $|\lambda_1| > |\lambda_2|$, then it should also be expected from (3) that the rate of convergence of the power method is governed by the size of λ_2/λ_1. If the ratio is small, convergence is rapid since (4) is now a very good approximation. To quantify these statements, suppose that the power method with scaling is employed and that $\beta_k = \mathbf{v}_k^T A \mathbf{v}_k$ denotes the kth Rayleigh quotient. It is not too hard to show that

$$\lim_{k \to \infty} \frac{\beta_{k+1} - \lambda_1}{\beta_k - \lambda_1} = \frac{\lambda_2}{\lambda_1}.$$

That is, $\beta_{k+1} - \lambda_1 \simeq (\lambda_2/\lambda_1)(\beta_k - \lambda_1)$, and this result can be interpreted to mean that the error $\beta_k - \lambda_1$ is decreased by about a factor of λ_2/λ_1 with each step of the power method.

EXAMPLE 3 As we mentioned at the end of Example 2, scaling was used in the program that produced the results in Table 7.2. The initial vector $\mathbf{v}_0$ was chosen as

$$\mathbf{v}_0 = \frac{1}{\sqrt{3}} \begin{bmatrix} 1 \\ 1 \\ 1 \end{bmatrix},$$

so $\|\mathbf{v}_0\| = 1$. When scaling is used, the vectors $\mathbf{v}_k$ are converging to $\mathbf{u}_1/\|\mathbf{u}_1\|$, where (to six places)

$$\frac{\mathbf{u}_1}{\|\mathbf{u}_1\|} = \begin{bmatrix} .203279 \\ .650493 \\ .731804 \end{bmatrix}.$$

Some of the vectors $\mathbf{v}_k$ produced in Example 2 were

$$\mathbf{v}_6 = \begin{bmatrix} .205821 \\ .642018 \\ .738546 \end{bmatrix}, \quad \mathbf{v}_{12} = \begin{bmatrix} .203392 \\ .650114 \\ .732109 \end{bmatrix},$$

$$\mathbf{v}_{18} = \begin{bmatrix} .203284 \\ .650475 \\ .731818 \end{bmatrix}, \quad \mathbf{v}_{24} = \begin{bmatrix} .203279 \\ .650492 \\ .731805 \end{bmatrix}.$$

Finally, to demonstrate the rate of convergence, we note that $\lambda_2/\lambda_1 = .6$. If we let $\beta_k = \mathbf{v}_k A \mathbf{v}_k$ be the kth estimate of λ_1, then a check of Table 7.2 will show that

$$\beta_{k+1} - 5 \simeq .6(\beta_k - 5).$$

Equivalently, the $(k + 1)$st error is approximately .6 times the kth error.

1. For the matrix A in Example 2 and for the starting vector

$$\mathbf{v}_0 = \begin{bmatrix} 1 \\ 1 \\ 1 \end{bmatrix},$$

calculate $\mathbf{v}_1, \mathbf{v}_2, \mathbf{v}_3, \mathbf{v}_4,$ and $\mathbf{v}_5$, where $\mathbf{v}_{i+1} = A\mathbf{v}_i$. Form the Rayleigh quotients as in (6) to estimate the dominant eigenvalue of A.

2. Let A be the (2×2) matrix

$$A = \begin{bmatrix} 3 & -4 \\ 2 & -3 \end{bmatrix}.$$

a) Verify that A has eigenvalues $\lambda = 1$ and $\lambda = -1$.
b) For $\mathbf{v}_0 = \mathbf{e}_1$, calculate $\mathbf{v}_1, \mathbf{v}_2, \mathbf{v}_3, \mathbf{v}_4,$ and $\mathbf{v}_5$.
c) Form the Rayleigh quotients and verify that the power method cannot work for A.

3. Verify that the characteristic polynomial of $H + E$ in Example 1 is $q(t) = (1 - t)^n + (-1)^{n+1}\varepsilon$. [Hint: Calculate $\det(H + E - tI)$.]

4. Suppose that A is an $(n \times n)$ matrix with eigenvalues $\lambda_1, \lambda_2, \ldots, \lambda_n$, where $|\lambda_1| > |\lambda_2| \geq \cdots \geq |\lambda_n|$, and that $\mathbf{v}_0$ is given by (1), where $a_1 \neq 0$. Suppose that $\mathbf{w}$ is a fixed vector in R^n such that $\mathbf{w}^T\mathbf{u}_1 \neq 0$, and for $k = 0, 1, \ldots$ that

$$\beta_k = \frac{\mathbf{w}^T\mathbf{v}_{k+1}}{\mathbf{w}^T\mathbf{v}_k},$$

where $\mathbf{v}_k = A\mathbf{v}_{k-1}, k = 1, 2, \ldots$. [Thus $\mathbf{v}_k$ has the form of (3).] Prove that

$$\lim_{k\to\infty} \beta_k = \lambda_1.$$

5. Using the same hypotheses as Exercise 4, prove that

$$\lim_{k\to\infty} \frac{\beta_{k+1} - \lambda_1}{\beta_k - \lambda_1} = \frac{\lambda_2}{\lambda_1}$$

when $a_2 \neq 0$ and $|\lambda_2| > |\lambda_3|$.

6. If A is an $(n \times n)$ symmetric matrix, we can assume that A has n linearly independent eigenvectors $\mathbf{u}_1, \mathbf{u}_2, \ldots, \mathbf{u}_n$, where $\mathbf{u}_i^T\mathbf{u}_j = 0$ when $i \neq j$ and $\mathbf{u}_i^T\mathbf{u}_i = 1$, $1 \leq i, j \leq n$. Calculate the Rayleigh quotient (6) under this assumption, and convince yourself that convergence should be

$$\lim_{k\to\infty} \frac{\beta_{k+1} - \lambda_1}{\beta_k - \lambda_1} = \left(\frac{\lambda_2}{\lambda_1}\right)^2.$$

7. Let A be an $(n \times n)$ symmetric matrix and let $\mathbf{x}$ be any nonzero vector in R^n. As in Exercise 6, we can assume that $\mathbf{x} = a_1\mathbf{u}_1 + a_2\mathbf{u}_2 + \cdots + a_n\mathbf{u}_n$, where the eigenvectors $\mathbf{u}_1, \mathbf{u}_2, \ldots, \mathbf{u}_n$ are orthonormal.
a) Calculate $\mathbf{x}^T\mathbf{x}$.
b) Calculate $\mathbf{x}^TA\mathbf{x}$.
c) Suppose that the eigenvalues of A are $\lambda_1, \lambda_2, \ldots, \lambda_n$, where $a \leq \lambda_i \leq b, 1 \leq i \leq n$. Prove that

$$a \leq \frac{\mathbf{x}^TA\mathbf{x}}{\mathbf{x}^T\mathbf{x}} \leq b.$$

7.4 The Inverse Power Method for the Eigenvalue Problem

The inverse power method is nothing more than the power method applied to the matrix $(A - \alpha I)^{-1}$. If α is a reasonably good estimate to an eigenvalue λ of A, then several steps of the inverse power method will give a very accurate estimate to λ and a corresponding eigenvector. To develop the theoretical basis of the inverse power method, we suppose that α is any scalar that is not an eigenvalue of A and that $A\mathbf{u} = \lambda\mathbf{u}, \mathbf{u} \neq \mathbf{0}$. Since $A\mathbf{u} = \lambda\mathbf{u}$, it is clear that

$$(A - \alpha I)\mathbf{u} = (\lambda - \alpha)\mathbf{u}.$$

Next, since α is not an eigenvalue of A, $A - \alpha I$ is nonsingular; and we can write

$$(A - \alpha I)^{-1}\mathbf{u} = \frac{1}{\lambda - \alpha}\mathbf{u}. \tag{1}$$

Clearly, (1) states that $1/(\lambda - \alpha)$ is an eigenvalue of $(A - \alpha I)^{-1}$ and that $\mathbf{u}$ is a corresponding eigenvector.

Suppose that A has eigenvalues $\lambda_1, \lambda_2, \ldots, \lambda_n$ and that α_1 is a good estimate to λ_1. By (1), the eigenvalues of $(A - \alpha_1 I)^{-1}$ are $\mu_1, \mu_2, \ldots, \mu_n$, where

$$\mu_1 = \frac{1}{\lambda_1 - \alpha_1}, \qquad \mu_2 = \frac{1}{\lambda_2 - \alpha_1}, \ldots, \qquad \mu_n = \frac{1}{\lambda_n - \alpha_1}. \tag{2}$$

If $\alpha_1 \simeq \lambda_1$, then we can expect that μ_1 is the dominant eigenvalue of $(A - \alpha_1 I)^{-1}$; that is, if the eigenvalues $\lambda_1, \lambda_2, \ldots, \lambda_n$ are reasonably separated, and if α_1 is a good estimate to λ_1, then we can anticipate that $|\mu_1| \gg |\mu_i|$ for $2 \le i \le n$. From Section 7.3, we know that if $|\mu_1| > |\mu_i|$ for $2 \le i \le n$, then the power method applied to $(A - \alpha_1 I)^{-1}$ will converge to μ_1; and the rate of convergence will be governed by the largest number among

$$\left| \frac{\mu_2}{\mu_1} \right|, \left| \frac{\mu_3}{\mu_1} \right|, \ldots, \left| \frac{\mu_n}{\mu_1} \right|.$$

Furthermore, we would expect rapid convergence when α_1 is close to λ_1 since $|\mu_1|$ is then quite large with respect to $|\mu_2|, |\mu_3|, \ldots, |\mu_n|$.

As an example to fix these ideas, suppose that A is a (3×3) matrix with eigenvalues $\lambda_1 = 3$, $\lambda_2 = -3$, and $\lambda_3 = 1$. Since $|\lambda_1| = |\lambda_2|$ and $\lambda_1 \ne \lambda_2$, the power method will not converge when applied to A. However, for $\alpha_1 = 2.9$, the matrix $(A - \alpha_1 I)^{-1}$ has approximate eigenvalues

$$\mu_1 = 10., \qquad \mu_2 = -.169, \qquad \mu_3 = -.526,$$

and the power method applied to $(A - \alpha_1 I)^{-1}$ will converge to $\mu_1 = 10$. Moreover, convergence will be extremely rapid, governed by the ratio $|\mu_3/\mu_1| = 1/19 \simeq .0526$. That is, the errors decrease by about a factor of $1/19$ per step, and so k steps will decrease the initial error by a factor of $(1/19)^k$. If $k = 3$, for instance, then $(1/19)^3 \simeq .000146$.

In general, if A is $(n \times n)$ and has eigenvalues $\lambda_1, \lambda_2, \ldots, \lambda_n$, and if we have good estimates $\alpha_1, \alpha_2, \ldots, \alpha_n$ to $\lambda_1, \lambda_2, \ldots, \lambda_n$, then we can expect that the power method applied to $(A - \alpha_j I)^{-1}$ will converge to μ_j, where

$$\mu_j = \frac{1}{\lambda_j - \alpha_j}.$$

Having μ_j, we can then recover λ_j from

$$\lambda_j = \frac{1}{\mu_j} + \alpha_j.$$

We wish to emphasize the eigenvalue "refinement" aspect of this procedure. When we transform A to a Hessenberg matrix H, we do not expect (in the presence of roundoff error) that the eigenvalues of H and A will be exactly the same. However, if the eigenvalues of H are $\alpha_1, \alpha_2, \ldots, \alpha_n$, then we can regard them as estimates to the eigenvalues of A. The estimates $\alpha_1, \alpha_2, \ldots, \alpha_n$ can be quickly corrected to $\lambda_1, \lambda_2, \ldots, \lambda_n$ when the power method is applied to $(A - \alpha_j I)^{-1}$ for $1 \le j \le n$. (The matrix A, of course, has not been contaminated by the roundoff errors made in Hessenberg reduction.) As we have tried to point

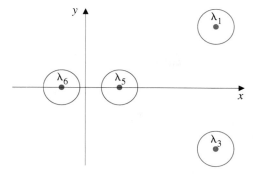

Figure 7.4
A distribution of
eigenvalues in the
complex plane

out above, if the estimates α_j are good relative to the separation of $\lambda_1, \lambda_2, \ldots, \lambda_n$, then the inverse power method will converge since the eigenvalue $\mu_j = 1/(\lambda_j - \alpha_j)$ will dominate all the other eigenvalues of $(A - \alpha_j I)^{-1}$.

As another example, suppose that A is a (6×6) real matrix with eignvalues $\lambda_1, \lambda_2, \ldots, \lambda_6$. Suppose further that λ_1 is complex, $\lambda_1 = a + ib$, that $\lambda_2 = \lambda_1$, $\lambda_3 = \lambda_4 = \bar{\lambda}_1$, and that λ_5 and λ_6 are real, where the ordering is

$$|\lambda_1| = |\lambda_2| = |\lambda_3| = |\lambda_4| > |\lambda_5| > |\lambda_6|.$$

This hypothetical situation is diagrammed in Fig. 7.4 and represents a case that the power method cannot deal with. [In Fig. 7.4, we think of a complex number $z = x + iy$ as being "identified" with the coordinates (x, y).] Given the situation illustrated in Fig. 7.4, suppose we have estimates α_1 to λ_1, α_3 to λ_3, α_5 to λ_5, and α_6 to λ_6 where α_j is in the circle centered at λ_j for $j = 1, 3, 5, 6$. Clearly the power method applied to $(A - \alpha_j I)^{-1}$ will converge to $1/(\lambda_j - \alpha_j)$, $j = 1, 3, 5, 6$. [Recall that a multiple eigenvalue, such as $1/(\lambda_1 - \alpha_1)$ or $1/(\lambda_3 - \alpha_3)$, does not affect the convergence of the power method.]

One remaining point must be made about the inverse power method: Since the inverse power method amounts to applying the power method to $(A - \alpha I)^{-1}$, we might feel that we should calculate the sequence

$$\mathbf{v}_k = (A - \alpha I)^{-1}\mathbf{v}_{k-1}, \qquad k = 1, 2, \ldots,$$

where $\mathbf{v}_0$ is some starting vector. However since an inverse matrix should not usually be calculated, we rewrite this iteration as

$$(A - \alpha I)\mathbf{v}_k = \mathbf{v}_{k-1}, \qquad k = 1, 2, \ldots. \tag{3}$$

Thus given $\mathbf{v}_0$, we solve $(A - \alpha I)\mathbf{x} = \mathbf{v}_0$ to find $\mathbf{v}_1$, and then solve $(A - \alpha I)\mathbf{x} = \mathbf{v}_1$ to find $\mathbf{v}_2$, and so on. In an algorithmic form, the inverse power method with scaling proceeds as follows. Let $\mathbf{v}_0$ be a starting vector where $\|\mathbf{v}_0\| = 1$ and for $k = 0, 1, 2, \ldots$.

1. Find $\mathbf{z}_{k+1}$ such that $(A - \alpha I)\mathbf{z}_{k+1} = \mathbf{v}_k$.
2. Set $\beta_k = \mathbf{z}_{k+1}^T \mathbf{v}_k$.
3. Let $\eta_{k+1} = \|\mathbf{z}_{k+1}\|$.
4. Set $\mathbf{v}_{k+1} = \mathbf{z}_{k+1}/\eta_{k+1}$, and return to step (1).

If we terminate this iteration after r steps with β_r and $\mathbf{v}_r$, then

$$\lambda = \frac{1}{\beta_r} + \alpha$$

is an estimate to the eigenvalue of A that is nearest to α, and $\mathbf{v}_r$ is an estimate to a corresponding eigenvector. If A is $(n \times n)$, then a reasonable starting vector $\mathbf{v}_0$ of norm 1 is

$$\mathbf{v}_0 = \frac{1}{\sqrt{n}} \begin{bmatrix} 1 \\ 1 \\ 1 \\ \vdots \\ 1 \end{bmatrix}.$$

As a final observation, we recall that our discussion of the power method in Section 7.3 presupposed that A had a set of n linearly independent eigenvectors. In Exercises 6 and 7, the reader is asked to show that this assumption is not necessary by using the fact that every $(n \times n)$ matrix A has a set of n linearly independent generalized eigenvectors. Thus the power method always converges in theory when $|\lambda_1| > |\lambda_2| \geq |\lambda_3| \geq \cdots \geq |\lambda_n|$ and when $a_1 \neq 0$; see (1) in Section 7.3. In practice, the restriction $a_1 \neq 0$ can be ignored, and the inverse power method satisfies the first restriction when sufficiently good estimates are available.

Figure 7.5 lists a FORTRAN program, Subroutine INVPOW, that implements the inverse power method as outlined above. The required inputs are the following:

A	An $(N \times N)$ matrix.
ALPHA	An estimate to an eigenvalue of A.
N	The size of A.
TOL	A convergence criterion. Control will return to the calling program when $|\beta_{k+1} - \beta_k| \leq \text{TOL}$.
MAXITR	The maximum number of iterations to be executed in INVPOW.
MAINDM	The declared dimension of the array containing A in the calling program.

The outputs from Subroutine INVPOW are the following:

EIGNVL	The machine estimate of the eigenvalue of A that is nearest ALPHA.
EIGNVC	The machine estimate of an eigenvector of A corresponding to EIGNVL.
IFLAG	A flag set equal to 1 if the convergence test is met. IFLAG is set equal to 2 or 3 to signal an abnormal return.

Figure 7.6 gives an example of a simple program that uses Subroutine INVPOW to find the eigenvalue of A that is nearest to α. As an example, we input the matrix

$$A = \begin{bmatrix} 4 & -5 & 0 & 3 \\ 0 & 4 & -3 & -5 \\ 5 & -3 & 4 & 0 \\ 3 & 0 & 5 & 4 \end{bmatrix},$$

```
      SUBROUTINE INVPOW(A,EIGNVC,ALPHA,EIGNVL,N,TOL,MAXITR,MAINDM,IFLAG)
      DIMENSION A(MAINDM,MAINDM),EIGNVC(MAINDM)
      DIMENSION AWORK(50,50),OLDVCT(50),REFVCT(50)
C
C     INITIALIZE THE ITERATION COUNTER AND SET UP THE WORKING
C     MATRIX AWORK=A-ALPHA*I.
C
      ITR=0
        DO 2 I=1,N
          DO 1 J=1,N
          AWORK(I,J)=A(I,J)
    1     CONTINUE
        AWORK(I,I)=AWORK(I,I)-ALPHA
    2   CONTINUE
C
C     CHOOSE A STARTING VECTOR OF NORM 1.
C
      AN=N
      Q=1./SQRT(AN)
        DO 3 I=1,N
        OLDVCT(I)=Q
    3   CONTINUE
C
C     CARRY OUT THE INVERSE POWER METHOD ITERATION.  REFVAL AND REFVCT
C     ARE REFINED ESTIMATES TO AN EIGENVALUE OF (A-ALPHA*I)**(-1)
C     AND A CORRESPONDING EIGENVECTOR.
C
      OLDVAL=ALPHA
    4 ITR=ITR+1
      CALL GAUSS(AWORK,OLDVCT,REFVCT,N,50,IERROR,RNORM)
      IF(IERROR.EQ.2)  GO TO 10
C
C     CALCULATE THE RAYLEIGH QUOTIENT AND THE NORM OF REFVCT.
C
      RAYQTN=0.
      SCALSQ=0.
        DO 5 I=1,N
        RAYQTN=RAYQTN+OLDVCT(I)*REFVCT(I)
        SCALSQ=SCALSQ+REFVCT(I)**2
    5   CONTINUE
C
C     CALCULATE A REFINED ESTIMATE OF AN EIGENVALUE OF A AND
C     SCALE REFVCT TO HAVE NORM 1.
C
      REFVAL=(1./RAYQTN)+ALPHA
      SCALE=SQRT(SCALSQ)
        DO 6 I=1,N
        REFVCT(I)=REFVCT(I)/SCALE
    6   CONTINUE
C
C     TEST FOR CONVERGENCE AND EXCESSIVE ITERATIONS.
C
      IF(ABS(OLDVAL-REFVAL).LE.TOL)  GO TO 8
      IF(ITR.GE.MAXITR)  GO TO 11
      OLDVAL=REFVAL
        DO 7 I=1,N
        OLDVCT(I)=REFVCT(I)
    7   CONTINUE
      GO TO 4
C
C     SUCCESSFUL RETURN.
C
    8 IFLAG=1
      EIGNVL=REFVAL
        DO 9 I=1,N
        EIGNVC(I)=REFVCT(I)
    9   CONTINUE
      RETURN
C
C     ABNORMAL RETURN --- AWORK MAY BE SINGULAR.
C
   10 IFLAG=2
      RETURN
C
C     ABNORMAL RETURN --- LIMIT ON NUMBER OF ITERATIONS IS EXCEEDED.
C
   11 IFLAG=3
      RETURN
      END
```

Figure 7.5

```
      DIMENSION A(20,20),EIGNVC(20)
      MAINDM=20
    1 READ 100,N
      IF(N.LT.2)  STOP
         DO 2 I=1,N
         READ 101,(A(I,J),J=1,N)
    2    CONTINUE
      READ 101,ALPHA
      TOL=.1E-06
      MAXITR=10
      CALL INVPOW(A,EIGNVC,ALPHA,EIGNVL,N,TOL,MAXITR,MAINDM,IFLAG)
      PRINT 102,IFLAG,EIGNVL
      PRINT 103,(EIGNVC(K),K=1,N)
      GO TO 1
  100 FORMAT(I2)
  101 FORMAT(10F4.0)
  102 FORMAT(1H0,6H FLAG=,I3,5X,36HREFINED ESTIMATE OF AN EIGENVALUE IS,
     1E15.5)
  103 FORMAT(1H0,30HA CORRESPONDING EIGENVECTOR IS,/,6E18.5)
      END
```

Figure 7.6

which has eigenvalues

$$\lambda_1 = 12, \qquad \lambda_2 = 1 + 5i,$$
$$\lambda_3 = 1 - 5i, \qquad \lambda_4 = 2$$

and corresponding eigenvectors

$$\mathbf{u}_1 = \begin{bmatrix} 1 \\ -1 \\ 1 \\ 1 \end{bmatrix}, \qquad \mathbf{u}_2 = \begin{bmatrix} 1 \\ -i \\ -i \\ -1 \end{bmatrix}, \qquad \mathbf{u}_3 = \begin{bmatrix} 1 \\ i \\ i \\ -1 \end{bmatrix}, \qquad \mathbf{u}_4 = \begin{bmatrix} 1 \\ 1 \\ -1 \\ 1 \end{bmatrix}.$$

We input $\alpha = 2.1$ as an estimate to λ_4, and Subroutine INVPOW returned $\lambda = .20000E\ 01$ and an estimated eigenvector

$$\begin{bmatrix} -.50000 \\ -.50000 \\ .50000 \\ -.50000 \end{bmatrix}.$$

Exercises 7.4

1. Use Subroutine INVPOW to refine the given eigenvalue estimates of A, where A is the (4×4) matrix at the end of this section. Include a print statement in INVPOW after the convergence test to print the array OLDVCT and the constant OLDVAL. Use the following values for α:

a) $\alpha = 2.1$ b) $\alpha = 2.01$
c) $\alpha = 12.1$ d) $\alpha = 12.01$.

2. Use INVPOW to refine the eigenvalue estimates of

$$A = \begin{bmatrix} 33 & 16 & 72 \\ -24 & -10 & -57 \\ -8 & -4 & -17 \end{bmatrix}.$$

a) $\alpha = 1.1$ b) $\alpha = 1.4$
c) $\alpha = 1.9$ d) $\alpha = 3.02$

3. The eigenvalues of

$$A = \begin{bmatrix} 6 & 4 & 4 & 1 \\ 4 & 6 & 1 & 4 \\ 4 & 1 & 6 & 4 \\ 1 & 4 & 4 & 6 \end{bmatrix}$$

are $\lambda_1 = 15$, $\lambda_2 = \lambda_3 = 5$, and $\lambda_4 = -1$. An eigenvector corresponding to λ_1 is

$$\mathbf{u}_1 = \begin{bmatrix} .5 \\ .5 \\ .5 \\ .5 \end{bmatrix}.$$

Since INVPOW uses $\mathbf{u}_1$ as its starting vector, we can expect this "textbook" eigenvalue problem to cause the subroutine difficulty. Try estimates of $\alpha = 5.1$ and $\alpha = -.9$ in INVPOW. If the program converges to $\lambda = 15$, modify the starting vector in INVPOW by setting the initial vector, OLDVCT, equal to a unit vector $\mathbf{e}_j$ (this can be done in the loop that terminates at the statement 3 CONTINUE). Run the program again with the estimates above.

4. The eigenvalues of

$$A = \begin{bmatrix} 2 & 1 & 3 & 4 \\ 1 & -3 & 1 & 5 \\ 3 & 1 & 6 & -2 \\ 4 & 5 & -2 & -1 \end{bmatrix}$$

are $\lambda = 7.9329$, $\lambda = -8.0286$, $\lambda = 5.6689$, and $\lambda = -1.5732$ to the places given. Run INVPOW with $\alpha = 8$, $\alpha = -8$, $\alpha = 5.5$, and $\alpha = -1.6$.

5. Write a calling program for INVPOW that checks the output by calculating $\|A\mathbf{u} - \lambda\mathbf{u}\|$, where λ and $\mathbf{u}$ are the eigenvalue and eigenvector estimates returned by INVPOW.

6. In Chapter 6, we saw that any $(n \times n)$ matrix A has a set of n linearly independent eigenvectors and generalized eigenvectors. As a simple example, suppose that A is a (3×3) matrix with two distinct eigenvalues, λ and γ, with corresponding eigenvectors $\mathbf{u}$ and $\mathbf{v}$. Thus

$$A\mathbf{u} = \lambda\mathbf{u}, \qquad \mathbf{u} \neq \mathbf{0},$$
$$A\mathbf{v} = \gamma\mathbf{v}, \qquad \mathbf{v} \neq \mathbf{0}.$$

Suppose that $\mathbf{q}$ is a generalized eigenvector corresponding to λ so that
$$(A - \lambda I)\mathbf{q} = \mathbf{u}, \qquad \mathbf{q} \neq \mathbf{0}.$$
Prove that $A^k\mathbf{q} = \lambda^k\mathbf{q} + k\lambda^{k-1}\mathbf{u}, k = 1, 2, \ldots$.

7. With the same hypotheses as in Exercise 6, suppose that $|\lambda| > |\gamma|$; and suppose $\mathbf{x}_0$ is a starting vector such that

$$\mathbf{x}_0 = a_1\mathbf{q} + a_2\mathbf{u} + a_3\mathbf{v}.$$

Suppose that $\mathbf{x}_k = A\mathbf{x}_{k-1}$, $k = 1, 2, \ldots$; and suppose $\mathbf{w}$ is a fixed vector such that either $\mathbf{w}^T\mathbf{q} \neq 0$ or $\mathbf{w}^T\mathbf{u} \neq 0$. Prove that

$$\lim_{k \to \infty} \frac{\mathbf{w}^T\mathbf{x}_{k+1}}{\mathbf{w}^T\mathbf{x}_k} = \lambda.$$

8. Modify Subroutine INVPOW so that complex estimates α can be employed. Using the complex version of GAUSS, test your modification on the matrix in Exercise 1 with $\alpha = 1.1 + 4.9i$ as an estimate.

7.5 Reduction to Hessenberg Form

In Section 6.5, we saw how to reduce an arbitrary $(n \times n)$ matrix A to Hessenberg form. That is, we saw how to construct Householder matrices $Q_1, Q_2, \ldots, Q_{n-2}$ such that

$$QAQ^{-1} = H, \tag{1}$$

where H is a Hessenberg matrix, and

$$Q = Q_{n-2} \cdots Q_2 Q_1. \tag{2}$$

Furthermore, we saw that if A is symmetric, $A^T = A$, then H is also symmetric, and hence H is tridiagonal. For example, a (5×5) symmetric Hessenberg matrix has the form

$$H = \begin{bmatrix} d_1 & b_1 & 0 & 0 & 0 \\ b_1 & d_2 & b_2 & 0 & 0 \\ 0 & b_2 & d_3 & b_3 & 0 \\ 0 & 0 & b_3 & d_4 & b_4 \\ 0 & 0 & 0 & b_4 & d_5 \end{bmatrix}. \tag{3}$$

Note that if any subdiagonal entry of H in (3) is zero, then H is a block diagonal matrix where each block is itself tridiagonal.

In this short section, we present the FORTRAN code for Subroutine SIMTRN, which will reduce an $(n \times n)$ matrix A to Hessenberg form by using Householder transformations.

Subroutine SIMTRN, listed in Fig. 7.7, uses Householder transformations to produce a Hessenberg matrix H, which is similar to A. Subroutine SIMTRN

```
      SUBROUTINE SIMTRN(A,H,N,MAINDM)
      DIMENSION A(MAINDM,MAINDM),H(MAINDM,MAINDM)
      DIMENSION U(50)
      NM2=N-2
C
C   INITIALIZE H--H WILL CONTAIN THE MACHINE ESTIMATE TO A HESSENBERG
C   MATRIX SIMILAR TO A WHEN CONTROL IS RETURNED TO THE CALLING PROGRAM.
C
      DO 2 I=1,N
        DO 1 J=1,N
        H(I,J)=A(I,J)
    1   CONTINUE
    2 CONTINUE
C
C   USE HOUSEHOLDER TRANSFORMATIONS TO ZERO ENTRIES BELOW THE
C   SUBDIAGONAL IN COLUMN J, J=1,2,...,N-2.
C
      DO 6 J=1,NM2
      JP1=J+1
      JP2=J+2
C
C   CALCULATE THE VECTOR U IN ALGORITHM 2, SECTION 6.5, WHERE K=J+1.
C
      SSQ=0.
        DO 3 I=JP1,N
        SSQ=SSQ+H(I,J)**2
    3   CONTINUE
      IF(SSQ.EQ.0.)  GO TO 6
      S=SQRT(SSQ)
      IF(H(JP1,J).GE.0.)  S=-S
      B=1./(SSQ-H(JP1,J)*S)
        DO 4 L=1,J
        U(L)=0.
    4   CONTINUE
      U(JP1)=H(JP1,J)-S
      H(JP1,J)=S
        DO 5 L=JP2,N
        U(L)=H(L,J)
        H(L,J)=0.
    5   CONTINUE
C
C   CALCULATE QHQ, WHERE Q IS THE HOUSEHOLDER TRANSFORMATION
C   DEFINED BY THE VECTOR U.
C
      CALL PRODCT(H,U,B,JP1,N,MAINDM)
    6 CONTINUE
      RETURN
      END
```

Figure 7.7
Subroutine SIMTRN

proceeds in stages in the obvious fashion. That is, given A, SIMTRN calculates a Householder transformation Q_1 such that $H_1 = Q_1 A Q_1$ has zeros in the $(3, 1)$, $(4, 1), \ldots, (N, 1)$ positions. Given H_1, SIMTRN finds Q_2 so that $H_2 = Q_2 H_1 Q_2$ has zeros in the $(4, 2), (5, 2), \ldots, (N, 2)$ positions, and so on. For clarity, SIMTRN calls Subroutine PRODCT to calculate $Q_i H_{i-1} Q_i$. In addition, Subroutine PRODCT calls Subroutine SCPROD to calculate the required scalar products $\mathbf{u}^T \mathbf{v}$. The necessary inputs are these:

A   An $(N \times N)$ matrix.
N   The size of A.
MAINDM The declared dimension of the array containing A in the calling program.

The output from Subroutine SIMTRN is this:

H   A machine estimate to an $(N \times N)$ Hessenberg matrix that is similar to A.

The two subroutines called by SIMTRN and a simple program that uses SIMTRN to transform A to Hessenberg form are listed in Figs. 7.8 and 7.9.

```
      SUBROUTINE PRODCT(H,U,B,JP1,N,MAINDM)
      DIMENSION H(MAINDM,MAINDM)
      DIMENSION U(50),V(50)
C
C     CALCULATE QH.
C
      DO 3 K=JP1,N
      DO 1 I=1,N
      V(I)=H(I,K)
   1  CONTINUE
      CALL SCPROD(U,V,UDOTV,N)
      DO 2 I=JP1,N
      H(I,K)=H(I,K)-UDOTV*B*U(I)
   2  CONTINUE
   3  CONTINUE
C
C     CALCULATE (QH)Q.
C
      DO 6 K=1,N
      DO 4 I=1,N
      V(I)=H(K,I)
   4  CONTINUE
      CALL SCPROD(U,V,UDOTV,N)
      DO 5 I=JP1,N
      H(K,I)=H(K,I)-UDOTV*B*U(I)
   5  CONTINUE
   6  CONTINUE
      RETURN
      END
      SUBROUTINE SCPROD(U,V,UDOTV,N)
      DIMENSION U(50),V(50)
C
C     THIS SUBPROGRAM CALCULATES UDOTV, WHERE UDOTV IS THE
C     SCALAR PRODUCT OF TWO N-DIMENSIONAL VECTORS, U AND V.
C
      UDOTV=0.
      DO 1 I=1,N
      UDOTV=UDOTV+U(I)*V(I)
   1  CONTINUE
      RETURN
      END
```

Figure 7.8
Subroutines PRODCT and SCPROD

```
      DIMENSION A(20,20),H(20,20)
      MAINDM=20
   1  READ 100,N
      IF(N.LT.3)   STOP
      DO 2 I=1,N
      READ 101,(A(I,J),J=1,N)
   2  CONTINUE
      CALL SIMTRN(A,H,N,MAINDM)
      DO 3 I=1,N
      PRINT 102,(H(I,J),J=1,N)
   3  CONTINUE
      GO TO 1
 100  FORMAT(I2)
 101  FORMAT(10F4.0)
 102  FORMAT(1H0,8E12.4)
      END
```

Figure 7.9

EXAMPLE 1   The (4 × 4) symmetric matrix

$$A = \begin{bmatrix} 6 & 4 & 4 & 1 \\ 4 & 6 & 1 & 4 \\ 4 & 1 & 6 & 4 \\ 1 & 4 & 4 & 6 \end{bmatrix}$$

has eigenvalues $\lambda_1 = 15$, $\lambda_2 = \lambda_3 = 5$, and $\lambda_4 = -1$. Corresponding eigenvectors are

$$\mathbf{u}_1 = \begin{bmatrix} 1 \\ 1 \\ 1 \\ 1 \end{bmatrix}, \quad \mathbf{u}_2 = \mathbf{u}_3 = \begin{bmatrix} -1 \\ -1 \\ 1 \\ 1 \end{bmatrix}, \quad \text{and} \quad \mathbf{u}_4 = \begin{bmatrix} 1 \\ -1 \\ -1 \\ 1 \end{bmatrix}.$$

This matrix was read in by the program in Fig. 7.9 and produced this output:

$$H = \begin{bmatrix} 0.6000\text{E }01 & -0.5745\text{E }01 & -0.2271\text{E}-17 & 0.2225\text{E}-15 \\ -0.5745\text{E }01 & 0.8909\text{E }01 & -0.5143\text{E }01 & -0.6661\text{E}-15 \\ 0.0 & -0.5143\text{E }01 & 0.4091\text{E }01 & 0.1998\text{E}-14 \\ 0.0 & 0.0 & 0.4441\text{E}-15 & 0.5000\text{E }01 \end{bmatrix}.$$

Although the input matrix A is symmetric, the output H is not quite symmetric and reflects the effect of roundoff error. (We expected H to be tridiagonal, but h_{13}, h_{14}, and h_{24} are not quite zero. The entries below the subdiagonal were set to zero by SIMTRN.) Interpreting the small entries as being zero, the machine approximation to a Hessenberg matrix that is similar to A is to four places:

$$H = \begin{bmatrix} 6.000 & -5.745 & 0 & 0 \\ -5.745 & 8.909 & -5.143 & 0 \\ 0 & -5.143 & 4.091 & 0 \\ 0 & 0 & 0 & 5.000 \end{bmatrix}.$$

Note that H is block diagonal, and we see that $\lambda = 5.000$ is probably one of the eigenvalues of A. We will return to this example in Section 7.6.

Exercises 7.5

1. Use Subroutine SIMTRN to reduce the following matrices to Hessenberg form:

$$A = \begin{bmatrix} 1 & 2 & 1 & 3 \\ 2 & 4 & 1 & 7 \\ 1 & 1 & 0 & 1 \\ 3 & 7 & 1 & 3 \end{bmatrix}, \quad B = \begin{bmatrix} 2 & 0 & 1 & 0 \\ 0 & 2 & 3 & 1 \\ 1 & 3 & 5 & 2 \\ 0 & 1 & 2 & 1 \end{bmatrix},$$

$$C = \begin{bmatrix} 1 & 1 & 1 & 1 \\ 1 & 1 & 1 & 1 \\ 1 & 1 & 1 & 1 \\ 1 & 1 & 1 & 1 \end{bmatrix}$$

2. Use SIMTRN to reduce the (12×12) matrix A to Hessenberg form:

$$A = \begin{bmatrix} 12 & 11 & 10 & 9 & \cdots & 2 & 1 \\ 11 & 11 & 10 & 9 & \cdots & 2 & 1 \\ 10 & 10 & 10 & 9 & \cdots & 2 & 1 \\ \vdots & & & & & & \vdots \\ 2 & 2 & 2 & 2 & \cdots & 2 & 1 \\ 1 & 1 & 1 & 1 & \cdots & 1 & 1 \end{bmatrix}$$

3. Reduce the following matrices to Hessenberg form using SIMTRN:

$$A = \begin{bmatrix} 1 & 2 & 1 & 3 \\ 0 & 4 & 1 & 5 \\ 2 & 1 & 1 & 2 \\ 0 & 1 & 4 & 1 \end{bmatrix}, \quad B = \begin{bmatrix} 3 & 6 & 4 & 2 \\ 1 & 5 & 3 & 0 \\ 2 & 1 & 8 & 4 \\ 1 & 3 & 9 & 6 \end{bmatrix}$$

4. Write a subroutine that calculates the characteristic polynomial of an unreduced Hessenberg matrix H. Test your program on the Hessenberg matrices obtained in Exercises 1, 2, and 3.

5. Most computing facilities have a routine that finds the roots of $p(t) = 0$, where $p(t)$ is a polynomial. Using SIMTRN, the subroutine from Exercise 4, and the root finder, write a program that estimates the eigenvalues of the matrices in Exercises 1, 2, and 3. Then use INVPOW to refine these estimates in the *original* matrices. Check your results by calculating $\|A\mathbf{u} - \lambda\mathbf{u}\|$, where λ and $\mathbf{u}$ are the returned eigenvalue and eigenvector. (*Caution:* Any procedure that estimates eigenvalues by finding the roots of the characteristic equation has a number of computational pitfalls. In particular, although the zeros of the calculated characteristic polynomial are refined to be accurate eigenvalues by INVPOW (and these answers can be checked directly), we cannot be certain that we have found all the eigenvalues unless we obtain n distinct eigenvalues when A is $(n \times n)$.)

7.6   Estimating the Eigenvalues of Hessenberg Matrices

Given an $(n \times n)$ matrix A, we have shown how Householder transformations can be used to reduce A to a Hessenberg matrix H. Now we come to the hard part—estimating the eigenvalues of H. Once we obtain estimates $\alpha_1, \alpha_2, \ldots, \alpha_n$ to the eigenvalues of H, we can refine them with the inverse power method and in the process find the corresponding eigenvectors for A. If we are confronted with the problem of finding the eigenvalues of H, our first impulse is to calculate the characteristic polynomial, $p(t)$, for H and then find the roots of $p(t) = 0$. This impulse has both reasonable and unreasonable aspects in terms of computation. The bad aspect is that because of roundoff errors, we cannot expect to find the coefficients of $p(t)$ exactly. Suppose, for example, that the characteristic polynomial of H is $p(t)$, but our machine calculations produce a slightly different polynomial $q(t)$:

$$\begin{aligned} p(t) &= t^n + a_{n-1}t^{n-1} + \cdots + a_1 t + a_0 \\ q(t) &= t^n + b_{n-1}t^{n-1} + \cdots + b_1 t + b_0. \end{aligned} \tag{1}$$

It is an unfortunate fact that even though the coefficients of $p(t)$ and $q(t)$ might be quite close, the roots of $p(t) = 0$ and $q(t) = 0$ may not be so close. We already have an illustration at hand in Example 1 of Section 7.3, where

$$p(t) = (1 - t)^{10}$$
$$q(t) = (1 - t)^{10} - 2^{-10}.$$

In terms of (1), the coefficients of $p(t)$ and $q(t)$ satisfy

$$b_9 = a_9, \qquad b_8 = a_8, \qquad \ldots, \qquad b_1 = a_1, \qquad b_0 = a_0 - 2^{-10};$$

so $p(t)$ and $q(t)$ are polynomials with nearly equal coefficients. However, $t = 1/2$ and $t = 3/2$ are roots of $q(t) = 0$, whereas $t = 1$ is the only root of $p(t) = 0$. [The other eight roots of $q(t) = 0$ are complex. Of course, $p(1/2) = p(3/2) = 2^{-10}$, which is quite small; but still $t = 1/2$ and $t = 3/2$ are not roots of $p(t) = 0$.]

The problem of estimating eigenvalues of Hessenberg matrices can be split in terms of computational difficulty into two classes—the eigenvalue problem is relatively easy when the Hessenberg matrix is symmetric, but the problem may be hard for a nonsymmetric matrix. We discuss the easy case first. Suppose H is an $(n \times n)$ symmetric Hessenberg matrix. Then H is tridiagonal and has the form

$$H = \begin{bmatrix} d_1 & b_1 & 0 & 0 & \cdots & 0 & 0 \\ b_1 & d_2 & b_2 & 0 & \cdots & 0 & 0 \\ 0 & b_2 & d_3 & b_3 & \cdots & 0 & 0 \\ 0 & 0 & b_3 & d_4 & \cdots & 0 & 0 \\ \vdots & & & & & & \vdots \\ 0 & 0 & 0 & 0 & \cdots & b_{n-1} & d_n \end{bmatrix}.$$

Suppose we define the sequence of polynomials

$$\begin{aligned}
p_0(t) &= 1 \\
p_1(t) &= d_1 - t \\
p_2(t) &= (d_2 - t)p_1(t) - b_1^2 p_0(t) \\
&\vdots \\
p_i(t) &= (d_i - t)p_{i-1}(t) - b_{i-1}^2 p_{i-2}(t) \\
&\vdots \\
p_n(t) &= (d_n - t)p_{n-1}(t) - b_n^2 p_{n-2}(t).
\end{aligned} \qquad (2)$$

It is an easy exercise to show (Exercise 3) that $p_n(t)$ is the characteristic polynomial for H.

If the subdiagonal entries $b_1, b_2, \ldots, b_{n-1}$ are all nonzero, then a very effective algorithm due to Givens can be used to isolate the roots of $p_n(t) = 0$. The algorithm proceeds as follows:

1. Let c be some real number.
2. Calculate the numbers $p_0(c), p_1(c), \ldots, p_n(c)$.
3. Let $N(c)$ be the number of agreements in sign in the sequence $p_0(c)$, $p_1(c), \ldots, p_n(c)$.
4. $N(c)$ is equal to the number of roots of $p_n(t) = 0$ that are in the interval $[c, \infty)$.

In the event that $p_k(c) = 0$ for some k, we take the sign of $p_k(c)$ to be that of $p_{k-1}(c)$. [Note that two successive terms in (2) cannot both vanish at $t = c$ unless $p_i(c) = 0$ for all i, $0 \leq i \leq n$. Thus if $p_k(c) = 0$, then $p_{k-1}(c) \neq 0$; and $p_{k-1}(c)$ has a well-defined sign.]

It is not too hard to show that assertion 4 in the algorithm is true, but it is not appropriate to include the proof here. As an example, if $n = 4$ and if $\{p_0(c), p_1(c), p_2(c), p_3(c), p_4(c)\}$ has the sign pattern

$$\{+, +, +, -, -\},$$

then $N(c) = 3$; there are three roots of $P_4(t) = 0$ in the interval $[c, \infty)$. Similarly, the pattern $\{+, -, +, -, -\}$ leads to $N(c) = 1$, and the pattern $\{+, -, 0, +, +\}$ gives $N(c) = 2$.

To use the Givens algorithm for computational purposes, we would first determine an interval $[a, b]$ that contains all the roots of $p_n(t) = 0$; that is, $[a, b]$ contains all the eigenvalues of H. (Since H is symmetric, all the eigenvalues of H are real.) Next, let c be the midpoint of $[a, b]$. If $N(c) > N(b)$, then there is at least one eigenvalue in $[c, b]$. Let d be the midpoint of $[c, b]$. If $N(d) > N(b)$, there is at least one eigenvalue in $[d, b]$; on the other hand, if $N(d) = N(b)$, then any eigenvalue in $[c, b]$ must be in $[c, d]$. In this fashion, by repeatedly halving and testing subintervals we can determine a small subinterval $[r, s]$ that contains $N(r) - N(s)$ eigenvalues of H. The remaining eigenvalues of H must be in $[a, r]$, and the bisection procedure described above can be applied to the interval $[a, r]$.

This process can be terminated when we have determined k small subintervals, $I_1, I_2, \ldots, I_k$, whose union contains all the eigenvalues of H. The midpoint of an interval, I_j, will serve as an estimate, α_j, to an eigenvalue of H. This estimate, α_j, can then be passed to the inverse power method and corrected to be a good estimate of an eigenvalue of A.

Several points must be made about the practical implementation of the Givens algorithm. First of all, we need not demand that the intervals I_j be terribly small, since the midpoint of I_j will be used only as an initial guess for an eigenvalue of A. Second, we note that the polynomials $p_i(t)$ in (2) are not actually calculated, since all we need is the value of $p_i(c)$ for various numbers c, $i = 0, 1, \ldots, n$. To calculate $\{p_0(c), p_1(c), \ldots, p_n(c)\}$, we merely compute the numbers

$$P_0 = 1$$
$$P_1 = d_1 - c$$
$$P_i = (d_i - c)P_{i-1} - b_{i-1}^2 P_{i-2}$$
$$\vdots$$

for $i = 2, 3, \ldots, n$; and we note that the number P_k is in fact $p_k(c)$. Finally, to start the algorithm, we need an initial interval $[a, b]$, which we know contains all the eigenvalues of H. Now if $C = (c_{ij})$ is an $(n \times n)$ real matrix, then it can be shown (see Exercise 4) that every eigenvalue λ of C satisfies the inequality

$$|\lambda| \leq \|C\|,$$

where

$$\|C\| = \left(\sum_{i=1}^{n} \sum_{j=1}^{n} c_{ij}^2 \right)^{1/2}. \tag{3}$$

Thus for the symmetric Hessenberg matrix H, we know that every eigenvalue of H is in the interval $[-\|H\|, \|H\|]$, where

$$\|H\| = \left(\sum_{i=1}^{n} d_i^2 + 2 \sum_{i=1}^{n-1} b_i^2 \right)^{1/2}.$$

Subroutine GIVENS listed in Fig. 7.10 applies the Givens algorithm to a symmetric Hessenberg matrix H. The required inputs are the following:

D	An (N × 1) array containing the diagonal entries of H.
B	An ((N − 1) × 1) array containing the subdiagonal entries of H. These entries are all presumed to be nonzero.
RADIUS	A scalar defining the desired radius of a subinterval containing an eigenvalue of H.
N	The size of H.
MAINDM	The declared dimension of the array containing H in the calling program.

The outputs from Subroutine GIVENS are the following:

ESTEIG	An array containing machine estimates to the eigenvalues of H. If ESTEIG $(k) = \alpha_k$, then there is at least one eigenvalue of H in the interval $[\alpha_k - r, \alpha_k + r]$, where $r = $ RADIUS.
NEIG	An array of integers. If NEIG(K) = j, there are j eigenvalues of H in $[\alpha_k - r, \alpha_k + r]$.
LSTEIG	An integer set equal to the number of subintervals found.

In summary,

$$\text{NEIG}(1) + \text{NEIG}(2) + \cdots + \text{NEIG}(\text{LSTEIG}) = \text{N},$$

and NEIG(K) eigenvalues of H are between ESTEIG(K) − RADIUS and ESTEIG(K) + RADIUS. Subroutine GIVENS calls a subprogram, Subroutine SGNCTR, which evaluates the quantity $N(c)$ required by the Givens algorithm; this subprogram is also listed in Fig. 7.10.

Figure 7.11 lists a simple program that uses Subroutine GIVENS. The program reads in a *symmetric* matrix A, uses SIMTRN to reduce A to a symmetric Hessenberg matrix, and then calls GIVENS to isolate the eigenvalues of H. When the program in Fig. 7.11 was given as input the (4 × 4) matrix A from Example 1 of Section 7.5, the program found the following:

Number of eigenvalues within .1 of 15.02 is 1.

Number of eigenvalues within .1 of 5.022 is 2.

Number of eigenvalues within .1 of −.9585 is 1.

As we noted in Example 1, the eigenvalues of A are $\lambda_1 = 15$, $\lambda_2 = \lambda_3 = 5$, and $\lambda_4 = 1$. The combination of Householder transformations and the Givens algorithm has worked well on this example.

The inverse power method can be incorporated into the program in Fig. 7.11 to produce a complete package for finding the eigenvalues and eigenvectors of a symmetric matrix. Such a program is listed in Fig. 7.12.

```
      SUBROUTINE GIVENS(D,B,ESTEIG,NEIG,RADIUS,N,MAINDM,LSTEIG)
C
C  INITIALIZATION PHASE--K IS THE INDEX OF A SUBINTERVAL OF LENGTH
C  2*RADIUS WHICH CONTAINS NEIG(K) EIGENVALUES OF H.  ALL EIGENVALUES OF
C  H ARE BETWEEN A1=-HNORM AND A2=HNORM.  NTOGO IS THE NUMBER OF
C  EIGENVALUES YET TO BE FOUND AT STAGE K.  SGNCTR IS A SUBROUTINE THAT
C  CALCULATES THE NUMBER OF SIGN AGREEMENTS, N(C).
C
      DIMENSION D(MAINDM),B(MAINDM),ESTEIG(MAINDM),NEIG(MAINDM)
      NM1=N-1
      K=1
      NTOGO=N
      HNRMSQ=0.
        DO 1 I=1,NM1
        HNRMSQ=HNRMSQ+2.*B(I)**2+D(I)**2
    1   CONTINUE
      HNRMSQ=HNRMSQ+D(N)**2
      HNORM=SQRT(HNRMSQ)
      A1=-HNORM
      A2=HNORM
      CALL SGNCTR(D,B,A2,N,NOFA2,MAINDM)
C
C  BASIC LOOP--BEGINS WITH AN INTERVAL [A1,A2] AND DETERMINES A
C  SUBINTERVAL OF LENGTH 2*RADIUS CONTAINING NEIG(K) EIGENVALUES OF
C  H, FOR K=1,2,...,LSTEIG.  NBISCT IS THE NUMBER OF BISECTIONS
C  NECESSARY TO PRODUCE THE K-TH SUBINTERVAL.
C
    2 NBISCT=ALOG(2.*(A2-A1)/RADIUS)/ALOG(2.)
      ICTR=0
    3 EVALPT=(A1+A2)/2.
      CALL SGNCTR(D,B,EVALPT,N,NOFEPT,MAINDM)
      IF(NOFEPT.EQ.NOFA2)  GO TO 4
      A1=EVALPT
      GO TO 5
    4 A2=EVALPT
    5 ICTR=ICTR+1
      IF(ICTR.LT.NBISCT)  GO TO 3
C
C  SET K-TH ESTIMATE OF AN EIGENVALUE EQUAL TO MID-POINT OF
C  K-TH SUBINTERVAL, CALCULATE NEIG(K) AND RETURN IF ALL THE
C  EIGENVALUES OF H HAVE BEEN ISOLATED.
C
      ESTEIG(K)=(A1+A2)/2.
      CALL SGNCTR(D,B,A1,N,NOFA1,MAINDM)
      CALL SGNCTR(D,B,A2,N,NOFA2,MAINDM)
      NEIG(K)=NOFA1-NOFA2
      LSTEIG=K
      NTOGO=NTOGO-NEIG(K)
      IF(NTOGO.LE.0)  RETURN
      K=K+1
      A2=A1
      NOFA2=NOFA1
      A1=-HNORM
      GO TO 2
      END
      SUBROUTINE SGNCTR(D,B,EVALPT,N,NOFEPT,MAINDM)
      DIMENSION D(MAINDM),B(MAINDM)
C
C  THIS SUBPROGRAM CALCULATES THE NUMBER OF SIGN AGREEMENTS, N(C),
C  AT C=EVALPT.  THE NUMBER N(EVALPT) IS RETURNED AS NOFEPT.
C
      NOFEPT=0
      PIM2=1.
      PIM1=D(1)-EVALPT
      IF(PIM2*PIM1.GE.0.)  NOFEPT=1
        DO 1 I=2,N
        PI=(D(I)-EVALPT)*PIM1-B(I-1)**2*PIM2
        IF(PI*PIM1.GT.0.)  NOFEPT=NOFEPT+1
        IF(PI.EQ.0.)  NOFEPT=NOFEPT+1
        PIM2=PIM1
        PIM1=PI
    1   CONTINUE
      RETURN
      END
```

Figure 7.10
Subroutine GIVENS and SGNCTR

```
      DIMENSION A(20,20),H(20,20),D(20),B(20),ESTEIG(20),NEIG(20)
      MAINDM=20
    1 READ 100,N
      IF(N.LT.3)  STOP
         DO 2 I=1,N
         READ 101,(A(I,J),J=1,N)
    2    CONTINUE
      CALL SIMTRN(A,H,N,MAINDM)
         DO 3 I=1,N
         PRINT 102,(H(I,J),J=1,N)
    3    CONTINUE
      NM1=N-1
         DO 4 I=1,NM1
         D(I)=H(I,I)
         B(I)=H(I+1,I)
         IF(B(I).EQ.0.)  GO TO 1
    4    CONTINUE
      D(N)=H(N,N)
      RADIUS=.1
      CALL GIVENS(D,B,ESTEIG,NEIG,RADIUS,N,MAINDM,LSTEIG)
         DO 5 I=1,LSTEIG
         PRINT 103,RADIUS,ESTEIG(I),NEIG(I)
    5    CONTINUE
      GO TO 1
  100 FORMAT(I2)
  101 FORMAT(10F4.0)
  102 FORMAT(1H0,8E12.4)
  103 FORMAT(1H0,28HNUMBER OF EIGENVALUES WITHIN,F8.4,2X,2HOF,
     1E16.4,4H  IS,I3)
      END
```

Figure 7.11
A program to isolate the eigenvalues of a symmetric matrix A

```
      DIMENSION A(20,20),H(20,20),D(20),B(20),ESTEIG(20),NEIG(20)
      DIMENSION EIGNVC(20)
      MAINDM=20
    1 READ 100,N
      IF(N.LT.3)  STOP
         DO 2 I=1,N
         READ 101,(A(I,J),J=1,N)
    2    CONTINUE
      CALL SIMTRN(A,H,N,MAINDM)
         DO 3 I=1,N
         PRINT 102,(H(I,J),J=1,N)
    3    CONTINUE
      NM1=N-1
         DO 4 I=1,NM1
         D(I)=H(I,I)
         B(I)=H(I+1,I)
         IF(B(I).EQ.0.)  GO TO 1
    4    CONTINUE
      D(N)=H(N,N)
      RADIUS=.1
      CALL GIVENS(D,B,ESTEIG,NEIG,RADIUS,N,MAINDM,LSTEIG)
         DO 5 I=1,LSTEIG
         PRINT 103,RADIUS,ESTEIG(I),NEIG(I)
    5    CONTINUE
      TOL=.1E-06
      MAXITR=10
         DO 6 I=1,LSTEIG
         ALPHA=ESTEIG(I)
         CALL INVPOW(A,EIGNVC,ALPHA,EIGNVL,N,TOL,MAXITR,MAINDM,IFLAG)
         PRINT 104,IFLAG,EIGNVL
         PRINT 105,(EIGNVC(K),K=1,N)
    6    CONTINUE
      GO TO 1
  100 FORMAT(I2)
  101 FORMAT(10F4.0)
  102 FORMAT(1H0,8E12.4)
  103 FORMAT(1H0,28HNUMBER OF EIGENVALUES WITHIN,F8.4,2X,2HOF,
     1E16.4,4H  IS,I3)
  104 FORMAT(1H0,6H FLAG=,I3,5X,36HREFINED ESTIMATE OF AN EIGENVALUE IS,
     1E15.5)
  105 FORMAT(1H0,30HA CORRESPONDING EIGENVECTOR IS,/,6E18.5)
      END
```

Figure 7.12
A program to find the eigenvalues and the eigenvectors of a
symmetric matrix A

EXAMPLE 1 As a simple test of the program in Fig. 7.12, we input the matrix

$$A = \begin{bmatrix} 2 & 4 & -6 \\ 4 & 2 & -6 \\ -6 & -6 & -15 \end{bmatrix},$$

which has eigenvalues $\lambda_1 = -18$, $\lambda_2 = 9$, and $\lambda_3 = -2$ and corresponding eigenvectors

$$\mathbf{u}_1 = \begin{bmatrix} 1 \\ 1 \\ 4 \end{bmatrix}, \quad \mathbf{u}_2 = \begin{bmatrix} 2 \\ 2 \\ -1 \end{bmatrix}, \quad \text{and} \quad \mathbf{u}_3 = \begin{bmatrix} 1 \\ -1 \\ 0 \end{bmatrix}.$$

The program returned eigenvalues correct to the five places printed and eigenvectors (of length 1):

$$\mathbf{v}_1 = \begin{bmatrix} -.23570 \\ -.23570 \\ -.94281 \end{bmatrix}, \quad \mathbf{v}_2 = \begin{bmatrix} .66667 \\ .66667 \\ -.33333 \end{bmatrix}, \quad \text{and} \quad \mathbf{v}_3 = \begin{bmatrix} .70711 \\ -.70711 \\ -.14343E{-}10 \end{bmatrix}. \quad \square$$

EXAMPLE 2 It is possible (especially when running a contrived problem rather than a real-world problem) that an input symmetric matrix might have an eigenvector each of whose entries is 1. This sort of matrix might fool the inverse power method, and Subroutine INVPOW may have to be modified to accommodate this. For example, when we input the matrix A from Example 1 of Section 7.5 to the program in Fig. 7.12, the program found estimates $\lambda_1 = 15.02$, $\lambda_2 = \lambda_3 = 5.022$, and $\lambda_4 = -.9585$, as we noted above. When these are passed to the inverse power method, convergence is to $\lambda = 15$ for each of these estimates because the eigenvector $\mathbf{u}_1$ corresponding to $\lambda_1 = 15$ is

$$\mathbf{u}_1 = \begin{bmatrix} 1 \\ 1 \\ 1 \\ 1 \end{bmatrix}.$$

To get the inverse power method to work properly, we changed the definition of the starting vector (OLDVCT in Subroutine INVPOW) to

$$\begin{bmatrix} 0 \\ 1/\sqrt{3} \\ 1/\sqrt{3} \\ 1/\sqrt{3} \end{bmatrix}.$$

With this modification all three eigenvectors and eigenvalues were found correctly. (Note that the output from the Givens algorithm alerted us that the inverse power method had been fooled.)

As we mentioned earlier, the eigenvalue problem for a nonsymmetric matrix may be quite difficult. If A is not symmetric, the currently accepted practice is to reduce A to Hessenberg form and then to employ some version of a procedure called the QR algorithm. Since the QR algorithm is rather hard to explain thoroughly, we refer the reader to a more advanced reference.

Exercises 7.6

1. Use the program in Fig. 7.12 to find the eigenvalues and the eigenvectors of the matrices in Exercises 1 and 2 of Section 7.5.

2. It can be shown that the eigenvalue problem for symmetric matrices is well behaved in that small changes in a symmetric matrix produce proportionately small changes in the eigenvalues. To illustrate this, choose one of the matrices in Exercise 1, add .01 to the (1, 4) and (4, 1) entries, and calculate the eigenvalues and the eigenvectors of the new matrix.

3. Prove that $p_n(t)$ as defined in (2) is the characteristic polynomial of the associated symmetric tridiagonal matrix H.

4. Let $C = (c_{ij})$ be an $(n \times n)$ matrix and let λ be an eigenvalue of C. Prove that $|\lambda| \leq \|C\|$, where $\|C\|$ is given in (3). To prove this, suppose that $Cx = \lambda x$, where $\|x\| = 1$, and show that $\|Cx\| = |\lambda|$. Use the Cauchy–Schwarz inequality and the properties of norms to show that

$$\|Cx\| \leq |x_1| \|\mathbf{C}_1\| + |x_2| \|\mathbf{C}_2\| + \cdots + |x_n| \|\mathbf{C}_n\|$$
$$\leq \|\mathbf{u}\| \|\mathbf{v}\|,$$

where

$$\mathbf{u} = \begin{bmatrix} |x_1| \\ |x_2| \\ \vdots \\ |x_n| \end{bmatrix} \quad \text{and} \quad \mathbf{v} = \begin{bmatrix} \|\mathbf{C}_1\| \\ \|\mathbf{C}_2\| \\ \vdots \\ \|\mathbf{C}_n\| \end{bmatrix}.$$

Finally, show that $\|\mathbf{u}\| = 1$ and $\|\mathbf{v}\| = \|C\|$.

5. *Hyman's method.* Let H be a (4×4) unreduced Hessenberg matrix and α a scalar, and consider

$$(h_{11} - \alpha)x_1 + h_{12}x_2 + h_{13}x_3 + h_{14}x_4 = g(\alpha)$$
$$h_{21}x_1 + (h_{22} - \alpha)x_2 + h_{23}x_3 + h_{24}x_4 = 0$$
$$h_{32}x_2 + (h_{33} - \alpha)x_3 + h_{34}x_4 = 0$$
$$h_{43}x_3 + (h_{44} - \alpha)x_4 = 0.$$

In this system, set $x_4 = 1$; determine x_3, x_2, and x_1 from the last three equations; and then let $g(\alpha)$ be defined by the first equation. Prove that $g(\alpha) = 0$ if and only if α is an eigenvalue of H. (*Note:* Half of this proof is easy; if $g(\alpha) = 0$, then you have generated a nontrivial solution of $(H - \alpha I)x = \mathbf{0}$. To complete the proof, show that if α is an eigenvalue of H, then x_4 can always be chosen to be equal to 1. Also note that this procedure can clearly be extended to $(n \times n)$ matrices H.)

6. If H is any matrix and λ is any real eigenvalue of H, then $-\|H\| \leq \lambda \leq \|H\|$ by Exercise 4. If H is an unreduced Hessenberg matrix, we can attempt to isolate the real eigenvalues of H by evaluating $g(\alpha)$ for various α in $[-\|H\|, \|H\|]$, where $g(\alpha)$ is as in Exercise 5. Write a subroutine that evaluates $g(\alpha)$ for various α, and test your program on

$$A = \begin{bmatrix} 15 & 11 & 6 & -9 & -15 \\ 1 & 3 & 9 & -3 & -8 \\ 7 & 6 & 6 & -3 & -11 \\ 7 & 7 & 5 & -3 & -11 \\ 17 & 12 & 5 & -10 & -16 \end{bmatrix}.$$

(*Note:* Matrix A has one real eigenvalue and of course must be reduced to Hessenberg form before Hyman's method can be implemented.)

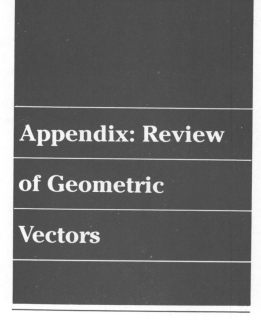

Appendix: Review of Geometric Vectors

Many physical quantities cannot be described mathematically by a single real number; an example is force, which has both magnitude and direction. Such physical quantities are called vectors, and we can represent them in two-space or three-space as directed line segments. Such a line segment, $\overrightarrow{AB}$, directed from A to B, is called a *geometric vector*. We note that a force vector is completely determined by its length and direction; so the initial point A in a representation $\overrightarrow{AB}$ is irrelevant. Similarly, any two geometric vectors having the same length and direction are called *equivalent vectors* and can be regarded as being equal. In Fig. A.1, the geometric vectors $\overrightarrow{OP}$, AB, and $\overrightarrow{CD}$ are all equivalent.

Given an arbitrary geometric vector $\mathbf{v} = \overrightarrow{AB}$, there is a unique geometric vector $\overrightarrow{OP}$ (called the position vector) that is equivalent to $\mathbf{v}$ and has its initial point at the origin of the coordinate system (see Fig. A.1.). If the coordinates of the terminal point P are (v_1, v_2), then we say that (v_1, v_2) are the *components* of $\mathbf{v}$, and we write this designation as

$$\mathbf{v} = (v_1, v_2). \tag{1}$$

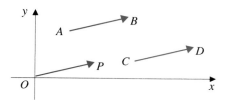

Figure A.1
Equivalent vectors

Similarly, for a geometric vector in three-space we would write $\mathbf{v} = (v_1, v_2, v_3)$, where the components of $\mathbf{v}$ are the coordinates of the terminal point of its equivalent position vector. If $\mathbf{v} = \overrightarrow{AB}$ is any geometric vector, and if A and B have coordinates (a_1, a_2) and (b_1, b_2), respectively, then clearly $\overrightarrow{OP}$ is equivalent to $\overrightarrow{AB}$, where the point P has coordinates $(b_1 - a_1, b_2 - a_2)$; that is, the line segment $\overrightarrow{OP}$ has the same length and direction as $\overrightarrow{AB}$. Therefore, the components of $\mathbf{v} = \overrightarrow{AB}$ are

$$\mathbf{v} = (b_1 - a_1, b_2 - a_2). \tag{2}$$

From Eq. (2) one sees how to find the components of any geometric vector $\mathbf{v} = \overrightarrow{AB}$. Conversely, if we are given the components of a geometric vector $\mathbf{v}$ as in Eq. (1), then we can construct the equivalent geometric vector $\overrightarrow{AB}$ whose initial point is at A; the coordinates of the terminal point B are $(a_1 + v_1, a_2 + v_2)$.

If $\mathbf{u}$ and $\mathbf{v}$ are any two geometric vectors, then the sum $\mathbf{u} + \mathbf{v}$ is determined as follows: Position $\mathbf{v}$ so that the initial point of $\mathbf{v}$ coincides with the terminal point of $\mathbf{u}$. Then $\mathbf{u} + \mathbf{v}$ is the geometric vector represented by the line segment from the initial point of $\mathbf{u}$ to the terminal point of $\mathbf{v}$ (see Fig. A.2). The components of $\mathbf{u} + \mathbf{v}$ are easy to calculate given the components $\mathbf{u} = (u_1, u_2)$ and $\mathbf{v} = (v_1, v_2)$. In particular, if the initial point of $\mathbf{u}$ is at $A = (a_1, a_2)$, then the terminal point of $\mathbf{u}$ is at $(a_1 + u_1, a_2 + u_2)$. Similarly, the terminal point of $\mathbf{v}$ is at $(a_1 + u_1 + v_1, a_2 + u_2 + v_2)$; so the sum $\mathbf{u} + \mathbf{v}$ has its initial point at (a_1, a_2) and its terminal point at $(a_1 + u_1 + v_1, a_2 + u_2 + v_2)$. Thus by Eq. (2) the components of $\mathbf{u} + \mathbf{v}$ are $(u_1 + v_1, u_2 + v_2)$; see Fig. A.3. Similarly, in three-space if $\mathbf{u} = (u_1, u_2, u_3)$ and $\mathbf{v} = (v_1, v_2, v_3)$, then $\mathbf{u} + \mathbf{v} = (u_1 + v_1, u_2 + v_2, u_3 + v_3)$.

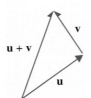

Figure A.2

EXAMPLE 1   As a numerical example to fix these ideas, suppose that $\mathbf{u} = \overrightarrow{AB}$ and $\mathbf{v} = \overrightarrow{CD}$, where $A = (7, 2)$, $B = (4, 4)$, $C = (1, 3)$, and $D = (2, 4)$. In terms of components, we have $\mathbf{u} = (-3, 2)$ and $\mathbf{v} = (1, 1)$. Note that $\mathbf{u}$ is equivalent to $\overrightarrow{OP}$ and $\mathbf{v}$ is equivalent to $\overrightarrow{OQ}$, where $P = (-3, 2)$ and $Q = (1, 1)$; see Fig. A.4, in which it is clear that $\overrightarrow{OP}$ and $\overrightarrow{AB}$ have the same length and direction as do $\overrightarrow{OQ}$ and $\overrightarrow{CD}$. In Fig. A.5, the vector $\mathbf{u} + \mathbf{v}$ is illustrated. Since $\mathbf{u} = (-3, 2)$ and $\mathbf{v} = (1, 1)$, it follows that $\mathbf{u} + \mathbf{v} = (-2, 3)$, or $\mathbf{u} + \mathbf{v}$ is equivalent to OR. $\square$

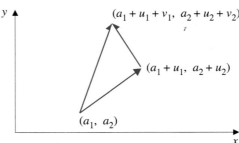

Figure A.3

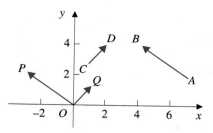

Figure A.4

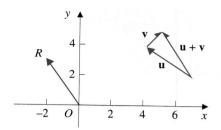

Figure A.5

If $\mathbf{v} = (v_1, v_2)$ is a geometric vector, then the **length** or **magnitude** of $\mathbf{v}$ is defined to be $\|\mathbf{v}\| = \sqrt{v_1^2 + v_2^2}$. [If $\mathbf{v} = \overrightarrow{AB}$, then Eq. (2) states that $\|\mathbf{v}\|$ is just the length of the line segment from A to B.] Similarly, if $\mathbf{v} = (v_1, v_2, v_3)$ is a geometric vector in three-space, then $\|\mathbf{v}\| = \sqrt{v_1^2 + v_2^2 + v_3^2}$. If c is a scalar, we define $c\mathbf{v}$ to be the geometric vector whose magnitude is $|c|\,\|\mathbf{v}\|$ and whose direction is the same as that of $\mathbf{v}$ when $c > 0$, and is opposite that of $\mathbf{v}$ when $c < 0$. For $c = 0$, the vector $c\mathbf{v}$ is defined to be the zero vector $\boldsymbol{\theta}$ (the direction of $\boldsymbol{\theta}$ is not defined, but the magnitude of $\boldsymbol{\theta}$ is 0). If $\mathbf{v}$ has components (v_1, v_2), it is an easy exercise to show that $c\mathbf{v}$ has components (cv_1, cv_2). For the special case $c = -1$, we usually write $-\mathbf{v}$ instead of $(-1)\mathbf{v}$; and we observe that if $\mathbf{u} = (u_1, u_2)$ and $\mathbf{v} = (v_1, v_2)$, then $\mathbf{u} - \mathbf{v} = (u_1 - v_1, u_2 - v_2)$.

In two-space and three-space, we define the angle between two vectors $\mathbf{u}$ and $\mathbf{v}$ in a natural way, using components. That is, if $\mathbf{u} = (u_1, u_2, u_3)$ and $\mathbf{v} = (v_1, v_2, v_3)$ are two vectors in three-space, then the angle φ between $\mathbf{u}$ and $\mathbf{v}$ is the angle AOB, where $A = (u_1, u_2, u_3)$, $O = (0, 0, 0)$, and $B = (v_1, v_2, v_3)$. (*Note:* If $\mathbf{u} = c\mathbf{v}$, $c > 0$, then $\varphi = 0$; if $c < 0$, then $\varphi = \pi$.) Figure A.6 illustrates the definition in two-space; the angle between $\mathbf{u}$ and $\mathbf{v}$ is φ, where $\mathbf{u}$ is equivalent to $\overrightarrow{OA}$ and $\mathbf{v}$ is equivalent to $\overrightarrow{OB}$. For $\mathbf{u} = (u_1, u_2, u_3)$ and $\mathbf{v} = (v_1, v_2, v_3)$, we define the **dot product** of $\mathbf{u}$ and $\mathbf{v}$, denoted by $\mathbf{u} \cdot \mathbf{v}$, as

$$\mathbf{u} \cdot \mathbf{v} = u_1 v_1 + u_2 v_2 + u_3 v_3. \tag{3}$$

Similarly, if $\mathbf{u} = (u_1, u_2)$ and $\mathbf{v} = (v_1, v_2)$ are in two-space, then $\mathbf{u} \cdot \mathbf{v} = u_1 v_1 + u_2 v_2$. There is a useful connection between $\mathbf{u} \cdot \mathbf{v}$ and the angle between $\mathbf{u}$ and $\mathbf{v}$.

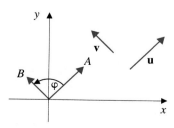

Figure A.6

THEOREM 1

If φ is the angle between two geometric vectors **u** and **v**, then

$$\mathbf{u} \cdot \mathbf{v} = \|\mathbf{u}\|\|\mathbf{v}\| \cos \varphi.$$

PROOF   If $\mathbf{u} = c\mathbf{v}$, then $\varphi = 0$ or $\varphi = \pi$, and the proof is immediate. For $0 < \varphi < \pi$, we can use the law of cosines in Fig. A.6 to deduce

$$\|\overrightarrow{AB}\|^2 = \|\overrightarrow{OA}\|^2 + \|\overrightarrow{OB}\|^2 - 2\|\overrightarrow{OA}\|\|\overrightarrow{OB}\| \cos \varphi. \tag{4}$$

With reference to Fig. A.6, we see that $\overrightarrow{AB}$ is equivalent to $\mathbf{v} - \mathbf{u}$; so if $\mathbf{u} = (u_1, u_2)$ and $\mathbf{v} = (v_1, v_2)$, then $\|\overrightarrow{AB}\|^2 = (v_1 - u_1)^2 + (v_2 - u_2)^2$. Similarly, $\|\overrightarrow{OA}\|^2 = u_1^2 + u_2^2$ and $\|\overrightarrow{OB}\|^2 = v_1^2 + v_2^2$; so (4) reduces to

$$-2u_1v_1 - 2u_2v_2 = -2\|\overrightarrow{OA}\|\|\overrightarrow{OB}\| \cos \varphi.$$

Since $\|\overrightarrow{OA}\| = \|\mathbf{u}\|$ and $\|\overrightarrow{OB}\| = \|\mathbf{v}\|$, Theorem 1 is proved in two-space and similarly follows in three-space. ☐

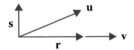

Figure A.7

From Theorem 1, if **p** and **q** are *nonzero* geometric vectors, then **p** and **q** are perpendicular if and only if $\mathbf{p} \cdot \mathbf{q} = 0$. Given vectors **u** and **v**, it is frequently necessary to express **u** as the sum of two vectors **r** and **s**, where **r** is a scalar multiple of **v** and **s** is perpendicular to **v**. That is, we want $\mathbf{u} = \mathbf{r} + \mathbf{s}$, where $\mathbf{r} = c\mathbf{v}$ and $\mathbf{s} \cdot \mathbf{v} = 0$ (see Fig. A.7). To determine **r** and **s**, we set $\mathbf{u} = \mathbf{r} + \mathbf{s} = c\mathbf{v} + \mathbf{s}$, and we calculate $\mathbf{u} \cdot \mathbf{v} = c(\mathbf{v} \cdot \mathbf{v}) + (\mathbf{s} \cdot \mathbf{v}) = c\|\mathbf{v}\|^2$. Thus, $\mathbf{r} = c\mathbf{v}$ is found from $c = \mathbf{u} \cdot \mathbf{v}/\|\mathbf{v}\|^2$; and once we have **r**, **s** is given by $\mathbf{s} = \mathbf{u} - \mathbf{r}$.

If **u** and **v** are any two nonzero vectors in three-space, it is often important to be able to construct a vector **x** that is perpendicular to both **u** and **v**. If $\mathbf{u} = (u_1, u_2, u_3)$ and $\mathbf{v} = (v_1, v_2, v_3)$, and if **u** is not a multiple of **v**, then Theorem 1 shows that we want a vector $\mathbf{x} = (x_1, x_2, x_3)$ such that $\mathbf{u} \cdot \mathbf{x} = 0$ and $\mathbf{v} \cdot \mathbf{x} = 0$. Writing these two conditions out, we see that **x** will be perpendicular to both **u** and **v** if

$$\begin{aligned} u_1x_1 + u_2x_2 + u_3x_3 &= 0 \\ v_1x_1 + v_2x_2 + v_3x_3 &= 0. \end{aligned} \tag{5}$$

Solving the system of equations in (5) shows that one vector **x** that satisfies $\mathbf{u} \cdot \mathbf{x} = 0$ and $\mathbf{v} \cdot \mathbf{x} = 0$ is

$$\mathbf{x} = (u_2v_3 - u_3v_2, u_3v_1 - u_1v_3, u_1v_2 - u_2v_1). \tag{6}$$

We define the **cross product** of **u** and **v**, denoted by $\mathbf{u} \times \mathbf{v}$, to be the geometric vector **x** in (6). For example, if $\mathbf{u} = (1, 2, 1)$ and $\mathbf{v} = (6, 0, 3)$, then

$$\mathbf{u} \times \mathbf{v} = (6, 3, -12);$$

and we see that $\mathbf{u} \cdot (\mathbf{u} \times \mathbf{v}) = 0$ and $\mathbf{v} \cdot (\mathbf{u} \times \mathbf{v}) = 0$. Theorem 2 summarizes some of the properties of the cross product.

> ## THEOREM 2
>
> Let **u**, **v** and **w** be geometric vectors in three-space. Then:
>
> 1. $\mathbf{u} \times \mathbf{v} = -(\mathbf{v} \times \mathbf{u})$.
> 2. $(\mathbf{u} \times \mathbf{v}) \cdot \mathbf{w} = \mathbf{u} \cdot (\mathbf{v} \times \mathbf{w})$ (the "box product").
> 3. $\mathbf{u} \times (\mathbf{v} \times \mathbf{w}) = (\mathbf{u} \cdot \mathbf{w})\mathbf{v} - (\mathbf{u} \cdot \mathbf{v})\mathbf{w}$ (the "triple product").
> 4. $\|\mathbf{u} \times \mathbf{v}\|^2 = \|\mathbf{u}\|^2\|\mathbf{v}\|^2 - (\mathbf{u} \cdot \mathbf{v})^2$.
> 5. $\mathbf{u} \cdot (\mathbf{u} \times \mathbf{v}) = \mathbf{v} \cdot (\mathbf{u} \times \mathbf{v}) = 0$.

It is conventional to let **i**, **j**, and **k** denote the standard unit vectors

$$\mathbf{i} = (1,0,0), \qquad \mathbf{j} = (0,1,0), \qquad \mathbf{k} = (0,0,1);$$

and we observe that if $\mathbf{v} = (v_1, v_2, v_3)$, then $\mathbf{v} = v_1\mathbf{i} + v_2\mathbf{j} + v_3\mathbf{k}$. These unit vectors can be used also to remember the form of $\mathbf{u} \times \mathbf{v}$. That is, if $\mathbf{u} = (u_1, u_2, u_3)$ and $\mathbf{v} = (v_1, v_2, v_3)$, then $\mathbf{u} \times \mathbf{v}$ can be represented symbolically as a (3×3) determinant:

$$\mathbf{u} \times \mathbf{v} = \begin{vmatrix} \mathbf{i} & \mathbf{j} & \mathbf{k} \\ u_1 & u_2 & u_3 \\ v_1 & v_2 & v_3 \end{vmatrix} = \mathbf{i}\begin{vmatrix} u_2 & u_3 \\ v_2 & v_3 \end{vmatrix} - \mathbf{j}\begin{vmatrix} u_1 & u_3 \\ v_1 & v_3 \end{vmatrix} + \mathbf{k}\begin{vmatrix} u_1 & u_2 \\ v_1 & v_2 \end{vmatrix}.$$

Finally, if we use Theorem 1 and property (4) of Theorem 2, we obtain the important formula

$$\|\mathbf{u} \times \mathbf{v}\| = \|\mathbf{u}\|\,\|\mathbf{v}\|\sin\varphi;$$

and so $\|\mathbf{u} \times \mathbf{v}\|$ is the area of the parallelogram determined by **u** and **v**.

We conclude by showing how geometric vectors can be used to express lines and planes in three-space. Consider the line L in three-space that passes through the point $P_0 = (x_0, y_0, z_0)$ and is parallel to the nonzero geometric vector $\mathbf{v} = (v_1, v_2, v_3)$. If $P = (x, y, z)$ is an arbitrary point on L, then the geometric vector $\overrightarrow{PP_0}$ is parallel to **v**, and so $\overrightarrow{PP_0} = t\mathbf{v}$. Componentwise, this equation is

$$x - x_0 = tv_1, \qquad y - y_0 = tv_2, \qquad z - z_0 = tv_3. \tag{7}$$

Since the point P is on L if and only if (7) is satisfied for some value of t, (7) gives the parametric equations of the line L. If v_1, v_2, and v_3 are all nonzero, we can eliminate the parameter t in (7) and obtain the symmetric equations of the line L as

$$\frac{x - x_0}{v_1} = \frac{y - y_0}{v_2} = \frac{z - z_0}{v_3}. \tag{8}$$

[There are obvious modifications to (8) if any of v_1, v_2, or v_3 is zero.]

Finally, we consider the equation of the plane Π that passes through the point $P_0 = (x_0, y_0, z_0)$ and has the nonzero geometric vector $\mathbf{v} = (v_1, v_2, v_3)$ perpendicular to it; that is, **v** is "normal" to Π. If $P = (x, y, z)$ is any point in Π, then the geometric vector $\overrightarrow{PP_0}$ is perpendicular to **v**. Setting $\overrightarrow{PP_0} \cdot \mathbf{v} = 0$ yields

$$v_1(x - x_0) + v_2(y - y_0) + v_3(z - z_0) = 0,$$

which is called the point-normal equation of the plane Π.

Answers to Selected Odd-Numbered Exercises*

Exercises 1.1

1. Linear

3. Linear

5. Nonlinear

7. $\begin{aligned} x_1 + 3x_2 &= 7 \\ 4x_1 - x_2 &= 2 \end{aligned}$

9. $\begin{aligned} x_1 + x_2 &= 0 \\ 3x_1 + 4x_2 &= -1 \\ -x_1 + 2x_2 &= -3 \end{aligned}$

11. $\begin{aligned} x_1 + x_2 &= 3 \\ 9x_2 &= 15 \end{aligned}$

13. $\begin{aligned} x_1 + 2x_2 - x_3 &= 1 \\ -x_2 + 3x_3 &= 1 \\ 5x_2 - 2x_3 &= 6 \end{aligned}$

15. $\begin{aligned} x_1 + x_2 &= 9 \\ -2x_2 &= -2 \\ -2x_2 &= -21 \end{aligned}$

17. $\begin{aligned} x_1 + 2x_2 - x_3 + x_4 &= 1 \\ x_2 + x_3 - x_4 &= 3 \\ 3x_2 + 6x_3 &= 1 \end{aligned}$

19. $x_1 = 1, x_2 = -3$

21. $x_1 = 0, x_2 = 0$

23. $x_1 = 0, x_2 = 1, x_3 = 1$

25. $x_1 = 1, x_2 = 0, x_3 = 1$

31. $a = 2$

33. All values of a except $a = 8$

35. $a = 3$ or $a = -3$

37. $a_{12} = 1, a_{22} = -1, x_1 = 1$

41. The parabola is $y = 2x^2 - x + 3$.

Exercises 1.2

1.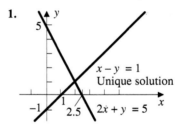
$x - y = 1$
Unique solution
$2x + y = 5$

3.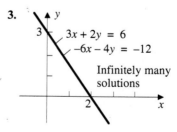
$3x + 2y = 6$
$-6x - 4y = -12$
Infinitely many solutions

---

* Many of the problems have answers that contain parameters or answers that can be written in a variety of forms. For problems of this sort we have presented one possible form of the answer. Your solution may have a different form and still be correct. You can frequently check your solution by inserting it in the original problem or by showing that two different forms for the answer are equivalent.

5. $x_1 = -3 - 2x_3$, $x_2 = 7 + x_3$, x_3 arbitrary

7. $x_1 = (5 - 3x_4)/2$, $x_2 = 2 + x_3 + 2x_4$, x_3 and x_4 arbitrary

9. $x_1 = -9 + 2x_4$, $x_2 = 6$, $x_3 = 1 + x_4$, x_4 arbitrary

11. (a) $x_1 = -3$, $x_2 = 7$, $x_3 = 0$;
(b) $x_1 = -1$, $x_2 = 6$, $x_3 = -1$;
(c) $x_1 = -5$, $x_2 = 8$, $x_3 = 1$

13. For Ex. 5: $x_1 = 11 - 2x_2$, x_2 arbitrary,
$x_3 = -7 + x_2$; $x_1 = 5$,
$x_2 = 3$, $x_3 = -4$
For Ex. 6: $x_1 = 2x_2$, x_2 arbitrary, $x_3 = (1 - x_2)/3$;
$x_1 = 6$, $x_2 = 3$, $x_3 = -2/3$

15. For Ex. 7: $x_1 = 4 - 3(x_2 - x_3)/4$, x_2 and x_3 arbitrary,
$x_4 = -1 + (x_2 - x_3)/2$;
$x_1 = 1$, $x_2 = 4$, $x_3 = 0$, $x_4 = 1$;
$x_1 = -2$, $x_2 = 0$, $x_3 = -8$, $x_4 = 3$
For Ex. 8: $x_1 = 2 + 2x_2 - 3x_3$, x_2 and x_3 arbitrary,
$x_4 = 1 - x_2$; $x_1 = 10$, $x_2 = 4$,
$x_3 = 0$, $x_4 = -3$; $x_1 = 26$, $x_2 = 0$,
$x_3 = -8$, $x_4 = 1$

17. $x_1 = (7 - x_3 - 2x_4)/3$, $x_2 = x_3$, x_3 and x_4 arbitrary,
$x_5 = -1 + x_3 + x_4$; $x_1 = 10/3$, $x_2 = 3$, $x_3 = 3$,
$x_4 = -3$, $x_5 = -1$

19. Inconsistent

21. $x_1 = 2 - x_2$, x_2 arbitrary

23. $x_1 = 2 - x_2 + x_3$, x_2 and x_3 arbitrary

25. $x_1 = 3 - 2x_3$, $x_2 = -2 + 3x_3$, x_3 arbitrary

27. $x_1 = 3 - (7x_4 - 16x_5)/2$, $x_2 = (x_4 + 2x_5)/2$,
$x_3 = -2 + (5x_4 - 12x_5)/2$, x_4 and x_5 arbitrary

29. Inconsistent

31. Inconsistent

33. $\alpha = \pi/3$ or $\alpha = 5\pi/3$; $\beta = \pi/6$ or $\beta = 5\pi/6$

35. Center at $(9/2, 1/2)$, radius is $5/\sqrt{2}$

37. Let A denote the number of adults, S the number of students, and C the number of children. Possible solutions are: $A = 5k$, $S = 67 - 11k$, $C = 12 + 6k$, where $k = 0, 1, \ldots, 6$.

39. The old system has solution $x_1 = 1$ and $x_2 = -3$. The new system has an additional solution of $x_1 = 0$ and $x_2 = -5$.

41. The solution set is the line $x = (1 - t)/2$, $y = 2$, $z = t$.

43. Coincident planes

Exercises 1.3

1. $A = \begin{bmatrix} 2 & 1 & 6 \\ 4 & 3 & 8 \end{bmatrix}$ **3.** $Q = \begin{bmatrix} 1 & 4 & -3 \\ 2 & 1 & 1 \\ 3 & 2 & 1 \end{bmatrix}$

5. $\begin{aligned} 2x_1 + x_2 &= 6 \\ 4x_1 + 3x_2 &= 8 \end{aligned}$ and $\begin{aligned} x_1 + 4x_2 &= -3 \\ 2x_1 + x_2 &= 1 \\ 3x_1 + 2x_2 &= 1 \end{aligned}$

7. $A = \begin{bmatrix} 1 & 1 & -1 \\ 2 & 0 & -1 \end{bmatrix}$, $B = \begin{bmatrix} 1 & 1 & -1 & 2 \\ 2 & 0 & -1 & 1 \end{bmatrix}$

9. $A = \begin{bmatrix} 1 & 1 & 2 \\ 3 & 4 & -1 \\ -1 & 1 & 1 \end{bmatrix}$, $B = \begin{bmatrix} 1 & 1 & 2 & 6 \\ 3 & 4 & -1 & 5 \\ -1 & 1 & 1 & 2 \end{bmatrix}$

11. $A = \begin{bmatrix} 1 & 1 & 1 \\ 2 & 3 & 1 \\ 1 & -1 & 3 \end{bmatrix}$, $B = \begin{bmatrix} 1 & 1 & 1 & 1 \\ 2 & 3 & 1 & 2 \\ 1 & -1 & 3 & 2 \end{bmatrix}$

13. (a) The matrix is in echelon form.
(b) The operation $R_1 - 2R_2$ yields reduced echelon form $\begin{bmatrix} 1 & 0 \\ 0 & 1 \end{bmatrix}$.

15. (a) The operation $R_2 - 2R_1$ yields echelon form $\begin{bmatrix} 2 & 3 & 1 \\ 0 & -5 & -2 \end{bmatrix}$.

17. (a) The operation $R_1 \leftrightarrow R_2$ yields echelon form $\begin{bmatrix} 2 & 0 & 1 & 4 \\ 0 & 0 & 2 & 3 \end{bmatrix}$.

19. (a) The matrix is in echelon form.
(b) The operations $R_1 - 2R_3$, $R_2 - 4R_3$, $R_1 - 3R_2$ yield reduced echelon form $\begin{bmatrix} 1 & 0 & 0 & 5 \\ 0 & 1 & 0 & -2 \\ 0 & 0 & 1 & 1 \end{bmatrix}$.

21. (a) The matrix is in echelon form.
(b) The operations $R_1 + 2R_3$, $R_2 + 3R_3$, $R_1 - R_2$ yield reduced echelon form $\begin{bmatrix} 1 & 0 & 1 & 0 \\ 0 & 2 & -2 & 0 \\ 0 & 0 & 0 & 1 \end{bmatrix}$.

23. Exercise 7: $x_1 = 1/2 + (1/2)x_3$; $x_2 = 3/2 + (1/2)x_3$; x_3 arbitrary
Exercise 9: $x_1 = 1$, $x_2 = 1$, $x_3 = 2$
Exercise 11: The system is inconsistent.

25. $x_1 = -1/2 + (1/2)x_3$, $x_2 = 1/3 - (2/3)x_3$, x_3 arbitrary

27. $x_1 = 5/2, x_2 = -1, x_3 = 1/3$

29. $x_1 = 0, x_2 = 0$

31. $x_1 = -2 + 5x_3, x_2 = 1 - 3x_3, x_3$ arbitrary

33. The system is inconsistent.

35. $x_1 = x_3 = x_4 = 0; x_2$ arbitrary

37. The system is inconsistent.

39. $x_1 = -1 - (1/2)x_2 + (1/2)x_4, x_3 = 1 - x_4, x_2$ and x_4 arbitrary, $x_5 = 0$

41. $(2n + 1)(n + 1)n/6$

43. $\dfrac{n^2}{12}[2n^4 + 6n^3 + 5n^2 - 1]$

45. $\begin{bmatrix} * & \times & \times \\ 0 & * & \times \end{bmatrix}, \begin{bmatrix} * & \times & \times \\ 0 & 0 & * \end{bmatrix},$

$\begin{bmatrix} * & \times & \times \\ 0 & 0 & 0 \end{bmatrix}, \begin{bmatrix} 0 & * & \times \\ 0 & 0 & * \end{bmatrix},$

$\begin{bmatrix} 0 & * & \times \\ 0 & 0 & 0 \end{bmatrix}, \begin{bmatrix} 0 & 0 & * \\ 0 & 0 & 0 \end{bmatrix}, \begin{bmatrix} 0 & 0 & 0 \\ 0 & 0 & 0 \end{bmatrix}$

47. The operations $R_2 - 2R_1, R_1 + 2R_2, -R_2$ transform B to I. The operations $R_2 - 3R_1, R_1 + R_2, (-1/2)R_2$ reduce C to I, so the operations $-2R_2, R_1 - R_2, R_2 + 3R_1$ transform I to C. Thus the operations $R_2 - 2R_1, R_1 + 2R_2, -R_2, -2R_2, R_1 - R_2, R_2 + 3R_1$ transform B to C.

Exercises 1.4

1. $\begin{bmatrix} 2 & 2 & -1 & 1 \\ 0 & 0 & 3 & 2 \\ 0 & 0 & 0 & 0 \\ 0 & 0 & 0 & 0 \end{bmatrix}$ $\begin{matrix} n = 3 \\ r = 2 \\ x_2 \\ {} \end{matrix}$

3. $\begin{bmatrix} 1 & 2 & 2 & -1 & 3 \\ 0 & -1 & 1 & 1 & 2 \\ 0 & 0 & 0 & 2 & 1 \end{bmatrix}$ $\begin{matrix} n = 4 \\ r = 3 \\ x_3 \end{matrix}$

5. $r = 2, r = 1, r = 0$

7. Infinitely many solutions or no solutions

9. Infinitely many solutions

11. Infinitely many solutions, a unique solution, or no solution

13. A unique solution or infinitely many solutions

15. Infinitely many solutions

17. A unique solution or infinitely many solutions

19. Infinitely many solutions

21. There are nontrivial solutions.

23. There is only the trivial solution.

25. (a) $\begin{bmatrix} * & \times & \times \\ 0 & * & \times \\ 0 & 0 & * \\ 0 & 0 & 0 \end{bmatrix}$

Exercises 1.5

1. (a) $\begin{aligned} x_1 \qquad\qquad + x_4 &= 1200 \\ x_1 + x_2 \qquad\quad &= 1000 \\ x_3 + x_4 &= 600 \\ x_2 + x_3 \qquad &= 400 \end{aligned}$
(b) $x_1 = 1100, x_2 = -100, x_3 = 500;$
(c) The minimum value is $x_1 = 600$ and the maximum value is $x_1 = 1000$.

3. $x_2 = 800, x_3 = 400, x_4 = 200$

5. $I_1 = 0.05, I_2 = 0.6, I_3 = 0.55$

Exercises 1.6

1. (a) $\begin{bmatrix} 2 & 0 \\ 2 & 6 \end{bmatrix}$; (b) $\begin{bmatrix} 0 & 4 \\ 2 & 4 \end{bmatrix}$;

(c) $\begin{bmatrix} 0 & -6 \\ 6 & 18 \end{bmatrix}$; (d) $\begin{bmatrix} -6 & 8 \\ 4 & 6 \end{bmatrix}$

3. $\begin{bmatrix} -2 & -2 \\ 0 & 0 \end{bmatrix}$ **5.** $\begin{bmatrix} -1 & -1 \\ 0 & 0 \end{bmatrix}$

7. (a) $\begin{bmatrix} 3 \\ -3 \end{bmatrix}$; (b) $\begin{bmatrix} 3 \\ 4 \end{bmatrix}$; (c) $\begin{bmatrix} 0 \\ 0 \end{bmatrix}$

9. (a) $\begin{bmatrix} 2 \\ 1 \end{bmatrix}$; (b) $\begin{bmatrix} 0 \\ 1 \end{bmatrix}$; (c) $\begin{bmatrix} 17 \\ 14 \end{bmatrix}$

11. (a) $\begin{bmatrix} 2 \\ 3 \end{bmatrix}$; (b) $\begin{bmatrix} 20 \\ 16 \end{bmatrix}$

13. $a_1 = 11/3, a_2 = -4/3$

15. $a_1 = -2, a_2 = 0$ **17.** No solution

19. $a_1 = 4, a_2 = -3/2$ **21.** $\mathbf{w}_2 = \begin{bmatrix} 1 \\ 3 \end{bmatrix}$

23. $\mathbf{w}_3 = \begin{bmatrix} -1 \\ 2 \end{bmatrix}$ **25.** $\begin{bmatrix} -4 & 6 \\ 2 & 12 \end{bmatrix}$

27. $\begin{bmatrix} 4 & 12 \\ 4 & 10 \end{bmatrix}$ **29.** $\begin{bmatrix} 0 & 0 \\ 0 & 0 \end{bmatrix}$

31. $AB = \begin{bmatrix} 5 & 16 \\ 5 & 18 \end{bmatrix}$, $BA = \begin{bmatrix} 4 & 11 \\ 6 & 19 \end{bmatrix}$

33. $A\mathbf{u} = \begin{bmatrix} 11 \\ 13 \end{bmatrix}$, $\mathbf{v}A = [8, 22]$

35. 66   **37.** $\begin{bmatrix} 5 & 10 \\ 8 & 12 \\ 15 & 20 \\ 8 & 17 \end{bmatrix}$ **39.** $\begin{bmatrix} 27 \\ 28 \\ 43 \\ 47 \end{bmatrix}$

41. $(BA)\mathbf{u} = B(A\mathbf{u}) = \begin{bmatrix} 37 \\ 63 \end{bmatrix}$

43.

$C(A(B\mathbf{u}))$	$(CA)(B\mathbf{u})$	$(C(AB))\mathbf{u}$	$C((AB)\mathbf{u})$
12	16	20	16

45. (a) AB is (2×4); BA is undefined.
  (b) Neither is defined.
  (c) AB is undefined; BA is (6×7).
  (d) AB is (2×2); BA is (3×3).
  (e) AB is (3×1); BA is undefined.
  (f) Both are (2×4).
  (g) AB is (4×4), BA is (1×1).

53. (a) For (i), $A = \begin{bmatrix} 2 & -1 \\ 1 & 1 \end{bmatrix}$, $\mathbf{x} = \begin{bmatrix} x_1 \\ x_2 \end{bmatrix}$, $\mathbf{b} = \begin{bmatrix} 3 \\ 3 \end{bmatrix}$;

for (ii), $A = \begin{bmatrix} 1 & -3 & 1 \\ 1 & -2 & 1 \\ 0 & 1 & -1 \end{bmatrix}$, $\mathbf{x} = \begin{bmatrix} x_1 \\ x_2 \\ x_3 \end{bmatrix}$,

$\mathbf{b} = \begin{bmatrix} 1 \\ 2 \\ -1 \end{bmatrix}$.

(b) For (i), $x_1 \begin{bmatrix} 2 \\ 1 \end{bmatrix} + x_2 \begin{bmatrix} -1 \\ 1 \end{bmatrix} = \begin{bmatrix} 3 \\ 3 \end{bmatrix}$;

for (ii), $x_1 \begin{bmatrix} 1 \\ 1 \\ 0 \end{bmatrix} + x_2 \begin{bmatrix} -3 \\ -2 \\ 1 \end{bmatrix} + x_3 \begin{bmatrix} 1 \\ 1 \\ -1 \end{bmatrix} = \begin{bmatrix} 1 \\ 2 \\ -1 \end{bmatrix}$.

(c) For (i), $\mathbf{b} = 2A_1 + A_2$;
  for (ii), $\mathbf{b} = 2A_1 + A_2 + 2A_3$.

55. $B = \begin{bmatrix} 2 & -1 \\ -1 & 1 \end{bmatrix}$

57. (a) $B = \begin{bmatrix} -1 & 6 \\ 1 & 0 \end{bmatrix}$; (b) Not possible;

(c) $B = \begin{bmatrix} -2a & -2b \\ a & b \end{bmatrix}$, a and b arbitrary

Exercises 1.7

1. $(DE)F = D(EF) = \begin{bmatrix} 23 & 23 \\ 29 & 29 \end{bmatrix}$

3. $DE = \begin{bmatrix} 8 & 15 \\ 11 & 18 \end{bmatrix}$, $ED = \begin{bmatrix} 12 & 27 \\ 7 & 14 \end{bmatrix}$

5. $F\mathbf{u} = F\mathbf{v} = \begin{bmatrix} 0 \\ 0 \end{bmatrix}$ **7.** $\begin{bmatrix} 3 & 4 & 2 \\ 1 & 7 & 6 \end{bmatrix}$

9. $\begin{bmatrix} 5 & 5 \\ 9 & 9 \end{bmatrix}$ **11.** $[0 \quad 0]$ **13.** -6

15. 36   **17.** 2   **19.** $\sqrt{2}$

21. $\sqrt{29}$ **23.** 0   **25.** $2\sqrt{5}$

29. D and F are symmetric.

31. AB is symmetric if and only if $AB = BA$.

33. $\mathbf{x}^T D\mathbf{x} = x_1^2 + 3x_2^2 + (x_1 + x_2)^2 > 0$

35. $\begin{bmatrix} -3 & 3 \\ 3 & -3 \end{bmatrix}$ **37.** $\begin{bmatrix} -27 & -9 \\ 27 & 9 \end{bmatrix}$

39. $\begin{bmatrix} -12 & 18 & 24 \\ 18 & -27 & -36 \\ 24 & -36 & -48 \end{bmatrix}$

41. (a) $\mathbf{x} = \begin{bmatrix} -10 \\ 8 \end{bmatrix}$; (b) $\mathbf{x} = \begin{bmatrix} 6 \\ -2 \end{bmatrix}$

Exercises 1.8

1. Linearly independent

3. Linearly dependent, $\mathbf{v}_5 = 3\mathbf{v}_1$

5. Linearly dependent, $\mathbf{v}_3 = 2\mathbf{v}_1$

7. Linearly dependent, $\mathbf{u}_4 = 4\mathbf{u}_5$

9. Linearly independent

11. Linearly dependent, $\mathbf{u}_4 = 4\mathbf{u}_5$

13. Linearly dependent, $\mathbf{u}_4 = \dfrac{16}{5}\mathbf{u}_0 + \dfrac{12}{5}\mathbf{u}_1 - \dfrac{4}{5}\mathbf{u}_2$

15. Those in Exercises 5, 6, 13, and 14

17. Singular; $x_1 = -2x_2$ **19.** Singular; $x_1 = -2x_2$

21. Singular; $x_1 = x_2 = 0$, x_3 arbitrary

23. Nonsingular

25. Singular; $x_2 = x_3 = 0$, x_1 arbitrary

27. Nonsingular   **29.** $a = 6$

31. $b(a - 2) = 4$

33. $c - ab = 0$

35. $\mathbf{v}_3 = \mathbf{A}_2$

37. $\mathbf{v}_2 = (\mathbf{C}_1 + \mathbf{C}_2)/2$

39. $\mathbf{u}_3 = (-8\mathbf{F}_1 - 2\mathbf{F}_2 + 9\mathbf{F}_3)/3$

41. $\mathbf{b} = -11\mathbf{v}_1 + 7\mathbf{v}_2$

43. $\mathbf{b} = 0\mathbf{v}_1 + 0\mathbf{v}_2$

45. $\mathbf{b} = -3\mathbf{v}_1 + 2\mathbf{v}_2$

Exercises 1.9

1. $p(t) = (-1/2)t^2 + (9/2)t - 1$

3. $p(t) = 2t + 3$

5. $p(t) = 2t^3 - 2t^2 + 3t + 1$

7. $y = 2e^{2x} + e^{3x}$

9. $y = 3e^{-x} + 4e^x + e^{2x}$

11. $\displaystyle\int_0^{3h} f(t)dt \approx \frac{3h}{2}[f(h) + f(2h)]$

13. $\displaystyle\int_0^{3h} f(t)dt \approx \frac{3h}{8}[f(0) + 3f(h) + 3f(2h) + f(3h)]$

15. $\displaystyle\int_0^{h} f(t)dt \approx \frac{h}{2}[-f(-h) + 3f(0)]$

17. $f'(0) \approx [-f(0) + f(h)]/h$

19. $f'(0) \approx [-3f(0) + 4f(h) - f(2h)]/(2h)$

21. $f''(0) \approx [f(-h) - 2f(0) + f(h)]/h^2$

27. $p(t) = t^3 + 2t^2 + 3t + 2$

29. $p(t) = t^3 + t^2 + 4t + 3$

35. $f'(a) \approx \dfrac{1}{12h}$
$\times [f(a - 2h) - 8f(a - h) + 8f(a + h) - f(a + 2h)]$

Exercises 1.10

1. $n = 2, m = 3$

3. $n = 4, m = 4$

5. $n = 4, m = 2$

7. $A^{-1} = \begin{bmatrix} 3 & -1 \\ -5 & 2 \end{bmatrix}$

9. $A^{-1} = (1/10)\begin{bmatrix} 3 & 2 \\ -2 & 2 \end{bmatrix}$

11. A has no inverse

13. $x_1 = 6, x_2 = -8$

15. $x_1 = 18, x_2 = 13$

17. $x_1 = x_2 = 5/2$

19. If $B = (b_{ij})$ is a (3×3) matrix such that $AB = I$, then $0b_{11} + 0b_{21} + 0b_{31} = 1$. Since this is impossible, no such matrix exists.

21. If $B = (b_{ij})$ is a (3×3) matrix such that $AB = I$, then $2b_{11} + 2b_{21} + 4b_{31} = 1$, $b_{11} + b_{21} + 7b_{31} = 0$, and $3b_{11} + 3b_{21} + 9b_{31} = 0$. This system of equations is inconsistent.

25. $Q^{-1} = C^{-1}A^{-1} = \begin{bmatrix} -3 & 1 \\ 3 & 5 \end{bmatrix}$

27. $Q^{-1} = (A^{-1})^T = \begin{bmatrix} 3 & 0 \\ 1 & 2 \end{bmatrix}$

29. $Q^{-1} = (A^{-1})^T(C^{-1})^T = \begin{bmatrix} -3 & 3 \\ 1 & 5 \end{bmatrix}$

31. $Q^{-1} = BC^{-1} = \begin{bmatrix} 1 & 5 \\ -1 & 4 \end{bmatrix}$

33. $Q^{-1} = (1/2)A^{-1} = \begin{bmatrix} 3/2 & 1/2 \\ 0 & 1 \end{bmatrix}$

35. $Q^{-1} = B(C^{-1}A^{-1}) = \begin{bmatrix} 3 & 11 \\ -3 & 7 \end{bmatrix}$

53. $b_0 = -5, b_1 = 2$

55. $b_0 = -7, b_1 = 0$

Exercises 1.11

1. $A^{-1} = \begin{bmatrix} -5 & 2 \\ 3 & -1 \end{bmatrix}$

3. $\begin{bmatrix} 3 & -1 \\ -2 & 1 \end{bmatrix}$

5. $\begin{bmatrix} -1/3 & 2/3 \\ 2/3 & -1/3 \end{bmatrix}$

7. $\begin{bmatrix} 1 & 0 & 0 \\ -2 & 1 & 0 \\ 5 & -4 & 1 \end{bmatrix}$

9. $\begin{bmatrix} 1 & -2 & 0 \\ 3 & -3 & -1 \\ -6 & 7 & 2 \end{bmatrix}$

11. $\begin{bmatrix} -1/2 & -2/3 & -1/6 & 7/6 \\ 1 & 1/3 & 1/3 & -4/3 \\ 0 & -1/3 & -1/3 & 1/3 \\ -1/2 & 1 & 1/2 & 1/2 \end{bmatrix}$

13. (a) $x_1 = -1/4, x_2 = 2/4$;   (b) $x_1 = -1, x_2 = 1$;
(c) $x_1 = -5, x_2 = 4$

15. (a) $x_1 = 3, x_2 = 4, x_3 = -9$;   (b) $x_1 = 7, x_2 = 14$,
$x_3 = -30$;   (c) $x_1 = -6, x_2 = -8, x_3 = 19$

17. $[A : \mathbf{c}_1, \mathbf{c}_2, \mathbf{c}_3]$ reduces to
$$\begin{bmatrix} 1 & 3 & -4 & 1 & 2 & 4 \\ 0 & -7 & 11 & -1 & -6 & -9 \\ 0 & 0 & 0 & 0 & 0 & 0 \end{bmatrix}.$$

Solutions to $A\mathbf{x} = \mathbf{c}_i$, $1 \le i \le 3$ are:

$$\mathbf{x} = \frac{1}{7}\begin{bmatrix} 4 - 5x_3 \\ 1 + 11x_3 \\ 7x_3 \end{bmatrix}, \quad \mathbf{x} = \frac{1}{7}\begin{bmatrix} -4 - 5x_3 \\ 6 + 11x_3 \\ 7x_3 \end{bmatrix},$$

$$\mathbf{x} = \frac{1}{7}\begin{bmatrix} 1 - 5x_3 \\ 9 + 11x_3 \\ 7x_3 \end{bmatrix};$$

$$B = \frac{1}{7}\begin{bmatrix} 4 & -4 & 1 \\ 1 & 6 & 9 \\ 0 & 0 & 0 \end{bmatrix} + \frac{x_3}{7}\begin{bmatrix} -5 & -5 & -5 \\ 11 & 11 & 11 \\ 7 & 7 & 7 \end{bmatrix}.$$

19. $B = \begin{bmatrix} 1 & 10 \\ 15 & 12 \\ 3 & 3 \end{bmatrix}$; $C = \begin{bmatrix} 13 & 12 & 8 \\ 2 & 3 & 5 \end{bmatrix}$

21. $(AB)^{-1} = B^{-1}A^{-1} = \begin{bmatrix} 2 & 35 & 1 \\ 14 & 35 & 34 \\ 23 & 12 & 70 \end{bmatrix}$,

$$(3A)^{-1} = \frac{1}{3}A^{-1} = \begin{bmatrix} 1/3 & 2/3 & 5/3 \\ 1 & 1/3 & 2 \\ 2/3 & 8/3 & 1/3 \end{bmatrix},$$

$$(A^T)^{-1} = (A^{-1})^T = \begin{bmatrix} 1 & 3 & 2 \\ 2 & 1 & 8 \\ 5 & 6 & 1 \end{bmatrix}$$

Exercises 2.1

1.

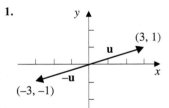

3.

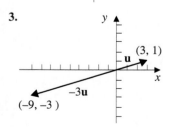

5.

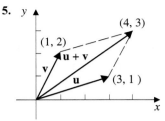

7.

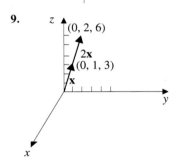

9.

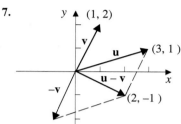

11.

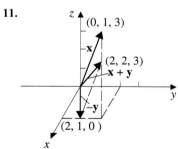

13.

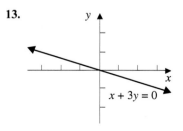

15.

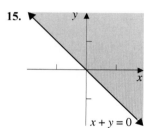

17.

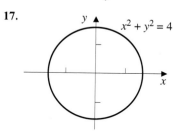

19. W is the plane with equation $x + y + 2z = 0$.

21. W is the set of points on the upper half of the sphere $x^2 + y^2 + z^2 = 1$.

23. $W = \left\{ \begin{bmatrix} a \\ 0 \end{bmatrix} : a \text{ any real number} \right\}$

25. $W = \left\{ \begin{bmatrix} a \\ 2 \end{bmatrix} : a \text{ any real number} \right\}$

27. $W = \left\{ \begin{bmatrix} x_1 \\ x_2 \\ x_3 \end{bmatrix} : x_1 + x_2 - 2x_3 = 0 \right\}$

29. $W = \left\{ \begin{bmatrix} 0 \\ x_2 \\ x_3 \end{bmatrix} : x_2, x_3 \text{ any real numbers} \right\}$

Exercises 2.2

1. W is a subspace. W is the set of points on the line with equation $x = 2y$.

3. W is not a subspace.

5. W is the subspace consisting of the points on the y-axis.

7. W is not a subspace.

9. W is the subspace consisting of the points on the plane $2x - y - z = 0$.

11. W is not a subspace.

13. W is not a subspace.

15. W is the subspace consisting of the points on the line with parametric equations $x = 2t$, $y = -t$, $z = t$.

17. W is the subspace consisting of the points on the x-axis.

19. W is the set of points on the plane $x + 2y + 3z = 0$.

23. W is the line formed by the two intersecting planes $x + 2y + 2z = 0$ and $x + 3y = 0$. The line has parametric equations $x = -6t$, $y = 2t$, $z = t$.

25. W is the set of points on the plane $x - z = 0$.

Exercises 2.3

1. $\text{Sp}(S) = \{\mathbf{x}: x_1 + x_2 = 0\}$; $\text{Sp}(S)$ is the line with equation $x + y = 0$.

3. $\text{Sp}(S) = \{\mathbf{e}\}$; $\text{Sp}(S)$ is the point $(0, 0)$.

5. $\text{Sp}(S) = R^2$

7. $\text{Sp}(S) = \{\mathbf{x}: 3x_1 + 2x_2 = 0\}$; $\text{Sp}(S)$ is the line with equation $3x + 2y = 0$.

9. $\text{Sp}(S) = R^2$

11. $\text{Sp}(S) = \{\mathbf{x}: x_1 + x_2 = 0\}$; $\text{Sp}(S)$ is the line with equation $x + y = 0$.

13. $\text{Sp}(S) = \{\mathbf{x}: x_2 + x_3 = 0 \text{ and } x_1 = 0\}$; $\text{Sp}(S)$ is the line through $(0, 0, 0)$ and $(0, -1, 1)$. The parametric equations for the line are $x = 0$, $y = -t$, $z = t$.

15. $\text{Sp}(S) = \{\mathbf{x}: 2x_1 - x_2 + x_3 = 0\}$; $\text{Sp}(S)$ is the plane with equation $2x - y + z = 0$.

17. $\text{Sp}(S) = R^3$

19. $\text{Sp}(S) = \{\mathbf{x}: x_2 + x_3 = 0\}$; $\text{Sp}(S)$ is the plane with equation $y + z = 0$.

21. The vectors $\mathbf{u}$ in (b), (c), and (e) are in $\text{Sp}(S)$; for (b), $\mathbf{u} = \mathbf{x}$; for (c), $\mathbf{u} = \mathbf{v}$; for (e), $\mathbf{u} = 3\mathbf{v} - 4\mathbf{x}$.

23. **d** and **e**

25. **x** and **y**

27. $\mathcal{N}(A) = \{\mathbf{x} \text{ in } R^2: -x_1 + 3x_2 = 0\}$; $\mathcal{R}(A) = \{\mathbf{x} \text{ in } R^2: 2x_1 + x_2 = 0\}$

29. $\mathcal{N}(A) = \{\mathbf{0}\}$; $\mathcal{R}(A) = R^2$

31. $\mathcal{N}(A) = \{\mathbf{x} \text{ in } R^3: x_1 + 2x_2 = 0 \text{ and } x_3 = 0\}$; $\mathcal{R}(A) = R^2$

33. $\mathcal{N}(A) = \{\mathbf{x} \text{ in } R^2: x_2 = 0\}$; $\mathcal{R}(A) = \{\mathbf{x} \text{ in } R^3: x_2 = 2x_1 \text{ and } x_3 = 3x_1\}$

35. $\mathcal{N}(A) = \{\mathbf{x} \text{ in } R^3: x_1 = -7x_3 \text{ and } x_2 = 2x_3\}$; $\mathcal{R}(A) = \{\mathbf{x} \text{ in } R^3: -4x_1 + 2x_2 + x_3 = 0\}$

37. $\mathcal{N}(A) = \{\boldsymbol{\theta}\}; \ \mathcal{R}(A) = R^3$

39. (a) The vectors **b** in (ii), (v), and (vi) are in $\mathcal{R}(A)$.
(b) For (ii), $\mathbf{x} = [1,0]^T$ is one choice; for (v), $\mathbf{x} = [0,1]^T$ is one choice; for (vi), $\mathbf{x} = [0,0]^T$ is one choice.
(c) For (ii), $\mathbf{b} = \mathbf{A}_1$; for (v), $\mathbf{b} = \mathbf{A}_2$, for (vi), $\mathbf{b} = 0\mathbf{A}_1 + 0\mathbf{A}_2$.

41. (a) The vectors **b** in (i), (iii), (v), and (vi) are in $\mathcal{R}(A)$.
(b) For (i), $\mathbf{x} = [-1,1,0]^T$ is one choice; for (iii), $\mathbf{x} = [-2,3,0]^T$ is one choice; for (v), $\mathbf{x} = [-2,1,0]^T$ is one choice; for (vi), $\mathbf{x} = [0,0,0]^T$ is one choice.
(c) For (i), $\mathbf{b} = -\mathbf{A}_1 + \mathbf{A}_2$; for (iii), $\mathbf{b} = -2\mathbf{A}_1 + 3\mathbf{A}_2$; for (v), $\mathbf{b} = -2\mathbf{A}_1 + \mathbf{A}_2$; for (vi), $\mathbf{b} = 0\mathbf{A}_1 + 0\mathbf{A}_2 + 0\mathbf{A}_3$.

47. $\mathbf{w}_1 = [-2,1,3]^T$, $\mathbf{w}_2 = [0,3,2]^T$

49. $\mathbf{w}_1 = [1,2,2]^T$, $\mathbf{w}_2 = [0,3,1]^T$

Exercises 2.4

1. $\{[1,0,1,0]^T, [-1,1,0,1]^T\}$

3. $\{[1,1,0,0]^T, [-1,0,1,0]^T, [3,0,0,1]^T\}$

5. $\{[-1,1,0,0]^T, [0,0,1,0]^T, [0,0,0,1]^T\}$

7. $\{[2,1,-1,0]^T, [-1,0,0,1]^T\}$

9. (a) $\mathbf{x} = 2\begin{bmatrix}1\\0\\1\\0\end{bmatrix} + \begin{bmatrix}-1\\1\\0\\1\end{bmatrix}$; (b) **x** is not in W.

(c) $\mathbf{x} = -3\begin{bmatrix}-1\\1\\0\\1\end{bmatrix}$; (d) $\mathbf{x} = 2\begin{bmatrix}1\\0\\1\\0\end{bmatrix}$

11. (a) $B = \begin{bmatrix}1 & 2 & 3 & -1\\0 & -1 & -1 & 1\\0 & 0 & 0 & 0\end{bmatrix}$

(b) A basis for $\mathcal{N}(A)$ is $\{[-1,-1,1,0]^T, [-1,1,0,1]^T\}$.
(c) $\{\mathbf{A}_1, \mathbf{A}_2\}$ is a basis for the column space of A; $\mathbf{A}_3 = \mathbf{A}_1 + \mathbf{A}_2$ and $\mathbf{A}_4 = \mathbf{A}_1 - \mathbf{A}_2$.
(d) $\{[1,2,3,-1], [0,-1,-1,1]\}$ is a basis for the row space of A.

13. (a) $B = \begin{bmatrix}1 & 2 & 1 & 0\\0 & 1 & 1 & -1\\0 & 0 & 0 & 0\\0 & 0 & 0 & 0\end{bmatrix}$

(b) A basis for $\mathcal{N}(A)$ is $\{[1,-1,1,0]^T, [-2,1,0,1]^T\}$.
(c) $\{\mathbf{A}_1, \mathbf{A}_2\}$ is a basis for the column space of A; $\mathbf{A}_3 = -\mathbf{A}_1 + \mathbf{A}_2$ and $\mathbf{A}_4 = 2\mathbf{A}_1 - \mathbf{A}_2$.
(d) $\{[1,2,1,0], [0,1,1,-1]\}$ is a basis for the row space of A.

15. (a) $B = \begin{bmatrix}1 & 2 & 1\\0 & 0 & -1\\0 & 0 & 0\end{bmatrix}$

(b) A basis for $\mathcal{N}(A)$ is $\{[-2,1,0]^T\}$.
(c) $\{\mathbf{A}_1, \mathbf{A}_3\}$ is a basis for the column space of A; $\mathbf{A}_2 = 2\mathbf{A}_1$.
(d) $\{[1,2,1], [0,0,-1]\}$ is a basis for the row space of A.

17. $\{[1,3,1]^T, [0,-1,-1]^T\}$ is a basis for $\mathcal{R}(A)$.

19. $\{[1,2,2,0]^T, [0,1,-2,1]^T\}$ is a basis for $\mathcal{R}(A)$.

21. (a) $\{[1,2]^T\}$; (b) $\{[1,2]^T\}$

23. (a) $\{[1,2,1]^T, [2,5,0]^T\}$;
(b) $\{[1,2,1]^T, [0,1,-2]^T\}$

25. (a) $\{[0,1,0]^T\}$; (b) $\{[-1,1,0]^T, [0,0,1]^T\}$;
(c) $\{[-1,1,0]^T\}$

27. $-2\mathbf{v}_1 - 3\mathbf{v}_2 + \mathbf{v}_3 = \boldsymbol{\theta}$, so S is linearly dependent. Since $\mathbf{v}_3 = 2\mathbf{v}_1 + 3\mathbf{v}_2$, if $\mathbf{v} = a_1\mathbf{v}_1 + a_2\mathbf{v}_2 + a_3\mathbf{v}_3$ is in $Sp\{\mathbf{v}_1, \mathbf{v}_2, \mathbf{v}_3\}$, then $\mathbf{v} = (a_1 + 2a_3)\mathbf{v}_1 + (a_2 + 3a_3)\mathbf{v}_2$. Therefore **v** is in $Sp\{\mathbf{v}_1, \mathbf{v}_2\}$.

29. The subsets are $\{\mathbf{v}_1, \mathbf{v}_2, \mathbf{v}_3\}, \{\mathbf{v}_1, \mathbf{v}_2, \mathbf{v}_4\}, \{\mathbf{v}_1, \mathbf{v}_3, \mathbf{v}_4\}$.

33. S is not a basis.

35. S is not a basis.

Exercises 2.5

1. S does not span R^2.

3. S is linearly dependent.

5. S is linearly dependent and does not span R^2.

7. S does not span R^3.

9. S is linearly dependent.

11. S is a basis.   **13.** S is not a basis.

15. $\dim(W) = 3$ **17.** $\dim(W) = 2$

19. $\dim(W) = 1$

21. $\{[-2,1]^T\}$ is a basis for $\mathcal{N}(A)$; nullity$(A) = 1$; rank$(A) = 1$.

23. $\{[-5,-2,1]^T\}$ is a basis for $\mathcal{N}(A)$; nullity$(A) = 1$; rank$(A) = 2$.

25. $\{[1,-1,1]^T, [0,2,3]^T\}$ is a basis for $\mathcal{R}(A)$; rank$(A) = 2$; nullity$(A) = 1$.

27. (a) $\{[1,1,-2]^T, [0,-1,1]^T, [0,0,1]^T\}$ is a basis for W; dim$(W) = 3$.

  (b) $\{[1,2,-1,1]^T, [0,1,-1,1]^T, [0,0,-1,4]^T\}$ is a basis for W; dim$(W) = 3$.

29. dim$(W) = 2$

33. (a) rank$(A) \le 3$ and nullity$(A) \ge 0$.
  (b) rank$(A) \le 3$ and nullity$(A) \ge 1$.
  (c) rank$(A) \le 4$ and nullity$(A) \ge 0$.

Exercises 2.6

5. $\mathbf{u}_1^T\mathbf{u}_3 = 0$ requires $a + b + c = 0$; $\mathbf{u}_2^T\mathbf{u}_3 = 0$ requires $2a + 2b - 4c = 0$; therefore $c = 0$ and $a + b = 0$.

7. $\mathbf{u}_1^T\mathbf{u}_2 = 0$ forces $a = 3$; then $\mathbf{u}_2^T\mathbf{u}_3 = 0$ requires $-8 - b + 3c = 0$, while $\mathbf{u}_1^T\mathbf{u}_3 = 0$ requires $4 + b + c = 0$; therefore $b = -5$, $c = 1$.

9. $\mathbf{v} = (2/3)\mathbf{u}_1 - (1/2)\mathbf{u}_2 + (1/6)\,\mathbf{u}_3$

11. $\mathbf{v} = 3\mathbf{u}_1$

13. $\mathbf{u}_1 = \begin{bmatrix} 0 \\ 0 \\ 1 \\ 0 \end{bmatrix}$, $\mathbf{u}_2 = \begin{bmatrix} 1 \\ 1 \\ 0 \\ 1 \end{bmatrix}$, $\mathbf{u}_3 = \begin{bmatrix} 1/3 \\ -2/3 \\ 0 \\ 1/3 \end{bmatrix}$

15. $\mathbf{u}_1 = \begin{bmatrix} 1 \\ 1 \\ 0 \end{bmatrix}$, $\mathbf{u}_2 = \begin{bmatrix} 1 \\ -1 \\ -1 \end{bmatrix}$, $\mathbf{u}_3 = \begin{bmatrix} 2 \\ -2 \\ 4 \end{bmatrix}$

17. $\mathbf{u}_1 = \begin{bmatrix} 0 \\ 1 \\ 0 \\ 1 \end{bmatrix}$, $\mathbf{u}_2 = \begin{bmatrix} -1 \\ -1 \\ 0 \\ 1 \end{bmatrix}$, $\mathbf{u}_3 = \begin{bmatrix} -2/3 \\ 1/3 \\ 1 \\ -1/3 \end{bmatrix}$

19. For the null space: $\begin{bmatrix} -3 \\ -1 \\ 1 \\ 0 \end{bmatrix}$, $\begin{bmatrix} 7/11 \\ -27/11 \\ -6/11 \\ 1 \end{bmatrix}$;

  for the range space: $\begin{bmatrix} 1 \\ 2 \\ 1 \end{bmatrix}$, $\begin{bmatrix} -11/6 \\ 8/6 \\ -5/6 \end{bmatrix}$

Exercises 2.7

1. (a) $\begin{bmatrix} 0 \\ 0 \end{bmatrix}$; (b) $\begin{bmatrix} -1 \\ 0 \end{bmatrix}$; (c) $\begin{bmatrix} 1 \\ -1 \end{bmatrix}$; (d) $\begin{bmatrix} -2 \\ 1 \end{bmatrix}$

3. (c) is not but (a), (b), and (d) are.

9. F is a linear transformation.

11. F is not a linear transformation.

13. F is a linear transformation.

15. F is a linear transformation.

17. F is not a linear transformation.

19. (a) $\begin{bmatrix} 3 \\ 1 \\ -1 \end{bmatrix}$; (b) $\begin{bmatrix} 0 \\ -1 \\ -2 \end{bmatrix}$; (c) $\begin{bmatrix} 7 \\ 2 \\ -3 \end{bmatrix}$

21. $T\left(\begin{bmatrix} x_1 \\ x_2 \end{bmatrix}\right) = \begin{bmatrix} x_1 + x_2 \\ x_1 - 2x_2 \end{bmatrix}$

23. $T\left(\begin{bmatrix} x_1 \\ x_2 \\ x_3 \end{bmatrix}\right) = \begin{bmatrix} -\dfrac{1}{2}x_1 - \dfrac{1}{2}x_2 + \dfrac{1}{2}x_3 \\ \dfrac{1}{2}x_1 + \dfrac{1}{2}x_2 + \dfrac{1}{2}x_3 \end{bmatrix}$

25. $A = \begin{bmatrix} 1 & 3 \\ 2 & 1 \end{bmatrix}$; $\mathcal{N}(T) = \{\boldsymbol{\theta}\}$; $\mathcal{R}(T) = R^2$; rank$(T) = 2$; nullity$(T) = 0$

27. $A = \begin{bmatrix} 3 & 2 \end{bmatrix}$; $\mathcal{N}(T) = \{\mathbf{x}$ in R^2: $3x_1 + 2x_2 = 0\}$; $\mathcal{R}(T) = R^1$; rank$(T) = 1$; nullity$(T) = 1$

29. $A = \begin{bmatrix} 1 & -1 & 0 \\ 0 & 1 & -1 \end{bmatrix}$; $\mathcal{N}(T) = \{\mathbf{x}$ in R^3: $x_1 = x_3$ and $x_2 = x_3\}$; $\mathcal{R}(T) = R^2$; rank$(T) = 2$; nullity$(T) = 1$

Exercises 2.8

1. $\mathbf{w}^* = \begin{bmatrix} 1/2 \\ 3 \\ 11/2 \end{bmatrix}$

3. $\mathbf{w}^* = \begin{bmatrix} 1 \\ 1 \\ 1 \end{bmatrix}$

5. $\mathbf{w}^* = \begin{bmatrix} 4 \\ 2 \\ 2 \end{bmatrix}$

7. $\mathbf{w}^* = \begin{bmatrix} 3 \\ -1 \\ 2 \end{bmatrix}$

9. $\mathbf{w}^* = \begin{bmatrix} 0 \\ 1 \\ -1 \end{bmatrix}$

11. $\mathbf{w}^* = \dfrac{4}{5}\begin{bmatrix} 2 \\ 1 \\ 0 \end{bmatrix} + \dfrac{11}{2}\begin{bmatrix} -1/5 \\ 2/5 \\ 1 \end{bmatrix}$

13. $\mathbf{w}^* = \begin{bmatrix} 1 \\ 1 \\ 0 \end{bmatrix} + 4\begin{bmatrix} 1/2 \\ -1/2 \\ 1 \end{bmatrix}$

15. $\mathbf{w}^* = 2\begin{bmatrix} 1 \\ -1 \\ 1 \end{bmatrix} + \begin{bmatrix} 1 \\ 1 \\ 0 \end{bmatrix}$

17. $P_W = \dfrac{1}{6}\begin{bmatrix} 5 & 2 & -1 \\ 2 & 2 & 2 \\ -1 & 2 & 5 \end{bmatrix}, \quad \mathbf{w}^* = \begin{bmatrix} 9/2 \\ 3 \\ 3/2 \end{bmatrix}$

19. $P_W = \dfrac{1}{6}\begin{bmatrix} 5 & 1 & 2 \\ 1 & 5 & -2 \\ 2 & -2 & 2 \end{bmatrix}, \quad \mathbf{w}^* = \begin{bmatrix} 2 \\ 0 \\ 1 \end{bmatrix}$

Exercises 2.9

1. $\mathbf{x}^* = \begin{bmatrix} -5/13 \\ 7/13 \end{bmatrix}$

3. $\mathbf{x}^* = \begin{bmatrix} (28/74) - 3x_3 \\ (27/74) + x_3 \\ x_3 \end{bmatrix}$, x_3 arbitrary

5. $\mathbf{x}^* = \begin{bmatrix} 2x_2 + 26/7 \\ -x_2 \end{bmatrix}$, x_2 arbitrary

7. $y = 1.3t + 1.1$

9. $y = 2t + 1$

11. $y = (1/3)t^2 + 2/3$

13. $y = 2t + 1$

Exercises 3.1

1. $\lambda = 1, \mathbf{x} = a\begin{bmatrix} -1 \\ 1 \end{bmatrix}, a \neq 0;$

$\lambda = 3, \mathbf{x} = a\begin{bmatrix} 0 \\ 1 \end{bmatrix}, a \neq 0$

3. $\lambda = 1, \mathbf{x} = a\begin{bmatrix} 1 \\ 1 \end{bmatrix}, a \neq 0;$

$\lambda = 3, \mathbf{x} = a\begin{bmatrix} -1 \\ 1 \end{bmatrix}, a \neq 0$

5. $\lambda = 1, \mathbf{x} = a\begin{bmatrix} -1 \\ 1 \end{bmatrix}, a \neq 0;$

$\lambda = 3, \mathbf{x} = a\begin{bmatrix} 1 \\ 1 \end{bmatrix}, a \neq 0$

7. $\lambda = 1, \mathbf{x} = a\begin{bmatrix} 0 \\ 1 \end{bmatrix}, a \neq 0$

9. $\lambda = 0, \mathbf{x} = a\begin{bmatrix} -1 \\ 1 \end{bmatrix}, a \neq 0;$

$\lambda = 5, \mathbf{x} = a\begin{bmatrix} 2 \\ 3 \end{bmatrix}, a \neq 0$

11. $\lambda = 2, \mathbf{x} = a\begin{bmatrix} -1 \\ 1 \end{bmatrix}, a \neq 0$

Exercises 3.2

1. $M_{11} = \begin{bmatrix} 1 & 3 & -1 \\ 2 & 4 & 1 \\ 2 & 0 & -2 \end{bmatrix}; A_{11} = 18$

3. $M_{31} = \begin{bmatrix} -1 & 3 & 1 \\ 1 & 3 & -1 \\ 2 & 0 & -2 \end{bmatrix}; A_{31} = 0$

5. $M_{34} = \begin{bmatrix} 2 & -1 & 3 \\ 4 & 1 & 3 \\ 2 & 2 & 0 \end{bmatrix}; A_{34} = 0$

7. $\det(A) = 0$

9. $\det(A) = 0;$ A is singular.

11. $\det(A) = -1;$ A is nonsingular.

13. $\det(A) = 6;$ A is nonsingular.

15. $\det(A) = 20;$ A is nonsingular.

17. $\det(A) = 6;$ A is nonsingular.

19. $\det(A) = 36;$ A is nonsingular.

21. $y = 2x - 1$

27. 5   **29.** 3/5

Exercises 3.3

1. $\det(A) = \begin{vmatrix} 1 & 2 & 1 \\ 3 & 0 & 2 \\ -1 & 1 & 3 \end{vmatrix} \overset{\substack{R_2 - 3R_1 \\ R_3 + R_1}}{=} \begin{vmatrix} 1 & 2 & 1 \\ 0 & -6 & -1 \\ 0 & 3 & 4 \end{vmatrix}$

$= \begin{vmatrix} -6 & -1 \\ 3 & 4 \end{vmatrix} = -21$

3. $\det(A) = \begin{vmatrix} 3 & 6 & 9 \\ 2 & 0 & 2 \\ 1 & 2 & 0 \end{vmatrix} = (3)(2)\begin{vmatrix} 1 & 2 & 3 \\ 1 & 0 & 1 \\ 1 & 2 & 0 \end{vmatrix}$

$\overset{\substack{R_2 - R_1 \\ R_3 - R_1}}{=} 6\begin{vmatrix} 1 & 2 & 3 \\ 0 & -2 & -2 \\ 0 & 0 & -3 \end{vmatrix} = 6\begin{vmatrix} -2 & -2 \\ 0 & -3 \end{vmatrix} = 36$

5. $\det(A) = \begin{vmatrix} 2 & 4 & -3 \\ 3 & 2 & 5 \\ 2 & 3 & 4 \end{vmatrix} = (\frac{1}{2}) \begin{vmatrix} 2 & 4 & -3 \\ 6 & 4 & 10 \\ 2 & 3 & 4 \end{vmatrix}$

$\underset{\substack{R_2 - 3R_1 \\ R_3 - R_1}}{=} (\frac{1}{2}) \begin{vmatrix} 2 & 4 & -3 \\ 0 & -8 & 19 \\ 0 & -1 & 7 \end{vmatrix}$

$= \begin{vmatrix} -8 & 19 \\ -1 & 7 \end{vmatrix} = -37$

7. $\begin{vmatrix} 1 & 0 & 0 & 0 \\ 2 & 0 & 0 & 3 \\ 1 & 1 & 0 & 1 \\ 1 & 4 & 2 & 2 \end{vmatrix} = (-1) \begin{vmatrix} 1 & 0 & 0 & 0 \\ 2 & 3 & 0 & 0 \\ 1 & 1 & 0 & 1 \\ 1 & 2 & 2 & 4 \end{vmatrix}$

$= \begin{vmatrix} 1 & 0 & 0 & 0 \\ 2 & 3 & 0 & 0 \\ 1 & 1 & 1 & 0 \\ 1 & 2 & 4 & 2 \end{vmatrix} = 6$

9. $\begin{vmatrix} 0 & 0 & 2 & 0 \\ 0 & 0 & 1 & 3 \\ 0 & 4 & 1 & 3 \\ 2 & 1 & 5 & 6 \end{vmatrix} = (-1) \begin{vmatrix} 2 & 0 & 0 & 0 \\ 1 & 0 & 0 & 3 \\ 1 & 4 & 0 & 3 \\ 5 & 1 & 2 & 6 \end{vmatrix} = \begin{vmatrix} 2 & 0 & 0 & 0 \\ 1 & 3 & 0 & 0 \\ 1 & 3 & 0 & 4 \\ 5 & 6 & 2 & 1 \end{vmatrix}$

$= (-1) \begin{vmatrix} 2 & 0 & 0 & 0 \\ 1 & 3 & 0 & 0 \\ 1 & 3 & 4 & 0 \\ 5 & 6 & 1 & 2 \end{vmatrix} = -48$

11. $\begin{vmatrix} 0 & 0 & 1 & 0 \\ 0 & 2 & 6 & 3 \\ 2 & 4 & 1 & 5 \\ 0 & 0 & 0 & 4 \end{vmatrix} = (-1) \begin{vmatrix} 2 & 4 & 1 & 5 \\ 0 & 2 & 6 & 3 \\ 0 & 0 & 1 & 0 \\ 0 & 0 & 0 & 4 \end{vmatrix} = -16$

13. $\det(B) = 3\det(A) = 6$

15. $\det(B) = -\det(A) = -2$

17. $\det(B) = -2\det(A) = -4$

19. $R_4 - \frac{1}{2}R_1$ gives $\begin{vmatrix} 2 & 4 & 2 & 6 \\ 1 & 3 & 2 & 1 \\ 2 & 1 & 2 & 3 \\ 0 & 0 & 0 & -2 \end{vmatrix} = (-2) \begin{vmatrix} 2 & 4 & 2 \\ 1 & 3 & 2 \\ 2 & 1 & 2 \end{vmatrix}$

$= (-2) \begin{vmatrix} 2 & 4 & 2 \\ 0 & 1 & 1 \\ 0 & -3 & 0 \end{vmatrix} = (-2)(2) \begin{vmatrix} 1 & 1 \\ -3 & 0 \end{vmatrix} = -12.$

21. $R_4 - 2R_3$ gives $\begin{vmatrix} 0 & 4 & 1 & 3 \\ 0 & 2 & 2 & 1 \\ 1 & 3 & 1 & 2 \\ 0 & -4 & -1 & 0 \end{vmatrix} = \begin{vmatrix} 4 & 1 & 3 \\ 2 & 2 & 1 \\ -4 & -1 & 0 \end{vmatrix}$

$= \begin{vmatrix} 4 & 1 & 3 \\ 2 & 2 & 1 \\ 0 & 0 & 3 \end{vmatrix} = 3 \begin{vmatrix} 4 & 1 \\ 2 & 2 \end{vmatrix} = 18.$

Exercises 3.4

1. $p(t) = (1 - t)(3 - t); \quad \lambda = 1, \lambda = 3$

3. $p(t) = t^2 - 4t + 3 = (t - 3)(t - 1); \quad \lambda = 1, \lambda = 3$

5. $p(t) = t^2 - 4t + 4 = (t - 2)^2; \quad \lambda = 2$, algebraic multiplicity 2

7. $p(t) = -t^3 + t^2 + t - 1 = -(t - 1)^2(t + 1);$ $\lambda = 1$, algebraic multiplicity 2; $\lambda = -1$, algebraic multiplicity 1

9. $p(t) = -t^3 + 2t^2 + t - 2 = -(t - 2)(t - 1)(t + 1);$ $\lambda = 2, \lambda = 1, \lambda = -1$

11. $p(t) = -t^3 + 6t^2 - 12t + 8 = -(t - 2)^3; \quad \lambda = 2$, algebraic multiplicity 3

13. $p(t) = t^4 - 18t^3 + 97t^2 - 180t + 100 =$ $(t - 1)(t - 2)(t - 5)(t - 10);$ $\lambda = 1, \lambda = 2, \lambda = 5, \lambda = 10$

Exercises 3.5

1. $\mathbf{x}$ is an eigenvector if and only if $\mathbf{x} = \begin{bmatrix} -x_2 \\ x_2 \end{bmatrix}$, $x_2 \neq 0$; basis consists of $\begin{bmatrix} -1 \\ 1 \end{bmatrix}$; algebraic and geometric multiplicities are 1.

3. $\mathbf{x}$ is an eigenvector if and only if $\mathbf{x} = \begin{bmatrix} -x_2 \\ x_2 \end{bmatrix}$, $x_2 \neq 0$; basis consists of $\begin{bmatrix} -1 \\ 1 \end{bmatrix}$; algebraic multiplicity is 2 and geometric multiplicity is 1.

5. $\mathbf{x}$ is an eigenvector if and only if $\mathbf{x} = \begin{bmatrix} a \\ -a \\ 2a \end{bmatrix}$, $a \neq 0$; basis consists of $\begin{bmatrix} 1 \\ -1 \\ 2 \end{bmatrix}$; algebraic and geometric multiplicities are 1.

7. $\mathbf{x}$ is an eigenvector if and only if $\mathbf{x} = \begin{bmatrix} x_4 \\ -x_4 \\ -x_4 \\ x_4 \end{bmatrix}, x_4 \neq 0$;

basis consists of $\begin{bmatrix} 1 \\ -1 \\ -1 \\ 1 \end{bmatrix}$; algebraic and geometric multiplicities are 1.

9. $\mathbf{x}$ is an eigenvector if and only if $\mathbf{x} = \begin{bmatrix} x_4 \\ x_4 \\ x_4 \\ x_4 \end{bmatrix}, x_4 \neq 0$;

basis consists of $\begin{bmatrix} 1 \\ 1 \\ 1 \\ 1 \end{bmatrix}$; algebraic and geometric multiplicities are 1.

11. $\mathbf{x}$ is an eigenvector if and only if

$\mathbf{x} = \begin{bmatrix} -x_2 -x_3 -x_4 \\ x_2 \\ x_3 \\ x_4 \end{bmatrix}$; basis consists of $\begin{bmatrix} -1 \\ 1 \\ 0 \\ 0 \end{bmatrix}$,

$\begin{bmatrix} -1 \\ 0 \\ 1 \\ 0 \end{bmatrix}, \begin{bmatrix} -1 \\ 0 \\ 0 \\ 1 \end{bmatrix}$; algebraic and geometric multiplicities are 3.

13. For $\lambda = 2$, $\mathbf{x} = \begin{bmatrix} x_1 \\ -2x_3 \\ x_3 \end{bmatrix} = x_1 \begin{bmatrix} 1 \\ 0 \\ 0 \end{bmatrix} + x_3 \begin{bmatrix} 0 \\ -2 \\ 1 \end{bmatrix}$; for

$\lambda = 3$, $\mathbf{x} = \begin{bmatrix} x_2 \\ x_2 \\ 0 \end{bmatrix}$; the matrix is not defective.

15. For $\lambda = 2$, $\mathbf{x} = \begin{bmatrix} x_1 \\ x_2 \\ 0 \end{bmatrix} = x_1 \begin{bmatrix} 1 \\ 0 \\ 0 \end{bmatrix} + x_2 \begin{bmatrix} 0 \\ 1 \\ 0 \end{bmatrix}$; for $\lambda = 1$,

$\mathbf{x} = \begin{bmatrix} -3x_3 \\ -x_3 \\ x_3 \end{bmatrix}$; the matrix is not defective.

17. For $\lambda = 2$, $\mathbf{x} = \begin{bmatrix} x_1 \\ -x_1 \\ 2x_1 \end{bmatrix}$; for $\lambda = 1$, $\mathbf{x} = \begin{bmatrix} -3x_2 \\ x_2 \\ -7x_2 \end{bmatrix}$; for

$\lambda = -1$; $\mathbf{x} = \begin{bmatrix} x_1 \\ 2x_1 \\ 2x_1 \end{bmatrix}$; the matrix is not defective.

19. For $\lambda = 1$, eigenvectors are $\mathbf{u}_1 = \begin{bmatrix} 1 \\ 0 \\ 0 \end{bmatrix}$, $\mathbf{u}_2 = \begin{bmatrix} 0 \\ 1 \\ 2 \end{bmatrix}$; for

$\lambda = 2$, $\mathbf{u}_3 = \begin{bmatrix} 1 \\ 2 \\ 3 \end{bmatrix}$. Therefore $\mathbf{x} = \mathbf{u}_1 + 2\mathbf{u}_2 + \mathbf{u}_3$, and

thus $A^{10}\mathbf{x} = \mathbf{u}_1 + 2\mathbf{u}_2 + 2^{10}\mathbf{u}_3 = \begin{bmatrix} 1025 \\ 2050 \\ 3076 \end{bmatrix}$.

Exercises 3.6

1. $3 + 2i$ 3. $7 - 3i$

5. 6   7. 17

9. $-5 + 5i$ 11. $17 - 6i$

13. $(10 - 11i)/17$ 15. $(3 + i)/2$

17. 1

19. $\lambda = 4 + 2i$, $\mathbf{x} = a\begin{bmatrix} 4 \\ -1 + i \end{bmatrix}$;

$\lambda = 4 - 2i$, $\mathbf{x} = a\begin{bmatrix} 4 \\ -1 - i \end{bmatrix}$

21. $\lambda = i$, $\mathbf{x} = a\begin{bmatrix} -2 + i \\ 5 \end{bmatrix}$;

$\lambda = -i$, $\mathbf{x} = a\begin{bmatrix} -2 - i \\ 5 \end{bmatrix}$

23. $\lambda = 2$, $\mathbf{x} = a\begin{bmatrix} -1 \\ 0 \\ 1 \end{bmatrix}$; $\lambda = 2 + 3i$, $\mathbf{x} = a\begin{bmatrix} -5 + 3i \\ 3 + 3i \\ 2 \end{bmatrix}$;

$\lambda = 2 - 3i$, $\mathbf{x} = a\begin{bmatrix} -5 - 3i \\ 3 - 3i \\ 2 \end{bmatrix}$

25. $x = 2 - i$, $y = 3 - 2i$

27. $\sqrt{6}$ 29. 4

Exercises 3.7

1. A is symmetric, so A is diagonalizable. For
$$S = \begin{bmatrix} 1 & -1 \\ 1 & 1 \end{bmatrix}, S^{-1}AS = \begin{bmatrix} 1 & 0 \\ 0 & 3 \end{bmatrix} \text{ and}$$
$$S^{-1}A^5S = \begin{bmatrix} 1 & 0 \\ 0 & 243 \end{bmatrix}; \text{ therefore}$$
$$A^5 = \begin{bmatrix} 122 & -121 \\ -121 & 122 \end{bmatrix}.$$

3. A is not diagonalizable; $\lambda = -1$ is the only eigenvalue and $\mathbf{x} = a\begin{bmatrix} 1 \\ 1 \end{bmatrix}$, $a \neq 0$, are the only eigenvectors.

5. A is diagonalizable since A has distinct eigenvalues.
For $S = \begin{bmatrix} -1 & 0 \\ 10 & 1 \end{bmatrix}, S^{-1}AS = \begin{bmatrix} 1 & 0 \\ 0 & 2 \end{bmatrix}$ and
$$S^{-1}A^5S = \begin{bmatrix} 1 & 0 \\ 0 & 32 \end{bmatrix}; \text{ therefore}$$
$$A^5 = \begin{bmatrix} 1 & 0 \\ 310 & 32 \end{bmatrix}.$$

7. A is not diagonalizable; $\lambda = 1$ is the only eigenvalue and it has geometric multiplicity 2. A basis for the eigenspace consists of $[1, 1, 0]^T$ and $[2, 0, 1]^T$.

9. A is diagonalizable since A has distinct eigenvalues.
For $S = \begin{bmatrix} -3 & -1 & 1 \\ 1 & 1 & 2 \\ -7 & -2 & 2 \end{bmatrix}, S^{-1}AS = \begin{bmatrix} 1 & 0 & 0 \\ 0 & 2 & 0 \\ 0 & 0 & -1 \end{bmatrix}.$

Therefore $A^5 = \begin{bmatrix} 163 & -11 & -71 \\ -172 & 10 & 75 \\ 324 & -22 & -141 \end{bmatrix}.$

11. A is not diagonalizable; $\lambda = 1$ has algebraic multiplicity 2 and geometric multiplicity 1.

13. Q is orthogonal.

15. Q is not orthogonal since the columns are not orthonormal.

17. Q is orthogonal.

19. $\alpha = 1/\sqrt{2}, \beta = 1/\sqrt{6}, a = -1/\sqrt{3}, b = 1/\sqrt{3},$
$c = 1/\sqrt{3}$

33. $\lambda = 2, \mathbf{u} = \dfrac{1}{\sqrt{2}}\begin{bmatrix} 1 \\ -1 \end{bmatrix}, \mathbf{v} = \dfrac{1}{\sqrt{2}}\begin{bmatrix} 1 \\ 1 \end{bmatrix}, Q = [\mathbf{u}, \mathbf{v}],$
$$Q^TAQ = \begin{bmatrix} 2 & -2 \\ 0 & 2 \end{bmatrix}$$

35. $\lambda = 1, \mathbf{u} = \dfrac{1}{\sqrt{2}}\begin{bmatrix} 1 \\ 1 \end{bmatrix}, \mathbf{v} = \dfrac{1}{\sqrt{2}}\begin{bmatrix} 1 \\ -1 \end{bmatrix}, Q = [\mathbf{u}, \mathbf{v}],$
$$Q^TAQ = \begin{bmatrix} 1 & 0 \\ 0 & 3 \end{bmatrix}$$

Exercises 3.8

1. $\mathbf{x}_1 = \begin{bmatrix} 4 \\ 2 \end{bmatrix}, \mathbf{x}_2 = \begin{bmatrix} 2 \\ 4 \end{bmatrix}, \mathbf{x}_3 = \begin{bmatrix} 4 \\ 2 \end{bmatrix}, \mathbf{x}_4 = \begin{bmatrix} 2 \\ 4 \end{bmatrix}$

3. $\mathbf{x}_1 = \begin{bmatrix} 80 \\ 112 \end{bmatrix}, \mathbf{x}_2 = \begin{bmatrix} 68 \\ 124 \end{bmatrix}, \mathbf{x}_3 = \begin{bmatrix} 65 \\ 127 \end{bmatrix},$
$\mathbf{x}_4 = \begin{bmatrix} 64.25 \\ 127.75 \end{bmatrix}$

5. $\mathbf{x}_1 = \begin{bmatrix} 7 \\ 1 \end{bmatrix}, \mathbf{x}_2 = \begin{bmatrix} 11 \\ 8 \end{bmatrix}, \mathbf{x}_3 = \begin{bmatrix} 43 \\ 19 \end{bmatrix}, \mathbf{x}_4 = \begin{bmatrix} 119 \\ 62 \end{bmatrix}$

7. $\mathbf{x}_k = 3(1)^k\begin{bmatrix} 1 \\ 1 \end{bmatrix} + (-1)^k\begin{bmatrix} -1 \\ 1 \end{bmatrix} = \begin{bmatrix} 3 + (-1)^{k+1} \\ 3 + (-1)^k \end{bmatrix};$
$\mathbf{x}_4 = \begin{bmatrix} 2 \\ 4 \end{bmatrix}, \mathbf{x}_{10} = \begin{bmatrix} 2 \\ 4 \end{bmatrix};$ the sequence $\{\mathbf{x}_k\}$ has no limit, but $\|\mathbf{x}_k\| \leq \sqrt{20}.$

9. $\mathbf{x}_k = 64(1)^k\begin{bmatrix} 1 \\ 2 \end{bmatrix} - 64(1/4)^k\begin{bmatrix} -1 \\ 1 \end{bmatrix} = 64\begin{bmatrix} 1 + (1/4)^k \\ 2 - (1/4)^k \end{bmatrix};$
$\mathbf{x}_4 = \begin{bmatrix} 64.25 \\ 127.75 \end{bmatrix}, \mathbf{x}_{10} = \begin{bmatrix} 64.00006 \\ 127.99994 \end{bmatrix};$ the sequence $\{\mathbf{x}_k\}$ converges to $[64, 128]^T.$

11. $\mathbf{x}_k = (3/4)(3)^k\begin{bmatrix} 2 \\ 1 \end{bmatrix} + (5/4)(-1)^k\begin{bmatrix} -2 \\ 1 \end{bmatrix} = $
$\dfrac{1}{4}\begin{bmatrix} 6(3)^k - 10(-1)^k \\ 3(3)^k + 5(-1)^k \end{bmatrix}; \mathbf{x}_4 = \begin{bmatrix} 119 \\ 62 \end{bmatrix};$
$\mathbf{x}_{10} = \begin{bmatrix} 88571 \\ 44288 \end{bmatrix};$ the sequence $\{\mathbf{x}_k\}$ has no limit and $\|\mathbf{x}_k\| \to \infty.$

13. $\mathbf{x}_k = -2(1)^k\begin{bmatrix} -3 \\ 1 \\ -7 \end{bmatrix} - 2(2)^k\begin{bmatrix} -1 \\ 1 \\ -2 \end{bmatrix} - 5(-1)^k\begin{bmatrix} 1 \\ 2 \\ 2 \end{bmatrix} = $
$\begin{bmatrix} 6 + 2(2)^k - 5(-1)^k \\ -2 - 2(2)^k - 10(-1)^k \\ 14 + 4(2)^k - 10(-1)^k \end{bmatrix}; \mathbf{x}_4 = \begin{bmatrix} 33 \\ -44 \\ 68 \end{bmatrix},$
$\mathbf{x}_{10} = \begin{bmatrix} 2049 \\ -2060 \\ 4100 \end{bmatrix};$ the sequence $\{\mathbf{x}_k\}$ has no limit and $\|\mathbf{x}_k\| \to \infty.$

15. $\mathbf{x}(t) = 3e^{2t}\begin{bmatrix} 2 \\ 1 \end{bmatrix} - 2e^{-t}\begin{bmatrix} 1 \\ 1 \end{bmatrix}$

17. $\mathbf{x}(t) = \begin{bmatrix} 0 \\ -2 \\ 2 \end{bmatrix} - e^{2t}\begin{bmatrix} -2 \\ -3 \\ 1 \end{bmatrix} + e^{3t}\begin{bmatrix} 1 \\ 2 \\ 0 \end{bmatrix}$

21. $\alpha = -.18$; $\mathbf{x}_k = \dfrac{16}{118}(1)^k\begin{bmatrix} 3 \\ 10 \end{bmatrix} + \dfrac{7}{118}(-.18)^k\begin{bmatrix} 10 \\ -6 \end{bmatrix}$;

the limit is $\dfrac{16}{118}\begin{bmatrix} 3 \\ 10 \end{bmatrix}$.

Exercises 4.2

1. $\begin{bmatrix} 0 & -7 & 5 \\ -11 & -3 & -12 \end{bmatrix}$, $\begin{bmatrix} 12 & -22 & 38 \\ -50 & -6 & -15 \end{bmatrix}$,
$\begin{bmatrix} 7 & -21 & 28 \\ -42 & -7 & -14 \end{bmatrix}$

3. $e^x - 2\sin x$, $e^x - 2\sin x + 3\sqrt{x^2 + 1}$,
$-2e^x - \sin x + 3\sqrt{x^2 + 1}$

5. $c_1 = -2 + c_3$, $c_2 = 3 - c_3$, c_3 arbitrary

7. Not a vector space

9. Not a vector space

11. Not a vector space

13. A vector space

15. Not a vector space

25. A vector space

27. Not a vector space

29. A vector space

Exercises 4.3

1. Not a subspace

3. A subspace

5. A subspace

7. Not a subspace

9. A subspace

11. Not a subspace

13. A subspace

15. Not a subspace

17. $p(x) = -p_1(x) + 3p_2(x) - 2p_3(x)$

19. $A = (-1 - 2x)B_1 + (2 + 3x)B_2 + xB_3 - 3B_4$,
x arbitrary

21. $\cos 2x = -\sin^2 x + \cos^2 x$

23. $W = \mathrm{Sp}\{1, x^2\}$

25. In Exercise 2, $W = \mathrm{Sp}\left\{\begin{bmatrix} 1 & 1 & 0 \\ 0 & 0 & 0 \end{bmatrix}, \begin{bmatrix} -2 & 0 & 1 \\ 0 & 0 & 0 \end{bmatrix}, \right.$
$\left.\begin{bmatrix} 0 & 0 & 0 \\ 1 & 0 & 0 \end{bmatrix}, \begin{bmatrix} 0 & 0 & 0 \\ 0 & 1 & 0 \end{bmatrix}, \begin{bmatrix} 0 & 0 & 0 \\ 0 & 0 & 1 \end{bmatrix}\right\}$;
in Exercise 3, $W = \mathrm{Sp}\left\{\begin{bmatrix} -1 & -1 & 1 \\ 0 & 0 & 0 \end{bmatrix}, \right.$

$\begin{bmatrix} 0 & 0 & 0 \\ 1 & 0 & 0 \end{bmatrix}, \begin{bmatrix} 0 & 0 & 0 \\ 0 & 1 & 0 \end{bmatrix}\right\}$; in Exercise 5,
$W = \mathrm{Sp}\{-1 + x, -2 + x^2\}$; in Exercise 6,
$W = \mathrm{Sp}\{1, -4x + x^2\}$; in Exercise 8,
$W = \mathrm{Sp}\{1 - x^2, x\}$

27. $W = \mathrm{Sp}\{B_1, B_2, E_{12}, E_{13}, E_{21}, E_{23}, E_{31}, E_{32}\}$, where
$B_1 = \begin{bmatrix} -1 & 0 & 0 \\ 0 & 1 & 0 \\ 0 & 0 & 0 \end{bmatrix}$ and $B_2 = \begin{bmatrix} -1 & 0 & 0 \\ 0 & 0 & 0 \\ 0 & 0 & 1 \end{bmatrix}$

29. $A = B + C$, where $B = (A + A^T)/2$ and
$C = (A - A^T)/2$

31. (a) $W = \mathrm{Sp}\{E_{12}, E_{13}, E_{23}\}$;

(b) $W = \mathrm{Sp}\left\{\begin{bmatrix} -1 & 0 & 0 \\ 0 & 1 & 0 \\ 0 & 0 & 0 \end{bmatrix}, \begin{bmatrix} -1 & 0 & 0 \\ 0 & 0 & 0 \\ 0 & 0 & 1 \end{bmatrix}, \right.$
$\left.\begin{bmatrix} 0 & -1 & 0 \\ 0 & 0 & 1 \\ 0 & 0 & 0 \end{bmatrix}, \begin{bmatrix} 0 & 0 & 1 \\ 0 & 0 & 0 \\ 0 & 0 & 0 \end{bmatrix}\right\}$;

(c) $W = \mathrm{Sp}\left\{\begin{bmatrix} 1 & 1 & 0 \\ 0 & 0 & 0 \\ 0 & 0 & 0 \end{bmatrix}, \begin{bmatrix} 0 & 0 & 1 \\ 0 & 0 & 1 \\ 0 & 0 & 0 \end{bmatrix}, \right.$
$\left.\begin{bmatrix} 0 & 0 & 0 \\ 0 & 1 & 0 \\ 0 & 0 & 1 \end{bmatrix}\right\}$;

(d) $W = \mathrm{Sp}\left\{\begin{bmatrix} 1 & 0 & 0 \\ 0 & 1 & 0 \\ 0 & 0 & 1 \end{bmatrix}, \begin{bmatrix} 0 & -1 & 0 \\ 0 & 0 & 1 \\ 0 & 0 & 0 \end{bmatrix}, \right.$
$\left.\begin{bmatrix} 0 & 0 & 1 \\ 0 & 0 & 0 \\ 0 & 0 & 0 \end{bmatrix}\right\}$

33. $x_1 = -6a + 5b + 37c + 15d$;
$x_2 = 3a - 2b - 17c - 7d$;
$x_3 = -a + b + 5c + 2d$; $x_4 = 2c + d$;
$C = -12B_1 + 6B_2 - B_3 - B_4$;
$D = 8B_1 - 3B_2 + B_3 + B_4$

Exercises 4.4

1. $\left\{\begin{bmatrix} -1 & 1 \\ 0 & 0 \end{bmatrix}, \begin{bmatrix} -1 & 0 \\ 1 & 0 \end{bmatrix}, \begin{bmatrix} -1 & 0 \\ 0 & 1 \end{bmatrix}\right\}$

3. $\{E_{12}, E_{21}, E_{22}\}$

5. $\{1 + x^2, x - 2x^2\}$

7. $\{x, x^2\}$

9. $\{-9x + 3x^2 + x^3, 8x - 6x^2 + x^4\}$

13. (a) $[2 \ -1 \ \ 3 \ \ 2]^T$;   (b) $[1 \ \ 0 \ -1 \ \ 1]^T$;
(c) $[2 \ \ 3 \ \ 0 \ \ 0]^T$

15. Linearly independent

17. Linearly dependent

19. Linearly dependent

21. Linearly independent

23. $\{p_1(x), p_2(x)\}$ **25.** $\{A_1, A_2, A_3\}$

27. $[-4 \ \ 11 \ -3]^T$

31. $[a + b - 2c + 7d, -b + 2c - 4d, c - 2d, d]^T$

38. The set $\{\mathbf{u}, \mathbf{v}\}$ is linearly dependent if and only if one of the vectors is a scalar multiple of the other.
(a) Linearly independent;
(b) Linearly independent;   (c) Linearly dependent;
(d) Linearly dependent;   (e) Linearly dependent

Exercises 4.5

1. (b) $\{E_{11}, E_{21}, E_{22}, E_{31}, E_{32}, E_{33}\}$ is a basis for V_1.
$\{E_{11}, E_{12}, E_{13}, E_{22}, E_{23}, E_{33}\}$ is a basis for V_2.
(c) $\dim(V) = 9, \dim(V_1) = 6, \dim(V_2) = 6$

3. $V_1 \cap V_2 = \left\{ \begin{bmatrix} a_{11} & 0 & 0 \\ 0 & a_{22} & 0 \\ 0 & 0 & a_{33} \end{bmatrix} : a_{11}, a_{22}, a_{33} \right.$
arbitrary real numbers$\}$; $\dim(V_1 \cap V_2) = 3$

5. $\dim(W) = 3$ **7.** $\dim(W) = 3$

9. (iii) The set S is linearly dependent.

11. (ii) The set S does not span V.

13. (iii) The set S is linearly dependent.

21. (a) $A = \begin{bmatrix} 1 & -1 & 1 \\ 0 & 1 & -2 \\ 0 & 0 & 1 \end{bmatrix}$;   (b) $[5 \ \ 2 \ \ 1]^T$

23. $A^{-1} = \begin{bmatrix} 1 & 1 & 1 \\ 0 & 1 & 2 \\ 0 & 0 & 1 \end{bmatrix}$ (a) $p(x) = 6 + 11x + 7x^2$;
(b) $p(x) = 4 + 2x - x^2$;   (c) $p(x) = 5 + x$;
(d) $p(x) = 8 - 2x - x^2$

Exercises 4.6

9. $\langle \mathbf{x}, \mathbf{y} \rangle = -3, \ \|\mathbf{x}\| = \sqrt{5}, \ \|\mathbf{y}\| = \sqrt{2}$,
$\|\mathbf{x} - \mathbf{y}\| = \sqrt{13}$

11. $\langle p, q \rangle = 52, \ \|p\| = 3\sqrt{6}, \ \|q\| = 3\sqrt{6}$,
$\|p - q\| = 2$

13. For $\langle \mathbf{x}, \mathbf{y} \rangle = \mathbf{x}^T\mathbf{y}$, the graph of S is the circle with equation $x^2 + y^2 = 1$. For $\langle \mathbf{x}, \mathbf{y} \rangle = 4x_1 y_1 + x_2 y_2$, the graph of S is the ellipse with equation $4x^2 + y^2 = 1$.

15. $a_1 = 7, a_2 = 4$

17. $q = (-5/3)p_0 - 5p_1 - 4p_2$

19. $p_0 = 1, p_1 = x, p_2 = x^2 - 2, p_3 = x^3 - (17/5)x$,
$p_4 = x^4 - (31/7)x^2 + 72/35$

25. $p^*(x) = (3/2)x^2 - (3/5)x + 1/20$

27. $p^*(x) \cong 0.841471p_0(x) - 0.467544p_1(x) -$
$0.430920p_2(x) + 0.07882p_3(x)$

29. (d) $T_2(x) = 2x^2 - 1, T_3(x) = 4x^3 - 3x$,
$T_4(x) = 8x^4 - 8x^2 + 1, T_5(x) = 16x^5 - 20x^3 + 5x$

Exercises 4.7

1. Not a linear transformation

3. A linear transformation

5. A linear transformation

7. Not a linear transformation

9. (a) $11 + x^2 + 6x^3$;
(b) $T(a_0 + a_1 x + a_2 x^2) =$
$(a_0 + 2a_2) + (a_0 + a_1)x^2 + (-a_1 + a_2)x^3$

11. (a) $8 + 14x - 9x^2$;
(b) $T\left(\begin{bmatrix} a & b \\ c & d \end{bmatrix}\right) = (a + b + 2d) +$
$(-a + b + 2c + d)x + (b - c - 2d)x^2$

13. (a) $\{2, 6x, 12x^2\}$ is a basis for $\mathcal{R}(T)$.
(b) Nullity$(T) = 2$
(c) $T[(a_0/2)x^2 + (a_1/6)x^3 + (a_2/12)x^4] =$
$a_0 + a_1 x + a_2 x^2$

15. $\mathcal{N}(T) = \{a_0 + a_1 x + a_2 x^2 : a_0 + 2a_1 + 4a_2 = 0\}$;
$\mathcal{R}(T) = R^1$

17. (b) $\mathcal{N}(I) = \{\mathbf{0}\}; \mathcal{R}(I) = V$

19.

rank(T)	3	2	1	0
nullity(T)	2	3	4	5

T cannot be one-to-one.

21.

rank(T)	3	2	1	0
nullity(T)	0	1	2	3

$\mathcal{R}(T) = \mathcal{P}_3$ is not possible.

27. (b) Nullity$(T) = 0$; rank$(T) = 4$

Exercises 4.8

1. $(S + T)(p) = p'(0) + (x + 2)p(x)$;
$(S + T)(x) = 1 + 2x + x^2$; $(S + T)(x^2) = 2x^2 + x^3$

3. $(H \circ T)(p) = p(x) + (x + 2)p'(x) + 2p(0)$;
$(H \circ T)(x) = 2x + 2$

5. (b) There is no polynomial p in $\mathscr{P}_3$ such that $T(p) = x$, so $T^{-1}(x)$ is not defined.

7. $T^{-1}(e^x) = e^x$; $T^{-1}(e^{2x}) = (1/2)e^{2x}$;
$T^{-1}(e^{3x}) = (1/3)e^{3x}$;
$T^{-1}(ae^x + be^{2x} + ce^{3x}) = ae^x + (b/2)e^{2x} + (c/3)e^{3x}$

9. $T^{-1}(A) = A^T$

11. (c) $T\left(\begin{bmatrix} a & b \\ c & d \end{bmatrix}\right) = a + bx + cx^2 + dx^3$

Exercises 4.9

1. $\begin{bmatrix} 0 & 1 & 0 & 0 \\ 0 & 0 & 0 & 0 \\ 0 & 0 & 0 & 0 \\ 0 & 0 & 0 & 0 \\ 0 & 0 & 0 & 0 \end{bmatrix}$

3. $\begin{bmatrix} 2 & 1 & 0 & 0 \\ 1 & 2 & 0 & 0 \\ 0 & 1 & 2 & 0 \\ 0 & 0 & 1 & 2 \\ 0 & 0 & 0 & 1 \end{bmatrix}$

5. $\begin{bmatrix} 1 & 1 & 0 & 0 & 0 \\ 0 & 0 & 2 & 0 & 0 \\ 0 & 0 & 0 & 3 & 0 \\ 0 & 0 & 0 & 0 & 4 \end{bmatrix}$

7. $\begin{bmatrix} 2 & 2 & 0 & 0 & 0 \\ 1 & 1 & 4 & 0 & 0 \\ 0 & 0 & 2 & 6 & 0 \\ 0 & 0 & 0 & 3 & 8 \\ 0 & 0 & 0 & 0 & 4 \end{bmatrix}$

9. (a) $[p]_B = \begin{bmatrix} a_0 \\ a_1 \\ a_2 \\ a_3 \end{bmatrix}$, $[T(p)]_C = \begin{bmatrix} 2a_0 \\ a_0 + 2a_1 \\ a_1 + 2a_2 \\ a_2 + 2a_3 \\ a_3 \end{bmatrix}$

11. (a) $Q = \begin{bmatrix} 1 & 0 & 0 \\ 0 & 2 & 0 \\ 0 & 0 & 3 \end{bmatrix}$; (b) $P = \begin{bmatrix} 1 & 0 & 0 \\ 0 & 1/2 & 0 \\ 0 & 0 & 1/3 \end{bmatrix}$

13. (a) $Q = \begin{bmatrix} 1 & 0 & 0 & 0 \\ 0 & 0 & 1 & 0 \\ 0 & 1 & 0 & 0 \\ 0 & 0 & 0 & 1 \end{bmatrix}$

15. $\begin{bmatrix} 3 & 6 & 0 \\ 3 & 3 & 0 \\ -1 & -1 & 3 \\ 0 & 0 & 0 \end{bmatrix}$

17. $\begin{bmatrix} 1 & 0 & 0 \\ 0 & 3 & 6 \\ 0 & 1 & 4 \end{bmatrix}$

19. $\begin{bmatrix} 0 & 0 & 1 & 1 \\ 1 & 0 & 1 & 0 \\ 0 & 1 & 0 & 0 \\ 0 & 0 & 0 & 3 \end{bmatrix}$

21. $\begin{bmatrix} -4 & -2 & 0 \\ 3 & 3 & 0 \\ -1 & 2 & 3 \end{bmatrix}$

23. $\begin{bmatrix} 2 & 0 & 0 \\ 0 & -3 & 0 \\ 0 & 0 & 3 \end{bmatrix}$

31. $T(a_0 + a_1x + a_2x^2 + a_3x^3) = (a_0 + 2a_2) + (a_1 + a_3)x + (-a_0 + a_1 - a_3)x^2$

Exercises 4.10

1. $T(\mathbf{u}_1) = \mathbf{u}_1$, $T(\mathbf{u}_2) = 3\mathbf{u}_2$, $\begin{bmatrix} 1 & 0 \\ 0 & 3 \end{bmatrix}$

3. $T(A_1) = 2A_1$, $T(A_2) = -2A_2$, $T(A_3) = 3A_3$,
$T(A_4) = -3A_4$, $\begin{bmatrix} 2 & 0 & 0 & 0 \\ 0 & -2 & 0 & 0 \\ 0 & 0 & 3 & 0 \\ 0 & 0 & 0 & -3 \end{bmatrix}$

5. $\begin{bmatrix} 1 & -1 & -1 \\ 1 & -1 & 0 \\ -1 & 2 & 1 \end{bmatrix}$;

$p(x) = (1 + x - x^2) + (1 + x^2)$;
$q(x) = -5(1 + x - x^2) - 3(1 + x^2) + 7(1 + x)$;
$s(x) = -2(1 + x - x^2) - (1 + x^2) + 2(1 + x)$;
$r(x) = (a_0 - a_1 - a_2)(1 + x - x^2)$
$\qquad + (a_0 - a_1)(1 + x^2)$
$\qquad + (-a_0 + 2a_1 + a_2)(1 + x)$

7. $\begin{bmatrix} 1/3 & 5/3 \\ 1/3 & -1/3 \end{bmatrix}$

9. $\begin{bmatrix} -1 & 1 & 2 & 3 \\ 1 & 0 & 0 & -3 \\ 0 & 0 & 1 & 0 \\ 0 & 0 & 0 & 1 \end{bmatrix}$;

$p(x) = -7x + 2(x + 1) + (x^2 - 2x)$;
$q(x) = 13x - 4(x + 1) + (x^3 + 3)$;
$r(x) = -7x + 3(x + 1) - 2(x^2 - 2x) + (x^3 + 3)$

11. The matrix of T with respect to B is
$Q_1 = \begin{bmatrix} 2 & 1 \\ 1 & 2 \end{bmatrix}$. The transition matrix from C to B is
$P = \begin{bmatrix} -1 & 1 \\ 1 & 1 \end{bmatrix}$. The matrix of T with respect to C is Q_2, where $Q_2 = P^{-1}Q_1P = \begin{bmatrix} 1 & 0 \\ 0 & 3 \end{bmatrix}$.

13. The matrix of T with respect to B is

$$Q_1 = \begin{bmatrix} -3 & 0 & 0 & 5 \\ 0 & 3 & -5 & 0 \\ 0 & 0 & -2 & 0 \\ 0 & 0 & 0 & 2 \end{bmatrix}.$$ The transition matrix

from C to B is $P = \begin{bmatrix} 1 & 0 & 0 & 1 \\ 0 & 1 & 1 & 0 \\ 0 & 1 & 0 & 0 \\ 1 & 0 & 0 & 0 \end{bmatrix}$. The matrix of

T with respect to C is Q_2, where

$$Q_2 = P^{-1}Q_1P = \begin{bmatrix} 2 & 0 & 0 & 0 \\ 0 & -2 & 0 & 0 \\ 0 & 0 & 3 & 0 \\ 0 & 0 & 0 & -3 \end{bmatrix}.$$

15. (a) $Q = \begin{bmatrix} 1 & 1 & 0 \\ 0 & 2 & 4 \\ 0 & 0 & 3 \end{bmatrix}$;

(b) $S = \begin{bmatrix} 1 & 1 & 2 \\ 0 & 1 & 4 \\ 0 & 0 & 1 \end{bmatrix}$; $R = \begin{bmatrix} 1 & 0 & 0 \\ 0 & 2 & 0 \\ 0 & 0 & 3 \end{bmatrix}$;

(c) $C = \{1, 1 + x, 2 + 4x + x^2\}$;

(d) $P = \begin{bmatrix} 1 & -1 & 2 \\ 0 & 1 & -4 \\ 0 & 0 & 1 \end{bmatrix}$;

(e) $T(\mathbf{w}_1) = -1 + 18x + 3x^2$;
$T(\mathbf{w}_2) = 5 + 4x + 3x^2$;
$T(\mathbf{w}_3) = 1 + 2x + 6x^2$

Exercises 5.2

1. -5

3. 0, $\mathbf{x} = \begin{bmatrix} -2 \\ 1 \end{bmatrix}$

5. 25

7. 6

9. $A_{11} = -2, A_{12} = 6, A_{13} = -2, A_{33} = 1$

11. $A_{11} = -2, A_{12} = 7, A_{13} = -8, A_{33} = 3$

13. $A_{11} = 3, A_{12} = -6, A_{13} = 2, A_{33} = -3$

15. 8

17. -35

19. -11

21. -9

23. 22

29. $C = \begin{bmatrix} 10 & 5 & -10 \\ -5 & -1 & 4 \\ -5 & -3 & 7 \end{bmatrix}$

35. (a) $H(n) = n!/2$;
(b) 3 seconds for $n = 2$; 180 seconds for $n = 5$;
5,443,200 seconds for $n = 10$

Exercises 5.3

1. $\begin{vmatrix} 1 & 2 & 1 \\ 2 & 0 & 1 \\ 1 & -1 & 1 \end{vmatrix} = \begin{vmatrix} 1 & 0 & 0 \\ 2 & -4 & -1 \\ 1 & -3 & 0 \end{vmatrix} = \begin{vmatrix} -4 & -1 \\ -3 & 0 \end{vmatrix} = -3$

3. $\begin{vmatrix} 0 & 1 & 2 \\ 3 & 1 & 2 \\ 2 & 0 & 3 \end{vmatrix} = -\begin{vmatrix} 1 & 0 & 2 \\ 1 & 3 & 2 \\ 0 & 2 & 3 \end{vmatrix} = -\begin{vmatrix} 1 & 0 & 0 \\ 1 & 3 & 0 \\ 0 & 2 & 3 \end{vmatrix}$

$= -\begin{vmatrix} 3 & 0 \\ 2 & 3 \end{vmatrix} = -9$

5. $\begin{vmatrix} 0 & 1 & 3 \\ 2 & 1 & 2 \\ 1 & 1 & 2 \end{vmatrix} = -\begin{vmatrix} 1 & 0 & 3 \\ 1 & 2 & 2 \\ 1 & 1 & 2 \end{vmatrix} = -\begin{vmatrix} 1 & 0 & 0 \\ 1 & 2 & -1 \\ 1 & 1 & -1 \end{vmatrix}$

$= -\begin{vmatrix} 2 & -1 \\ 1 & -1 \end{vmatrix} = 1$

7. -6

9. 3

11. 3

13. Use the column interchanges: $[\mathbf{C}_1, \mathbf{C}_2, \mathbf{C}_3, \mathbf{C}_4] \to$ $[\mathbf{C}_1, \mathbf{C}_4, \mathbf{C}_3, \mathbf{C}_2] \to [\mathbf{C}_1, \mathbf{C}_4, \mathbf{C}_2, \mathbf{C}_3]$; the determinant is 6.

15. Use the column interchanges: $[\mathbf{C}_1, \mathbf{C}_2, \mathbf{C}_3, \mathbf{C}_4] \to$ $[\mathbf{C}_2, \mathbf{C}_1, \mathbf{C}_3, \mathbf{C}_4] \to [\mathbf{C}_2, \mathbf{C}_4, \mathbf{C}_3, \mathbf{C}_1] \to$ $[\mathbf{C}_2, \mathbf{C}_4, \mathbf{C}_1, \mathbf{C}_3]$; the determinant is -12.

17. $\begin{vmatrix} 2 & 4 & -2 & -2 \\ 1 & 3 & 1 & 2 \\ 1 & 3 & 1 & 3 \\ -1 & 2 & 1 & 2 \end{vmatrix} = \begin{vmatrix} 2 & 0 & 0 & 0 \\ 1 & 1 & 2 & 3 \\ 1 & 1 & 2 & 4 \\ -1 & 4 & 0 & 1 \end{vmatrix}$

$= 2\begin{vmatrix} 1 & 2 & 3 \\ 1 & 2 & 4 \\ 4 & 0 & 1 \end{vmatrix} = 2\begin{vmatrix} 1 & 0 & 0 \\ 1 & 0 & 1 \\ 4 & -8 & -11 \end{vmatrix}$

$= 2\begin{vmatrix} 0 & 1 \\ -8 & -11 \end{vmatrix} = 16$

19. $\begin{vmatrix} 1 & 2 & 0 & 3 \\ 2 & 5 & 1 & 1 \\ 2 & 0 & 4 & 3 \\ 0 & 1 & 6 & 2 \end{vmatrix} = \begin{vmatrix} 1 & 2 & 0 & 3 \\ 0 & 1 & 1 & -5 \\ 0 & -4 & 4 & -3 \\ 0 & 1 & 6 & 2 \end{vmatrix}$

$= \begin{vmatrix} 1 & 1 & -5 \\ -4 & 4 & -3 \\ 1 & 6 & 2 \end{vmatrix} = \begin{vmatrix} 1 & 1 & -5 \\ 0 & 8 & -23 \\ 0 & 5 & 7 \end{vmatrix}$

$= \begin{vmatrix} 8 & -23 \\ 5 & 7 \end{vmatrix} = 171$

21.
$$\begin{vmatrix} 1 & 1 & 2 & 1 \\ 0 & 1 & 4 & 1 \\ 2 & 1 & 3 & 0 \\ 2 & 2 & 1 & 2 \end{vmatrix} = \begin{vmatrix} 1 & 1 & 2 & 1 \\ 0 & 1 & 4 & 1 \\ 0 & -1 & -1 & -2 \\ 0 & 0 & -3 & 0 \end{vmatrix}$$

$$= \begin{vmatrix} 1 & 4 & 1 \\ -1 & -1 & -2 \\ 0 & -3 & 0 \end{vmatrix} = \begin{vmatrix} 1 & 4 & 1 \\ 0 & 3 & -1 \\ 0 & -3 & 0 \end{vmatrix}$$

$$= \begin{vmatrix} 3 & -1 \\ -3 & 0 \end{vmatrix} = -3$$

Exercises 5.4

1. $A \to \begin{bmatrix} 1 & 0 & 3 \\ 2 & 1 & 1 \\ 4 & 3 & 1 \end{bmatrix} \to \begin{bmatrix} 1 & 0 & 0 \\ 2 & 1 & -5 \\ 4 & 3 & -11 \end{bmatrix} \to \begin{bmatrix} 1 & 0 & 0 \\ 2 & 1 & 0 \\ 4 & 3 & 4 \end{bmatrix};$
$\det(A) = -4.$

3. $A \to \begin{bmatrix} 2 & 0 & 0 \\ 1 & 2 & 2 \\ -1 & 3 & 3 \end{bmatrix} \to \begin{bmatrix} 2 & 0 & 0 \\ 1 & 2 & 0 \\ -1 & 3 & 0 \end{bmatrix}; \det(A) = 0$

5. $A \to \begin{bmatrix} 1 & 0 & 0 \\ 3 & 1 & 9 \\ 0 & 1 & 2 \end{bmatrix} \to \begin{bmatrix} 1 & 0 & 0 \\ 3 & 1 & 0 \\ 0 & 1 & -7 \end{bmatrix} \to \begin{bmatrix} 1 & 0 & 0 \\ 0 & 1 & 0 \\ 0 & 0 & -7 \end{bmatrix} \to I;$
$\det(A) = -7$

7. (a) 6; (b) 18; (c) 3/2 (d) 4; (e) 1/16

9. $\det[B(\lambda)] = -\lambda^2 + 2\lambda$; $\lambda = 0$ and $\lambda = 2$

11. $\det[B(\lambda)] = 4 - \lambda^2$; $\lambda = 2$ and $\lambda = -2$

13. $\det[B(\lambda)] = (\lambda + 2)(\lambda - 1)^2$; $\lambda = -2$ and $\lambda = 1$

15. $\det(A) = -2$, $\det(B_1) = -2$, $\det(B_2) = -4$; $x_1 = 1$, $x_2 = 2$

17. $\det(A) = -2$, $\det(B_1) = -8$, $\det(B_2) = -4$, $\det(B_3) = 2$; $x_1 = 4$, $x_2 = 2$, $x_3 = -1$

19. $\det(A) = \det(B_1) = \det(B_2) = \det(B_3) = \det(B_4) = 3$; $x_1 = x_2 = x_3 = x_4 = 1$

21. $\det(A) = 1$, $\det(B_1) = a-b$, $\det(B_2) = b - c$, $\det(B_3) = c$; $x_1 = a - b$, $x_2 = b - c$, $x_3 = c$

27. $\det(A^5) = [\det(A)]^5 = 3^5 = 243$

Exercises 5.5

1. $\begin{vmatrix} 1 & 2 & 1 \\ 2 & 3 & 2 \\ -1 & 4 & 1 \end{vmatrix} = \begin{vmatrix} 1 & 2 & 1 \\ 0 & -1 & 0 \\ 0 & 6 & 2 \end{vmatrix} = \begin{vmatrix} 1 & 2 & 1 \\ 0 & -1 & 0 \\ 0 & 0 & 2 \end{vmatrix} = -2$

3. $\begin{vmatrix} 0 & 1 & 3 \\ 1 & 2 & 2 \\ 3 & 1 & 0 \end{vmatrix} = -\begin{vmatrix} 1 & 2 & 2 \\ 0 & 1 & 3 \\ 3 & 1 & 0 \end{vmatrix} = -\begin{vmatrix} 1 & 2 & 2 \\ 0 & 1 & 3 \\ 0 & -5 & -6 \end{vmatrix}$

$$= -\begin{vmatrix} 1 & 2 & 2 \\ 0 & 1 & 3 \\ 0 & 0 & 9 \end{vmatrix} = -9$$

5. $\mathrm{Adj}(A) = \begin{bmatrix} 4 & -2 \\ -3 & 1 \end{bmatrix}$; $A^{-1} = -\dfrac{1}{2}\mathrm{Adj}(A)$

7. $\mathrm{Adj}(A) = \begin{bmatrix} 0 & 1 & -1 \\ -2 & 1 & 0 \\ 1 & -1 & 1 \end{bmatrix}$; $A^{-1} = \mathrm{Adj}(A)$

9. $\mathrm{Adj}(A) = \begin{bmatrix} -4 & 2 & 0 \\ 1 & 0 & -1 \\ 1 & -2 & 1 \end{bmatrix}$; $A^{-1} = -\dfrac{1}{2}\mathrm{Adj}(A)$

11. For all x, $w(x) = 2$; therefore, the set is linearly independent.

13. For all x, $w(x) = 0$; the Wronskian gives no information; the set is linearly dependent since $\cos^2 x + \sin^2 x = 1$.

15. For all x, $w(x) = 0$; the Wronskian gives no information; the set is linearly independent.

17. $L = \begin{bmatrix} 1 & 0 & 0 \\ 2 & 1 & 0 \\ 2 & 2 & -1 \end{bmatrix}$, $E_1 = \begin{bmatrix} 0 & 1 & 0 \\ 1 & 0 & 0 \\ 0 & 0 & 1 \end{bmatrix}$,

$E_2 = \begin{bmatrix} 1 & 0 & -3 \\ 0 & 1 & 0 \\ 0 & 0 & 1 \end{bmatrix}$, $E_3 = \begin{bmatrix} 1 & 0 & 0 \\ 0 & 1 & 2 \\ 0 & 0 & 1 \end{bmatrix}$

19. $L = \begin{bmatrix} 1 & 0 & 0 \\ 3 & -1 & 0 \\ 4 & -8 & -26 \end{bmatrix}$, $E_1 = \begin{bmatrix} 1 & -2 & 0 \\ 0 & 1 & 0 \\ 0 & 0 & 1 \end{bmatrix}$,

$E_2 = \begin{bmatrix} 1 & 0 & 1 \\ 0 & 1 & 0 \\ 0 & 0 & 1 \end{bmatrix}$, $E_3 = \begin{bmatrix} 1 & 0 & 0 \\ 0 & 1 & 4 \\ 0 & 0 & 1 \end{bmatrix}$

21. $\det[A(x)] = x^2 + 1$; $[A(x)]^{-1} = \dfrac{1}{x^2 + 1}\begin{bmatrix} x & -1 \\ 1 & x \end{bmatrix}$

23. $\det[A(x)] = 4(2 + x^2)$;

$[A(x)]^{-1} = \dfrac{1}{4(2 + x^2)}\begin{bmatrix} 4 + x^2 & -2x & x^2 \\ 2x & 4 & -2x \\ x^2 & 2x & 4 + x^2 \end{bmatrix}$

Exercises 6.1

1. $A = \begin{bmatrix} 2 & 2 \\ 2 & -3 \end{bmatrix}$

3. $A = \begin{bmatrix} 1 & 1 & -3 \\ 1 & -4 & 4 \\ -3 & 4 & 3 \end{bmatrix}$ **5.** $A = \begin{bmatrix} 2 & 2 \\ 2 & 1 \end{bmatrix}$

7. $Q = \dfrac{1}{\sqrt{2}} \begin{bmatrix} 1 & 1 \\ 1 & -1 \end{bmatrix}$; the form is indefinite with eigenvalues $\lambda = 5$ and $\lambda = -1$.

9. $Q = \dfrac{1}{\sqrt{6}} \begin{bmatrix} \sqrt{2} & \sqrt{3} & -1 \\ \sqrt{2} & -\sqrt{3} & -1 \\ \sqrt{2} & 0 & 2 \end{bmatrix}$; the form is indefinite with eigenvalues $\lambda = 5$ and $\lambda = -1$.

11. $Q = \dfrac{1}{\sqrt{2}} \begin{bmatrix} 1 & 1 \\ 1 & -1 \end{bmatrix}$; the form is positive definite with eigenvalues $\lambda = 2$ and $\lambda = 4$.

13. $Q = \begin{bmatrix} 1/2 & \sqrt{3}/2 \\ -\sqrt{3}/2 & 1/2 \end{bmatrix}$; the graph corresponds to the ellipse $\dfrac{u^2}{20} + \dfrac{v^2}{4} = 1$.

15. $Q = \dfrac{1}{\sqrt{10}} \begin{bmatrix} -1 & 3 \\ 3 & 1 \end{bmatrix}$; the graph corresponds to the hyperbola $\dfrac{v^2}{4} - u^2 = 1$.

17. $Q = \dfrac{1}{\sqrt{2}} \begin{bmatrix} 1 & -1 \\ 1 & 1 \end{bmatrix}$; the graph corresponds to the hyperbola $\dfrac{u^2}{4} - \dfrac{v^2}{4} = 1$.

19. $Q = \dfrac{1}{\sqrt{2}} \begin{bmatrix} 1 & 1 \\ -1 & 1 \end{bmatrix}$; the graph corresponds to the ellipse $\dfrac{u^2}{4} + \dfrac{v^2}{8} = 1$.

Exercises 6.2

1. $\mathbf{x}'(t) = A\mathbf{x}(t)$, $A = \begin{bmatrix} 5 & -2 \\ 6 & -2 \end{bmatrix}$, $\mathbf{x}(t) = \begin{bmatrix} u(t) \\ v(t) \end{bmatrix}$;

$\mathbf{x}(t) = b_1 e^t \begin{bmatrix} 1 \\ 2 \end{bmatrix} + b_2 e^{2t} \begin{bmatrix} 2 \\ 3 \end{bmatrix}$;

$\mathbf{x}(t) = e^t \begin{bmatrix} 1 \\ 2 \end{bmatrix} + 2e^{2t} \begin{bmatrix} 2 \\ 3 \end{bmatrix} = \begin{bmatrix} e^t + 4e^{2t} \\ 2e^t + 6e^{2t} \end{bmatrix}$

3. $\mathbf{x}'(t) = A\mathbf{x}(t)$, $A = \begin{bmatrix} 1 & 1 \\ 2 & 2 \end{bmatrix}$, $\mathbf{x}(t) = \begin{bmatrix} u(t) \\ v(t) \end{bmatrix}$;

$\mathbf{x}(t) = b_1 \begin{bmatrix} 1 \\ -1 \end{bmatrix} + b_2 e^{3t} \begin{bmatrix} 1 \\ 2 \end{bmatrix}$;

$\mathbf{x}(t) = 3 \begin{bmatrix} 1 \\ -1 \end{bmatrix} + 2e^{3t} \begin{bmatrix} 1 \\ 2 \end{bmatrix}$

$= \begin{bmatrix} 3 + 2e^{3t} \\ -3 + 4e^{3t} \end{bmatrix}$

5. $\mathbf{x}'(t) = A\mathbf{x}(t)$, $A = \begin{bmatrix} .5 & .5 \\ -.5 & .5 \end{bmatrix}$, $\mathbf{x}(t) = \begin{bmatrix} u(t) \\ v(t) \end{bmatrix}$;

$\mathbf{x}(t) = b_1 e^{(1+i)t/2} \begin{bmatrix} 1 \\ i \end{bmatrix} + b_2 e^{(1-i)t/2} \begin{bmatrix} 1 \\ -i \end{bmatrix}$;

$b_1 = 2 - 2i$ and

$b_2 = 2 + 2i$, or $\mathbf{x}(t) = 4e^{t/2} \begin{bmatrix} \cos(t/2) + \sin(t/2) \\ \cos(t/2) - \sin(t/2) \end{bmatrix}$

7. $\mathbf{x}'(t) = A\mathbf{x}(t)$, $A = \begin{bmatrix} 4 & 0 & 1 \\ -2 & 1 & 0 \\ -2 & 0 & 1 \end{bmatrix}$, $\mathbf{x}(t) = \begin{bmatrix} u(t) \\ v(t) \\ w(t) \end{bmatrix}$;

$\mathbf{x}(t) = b_1 e^t \begin{bmatrix} 0 \\ 1 \\ 0 \end{bmatrix} + b_2 e^{2t} \begin{bmatrix} -1 \\ 2 \\ 2 \end{bmatrix} + b_3 e^{3t} \begin{bmatrix} -1 \\ 1 \\ 1 \end{bmatrix}$;

$\mathbf{x}(t) = e^t \begin{bmatrix} 0 \\ 1 \\ 0 \end{bmatrix} - e^{2t} \begin{bmatrix} -1 \\ 2 \\ 2 \end{bmatrix} + 2e^{3t} \begin{bmatrix} -1 \\ 1 \\ 1 \end{bmatrix}$

$= \begin{bmatrix} e^{2t} - 2e^{3t} \\ e^t - 2e^{2t} + 2e^{3t} \\ -2e^{2t} + 2e^{3t} \end{bmatrix}$

9. (a) $\mathbf{x}'(t) = A\mathbf{x}(t)$, $A = \begin{bmatrix} 1 & -1 \\ 1 & 3 \end{bmatrix}$, $\mathbf{x}(t) = \begin{bmatrix} u(t) \\ v(t) \end{bmatrix}$;

$\mathbf{x}_1(t) = b_1 e^{2t} \begin{bmatrix} 1 \\ -1 \end{bmatrix}$

(b) The vector $\mathbf{y}_0$ is determined by the equation $(A - 2I)\mathbf{y} = \mathbf{u}$, where $\mathbf{u} = \begin{bmatrix} 1 \\ -1 \end{bmatrix}$. One choice is $\mathbf{y}_0 = \begin{bmatrix} -2 \\ 1 \end{bmatrix}$. Thus, $\mathbf{x}_2(t) = te^{2t} \begin{bmatrix} 1 \\ -1 \end{bmatrix} + e^{2t} \begin{bmatrix} -2 \\ 1 \end{bmatrix}$ is another solution of $\mathbf{x}'(t) = A\mathbf{x}(t)$.

(c) Note that $\mathbf{x}_1(0) = \begin{bmatrix} 1 \\ -1 \end{bmatrix}$ and $\mathbf{x}_2(0) = \begin{bmatrix} -2 \\ 1 \end{bmatrix}$.

Thus, $\{\mathbf{x}_1(0), \mathbf{x}_2(0)\}$ is a basis for R^2.

Exercises 6.3

1. $H = Q_1 A Q_1^{-1} = \begin{bmatrix} -7 & 16 & 3 \\ 8 & 9 & 3 \\ 0 & 1 & 1 \end{bmatrix}$; $Q_1 = \begin{bmatrix} 1 & 0 & 0 \\ 0 & 1 & 0 \\ 0 & -4 & 1 \end{bmatrix}$

3. $H = Q_1 A Q_1^{-1} = \begin{bmatrix} 1 & 1 & 3 \\ 1 & 3 & 1 \\ 0 & 4 & 2 \end{bmatrix}$; $Q_1 = \begin{bmatrix} 1 & 0 & 0 \\ 0 & 0 & 1 \\ 0 & 1 & 0 \end{bmatrix}$

5. $H = Q_1 A Q_1^{-1} = \begin{bmatrix} 3 & 2 & -1 \\ 4 & 5 & -2 \\ 0 & 20 & -6 \end{bmatrix}$; $Q_1 = \begin{bmatrix} 1 & 0 & 0 \\ 0 & 1 & 0 \\ 0 & 3 & 1 \end{bmatrix}$

7. $H = Q_1 A Q_1^{-1} = \begin{bmatrix} 1 & -3 & -1 & -1 \\ -1 & -1 & -1 & -1 \\ 0 & 0 & 2 & 0 \\ 0 & 0 & 0 & 2 \end{bmatrix}$;

$Q_1 = \begin{bmatrix} 1 & 0 & 0 & 0 \\ 0 & 1 & 0 & 0 \\ 0 & -1 & 1 & 0 \\ 0 & -1 & 0 & 1 \end{bmatrix}$

9. $H = Q_2 Q_1 A Q_1^{-1} Q_2^{-1} = \begin{bmatrix} 1 & 3 & 5 & 2 \\ 1 & 2 & 4 & 2 \\ 0 & 1 & 7 & 3 \\ 0 & 0 & -11 & -5 \end{bmatrix}$;

$Q_1 = \begin{bmatrix} 1 & 0 & 0 & 0 \\ 0 & 0 & 0 & 1 \\ 0 & 0 & 1 & 0 \\ 0 & 1 & 0 & 0 \end{bmatrix}$, $Q_2 = \begin{bmatrix} 1 & 0 & 0 & 0 \\ 0 & 1 & 0 & 0 \\ 0 & 0 & 1 & 0 \\ 0 & 0 & -2 & 1 \end{bmatrix}$

13. The characteristic polynomial is $p(t) = (t+2)(t-2)^3$.

15. $[\mathbf{e}_1, \mathbf{e}_2, \mathbf{e}_3]$, $[\mathbf{e}_1, \mathbf{e}_3, \mathbf{e}_2]$, $[\mathbf{e}_2, \mathbf{e}_1, \mathbf{e}_3]$, $[\mathbf{e}_2, \mathbf{e}_3, \mathbf{e}_1]$, $[\mathbf{e}_3, \mathbf{e}_1, \mathbf{e}_2]$, $[\mathbf{e}_3, \mathbf{e}_2, \mathbf{e}_1]$

17. $n!$

Exercises 6.4

1. The system (4) is

$\begin{aligned} a_0 + 2a_1 &= -4 \\ a_1 &= -3 \end{aligned}$; $p(t) = t^2 - 3t + 2$

3. The system (4) is

$\begin{aligned} a_0 + a_1 + a_2 &= -3 \\ 2a_1 + 4a_2 &= -6; \quad p(t) = t^3 - 4t^2 + 5t - 4 \\ 2a_2 &= -8 \end{aligned}$

5. The system (4) is

$\begin{aligned} a_0 + 2a_1 + 8a_2 &= -29 \\ a_1 + 3a_2 &= -14; \quad p(t) = t^3 - 8t^2 + 10t + 15 \\ a_2 &= -8 \end{aligned}$

7. The system (4) is

$\begin{aligned} a_0 \quad\quad + a_2 + 2a_3 &= -8 \\ a_1 + 2a_2 + 6a_3 &= -18 \\ a_2 + 2a_3 &= -8 \\ 2a_3 &= -6 \end{aligned}$

$p(t) = t^4 - 3t^3 - 2t^2 + 4t$

9. The blocks are $B_{11} = \begin{bmatrix} 1 & -1 \\ 1 & 3 \end{bmatrix}$ and $B_{22} = \begin{bmatrix} 2 & -1 \\ -1 & 2 \end{bmatrix}$. The only eigenvalue of B_{11} is $\lambda = 2$; the eigenvalues of B_{22} are $\lambda = 1$ and $\lambda = 3$. The eigenvectors are $\lambda = 2$, $\mathbf{u} = [1, -1, 0, 0]^T$; $\lambda = 1$, $\mathbf{u} = [-9, 5, 1, 1]^T$; and $\lambda = 3$, $\mathbf{u} = [3, -9, 1, -1]^T$.

11. The blocks are $B_{11} = \begin{bmatrix} -2 & 0 & -2 \\ -1 & 1 & -2 \\ 0 & 1 & -1 \end{bmatrix}$ and $B_{22} = [2]$. The eigenvalues of B_{11} are $\lambda = 0$ and $\lambda = -1$; the eigenvalue of B_{22} is $\lambda = 2$. The eigenvectors are $\lambda = 0$, $\mathbf{u} = [-1, 1, 1, 0]^T$; $\lambda = -1$, $\mathbf{u} = [2, 0, -1, 0]^T$; and $\lambda = 2$, $\mathbf{u} = [1, 15, 1, 6]^T$.

15. $P = [\mathbf{e}_2, \mathbf{e}_3, \mathbf{e}_1]$

Exercises 6.5

1. $Q\mathbf{x} = \mathbf{x} + \mathbf{u} = [4, 1, 6, 7]^T$

3. $Q\mathbf{A}_1 = \mathbf{A}_1 + \mathbf{u}$, $Q\mathbf{A}_2 = \mathbf{A}_2 + 2\mathbf{u}$; therefore,

$QA = \begin{bmatrix} 3 & 3 \\ 5 & 1 \\ 5 & 4 \\ 1 & 2 \end{bmatrix}$

5. $\mathbf{x}^T Q = \mathbf{x}^T + \mathbf{u}^T = [4, 1, 3, 4]$

7. $AQ = \begin{bmatrix} 1 & 2 & 1 & 2 \\ 2 & -1 & 2 & 3 \end{bmatrix}$

9. $\mathbf{u} = \begin{bmatrix} 0 \\ 5 \\ 2 \\ 1 \end{bmatrix}$

11. $\mathbf{u} = \begin{bmatrix} 0 \\ 0 \\ 9 \\ 3 \end{bmatrix}$

13. $\mathbf{u} = \begin{bmatrix} 0 \\ 0 \\ 0 \\ -8 \\ 4 \end{bmatrix}$

15. $\mathbf{u} = \begin{bmatrix} 0 \\ 8 \\ 4 \end{bmatrix}$

17. $\mathbf{u} = \begin{bmatrix} 0 \\ -9 \\ 3 \end{bmatrix}$

19. $\mathbf{u} = \begin{bmatrix} 0 \\ 0 \\ -8 \\ 4 \end{bmatrix}$

Exercises 6.6

1. $\mathbf{x}^* = \begin{bmatrix} 1 \\ 1 \end{bmatrix}$

3. $\mathbf{x}^* = \begin{bmatrix} 2 \\ 1 \\ 2 \end{bmatrix}$

5. $R = \begin{bmatrix} -5 & -11 \\ 0 & 2 \end{bmatrix}$, $\mathbf{u} = \begin{bmatrix} 8 \\ 4 \end{bmatrix}$

7. $R = \begin{bmatrix} -4 & -6 \\ 0 & -2 \end{bmatrix}$, $\mathbf{u} = \begin{bmatrix} 4 \\ 4 \end{bmatrix}$

9. $R = \begin{bmatrix} 1 & 2 & 1 \\ 0 & -1 & -8 \\ 0 & 0 & -6 \end{bmatrix}$, $\mathbf{u} = \begin{bmatrix} 0 \\ 1 \\ 1 \end{bmatrix}$

11. $Q_2 Q_1 A = \begin{bmatrix} -5 & -59/3 \\ 0 & -7 \\ 0 & 0 \\ 0 & 0 \end{bmatrix}$, where $\mathbf{u}_1 = \begin{bmatrix} 6 \\ 2 \\ 2 \\ 4 \end{bmatrix}$,

$\mathbf{u}_2 = \begin{bmatrix} 0 \\ 13 \\ 2 \\ 3 \end{bmatrix}$

13. $Q_1 A = \begin{bmatrix} 2 & 4 \\ 0 & -5 \\ 0 & 0 \\ 0 & 0 \end{bmatrix}$, where $\mathbf{u}_1 = \begin{bmatrix} 0 \\ 8 \\ 0 \\ 4 \end{bmatrix}$

15. $\mathbf{x}^* = \begin{bmatrix} -38/21 \\ 15/21 \end{bmatrix}$

17. $\mathbf{x}^* = \begin{bmatrix} -87/25 \\ 56/25 \end{bmatrix}$

Exercises 6.7

1. $q(A) = \begin{bmatrix} -1 & 0 \\ 0 & -1 \end{bmatrix}$; $q(B) = \begin{bmatrix} 0 & 0 \\ 0 & 0 \end{bmatrix}$;

$q(C) = \begin{bmatrix} 15 & -2 & 14 \\ 5 & -2 & 10 \\ -1 & -4 & 6 \end{bmatrix}$

3. (a) $q(t) = (t^3 + t - 1)p(t) + t + 2$;

(b) $q(B) = B + 2I = \begin{bmatrix} 4 & -1 \\ -1 & 4 \end{bmatrix}$

Exercises 6.8

1. (a) $p(t) = (t - 2)^2$; $\mathbf{v}_1 = \begin{bmatrix} 1 \\ -1 \end{bmatrix}$, $\mathbf{v}_2 = \begin{bmatrix} 1 \\ -2 \end{bmatrix}$

(b) $p(t) = t(t + 1)^2$; for $\lambda = -1$, $\mathbf{v}_1 = \begin{bmatrix} -2 \\ 0 \\ 1 \end{bmatrix}$,

$\mathbf{v}_2 = \begin{bmatrix} 0 \\ 1 \\ 1 \end{bmatrix}$; for $\lambda = 0$, $\mathbf{v}_1 = \begin{bmatrix} -1 \\ 1 \\ 1 \end{bmatrix}$

(c) $p(t) = (t - 1)^2(t + 1)$; for $\lambda = 1$,

$\mathbf{v}_1 = \begin{bmatrix} -2 \\ 0 \\ 1 \end{bmatrix}$,

$\mathbf{v}_2 = \begin{bmatrix} 5/2 \\ 1/2 \\ 0 \end{bmatrix}$; for $\lambda = -1$, $\mathbf{v}_1 = \begin{bmatrix} -9 \\ -1 \\ 1 \end{bmatrix}$

3. (a) $QAQ^{-1} = H$, where $Q = \begin{bmatrix} 1 & 0 & 0 \\ 0 & 1 & 0 \\ 0 & 3 & 1 \end{bmatrix}$,

$H = \begin{bmatrix} 8 & -69 & 21 \\ 1 & -10 & 3 \\ 0 & -4 & 1 \end{bmatrix}$, $Q^{-1} = \begin{bmatrix} 1 & 0 & 0 \\ 0 & 1 & 0 \\ 0 & -3 & 1 \end{bmatrix}$

The characteristic polynomial for H is $p(t) = (t + 1)^2(t - 1)$; and the eigenvectors and generalized eigenvectors are

$\lambda = -1$, $\mathbf{v}_1 = \begin{bmatrix} 3 \\ 1 \\ 2 \end{bmatrix}$, $\mathbf{v}_2 = \begin{bmatrix} -7/2 \\ -1/2 \\ 0 \end{bmatrix}$;

$\lambda = 1$, $\mathbf{w}_1 = \begin{bmatrix} -3 \\ 0 \\ 1 \end{bmatrix}$.

The general solution of $\mathbf{y}' = H\mathbf{y}$ is $\mathbf{y}(t) = c_1 e^{-t} \mathbf{v}_1 + c_2 e^{-t}(\mathbf{v}_2 + t\mathbf{v}_1) + c_3 e^t \mathbf{w}_1$; and the initial condition $\mathbf{y}(0) = Q\mathbf{x}_0$ can be met by choosing $c_1 = 0, c_2 = 2, c_3 = -2$. Finally $\mathbf{x}(t) = Q^{-1}\mathbf{y}(t)$, or

$$\mathbf{x}(t) = \begin{bmatrix} e^{-t}(6t - 7) & + 6e^t \\ e^{-t}(2t - 1) \\ e^{-t}(-2t + 3) - 2e^t \end{bmatrix}.$$

(b) $QAQ^{-1} = H$, where

$$Q = \begin{bmatrix} 1 & 0 & 0 \\ 0 & 1 & 0 \\ 0 & 3 & 1 \end{bmatrix}, \quad H = \begin{bmatrix} 2 & 4 & -1 \\ -3 & -4 & 1 \\ 0 & 3 & -1 \end{bmatrix},$$

$$Q^{-1} = \begin{bmatrix} 1 & 0 & 0 \\ 0 & 1 & 0 \\ 0 & -3 & 1 \end{bmatrix}.$$

The characteristic polynomial is $p(t) = (t + 1)^3$; and

$$\mathbf{v}_1 = \begin{bmatrix} 1 \\ 0 \\ 3 \end{bmatrix}, \quad \mathbf{v}_2 = \begin{bmatrix} 0 \\ 1 \\ 3 \end{bmatrix}, \quad \mathbf{v}_3 = \begin{bmatrix} 0 \\ 1 \\ 4 \end{bmatrix}.$$

The general solution of $\mathbf{y}' = H\mathbf{y}$ is

$$\mathbf{y}(t) = e^{-t}\left[c_1 \mathbf{v}_1 + c_2(\mathbf{v}_2 + t\mathbf{v}_1) + c_3\left(\mathbf{v}_3 + t\mathbf{v}_2 + \frac{t^2}{2}\mathbf{v}_1 \right) \right];$$

and the initial condition $\mathbf{y}(0) = Q\mathbf{x}_0$ is met with $c_1 = -1, c_2 = -5, c_3 = 4$. Finally $Q^{-1}\mathbf{y}(t)$ solves $\mathbf{x}' = A\mathbf{x}, \mathbf{x}(0) = \mathbf{x}_0$.

(c) $QAQ^{-1} = H$, where

$$Q = \begin{bmatrix} 1 & 0 & 0 \\ 0 & 1 & 0 \\ 0 & 3 & 1 \end{bmatrix}, \quad H = \begin{bmatrix} 1 & 4 & -1 \\ -3 & -5 & 1 \\ 0 & 3 & -2 \end{bmatrix},$$

$$Q^{-1} = \begin{bmatrix} 1 & 0 & 0 \\ 0 & 1 & 0 \\ 0 & -3 & 1 \end{bmatrix}.$$

The characteristic polynomial is $p(t) = (t + 2)^3$; and

$$\mathbf{v}_1 = \begin{bmatrix} 1 \\ 0 \\ 3 \end{bmatrix}, \quad \mathbf{v}_2 = \begin{bmatrix} 0 \\ 1 \\ 3 \end{bmatrix}, \quad \mathbf{v}_3 = \begin{bmatrix} 0 \\ 1 \\ 4 \end{bmatrix}.$$

The general solution of $\mathbf{y}' = H\mathbf{y}$ is

$$\mathbf{y}(t) = e^{-2t}\left[c_1 \mathbf{v}_1 + c_2(\mathbf{v}_2 + t\mathbf{v}_1) + c_3\left(\mathbf{v}_3 + t\mathbf{v}_2 + \frac{t^2}{2}\mathbf{v}_1 \right) \right];$$

and the initial condition $\mathbf{y}(0) = Q\mathbf{x}_0$ is met with $c_1 = -1, c_2 = -5, c_3 = 4$. Finally $Q^{-1}\mathbf{y}(t)$ solves $\mathbf{x}' = A\mathbf{x}, \mathbf{x}(0) = \mathbf{x}_0$.

5. $\mathbf{x}(t) = c_1 e^t \begin{bmatrix} -2 \\ 0 \\ 1 \end{bmatrix} + c_2 e^t \left(\begin{bmatrix} 5/2 \\ 1/2 \\ 0 \end{bmatrix} + t \begin{bmatrix} -2 \\ 0 \\ 1 \end{bmatrix} \right)$

$\quad + c_3 e^{-t} \begin{bmatrix} -9 \\ -1 \\ 1 \end{bmatrix}$

Exercises 7.1

1. $a = (1/2 + 1/8)8 = 5, b = 6.75, c = 38$

3. $a = 2857, b = 145.75, c = 51.1875$

5. $a = .111 \times 2^4, b = .111011 \times 2^5, c = .11011 \times 2^3,$
$d = .10000110111 \times 2^8$

7. $a = .E \times 16, b = .1D8 \times 16^2, c = .6C \times 16,$
$d = .86E \times 16^2$

11. $a + b = .20242 \times 10^2, ab = .63306 \times 10^0,$
$a(b + 3c) = .30379 \times 10^3, c + (d + d) = .50001 \times 10^1,$
$(c + d) + d = .50000 \times 10^1$

GLOSSARY OF SYMBOLS

Below is a list of frequently used symbols and the pages on which they are defined.

$A = (a_{ij})$	matrix, 22
$[A\,|\,b]$	augmented matrix, 23
R^n	Euclidean n-space, 53
$A = [\mathbf{A}_1, \mathbf{A}_2, \ldots, \mathbf{A}_n]$	column form of a matrix, 58
$\mathcal{O}$	zero matrix, 65
A^T	matrix transpose, 67
$\|\mathbf{x}\|$	norm of a vector, 70
$\boldsymbol{\theta}$	zero vector, 74
I	identity matrix, 95
$\mathbf{e}_k$	natural unit vectors in R^n, 77
A^{-1}	inverse of the matrix A, 98
$\mathcal{N}(A)$	null space of A, 132
$\mathcal{R}(A)$	range space of A, 134
$\dim(W)$	dimension of W, 157
$\det(A)$	determinant of A, 228
$\mathscr{P}_n$	set of polynomials of degree n or less, 304
$\mathrm{Sp}(Q)$	span of Q, 130
$[\mathbf{w}]_B$	coordinate vector, 321
$\langle \mathbf{u}, \mathbf{v} \rangle$	inner product, 335
$T \circ S$	composition of transformations, 355

INDEX